Estuaries of the World

Series Editor

Jean-Paul Ducrotoy, The University of Hull, Hull, UK

Estuaries are amongst the most endangered areas in the world. Pollution, eutrophication, urbanization, land reclamation; over fishing and exploitation continuously threaten their future. The major challenge that humans face today is managing their use, so that future generations can also enjoy the fantastic visual, cultural and edible products that they provide. Such an approach presupposes that all users of the environment share views and are able to communicate wisely on the basis of robust science. The need for robust science is pressing. Over the last decade there have been numerous advances in both understanding and approach to estuaries and more and more multidisciplinary studies are now available. The available scientific information has come from a multiplicity of case studies and projects local and national levels. Regional and global programs have been developed; some are being implemented and some are in evolution. However, despite the rapidly increasing knowledge about estuarine ecosystems, crucial questions on the causes of variability and the effects of global change are still poorly understood. Although the perception of politicians and managers of coasts is slowly shifting from a mainly short-term economic approach towards a long-term economic – ecological perspective, there is a need to make existing scientific information much more manageable by non-specialists, without compromising the quality of the information. The book series includes volumes of selected invited papers and is intended for researchers, practitioners, undergraduate and graduate students in all disciplines who are dealing with complex problems and looking for cutting-edge research as well as methodological tools to set up truly transversal science and technology projects, such as the restoration of damaged habitats.

More information about this series at http://www.springer.com/series/11705

Benjamin E. Cuker
Editor

Diet for a Sustainable Ecosystem

The Science for Recovering the Health of the Chesapeake Bay and its People

Springer

Editor
Benjamin E. Cuker
Department of Marine and Environmental Science
Hampton University
Hampton, Virginia, USA

ISSN 2214-1553 ISSN 2214-1561 (electronic)
Estuaries of the World
ISBN 978-3-030-45480-7 ISBN 978-3-030-45481-4 (eBook)
https://doi.org/10.1007/978-3-030-45481-4

This Springer imprint is published by the registered company Springer Nature Switzerland AG.
The registered company address is: Gewerbestrasse 11, 6330 Cham, Switzerland

This work is dedicated to Dawn, Aaron, Lydia, Lindsay, Ava, Slocum, and Ainslie, and to the memory of Mina, Sy, Sam, Vladimir, Tovi, Tasha, and Scupper.

Preface

This book emerged from the merging of four of my passions.

Passion one, aquatic sciences and sustainability I came of age in the 1960s before the passage of the Clean Water Act of 1972. I was a sophomore in high school when we first celebrated Earth Day in 1970. It was a time of contradictions. One day the media showed images of Ohio's Cayuga River alight, hydrocarbons on its surface fed smoky fires. The next evening Jacques Cousteau would light up the family's first color television with beautiful images of pristine tropical coral reefs teeming with the bright palette of life. Growing up, my family vacations often meant a week in a waterside cabin with a rowboat to explore the lake or coastal estuary. All this fostered my love of the watery world and the desire to study it with an eye to preserving the natural wonders for future generations.

As a graduate student I studied lakes in Arctic Alaska and was there to take bottom samples a day after the first pond on the North Slope was spoiled by a blown valve of the newly opened Trans-Alaska Pipeline.

In 1981, my first academic job brought me to Shaw University, in Raleigh, NC. There I studied the effects of sediment pollution on the ecology of a small reservoir. In 1988, I moved to Hampton University on the shores of Hampton Roads, near the mouth of the Chesapeake Bay. I studied various aspects of the Bay, but spent much of my effort on understanding its eutrophication (response to excess nutrients, primarily from the food system).

Passion two, diversity I had the good fortune of spending my teenage years in Detroit, Michigan. My family moved into that city a week after the Detroit uprising of 1967. We were one of the few white families to resist the flight to the suburbs in the ensuing years. I was one of six white students to graduate from Mumford High School in 1972, the rest of my classmates being African American. This gave me some insight into issues of social and environmental justice as well as the importance of diversity, which went on to inform my career. I have enjoyed 38 years of teaching at historically black universities. I am best known in the world of aquatic sciences for founding the Association for the Sciences of Limnology and Oceanography's Multicultural Program (ASLOMP) in 1990. This National Science Foundation funded program continues today, each year bringing 70 or so diverse students to the annual meetings of the society. The success of ASLOMP led to the honor of being named a Pew Fellow in Marine Conservation in 1999. That award funded what became a nine-year long program called Multicultural Students at Sea Together (MAST). Between 2000 and 2008, every summer I took a dozen diverse students from across the USA on a month-long exploration of the Chesapeake Bay aboard a ketch-rigged sailboat. This included a rigorous study of the distribution of oxygen and chlorophyll levels, marine policy, seamanship, and the role of African Americans and Native Americans on the Bay.

Passion three, food for the health of people and the planet When I started my undergraduate studies at the University of Michigan in 1972, I heard the echoes of a speech delivered there by Frances Moore Lappé the preceding year, where she introduced her revolutionary book, *Diet* *for a Small Planet*. While never formally addressed in courses of that era, classmates in the School of Natural Resources talked a lot about how food informed our lives and the environment. I joined the People's Food Co-op that serviced the Ann Arbor community. At Michigan I also met my future wife, Dawn Gerbing, who was a vegetarian. It took an elevated blood pressure reading for me to also become a vegetarian in 2002, yet I continued to eat dairy, seafood, and occasionally eggs. It was not till I began teaching an honor's seminar in 2009 called, Eating for a Healthy You and a Sustainable Planet, that I took a serious look at all the new science surrounding dietary choices. It became clear that my dozen years as a pesca-vegetarian was not sufficient to sustain my own health or that of the environment, and I adopted a whole plant-based diet in 2014, at the age of 59. This brought me into my sixth decade with a new sense of health and vitality. I lost weight, and achieved low blood cholesterol levels. I felt the best I ever had in my life. Becoming vegan also opened my eyes to the systematic misery visited upon the animals raised for food, something necessary to ignore for those eating meat and dairy.

Passion four I discovered the joys of sailboats during my Ph.D. studies at North Carolina State University. Upon relocating to Hampton University in 1988, I found myself on the shore of the Chesapeake Bay, with all of its possibilities for exploration. My family and I became ardent cruisers and racers. This allowed me to see the ecosystem and its food system in varied moods and seasons, gaining experiences beyond what I could get from sampling particular stations or reading papers about the Bay. While sailing the Bay I regularly share the waters with crabbers, oyster tongers, and fishers working gill nets. More than once we found ourselves tangled in an unlit pound net while making passage on a dark night. And of course, like every other sailor on the Bay, we have snagged our share of tethered crab pot buoys. We have smelled the sugar ships long before we passed them offloading their sweet brown granules in Baltimore's Inner Harbor. And we have experienced the stench of menhaden rendering while sailing into Reedville more times than we care to recall. On one cruise to the Eastern Shore we brought our bicycles to explore the landscape, only to be sprayed with herbicide by a low-flying crop duster working the cornfields through which we passed. Too often we have sailed through algal blooms but must admit to taking delight in their bioluminescent glow during night passages.

Combining the passions into this book My four passions led to the creation of this book. I clearly saw how our food system shaped the Chesapeake Bay ecosystem and that remaking our relationship with nutrition was required if we were to return the estuary to its former majesty. I also witnessed the decline of the health of the people living around the Bay over the last three decades. So many of my fellow Bay residents suffer obesity, diabetes, and related diseases brought on by the meat-sweet Standard American Diet. This is particularly true for the region's most vulnerable residents, the poor and people of color. And people of color do much of the essential and difficult work that keeps the food system running, often with low pay and poor working conditions. In this book I note the special relationship between people of color and the food system.

To help me create this work I assembled a diverse team of authors and coauthors. The 20 collaborators includes 15 women, 8 people of color, well-established scientists, mid and early career researchers, retired experts, and several students. I particularly sought women as collaborators, in keeping with the theme of an eco-feminist world view.

I hope that this volume helps readers understand the Chesapeake Bay and the food system in a new way that enables them to imagine a healthier and more sustainable approach for the future. The book will challenge its readers to consider their own relationship with food and the environment. Although the Chesapeake Bay is at the heart of this narrative, the story should inform the understanding of the nexus of food systems, society, and aquatic ecosystems around the planet.

Hampton, VA, USA
September 20, 2019

Benjamin E. Cuker

Acknowledgments

Many colleagues and friends made possible the creation of this book. I thank the funders of my professional activities over the years that provided the experience essential for assembling this volume. This includes the National Science Foundation, The Pew Fellows Program in Marine Conservation, the US Army Corps of Engineers, and The National Oceanic and Atmospheric Administration. I owe my gratitude to the coauthors (named below) who contributed to this volume. Much thanks to the colleagues who provided reviews of the chapters, including Greg Cutter, Nicholas DiPasquale, Michael Fine, Ted Henifin, Vic Kennedy, Rom Lipcius, Terran Niedrauer, Pete Sheridan, Leila Rice and Rosemary Ziemba. Dawn Gerbing, Ruth Belcastro and Alison Smith contributed copyediting and research assistance. Dawn Gerbing and Andrij Horodysky provided stimulating discussion of the material. Rosemary Ziemba, a friend from my high school years in Detroit helped inspire this effort after she invited me to join her as a coauthor on a book chapter addressing water pollution, human health, and the role of health care providers. From Springer, Eva Loerinczi provided kind guidance and encouragement. S. Padmashri and Abinay Subramaniam proved essential for turning the manuscript into a published volume. The authors of the oyster chapter thank Andrew Button, Karen Hudson, Gail Scott and Clara Robison for their assistance and note that their effort is VIMS contributions #3920. Finally, much thanks to my wife and life partner Dawn Gerbing for facilitating the long hours devoted to this effort, for encouraging the work, and for providing the intellectual stimulation that helped nurture the project.

Contents

Editor and Contributors

About the Editor

Benjamin E. Cuker, Ph.D., Professor of Marine and Environmental Science, Hampton University, Hampton, VA, USA. Dr. Cuker is the main author and editor of this volume. He earned his BS and MS degrees in Natural Resources from the University of Michigan and a Ph. D. in Zoology (Aquatic Ecology) from North Carolina State University. He is a Professor of Marine and Environmental Science at Hampton University. His research includes arctic limnology, the ecology of turbid reservoirs, and studies of the Chesapeake Bay. He was the first recipient of two awards from the Association for the Sciences of Limnology and Oceanography (ASLO), the Distinguished Service Award in 1995, and the Excellence in Education Award in 2009. He was named an ASLO fellow in 2017. Cuker was named a Pew Fellow in Marine Conservation in 1999. He has published in various journals, including *BioScience*, *Conservation Biology*, *Ecology*, *Estuaries*, and *Limnology and Oceanography*. Cuker holds a certificate in Plant-Based Nutrition from Cornell University. He has twice been the overall winner of the annual Annapolis to Hampton, Down the Bay Race, in his Cal 3-30 sailboat, *Callinectes*.

List of Contributors

Matthew Balazik Research Development Center, Vicksburg, MS, USA

Eugene M. Burreson Emeritus Faculty, Virginia Institute of Marine Science, Gloucester Point, VA, USA

Victoria Coles University of Maryland Center for Environmental Science, Horn Point Laboratory, Cambridge, MD, USA

Jeffery Cornwell University of Maryland Center for Environmental Science, Horn Point Laboratory, Cambridge, MD, USA

Carie Curry-Cutter Hampton Roads Sanitation District, Virginia Beach, VA, USA

Karen Davis United Poultry Concerns, Machipongo, VA, USA

Kendra Dorsey Department of Marine and Environmental Science, Hampton University, Hampton, Virginia, USA

Ileana Fenwick Department of Marine and Environmental Science, Hampton University, Hampton, VA, USA

Dawn D. Gerbing Lactation Consultant, Fairfax, VA, USA

Deidre M. Gibson Department of Marine and Environmental Science, Hampton University, Hampton, VA, USA

Sierra S. Hildebrandt Hampton University, Hampton, VA, USA

Andrew Justice Department of Marine and Environmental Science, Hampton University, Hampton, Virginia, USA

Martin Kaufman University of Michigan, Flint, MI, USA

Kathryn MacCormick Biology Faculty, Rappahannock Community College, Glenns, VA, USA

Bill Matuszeski Chesapeake Bay Program, Washington, DC, USA

Michelle Penn-Marshall The Graduate College and V.P. for Research, Hampton University, Hampton, VA, USA

Kimberly S. Reece Virginia Institute of Marine Science, Gloucester Point, VA, USA

Meredith Evans Seeley Virginia Institute of Marine Science, Gloucester Point, VA, USA

Rochelle D. Seitz Virginia Institute of Marine Science, Gloucester Point, VA, USA

Indu Sharma Department of Biological Sciences, Hampton University, Hampton, VA, USA

Nefertiti Smith Department of Marine and Environmental Science, Hampton University, Hampton, Virginia, USA

Kari St. Laurent Delaware National Estuarine Research Reserve, Townsend, DE, USA

Olivera Stojilovic Department of Marine and Environmental Science, Hampton, Virginia, USA

Part I

Introduction and Background

Introduction: Starting the Journey to a Sustainable Ecosystem and Healthy People

Benjamin E. Cuker and Kathryn MacCormick

The Bay and the land around it were once so fair,
that people could find and grow their food without much care.
Then came the colonists who didn't understand,
how to manage the sacred land.
They built dams that blocked the flow of fish,
and took from the Bay too much food for their dish.
Meat and sweets became the order of the day,
this made sick the people and the Bay.
To bring both the people and ecosystem back to health,
from better foods, farmers must look to for wealth.

Abstract

This book is about the intersection between the food system, the condition of the Chesapeake Bay ecosystem, and the health of the people that live there. It takes a historical approach examining how the food system developed from early colonial times to the present. The book introduces the reader to the basic science governing the estuary and human health. It then examines the issues of exploitation of key animal populations, effects of industrial/chemical agriculture on the ecosystem, diet driven human health outcomes, pollution from the food system, economics and ethics of the food system, current policy environment, and the nexus between the food system and climate change in the Chesapeake Bay Region. The Standard American Diet (SAD) is examined from the perspectives of the impacts on the health of the ecosystem that produces it and the people that consume it. The authors introduce the concept that the food system drives eutrophication of people as well as ecosystems. The book presents evidence for the efficacy of replacing the SAD with a food system restructured to produce a whole foods plant based diet. It is written with social context that addresses issues of gender and race as they pertain to the food system and health outcomes. It also notes particular contribution made by women and people of color.

Keywords

Chesapeake Bay · Food system · Ecosystem health · Ecosystem restoration · Standard American diet

The English Meet the Chesapeake Bay and its People On a quiet spring day in 1607, indigenous women on the Chesapeake's Eastern Shore were likely gathering seasonal snacks from the water's edge when the sails of Captain Newport's fleet approached the shore. After a 4-month journey, 104 English men disembarked from the cramped quarters of the *Discovery*, *Susan Constant*, and *Godspeed* and came ashore with dreams of finding gold and new trade routes leading to the storied riches of the Orient. These men were soldiers, explorers, and tradesmen who brought guns and trade goods but not one among them was a farmer or fisher. The Virginia Company planned for colonists to acquire food from local tribes. The ships contained only 14 weeks supply of victuals, necessitating dependence on Algonquin food (Rountree et al. 2007). But there was no understanding or planning for how seasonal and climactic

B. E. Cuker (✉)
Department of Marine and Environmental Science, Hampton University, Hampton, Virginia, USA
e-mail: Benjamin.cuker@hamptonu.edu

K. MacCormick
Biology Faculty Rappahannock Community College, Warsaw, VA, USA
e-mail: Kathryn.e.Maccormick@dominionenergy.com

B. E. Cuker (ed.), *Diet for a Sustainable Ecosystem*, Estuaries of the World, https://doi.org/10.1007/978-3-030-45481-4_1

constraints may affect the hospitality of their indigenous hosts. Early Spring is a food scarce time for tribes living on a seasonal diet and the entire region had been suffering the worst drought in 800 years (Blanton 2000). The false assumption that the natural resources of the Chesapeake Bay inherently belong to and will always be available for men with economic power and disposable human capitol led to disastrous results for early English settlers. Of the 6000 people sent to Jamestown before 1624, only 1200 were still alive in 1625 and the majority of the 4800 deaths were caused by starvation (Kupperman 1979). A 20% survival rate is an unacceptable trade of human life on behalf of corporate gain to the eye of most modern beholders, but the Virginia Company and other corporations like it went on to form the foundation for the modern American diet. Food, its abundance or absence, is central to the earliest stories of Colonial America and, as we will see throughout this book, continues to define the relationship between people, their health, and the well-being of the Chesapeake Bay ecosystem.

The Virginia Company of London chose Captain John Smith to lead the new colony. He was one of those first Englishmen who splashed through pristine waters near the mouth of the Chesapeake Bay on April 27, 1607 to plant an English flag and name the place Cape Henry in honor of their monarch far across the ocean. No doubt the indigenous people who first saw the ships had raised the alarm and so five Algonquin speaking warriors approached the English camp on the first night. As Smith tells it, the Indians attacked with arrows and the English shot back with pistols (Smith 1608). However, there is some mystery surrounding this initial conflict; the Virginia Company men had orders to build trading partnerships with locals and plenty of other tribes welcomed foreigners with food and not arrows. Is it possible that memories of conflict between Powhatans and Spanish explorers from a half century earlier were still fresh with this tribe? Even the English were wary of encounters with the Spanish, so they sailed up the southernmost tributary river to the Chesapeake Bay, named it the James River after their King, and founded Jamestown (Fig. 1) on an inhospitable island that provided a good defensive position and a deep-water port for trade but lacked fresh water during the summer and did not provide easy access to the resources of the mainland. Dismissing indigenous names, focusing their attention on countering potential Spanish influence in the region, and ignoring basic aspects of the ecosystem crucial for their survival surely was all meant to please the English authorities but it reveals a lack of appreciation for the local politics of the Chesapeake and it lead to conflicts with indigenous people that contribute heavily to the shortage of food among Jamestown colonists.

The Jamestown residents certainly needed to worry about conflict. The territory on which they landed and which was explored later by John Smith was under control of the Powhatan Confederacy. Unlike the small, simple groups the English assumed they would encounter, the Powhatan Confederacy was an alliance of 36 tribes of Algonquin-speaking peoples stretching across the tidewater region of what is now Virginia. The alliance of Algonquin speaking tribes in the region had arisen only a generation before the English arrival and while it served as defense against enemies to the North, the Iroquois, the Confederacy also contained internal struggles among member tribes. It's leader, Paramount Chief Powhatan (Wahunsenacawh), had complex political and economic reasons to both show the English the strength of his empire and to offer them friendship. Local tribes appreciated the various manufactured goods that the colonists brought to trade but remained suspicious of the English (Rountree et al. 2007). At times, Powhatan warriors attacked the parties of colonists who ventured outside of the fortified Jamestown settlement. It wasn't long before the English colonists realized that restricted access to the mainland meant it was easy for the Powhatan warriors to limit their ability to forage for food away from the settlement site.

In the initial months of quasi peace with Powhatan, Capt. Smith found safe passage out of Jamestown with a small band of men for two extensive voyages of exploration over the summer of 1608 (Rountree et al. 2007). They used a small vessel called a shallop. This boat featured shallow draft, a single mast with sail and oars, making it useful for exploring the shoal waters of the various tributaries entering the Bay. The summer of voyaging in the Bay, along with land-based excursions, formed the basis of Capt. Smith's *A MAP OF VIRGINIA with a Description of the Countrey, the Commodities, People, Government and Religion*, which he published in 1612 (Smith 1612; Fig. 2). This publication contains the map promised in the title and many detailed observations on the Bay, the land, and its people. While off in scale and proportion, Smith's map compares well to modern charts of the Bay. His comments on the land and its people provide the modern reader a trip back in time to glimpse a very different Chesapeake than what we know today.

During his two voyages of the Chesapeake Bay Smith encountered native people almost every day. Sometimes the English were met with arrows but more often the tribes treated then as honored guests bearing goods to trade. The chiefs expressed their hospitality through food. Smith writes in detail about the various feasts put on to honor the band of explorers (Smith 1608). These accounts help create a picture of the pre-colonial food system for the region (Smith 1612), which will be explored in the Chapter "The Algonquin Food System and How It Shaped the Ecosystem and Interactions with the English Colonists of the Chesapeake Bay".

Fig. 1 Print of Captain John Smith landing in Jamestown, Virginia, 1607. From, *The Story of Pocahontas and Captain John*. Courtesy of the New York Public Library. Source=http://digitalgallery.nypl.org/nypldigital/id?808046 |A https://upload.wikimedia.org/wikipedia/commons/thumb/f/f4/Captain_John_Smith_landing_in_Jamestown.jpeg/745px-Captain_John_Smith_landing_in_Jamestown.jpeg

Smith described the clear water of the Bay, thriving reefs of oysters, and massive schools of fish. In July of 1608, he encountered a school of fish (probably menhaden) so dense that his men tried to scoop them up with frying pans (Rountree et al. 2007). Smith noted beds of submerged grasses, marshes teaming with wildlife, an abundance of fowl, and extensive forests (Smith 1612).

The health of the peoples Capt. Smith met on the shores of the Chesapeake mirrored that of the Bay ecosystem. He admired their physical appearance, noting their height and vigor. Native American men sported the trim bodies of quick runners, well suited to their occupations of hunter and warrior. Native women presented strong musculature well adapted to their daily routine of heavy work (Smith 1612).

A Return Voyage: Discovering a Stricken Bay and its Sickened People Anyone retracing John Smith's voyaging today would encounter a very different Chesapeake Bay and population of human inhabitants. Cloudy green water laps the shores and obscures from view the features of the shallow bottom. Urban and suburban tributaries feature shores hardened with concrete, stone and steel, eliminating the muddy and marshy shoreline favored by birds and other wildlife. The once abundant oyster reefs that protruded the surface at low tide are long gone, victims of a collapsed shell fishing industry. Current harvest rates are 1% of the peak landings of the 1880s (Couranz 2018). The greatly reduced beds of sea grass cling to life, restricted to shallow depths where sufficient sunlight can penetrate the murky water. Industrial, urban, suburban and agricultural development replaced much of the original maritime forests and marsh lands. From late spring to early fall a dead zone of low oxygen water excludes larger forms of life from the deeper reaches of the Bay. Discarded plastic wrappers, bottles, and food containers dot the open waters and mingle with dislodged sea grasses and shells of the strandline that inscribes the shores. Unseen are the contaminants that impair 70% of the Bay (Phillips 2018). Due to contamination, the State of Maryland advises against eating more than four meals per month of almost any fish caught in the Bay. For the most common catch, striped bass (over 28 inches), that falls to once per month for adults, and once every other month for children (MDE 2014).

As the physical and biological environment of the Bay has changed, the same is true for its people. Capt. Smith recorded the number of residents he observed throughout his explorations. The populations of most villages numbered less than 200. He estimated some 5000 people lived within 60 miles (97 km) of Jamestown (Smith 1612). That would put

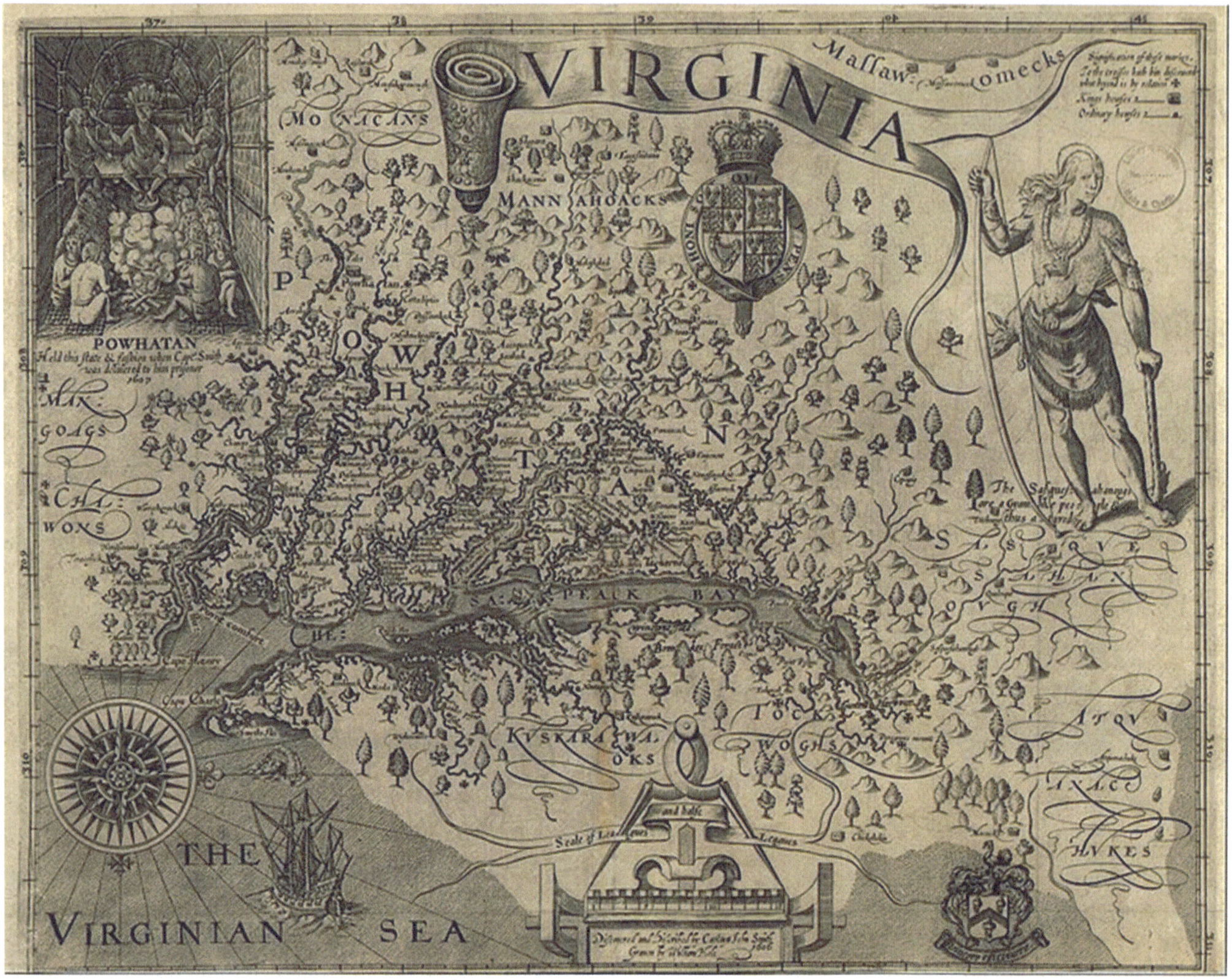

Fig. 2 Capt. John Smith's Map of Virginia from Smith (1612)

the native population of the entire watershed somewhere in the tens of thousands. From A Map of Virginia, "*The land is not populous, for the men be fewe; their far greater number is of women and children. Within 60 miles of James Towne there are about some 5000 people, but of able men fit for their warres scarse 1500. To nourish so many together they have yet no means because they make so smal a benefit of their land, be it never so fertill or 700 have beene the most hath beene seene together, when they gathered themselves to have surprised Captaine Smyth at Pamaunke, having but 15 to withstand the worst of their furie (Smith 1612).*" This passage on the population size also notes the uneven demographics for the Indians, with reduced numbers of men, presumably from loss to war. It also speculates that a lack of food restrains their population size.

Today 18 million people live in the watershed of the Chesapeake Bay (CBP 2018, Fig. 3). The sheer numbers of people living on the drainage means that no matter how effective, attempts to recover the health of the Bay will never return it to the condition of the time when Powhatan and the colonists shared its wealth and beauty. Urban sprawl, deforestation, draining of wetlands, and all the rest that comprises the footprint of the modern inhabitants of the region are now fixed features of the system. Yet, reshaping the food system can reverse much of the damage done by destructive agricultural practices and overharvesting of marine food resources.

Four centuries later, people living around the Chesapeake Bay appear quite different from the robust native inhabitants encountered by Capt. John Smith. Many of today's residents seem unable or unwilling to walk more than a few steps to get between an automobile and a destination. Most appear fated to spend their later years on a path of decline, kept functioning only by a plethora of medications and medical procedures. Nationally, 17% of the children are obese and 71% of adults are obese or over weight (CDC 2017a). For the states of the Chesapeake Bay watershed its worse, with

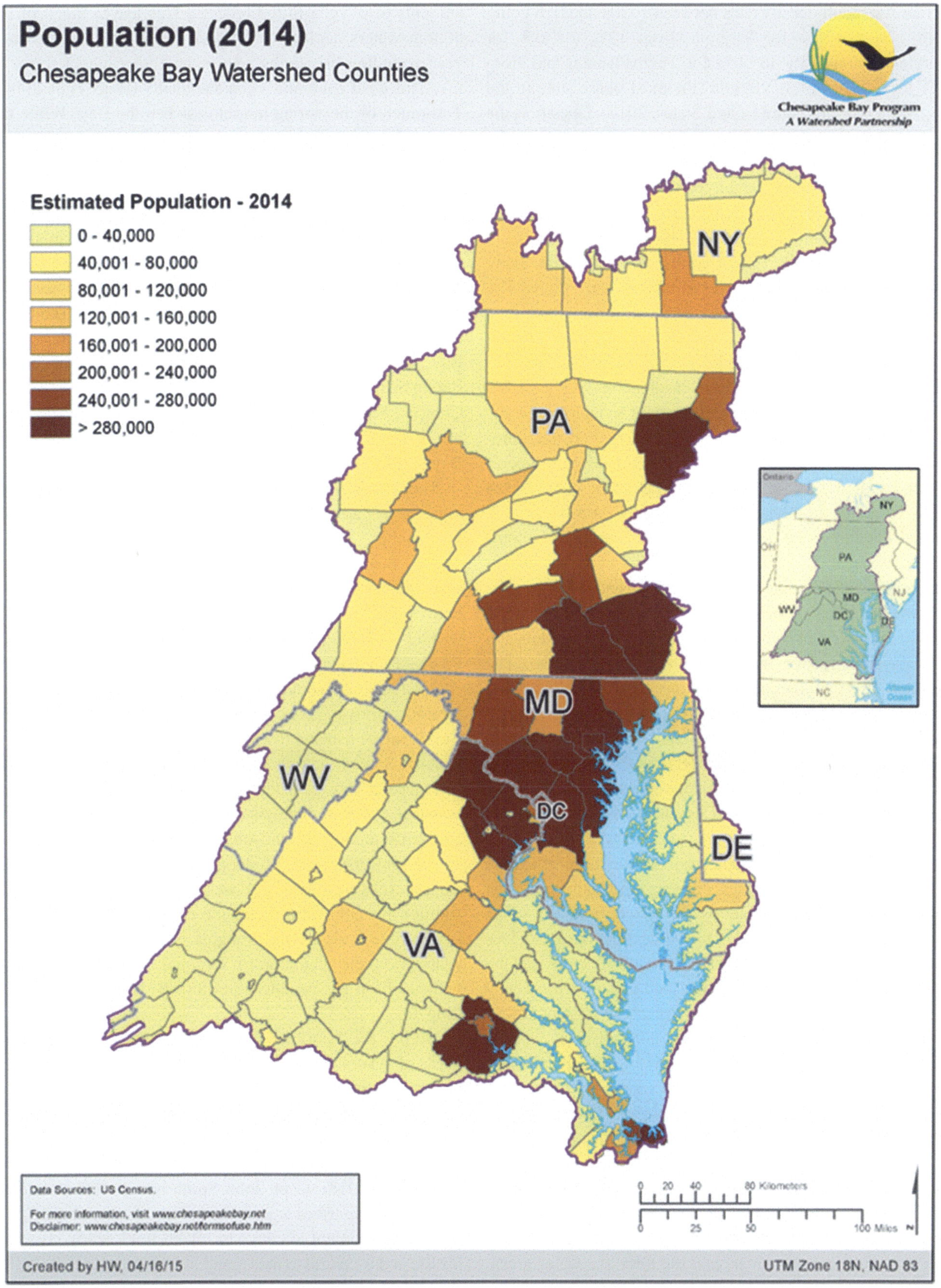

Fig. 3 Population distribution in the Chesapeake Bay Watershed (CWB). Presently about 18 million people live in the CBW, with highest densities in the urban areas of Washington, DC. Maryland, and southern Pennsylvania. https://www.chesapeakebay.net/images/maps/population2014.pdf

increases in adult obesity rates between 1990 and 2016 going from 12% to 28% for Virginia, from 12% to 30% for Maryland, from 13% to 30% for Pennsylvania, and from 13% to 37% for West Virginia (the most obese state in the US, Adult Obesity in the United States, 2017). Obesity is the most obvious expression of how the food system is failing the public. Other more cryptic markers of damaged health, such as heart disease, cancers, and hypertension will be discussed later. But for now know that every day even adults that appear trim die from preventable diet-driven heart attacks.

For the US overall, 57% of adults take at least one daily prescription medication and 11% take five or more. The situation worsens with age, as 66% of those older than 45 years take at least one prescription drug daily, jumping to 92% for those 65 years and older. For the 65+ set, 43% take five different prescription medications daily (CDC 2017b). Most of these medications treat symptoms of the Standard American Diet (SAD) such as hypertension, coronary artery disease, elevated cholesterol levels, type-2-diabetes, and erectile dysfunction. Doctors either don't know or bother to advise their patients to eat a better diet, instead they prescribe medications to manage the symptoms of SAD (Greger 2015). The overmedication of the population poses problems for the Chesapeake Bay and other aquatic ecosystems. Many of these drugs pass through the patient and into sewage systems, where wastewater treatment plants and septic systems prove ineffective in destroying them. The excreted drugs contaminate receiving waters. The displaced drugs may then move into the aquatic food chain, disrupting the biology of various species (Boxall 2004, USEPA et al. 2012). This will be addressed in detail in Chapter "Finishing the Journey: Wastewater as a misplaced resource".

The thing that sickens the Chesapeake Bay is also what sickens its people and the food system. Overharvesting of oysters, fish, and other seafood from the Bay diminished the ecosystems as do the chicken growing operations along its eastern shore. Runoff from overfertilized fields of crops destined for animal feed adds to the destruction. Pesticides applied in the battle against nature contaminate crops and the Bay. These chemicals move up the food chain to chronically poison fish and fowl. People consuming the food produced by this system suffer for it. Obesity, heart disease, cancers, diabetes, and a host of other afflictions of misapplied affluence plague today's people of the Chesapeake region.

As will be discussed in detail in Chapter "Eutrophication: Obesity of the Bay and Its People", the Chesapeake Bay suffers from cultural eutrophication. Plant nutrients from artificial fertilizers, chicken growing factories, and wastewater treatment facilities stimulate excessive growth of the tiny algae cells called phytoplankton. From early spring to late fall this overabundance of phytoplankton turns the water green, red or brown. Most of the algae eventually settles to the bottom where bacteria consume it along with massive amounts of oxygen in the process. This leaves much of the bottom waters depleted of oxygen for several months each year excluding fish, crabs, oysters and other higher forms of life. This dead zone now occurs annually (Hagy et al. 2004). Too much of the wrong nutrients sicken the Bay. While the term cultural eutrophication describes the consequences of overfeeding water bodies, the concept also illustrates what happens with overfeeding human bodies. Consuming too much of the wrong foods and too little of the right ones leads to cultural eutrophication of our own bodies, including reduced ability to bring oxygen to our tissues. The miss-feeding of the people of the Bay and its consequences will be addressed in Chapters "Livestock and Poultry: The Other Colonists Who Changed the Food System of the Chesapeake Bay, Instead of Eating Fish: The Health Consequences of Eating Seafood from the Chesapeake Bay Compared to Other Choices, Sugar Twice Enslaves: Consequences for the People of the Chesapeake, and Eutrophication: Obesity of the Bay and Its People".

The founders of the Virginia Colony didn't set sail on such a dangerous adventure simply to find a place of sustenance and shelter. They wanted material wealth and took for granted that they would find or trade for sufficient food to support their needs while in pursuit of valuables (Fig. 4). Poor organization, insufficient agricultural skills, lousy weather, tainted drinking water, bad relations with the Powhatan Indians, and distraction with growing tobacco for commerce all conspired to teach the colonists the primacy of food for survival. As noted above, Capt. Smith described the Bay as full of fish and fowl and surrounded by lush forest and marsh. Surely getting food would be easy. Yet the Chesapeake Bay is a temperate location experiencing the seasonal cycle of biological abundance and scarcity. Fish and birds migrate in and out of the Bay, berries and nuts appear and disappear. Winters, especially those of the little ice age (circa 1400–1700) were severe (White 2017). In addition to temperature, climatic conditions also vary over longer time scales, with droughts often lasting years. Two years into their habitation of Virginia, the English colonists suffered from the confluence of a cold winter and extended drought.

During the "starving time" (winter of 1609–1610) two thirds of the colony died from lack of food or disease enabled by malnutrition. One other food related death was that of an unfortunate pregnant wife, killed, salted and consumed by her husband. Add to that the husband's execution for his crime, ordered by George Percy, President of the colony (Stromberg 2018). Capt. John Smith bought and stole corn from the native tribes to save the colony from extinction. His clumsy and violent attempts to obtain food nearly cost him his life and is central to the legend of Pocahontas (NPS 2018). The moral of the story of Jamestown is neglect the importance of food at your own peril. The residents of the modern

Fig. 4 John Smith trying to get more food for the settlers from the Powhatan. National Park Service Image https://www.nps.gov/jame/learn/historyculture/pocahontas-her-life-and-legend.htm

Chesapeake Bay don't experience the food deprivations of 1609, but they unwittingly do suffer the consequences of a food system that robs them of their health with abundance. The current food system yields a SAD (Standard American Diet) loaded with calories and disease-promoting agents, while short in fiber, antioxidants, vitamins, and other health-promoting ingredients (Greger 2015).

While this book focuses on the food system, the environment, and people's health, it does so in the context of the greater agricultural enterprise. As mentioned above, tobacco, a nonfood crop played a part in the Jamestown famine. John Rolfe (who later married Pocahontas and took her to her death in England) promoted the idea of planting tobacco as a cash crop for export to England. Although Rolfe's tobacco did eventually dominate the economy of colonial Virginia, its cultivation by Jamestown settlers proved fatal long before the noxious plant caused its first case of lung cancer in English smokers. Had those first colonists planted beans, corn, squash, or other food crops on the plots reserved for tobacco, more of them might have survived the famine. So it is difficult to discuss the food system without also understanding the role of nonfood crops in the overall agricultural scene.

The Jamestown colonists also planted cotton hoping to use and sell the fiber for producing textiles. And cotton eventually did join tobacco as a leading export product (Cotton 2018). As with tobacco, cotton competed with food crops for the land and attention of later colonial farmers. That competition occurred in the short term as land devoted to cotton meant land not planted for food that season. Not understood by the early farmers, the consequences of cotton planting extended beyond the annual agricultural cycle. The cotton plants exhausted the land of nitrogen and other essential nutrients needed to support subsequent plantings of food crops. The role of soil fertility will be addressed in Chapters "The Journey from Peruvian Guano to Artificial Fertilizer Ends with too Much Nitrogen in the Chesapeake Bay and An Organic-Based Food System: A Voyage Back and Forward in Time".

The Jamestown colonists and their Powhatan neighbors used biofuel in the form of wood to cook their food, warm their dwellings, and fire their pottery. That wood came from natural forests managed by Indian practices discussed in Chapter "The Algonquin Food System and How it Shaped the Ecosystem and Interactions with the English Colonists of the Chesapeake Bay". Today's biofuel production centers on producing ethanol from corn to help fuel motor vehicles. For the US, about 40% of the corn crop is funneled through the gas pump. So one must consider the consequences of corn as car fuel when understanding it as cattle fuel or people fuel (Foley and Foley 2013).

Corn and Soy dominate US crop plantings, followed by wheat (USDA 2017). Yet about thrice the irrigated land used for these crops is planted in lawns (Milesia et al. 2005). This is significant for our food system in two ways. First, lawns compete with food crops for space and water. Second, the chemical agriculture industry promotes the use of the same herbicides, pesticides, and artificial fertilizers for use on lawns that it does for food crops.

Sadly, the Chesapeake's tale of food system driven degradation of place and people is the current story of ecosystems and their human inhabitants worldwide. Yet all is not doom and gloom. While the Chesapeake Bay ecosystem continues to suffer in many ways some aspects have improved in response to directed efforts over recent years. Curtailing the use of DDT and other similar pesticides in the 1970s saved from extinction bald eagles, cormorants, osprey, brown pelican, and other water birds that now thrive on the Bay. Better regulations of fishing have recovered the population of striped bass (*Morone saxatilis*) and sustains other species (Richards and Rago 1999). Reduction in nitrogen pollution drives extensive recovery of beds of submerged aquatic vegetation (Lefcheck et al. 2018). And not every person who occupies the Chesapeake Bay region suffers so completely from the food system. Increasing numbers of residents of the area now recognize that they can vastly improve their health and reduce their impact on the environment by changing the way they eat. Many are adopting a lifestyle built around a Whole Food Plant Based Diet (WFPBD), rejecting the Standard American Diet (SAD) and its associated food system.

Race and Gender Race and race based conflict informs the food system of the Chesapeake Bay region. As already noted, the need for food and croplands motivated the white colonist's suppression and displacement of their red neighbors. The importation of black slave labor to work the newly acquired land and adjacent marine resources soon followed. The Civil War ended chattel slavery and for the Chesapeake Bay it created a new economic order in which the labor of former slaves teamed with capitol from northern entrepreneurs to create vast wealth through the exploitation of oyster, crabs (Fig. 5) and the oily menhaden fish. Black labor filled the floors of slaughter and packing houses and accounted for much of the agricultural production on family owned farms. By the end of the twentieth century, labor in brown and yellow skin joined the food system work force. Migrants from Mexico and Central America came in seasonal waves to harvest fruit and vegetable crops and to process seafood. Some obtained longer lasting jobs in swine and poultry processing factories. Fishers displaced by the US war in their home country of Vietnam migrated to the US and took up their trade in the waters of the Chesapeake. Chapter "Ethics and Economics of Building a Food System to Recover the Health of the Chesapeake Bay and Its People" includes discussion of aspects of race and gender in the food system.

Race also determines the health response of people on the receiving end of the food system. African Americans suffer disproportionately higher levels of disease driven by the SAD. In Virginia the 2016 obesity rate was 27.3% for whites and 39.4% for African Americans. The rates of obesity were essentially the same for Maryland, 27.8% for whites, and 38.1%. For West Virginia 36.0% of whites are obese compared to 44.7% of African Americans. For the District of Columbia, the racial disparity was even more pronounced, 9.7% of whites and 35.5% of African Americans suffered obesity (CDC 2017a). It is similar for diabetes. In Virginia 9.6% of whites and 13.3% of blacks suffer from diabetes, with hospitalization rates being three times higher for blacks than whites (Diabetes in Virginia 2018).

Gender informed the Chesapeake food system long before European contact. Native American society cast women as the cultivators of domesticated crops, gatherers of natural plant foods, and cooks. Men hunted and fished. European colonists brought their own system of gender-based labor differentiation. Notably, men joined the agricultural labor force and controlled the activities on most farms. The expanding exploitation of marine resources in the nineteenth and twentieth centuries meant more men working on the water and women dominating the hand labor force processing their catch of crabs, oysters, and fish. Gender interacted with race as most of the crab picking and oyster shucking jobs went mostly to women of color, first black and later brown.

The Food-Ecosystem-Health-Problem (FEHP) Think of this book as a second voyage of exploration some four centuries since Capt. John Smith's. This voyage searches for understanding of how our food system transformed the healthy Chesapeake of Smith's time to a stricken ecosystem that struggles to breathe every summer. And we set sail to learn how our food system sickens the people too many of whom also struggle to catch their breath. Our destination is more than explaining what went wrong. We finish the voyage with a vision for the future when a new sustainable food system restores the health of the Chesapeake and its People.

The health of the Earth and humanity are suffering from the way we eat. Scientists began to understand this in the 1960s and now can quantify such with enough detail to point to a common cure for people and the planet. The only way to reverse the epidemic of the diseases of affluence plaguing the wealthier portion of the population, provide enough to eat for everyone, and simultaneously rebuild the integrity of the Earth's ecosystems is to change our food system. This means altering what we eat and how we produce our food. In short, we need to switch to a diet based upon whole plant foods, and a food system that produces them with sustainable organic techniques. This is a global problem and other books and papers have addressed it at that level (Tilman and Clark 2014). This volume uses the Chesapeake Bay ecosystem, its people, and their food to tell the story. We will look back at the history of the Bay to understand how our food system came to sicken people and the ecosystem. We will look forward to examine the common solution to restore the health of the Bay and its people.

Fig. 5 African American women dominated the crab picking workforce of the Chesapeake Bay, as shown in this 1940 image from Crisfield, MD, Maryland State Archives, https://www.google.com/search?q=Black+crab+pickers&tbm=isch&source=iu&ictx=1&fir=C8X3-b8YnoncrM%253A%252CTc-PNlKKlMIFsM%252C_&usg=__TQpoBYD-rYsxzMxYrZ0Rk7sSxbc%3D&sa=X&ved=0ahUKEwi4pI7wuY3aAhUN62MKHeTEA9cQ9QEISzAF#imgrc=C8X3-b8YnoncrM:

This book uses the Chesapeake Bay ecosystem and its human society to reveal universal truths and solutions applicable around the globe. Many other ecosystems could be treated as we will the Chesapeake Bay. However, the Chesapeake offers several advantages for exploring the Food Ecosystem Health Problem (FEHP). The Bay is a large and open ecosystem informed by an extensive watershed with a long history of agriculture. Its waters fed the earliest Native Americans and continue to support extensive commercial and recreational fisheries. The Chesapeake communicates directly with the Atlantic Ocean. Many marine species split their lives between time spent in the Bay and the ocean. Nutrients and toxins freely flow in and out of the Bay. Its shores provided the first permanent English New World settlement in 1607, and a decade later the landing of the first enslaved Africans. The Bay and its extensive network of tributaries were the highways of commerce for its original inhabitants and the colonists who displaced them. Although today paved roads and trains transport most of the food around the states bordering the Chesapeake, the Bay still serves as the port of entrance and exit for international trade in agriculture. The Chesapeake Bay ecosystem encompasses most of the important themes around the development of the FEHP and offers an excellent venue for implementing sustainable solutions.

Changing Worldviews Our voyage of exploration approaches the Food Environment Health Problem and the Chesapeake Bay with a world view shaped by the environmental movement that emerged in the last half of the twentieth century. But first let's understand the world view that characterized the development of modern civilization and brought us to the FEHP. In the broadest terms, European colonization, the Industrial Revolution and human-centric religious philosophy drove western society to view nature as a body of resources waiting for exploitation by men. This human-centered worldview is a masculine approach that says man is above and separate from nature and it is his job to conquer nature so it can best serve his needs. For those in power this worldview worked well enough in the short term until human populations grew, resources became scarce, and ecosystems were degraded by habitat destruction and pollution. For many of those not in power, this approach failed them immediately, losing access to resources rendered artificially scarce by colonizers and landlords who monopolized the common wealth.

In response to the obvious environmental destruction and resource scarcity wrought by the human-centered worldview, this philosophy evolved to include the beliefs that free-markets, continued economic growth, technological advances, and effective stewardship would solve the problems. For example, consider pesticides. Industrialization of agriculture created the perfect conditions for the eruption of populations of insects that would consume the crops before farmers could harvest. A free-market and technological solution was found in the application of toxic chemicals to kill the insects. This attempt to conquer nature did yield great profits for the chemical industry but it burdened the farmer and environment with a host of unintended consequences such as poisoning ground and surface waters, destroying predators that could control the pests, killing pollinators, devastating bird populations, sickening workers, tainting the crop, and producing super-races of resistant insects to eat our food. Societies guilty of violating laws of nature eventually pay stiff penalties for their transgressions.

The environmental wisdom worldview (Miller and Spoolman 2017) that informs this book stands in stark contrast to the human centered philosophy. The environmental wisdom worldview is built on these principles: We need to understand nature; we are a part of nature, not apart from it; we are subject to nature's laws; human economies are subsystems of the earth's life support systems; earth's natural capitol sustains us; we need to work with nature rather than conquer it; caring for the earth is caring for ourselves; we are responsible for leaving the earth in better condition for future generations. Although difficult to find in actual practice, the environmental wisdom worldview is at the core of the Earth Charter adopted by the United Nations in the year 2000 (Earth Charter 2020).

The Pioneering Women Navigators of the Food Environment Health Problem Three women shaped the course for understanding the FEHP: Rachel Carson, Frances Moore Lappé, and Vandama Shiva. While many other authors, scientists, and journalists contributed vastly to recognizing and offering solutions to the FEHP, Carson, Lappé and Shiva each navigated their ships of inquiry into uncharted waters. They are the founding mothers of FEHP understanding. Considering the male dominance of science during their era of emergence (1950s–1980s), it is no accident or coincidence that each of these pioneers were women. They brought a feminine perspective to understanding a problem emerging from a western-masculine approach to nature. They each shared the environmental wisdom worldview that said humans are part of nature, and that our survival depends upon our ability to work in harmony with the principles that govern the natural world. Dr. Shiva argues that this perception emerges directly from a feminine understanding of the world (Mies and Shiva 2014).

Let's review the remarkable contributions to understanding the FEHP made by these three women. Rachel Carson (1907–1964) did much of her science training on the Chesapeake and wrote about the Bay for the Baltimore Sun newspaper. She joined the U.S. Fish and Wildlife Service in 1937 and wrote two best selling books in the 1950s, *The Sea Around Us* and *The Edge of The Sea*. In 1962, Carson published the book that began the environmental movement, *Silent Spring*. She stood up to the powerful chemical/agricultural industry, documenting the consequences of pesticide use on the environment and human health. Carson also wrote about pesticide applications for purposes other than foodcrops, such as control of insects as vectors of disease and those that attack fiber producing plants, but the food system was central to her book. Anyone reading the Silent Spring today will be struck by how much she and science had figured out about the consequences of pesticides nearly six decades ago. Chapter seven of Silent Spring begins, "As man proceeds toward his announced goal of the conquest of nature, he has written a depressing record of destruction, directed not only against the earth he inhabits, but against the life that share it with him (Carson 2002)." While Silent Spring focused on pesticides, and naturally evoked the wrath of the chemical industry, it also challenged the entire human centered worldview, threatening the profits of the myriad of industries in the business of conquering nature. Sadly, Carson died 2 years after publication of Silent Spring. She succumbed to breast cancer at the age of 56. It is ironic that while suffering from the disease, she wrote a book that exposed the link between cancer and pesticides.

While Rachel Carson is best remembered for exposing the toxic consequences of industrial agriculture, she also made a fundamental observation that the real crisis of agriculture in the US is overproduction of food. She said, in essence, that the need for pesticides to sustain sufficient agricultural production is a myth. After all, the basic law of supply and demand says that overproduction will saturate a market and reduce prices, reducing profits for the farmers. Overproduction also destroys fisheries, as will be addressed later.

A decade after *Silent Spring* revolutionized thinking about our food system, Frances Moore Lappé (1944–) published *Diet for a Small Planet* in 1971 (Lappé 2011). While Silent Spring was the capstone of Carson's career, *Diet for a Small Plant* launched the 26 year old Lappe' on her path. As with Carson, Lappé matriculated in a man's world. After college she worked as a community organizer for the Welfare Rights Organization in Philadelphia. That prompted a try at a graduate degree in Social Work at the University of California, Berkeley. Unsure of her direction in life, she left that program and on her own without the benefit (or inhibition) of formal guidance she studied food and world hunger. To her surprise she discovered that the world was producing enough food to feed its population. The real issues were the distribution of that food and the kind of food produced. Hours spent pouring over agricultural records in the UC Berkeley library and doing calculations with a slide rule documented the myth of underproduction of food as the cause of hungry people. This was consistent with Carson's observation of overproduction in the US. Lappé revealed several other fundamental truths. One was that the production of meat is inherently inefficient at meeting the nutritional needs of people, wastes water, and destroys ecosystems. She also exploded the myth that meat consumption was needed to provide the world with sufficient protein and showed that a balanced whole food plant- based diet would do so without the environmental and economic damage wrought by the consumption of animals. Lappé also addressed the issue of subsidies used to sustain meat production. These included, money, energy, and artificial fertilizers.

Diet for a Small Planet gained popularity as much for its recipes as for its critique of the food-system. This innovation to tie analytical science to how individuals could address the FEHP with instructions for cooking sustainable meals was revolutionary, empowering citizens to make better decisions for their health and the environment. As with Carson, Lappé challenged the dominant worldview threatening the profits of corporations vested in overproduction to feed animals rather than people. Lappé continues to pursue and expand upon the ideas in *Diet for a Small Planet*, through publication, speaking and developing organizations.

Dr. Vandana Shiva (1952–) advanced the understanding of the FEHP by tying together traditional ecological and farming knowledge in her native India with the teachings of Carson and Lappé. She earned a degree in nuclear physics, but soon realized that her energies were better spent addressing the human relationship with the environment. Beginning with work protecting forests (she was one of the original "tree-huggers," physically defending trees from harvest) she moved on to the global food system with a focus on her native India. Publishing 24 books since 1981, Shiva became one of the leading environmentalists on the world stage. Much of her work addresses the consequences of the so called "Green Revolution" that transformed India's food system from one steeped in sustainable, time-tested and culturally relevant practice, to a "modern" approach that relied on massive inputs of capitol, fossil fuels, artificial fertilizers, ground water, and inappropriate super-producing breeds of plants (Shiva 1991). Shiva demonstrated the fallacy of using yield per unit of area of a particular crop as the measure of successful agriculture.

Shiva campaigns for protecting biodiversity and the right of farmers to save their seeds in the face of corporate efforts to patent life. This includes thwarting the introduction of GMO's (Genetically Modified Organisms) into the world food system. Dr. Shiva established the Navdanya demonstration farm in India that uses biodiversity and traditional organic methods to outperform the industrial/chemical farms of the "Green Revolution." Shiva also advocates that the world consists of various commons, such as forest, field, water and air. She says no one should own the commons nor have the right to pollute the commons. This is contrary to the cap and trade schemes used to control pollution by selling the right to dirty the common.

The contributions of Carson, Lappé and Shiva are legion and would consume an entire book and then some to explore. Less famous than this trio is another women scientist who must be recognized for her work on the Chesapeake Bay. Dr. Grace Brush (1932–) earned a Ph.D. in Biology from Harvard University in 1956. In 1969, she joined the faculty of Johns Hopkins University (where Rachel Carson earned her MS degree years before). Dr. Brush produced 61 scientific papers and five books on the ecology of the Chesapeake Bay. She used pollen, seeds, microfossils of algae, and mineral particles deposited in the sediments of the Bay to decipher the paleoecology of the region. Brush uncovered the information buried in the sediments to reconstruct the changing ecology of the Bay over thousands of years (Brush 2017). Her studies transport us back in time well before the observations of Capt. John Smith. This work is critical to understanding how the FEHP developed in the Chesapeake. She also published extensively on the nitrogen cycle and soil processes related to agriculture and land use, topics to be visited in later chapters. The story of this book couldn't be told without Dr. Brush's contributions.

Expanding the Human Health Component *Silent Spring* and *Diet for a Small Planet* each spoke about the human health component of the FEHP in their own important but limited way. The first focused on the toxicity of pesticides to workers and consumers. The second stressed food sufficiency to prevent malnutrition. In 1987, John Robbins (1947–) published "Diet for a New America (Robbins 2012)." Its title is clearly homage to Lapee's book. Robbins expanded the discussion to include the role of animal-based diets in promoting diseases of civilization and the unethical treatment of animals by the food-system. This work showed the direct connections between animal- based diets and the main maladies plaguing modern civilization such as heart disease, cancer, diabetes, and obesity. Robbins also forces his readers to open their minds to the suffering of animals caused by the meat industry. This is a less obvious, but a real part of the health component of the FEHP. For anyone familiar with the suffering that goes into the production of meat, yet who still consumes it, must either suppress emotions of empathy or cope through denial. Such denial allows people to continue to make food choices which are bad for their health. Robbins followed-up *Diet for a New America* with *The Food Revolution* (Robbins 2010). The second book expanded the themes of the first and importantly included a new focus on the benefits of eating plants in *preventing* and *reversing* disease. He essentially added information about food as medicine.

Robbins continues to publish on the FEHP. Note that Robbins is the son of Irv Robbins, the co-founder of Baskin and Robbins, the world's largest ice cream company. He walked away from his father's business, rejecting a life of financial wealth to pursue one of promoting human and world health.

In 2005, Dr. T. Colin Campbell published the *China Study* and updated it in 2016 (Campbell and Campbell 2016). This book uses epidemiological, clinical, and laboratory- based evidence to demonstrate the links between the SAD (Standard American Diet) and the diseases of affluence including, heart disease, cancer, stroke, Alzheimer's, obesity and diabetes. His work demonstrates that the proper diet both reduces the risk of getting cancer and can reverse the disease. In a radical departure from dogma, Campbell showed that certain animal foods are unto themselves powerful carcinogens and often more so than many known synthetic carcinogens. Campbell grew-up on a dairy farm and began his academic career focused on increasing meat production. His scientific journey included pioneering studies that brought him to being a proponent for the WFPBD (Whole Food Plant Based Diet). His Center for Nutrition Studies (https://nutritionstudies.org/) provides WFPBD recipes, articles, and a portal to a formal web-based course on whole plant nutrition through Cornell University.

Michael Greger, MD built a career in lifestyle medicine focused on the prevention and cure of disease through science based dietary decisions. He published *How Not to Die* (Greger 2015) to inform the public about the most current nutrition science and its relationship to health. He is an active proponent of WFPBD and provides the public with direct access to the scientific literature through his website Nutritionfacts.org.

Dr. John McDougall authored several books making a compelling argument for adhering to grains, tubers and other starches as the center of a WFPBD, and promotes this thesis on his website (https://www.drmcdougall.com/). While Robbins, Campbell, Greger, and McDougall are some of the best known advocates for WFPBD, certainly others have also helped shape the course to better human health though this approach to eating.

Sailing to a Better Destination It is easy to get overwhelmed and depressed about the state of our environment and our health. Yet one's everyday choices can turn despair into optimism and empowerment. Dietary choices can improve the environment and one's health. People cast a vote for the health of the planet and themselves with each dollar they spend to eat. According to a 2017 report from a food-industry research group, 6% of the US population describes itself as vegan (eating only plants and fungi). This is up from just 0.5% in 2009 (Report Buyer 2017; Veg. Times 2018). The US Department of Agriculture reports that the sale of organically produced food increased from $13 billion in 2005 to $34 billion in 2015 (ERSUSDA 2017). More people are also growing their own food in the US. Home food gardening increased from 36 million households in 2008 to 42 million in 2013. Most encouraging is that younger people (age 18–34) increased participation in gardening 63% (Garden.org 2014). Chapters "A New Food System for the Chesapeake Bay Region and a Changing Climate, An Organic-Based Food System: A Voyage Back and Forward in Time, and Ethics and Economics of Building a Food System to Recover the Health of the Chesapeake Bay and Its People" explore these themes.

Each of Capt. John Smith's two voyages of discovery in the summer of 1608 consisted of many connected smaller expeditions that explored distinct tributary rivers flowing into the Chesapeake Bay. The organization of this book takes a similar approach, each chapter exploring a river of knowledge that flows into a greater understanding of the Chesapeake Bay and its people in the context of the food system.

Part I consists of three chapters. The introduction is followed by Chapter "The Bay and Its Watershed: A Voyage Back in Time" which explores the geological and ecological history of the Chesapeake Bay ecosystem. This will aid in understanding how food systems have transformed the ecosystem. Chapter "Scientific Concepts for Understanding the Health of the Chesapeake Bay and Its People" summarizes the basic science needed to understand the various phenomena that govern the functioning of the Chesapeake Bay ecosystem and the humans that reside there.

Part II explores ports of call in the food system, beginning with Chapter "The Algonquin Food System and How it Shaped the Ecosystem and Interactions with the English Colonists of the Chesapeake Bay", that covers the Native American ways of gathering, farming, hunting and fishing. Chapter "A Fishing Trip: Exploiting and Managing the Commons of the Chesapeake Bay" examines the science and economics of fishing and its impact on the Bay. Chapter "Menhaden, the Inedible Fish that Most Everyone Eats" explores the reduction industry that turns Atlantic menhaden into animal feed and dietary supplements. Chapter "Blue Crabs: Beautiful Savory Swimmers of the Chesapeake Bay" looks at the blue crab fishery and its role in the culture of the Bay. Chapter "The Chesapeake Bay Oyster: Cobblestone to Keystone" tells the story of the demise and possible recovery of the oyster fishery. Chapter "Passenger Pigeon and Waterfowl: Flights to Extinction and Not" examines how industrialized hunting drove to extinction the most common bird in North America and how other species of the Bay managed to escape that fate.

Part III is a terrestrial journey into the landbased food system of the Bay. Chapter "The Journey from Peruvian Guano to Artificial Fertilizer Ends with too Much Nitrogen in the Chesapeake Bay" examines fertilizers and soil fertility with focus on nitrogen. Chapter "Pesticides Bring the War on Nature to The Chesapeake Bay" recounts the consequences of pesticide use in the Chesapeake Bay Region. Chapter "Livestock and Poultry: The Other Colonists Who Changed the Food System of the Chesapeake Bay" appraises the livestock and poultry industries and their impact on the Bay and its people.

Part IV is a cruise through the health consequences of the Standard American Diet for the ecosystem and the people that live there. Chapter "Instead of Eating Fish: The Health Consequences of Eating Seafood from the Chesapeake Bay Compared to Other Choices" examines the health consequences associated with seafood. Chapter "Sugar Twice Enslaves: Consequences for the People of the Chesapeake" looks at the special history and role of refined plant sugar in shaping our diet and its consequences. Chapter "Eutrophication: Obesity of the Bay and Its People" examines the eutrophication of the Bay in analogy with the miss-feeding of the Bay's people. Chapter "Finishing the Journey: Wastewater as a Misplaced Resource" addresses sewage as a part of the food system and the wastewater industry that developed to protect the environment and human health. Chapter "Plastic Pollution and the Chesapeake Bay: The Food System and Beyond" investigates the role of the food system in creating pollution from plastics. Chapter "A New Food System for The Chesapeake Bay Region and a Changing Climate" explores the role of the food system in climate change and offers an alternative diet as a way to address the problem.

Part V looks to the future. Chapter "An Organic-Based Food System: A Voyage Back and Forward in Time" examines the rediscovery of organic farming in the Bay's watershed. Chapter "What Nature, Politics and Policy Demand of the Chesapeake Bay and Its Food System" delves into the policies developed to restore the health of the Bay. Chapter "Ethics and Economics of Building a Food System to Recover the Health of the Chesapeake Bay and Its People" looks at the ethics and economics of building a new sustainable food system for the Bay.

References

Blanton D (2000) Drought as a factor in the Jamestown colony, 1607–1612. Hist Archaeol 34:74–81. https://doi.org/10.1007/bf03374329

Boxall A (2004) The environmental side effects of medication. EMBO Rep 5(12):1110–1116. https://doi.org/10.1038/sj.embor.7400307

Brush G (2017) Decoding the deep sediments: the ecological history of the Chesapeake Bay. Sea Grant, College Park, MY

Campbell T, Campbell T (2016) The China study. BenBella Books, Dallas, TX

Carson R (1962/2002) Silent spring. Mariner Books (Houghton Mifflin), London

CDC (2017a) FastStats. National Center for Health Statistics. Centers for Disease Control and Prevention, Obesity and overweight [online]. Available at: https://www.cdc.gov/nchs/fastats/obesity-overweight.htm. Accessed 27 Mar 2018

CDC (2017b) FastStats. National Center for Health Statistics. Centers for Disease Control and Prevention. Therapeutic drug use [online]. Available at: www.cdc.gov/nchs/fastats/drug-use-therapeutic.htm. Accessed 1 Mar 2018

Chesapeake Bay Program (2018) Facts & figures. Chesapeake Bay program [online]. Available at: https://www.chesapeakebay.net/discover/facts. Accessed 20 Mar 2018

Cotton.org (2018) The story of cotton- history of cotton [online]. Available at: https://www.cotton.org/pubs/cottoncounts/story/. Accessed 27 Mar 2018

Couranz K (2018) Oysters. Blue water media [online]. Chesapeakebay.noaa.gov. Available at: https://chesapeakebay.noaa.gov/index.php?option=com_content&view=article&id=201&Itemid=200. Accessed 27 Mar 2018. Oysters, Chesapeake Bay Program 2018. https://chesapeakebay.noaa.gov/index.php?option=com_content&view=article&id=201&Itemid=200. Accessed 12 Mar 2018

Diabetes in Virginia (2018). http://www.vdh.virginia.gov/content/uploads/sites/75/2016/12/Diabetes-Burden-Report.pdf. Accessed 18 Mar 2018

Earth Charter (2020) Read the Earth Charter - Earth Charter. In: Earth Charter. https://earthcharter.org/read-the-earth-charter/. Accessed 15 May 2020

ERSUSDA (2017) USDA ERS – Organic market overview [online]. Available at: https://www.ers.usda.gov/topics/natural-resources-environment/organic-agriculture/organic-market-overview.aspx. Accessed 27 Mar 2018

Foley J and Foley J (2013) It's time to rethink America's corn system [online]. Scientific American. Available at: https://www.scientificamerican.com/article/time-to-rethink-corn/. Accessed 27 Mar 2018

Garden.org (2014) National gardening association special report: Garden to table. A -5 year look at food gardening in America [online]. Available at: https://garden.org/special/pdf/2014-NGA-Garden-to-Table.pdf. Accessed 27 Mar 2018

Greger M (2015) How not to die: discover the foods scientifically proven to prevent and reverse disease. Flatiron Books, New York

Hagy J, Boynton W, Keefe C et al (2004) Hypoxia in Chesapeake Bay, 1950–2001: Long-term change in relation to nutrient loading and river flow. Estuaries 27:634–658

Kupperman K (1979) Apathy and death in early Jamestown. J Am Hist 66:24–40

Lappé F (2011) Diet for a small planet. Random House Publishing Group, New York

Lefcheck J, Orth R, Dennison W, Wilcox D, Murphy R, Keisman J, Gurbisz C, Hannam M, Landry J, Moore K, Patrick C, Testa J, Weller D, Batiuk R (2018) Long-term nutrient reductions lead to the unprecedented recovery of a temperate coastal region. Proc Natl Acad Sci 115(14):3658–3662

MDE (2014) Guidelines for recreationally caught fish species in Maryland http://mde.maryland.gov/programs/Marylander/fishandshellfish/Documents/Fish%20Consumption%20Docs/Maryland_Fish_Advisories_2014_March17.pdf. Accessed 12 Mar 2018

Mies M, Shiva V (2014) Ecofeminism. Zed Books, London

Milesia C, Elvidgeb C, Dietzc J, et al (2005) A strategy for mapping and modeling the ecological effects of US lawns. http://www.isprs.org/proceedings/XXXVI/8-W27/milesi.pdf Accessed 27 Mar 2018

Miller G, Spoolman S (2017) Living in the environment, 19th edn. National Geographic Learning/Cengage, Boston, MA, pp 683–686

NPS (2018) Pocahontas: her life and legend – Historic Jamestowne Part of Colonial National Historical Park (U.S. National Park Service) [online]. Available at: https://www.nps.gov/jame/learn/historyculture/pocahontas-her-life-and-legend.htm. Accessed 27 Mar 2018

Phillips, S (2018) USGS Chesapeake Bay activities toxic contaminants [online]. Chesapeake.usgs.gov. Available at: https://chesapeake.usgs.gov/toxiccontaminants.html. Accessed 27 Mar 2018

Report Buyer (2017) Top Trends in Prepared Foods 2017: Exploring trends in meat, fish and seafood; pasta, noodles and rice; prepared meals; savory deli food; soup; and meat substitutes. In: Reportbuyer.com. https://www.reportbuyer.com/product/4959853/top-trends-in-prepared-foods-2017-exploring-trends-in-meat-fish-and-seafood-pasta-noodles-and-rice-prepared-meals-savory-deli-food-soup-and-meat-substitutes.html. Accessed 15 May 2020

Richards R, Rago P (1999) A case history of effective fishery management: Chesapeake Bay striped bass. N Am J Fish Manag 19 (2):356–375

Robbins J (2010) The food revolution. Conari Press, San Francisco, CA

Robbins J (2012) Diet for a new America. H J Kramer, Tiburon, CA

Rountree H, Clark W, Mountford K et al (2007) John Smith's Chesapeake voyages, 1607–1609. University of Virginia Press, Charlottesville, VA

Shiva V (1991) The violence of the green revolution: third world agriculture, ecology and politics. Zed Books, London

Smith J (1608) A true relation of such occurrences and accidents of noate as hath hapned in Virginia since the first planting of that collony, which is now resident in the south part thereof, till the last returne from thence written by Captaine Smith [Cor]one[ll] of the said collony, to a worshipfull friend of his in England. London: Printed for Iohn Tappe, and are to bee solde at the Greyhound in Paules-Church-yard, by W.W. In The complete works of Captain John Smith (1580–1631), edited by Philip L. Barbour, (1986) 3 vols., 1:3–118. University of North Carolina Press, Chapel Hill

Smith J (1612) A map of Virginia. Joseph Barnes, Oxford, pp 1–3. In The complete works of Captain John Smith (1580–1631), edited by Philip L. Barbour, (1986) 3 vols., 1:119–190, University of North Carolina Press, Chapel Hill. Available in Capt. John Smith's complete works. http://www.virtualjamestown.org/exist/cocoon/jamestown/fha-js/SmiWorks1

Stromberg J (2018) Starving settlers in Jamestown Colony resorted to cannibalism [online]. Smithsonian. Available at: https://www.smithsonianmag.com/history/starving-settlers-in-jamestown-colony-resorted-to-cannibalism-46000815/. Accessed 27 Mar 2018

Tilman D, Clark M (2014) Global diets link environmental sustainability and human health. Nature 515(7528):518–522

U.S. Environmental Protection Agency, U.S. Geological Survey, and U.S. Fish and Wildlife Service (2012) Toxic contaminants in the Chesapeake Bay and its watershed. Extent and severity of occurrence and potential biological effects. Technical report. December, 2012: Annapolis, Md., U.S. Environmental Protection Agency Chesapeake Bay Program Office, 175 p [online]. Available at: https://federalleadership.chesapeakebay.net/ChesBayToxics_finaldraft_11513b.pdf. Accessed 27 Mar 2018

USDA.mannlib.cornell.edu (2017) [online]. Available at: http://usda.mannlib.cornell.edu/usda/current/Acre/Acre-06-30-2017.pdf. Accessed 27 Mar 2018

Veg. Times (2018) Vegetarianism in America [online]. Vegetarian Times. Available at: https://www.vegetariantimes.com/uncategorized/vegetarianism-in-america. Accessed 27 Mar 2018

White S (2017) New worlds of climate change: the little ice age and the colonization of North America [online]. HistoricalClimatology.com. Available at: https://www.historicalclimatology.com/blog/new-worlds-of-climate-change-the-little-ice-age-and-the-colonization-of-north-america. Accessed 27 Mar 2018

The Bay and Its Watershed: A Voyage Back in Time

Benjamin E. Cuker

The ancient Susquehanna River once did flow directly to the sea.
Cycling climate and an extraterrestrial visitor made that to no longer be.

Now seven mighty rivers and many smaller others come together to form the Chesapeake Bay.
Mixed with salty flow from the Atlantic, these waters make an estuary that churns every night and day.

The moon and sun pull upon the water to create tides.
Upon these currents, nutrients, oxygen and life rides.

Yet people found the Bay as it was to be not enough.
So they built dams which meant for many fish, surviving would be tough.

Abstract

To understand the current state of the Chesapeake Bay, its people and food system, and to explore health restoring solutions, it is important to know something about the geological history and physical features of this magnificent ecosystem. This chapter examines how the Bay ecosystem was formed and its main physical features, including those reshaped by human activity. The history of the physical structure of the Chesapeake Bay is driven by geological and atmospheric processes that proceeded over a range of time scales going from thousands to millions of years. However, one seminal event that took place in a matter of minutes about 480 million years ago completely reshaped the Bay. A bolide (exploding meteor) left a crater that reconfigured the drainage of the three most southern tributary rivers and reshaped the mouth of the estuary. Thousands of years of cyclical climate change linked with rising and lowering sea levels transformed the ancient Susquehanna River bed into the modern Chesapeake Bay. This chapter calls on atmospheric science and evolutionary biology to help explain that process. The journey closes with a look at how humans reshaped the physical aspects of the Bay with dams and other features, often in response to the needs of their food systems.

Keywords

Bolide · Changing sealevels · Dams · Estuary formation

The Bay Arrives: Plate Tectonics, Rising Mountains and a Blast from the Past Understanding the physical structure of the Chesapeake Bay (Fig. 1) requires a look back through time. It is a story of slow geological and climatic processes, punctuated by an instantaneous cataclysm of extraterrestrial origin. Some 480 million years ago, continental plates were slowly sliding together on the way to forming the super continent called Pangea. The eastern margin of what is modern-day North America butted against the Iapetus ocean plate. The slow-motion collision of continental and ocean plates gave rise to the original Appalachian Mountain range. A zone of subduction drove the ocean plate under its continental neighbor, wrinkling the landscape with uplift and volcanoes. The resulting Appalachian range ran roughly from the north to the south between the modern era states of Maine and Georgia (Frank 2018). Rain and meltwaters filled rivers that flowed mostly perpendicular to the range, draining water and eroded minerals to the adjacent ocean. Yet some rivers existed before the Appalachian peaks reshaped the

B. E. Cuker (✉)
Department of Marine and Environmental Science, Hampton University, Hampton, Virginia, USA
e-mail: Benjamin.cuker@hamptonu.edu

B. E. Cuker (ed.), *Diet for a Sustainable Ecosystem*, Estuaries of the World, https://doi.org/10.1007/978-3-030-45481-4_2

Fig. 1 Landsat image of the Chesapeake Bay from 2009 U.S. Geological Survey, NASA. Public domain. https://prd-wret.s3-us-west-2.amazonaws.com/assets/palladium/production/s3fs-public/Chesapeakelandsat.jpg

landscape, persisting through the upheaval. The ancient Susquehanna River is one such flow that predated the rise of the Appalachian Mountains. It cut through the lifting landscape on its eastward path to the ocean. Draining an area that now includes parts of New York, Pennsylvania and Maryland, the ancient river of 300–500 thousand years ago flowed to the sea through a now-extinct channel located about midway up the present Delmarva Peninsula. The ancient Susquehanna River and its valley built the foundation for the modern Chesapeake Bay. It did so in the context of dynamic cycling of global climate.

The earth periodically warms and cools, shifting from one stable climatic regime to the other (Graham 2000). During cool periods, much of the precipitation that falls on land remains frozen for thousands of years. Since most of that precipitation originated as water evaporated from the ocean, this causes sea level to drop. Cooler temperature of the ocean also causes the water to contract, further reducing sea level. Conversely during warm periods, as is the case today, water locked up in glaciers and frozen soils melts, returning to the ocean. This transfer of water back to the ocean combined with thermal expansion of the warmer seawater raises sea levels.

Such climatic cycling helped shape the Chesapeake Bay. During periods of high sea level, the sediments eroded from the Appalachian Mountains were transported to the coastal ocean and deposited by long-shore currents along the eastern seaboard. During the cold ice ages, sea level dropped and exposed the former marine deposits, extending the Delmarva Peninsula southward. Each subsequent cycle of sea level rise and drop caused the ancient Susquehanna to cut a new and more southerly channel to the sea (Fig. 2). During this migration, the Susquehanna cut across the previously independent Potomac and Rappahannock River channels, diverting their flows into its own. Climate cycling proceeded over the last 500 thousand years to produce an exit channel for the Susquehanna River at

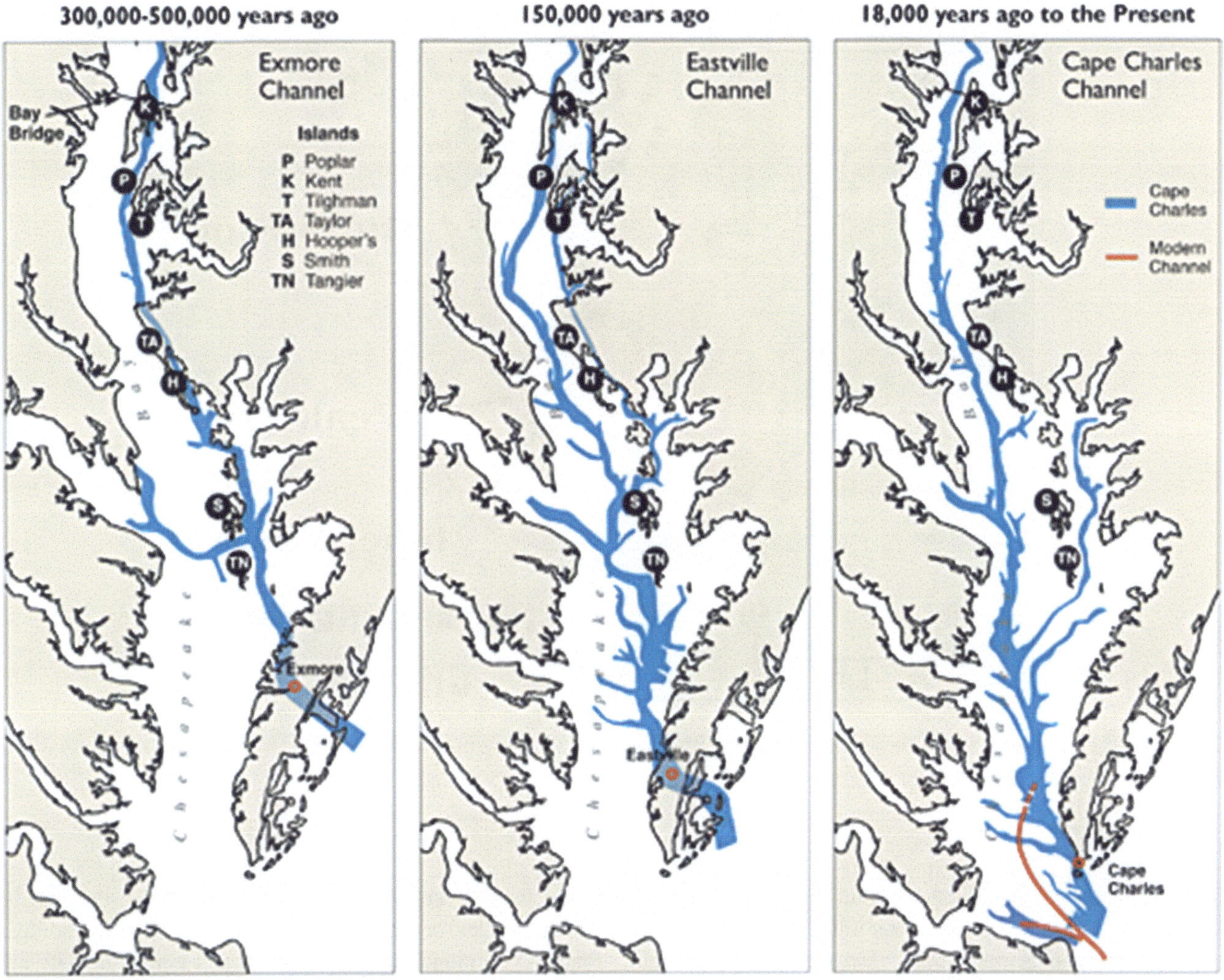

Fig. 2 Movement of the mouth of the Chesapeake Bay southward with the cycling between warm high-sea level periods and cold, low-sea level periods. https://www.google.com/search?tbm=isch&q=formation+of+chesapeake+bay&chips=q:formation+of+chesapeake+bayonline_chips:ancient&sa=X&ved=0ahUKEwjXgurElJLaAhVY12MKHQsdDBgQ4lYILigG&biw=950&bih=593&dpr=1.25#imgrc=EdcuNF-bZkkHMM. Used by permission from Maryland Sea Grant

the present Cape Charles (Rountree et al. 2008; Virginiaplaces 2018). This is the current northern cape of the Chesapeake Bay, where it meets the Atlantic Ocean.

The last ice age peaked around 21,000 years ago, with thick glaciers covering much of the northern watershed of the Susquehanna River, as was true of much of the world at mid-latitudes and higher. Sea level was 125 m (480 ft) lower than today (USGS 2018). As noted above, water freed from the melting glaciers flowed to the ocean, raising sea levels. The rising sea flooded river mouths and valleys around the world, creating new structures called estuaries. The present Chesapeake Bay estuary is the flooded ancient Susquehanna River valley. A nautical chart of the Bay reveals the outlines of the old river as a deep channel running the length of the Bay. Red and green navigation buoys demark the drowned river channel for deep-draft ships transiting the Bay.

Unlike the slow collision of continental plates or melting of glaciers, a sudden, shattering, heavenly event 36 million years ago also determined the shape of the present Chesapeake Bay. A bolide (meteor that explodes in the atmosphere) came hurling to earth from deep space (Poag et al. 2004). It hit hard on a span of continental shelf under the coastal sea of that warm period. Although submerged at the time, the margins of the zone of impact now include important cities of the southern Chesapeake Bay, including Newport News, Hampton and Norfolk (Fig. 3). Since it hit during a very warm period, the actual land lay under about 100 m of coastal sea. The blanket of water did little to soften the blow. The impact left a crater 85 km in diameter and 1.3 km deep. Its area is twice that of the state of Rhode Island. The physical force and heat from the collision vaporized life for miles around. Whatever survived on land some distance from the event, was quickly enveloped in a wall of water

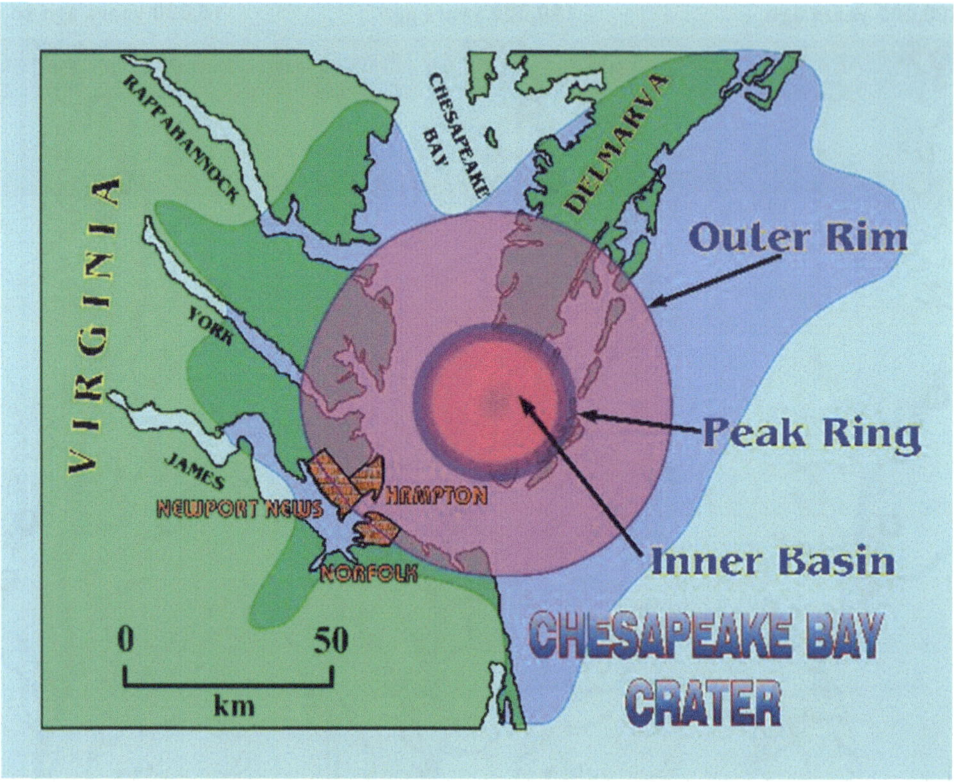

Fig. 3 Illustration of the impact carter formed by the bolide collision 36 million years ago, note how it cut across the mouths of several rivers, redirecting their drainage. https://commons.wikimedia.org/wiki/File:Chesapeake_Bay_impact_crater4.jpg

from a tsunami that washed the foothills of the Appalachian Mountains. The under-water crater filled first with debris from the collapsing crater walls, and then over millennia with sediment deposited from land and sea. So thoroughly did the sediments fill the crater, that it remained unknown until discovered in the early 1980s (Poag et al. 2004). Geologists cued by a variety of findings began to look for evidence of a large impact crater. They used seismic reflection profiles, drilling cores from petroleum prospectors, and other geological measures to piece together the story of the crater. Although the bolide event remained obscure until recently, it left two signatures evident today. Most obvious is the joining of the York and James Rivers to the Chesapeake Bay estuary. These rivers turn near their mouths to flow into the depression left from the ancient crater. The bolide added these two rivers to the Bay ecosystem. The second long-lasting consequence of the impact is the gradual but continuous subsidence of the land overlying the ancient impact crater. Time and pressure works to compress the sediments and debris, resulting in sinking lands. Coupled with rising sea level, this subsidence makes the southern shores of the Chesapeake Bay susceptible to regular flooding from storms and extreme tides.

From Icey River to Estuary As noted above, the earth's climate cycles between periods of warming and cooling. Planetary motion, geology, biology, and human activity all play parts in the climate cycle drama. The Milinkovic Theory attributes changes in the average temperature of the earth to the way it spins and orbits the sun. This includes changes in the relationship between the earth's angle of rotation to its orbital path around the sun, how elliptical or circular the orbit is, and wobbling of the axis of rotation. The earth cycles through periods when these motions coincide to result in a hotter planet followed by a period of cooling (Graham 2000). Geology also contributes to climate cycling. Rising and eroding mountains and volcanic activity alter concentrations of greenhouse gases, such as carbon dioxide and methane. Increasing concentrations of these gases raise the average temperature of the planet. In the distant past, episodic volcanic activity spewed vast quantities of ash, carbon dioxide, and methane into the atmosphere. The ash tended to cool the planet, blocking some of the sunlight. But the gases persisted longer than the heavier ash particles resulting in net warming of the planet. And it wasn't just volcanic peaks that changed the climate. Rain weathered the calcium carbonate-rich mountains of uplifted sedimentary limestone rock, causing it to dissolve and release carbon dioxide to the atmosphere (Berardelli 2009).

Biology is the oft neglected actor in the climate change drama. The earth is about 5 billion years old, and life on it first arose about 1.2 billion years after the planet formed. So life has characterized most of earth's existence. Early life consisting of singlecelled organisms hummed along for the first 500 million years, evolving greater complexity along the way. All of those first lifeforms derived their energy from the chemicals in the environment or from consuming other cells. At the half-billion-year anniversary of life something new occurred, some organisms evolved the ability to obtain energy from the sun in the new process of photosynthesis (covered in more detail in Chapter "Scientific Concepts for Understanding the Health of the Chesapeake Bay and Its People"). Able to tap a hitherto unavailable source of energy meant that photosynthetic cells could proliferate. Free oxygen (O_2) is a key waste product of photosynthesis and it slowly entered the primordial atmosphere. There is much debate about the composition of earth's early atmosphere, but it probably included methane, ammonia, carbon monoxide, hydrogen gas, and water vapor (Zahnle et al. 2010). Free oxygen from photosynthesis changed all that as it readily combined with those gases to produce our current atmosphere of 78% dinitrogen (N_2), 21% oxygen (O_2), 0.04% carbon dioxide (CO_2), water vapor, and a host of lesser volatile compounds.

In addition to completely changing the atmosphere, photosynthesis performed another spectacular feat. Over the eons, it translocated massive amounts of carbon from the atmosphere, ocean and rocks to buried stores of fossil fuels. The point of photosynthesis for the microbes and the green plants that were their descendants was not to produce oxygen, but to use the energy from sunlight to fix atmospheric carbon (CO_2) into biomolecules used for growth and energy storage. When these cells died they decomposed and returned much of that fixed carbon to the atmosphere, mostly by the action of bacteria and other microbes. Yet depending on local conditions, some of the carbon stored in plant biomolecules wasn't retuned to the atmosphere and instead was buried under sediments. Time, heat, and pressure transformed this interred residue of life into the familiar fossil fuels of petroleum, natural gas, and coal as well as lesser recognized substances such as bitumen, tar, and oil shales.

The problem faced by the Chesapeake Bay today and ecosystems worldwide, is that in their search for useable forms of energy humans learned how to quickly reverse what millions of years of photosynthesis and burial achieved. Presently the earth is warming at an unprecedented rate due to the release of greenhouse gases from industrialization of human life and agriculture (IPCC 2018a; USEPA 2018). Most of the problem gasses come from burning the fossil fuels, but others emanate from animal husbandry and abuse of soils (explored in Chapters "The Journey from Peruvian Guano to Artificial Fertilizer Ends with too Much Nitrogen in the Chesapeake Bay and Livestock and Poultry: The Other Colonists Who Changed the Food System of the Chesapeake Bay"). Humans have so altered the cycling of carbon over the last two centuries, driving such a major change in earth conditions, that scientists now recognize a whole new geological epoch, called the *Anthropocene*. As we shall see, today's Chesapeake Bay ecosystem illustrates the causes and consequences of the arrival of the Anthropocene.

Although the public is well aware of the connection between certain gases and global warming, many readers may not understand the mechanism involved. Let's explore the basic science. First consider that Earth and the moon are essentially the same distance from the Sun. As the moon orbits the Earth, it alternates between being somewhat closer and somewhat further away, but on average it is the same distance. Yet the Earth's average surface temperature is 15° C, much warmer than the -35° C for the moon (NASA 2018a; UCAR 2018). The difference in temperature is due to the robust atmosphere enveloping Earth compared to the gossamer layer of gases surrounding the moon. Why the difference in the atmospheres of the Earth and its moon? It is a matter of gravity. The earth is 81 times as massive as the moon (NASA 2018b). The greater mass of earth exerts a gravitational force sufficient to hold a thick blanket of gasses, while the lesser gravity of the moon can't do the same. Earth's atmosphere acts as the glass roof and walls of a garden greenhouse, retaining sufficient amounts of solar energy to make the planet habitable. As noted above, human activity since the Industrial Revolution has intensified the greenhouse effect, changing patterns of climate around the world.

What is in the air? The atmosphere is 78% nitrogen (N_2), and 21% oxygen (O_2). The other 1% consists of various gases found in trace amounts. Very important are the greenhouse gases of water vapor (variable in concentration, 10–50,000 ppb), carbon dioxide (CO_2, 0.04% or 400 ppm), methane (CH_4 1.8 ppm), nitrous oxide (N_2O, 2.8 ppm), and ozone (O_3, 5 ppm). There are other greenhouse gasses primarily of human origin and of lesser concentration. To make sense of the concentrations of the various greenhouse gasses let's review some basic units of concentration for liquids and gases (USEPA 2018). This review will also help the understanding of subsequent chapters addressing pollution of water, air, the food we eat, and our bodies.

Percentage (parts per hundred) is used when the fraction of the substance is relatively large. For matter in lower concentration, it is more convenient to use ppt (parts per thousand), ppm (parts per million), or ppb (parts per billion). Using ppm and ppb saves on writing lots of extra zeros when substances occur at very low percentages. Consider the present concentration of atmospheric CO_2. That is now in excess of 0.04%, 400 ppm, or 400,000 ppb.

The greenhouse gases alone don't warm the earth. They do so only in concert with the radiation received from the sun (Myhre and Shindell 2018). Sunlight consists primarily of shortwave radiation, visible light, and ultraviolet light. About half of the light reaching the earth is reflected at the top of the atmosphere which is why the planet is a glowing blue and white sphere when viewed from far away in space. The remainder penetrates the atmosphere. When it does, this shortwave radiation is intercepted by gases and particles in the atmosphere, the land, and the sea. When shortwave radiation is absorbed it changes to longer wave radiation called infrared. If you've worn a black shirt on a sunny day, you felt the heat generated as the dark cloth struck by shortwave light reradiates that energy at longer wavelengths. While the shortwave radiation easily penetrates the atmosphere, once it is converted to longer wave radiation it tends to be retained by the greenhouse gases. Consider a car with closed windows on a clear and cold winter day. The shortwave sunlight passes through the automobile's glass windows, being absorbed by the interior of the car and therefore being transformed to longer wave radiation. The windows block the escape of much of that longer wave radiation, resulting in significant warming. The greenhouse gases in the atmosphere act like the glass windows of the car.

Water vapor accounts for about half of the greenhouse effect, clouds 25%, CO_2 20%, the remaining 5% tracing to methane, nitrous oxide, ozone, and a few other lesser gases (IPCC 2018a). The simple concentration of the different greenhouse gasses tells only part of their story. Two other properties of the different gases also contribute to their importance as climate changers. First, the gasses differ in their ability to warm the planet. Second, the gasses spend differing amounts of time in the atmosphere. The two factors combine to determine their relative Global Warming Potential (GWP, IPCC 2018b). This system scores carbon dioxide (CO_2) as the standard, with a GWP of 1; methane (CH_4) at 28 and nitrous oxide (NO_2) at 228. The concentrations of CO_2 and CH_4 are 400 ppm and 1.8 ppm respectively, so there is 222 times more CO_2 than CH_4 in the atmosphere. Yet CH_4 is 100 times better at blocking the exit of longwave radiation than CO_2. However, CH_4 spends only about 12 years in the atmosphere, compared to 1000 years for CO_2. So the shorter atmospheric residence time of CH_4 mitigates its greater ability to retain heat over time. Nitrous oxide stays in the atmosphere 114 years, about ten times longer than methane and ten times shorter than carbon dioxide. While knowing that the greenhouse effect is important for understanding the development of the Chesapeake Bay over geological time, it is also critical to understanding how the foodsystem is contributing to climate change. As will be discussed later, elevated CO_2, CH_4, and NO_2 emissions are part of industrial agriculture.

During the cold Pleistocene (2.5 million to 11,700 years ago) ice sheets covered up to 30% of the land, forming glaciers 1500–3000 m (over a mile) thick. With all that water stored on land, sea level ran 100 m (330 feet) lower than today. At that time there was no Chesapeake Bay, only the ancient Susquehanna River laying in its floodplain. The warming that ended the Pleistocene meant the beginning of the Chesapeake Bay estuary. The warming temperatures melted the massive glaciers, releasing torrents of water that etched new rivers into the landscape and scoured out existing channels. The flow of water from glaciers to the ocean and thermal expansion of the water caused sea level to rise. The expanded ocean flooded the old Susquehanna River Valley and that of its tributaries, creating the Chesapeake Bay.

The flooding of the ancient Susquehanna River Valley transformed a unidirectional freshwater flow into a bi-directional brackish water *estuary* of many branches. The freshwater draining from the watershed mixes with the salty water of the Atlantic Ocean in a pattern of *estuarine circulation* (see Chapter "Scientific Concepts for Understanding the Health of the Chesapeake Bay and Its People"). Knowing the basics of estuarine circulation is critical to understanding the workings of the Bay, from the transport of larvae of blue crabs in the spring to the formation of oxygen depleted dead zones in the summer. But first a word about salinity, as this will help with understanding estuarine circulation, as discussed later, and the distribution of different species in the Bay.

The saltiness of water is expressed in units of *salinity*. Originally scientists used the units of part per thousand (ppt) to express saltiness. This is based upon how many grams of sodium chloride (NaCl) are dissolved in a liter of water. The salinity of the Atlantic Ocean is equivalent to 37 ppt (Garrison 2001), or 37 g of salt per liter of water (in truth a portion of the salinity comes from salts other than NaCl, so today marine scientists tend to use PSU [practical salinity units] instead of ppt, or simply say the Atlantic has a salinity of 37). The salinity of the freshwater entering the Bay is about 0.1–0.5. Because of all the salt dissolved in ocean water, it is more dense than freshwater (see Chapter "Scientific Concepts for Understanding the Health of the Chesapeake Bay and Its People"). So when the mass of low density fresh water from the land meets the higher density salt water from the sea, the fresh water will tend to form a lens that floats on top. Think of the less dense oil floating on top of the denser vinegar in a carafe of salad dressing.

The freshwater flowing into the Bay is powered by gravity as it slides off the land and floats atop the mass of marine waters from the Atlantic. Some of the energy of motion of the freshwater flowing out of the Bay is imparted to the marine waters below, actually driving them in the opposite direction, up the Bay. All along the way some mixing occurs at the boundary between the fresher surface waters and the saltier deeper waters. This creates a gradient of salinity, going from saltier near the mouth of the Bay, and fresher toward the headwaters of each tributary (Fig. 4). All of the saltier portions of the estuary and its tributaries experience the twice daily rhythm of tidal flow (see Chapter "Scientific Concepts for Understanding the Health of the Chesapeake Bay and Its People"). The freshwater reaches of the tributaries closest to the Bay also rise and fall with the tides, hence they are called *freshwater tidal* waters. Further upslope, the freshwater in the tributaries are unaffected by tides and hence called *non-tidal freshwater*. As noted in Chapter "Scientific Concepts for Understanding the Health of the Chesapeake Bay and Its People", the gradient in salinity means different habitats occupied by different biological communities adapted to each.

Dams and Impoundments When Capt. John Smith explored the Bay in the summer of 1608, he was able to sail and row his shallop well up to the headwaters of each tributary he visited, being stopped only by shallowing waters or encounters with hostile natives. As the colonial occupation expanded, the English built dams to create impoundments on many of the smaller branches of tributaries. The dams served as a barrier preventing the movement of brackish Bay water into the barricaded tributary. Fresh water filled the impoundments and the excess ran over the tops of the dams to continue on the journey to the Bay. These impoundments provided reservoirs of fresh water and served the developing colonial food system by providing hydropower to turn the paddle wheels of grist mills. The mills produced flour from the farmer's grain. While the dams proved essential for the production of one food they did so at the expense of another because they blocked the migration of shad and other species of fish used by native and colonist alike.

Closer to the Bay the slope of the floodplain proved too gentle to build impoundments to power traditional water mills. So the colonists adopted a medieval European technology of building mills run by tidal flow (McErlean 2018). The tide mill would be situated on a creek with suitable ability to store water at high tide. A dam would close the opening of the creek to the adjacent body of water, a larger river that communicated with the Bay. The builders placed a dam across the mouth of the creek with two openings. One opening served as an exit and entrance for water to move through a sluice where it would turn the mill's wheel. The other opening provided a second path for filling the temporary impoundment with the rising tide. Gates regulated the flow through both openings, allowing the miller to control the water level and the pressure driving the wheel. The entering and escaping water turned the wheel which rotated the grindstone for the production of flour. It is about 6.5 h between each high and low tide on the Bay (see Chapter "Scientific Concepts for Understanding the Health of the Chesapeake Bay and Its People"). Timing was everything as average difference between high and low tide on the Bay is only 1 m (39 inches) and the miller needed to maintain sufficient differences in water level between creek and river for gravity to do its work. It seems most communities in Tidewater Virginia feature roads and lanes named "Tide Mill," an indication of the importance of this food processing technology to the pre-industrial Chesapeake.

The Poplar Grove Tide Mill is the last remaining mill of this sort on Chesapeake Bay (Fig. 5). It was built by slave labor in the 1700s on the East River, a tributary to Mobjack Bay, on the western shore of the lower Chesapeake. The mill supplied flour to George Washington's troops during the Revolutionary War. It burned during the Civil War, but was quickly rebuilt and remained in operation until 1912. These days Chesapeake Bay sailors frequently anchor in the East River near the mill and use dinghies or kayaks to enter Poplar Grove creek, explore the historic structure, and contemplate the story it tells about the food system of an earlier time. The legendary musician John Lennon (the Beatles) and his wife Yoko Ono bought the Poplar Grove Mill and associated mansion in 1980. Sadly, he was killed later that year, so the couple never enjoyed the view of the East River nor the glimpse back in time (Swenson 2013).

Steam engines and electric motors eventually replaced the water mills of the Bay, but many of their impoundments remain as testimony to how the food system reshaped a portion of the estuary for hundreds of years. One such mill pond went on to also tell a story of the food system that predated the colonists. In 1999, the heavy rains of Hurricane Floyd burst the earthen colonial-era dam built on Mill Creek near Williamsburg, VA. This drained Lake Powell and revealed the remains of two Indian cites, allowing archeologists to discover much about the food system of precolonial times (Gallivan 2016). Those findings will be explored in Chapter "The Algonquin Food System and How it Shaped the Ecosystem and Interactions with the English Colonists of the Chesapeake Bay".

The Susquehanna River flows 747 km (464 miles) from its origin in Cooperstown, NY to Havre De Grace, Md, where it joins the upper Chesapeake Bay. While this important river delivers most of the freshwater to the Bay, a series of rocky rapids inhibited navigation beyond 16 km (10 miles) north of the confluence with the Chesapeake. To resolve this problem, in 1783 the State of Maryland authorized the construction of a canal on the eastern shore of the river. The canal proved too short and shallow to be effective and it closed in 1817. The Susquehanna and Tidewater canal opened in 1840 on the western side of the river as a replacement for the earlier effort. It ran 69 km (43 miles) between Havre De Grace and Wrightsville, PA. The barges went through 29 locks along the way to bridge the difference in elevation of 73 m (239 feet). Vendors of oysters and other food system products

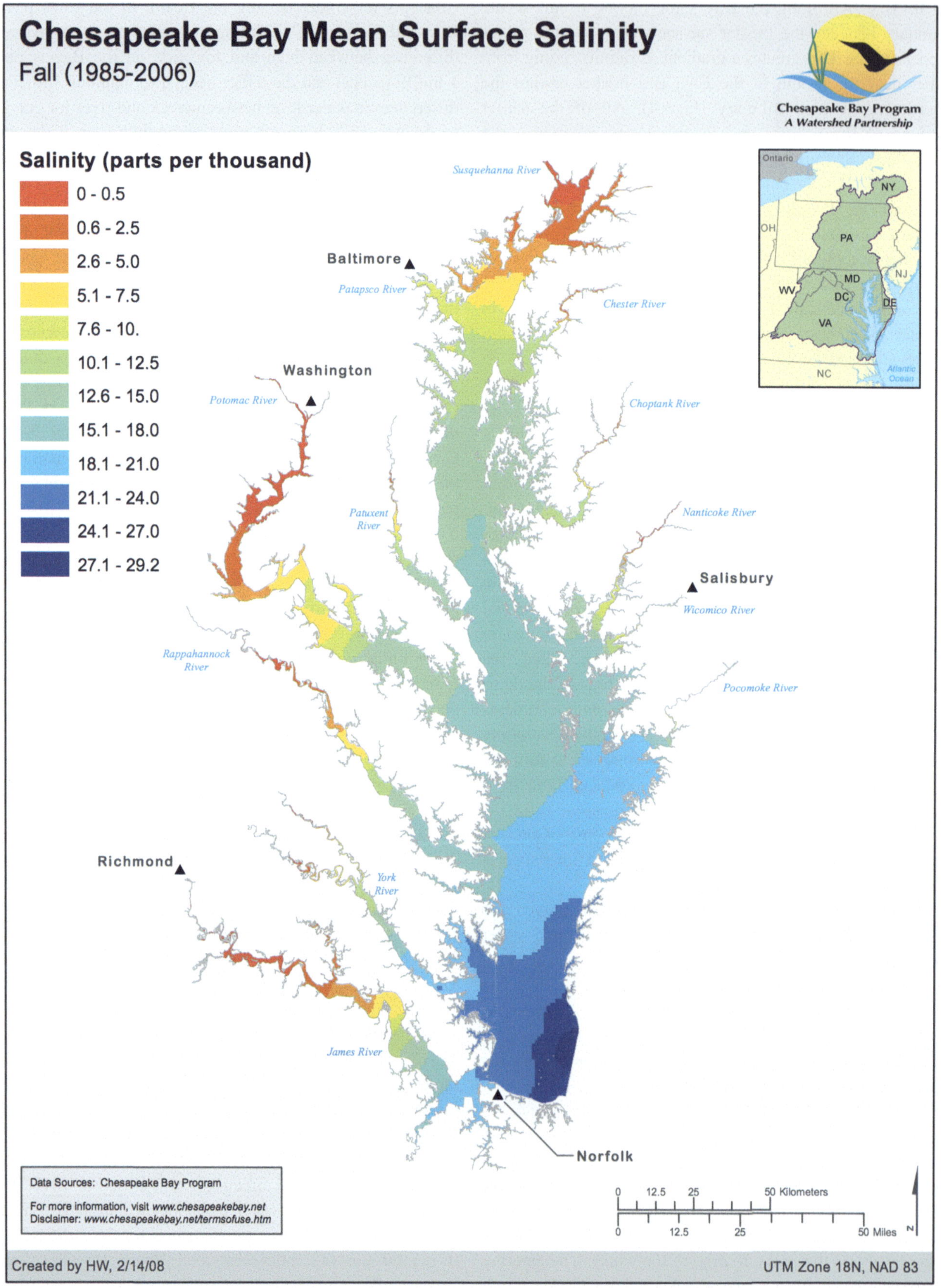

Fig. 4 Salinity zones in the Chesapeake Bay surface waters. https://www.chesapeakebay.net/what/maps/chesapeake_bay_mean_surface_salinity_fall_1985_2006

Fig. 5 The Poplar Grove tide mill circa 1935, located on the East River of Mathews County, VA off of Mobjack Bay on the western shore of the Chesapeake Bay, as seen from the creek side with the East River in the background. https://www.dhr.virginia.gov/historic-registers/057-0008/

used the canal to expand their markets. The canal closed in 1894, succumbing to competition from railroads (Msa. maryland 2018).

The rapids that proved an obstacle to navigation provided an ideal site for locating a hydroelectric dam. Finished in 1928, the Conowingo Dam spans the Susquehanna River some 16 km (10 miles) north of where the river meets the Bay. This is the largest dam on the Chesapeake. It and two other dams built upriver, blocked the migration of shad and other fish. However, it proved useful in trapping much of the sediment and its associated load of excess nutrients that washed from the rich agricultural fields of Pennsylvania and New York. Eventually the trapped sediments filled much of the impoundment behind the dam, so that today it no longer provides the service of pollution reduction to the Chesapeake ecosystem. Severe flooding resuspends some of the pollution laden sediments stored behind the dam transporting them downstream to the Bay (Krikstan 2012; CBP 2018b; Conowingodam 2018) (Fig. 6).

Physical Dimensions of the Bay and Its Watershed The Chesapeake Bay is the largest estuary in the US, and the third largest in the world. It is oriented approximately north-south, its long axis spanning 322 km (200 miles). The width varies from 48 km (30 miles) across its lower portion between Cape Charles and Hampton, Virginia, to 6.4 km (4 miles) between Aberdeen, MD., and the Eastern Shore. The Bay joins the Atlantic Ocean between Cape Charles (north) and Cape Henry (south), a span of 19.3 km (12 miles). The surface area of the Bay, including the tidal portion of its tributaries, is 30,261 km^2 (11,684 sq. miles). The average depth of the Bay at mean low water (when the tide is out) is 8 m (26 feet). The ancient Susquehanna River bed provides a deep channel that permits large ships to make passage from the mouth all the way to Baltimore and beyond to the Chesapeake and Delaware (C&D) Canal. Dredging by the U.S. Army Corps maintains a minimum depth of 15 m (50 ft.) in the shipping channel. Several depressions in excess of 30 m (99 ft.) dot the Bay, the deepest at 53 m (174 ft.) near Bloody Point on the Eastern Shore (Chesapeake Bay Program 2018a, b).

In addition to the dams discussed above, European colonization reshaped the physical Bay. Deforestation of the watershed, agriculture, and development meant massive erosion and runoff of soil into the Bay, filling shallows and channels and creating marshes over ancient shoals. Sediment so fills some Colonial era harbors that no ship of any size can now enter. Constant dredging keeps open a portion of the previously navigable waterways.

In a technological triumph of the industrial age, the C&D Canal opened in 1829 to connect the upper Chesapeake Bay with the Delaware Bay. The 22.5 km (14 mile) shipping way used a system of four locks filled by steam powered paddle

Fig. 6 The Conowingo hydroelectric dam spanning the Susquehanna River and flooding the former area of extensive rapids behind it. https://www.flickr.com/photos/aharrin1/13205232323. Image by Aaron Harrington 2013

wheels. By 1927, the C&D Canal was dug deep enough to allow a sea-level system free of locks with the ability to accommodate larger ships (US Army Corps 2018).

Today, water from 166,000 km^2 (64,000 sq. miles) of land drains in to the Chesapeake Bay. The watershed area is 14 times the area of the bay, the largest ratio for any coastal body of water in the world. Although ocean and freshwater mix to create the brackish Bay, what happens in the bay is mostly a product of what happens in the watershed. The watershed includes parts of Delaware, Maryland, New York, Pennsylvania, West Virginia and Virginia (Fig. 7). Only Maryland and Virginia feature extensive shoreline along the Bay. Naturally, those two states show the most interest in efforts to restore the Chesapeake.

The Susquehanna River is the largest tributary to the Bay, contributing about half of its freshwater. The Susquehanna watershed features a large area of intensely farmed land. Runoff from those farms supplies most of the excess nitrogen and phosphorus to the Bay (Hagy et al. 2004, Fig. 8). As noted above, the states producing the nutrient pollution from the farms of the Susquehanna (New York and Pennsylvania) don't border the Bay, so residents of those states don't live with the direct consequences from the pollution they produce. The other Western Shore major tributary rivers are the James, York, Rappahannock, Potomac, Patuxent, and Patapsco Rivers. Those on the Eastern Shore are the Pocomoke, Wicomico, Nanticoke, Choptank and Chester Rivers. The Chesapeake Bay and its tributaries present 18,800 km (11,684 miles) of shoreline, a testament to the vast number of small rivers, creeks, and bays that connect the watershed to this great body of water (CBP 2018a).

Being an estuary, water enters the Bay from both the watershed and the ocean, in a pattern of circulation explained in the next chapter. Gravity dictates that the net flow is from land to the sea. Residence time of water in the Bay ranges from 110 to 264 days, with an average of 180 days. Residence time is greatest during the summer stratification, and lowest in winter when it behaves more like a river (Du and Shen 2016). The months long time that water lingers in the Bay setsup the conditions for pollution from our food system to promote the growth of excessive algae, as explained in Chapter "Reversing the Eutrophication of the Chesapeake Bay and its People".

The Airshed The pattern of drainage from the surrounding land clearly defines the watershed of the Bay. Yet it's not just water that flows in to the system. Aerosols and particulates from the atmosphere fall on the watershed and directly into the Bay's waters. These include nutrients and pollutants, ranging from oxides of nitrogen and sulfur, to ash from combustion, and toxic herbicides and pesticides. Most of these substances arrive from an area traversed by the prevailing winds. That area constitutes the Bay's airshed (Fig. 9). The airshed of the Bay is more loosely defined

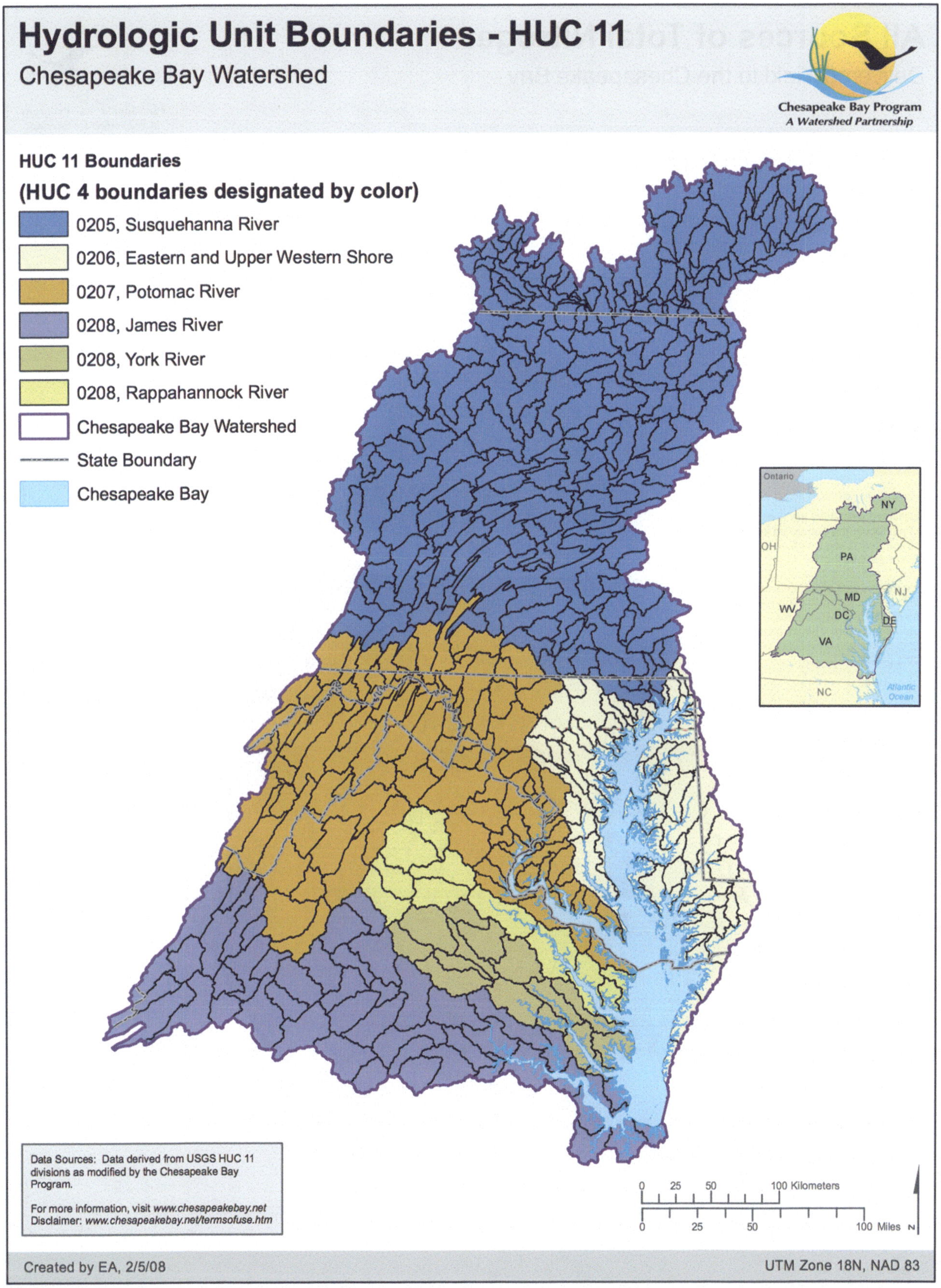

Fig. 7 Map of the Chesapeake Bay watershed divided into the drainage areas form the major tributaries. Note that the Susquehanna is the tributary with the largest drainage area. HUC = Hydrologic Unit Code. It includes tertiary streams draining into the main tributaries of the Chesapeake Bay. From the Chesapeake Bay Program: https://www.chesapeakebay.net/images/maps/cbp_18894.pdf

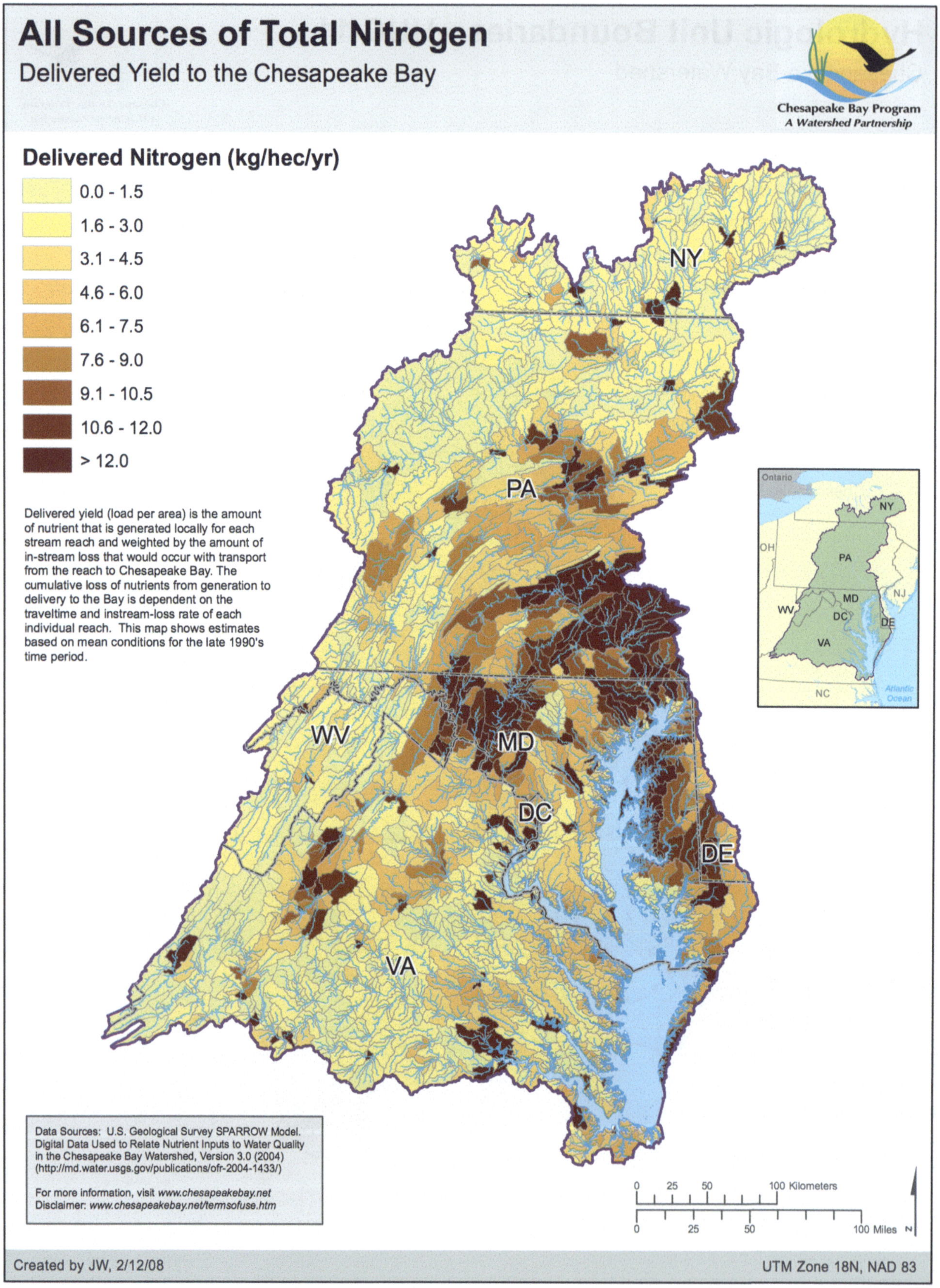

Fig. 8 Map showing the major sources of watershed nitrogen loading to the Chesapeake Bay, note that the Susquehanna River basin is the largest source of N. https://www.chesapeakebay.net/images/maps/cbp_20150.pdf

Fig. 9 The Chesapeake Bay airshed map with inset showing the watershed. Note that the airshed is much more extensive and includes jurisdictions well beyond the watershed. https://www.chesapeakebay.net/what/maps/chesapeake_bay_airshed

than the watershed. Certainly some aerosols and particles come to the Chesapeake from places quite distant, but most trace to nearer areas that constitute the airshed.

References

Berardelli P (2009) The mountains that froze the world [online]. Science. AAAS. Available at: http://www.sciencemag.org/news/2009/11/mountains-froze-world. Accessed 20 Mar 2018

CBP (2018a) Facts & figures. Chesapeake Bay program [online]. Available at: https://www.chesapeakebay.net/discover/facts. Accessed 20 Mar 2018

CBP (2018b) Conowingo Dam. Chesapeake Bay program [online]. Available at: https://www.chesapeakebay.net/issues/conowingo_dam. Accessed 23 Mar 2018

Conowingodam.org (2018) Conowingo Dam. About the River and Conowingo Dam [online]. Available at: http://www.conowingodam.org/issues/river-dam/. Accessed 23 Mar 2018

Du J, Shen J (2016) Water residence time in Chesapeake Bay for 1980–2012. J Mar Syst 164:101–111

Frank D (2018) USGS geology and geophysics [online]. Geomaps.wr.usgs.gov. Available at: https://geomaps.wr.usgs.gov/parks/province/appalach.html. Accessed 20 Mar 2018

Gallivan M (2016) The Powhatan landscape: an archaeological history of the Algonquian Chesapeake. University Press of Florida, Gainesville, pp 1–197

Garrison T (2001) Essentials of oceanography, 2nd edn. Brooks/Cole, Pacific Grove, CA

Graham S (2000) Milutin Milankovitch: feature articles [online]. Earthobservatory.nasa.gov. Available at: https://earthobservatory.nasa.gov/Features/Milankovitch/. Accessed 20 Mar 2018

Hagy J, Boynton W, Keefe C et al (2004) Hypoxia in Chesapeake Bay, 1950–2001: Long-term change in relation to nutrient loading and river flow. Estuaries 27:634–658

IPCC (2018a) 2.10.2 Direct global warming potentials – AR4 WGI Chapter 2: Changes in atmospheric constituents and in radiative forcing [online]. Available at: http://www.ipcc.ch/publications_and_data/ar4/wg1/en/ch2s2-10-2.html#table-2-14. Accessed 20 Mar 2018

IPCC (2018b) IPCC – Intergovernmental panel on climate change. Agriculture and energy cropping [online]. Available at: http://www.ipcc.ch/ipccreports/tar/wg3/index.php?idp=115. Accessed 20 Mar 2018

Krikstan C (2012) Sediment reservoirs in lower Susquehanna reach capacity, deliver more pollutants into bay. Chesapeake Bay Program [online]. Chesapeakebay.net. Available at: https://www.chesapeakebay.net/news/blog/sediment_reservoirs_in_lower_susquehanna_reach_capacity_deliver_more_pollut. Accessed 23 Mar 2018

McErlean T (2018) The World's earliest dated tide mill [online]. Medievalists.net. Available at: http://www.medievalists.net/2015/12/the-worlds-earliest-dated-tide-mill/. Accessed 23 Mar 2018

Msa.maryland.gov (2018) Maryland canals [online]. Available at: http://msa.maryland.gov/msa/mdmanual/01glance/html/canals.html. Accessed 23 Mar 2018

Myhre G, Shindell D (2018) Anthropogenic and natural radiative forcing [online]. Ipcc.ch. Available at: https://www.ipcc.ch/pdf/assessment-report/ar5/wg1/WG1AR5_Chapter08_FINAL.pdf. Accessed 20 Mar 2018

NASA (2018a). The moon [online]. Available at: https://www.nasa.gov/moon. Accessed 20 Mar 2018

NASA (2018b) Moon fact sheet [online]. Available at: https://nssdc.gsfc.nasa.gov/planetary/factsheet/moonfact.html. Accessed 20 Mar 2018

Poag C, Koeberl C, Reimold W (2004) The Chesapeake Bay crater. Springer, Heidelberg

Rountree H, Clark W, Mountford K et al (2008) John Smith's Chesapeake voyages, 1607–1609. University of Virginia Press, Charlottesville, VA

Swenson B (2013). Poplar grove; imagine There's no tide mill. Abandoned country [online]. Abandonedcountry.com. Available at: http://www.abandonedcountry.com/2013/10/21/poplar-grove-imagine-theres-no-tide-mill/. Accessed 23 Mar 2018

UCAR (2018) What is the average global temperature now? UCAR – University Corporation for Atmospheric Research [online]. Available at: https://www2.ucar.edu/climate/faq/what-average-global-temperature-now. Accessed 20 Mar 2018

US Army Corps (2018) Philadelphia District > Missions > Civil Works > Chesapeake & Delaware Canal > Canal History [online]. Available at: http://www.nap.usace.army.mil/Missions/Civil-Works/Chesapeake-Delaware-Canal/Canal-History/. Accessed 20 Mar 2018

USEPA (2018) Climate change indicators: atmospheric concentrations of greenhouse gases. US EPA [online]. Available at: https://www.epa.gov/climate-indicators/climate-change-indicators-atmospheric-concentrations-greenhouse-gases. Accessed 20 Mar 2018

USGS (2018) Glacier and landscape change in response to changing climate [online]. Available at: https://www2.usgs.gov/climate_landuse/glaciers/glaciers_sea_level.asp. Accessed 20 Mar 2018

Virginiaplaces.org (2018) Chesapeake Bay geology and sea level rise [online]. Available at: http://www.virginiaplaces.org/chesbay/chesgeo.html. Accessed 20 Mar 2018

Zahnle K, Schaefer L, Fegley B (2010) Earth's earliest atmospheres. Cold Spring Harb Perspect Biol 2(10):4895–4895

Scientific Concepts for Understanding the Health of the Chesapeake Bay and Its People

Benjamin E. Cuker

The same science speaks to the body and the Bay
It explains how in each, the components come together in an integrated way
Oxygen, nutrients, and energy flows
In the Chesapeake and down to our fingers and toes

Abstract

This chapter provides an introduction to the basic science needed for the reader to understand the ensuing chapters that explore the relationship between the health of the Chesapeake Bay, its food system, and the wellbeing of its people. For many readers, high school or college science courses may sit dim in their memories. This chapter will refresh some of those old facts and concepts, as well as introduce some important new ones. It focuses on the most important aspects of physics, chemistry, biology, and ecology needed to understand the subsequent material. Whenever possible the chapters apply the same scientific principles to explain phenomena that are common to human health and that of the Chesapeake Bay Ecosystem.

Keywords

Estuarine ecosystem · Human health · Circulation · Nutrition · Food system

Systems of Life Life scientists typically organize the living world with a system of hierarchical classification. The classification starts out with atoms that are organized in molecules. The molecules of life form the bits of living cells. Each cell may be a free living individual or organized with other cells into a more complex multicellular organism. In more advanced organisms (plants, animals, and fungi), the cells are further organized into tissues and organs with specialized functions. The organs work together in organ systems. Whether simple or complex, each organism belongs to its own species. Members of the same species in an area constitute a population. All the populations that interact in an area comprise the living community. The community is also the living portion of an ecosystem. The ecosystem includes non-living attributes too, such as minerals, gases, water, and temperature.

At first glance a body of water and the human body seem to share little in common. Yet each is like the other in many important ways. Both are living systems that share similar functions. Each requires inputs of energy and materials that are processed or metabolized, yielding various outputs. The Chesapeake Bay system is mostly water and so are people. The Bay uses energy and basic nutrients to build elaborate life forms. People use their food system to do much the same to build and maintain their complex bodies.

The Chesapeake Bay and its surrounding land is an ecosystem, comprised of many different species and the non-living aspects of the environments. Humans are but one of many populations of species that constitute the community of life in the Chesapeake Bay ecosystem. Traditional science classifies human beings as distinct individual organisms, organized in populations with other members of their same species (*Homo sapiens*). Yet, a closer look reveals that a human is actually more than just one individual organism. Each human harbors trillions of bacteria, particularly in the intestines. These microbes constitute the microbiome. The tiny bacterial cells of the microbiome outnumber human cells by ten to one and constitute 1–3% of the bodies mass (NIH 2013). Indeed,

B. E. Cuker (✉)
Department of Marine and Environmental Science, Hampton University, Hampton, Virginia, USA
e-mail: Benjamin.cuker@hamptonu.edu

B. E. Cuker (ed.), *Diet for a Sustainable Ecosystem*, Estuaries of the World, https://doi.org/10.1007/978-3-030-45481-4_3

human health depends upon a well-functioning microbiome in the gut. So, in a sense, an individual human is a bit of an ecosystem unto itself.

The Chesapeake Bay is an Estuary Estuaries are semi-enclosed bodies of water found where rivers enter the ocean (Cuker and Bugyi 2016). The Chesapeake Bay is neither a freshwater river nor a salty ocean, but a mix of both. The freshwater entering the bay drains from a watershed that includes six different states. That watershed is a patchwork of forest, agricultural lands, suburban and urban communities, industrial sites, towering mountains and marshy plains. The substances carried by water draining from this diverse landscape determine the health of the Bay. In many ways, the watershed and Atlantic Ocean feed the Bay in the same way that people drink and eat to feed themselves.
The Bay and its numerous tributaries are home to a variety of plants and animals that thrive in the special conditions provided by the estuarine environment. Subsequent chapters will explore the details of how our quest for food reshaped the Bay. For now, keep in mind that the Chesapeake Bay is a product of the interaction between land, atmosphere, sea, and life. And understand that the Bay ecosystem is much more than the water that laps its shores.

A Taste of Water Chemistry Since the Chesapeake Bay is foremost a body of water and the people that live there are foremost made of water (about 60% by weight), and our food system depends on water, it is important to understand something about the nature of water. To understand the special properties of water so important to the functioning of the Bay and its people, one needs to understand some basic information about the chemistry of water molecules. A molecule of water is comprised of two tiny atoms of hydrogen (H) and one bigger one of oxygen (O). The atoms are held together by sharing electrons that orbit around their respective nuclei. This is called a *covalent bond* which is very strong. Each of the two hydrogen atoms shares one electron with the oxygen atom, making for a very stable molecule of water. The shape of a water molecule resembles "Mickey Mouse ears," with an angle of 104.5° between the two "ears" of Hydrogen atoms attached to Mickey's head of Oxygen (Fig. 1) (Cuker and Bugyi 2016).

The atom of oxygen in a water molecule is much larger than each atom of hydrogen. Oxygen has eight neutrons (neutral charge) and eight protons (positive charge) in its nucleus. Hydrogen, the smallest of all atoms, only has one proton and no neutron in its nucleus. When oxygen and hydrogen are separated from each other, each atom is unstable. The oxygen needs two electrons (negative charge) to achieve a full and

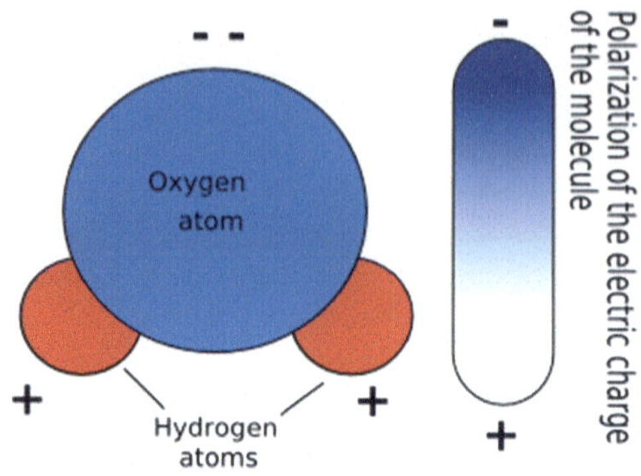

Fig. 1 Structure of a water molecule showing how one end carries a double negative charge while the other end presents two positive charges. https://commons.wikimedia.org/wiki/File:H2O_molecule_scheme_of_dipole.png by Oz Nahum

therefore stable outer shell of electrons. The hydrogen needs just one electron to achieve stability. For water, one oxygen atom achieves stability by bonding with two hydrogen atoms. But here is the interesting part, since oxygen has eight protons in its nucleus and hydrogen has only one, the shared negatively charged electrons spend most of their time near the oxygen atom. It is said that oxygen is more *electronegative* than hydrogen, its eight protons being more attractive to electrons. That means that the oxygen end of the water molecule has two negative charges associated with it, while each of the hydrogen atoms has one positive charge. This arrangement of charges causes water to be a *polar* substance, with a positive end and a negative end (Fig. 1). And this is essential as it is the basis for all of the very special properties of water.

Because water molecules are polar in the liquid and solid states, they interact with each other in a particular way. The negatively charged oxygen ends of one molecule of water will attract the positively charged hydrogen ends of two adjacent water molecules. That attractive force between the positive and negative aspects of adjacent water molecules is called a *hydrogen bond*. When in the liquid state, each molecule of water forms hydrogen bonds with about 3.4 other water molecules at a time, with four being the maximum (Armstrong 1983). So, if a thimble of water contains billons of water molecules, it also contains up to four times that many hydrogen bonds that join the water molecules together (Fig. 2).

Individually, hydrogen bonds are weak. Recall the covalent bonds holding the hydrogen atoms to the oxygen atom. The energy required to make or break covalent bonds between an H and O is 111 kcal/mole. In contrast, the energy of a hydrogen bond between adjacent water molecules is just one twentieth of the strength of the H-O covalent bond. Yet, water contains so many hydrogen bonds that there is real strength in numbers.

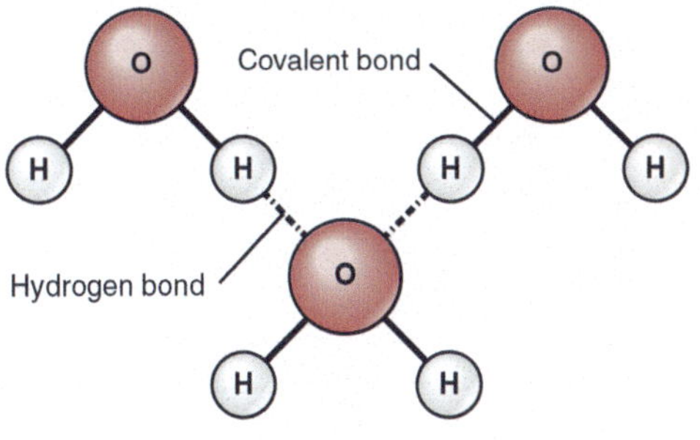

Fig. 2 Water molecules in a matrix showing hydrogen bonds, each water molecule can form hydrogen bonds with up to four other water molecules. https://upload.wikimedia.org/wikipedia/commons/0/0f/210_Hydrogen_Bonds_Between_Water_Molecules-01.jpg

All of those hydrogen bonds give water some very special properties. Water has a very high *specific heat*, meaning that it takes a lot of energy to add to those hydrogen bonds to cause the water to increase in temperature just one degree Celsius. Indeed, water is the defining substance for specific heat. At standard conditions, the specific heat of water is one calorie per gram. It takes one calorie to heat that water one degree Celsius. In contrast, the specific heat of quartz sand is only 0.19 cal/g (Engineering Tool Box 2016). Thus, the same about of heat that causes 1 g of water to rise in temperature 1 °C would cause the same mass of quartz sand to increase about 5 °C. It is the same story when water loses heat. It takes about five times more heat to be lost from water than it does quartz sand in order for the temperature to drop 1 °C. As such, the Chesapeake Bay takes time to warm-up every spring and cool down every fall. The large mass of water in the Bay also buffers changes in the temperature of the surrounding land. It never gets quite as cold or hot on the shores of the Chesapeake as it does inland 1 km (less than a mile) or so.

In addition to water's high specific heat, it also takes a lot of energy to change the state of water between solid and liquid, and even more between liquid and gas. The energy exchanged when water goes between ice and the liquid state is 80 cal/g. This is called the *heat of fusion*. Think of a gram of water at about room temperature, say 22 °C. It would have to give up 22 calories to reach a temperature of 0°C (*freezing point* of water). But for it to transform to ice it would need to lose another 80 cal.

The structure of ice differs from that of liquid water in two ways. First, the molecules of water in ice form a regular and highly rigid crystal pattern. Second, spaces between the molecules in ice are greater than that for liquid water (Fig. 3). The additional space between water molecules in ice means that it is less dense than liquid water, the same mass takes up more space. This means that solid ice floats on the denser liquid water. As winter settles on the Chesapeake Bay, one often sees ice floating on the surface of its small tributaries.

For 1 g of liquid water to change to gaseous water vapor, 540 calories must be added. This is called the heat of vaporization (540 cal/g). It takes so much energy to evaporate water because it means the complete breaking of nearly four hydrogen bonds for each molecule that becomes a gas. Warming up water requires heat to expand hydrogen bonds, while evaporating water requires much more heat to break those hydrogen bonds. Starting with that 1 g of water at room temperature of 22 °C, 78 calories must be added to bring it to 100 °C (boiling point of water). But it will take an additional 540 calories to evaporate the gram of liquid water into gaseous state. So, it is not just the high specific heat of water that cools adjacent lands during the summer. It is also the huge amount of heat dissipated by the Bay as it evaporates water from its surface. People of the Chesapeake region directly experience the power of evaporative cooling every summer, as they sweat their way through those hot months of the year. But when the air becomes saturated with water on the most humid days, people get less relief. The air can hold no more water so that the perspiration coalesces into drops that fail to evaporate, leaving folks both hot and wet.

Above it was noted that ice is less dense than liquid water and as such it floats at the surface of the wintery Chesapeake. But temperature also affects the density of liquid water and this is key to understanding important patterns of circulation in the Bay, particularly in the warmer months. For fresh water, maximum density occurs at 4 °C. As freshwater cools below 4 °C it begins to expand, becoming less dense. For temperatures above 4 °C, the warmer the water the less dense it is. Just like cold ice in the winter, the warmest and lowest density water floats on top of the cooler and denser water during the summer. This has profound implications for how the Chesapeake Bay functions, and it gets a bit more complicated when we consider the saltiness of the system, to be explained below.

That water is H_2O, is only mostly true. In fact, water appears in nature mostly as H_2O, but importantly a small portion of that water disassociates into ions (charged molecules or atoms) of H^+ and OH^-. Pure water is said to be a weak acid, since the small portion of the water that does disassociate produces more H^+ ions than it does OH^- ions (Fig. 4). A substance like ammonia (NH_3) disassociates partially and produces more negatively charged ions so it is called a weak base.

The pH scale is used to measure the acidity of a solution (Fig. 5). When the concentrations of H^+ and OH^- ions are equal, the solution is neutral and has a pH of seven. Why seven? The term pH is shorthand for the $^-$log of the hydrogen ion activity (think concentration). So, at pH 7, the concentration of H^+ is 10^{-7} moles/liter and the concentration of OH is also 10^{-7} moles/liter (a mole is the mass of a substance containing the same number of fundamental units as there are atoms in exactly 12 g of carbon 12). Water with a pH less

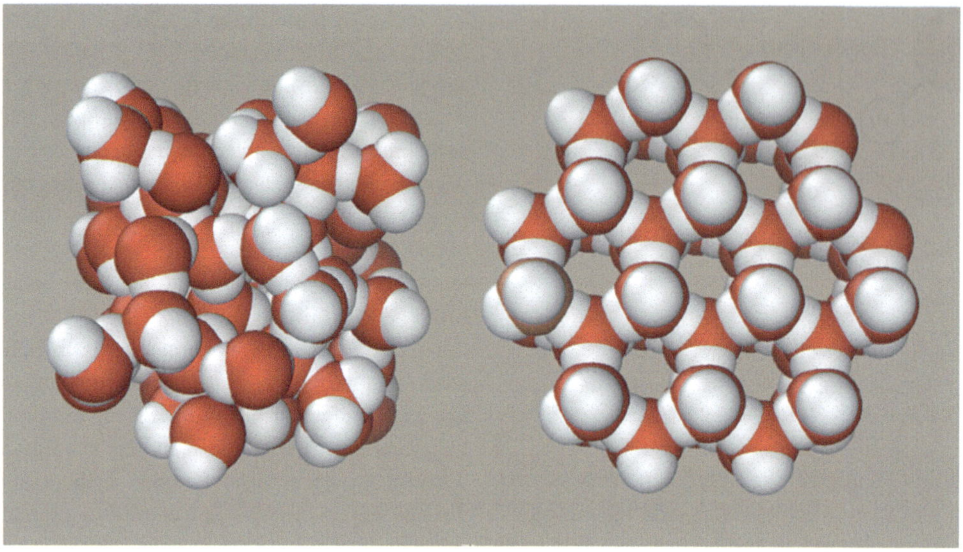

Fig. 3 The molecular arrangement of liquid water is shown on the left, and that of ice on the right. Note that there is more space between the molecules of ice than liquid water. This makes the ice less dense then liquid water and is why ice floats on the surface of frozen rivers and bays. https://commons.wikimedia.org/wiki/File:Liquid-water-and-ice.png

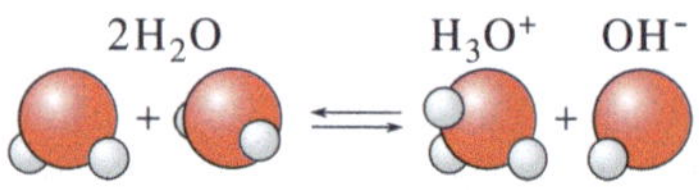

Fig. 4 The disassociation of water into ions. https://en.wikipedia.org/wiki/Self-ionization_of_water#/media/File:Autoprotolyse_eau.svg. By Cdang—Own work, Public Domain, https://commons.wikimedia.org/w/index.php?curid=7942156

than seven is acid and greater than seven is basic. The pH scale is deceptive, since it is logarithmic. So, water with a pH of five will have 100 times more H+ ions than water with pH of seven. Each pH unit represents a ten-fold change in H+ concentration (Cuker and Bugyi 2016).

Adding or subtracting H^+ or OH^- ions will change the pH of water. Yet the waters of the Chesapeake Bay and the human body contain dissolved salts that can absorb some of the excess ions. Such salts are called buffers. In the Chesapeake Bay, calcium carbonate ($CaCO_3$), and to a lesser extent magnesium carbonate ($MgCO_3$) serve as buffers. Human blood is also buffered by $CaCO_3$.

Buffers in the waters of the Bay prevent large swings in pH that would otherwise follow acid rain events (from air pollution). Despite these buffers, the Chesapeake Bay is becoming acidified by the increasing levels of carbon dioxide (CO_2) in the atmosphere that diffuses into the water (Fig. 6).

Oceanographers once considered the ocean and estuaries like the Chesapeake Bay to be so well buffered that they paid little attention to measuring pH. However, the extensive burning of fossil fuels since the Industrial Revolution combined with deforestation and animal husbandry has driven up atmospheric CO_2 levels by 56% (going from 180 ppm to 415 ppm as of 2019) over the last two centuries. Over recent years, the pH of the ocean has dropped from an average of 8.2 to 8.1, meaning a 30% increase in acidity (NOAA 2020). Acidification is worse in the Chesapeake Bay than the ocean due to lower availability of buffers than seawater and production of acids under seasonal anaerobic conditions (Cai et al. 2017). The primary buffer in the Bay, calcium carbonate, is also critical for building the shells of oysters (*Crassostrea virginica*) and other animals of the Bay. The acidification of the Bay uses up so much of the carbonate buffer, that it makes it harder for oysters to build their shells (Waldbusser et al. 2011).

The carbonate buffering system that helps regulate the pH of the Chesapeake Bay performs the same function in the blood of the people that reside on its shores. Yet, like in the Bay, the buffering is imperfect. And as with the Bay, its people have also become more acidic over recent years. This is not due to carbon dioxide in the atmosphere, but instead diets heavy in animal-based items and light in plant-based foods. Such acid-causing diets are linked to kidney disease and other medical problems (Lin et al. 2010; Deriemaeker et al. 2010; Sebastian et al. 2002). And as noted in later chapters, the production of acid-forming foods (animal flesh) is directly responsible for a large portion of the carbon dioxide that is acidifying the Bay.

What's in the Water? At this point the reader knows the essentials about the chemistry of pure water, and a bit about one type of constituent of water, buffers. The Chesapeake Bay is not pure water and quite far from it. It contains a diversity of non-living and living constituents that interact in

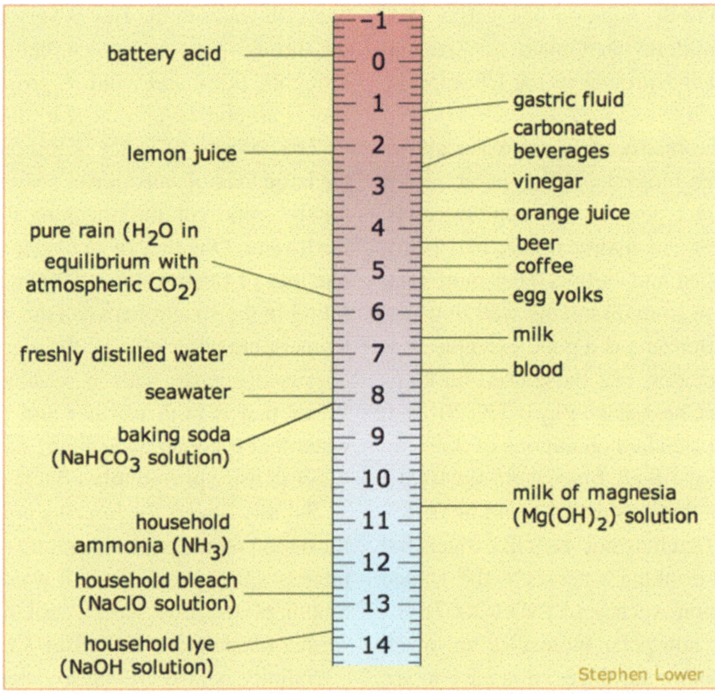

Fig. 5 The pH scale with examples of familiar liquids. Note that the pH of seawater and human blood are relatively close. http://chemwiki.ucdavis.edu/@api/deki/files/8232/639px-PH_scale.png?revision=1 by Stephen Lower

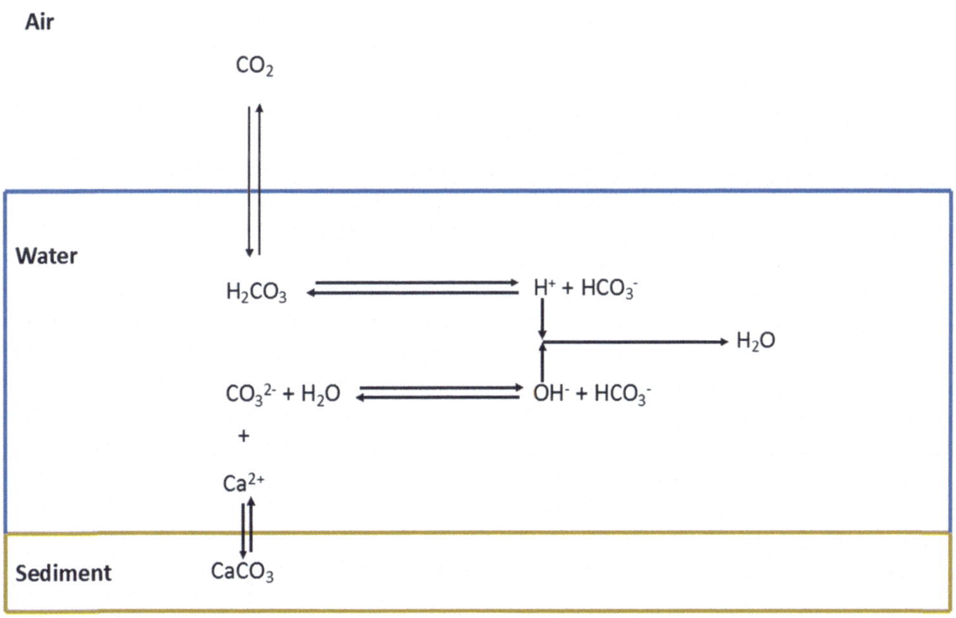

Fig. 6 The carbonate buffer system in natural waters. Additions of acids are countered by the disassociation of $CaCO_3$ and the subsequent reactions, which results in maintaining a near constant pH. If the addition of acids exceeds the buffering capacity of the water, the pH will drop. By B. Cuker

a myriad of ways to yield a vibrant system. The same is true of the water circulating in our bodies.

Water is the universal solvent. That means that most substances dissolve, at least to some degree, in water. Recall that water is polar and as such easily dissolves other polar substances. This includes *cations* (positively charged ions) such as Ca^{++}, Mg^{++}, Na^+ and K^+; *anions* (negatively charged ions) such as HCO_3^-, CO_3^-, SO_4^-, and Cl^- and atmospheric gases (O_2, N_2 and CO_2). The polar quality of these entities easily fit into the matrix of liquid polar water molecules. As will be seen later, excess levels of ionic forms of compounds of nitrogen and phosphorus originating from the agricultural system play a large role in the degradation of the Bay (See Chapter "The Journey from Peruvian Guano to Artificial Fertilizer Ends with too Much Nitrogen in the Chesapeake Bay").

The dynamic nature of the matrix of liquid water allows some solubility for non-polar molecules. Non-polar means that the molecules lack distinct positive and negative ends. Familiar non-polar substances are hydrocarbons like petroleum. They consist of hydrogen and carbon. Their non-polar charge distribution keeps them from dissolving well in water, but to some extent they do. Benzene is a good example. It is found in petroleum and pesticides, like those used on crops grown in the watershed of the Chesapeake Bay (CDC 2016). It is a toxin and carcinogen. At standard conditions of 25 °C in freshwater, it has a solubility of 1.8 g/l. Since there are 1000 g of water in a liter, this means that freshwater could be 0.18% contaminated with benzene. That may not seem like much, but it is 36 times the safe limit for drinking water set by the United States Environmental Protection Agency (ATSDR 2007).

Since benzene and other non-polar molecules are much more soluble in other non-polar substances, it is easy to see why they tend to accumulate in the non-polar fatty or oily tissues of animals in the Bay ecosystem, including people. Indeed, many of the toxic residues from the chemical food-system used to grow crops in the watershed are non-polar substances that accumulate in the bodies of bay animals. Now banned, the pesticide DDT was widely used on crops and for mosquito control from 1945–1972. The accumulation of this infamous toxin in the tissues of birds nearly drove to extinction the eagles, cormorants and osprey of the Bay (Watts et al. 2008, See Chapter "Pesticides Bring the War on Nature to the Chesapeake Bay").

Being an estuary, the water in the Chesapeake Bay has two main origins. Freshwater drains from the watershed and falls from the sky. Saltwater flows in from the ocean.

Freshwater tributaries to the Bay contain a variety of substances derived from the watershed. As rain or melting snow water moves through and across the landscape, it picks up dissolved and suspended particles. The list of cations and anions given above dominate freshwater lakes and rivers. The masses of all of those and other ions in the water total to what is called *salinity*. Traditionally salinity was expressed as *parts per thousand* (ppt). Ocean water is typically about 34–37 ppt, which means that there are 34–37 g of salts dissolved in 1000 g (or l) of water (Garrison 2001). Freshwaters are defined as those having salinity less than 0.5 ppt (Wetzel 2001). Note that in recent years many chemical oceanographers prefer to present the measure as simply *salinity*, without the units of concentration for technical reasons addressed in the previous chapter.

The freshwater from the land and the saltwater from the ocean mix along the length of the Chesapeake Bay, creating *brackish* water of intermediate salinity. The salinity of the brackish water in the Bay is very dynamic, constantly changing through time and space, being influenced by rainstorms, droughts, tides, and wind. A gross average of salinity for the Bay is around 14 ppt, close to that of human blood.

The salinity of water is important in several ways. First, the large load of salts makes seawater heavier than freshwater for the same volume, meaning that saltwater is denser than freshwater. One liter of saltwater weighs 1034 g compared to one liter of fresh water at 1000 g. All of those polar salts are filling in the spaces between the water molecules. Recall that temperature also affects the density of water. Freshwater density decreases with increasing temperatures above 4 °C. Water that is both warmer and fresher will float on top of water that is cooler and saltier. As will be seen, differences in water density profoundly effects circulation in the Bay.

Salinity lowers the freezing point of water such that sea ice formation requires a temperature of -2 °C. The added salt also increases the specific heat of water. So saltier waters near the mouth of the Bay won't cool off or freeze as quickly as fresher headwaters of the Bay's tributaries.

Salinity is also critical to determining the types of plants and animals that can inhabit a body of water. Water diffuses into and out of the cells of living organisms in a process called osmosis. Water will naturally move toward a saltier solution. Freshwater plants and animals have an internal salinity that is higher than their surroundings. The reverse is true for many forms of life in the ocean; they are internally less salty than the water in which they reside. This means that freshwater fauna and flora must fight the continual intrusion of water diffusing in. It is the opposite for life forms of low internal salinity living in the ocean as they must strive to keep from losing their water to the surrounding sea. However, some marine species survive in different external salinities by changing their internal salinities to match. These are called osmotic conformers, as opposed to the osmotic regulators that spend energy to control the flow of water between themselves and their surroundings.

As noted above, the salinity in the Bay is highly variable. The plants and animals of the Bay must be able to endure regular swings in salinity that may happen over hours. As such some familiar marine species, such as sea urchins, sea cucumbers, and sea stars are absent from the Chesapeake Bay and other estuaries. They are osmotic conformers and can't tolerate large swings in salinity that characterize the Bay. All resident and visiting forms of life in the Chesapeake Bay share in common their capacity to thrive in waters of variable salinity.

As with the creatures of the Chesapeake, the people on its shores also have challenges with salt. The Standard American Diet (SAD) includes high levels of salts of sodium chloride (NaCl), which is added to most processed foods. Excess sodium raises blood pressure and increases incidence of heart disease (Whelton et al. 2012; He et al. 2014).

The water of the Chesapeake Bay contains the same gases found in the atmosphere above, most notably nitrogen (N_2), oxygen (O_2), and carbon dioxide (CO_2). N_2 accounts for 78% of the atmosphere, O_2 21% and CO_2 only about 0.04%. Of these, oxygen is the most critical for higher forms of life in the Chesapeake, including the people that live there.

Oxygen is necessary for the process called aerobic respiration, the metabolic sequences used by cells to release energy from sugars and other organic molecules. A portion of this process is the electron transport chain. High energy electrons released from food, skip down this chain and as they do some of their energy is used to produce ATP (Adenosine Triphosphate). ATP is the direct energy source for most of the chemical reactions in cells. To keep the flow of electrons going down the chain there must be something to remove the electron as it finishes its journey. That something is free oxygen. It readily combines with the now low-energy electron along with hydrogen to form water. Without oxygen to take away the electrons, the transport chain stops, meaning there is no more production of ATP. For higher organisms from fish to humans, the absence of oxygen spells death in a matter of minutes (Armstrong 1983).

There are three ways oxygen gets into the Chesapeake Bay. First, oxygen diffuses from the atmosphere into the surface waters. Wind and waves speed the process and also help mix the newly oxygenated surface water to somewhat deeper depths. Second, water flowing into the Bay from its many tributaries and the ocean brings in its own load of oxygen. This transportation by flowing water is called advection (Garrison 2001).

The third source of oxygen to the Bay is photosynthesis. This is a metabolic process done by green plants and algae that contain the pigment chlorophyll. Photosynthesis uses energy from sunlight to split water molecules into atoms of oxygen and hydrogen. Recall that there is much energy in the covalent bonds that hold together the hydrogen and oxygen atoms of water molecules (Fig. 7). The splitting releases high energy electrons to power the production of ATP. The energy stored in ATP is then used to convert carbon dioxide into sugars, fats, and proteins, the building blocks of life. The oxygen produced by the splitting of water is a waste product, and it is released by the cells into the surrounding water. On a clear, still day along the shores of the Chesapeake, one can look down into a bed of submerged sea grass and see oxygen bubbles floating to the surface.

That oxygen is a waste product might seem strange to most readers. After all, oxygen is critical to the process of aerobic respiration that keeps all higher organisms alive. However, what makes free oxygen so good at picking up those low energy electrons at the end of the electron transport chain in aerobic respiration, also makes it good at disrupting important molecules in the cell by stealing their electrons. If

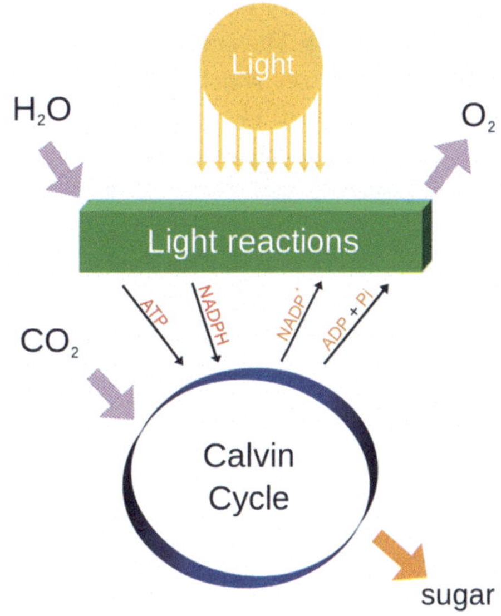

Fig. 7 Simplified diagram of events in photosynthesis. This process uses the energy from sunlight to make the basic building blocks of biomolecules (Calvin Cycle) and releases oxygen as a waste product (part of the light reactions). By Daniel Mayer. https://commons.wikimedia.org/wiki/File:Simple_photosynthesis_overview.svg

not quickly paired with a second oxygen atom to make O_2, the uncoupled oxygen atom is called a *free-radical*, ready to steal electrons away from other molecules. This is the process of *oxidation* and it causes havoc for the fats, proteins, and nucleic acids that are the structural and functional basis of cellular life. So, this benefits plant and algae cells to quickly expel excess oxygen. In addition, plant cells protect themselves from free-radicals with a class of compounds called *anti-oxidants*. Not surprisingly, it is plant foods that provide humans with dietary antioxidants. And it is the lack of fresh fruits and vegetables in the SAD (Standard American Diet) that leads to diseases associated with damage from cellular oxidation including, "…atherosclerosis, cancer, diabetes, rheumatoid arthritis, post-ischemic perfusion injury, myocardial infarction, cardiovascular diseases, chronic inflammation, stroke, septic shock, aging and other degenerative diseases in humans (Uttara et al. 2009)."

While humans don't photosynthesize and therefore don't have as much oxidation from excess oxygen in cells, they do produce large quantities of oxidants as a consequence of metabolizing their food. Recall that metabolism involves the shuffling of electrons between molecules. The process is messy, resulting in the production of some molecules that are missing an electron in their outer shell. These free-radicals cause damage to the cell's molecular structure. The more people eat, the more free-radicals they create. Since free-radicals are so damaging to human health, people should choose plant centered diets rich in antioxidants (Prior et al. 2007; Matés et al. 1999).

The physics and chemistry of water determines the solubility of oxygen and gases. The concentrations of oxygen in the Chesapeake Bay fluctuate widely with time and location, ranging from zero to about 14 mg/l. Salinity reduces the solubility for oxygen in water, as it does for all gasses. In essence, the salt molecules take up space otherwise used by dissolved gasses. Increasing temperature also reduces gas solubility in water. Water, and for that matter all substances above absolute zero (much colder than any place on Earth), vibrates. The warmer the water, the faster the water molecules vibrate, and that leaves less effective room for gasses to dissolve.

Now consider the combined effects of temperature and salinity. The cold and fresher waters found in the upper tributaries in spring can potentially contain twice the oxygen found in warm, salty waters near the mouth of the Bay in summer. Water is *saturated* with a gas when it can't hold any more over time, with the excess forming bubbles that float to the surface. Cold fresh water of 1 °C (just above the freezing point of water) entering the Bay in early spring will saturate at 14 mg/l of oxygen. The warm 20 °C and salty (30 ppt) water near the mouth of the Bay in summer will saturate with 5 mg/l, or at nearly two-thirds less oxygen than the cold fresh water (Wetzel 2001). So, it is tougher for a fish to use its gills to extract oxygen from warm salty water than cold fresher water. But it gets worse! The metabolic rate of fish (and all living creatures) is proportional to temperature. That means that they require more oxygen at higher temperatures. For example, the striped bass (aka rock fish, *Morone saxatilis*) may live in the Bay year-round. Following the general rule of a doubling of metabolic rate for every 10 °C increase in temperature, a striped bass will require nearly four times as much oxygen at 28 °C in summer as it will at 1 °C in winter (Clarke and Johnston 1999).

Often water in the Bay is either undersaturated or supersaturated with oxygen. Biology plays an important role here. Recall that one of the three ways oxygen gets into the Bay is by photosynthesis. Under the right conditions, the algae or sea grass in a portion of the Bay may be so stimulated with light and nutrients, that it supersaturates the water with oxygen. Once the rate of photosynthesis subsides, the oxygen concentration will return to saturation as the excess bubbles away. Rapidly warming surface waters, say during a heat wave in the late spring, can also supersaturate the water. Recall that the newly warmed water can no longer hold as much oxygen and it will bubble away over the course of a day or two.

Supersaturation with oxygen rarely presents problems for life in the Bay, but under-saturation can be devastating. There are four ways oxygen leaves the water. (1) It can move out by advection, dissolved in water flowing into the ocean. (2) It can diffuse or bubble into the atmosphere. (3) Oxygen may react with chemicals in the water, like iron or sulfur, to be locked-up in rust or bubbled away as sulfate. (4) And most importantly, oxygen may be used and even used up in the process of aerobic respiration conducted by living organisms, as discussed above. The loss of oxygen to respiring organisms (primarily bacteria) is indeed the biggest challenge faced by the Chesapeake Bay today. This will be explored in more detail below and in Chapter "Reversing the Eutrophication of the Chesapeake Bay and Its People".

The Chesapeake Bay and Its Watershed is an Ecosystem Recall the hierarchy of life presented at the beginning of the chapter. The Chesapeake Bay and its watershed constitute an *ecosystem*. The living (biotic) and nonliving components of the Bay and its watershed interact in a dynamic way to yield an ecosystem. The nonliving constituents of this system include sunlight, gases such as oxygen, nitrogen (N_2), and carbon dioxide; water; dissolved mineral salts; mineral particles (clays, silts and sands), and temperature. The living portion of the Chesapeake Bay ecosystem includes, viruses, bacteria, tiny cells of algae (phytoplankton), and various plants and animals of all sorts and sizes. Energy and materials from outside the system (inputs) are used and modified by these abiotic and biotic components of the ecosystem to produce various outputs. For example, the inputs of CO_2, water, plant nutrients, and sunlight result in the outputs of oxygen and blue crabs (*Callinectes sapidus*). Cleary a lot happens in the Bay to transform simple inputs into complex outputs like blue crabs!

All life in the Chesapeake Bay ecosystem depends directly or indirectly on photosynthesis to provide the energy that powers metabolism and builds the molecules needed to assemble cells. Sunlight intercepted by leaves of tall oak trees and fields of planted corn in the watershed, or by tiny algal cells drifting in the Bay, is used to power the process of photosynthesis. Only about 1% of the sun's energy is captured in photosynthesis (Odum and Barrett 2005), but that is enough to support an abundance of life in the Chesapeake Bay and its watershed.

Photosynthesis makes simple sugars that algae and green plants turn into the variety of biological molecules needed to support life. Some of those molecules like sugars, starch and lipids store energy used to run the metabolism of the plant. Others form the very stuff of life, biomolecules used to build cells and organisms. So, photosynthesis provides both the fuel of life and the building blocks of its construction.

The main biomolecules of life belong to four different groups. (1) *Carbohydrates* (contain Carbon [C], Hydrogen [H], Oxygen [O]) range from simple sugars and starch to cellulose (think wood) and other strong complex fibers that support tall trees. (2) *Lipids* (C, H, O) are fats, oils, and waxes that store energy, form cell membranes, and some play the role of chemical messengers (hormones) in organisms. (3) *Amino acids* (C, H, O, Nitrogen [N], Sulfur [S]) link together to form proteins. Proteins are the action molecules of life, involved in essentially all the chemical reactions in cells (metabolism) as *enzymes*, that regulate those reactions. (4) *nucleic acids* (C, H, O, N, P) are modified sugars that play two major roles. First, they are used for transferring energy in metabolism (Adenosine Diphosphate and Adenosine Triphosphate, ADP and ATP). Second, nucleic acids string together in chains of DNA (Deoxyribonucleic Acids) and RNA (Ribonucleic acids) that contain the genetic code. The DNA is the library of life, storing the directions for making the proteins that organize the lives of cells and organisms.

Sunlight provides the energy for photosynthesis, but not the raw materials needed to store that energy or build the molecules of life. The air, land, and water provide the nutrients used by plants for those purposes. Ecologists classify those nutrients by the relative quantity that is needed by plants or animals. *Macronutrients* are those needed in large amounts. CHNOPS, or carbon, hydrogen, nitrogen, oxygen, phosphorus, and sulfur are all required in large amounts. They form the backbone of the molecules of life (carbohydrates, lipids, proteins and nucleic acids). *Micronutrients* are needed in much smaller amounts. These include minerals such as potassium, copper, iron, magnesium, calcium, and zinc. Also required in small amounts and not manufactured by the cells are organic molecules called vitamins. For a plant or animal to grow, it needs to have a source of energy and all of its required macro and micronutrients. If one of these is in short supply, not enough for growth to occur, it is said to be the *limiting factor*. During the warmer months of the year the limiting factor for plant growth in the Bay is usually a macro nutrient nitrogen (N) or phosphorus (P). But in winter, it might be lack of sunlight or low temperature that is the limiting factor for plant growth.

All the molecules of life are built on a skeleton of C, H, and O atoms. Those three elements alone are sufficient to build carbohydrates and lipids. The addition of N and S is needed to produce amino acids for making proteins. Adding P provides the essentials for building nucleic acids that are used in making the genetic material and in energy transfer. The ratio of these atoms in the algae cells like those that form the base of the food chain in the Bay is roughly; 263 H: 106 C: 110 O: 16 N: 2 S: 1 P (Chen et al. 1996). So, a cell needs about 106 C atoms and 12 N atoms for every one atom of P. In unpolluted waters there is plenty of C available, but usually less N or P to satisfy the needed ratio for building algae cells. This means that unless purposely fertilized or accidently polluted, a lack of N or P typically limits plant growth during the warmer months for the Chesapeake Bay ecosystem.

Most commonly N is the limiting nutrient for the middle and lower portion of the Chesapeake Bay ecosystem. This is true for forests and most croplands, as it is for plants and algae living in the water. That means that if one adds nitrogen in an easily absorbed form, like ammonia or nitrate, plant growth will increase. Farmers in the region know this very well, and as such fertilize their crops with nitrogen. Prior to the mid-twentieth century, farmers mostly used animal and plant manures that we call organic fertilizers today. Those organic fertilizers tended to remain in place when applied to fields. Organic fertilizers slowly released N & P as they naturally decayed over time. Although introduced early in the twentieth century, it wasn't until after WWII, that artificial fertilizers began to replace the manures on a large scale. The new synthetic forms of nitrogen were easy to apply but tended to wash away with rain and irrigation. The wasted artificial nitrogen fertilizer finds its way through streams, rivers and groundwater into the Bay. There it stimulates the growth of algae, turning the waters of the Bay green, red or brown. The massive growths of algae are called blooms. As the blooms first form, they supersaturate the water with oxygen released from photosynthesis. But soon, the algae sink and die, being eaten and decomposed by bacteria. The rotting blooms fuel massive bacterial respiration that depletes the water of oxygen. This process of over-fertilizing the Bay with nutrients, the subsequent algal bloom, and ultimate depletion of oxygen is called *cultural eutrophication*. As detailed below and in subsequent chapters, cultural eutrophication is the main agent degrading the Chesapeake Bay.

Why is nitrogen so important to plant growth? It is one of the macronutrients, being essential to the construction of proteins and nucleic acids. And it is hard to acquire, despite N_2 (elemental nitrogen) being the dominant gas in the atmosphere (78%). N_2 is held together by a triple covalent bond, meaning that it takes a lot of energy to break it apart and reconfigure it to become NH_4 or NO_3, the forms easily taken-up by plants. Chapter "The Journey from Peruvian Guano to Artificial Fertilizer Ends with too Much Nitrogen in the Chesapeake Bay" will explain much more about the role of nitrogen in the Chesapeake Bay ecosystem.

Food Chains and Food Webs Energy flows through the Chesapeake Bay ecosystem along food chains and food webs. The food chain is a simple way of thinking about energy flow in ecosystems. Each of the steps in the food chain is called a trophic level. The chain starts with primary producers, the green plants and algae that use photosynthesis to change energy from sunlight into the chemical energy of biomolecules. Some of these primary producers are consumed by primary consumers, which in turn are eaten by secondary consumers, which are eaten by tertiary consumers etc., but only to the point when the energy runs out. Decomposers constitute a trophic level adjacent to all the consumer levels in the food chain.

Decomposers consume the dead bodies and fragments of producers and consumers. Such dead material is termed *detritus*, and the organisms that feed upon it are called detritivores. Detritus nourishes a wide range of organisms, from bacteria and fungi to worms, blue crabs and gulls. Decomposers play a critical role in recycling nutrients, making the basic elements (primarily N and P) of life available to sustain the growth of primary producers. That is, decomposers get their energy from breaking down biomolecules and in the process regenerate nutrients for use by primary producers.

As a general rule, only about 10% of the energy available at one trophic level moves up the food chain to become available at the next trophic level. Energy is lost between trophic levels for several reasons. First, the second law of thermodynamics deems that some useful energy is lost every time energy is transferred from place to another. Second, only a portion of what a consumer eats is digested and absorbed into its body. If consumers only dined on simple sugars and lipids, then almost all of what they ate would be absorbed. But food items often contain such things as hard to digest plant fibers or animal shells. For humans, think of the difference in available calories between 100 g of each; glazed doughnuts (426 kcal), ground beef (295 kcal), and kale (49 kcal) (USDA 2018). Energy is also lost between trophic levels because consumers must use a portion of what they eat to keep themselves alive and do such things as move, defend themselves, seek mates, and reproduce. Homeothermic (warm blooded) consumers like birds and mammals spend much of the energy they consume on producing heat to stay warm. Consumers also share a portion of their plate with the microbiome (bacteria) in their guts.

Consider the energy loss along one important food chain of the Chesapeake Bay assuming the 10% efficiency rule. Suppose that one kg of algae floating in the Bay is consumed by the suspension feeding (filter feeding) fish called menhaden (*Brevoortia tyrannus*). That would result in 0.1 kg of menhaden flesh. A striped bass that ate that 0.1 kg of menhaden would only add 0.01 kg to its own body mass. A bottle-nosed dolphin (*Tursiops truncates*) that ate 0.01 kg of the striped bass would only add 0.001 kg to its own body. By the time the 1 kg of algal biomass passed up the food chain to the dolphin, only 0.1% would be left. Another way of looking at this is that each 500 kg bottle-nosed dolphin would have to have eaten 5000 kg of striped bass, which would have to had consumed 50,000 kg of menhaden, which in turn ate 500,000 kg of algae! Alas, bottle-nosed dolphins tend to eat lower on the food chain, eating more menhaden than striped bass, but it is clear that food chains can only be four or five steps long before the energy is all but gone.

To be fair, aquatic food chains tend to be more efficient than their terrestrial counterparts, with closer to 20% efficiency of energy transfer between trophic levels. Neverless, the vast majority of energy stored at one trophic level will never reach the one above.

Eating lower on the food chain means more energy (food) is available to be consumed. This is as true for humans as it is for fish and dolphins. It takes about 10 kg of plant food to produce 1 kg of beef. So, it takes about the same amount of land to produce one beef burger as it would ten veggie burgers of the identical mass. Animal based foods account for about 30% of the calories consumed in the SAD (USDAERS 2016). Assuming all food was from land, this means that the animal food component accounts for about 3.3 times as much land use as is needed for the rest of the SAD which is based upon eating plants.

Food chains are handy ways of thinking about energy flow through the ecosystem. But in reality, most consumers are not restricted to eating just one particular plant or animal or at one particular trophic level. Omnivory, eating multiple types of species is the rule rather than the exception. Animals will choose different dietary items depending on relative availability, feeding preference, energetic reward, season, and life cycle. For example, striped bass appear to prefer menhaden to most other prey choices. The oily menhaden is a dense source of calories. Yet over-fishing of menhaden has driven a shift to eating the less nutritious blue crab (*Callinectes sapidus*) (Uphoff 2003). Thus, a food web better depicts the energy relationships between species than a simple food chain (Fig. 8).

There are different types of food-chains in the Bay. To understand them, it is important to grasp a bit of basic terminology that describes lifestyles of species. *Plankton* refers to the microbes, plants and animals that drift in the waters of the Bay. Some members of the plankton simply float along with the water currents. This is true of many of the algae that constitute the *phytoplankton* (plant-like plankton), as well as the microbes that make up the viroplankton and bacterioplankton. Others can swim and control their vertical distribution in the Bay, but are not strong enough to resist the horizontal currents. The Atlantic stinging nettle jellyfish (*Chrysaora quinquecirrha*), copepods, and the larvae of oysters are examples of such swimming zooplankton (animallike plankton). Almost all of the living creatures of the Bay spend at least the beginning portion of their life cycles in the plankton, even if they grow up to live elsewhere.

The *nekton* refers to the good swimmers of the Bay. Examples include fish, blue crabs, turtles, and dolphins. All

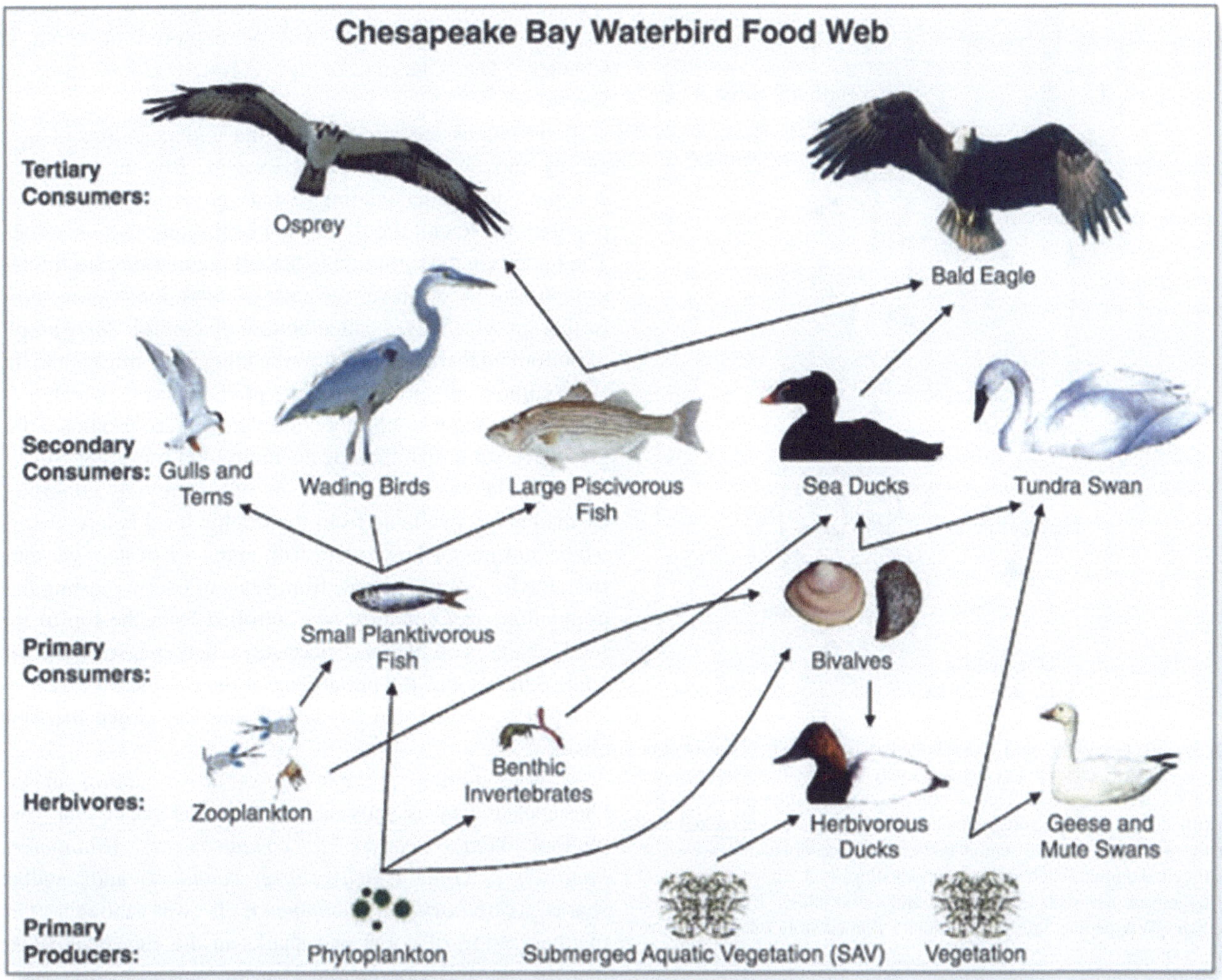

Fig. 8 A portion of the food web of the Chesapeake Bay emphasizing the role of waterbirds. Matthew C. Perry—US Geological Survey. "Chapter 14: Changes in Food and Habitats of Waterbirds." Figure 14.1. Synthesis of U.S. Geological Survey Science for the Chesapeake Bay Ecosystem and Implications for Environmental Management. USGS Circular 1316. https://pubs.usgs.gov/circ/circ1316/html/circ1316chap14.html

but the dolphins start their lives in plankton but mature into the nekton.

Benthic plants and animals live on the bottom of the Bay, from shallow marshes to the deepest channels. Sea lettuce (a leafy algae), oysters, blood worms, and snails are all members of the benthos as adults but begin life in the plankton.

The most familiar food chain goes like this: phytoplankton >zooplankton> small fish> big fish. One benthic food chain goes this way: phytoplankton>oyster> fish or bird. Another benthic food chain goes as follows: detritus> bacteria> worms> blue crabs or fish> big fish.

One critically important but often overlooked food chain in the Bay is called the microbial loop (Fig. 9). This one goes dissolved organic carbon (DOC)>bacteria> protozoans (single-celled animal like)> copepods> small fish> big fish. The microbial-loop food chain helps sustain the large populations of planktonic, benthic and necktonic creatures of the Bay. It utilizes the dissolved organic products of decomposition that wash into the Bay from the watershed and those created in the Bay proper. Algae, plants and animals all leak DOC into the water. Although one can't see it, DOC is the largest pool of biomolecules in the Bay. In other words, if one were to take a square area of the Bay and add up all organic matter in the plankton, nekton, and benthos, it would be less than the DOC in the water under that same patch! The millions of bacterial cells per milliliter (ml) feast on the abundant DOC, and thus making this invisible resource available to the rest of the food chain.

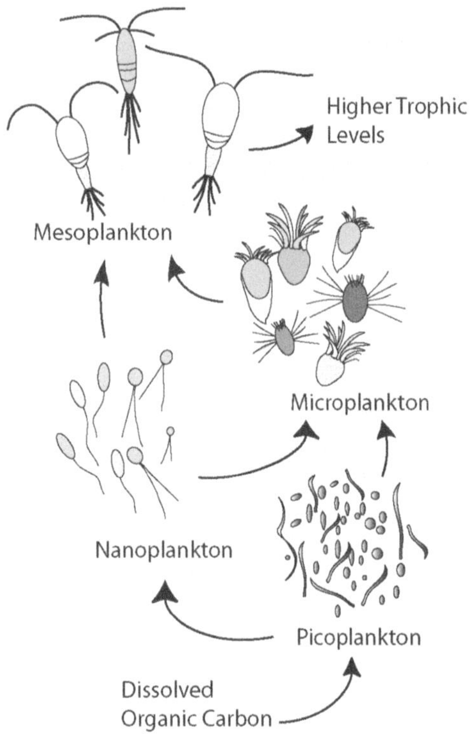

Fig. 9 The microbial loop food chain based upon dissolved organic carbon. The picoplankton refers to mostly bacteria that consume the dissolved organic carbon. The nanoplankton are smaller protozoa (single celled animals). Microplankton are mostly larger protozoa and they eat the picoplankton and nanoplankton. The mesoplankton are small multi-celled animals such as the copepods illustrated. They serve as food for larger animals such as small fish and comb jellies. By Umarina at English Wikipedia. https://commons.wikimedia.org/wiki/File:Microbial_Loop.jpg

The role of bacteria and the importance of the microbial loop were only recently understood, beginning in the 1970s. Likewise, physiologists are only now grasping the importance of the microbial community that lives in the guts of humans. For example, a healthy human microbiome in the large intestine converts dietary fiber to small chain fatty acids necessary to feed the cells in the walls of the colon (Possemiers et al. 2011; Sonnenburg and Sonnenburg 2014).

Since omnivory is the rule, the feeding relationships in the Bay are best thought of as a food web that links all plants and animals together to greater and lesser degrees. The complex food web of the Bay helps make for a stable ecosystem. If one species is diminished, say Atlantic croaker (*Micropogonias undulatus*) for example, it may be replaced in the diets of predators by a similar fish such as spot (*Leiostomus xanthurus*). However, the stability of the food web breaks-down with large fluctuations in the populations of certain key species. For example, the over exploitation of oysters for the food system during the twentieth century had dramatic effects on the quality of water in the Bay and the populations of other species that either feed on the oysters or used the oyster reefs as essential habitat. This will be explored in depth in Chapter "The Chesapeake Bay Oyster: Cobblestone to Keystone".

Food chains are important to the understanding of how energy moves through the ecosystem. But they are also essential to understanding what governs the size of populations. Recall the discussion of limiting factors earlier. The underlying assumption is that all populations are limited in their size by whatever resource or condition is preventing further growth. This is called bottom up control. An example of bottom up theory is as follows: adding more nitrogen to the Bay promotes the growth of more phytoplankton, which then causes an increase in zooplankton, which promotes the growth of small fish that means more food for larger fish.

If populations of plants and animals were only controlled by the energy available from the trophic level below, then it would not matter how many fish, crabs, or oysters we took from the Bay. Yet we know from years of overharvesting that populations can certainly be controlled from the top of the food chain. Top down theory says that a consumer can control the size of the population in the trophic level below, and that the effect can cascade all the way down the food chain.

A once touted example of top-down control in the Chesapeake Bay is provided in the food chain that goes phytoplankton> oysters > cownose rays (*Rhinoptera bonasus*) > cobia (*Rachycentron canadum*) and sandbar sharks (*Carcharhinus plumbeus*). It was thought that overharvesting of cobia and sharks in the Bay enabled an increase in the population size of cownose rays. The increased ray population supposedly meant more predation on oysters and other mollusks (Myers et al. 2007). Yet more recent research casts doubt on this cascade, exonerating cownose rays for failures of oyster recover efforts (Grubbs et al. 2016). Nevertheless, there is no doubt that humans exerted sufficient top down control through fishing as to drastically reduce key populations in the Bay.

The consequences of over-harvesting seafood from the Bay will be explored in Chapter "A Fishing Trip: Exploiting and Managing the Commons of the Chesapeake Bay."

One way to understand the health of the Chesapeake Bay is to compare the relative size of current populations with those of earlier times. The contemporary population size of oysters is about 1% of what it was in 1900. In contrast, the population size of various species of phytoplanktonic algae is many times that of earlier days (NOAA 2019).

Given unrestricted resources and the right conditions, all populations are capable of very rapid exponential growth. This is especially true of small species capable of quick reproduction. Consider phytoplanktonic algae in the Bay. During the long solar days of summer, the water is warm and, if provided with sufficient nutrients such as nitrogen and phosphorus, algal cells can divide once each day. One cell will give rise to two, then four, then eight, then 64 and so forth. After 20 days that population of a single algal cell would have grown to a size of 1,048,576 cells. Of course, in nature some of the algae will get eaten by zooplankton or oysters, and others will sink to the sediments, yet, it is clear that populations possess the ability to grow rapidly.

Biodiversity In his Map of Virginia, Capt. John Smith used the term "diverse" seven times in describing the natural wealth of the Chesapeake Bay and its environs (Smith 1612). This suggests that he understood the importance of variety in nature. Biodiversity is a hallmark of stable and productive ecosystems. Each different species of plant, animal, and microbe functions in the processing of energy and chemicals. They also work together in a complex set of interactions that sustains the integrity of the ecosystem over time, enabling it to endure predictable cycles of nature, such as seasonal change, and less frequent but intense situations such as fires, extended droughts, and severe storms.

Some species create living space and conditions for many others. The reefs built by generations of oysters (*Crassostrea virginica*) provide habitat and food for a variety of fish, birds, crabs, sponges, worms, and numerous other species (Gedan et al. 2014). The same is true for the beds of underwater plants created by eel grass (*Zostera marina*), that blanketed the shallows of the lower Chesapeake Bay during Capt. John Smith's time (Lefcheck et al. 2017). The oyster reefs and grass beds provide the important ecosystem function of cleaning the water. The oysters remove excess algae and sediments from the water. The eel grass beds slow water movement, causing particles to settle, and take up excess nutrients from the water. Since their presence is so important for the viability of the ecosystem, the oysters and eel grass are termed foundation species.

In addition to celebrating the natural biodiversity of the region, Capt. John Smith noted how the Indians applied diversity in planting their crops. "In May also amongst their corne they plant Pumpeons, and a fruit like unto a muske millen, but lesse and worse, which they call Macocks. These increase exceedingly, and ripen in the beginning of July, and continue until September. They plant also Maracocks a wild fruit like a lemmon, which also increase infinitely (Smith 1612)." The Algonquin mimicked nature with a system of polyculture (many species), planting a variety of species together. They knew the corn stalks provided support for their climbing and clinging beans, but probably didn't understand that the beans enriched their garden by providing the service of nitrogen fixation (See Chapters "The Algonquin Food System and How it Shaped the Ecosystem and Interactions with the English Colonists of the Chesapeake Bay and The Journey from Peruvian Guano to Artificial Fertilizer Ends with too Much Nitrogen in the Chesapeake Bay").

The decedents of the colonists, and others that followed later, seemed to have lost their faith in the variety of life revered by their Algonquin predecessors. This is evident in the chemical-industrial agriculture so widely practiced in the watershed since the mid-twentieth century.

Monocultures (single species, often genetically identical) now occupy great swaths of land, planted all at the same time in uniform rows to accommodate farm machinery, and all harvested at the same time. Monocultures are virtually absent in nature, as they are inherently unstable, and soon overtaken by insects, rodents and other consumers. They also lack the services of nutrient recycling and soil protection that would be provided by neighboring species. As such, agricultural monocultures require massive inputs of energy, fertilizers, pesticides and water to produce a useful harvest. Chapters "A New Food System for the Chesapeake Bay Region and a Changing Climate and An Organic-Based Food System: A Voyage Back and Forward in Time" address the problems of monoculture and alternative forms of agriculture that embrace biodiversity.

Pollution and Cultural Eutrophication of the Bay and Its People Pollution of the Chesapeake Bay ecosystem takes various forms, much of which traces directly or indirectly to our food system. Excess nutrients and organic matter drive cultural eutrophication. This process is a growing problem around the world, causing coastal dead zones of low oxygen in the Gulf of Mexico and over 400 other locations (Diaz and Rosenberg 2008). As noted above, excess nutrients from agriculture and human sewage promote the rapid growth of phytoplankton. When the algae die and sink to the bottom waters, they are decomposed by bacteria, which use all or most of the available oxygen in the process. Organic waste, such as those found in sewage, animal manure, or from paper mills, also get attacked by the oxygen consuming bacteria. Cultural eutrophication is responsible for the annual occurrence of the dead zone in the Chesapeake Bay, where oxygen levels are too low to support higher forms of life such as fish, crabs and oysters (Hagy et al. 2004, Chapter "Reversing the Eutrophication of the Chesapeake Bay and its People").

Plastic bottles, bags, and food wrappers are solid waste pollutants (See Chapter "Plastic Pollution and the Chesapeake Bay: The Food System and Beyond"). Agricultural pesticides are toxic pollutants, causing harm to the human and non-human inhabitants of the ecosystem (See Chapter "Pesticides Bring the War on Nature to the Chesapeake Bay").

We can think about pollutants as contaminates of water, land, and air, which is true. But we should also recognize that most pollutants are misplaced resources. For example, animal manures are a valuable resource when used properly on agricultural lands for restoring fertility to the soil. But when misplaced by our food system into the Chesapeake Bay, they create cultural eutrophication.

The process of cultural eutrophication by the addition of excess inorganic nutrients provides too much nutrition, and the wrong kind of nutrition for the Bay to be healthy. Since the extra nutrients that feed eutrophication come from the watershed, which is outside the Bay proper, they are called *allocanthous*. This is different from the *authocanthous* nutrients that are recycled within the Bay. For example, oysters secrete ammonia (NH_4) as a waste product from digesting algae they have eaten. That released nitrogen may then fertilize other algae, sustaining the ecosystem through natural recycling of nutrients.

Another type of allocanthous input originating from the watershed is organic matter (detritus). Some is in the form of dead and decaying plant matter, mostly leaves and pieces of wood. Decaying marsh plants are an important source of detritus in the Bay. Detritus is different from the inorganic nutrient pollution mentioned above. It is made mostly of plant fibers like cellulose and ligands that are difficult to digest and decompose. It is much like what constitutes a high fiber diet for humans. While such detritus does contain some nitrogen, it is in a very low proportion to the carbon in its molecules. It takes weeks or months for detritivores, bacteria, and fungi to decompose this fibrous material. So, the nitrogen in the plant detritus is released more slowly than the pulses of inorganic nutrients that are washed in from farmers' fields. The slow nutrient release from natural detritus doesn't cause the extensive algal blooms linked to inorganic nitrogen released from industrial agriculture.

Humans also add a big helping of processed sewage to the Bay. Native Americans and colonists traditionally urinated and defecated on the land. That kept the organic matter and nutrients in the soil. The growth of towns and cities in the watershed coupled with the adoption of the flush toilet in the mid-nineteenth century changed this, with human excrement carried in pipes to the nearest river. Prior to the advent of wastewater treatment plants in the second half of the twentieth century, sewage flowed directly into tributaries of the Bay. The raw sewage provided a feast for waterborne bacteria, their respiration drawing down the available oxygen. The sewage also contained pathogens that caused dysentery other diseases, and forced the closing of prime oyster grounds due to contamination.

These days, most of the organic matter of sewage is digested and collected as biosolids (sludge) at wastewater treatment facilities that dot the coastlines of the tributaries draining into the Bay. However, the decomposition process at the treatment plants, which do an excellent job of removing excess organic matter and killing pathogens, still releases large quantities of nitrogen and other nutrients to the Bay, adding to cultural eutrophication. Improved ways of treating sewage are addressed in Chapter "Finishing the Journey: Wastewater as a Misplaceed Resource".

It is not just human feces and urine that contributes to cultural eutrophication of the Bay. Our current food system relies on highly concentrated facilities to feed chicken, pigs, and cattle. These Concentrated Animal Feeding Operations (CAFOs) house thousands of animals on small plots of land that produce tons of waste daily. Much of the organic matter and nitrogen from those wastes wash into the Chesapeake Bay, contributing to cultural eutrophication. This is be covered in Chapters "The Journey from Peruvian Guano to Artificial Fertilizer Ends with too Much Nitrogen in the Chesapeake Bay and Livestock and Poultry: The Other Colonists Who Changed the Food System of the Chesapeake Bay".

The cultural eutrophication of the Chesapeake Bay is closely linked to our food system. That same food system which is driving cultural eutrophication of the Bay is having a similar effect on the human residents of the watershed. As with the Bay, too much of the wrong kinds of nutrition produced by the food system is making people sick. The Standard American Diet (SAD) drives obesity, heart disease, diabetes, and a whole host of other human diseases.

Toxins Pollution from nutrients and organic matter are detrimental, but generally not toxic. Indeed, they promote the growth of lifeforms such as algae and bacteria and as such imbalance the ecosystem. A consequence may be oxygen depletion, which is lethal to many animals. But excess nutrients and organic matter pollution aren't directly deadly.

In contrast, insecticides and herbicides are designed to kill insects and weeds, so naturally they can kill or sicken animals in the Bay. Thanks to water pollution laws, releases of toxins that outright kill fish and other animals in the Bay are rare events these days. However, the widespread use of a variety of pesticides in our food system still creates problems for many species, such as lower birthrates or changes in behaviors. Acute toxicity is generally deadly or highly damaging. Chronic toxicity does not kill outright but diminishes the health of affected populations. This is true for fish in the Bay such as striped bass, as well as for the people that eat the fish (MD Fish Consumption Advisory 2014). Mercury is a common chronic toxin in predatory fish of the Bay. Among other things, mercury diminishes the function of the nervous system, something to think about as long you still can, if you are eating too much fish. Mercury enters the Bay primarily as a byproduct of coal burning to produce electricity. A significant portion of that electrical energy is used to subsidize our food system, from production of fertilizers, to refrigeration and food processing.

Atrazine is a very widely used herbicide designed to kill broadleaf plants by disrupting their photosynthetic system. Since animals don't photosynthesize, this is not a problem for them. Yet, Atrazine disrupts the endocrine system in animals, interfering with the activity of hormones critical for development and organ system functioning. It also causes cancer in mammals, including humans (Fan et al. 2007). So, chemicals may be toxic in one way for one species, and in another way for a different species.

To understand the distribution of toxic substances in the Bay, one must appreciate the role of biological magnification of materials up the food chain. Recall that water is polar and that nonpolar substances can dissolve in it, but only to a limited degree. Many toxins have limited solubility in water and quickly move from water into living cells, be they cells of bacteria, crabs, or fish. Often the cells have limited ability to either degrade or excrete the toxin, so it tends to stay put in the living tissue. Consider mercury. It may enter the Chesapeake Bay in the elemental form from the atmosphere as a consequence of burning coal. Being heavy the mercury quickly sinks to the sediments. There, in the absence of oxygen, it is changed to methylmercury, which enters bacterial cells. Millions of those bacteria are consumed by a worm which ends up storing the mercury in its tissues. A small fish like an Atlantic croaker (*Micropogonias undulates*) might eat that worm and hundreds of others like it, retaining most of the mercury in its tissues (Gilmour and Riedel 2000). A striped bass that ate 100 croakers would retain most of the mercury from its meals. An angler that ate the striped bass would get its full load of concentrated mercury. This is why authorities warn against eating more than one or two servings of striped bass a month. To summarize, biological magnification concentrates toxins at ever increasing levels with each step in the food chain. The reader will learn more about toxic substances in the Bay in Chapter "Pesticides Bring the War on Nature to the Chesapeake Bay".

Biological Pollution Another form of pollution in the Bay is the accidental or purposeful introduction of new species that did not evolve there. Such exotic introduced species may disrupt the stability of the food web. Lacking local enemies in their new environment, exotics may display population explosions at the expense of native species. The Veined Rapa Whelk (*Rapana venosa*) is one such invasive species that appeared in the lower Bay in 1988, probably transported accidently in the ballast water of ships from Asia. It consumes local bivalves such as clams, mussels and oysters (Harding and Mann 2005).

Exotic parasites often devastate native species lacking any history of coevolution. MSX (*Haplosporidium nelsoni*) kills the native eastern oyster (*Crassostrea virginica*) and appears to have arrived in the region in plantings of the Pacific oyster (*C. gigas*) imported from Japan. If so, this is an unintended consequence of an attempt to alter the local food system (Hagan 2018).

Perhaps 90% of the Native American population in North America was wipedout by infectious diseases brought by Spanish explorers and colonists (Walbert 2016). Most of those pathogens came from the European food system that utilized livestock: birds, cattle, sheep, and pigs. Measles, smallpox, influenza, diphtheria, the common cold, and tuberculosis all trace to animal husbandry. The Europeans had coevolved with these zoonotic (of animal origin) diseases, but for the Native Americans, the pathogens were all newly introduced species for which their immune systems were unprepared. These diseases of the European food system were so potent that they spread faster than conquistadores could carry them. Some scientists suggest that the Algonquin of the Chesapeake escaped most of these devastating diseases due to their relative isolation and habit of migrating between agricultural settlements and hunting/fishing camps on a seasonal basis (Rountree et al. 2007). The Covid-19 pandemic of 2020 originated with exotic animals (bats) captured for food.

Circulation in the Bay and Human Bodies The circulation of fluids in humans is essential for good health. The same is true for the circulation of water and the health of the Chesapeake Bay. For both humans and the Bay, circulation provides nutrients and oxygen where needed and removes waste products.

The heart is the main driver of circulation in the human body (Fig. 10). The pumping heart ensures that the cells of the body are constantly supplied with oxygen, and food, while being relieved of waste products like CO_2 and ammonia (NH_4). The right side of the heart collects blood from all over the body and sends it to the lungs where it refreshes its supply of oxygen and releases much of its CO_2. The oxygen-rich blood returning from the lungs enters the left side of the heart where it is pumped back to the rest of the body.

Human bodies have sufficient reserves of nutrition to survive weeks without eating but starved of oxygen people die within minutes. A stopped heart leads to sudden death. Heart failure is most often a result of clogged cardiac arteries that supply the heart muscle. The clogging is from deposits of cholesterol originating from the consumption of animal-based foods, the center of the SAD. Cholesterol is essentially absent from plant-based food. While people do make cholesterol themselves, it is those consuming the extra dietary cholesterol that develop heart disease and consequently die of cardiac arrest.

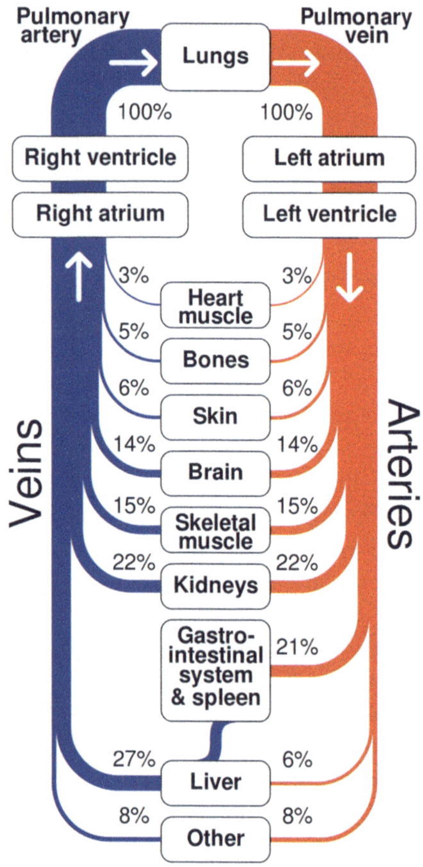

Fig. 10 The circulation of blood in the human body illustrating the pattern of relative distribution to various organs given as percent of flow. By CMG Lee. https://upload.wikimedia.org/wikipedia/commons/d/dd/Sankey_diagram_human_circulatory_system.svg

There is no heart to supply the force for circulating the waters of the Chesapeake Bay. Yet the SAD that interferes with the supply of oxygen in the human body is based on a food system that also interferes with the supply of oxygen in the Bay, as will be illustrated below.

Absent a pumping heart, gravity and wind do the work of moving the Bay's waters. Gravity circulates water in the Bay by two different means, driving tidal flow and as a component of the hydrological cycle.

First, consider the tides. The moon and the sun both exert gravitational force upon the Earth. It is most evident in the coastal sea as the gravitational forces from those celestial bodies cause the water to slosh back and forth in the various ocean basins. A flooding tide is one that moves water into the Chesapeake Bay from the Atlantic Ocean, resulting in a high tide. This is followed by the ebbing tide of water moving back to the ocean and resulting in a low tide. The brief periods of an hour or so between flooding and ebbing is called slack tide. The tidal rhythm of the Bay is called semi-diurnal, meaning that there are two high tides and two low tides each day of roughly equal tidal height. When the Sun, Moon, and Earth all fall in a line, the gravitational forces are additive, resulting in big swings in the tide, producing the highest high and the lowest low tides. These are called *spring tides* and happen twice a month, on the days around the new moon and the days around the full moon. When the moon, earth, and sun form a right triangle, the gravitational forces of the sun and moon are out of phase, producing *neap tides* of the least extreme between high and low. Neap tides occur twice a month around the times of the half-moons. A significant portion of the water in the Bay is moved on each tidal cycle, providing water circulation throughout the entire system. The time between a high and a low tide is about 6.5 h. For a portion of the Bay, where the average ebbing tidal velocity is 1 km per hour, a mass of water would move 6.5 km toward the ocean. The tidal flows also cause mixing on small scales, as along the bottom sediments and in marshes along the shores.

Now, consider the role of gravity in the hydrological cycle that feeds freshwater to the Bay. Energy from the sun evaporates water from land and sea. The gaseous water vapor easily floats into the atmosphere where it eventually condenses into clouds. The independent water vapor molecules collide together, forming new hydrogen bonds. The expanding molecules of water form liquid droplets, or when cold enough, crystals of ice. Gravity returns the water to the ground as precipitation (rain, snow, or hail). Some of that precipitation drains off or through the land and into the Bay via an extensive system of tributaries. The force of Earth's gravity keeps the water moving down and out of the Bay into the Atlantic Ocean.

Recall that the freshwater draining into the Bay is less dense than the marine waters filling the estuary from the Atlantic Ocean. The lower density freshwater draining from the land floats on top of the saltier marine water from the sea. This divides the bay into two distinct layers, the fresher surface layer and the saltier deep layer. This layering is most distinct in summer when the surface water density is reduced by elevated temperature as well as low salinity. The seasonal formation of distinct layers is called *stratification*, and this has profound consequences for circulation of water in the Bay. During the warm months of the year, the two layers fit together like matching wedges. The marine salt wedge narrows as it moves up the Bay, while the terrestrial freshwater wedge narrows as it extends to the ocean. This creates the conditions for the pattern of estuarine circulation explained next (Fig. 11).

As the fresher surface waters journey toward the mouth of the Bay, their gravity-driven flow rubs along the saltier and deeper layer. This generates an internal wave that drives the saltier marine waters in the opposite direction, up the estuary. The boundary between the fresher surface and saltier deeper waters provides minimal mixing between the two layers. This

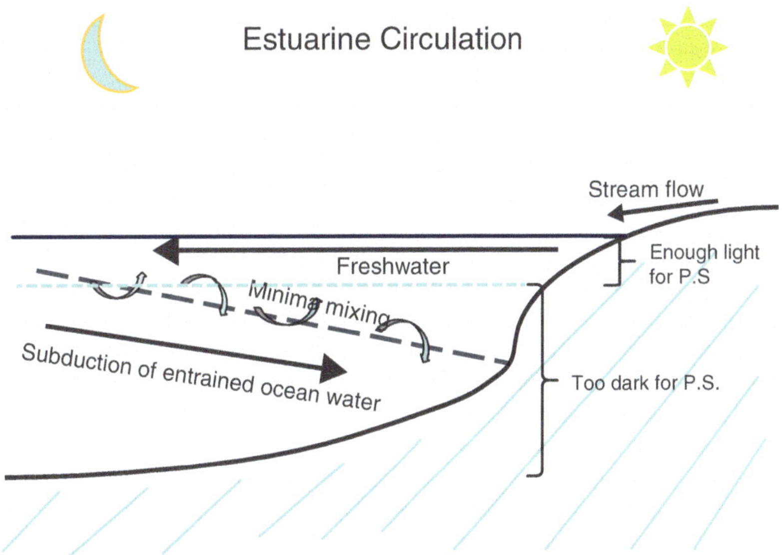

Fig. 11 The momentum of freshwater leaving the Bay transfers to the incoming water from the ocean, with some mixing at the boundary between the two water masses, note the surface waters have sufficient light to support photosynthesis (PS) during the day, but this is not the case for deeper waters. By B. Cuker

sets-up conditions for the deeper waters to be starved of oxygen by the time they complete their journey up the Bay.

The fresher surface water flowing out of the Bay is generally well supplied with oxygen. It has direct contact with the atmosphere, and it contains trillions of tiny algal cells performing photosynthesis during daylight hours, also adding oxygen to the water. But the deeper, cooler, saltier, and denser marine layer continues to sink on its way up the Bay. This means it is prevented from receiving oxygen from the atmosphere. In addition, little sunlight reaches the deeper water so there is essentially no photosynthesis to produce oxygen either. Despite arriving at the mouth of the Bay with a full load of oxygen, that supply is dissipated by bacteria and other decomposers eating dissolved and particulate organic matter. The food system that supplies the SAD makes the oxygen depletion worse by causing cultural eutrophication. The excess N and other nutrients from the food system that wash into the Bay stimulate blooms of algae in the surface waters. The algae then sink into the bottom waters of the Bay. Bacteria and other decomposers use oxygen in aerobic respiration as they feast on the rain of dead and dying algae. Now recall that the deeper marine waters are moving up the estuary, opposite to the freshwater layer above which is flowing out. This means that the residue of algal blooms fueling the oxygen-depleting bacteria is mostly retained in the Bay rather than being washed out to sea.

The cooling of the Chesapeake during fall and winter dissipates the strength of the stratification that forms in spring and summer. This allows for improved circulation of oxygen to the bottom waters and permits some flushing of excess organic matter into the Atlantic Ocean.

Winds also improve circulation of oxygen in the Bay. Winds create surface waves that increase the exchange of gasses with the atmosphere. Surface waves are the ones most people think of, the kind that make for a bumpy boat ride and that lap a shoreline. Sustained winds also create internal waves, ones not evident to the eye. These internal waves cause the different layers of the Bay to rock back and forth. With sufficient wind, the deep low-oxygen layer may rock so much that it is brought to the surface along the shore. That exposes the formerly deep water to the atmosphere where it can "catch a breath" and gain some oxygen (Scully 2010). However, when the low oxygen water first arrives at shore it stresses the animals living there. Blue crabs may crawl to the very edge of the water to escape the hypoxia. Coastal residents of the Bay call this a "crab jubilee" or "crab walk," as it makes for easy capture of one of their favorite foods (Mountford 2008).

This discussion of circulation focuses on blood and lymph vessels in humans and river basins in estuaries simply as the conduits for flow. However, each is more than just a structure to contain the moving liquid. Both the vessels of the body and the basins of the Bay are living structures that perform a variety of functions critical to the health of their respective systems. Endothelium cells that line blood vessels regulate the passage of materials to and from the adjacent tissues they serve. They also play important roles in the immune and inflammation response (Mai et al. 2013) and endocrine system (Ingami et al. 1995).

The sediments and structures that line the shores and bottom of the Bay and its numerous tributaries are as alive as the walls of blood vessels. They support benthic communities diverse with life. Intertidal and shallow areas with sufficient light feature photosynthetic microalgae, seaweeds and vascular plants. Suspension feeders such as oysters, mussels, barnacles, and sponges clean the water of excessive phytoplankton and detritus. These intermingle with the digestive system of the Bay: detritivores (worms, crabs, etc.), bacteria, and fungi. Fish and birds rely on the benthic community for sustenance and habitat.

The more circulation the better? More circulation means better delivery of oxygen and nutrients to human tissues and components of the Bay ecosystems. For humans, the consequence of reduced circulation include increased chances of heart attack, erectile dysfunction (Gupta et al. 2011), Alzheimer's disease (de la Torre 2002), peripheral arterial disease (Rajendran et al. 2013), and lower back pain (Kauppila 2009). For the Chesapeake Bay, poor circulation coupled with eutrophication creates areas of oxygen depletion and means less food for suspension feeders such as oysters, mussels, and sponges. Yet, excessive circulation may cause problems for both humans and the Bay. For humans, cancerous tumors require the proliferation of additional blood vessels (angiogenesis) to support their growth and spread to other organs. Type 2 diabetes, a disease caused by SAD, too often leads to diabetic retinopathy where abnormal blood vessels proliferate on the light sensitive portion of the eye. This diet driven excess circulation is the leading cause of adult blindness in the US (NEI 2018). Excessive circulation also causes problems in the Bay. The heavy precipitation from large storms erodes the soils of the watershed, causing vast quantities of sediments and excess nutrients to flood into the Chesapeake. The sediments clog channels and the nutrients drive eutrophication. Flooding and storm driven waves often destroy marsh and beach habitats.

Subsidies of Energy to the Chesapeake Bay Ecosystem and Its People We usually think of the Chesapeake Bay ecosystem as being powered by sunlight working through the biological process of photosynthesis. Producers in the form of green plants and algae use photosynthesis to turn sunlight into chemical energy. The producers provide energy to the trophic (feeding) levels above, including the decomposers. The solar energy harvested by algae and marsh plants supports a food chain leading all the way through striped bass up to osprey and eagles. The photosynthesis of terrestrial plants including grasses and trees powers a food chain leading to owls and foxes. These sun powered food chains also nourished the Powhatan and the original colonists.

In addition to what's provided by photosynthesis, the Chesapeake Bay ecosystem requires other sources of energy to function. These other non-photosynthetic sources are called *energy subsidies.*

Sunlight that isn't used for photosynthesis acts as an energy subsidy in several ways. The solar energy heats the water of the Bay, the soil, and the air, yielding a temperature regime suitable for the organisms of the ecosystem. The annual solar thermal cycle controls the activity of life on land and in the water. The metabolic rates of the plants, microbes, and most of the animals depend directly on the temperature. The warmth of spring awakens biological activity from winter's cold-induced dormancy. While mammals, birds and a few fish can generate heat from metabolizing their food, the vast majority of life depends on solar energy to provide the temperatures needed for active living. The annual timing of the solar energy subsidy not only regulates the year-round residents of the Chesapeake region, but also determines the patterns of migration for its visiting fish and fowl populations. The Powhatan and other tribes on the Chesapeake planted their crops on fields they knew would warm quickly in springtime, and sunshine, and timed their hunts to correspond with animal migrations (Rountree et al. 2007). See Chapter "The Algonquin Food System and How it Shaped the Ecosystem and Interactions with the English Colonists of the Chesapeake Bay".

Solar thermal energy works with gravity to provide another subsidy by powering the hydrological (water) cycle. Thermal energy from sunlight evaporates water from the salty ocean. This process produces essentially pure water vapor, cleaned of salts and other contaminants. When cooled, the vapor condenses into clouds. Eventually the clouds condense further, and the force of gravity returns the cleaned water to the earth in the form of rain, snow, or sleet.

Solar thermal energy and gravity also team to subsidize the ecosystem by creating wind. In the simplest of terms, wind is the flow of air from an area of high pressure to one of low pressure. This is true on the global scale as well as on the local level such as on the Chesapeake Bay. Gravity works on the mass of air to create pressure. Solar thermal energy warms a mass of air, causing it to expand and thus be of lower density. The lower density means fewer molecules of air, and therefore less atmospheric pressure. In contrast, air cooled such as by losing heat at night, will be of higher density and therefore of higher atmospheric pressure. Air will flow from areas of high pressure to area of low perssure, creating wind.

Capt. John Smith and subsequent sailors of the Chesapeake would all know well the predictable patterns of wind that blow across the Bay on clear days from spring to fall. The morning sun warms the land and the mass of air above it. By midday the warmed, low pressure air above the land begins to float skyward. It is replaced by a mass of cooler air coming from above the adjacent and colder Atlantic Ocean. This sea breeze flowing from the east appears most afternoons during the warmer months and persists till sundown. At night the land may cool enough to create a reverse flow from the west, called a land breeze.

Wind energy subsidies the ecosystem in several important ways. As mentioned above, it drives surface and internal waves that play a role in circulating water in the Bay. Anyone

Fig. 12 A soaring osprey (*Pandion haliaetus*) takes advantage of a subsidy from solar energy that warms rising air masses. The energy saved on flapping wings is available for growth, reproduction, and other body functions. By Pete Markham from Loretto, USA. https://upload.wikimedia.org/wikipedia/commons/thumb/d/db/Pandion_haliaetus_-Sanibel_Island%2C_Florida%2C_USA_-flying-8.jpg/1024px-Pandion_haliaetus_-Sanibel_Island%2C_Florida%2C_USA_-flying-8.jpg

parking a car in the Chesapeake watershed during the spring knows of the wind bourn pollen from pines and oaks that covers their vehicle in a dust of green. Wind is an energy subsidy for plants that lack the services of pollinators such as bees. The wind also disperses the winged seeds of maple and beach trees, and the hairy ones generated by willows (Andersen 1991).

While solar powered photosynthesis feeds the food chain leading to osprey (*Pandion haliaetus*) and bald eagles (*Haliaeetus leucocephalus*), it is the subsidy of solar thermal energy that keeps these birds aloft (Fig. 12). The sunlight heats the atmosphere, causing rising columns of lower-density air (thermals). The osprey and eagles ride the thermals, reducing the energy expended in flapping wings while hunting or migrating (Katzner et al. 2015).

The solar subsidy that helps osprey and other fishing birds stay aloft, also enables the raptors to maintain the integrity of their wings for flight and swimming. After plunging into the Bay in pursuit of a meal, osprey, cormorant and other birds will perch atop a tree limb, pilling, or any convenient structure to dry their wings in the sunshine (Fig. 13). The solar thermal energy and breeze evaporates away the water, leaving the birds lithe for the next flight, and sufficiently buoyant to survive their next swim (Lippson and Lippson 2006). This solar subsidy also reduces the burning of their own energy reserves to power thermogenesis (metabolic production of heat) otherwise needed to evaporate the water from their wet feathers.

In the previous discussion of circulation, the reader learned how gravity from the moon and sun powers tidal currents. These predictable tidal currents provide an energy subsidy to the planktonic larvae of blue crabs and other animals. The crab larvae swim up from the bottom to ride the flooding tidal currents up the Bay. The larvae migrate back down to the bottom to avoid being washed back down the Bay by the subsequent ebb tidal current. Over the course of some days they use the energy subsidy of the flooding tide to make the 200 km (124 miles) passage up the Bay (Welch et al. 1999). They could never swim that far on their own.

The human population of the Chesapeake Bay ecosystem also benefits from the natural energy subsidies of wind, solar heat, and gravity. The Algonquin and the colonists relied on wind to pollinate their corn. Both took advantage of gravity driven tides to quicken their navigation of the Bay and to help trap migrating fish (See Chapters "The Algonquin Food System and How it Shaped the Ecosystem and Interactions with the English Colonists of the Chesapeake Bay and A Fishing Trip: Exploiting and Managing the Commons of the Chesapeake Bay"). The colonists took the gravity subsidy a step further, using impounded water to drive grist mills to grind their grain. Wind subsidies brought Capt. John Smith and company across the Atlantic Ocean and powered much

Fig. 13 A double crested cormorant (*Phalacrocorax auritus*) drying its wings in the sun https://pixabay.com/en/fauna-birds-wild-cormorant-black-2613238/ Free use and no attribution required

of the navigation on the Bay and beyond until the age of steam, two centuries later.

In recent times the residents of the Chesapeake region got better at using the natural energy subsides provided by gravity, solar, and wind. As noted in Chapter "The Bay and Its Watershed: A Voyage Back in Time", in the early twentieth century they built the Conowingo Dam to produce electricity from gravity and falling water. Three generations later, the residents began to use sunlight to produce electricity from photovoltaic cells arrayed in abandoned farm fields or atop buildings. Currently the region is moving very slowly to building offshore turbines to take advantage of the wind subsidy.

The early twentieth century saw the introduction of a new and *unnatural* energy subsidy to the Chesapeake Bay ecosystem. Humans began using fossil fuels to subsidize the food system. Fossil fuels ultimately do come from photosynthesis, but their formation took place over a time span stretching back millions of years. So, the use of these long stored products of ancient photosynthesis is considered a subsidy to the agricultural portion of the ecosystem.

Upon their death, organisms start to decompose. Decomposers such as bacteria and fungus quickly consume the easily digestible components of the dead cells, leaving behind more resistant portions. If that residue is buried in terrestrial or aquatic sediments it may lose contact with free oxygen and exist in an acidic environment. The lack of oxygen and low pH is inhospitable to most decomposers. Over time, the carbohydrates, proteins, lipids, and nucleic acids from these dead organisms are transformed into hydrocarbons to produce the familiar fossil fuels of coal, petroleum and natural gas. With a source of ignition these fossil fuels will combine with free oxygen (burn), releasing the stored energy that eluded decomposers millions of years before. The burning also transforms the carbon previously stored in the fossil fuels into carbon dioxide (CO_2), a greenhouse gas that warms the planet and acidifies the Bay.

Food produced in the modern industrial-chemical food system represents much more energy from fossil fuel subsidy than it does from recent photosynthesis. For every calorie of food produced, it takes about 10 calories of fossil fuel subsidy. The energy from a dinner created by the industrial-chemical food system is about 10% from recent photosynthesis and 90% from fossil fuels! The fossil fuels are used to run farm machinery, process the food, package it, and transport it. Fossil fuel is also used to make artificial nitrogen fertilizers, herbicides, pesticides and pump water to irrigate the land. If a fast food serving of hamburger, fries, and a milkshake seems oily, it should, as most of the energy comes from petroleum and other hydrocarbons (Pimentel and Pimentel 2003; Pimentel et al. 2008; Canning et al. 2017). Chapter "A New Food System for The Chesapeake Bay Region and a Changing Climate" examines energy subsidies in the Chesapeake food system in more depth.

In a very real sense, people consuming the SAD (Standard American Diet) produced by industrial agriculture are like conventional automobiles, getting most of their energy from fossil fuels. People also use fossil fuel subsidies for transportation and heating buildings. This means that fewer of the calories they consume are used in walking or bike riding, and in thermoregulation (using metabolism to create heat). Fossil fuel subsidies are distorting the food system and reducing human physical activity, which contributes to morbidity and mortality (Gremeaux et al. 2012).

The basic science reviewed in this and the previous chapter will provision the reader with sufficient background to undertake a voyage to understanding the interactions between the food system, the Chesapeake Bay, and its people.

References

Andersen M (1991) Mechanistic models for the seed shadows of wind-dispersed plants. Am Nat 137(4):476–497

Armstrong F (1983) Biochemistry, 2nd edn. Oxford University Press, New York

ATSDR (2007) ATSDR - toxicological profile: Benzene. In: Atsdr.cdc.gov. https://www.atsdr.cdc.gov/toxprofiles/tp.asp?id=40&tid=14. Accessed 16 May 2020

Cai W, Huang W, Luther G et al (2017) Redox reactions and weak buffering capacity lead to acidification in the Chesapeake Bay. Nat Commun 8:369. https://doi.org/10.1038/s41467-017-00417-7

Canning P, Rehkamp, S, Waters A et al (2017) The role of fossil fuels in the U.S. food system and the American diet [online]. Ers.usda.gov. Available at: https://www.ers.usda.gov/webdocs/publications/82194/err-224.pdf?v=42804. Accessed 29 Mar 2018

CDC (2016) Facts about Benzene. Centers of Disease Control and Prevention. http://www.bt.cdc.gov/agent/benzene/basics/facts.asp

Chen C, Lin C, Huang B et al (1996) Stoichiometry of carbon, hydrogen, nitrogen, sulfur and oxygen in the particulate matter of the western North Pacific marginal seas. Mar Chem 54(1–2):179–190

Clarke A, Johnston M (1999) Scaling of metabolic rate with body mass and temperature in teleost fish. J Anim Ecol 68:893–905. https://doi.org/10.1046/j.1365-2656.1999.00337.x

Cuker B, Bugyi G (2016) The fundamentals of water and natural Waters. Practice, Progress, and proficiency in sustainability. In: McKeown A, Bugyi G (eds) Impact of water pollution on human health and environmental sustainability. IGI Global, Hershey, PA, pp 1–28. https://doi.org/10.4018/978-1-4666-9559-7.ch001

de la Torre JC (2002) Vascular basis of Alzheimer's pathogenesis. Ann N Y Acad Sci 977:196–215

Deriemaeker P, Aerenhouts D, Hebbelinck M et al (2010) Nutrient based estimation of acid-base balance in vegetarians and non-vegetarians. Plant Foods Hum Nutr 65(1):77–82

Diaz RJ, Rosenberg R (2008) Spreading dead zones and consequences for marine ecosystems. Science 321(5891):926–929. https://doi.org/10.1126/science.1156401

Fan W, Yanase T, Morinaga H et al (2007) Atrazine-induced aromatase expression is SF-1 dependent: implications for endocrine disruption in wildlife and reproductive cancers in humans. Environ Health Perspect 115(5):720–727. https://doi.org/10.1289/ehp.9758

Garrison T (2001) Essentials of oeanography, 2nd edn. Brooks/Cole, Pacific Grove, CA

Gedan K, Kellogg L, Breitburg D (2014) Accounting for multiple foundation species in oyster reef restoration benefits. Restor Ecol 22(4):517–524

Gilmour C, Riedel G (2000) A survey of size-specific mercury concentrations in game Fish from Maryland fresh and estuarine Waters environ. Contam Toxicol 39:53–59. https://doi.org/10.1007/s002440010079

Gremeaux V, Gayda M, Lepers R et al (2012) Exercise and longevity. Maturitas 73(4):312–317

Grubbs R, Carlson J, Romine J et al (2016) Critical assessment and ramifications of a purported marine trophic cascade. Sci Rep 6. https://doi.org/10.1038/srep20970

Gupta B, Murad M, Clifton M et al (2011) The effect of lifestyle modification and cardiovascular risk factor reduction on erectile dysfunction: a systematic review and meta-analysis. Arch Intern Med 171(20):1797–1803. https://doi.org/10.1001/archinternmed.2011.440

Hagan M (2018) NEMESIS: National Exotic Marine and Estuarine Species Information System. [online] Invasions.si.edu. Available at http://invasions.si.edu/nemesis/CH-TAX.jsp?Species_name=Haplosporidium%20nelsoni. Accessed 28 Mar 2018

Hagy J, Boynton W, Keefe C et al (2004) Hypoxia in Chesapeake Bay, 1950–2001: Long-term change in relation to nutrient loading and river flow. Estuaries 27:634–658

Harding J, Mann R (2005) Veined rapa whelk (*Rapana venosa*) range extensions in the Virginia waters of Chesapeake Bay, USA. J Shellfish Res 24(2):381–385. https://doi.org/10.2983/0730-8000(2005)24[381:VRWRVR]2.0.CO;2

He F, Pombo-Rodrigues S, MacGregor G (2014) Salt reduction in England from 2003 to 2011: its relationship to blood pressure, stroke and ischaemic heart disease mortality. BMJ Open 4(4):4549

Ingami T, Naruse M, Hover R (1995) Endothelium as endocrine organ. Annu Rev Physiol 57:171–189. https://doi.org/10.1146/annurev.ph.57.030195.001131

Katzner T, Turk P, Duerr A et al (2015) Use of multiple modes of flight subsidy by a soaring terrestrial bird, the golden eagle Aquila chrysaetos, when on migration. J R Soc Interface 12(112):20150530. https://doi.org/10.1098/rsif.2015.0530

Kauppila L (2009) Atherosclerosis and disc degeneration/low-back pain--a systematic review. Eur J Vasc Endovasc Surg 37 (6):661–670. https://doi.org/10.1016/j.ejvs.2009.02.006

Lefcheck J, Wilcox D, Murphy R et al (2017) Multiple stressors threaten the imperiled coastal foundation species eelgrass (*Zostera marina*) in Chesapeake Bay, USA. Glob Chang Biol 23(9):3474–3483

Lin J, Hu F, Curhan G (2010) Associations of diet with albuminuria and kidney function decline. Clin J Am Soc Nephrol 5(5):836–843

Lippson A, Lippson R (2006) Life in the Chesapeake Bay, 3rd edn. Johns Hopkins University Press, Baltimore, MD

Mai J, Virtue A, Shen J et al (2013) An evolving new paradigm: endothelial cells – conditional innate immune cells. J Hematol Oncol 6:61. https://doi.org/10.1186/1756-8722-6-61

Matés JM, Pérez-Gómeza C, De Castroa I (1999) Antioxidant enzymes and human diseases. Clin Biochem 32:595–603

MD Fish Consumption Advisory (2014) Guidelines for recreationally caught fish species in Maryland. http://mde.maryland.gov/programs/Marylander/fishandshellfish/Documents/Fish%20Consumption%20Docs/Maryland_Fish_Advisories_2014_March17.pdf. Accessed 12 Mar 2018

Mountford K (2008) It's called a jubilee, but the crabs aren't happy & we shouldn't be [online]. Bay Journal. Available at: https://www.bayjournal.com/article/its_called_a_jubilee_but_the_crabs_arent_happy_we_shouldnt_be. Accessed 12 Apr 2018

Myers R, Baum J, Shepherd T et al (2007) Cascading effects of the loss of apex predatory sharks from a Coastal Ocean. Science 315:1846–1850. https://doi.org/10.1126/science.1138657

NEI (2018) Diabetic eye disease. National Eye Institute. https://nei.nih.gov/health/diabetic/retinopathy. Accessed 18 March 2018

NIH (2013) NIH human microbiome project defines normal bacterial makeup of the body. National Institutes of Health [online]. Available at: https://www.nih.gov/news-events/news-releases/nih-human-microbiome-project-defines-normal-bacterial-makeup-body. Accessed 28 Mar 2018

NOAA (2019) Oyster reefs – oysters – chesapeakebay.noaa.gov. In: Chesapeakebay.noaa.gov. https://chesapeakebay.noaa.gov/oysters/oyster-reefs. Accessed 12 Aug 2019

NOAA (2020) Ocean acidification | National Oceanic and Atmospheric Administration. In: Noaa.gov. https://www.noaa.gov/education/resource-collections/ocean-coasts/ocean-acidification. Accessed 16 May 2020

Odum E, Barrett G (2005) Fundamental of ecology, 5th edn. Thomson, Belmont, CA, 598 p

Pimentel D, Pimentel M (2003) Sustainability of meat-based and plant-based diets and the environment. Am J Clin Nutr 78(3):660S–663S
Pimentel D, Williamson S, Alexander C et al (2008) Reducing energy inputs in the US food system. Hum Ecol 36(4):459–471
Possemiers S, Bolca S, Verstraete W et al (2011) The intestinal microbiome: a separate organ inside the body with the metabolic potential to influence the bioactivity of botanicals. Fitoterapia 82 (1):53–66
Prior R, Gu L, Wu X et al (2007) Plasma antioxidant capacity changes following a meal as a measure of the ability of a food to alter in vivo antioxidant status. J Am Coll Nutr 26(2):170–181
Rajendran P, Rengarajan T, Thangavel J et al (2013) The vascular endothelium and human diseases. Int J Biol Sci 9(10):1057–1069. https://doi.org/10.7150/ijbs.7502
Rountree H, Clark W, Mountford K et al (2007) John Smith's Chesapeake voyages, 1607–1609. University of Virginia Press, Charlottesville, VA
Scully ME (2010) Wind modulation of dissolved oxygen in Chesapeake Bay. Estuar Coasts 33:1164–1175
Sebastian A, Frassetto L, Sellmeyer D et al (2002) Estimation of the net acid load of the diet of ancestral preagricultural Homo sapiens and their hominid ancestors. Am J Clin Nutr 76(6):1308–1316
Smith J (1612) A map of Virginia. Oxford: Joseph Barnes, pp. 1–3. In: The complete works of captain John Smith (1580–1631), edited by Philip L. Barbour, (1986) 3 vols., 1:119–190. University of North Carolina Press, Chapel Hill. Available in Capt. John Smith's complete works http://www.virtualjamestown.org/exist/cocoon/jamestown/fha-js/SmiWorks1
Sonnenburg E, Sonnenburg J (2014) Starving our microbial self: the deleterious consequences of a diet deficient in microbiota-accessible carbohydrates. Cell Metab 20(5):779–786
Uphoff J (2003) Predator–prey analysis of striped bass and Atlantic menhaden in upper Chesapeake Bay. Fisheries Manag Ecol 10:313–322. https://doi.org/10.1046/j.1365-2400.2003.00366.x
USDA (2018) Food composition databases show nutrients list. Ndb.nal.usda.gov [online]. Available at: https://ndb.nal.usda.gov/ndb/nutrients/report/nutrientsfrm?max=25&offset=0&totCount=0&nutrient1=208&nutrient2=&nutrient3=&subset=0&sort=f&measureby=m. Accessed 29 Mar 2018
USDAERS (2016) A look at calorie sources in the American diet [online]. Available at: https://www.ers.usda.gov/amber-waves/2016/december/a-look-at-calorie-sources-in-the-american-diet/. Accessed 29 Mar 2018
Uttara B, Singh A, Zamboni P et al (2009) Oxidative stress and neurodegenerative diseases: a review of upstream and downstream antioxidant therapeutic options. Curr Neuropharmacol 7:65–74
Walbert D (2016) Disease and catastrophe. North Carolina Digital History. http://www.learnnc.org/lp/editions/nchist-twoworlds/1689
Waldbusser G, Voigt E, Bergschneider H et al (2011) Biocalcification in the eastern oyster (*Crassostrea virginica*) in relation to long-term trends in Chesapeake Bay pH. Estuar Coast 34:221–231. https://doi.org/10.1007/s12237-010-9307-0
Watts B, Therres G, Byrd M (2008) Recovery of the Chesapeake Bay bald eagle nesting population. J Wildl Manag 72:152–158
Welch J, Forward R, Howd P (1999) Behavioral responses of blue crab *Callinectes sapidus* postlarvae to turbulence: implications for selective tidal stream transport. Mar Ecol Prog Ser 179:135–143
Wetzel R (2001) Limnology, 2nd edn. Academic Press, San Diego, CA
Whelton P, Appel L, Sacco R et al (2012) Sodium, blood pressure, and cardiovascular disease: further evidence supporting the American Heart Association sodium reduction recommendations. Circulation 126(24):2880–28809

Part II

Foundations of the Chesapeake Bay Food System and the Consequences of Over-Extraction

The Algonquin Food System and How It Shaped the Ecosystem and Interactions with the English Colonists of the Chesapeake Bay

Benjamin E. Cuker and Kathryn MacCormick

Maize came from the west
Its two sisters, beans and squash took the same trip some years later to provide for the best
Together these siblings of food remade the way the Algonquin of the Chesapeake Bay lived and ate
Foraging, fishing, hunting and now farming as a new trait
Each new child that was born
Would live well nourished by corn
Then came the English from the east
Upon the Powhatan's maize Colonists chose to feast
These new-comers showed no respect
For the Chief and the people he needed to protect
After years of war
The English took away their access to food and made sure that the Indians would be no more
Yet the ghosts of the Powhatan sill haunt the sacred land
Those sprits cheer anyone that for a better and just food system takes a stand

Abstract

The pre-colonial residents of the Chesapeake Bay region developed a sophisticated food system based upon farming, gathering and hunting/fishing. That food system informed all aspects of their lives. The Mesoamerican "three sisters" system of companion farming began to arrive in the region circa 1100 AD. The introduced crops of corn (*Zea mays*), beans (*Phaseolus vulgaris*) and squash (*Cucurbita sp.*) transformed the Algonquin speaking tribes into part-time farmers. The advent of corn disrupted the preceding food system of hunting and gathering. By 1500 AD, corn accounted for half of the Algonquin diet, reducing their dependence on oysters (*Crassostrea virginica*) and other marine foods. The Algonquin food system relied on fire as a tool to clear lands for farming and hunting, as well as for cooking and making pottery. The widespread use of fire altered the landscape mosaic, leaving plots in various stages of succession. The English colonists that first arrived in 1607 suffered starvation in the winter of 1609–1610 through a combination of bad weather and inability to employ the Algonquin food system. After the colony recovered and was reinforced with new arrivals, it used denial of access to food as a weapon that ultimately rid the colony of all but a few Native American individuals.

Keywords

Native American · Three sisters farming · Hunting · Land management · Fire

The Foundational Food System In 1607 the three ships of English colonists sent to settle the shores of the Chesapeake Bay arrived with but a few months' worth of food in their holds. The Virginia Company of London knew from a previous failed attempt of colonization in nearby Roanoke (what is now North Carolina) in 1587 that the Algonquin speaking natives had a sophisticated food system that produced sufficient excess for trade. The Company instructed the colonists to use their supply of small manufactured trade goods to

B. E. Cuker (✉)
Department of Marine and Environmental Science, Hampton University, Hampton, Virginia, USA
e-mail: Benjamin.cuker@hamptonu.edu

K. MacCormick
Department of Biology, Rappahannock Community College, Warsaw, VA, USA
e-mail: Kathryn.e.Maccormick@dominionenergy.com

B. E. Cuker (ed.), *Diet for a Sustainable Ecosystem*, Estuaries of the World, https://doi.org/10.1007/978-3-030-45481-4_4

obtain food from the Indians during the transition to self-sustainability. Capt. John Smith and the band of colonists he led were expected to use the seeds and farming implements shipped with them to grow their own food in their settlement at Jamestown. Fishing and hunting was to supplement their farming. Alas, bad weather, incompetence, failed relationships with the Algonquin tribes and distraction with the search for wealth, left Jamestown hungry. The inability of the English to feed themselves during the winter of 1609–1610 nearly destroyed the colony through starvation and left a legacy of cycles of violent conflict with the Powhatans over access to the food the Indians produced.

To understand how the evolution of the modern food system shaped the health of the Bay and its residents, and to imagine a more sustainable one for the future, it's important to understand how the original peoples of the ecosystem fed themselves and the impact that left on the environment. Most of what is known of the precolonial Chesapeake Bay's food system comes from three types of information. The first is the chronicles left by Capt. John Smith and other early English writers. The second is archelogy that explores physical remains from that era. The third is paleoecological records left in the sediments deposited before the time of colonization.

The Algonquin speaking people of the Chesapeake Bay encountered by Capt. John Smith displayed a resilient culture that was evolving to better meet the challenges and opportunities offered by a dynamic ecosystem. The Algonquin of 1607 weren't the same people who lived there 15,000 or 500 hundred years before. Although sharing some genetics with the earlier residents, the Algonquin tribes were remarkably different from their ancestors. Waves of migration and ecosystem change informed the culture of the 1607 Algonquins. The Paleo-Indians (16,000–8500 BCE) carved a living out of the forest by subsisting on game and gathering wild fruits and vegetables during the brief summers and fall. As the climate warmed, nuts, fruits, tubers, wild rice, fish and marine invertebrates became more available. The Archaic Period Indians (8500–1000 BCE) took advantage of these new foods. But it would be the subsequent Woodland Period Indians (1000 BCE–1650 AD) that evolved the culture that greeted the English Colonists. The ancestors of the Powhatan and other tribes invented pottery to cook and store their foods. And they invented a new food system that leaned heavily upon farming crops introduced from Mesoamerica, thousands of km away (Roundtree and Roundtree 2014).

The Native People's Relationship to Food Unlike the English colonists, food was the central organizing agent of Algonquin and other native tribal cultures. These peoples certainly occupied their time with other pursuits such as creating art, tool making, storytelling, courtship, taking care of family and neighbors, building shelters, making clothing, religious rituals, enjoying each other's company, appreciating nature, and engaging in war. Yet all of these other activities were either secondary to, or at least tightly linked with food. Art included pottery for cooking and storing food. Toolmaking produced implements for cooking, hunting, fishing and cultivation. Rituals helped sort out decisions regarding time to hunt, plant and harvest. War making sometimes meant protecting or securing access to land for farming or hunting.

The English colonists no longer clung to the central focus on food that certainly informed their own ancestors. Those early colonists came to the Chesapeake not to find food, but fortune. They could have stayed in England and fed themselves. Instead they crossed an ocean in search of wealth. That attitude helped set the stage for the famine they suffered in 1609–1610. While colonization of the Chesapeake and the rest of what is now the US ultimately succeed due to the food-focused efforts of future generations of farmers, in many ways the attitude of the original English at Jamestown informs modern life in the US. The idea of a home-cooked family meal is on the wane, with half of the family food dollar used to buy meals prepared out of the kitchen (Saksena et al. 2019). Only 2% of the US population is directly involved in the production of food (USDA 2014). Half of the population is sick from eating too much of the wrong foods and too little of the right ones (Afshin et al. 2019; See chapters "Reversing the Eutrophication of the Chesapeake Bay and is People" and "Ethics and Economics of Building a Food System to Recover the Health of the Chesapeake Bay and Its People").

The Native Food System Through the Eyes of the English The relationship between the Indians of the Chesapeake Bay and the English fluctuated between times of war and peace. The Powhatan and other tribes saw the colonists as a source of desirable trade goods such as hatchets, sewing needles, mirrors, and ornaments. The tribes also wared with each other and perceived the English as potential well-armed allies. During intervals of peace, Capt. Smith explored much of the Chesapeake and its shores, taking careful notes on the native food system. This was more than observation at a distance. He attended many a feast put on by local chiefs to honor his presence and duly noted the menu. The Algonquin food system merged three modes of acquisition; foraging, hunting/fishing, and farming. Smith and other early English explores paint a picture of how the Native food system functioned.

The Algonquin's used a calendar with five seasons, all organized around their food system. *Cattapeuk* (early to midspring) was busy with planting and fishing the anadromous species (particularly shad, herring and striped bass) on their spawning runs up the tributaries of the Bay. They also harvested tubers of tuckahoe (*Peltandra virginica*) from freshwater marshes as source of starch. The foraging and fishing required the Indians to disperse from their villages.

Cohattayough (late spring to midsummer) meant the time to forage from the local environment for berries, greens and grains as well as to tend the emerging crops. Fishing continued with focus on the large Atlantic sturgeon that came up tributaries to spawn (*Acipenser oxyrhynchus*).

Nepinough (late summer), meaning earing of corn, was busy with weeding fields and the first harvests of corn, beans and squash. They also foraged for wild rice. The starchy foods gave the Indians a positive energy budget, allowing them to add adipose to their otherwise lean bodies.

Taquitock (autumn to early winter), brought the largest harvest of corn, which required processing for storage. But the work paused for extended days of feasting to celebrate food sufficiency for the coming months. This was also a time of food-strengthened bodies and free time, allowing the men to make war on other tribes. After the harvest and warring, the population dispersed from their towns to inland areas rich in game and forage.

Popanow (late winter to early spring) was the leanest time of the year, absent fruits, nuts and other high energy density forage. Fishing weirs provided some food, but the Indians had to rely on stores of corn and nuts from the preceding months to make it through the cold days (Roundtree and Roundtree 2014).

Roundtree and Roundtree (2014) notes that this food system calendar provided the Algonquin with a diet replete with phytonutrients, fiber, and omeg-3 fatty acids for most of the year. These are just the things missing from the Standard American Diet that wrongly nourishes most of the present residents of the Chesapeake Bay region.

Foods Taken from the Ecosystem by Native Tribes Smith published a list of foods he saw collected, hunted, or fished by Chesapeake Bay Indians in his *Map of Virginia* (Smith 1612). He used European names for the species that resembled what he knew from the Old World and Indian nomenclature for previously unfamiliar items. The following paragraphs annotate his list with the currently accepted common and scientific names for what he was probably describing. Some of the names he used refer to very diverse assemblages with multiple species, so in those cases only the genus, family or order is listed.

Smith noted the Indian's use of fruits and nuts gathered from trees including; Walnuts (*Juglans sp.*), acorns (*Quercus sp.*), plums (*Prunus americana*), cherries (*Prunus serotina*), mulberries (*Morus rubra*), grapes (*Vitis sp.*) chinquapin (*Castanea pumila*, beech and oak family), and passion flower fruit (*Passiflora incarnate*). Shrubs and herbs included strawberries (*Fragaria virginiana*), raspberries (Rubus sp.), tuckahoe tubers (*Peltandra virginica*), ground nuts and the associated tuber (*Apios americana*) and small onions (*Allium sp.*). For salads they used various greens including; violets (*Viola sp*), purslane (*Portulaca oleracea*) and sorrel (*Rumex acetosa*).

Mammalian game included; eastern white-tailed deer (*Odocoileus virginianus*), eastern grey squirrel (*Sciurus carolinensis*), flying squirrel (*Glaucomys sp.*) opossum (*Didelphis virginiana*), muskrat (*Ondatra zibethicus*), hare (*Sylvilagus floridanus*) black bear (*Ursus americanus*), beaver (*Castor canadensis*), and river otter (*Lontra canadensis*).

Birds taken for food included; partridge (bobwhite quail *Colinus virginianus*), wild turkey (*Meleagris gallopavo*), swan (*Cygnus columbianus*), brant (Canada goose, *Branta canadensis*), various ducks (*Anas sp., Aythya sp.*) and passenger pigeon (*Ectopistes migratorius,* now extinct, see chapter "Passenger Pigeon and Waterfowl: Flights to Extinction and Not").

Fish included; Atlantic sturgeon, (*Acipenser oxyrhynchus*) Atlantic stingray (*Hypanus sabina*), mullet (*Mugil sp.*), white salmon (Salmonidae), trout (*Salvelinus fontinalis*), sole (Achiridae) plaice (flounder, *Paralichthys sp.*) herring (*Clupea harengus, Alosa sapidissima, A. mediocris*) conyfish (*Cephalopholis fulva*), rockfish (*Morone saxatillis*), American eel (*Anguila rostrata*), sea lamprey (*Petromyzon marinus*), catfish (Ictaluridae), shad (*Dorosoma sp.*), perch (Perciformes), puffer fish (*Sphoeroides machlatus*), and sea robin (*Prionotus scitulus*).

He recorded the presence of several marine mammals; grampus (Risso's Dolphin, *Grampus griseus*) porpoise (harbor porpoise, *Phocoena phocoena* and probably bottlenose dolphin, *Tursiops truncatus*) and harbor seals (*Phoca vitulina*) (Blaylock 1985). Although Smith didn't document that Indians ate any of these. Colonial hunters eliminated harbor seals, which are presently returning to the mouth of the Bay in small numbers (NOAA 2018).

Marine invertebrates were the: eastern oyster (*Crassostrea virginica*) cockles (hard clam, *Mercenaria mercenaria*, macoma clam, *Macoma spp.*, soft shell clam, *Mya arenaria*) muscles (Atlantic ribbed mussel, *Geukensia demissa*, hooked mussel, *Ischadium recurvum*). Although not in the list from *Map of Virginia*, Smith writes on other occasions about blue crabs (*Callinectes sapidus*).

Algonquin Framing In a section of *Map of Virginia*, called, "Of their planted fruits in Virginia and how they use them," Capt. John Smith explained the Algonquin system of farming. Although not calling it such, it was the "three sisters" system typical of Native American agriculture across North America (Fig. 1, Landon 2008). The sisters were corn (*Zea mays*), beans (*Phaseolus vulgaris*) and squash (*Cucurbita sp.*). The following flows from that description and includes

Fig. 1 The three sisters system of companion planting used by the Powhattan and many other tribes across North America. The seeds of corn (*Zea mays*), and beans (*Phaseolus vulgaris*) were planted together. The beans wound around the corn stalk for support. Squash (*Cucurbita sp.*) of various types were planted between the corn stalks and their expansive leaves shaded the ground below, inhibiting growth of weeds. Digital and pencil drawing by Anna Juchnowicz. Content free to use: https://commons.wikimedia.org/wiki/File:Three_Sisters_companion_planting_technique.jpg

annotations based on authors to offer further understanding (McCary 1957; Gallivan 2018).

Gardens of 10–20 m^2 (100–200 ft^2) accompanied each house in a village. Indians also cultivated larger communal fields, with the produce belonging to the chief. John White painted an Algonquin village illustrating the corn agriculture (Fig. 2).

Although the Indians planted seeds in the spring, they began preparing the fields the year before. That involved bruising the bark of trees near the roots. This killed the cambium (the living tissue that surrounds the dead woody trunk). They then used fire to complete the kill. The following year a stout branch was used to break-up the fallen and charred wood, creating a mulch for planting.

If the field was used the preceding year it would be prepared for planting by loosening the soil with a long-handled hoe, weeding, and burning of old corn stalks.

The Algonquin women organized the fields in rows. Along the rows they used wooden rods to pierce holes in the soil, each about one meter from the next. They put four kernels of corn and two beans in each hole. In between the corn and beans were planted pumpkins, squash melons, peas, sunflowers, and passion fruit. These plants produce fast-growing leaves that overtop and shade-out most weeds.

When the corn reached about half its final height, the women hilled each stalk by pushing soil around its base. Women and children weeded the fields during the growing season. The growing bean stems found support in winding themselves around the stalks of corn. The beans were a source of protein for the Indians, and also enriched the soil with nitrogen vie their mutualistic root nodules that contained cyanobacteria. The bacteria preform nitrogen fixation, converting N_2 from the atmosphere to reactive nitrogen that could be used by the corn (See chapter The Journey from Peruvian Guano to Artificial Fertilizer ends with too Much Nitrogen in the Chesapeake Bay). Indians also planted peas, another nitrogen-fixing legume, in the same field. Farmers built small shaded platforms in the fields for guards to stand vigil against hungry birds and mammals.

Planting began in April, peaked in May, and continued into June. April's planting yielded harvest in August, May's in September, and June's in October. They planted three varieties of corn, each with its own purpose. (1) Small corn grew to about 1 m in height and matured early, allowing two plantings per season. This variety produced popcorn. (2) Hominy corn with large kernels produced in various colors. (3) Flour corn for making bread.

Fig. 2 A John White illustration of the Algonquin Secoton village in present day North Carolina, just south of the area of the Chesapeake Bay colonized by the Virginia Company. The upper right side of the painting illustrates fields of corn in three different periods of growth. Tall, near mature corn is on top, in the middle is green corn, and on the bottom newly emergent seedlings. Image is in the public domain due to age. https://commons.wikimedia.org/wiki/File:North_carolina_algonkin-dorf.jpg

Each corn stalk usually bore two ears, but occasionally up to four. Each ear had 200–500 kernels. The Indians also harvested the green corn stalks, which Smith said were, "...somewhat like a sugar cane... .."

Cooking and Storing Food The Algonquin ate a very diverse diet, more so than the English colonists (Roundtree et al. 2008). One food item ruled supreme, corn. The different varieties of corn received different processing. The fast growing popping corn was dried and the kernels kept for later popping over a fire. Cooks roasted or dried other maize in the husk over fire or in hot beds of coals. Some of the cooked corn went to making stews with beans and other vegetables (Fig. 3). Much of it was pounded with wooden pestles to make four for later baking into bread. Hominy was made from corn broken in a mortar and soaked in water. It included the fibrous cob and as such required 12 h of boiling

Fig. 3 A Theodor de Bry painting after a John White rendering of Algonquin Indians boiling a stew that included corn and fish. The large size of the earthenware vessel illustrates the level of skill possessed by the Indians in creating implements for their food system. The size of the pot suggest a communal meal. This image was created in 1590 and is in the open domain. https://www.flickr.com/photos/britishlibrary/12458854075

to make it eatable. Cobs were also cooked and pounded into powder, the meal later mixed into broth and breads.

The Indians used corn flour and meal, along with mashed; wild rice, sunflower seeds, chestnuts, walnuts and hickory nuts for making breads.

Animal flesh, including fish, was boiled in a pot, roasted, or broiled over the fire. The Algonquin used smoky fires to preserve oysters, shad, and other meats. These they stored in baskets for later consumption (Fig. 4).

Corn as the King of Crops When Capt. John Smith and his band of colonists arrived in the Chesapeake Bay in 1607, they encountered a culture dependent on corn (maize). It hadn't always been that way. Maize was native to Mesoamerica and domesticated by the Indians of that region. Trade brought corn to the Chesapeake Bay region about 1100 AD. Beans arrived around 1450 AD (Gallivan 2018).

The arrival of corn profoundly changed the cultures of Chesapeake Bay tribes, shifting their food system more toward farming with less dependence upon hunting, fishing, and gathering. The story is told in the bones and teeth of the ancestors of the Powhatan. To understand that story one must first appreciate the chemical signature left by corn in the remains of people that ate it. To do that requires a voyage into the chemistry of carbon and nitrogen isotopes and the biology of photosynthesis.

Carbon (C) and nitrogen (N) are two essential atoms of life. Carbon forms the skeleton of all biological molecules and nitrogen is an important part of proteins and nucleic acids. Carbon occurs in the atmosphere as carbon dioxide (CO_2) and nitrogen primarily as dinitrogen (N_2). Not all the carbon atoms are the same, which is also true of nitrogen.

Carbon appears in three different isotopes (elements with the same chemical properties but different numbers of neutrons). Most (nearly 99%) is ^{12}C, meaning that it has six protons and six neutrons in its nucleus. About 1% is ^{13}C, with six protons and seven neutrons. Both ^{12}C and ^{13}C are stable isotopes with little tendency to degrade. About 1 in a trillion carbon atoms is ^{14}C. This is an unstable or radioactive isotope that naturally degrades over time, the extra neutrons breaking into electron sized particles (beta particles). The half-life of ^{14}C is 5730 years, meaning that in that time a

Fig. 4 Drawing by John White of Algonquin men cooking and smoking fish. Note the large size of the fish which suggests a low level of exploitation of the fishery. Public domain image from 1590. https://upload.wikimedia.org/wikipedia/commons/4/48/How_they_cook_their_fish_%281590%29.jpg

given amount of ^{14}C will be cut in half by radioactive degradation. This is the basis for radio-carbon dating.

Now consider nitrogen. It appears in two stable isotopes. ^{14}N accounts for 99.6% of nitrogen and has seven neutrons. ^{15}N makes up almost all of the remaining 0.4% of N, with eight neutrons.

The ratio of ^{14}C to ^{12}C and that of ^{15}N to ^{14}N is fairly constant in the atmosphere over time. However, biological processes uses each isotope at a slightly different rate. Knowing that helps reveal what animals have been eating, as the isotopes become part of the animal's tissues.

The ratio of the lighter isotope to the heavier version is expressed $\delta X = [(R_{sample}/R_{standar}d)^{-1}] (10)^3$, where X is ^{13}C or ^{14}N and R is the corresponding ratio; $^{13}C/^{12}C$, $^{15}N/^{14}N$. The δ values increase when a greater proportion of heavy isotopes appear, and decrease with greater levels of the lighter isotope. A mass spectrophotometer is used to detect the tiny differences in the heavy and light isotopes. Biological processes often *fractionate* the isotopes, taking up or releasing one more than another. This is measured as *discrimination* (*D*). $D = \delta_{substrate} - \delta_{product}$. The units for *D* are parts per thousand, indicated by *0/00* or called *mil* (Peterson and Fry 1987).

Most plants that evolved for moister conditions use the C-3 system of photosynthesis. A carrier molecule called ribulose bisphosphate (RUBP) picks up carbon dioxide (CO_2) from the air that diffuses into the stomata (pores) in the leaves and brings it into the Calvin cycle. The Calvin cycle fixes the captured CO_2 into sugar, which is the basis for making many other biomolecules. This takes place in the chloroplasts in the cells of the leaves. The problem is that stomata open to let CO_2 enter, also lets water vapor out of the leaf in a process called transpiration. In dry conditions the plant may lose too much water and wilt.

Corn is a tropical grass adapted for periodic drier conditions. It uses the C-4 system of photosynthesis. To avoid too much water loss, corn reduces the opening of the stomata when the soil is dry. This cuts the amount of water lost by transpiration through the stomata, but also reduces the influx of carbon dioxide the plant needs for photosynthesis. To compensate for this, C-4 plants use an intermediary molecule with very high affinity for the scarce CO_2, called phosphoenolpyruvate (PEP). C-3 plants discriminate against ^{13}C while C-4 plants preferentially take up ^{13}C (Farquhar et al. 1982). C-3 plants have a D of 21 *0/00*, and that for C-4 plants is 6 *0/00* (Peterson and Fry 1987). Knowing the relative amounts of ^{13}C and ^{12}C in a person's tissue gives a clue to the diet that put the isotopes there.

Proteins from marine food chains have a different N isotope signature than those from terrestrial animals. The multiple steps of the marine food chain favors an increase in the fraction of ^{15}N in animal tissue. Cultures eating mostly marine animals will have a $\delta^{15}N$ of 13–15 *0/00* and those eating mostly grains and other terrestrial plants present $\delta^{15}N$ of 9–10 *0/00* (Schoeninger et al. 1983).

Stable isotopes of C and N found in human bones and teeth of the Chesapeake Bay Indians provide a clue to the source of their food. Paring that with radio-carbon dating of those skeletal remains reveals how their diets changed over time. A study of the remains of people in the Chickahominy tribe (part of the Powhatan confederation) showed the arrival of corn in their diets. After corn was first introduced to their food system in about 1100 AD, $\delta^{15}N$ values fell below 12 *0/00*. This indicated a shift away from dependence on the marine food chain with the consumption of corn. By the sixteenth Century, prior to the arrival of the English, $\delta^{13}C$ isotope ratios exceeded −14.4 *0/00*, indicating that half of the diet came from corn (Gallivan 2018).

Additional evidence of the arrival of corn circa 1100 AD shows-up in the teeth of the Chickahominy. The starchy diet of corn promoted tooth-decay, as seen with the increasing incidence of dental caries. Finally, the amount of charcoal in sediments increases after the introduction of corn. This indicates more burning of forests to prepare land for corn planting (Gallivan 2018).

Corn in Culture and War The harvest of corn in the fall occasioned the Great Annual Feast in Werowocomoco, the town that served as the political center of the Powhatan Confederation. The event celebrated harvest and storage of adequate food to get the Indians through the winter and freed them to make war if necessary. It also marked the time for the population to shift to interior hunting camps for the winter to exploit another part of their food system.

The Chickahominy tribe in the Powhatan Empire took their name directly from their relationship with maize. The name means either "Course-pounded corn people" or "Refined-corn people." The Chickahominy grew corn and crushed the kernels to produce a meal used to make bread. They traded the corn they produced with other tribes and supplied it to the Jamestown colonists. That corn proved essential to the colonists. It was Capt. John Smith's attempt to negotiate directly with the Chickahominy for corn that angered paramount chief Powhatan (Wahunsenacawh). Smith violated protocol by going around the paramount chief, which led to Powhatan holding Smith a prisoner for 4 weeks in December of 1607. During his captivity Smith observed a Powhatan ceremony that involved corn, indicating the spiritual and well as nutritional value placed on this crop.

Near the end of his captivity is when Smith reported that Pocahontas, the young daughter of Wahunsenacawh intervened to save his life. Wahunsenacawh made Smith an honorary son before setting him free with the expectation of copper, hatchets, and beads to be delivered later.

A year after winning his release from captivity, "Wahunsenacawh repeated his demand for guns and swords, adding sardonically that corn was more valuable than these items, since corn could be eaten (Gallivan 2018)." Gallivan notes that Wahunsenacawh attempted to control the English through dependence on food as well as the ceremonial kinship and the corn ceremony experienced by Smith while held captive.

Wahunsenacawh was right when he observed the value of corn over weaponry. During the "starving time" (winter 1609–1610) some of the residents ventured from Jamestown to look for food. They "...were found slain, their mouths stuffed full of bread in an act read by the colonists as a sign of contempt and scorn... .(Gallivan 2018)."

In 1610 new English colonists and supplies arrived at Jamestown. They rebuilt the settlement and went to war against the Powhatan. This began with a mock feast setup in the English-occupied village of Kecoughtan (now Hampton). The colonists used food, music, and dancing to lure the local Indians into the open. The English then massacred the lot (Vaughan 1978).

This first Anglo-Powhatan war waxed and waned over subsequent years. The colonists waged war on the Indians to take their corn. This was more than to satisfy the need of the English for food. It was also to drive the Indians from their lands. The raids for corn including the burning of houses and fields and killing of villagers. As such, the English weaponized the main staple food. Eventually the English forced the Chickahominy from their lands in order to plant tobacco, a crop they prized more than corn. Tobacco exported to England meant cash in their pockets.

The Location of Algonquin Villages in Time and Space Much of the following discussion is informed by Roundtree et al. (2008). The food system dictated where the Indians of the Chesapeake Bay built their villages and when they occupied them. The importance of the three sisters agriculture meant constructing the main villages near fertile lands that warmed early in the springtime to facilitate early planting. The best fields were somewhat raised above the adjacent waterway, making for a low water table and quick

Fig. 5 Engraving by Theodor de Bray (1590) after a John White watercolor (1585) originally published with the caption, "A weroan or great Lorde of Virginia." This image of an Algonquin chief illustrates several points. The physicality of the depicted body suggest a food system that produced strong and healthy individuals. The bow and arrow is an effective weapon for hunting, as shown in the background on the right. The image also depicts a large open field adjacent to a forest, suggestive of how the Indians used fire to manage the landscape. This image was created in 1590 and in the open domain. https://commons.wikimedia.org/wiki/File:De_Bry_Chief_Virginia.jpg

warming of the soil in the lengthening spring sunlight. The Indians sited their villages along waterways to facilitate travel and transport by canoe. Such locations also provided access to fishing, gathering and hunting of marine animals. Up-stream villages also utilized the adjacent freshwater marshes for harvesting tuckahoe. The waterways provided water for bathing, and drinking if up-stream. The more brackish locations relied upon freshwater springs for drinking. Typically some 40–50 people would occupy a village.

Locations of farming villages shifted regularly. Several years of cultivation would deplete the soil of nutrients. This prompted the burring and clearing of nearby forest for a campus of new fields and houses. The new locations weren't far from the preceding ones, allowing the Indians to return to their fallow fields to forage for edible greens, berries and little barley (*Hordeum pusillum*) that colonized the abandoned croplands. These fields also attracted game for hunting including, deer, racoon and possum.

Farming occupied most of the Algonquin calendar, but not all of it. After completing the harvest in late fall or early winter, the Algonquin would migrate to temporary upland fishing and hunting camps. There they pursued deer, turkey, racoon, possum, and black bear, all of which were foraging on acorns and other nuts. The Algonquin competed with the game for the same nuts. The nuts stored well and were used in various cooked dishes and breads. Women and children gathered the nuts while the men hunted. Hunting produced not only flesh for food, but bones for making farming, fishing and hunting implements. Deer scapula were fashioned into plows. Other bones provided fish hooks and tips for arrows and spears.

The Algonquin men were expert archers, using the longbow and arrow to hunt beast and fellow man. They used bows fashioned from hickory or ash, typically nearly 2 m in length. Arrows tipped in sharpened stone or bone could pass right through a deer (Fig. 5).

Winter was time for hunting waterfowl that flew in from the north (see chapter "Passenger Pigeon and Waterfowl: Flights to Extinction and Not"). Before the water cooled it was a time to catch fish in weirs, by hook or spear. The Native Americans pioneered a variety of efficient fishing techniques, many of which are still in use today on the Chesapeake Bay (see chapter "A Fishing Trip: Exploiting and managing the commons of the Chesapeake Bay"). It was also a good season for collecting oysters (see chapter "The Chesapeake Bay Oyster: Cobblestone to Keystone").

Oysters Oysters played a large role in the Algonquin diet. The Powhatan village of Kecoughtan (now the City of Hampton) hosted Capt. John Smith in the fall of 1607 and again in 1608. He clearly delighted in consuming the large oysters they provided (Smith 1608). The ancient native people of the Chesapeake tell their culinary story through piles of discarded oysters called middens. Colonists used the omnipresent oyster middens as sources of building material, paving paths and grinding shells to produce mortar for masonry. Today archeologists use the discarded oysters to peer back in time. One midden located in the ancient Algonquin town of Kiskiak (on the south bank of the York River) chronicles changes in both the oyster resource and the arrival of corn as an alternative food (Gallivan 2018). The amount of shell deposited in the midden peaked in the Middle Woodland II period (circa 200–900 AD). It declined during the Late Woodland period (900–1200) and increased again through the time of European colonization. The average size of the oysters found in the various strata of the midden varied in opposition to their number. The more numerous the shells, the smaller they were. One explanation is that periods with high numbers of shells indicated times of over-harvesting, leaving a population of only younger and therefore smaller oysters for exploitation. The decline of oysters in the midden between 900 and 1200 AD also corresponds to the introduction of corn-based agriculture, alleviating pressure on the oyster population. While the arrival of corn meant less dependence on oysters, they remained a staple in the Algonquin diet. They were particularly important in years of poor corn harvest.

From Food to Tobacco and More War Violent food-stealing assaults by colonists in the 1620s compelled the Kiskiak residents to relocate their town a safer distance away from Jamestown, abruptly ending the history of the oyster midden. Note that it was corn and not oysters for which the English looted and burned Kiskiak. But it wasn't that the English wanted to grow corn there themselves. The English forced the Chickahominy from their lands in order to plant tobacco. The colonists had shipped their first commercial crop of tobacco to England in 1613.

The Chickahominy and the other Powhattan tribes saw their food system being crushed by the English. The tribes were forced inland, away from the life-sustaining waterways (Roundtree and Roundtree 2014). Cut off from a large portion of their food supply, the Indians had to act to save their community. Consequently, in 1622 the Algonquians attacked the tobacco plantations, killing 347 of the 1240 colonists in Virginia. This included killing women and children, something previously avoided by the Indians. Opechancanough, the brother of the Powhatan Chief, Wahunsenacawh, led that revolt and a subsequent one in 1644. Each time the English took their retribution, attacking the Algonquin directly and making war on their food system. The English destroyed fish weirs, burnt villages along with crops in the field and stores of grain. This eventually expatriated most of the Algonquins from their lands.

Although the Powhatan were mostly gone from the Chesapeake by the end of the seventeenth Century, much of their food system persisted in the hands of the Colonists. The English learned the hard lesson that their survival depended upon adopting most of the farming, fishing and hunting techniques develop by the people they dispossessed. While industrialized agriculture erased three sister planting and many other aspects of the Algonquin food system, one feature still remains, the overwhelming importance of corn, the number one crop in the US. However, only about 5% of that corn goes to nourish people directly. The rest is split about evenly for feeding animals and making ethanol for automobile engines (USDA 2018). This means that the current residents of the Bay still eat lots of corn, but mostly indirectly through their consumption of chicken, beef, and pork, washed down with beverages sweetened with high fructose corn syrup.

Native Americans Used Fire to Manage the Food System, Altering the Ecosystem Of all the various ways Native Americans obtained food from their environment, the use of fire most left its mark on the Chesapeake Bay Ecosystem. The overall population of Indians living around the Chesapeake Bay first encountered by Capt. John Smith was only 17,000. Much less than the 14 million residents of today. Fishing and hunting by Indians may have depleted local populations of prey. But such a small human presence did little to reduce the ecosystem-wide numbers of fish and wildlife. Instead, Native Americans left their biggest mark on the Bay's ecosystem by their use of fire.

The Indians used fire to manage their food system in several ways. They burned forests to make way for planting of crops. The ashes also fertilized the soil. They used ground-fires to burn away the low ground-cover below mature forests canopies. This left a park-like forest-floor that made for easy travel and hunting of deer and other game. They also used fire to concentrate deer and other wildlife, exposing the prey to hunting. Large groups of hunters would ignite a 5-mile (8 km) wide ring in a forest. The burning brush and trees focused deer into the center of the ring where they were easily taken by the awaiting hunters.

Following the fires, small woody and herbaceous plants, including grasses would dominate the recovering landscape. Deer easily browsed upon the post-fire emergent plant community. The boon in food for deer concentrated these herbivores for easier hunting on the recently open lands. The increased food also meant a higher carrying capacity for deer on the land, and thus more deer to hunt. American bison (*Bison bison,* also called buffalo) migrated from western lands to take advantage of the grasses on the postfire landscape. The bison provided food for Indians in the western portion of the Bay's watershed. Left alone, the burned lands would revert to forests in one or two decades. So Indians would regularly reburn the fields to sustain the interest of deer and other hunted species.

Fire also changed the community structure of the forest. The Indians made good use of certain fire-survivors that grew well in the ash-fertilized soil and the absence of other trees that competed for light and water. These included red mulberry (*Morus rubra*), persimmon (*Diospyros virginiana*) Hickory (*Carya sp.*), black walnut (*Juglans nigra*) and plum (*Prunus Americana*). The open-grown form of these trees produce much more fruit than those in a closed forest with less access to light and nutrients. Capt. John Smith commented on the abundance of mulberry trees in Native villages and noted their fondness for the fruits (Brown 2000).

Pre-Colonial era use of fire for the Native American food system added diversity to the mosaic of landscapes comprising the Chesapeake Bay's Watershed. As said by Brown (2000), "...accounts by early European settlers and travelers, coupled with what we know about Virginia's climate in recent millennia, consistently point to one conclusion: that at least some of Virginia's ecosystems evolved with, and depended on, frequent burning by American Indians. Shaped and maintained to make the land livable, such ecosystems should not be confused with wilderness. Instead, they should be treated as what they were—a cultural imprint left on the land by Virginia's first inhabitants."

One of the sources used for the account above was an article by Forest Service employee, Hu Maxwell (1910). While his story is sound, it must be noted that Maxwell peppered his paper with profoundly racist comments against Native Americans, such as... "He was wasteful and destructive, as savages usually are, and the word economy had no place in his vocabulary. When he had abundance, he squandered like a pirate, and when want pinched, he stood it like a stoic." While the extensive use of fire by Native Americans did leave a large footprint on the watershed for such a small group of people, those effects pale in comparison to what the English and other colonists subsequently did to the Bay's watershed. Other than the clearings noted by Maxwell, forests dominated the pre-Colonial landscape. By the mid-seventeenth Century, the colonists removed about 20% of the original forest. By the end of the nineteenth century, 80% of the forest was gone (Brush 2017).

Pollen and charcoal deposited in sediments confirm that people of the Middle Woodland II period (200–900 AD) through the time of contact with the English used fire to manage forests of the Tidewater Region of Virginia. Charcoal provides direct evidence of the burning which is supported by a decrease in pollen from oak and hickory trees, accompanied by an increase in that from herbaceous or weedy species that populated the deforested areas (Gallivan 2018).

In addition to ecosystem management, the Powhattan and other tribes used fire for other food system purposes. Fire was essential for cooking game, corn, and tubers such as the staple tuckahoe. It also served to preserve foods for storage and later use. The Indians treated shad and other fish with baths of smoke that dried the flesh and reduced microbial attack. They created pottery needed for cooking and storing food by firing vessels shaped from clay. And they attracted fish at night with burning torches and fire platforms in canoes.

References

Afshin A, Sur P, Fay K et al (2019) Health effects of dietary risks in 195 countries, 1990–2017: a systematic analysis for the Global Burden of Disease Study 2017. The Lancet 393:1958–1972. https://doi.org/10.1016/s0140-6736(19)30041-8

Blaylock R (1985) The marine mammals of Virginia. The College of William and Mary, VIMS Education Series No. 35. Williamsburg

Brown H (2000) Wildland burning by American Indians in Virginia. Fire Management Today 60:29–39

Brush G (2017) Decoding the deep sediments, 1st edn. Maryland Sea Grant, College Park

Farquhar G, O'Leary M, Berry J (1982) On the relationship between carbon isotope discrimination and the intercellular carbon dioxide concentration in leaves. Aust J Plant Physiol 9(2):121–137. https://doi.org/10.1071/pp9820121

Gallivan M (2018) The Powhatan Landscape: an archaeological history of the Algonquin Chesapeake. University Press of Florida, Gainesville

Landon A (2008) The "How" of the three sisters: The origins of agriculture in Mesoamerica and the human niche. In: Digitalcommons.unl.edu. https://digitalcommons.unl.edu/cgi/viewcontent.cgi?article=1039&context=nebanthro. Accessed 19 Aug 2018

Maxwell H (1910) The use and abuse of forests by the Virginia Indians. The William and Mary Quarterly 19:73. https://doi.org/10.2307/1921261

McCary B (1957) Indians in seventeenth century Virginia. University of Virginia Press. Charlottesville, VA

NOAA (2018) Harbor seals: Aerial surveys over Maine, tagging in Virginia | NOAA Fisheries. In: Fisheries.noaa.gov. https://www.

fisheries.noaa.gov/feature-story/harbor-seals-aerial-surveys-over-maine-tagging-virginia. Accessed 1 Jul 2019

Peterson B, Fry B (1987) Stable isotopes in ecosystem studies. Annu Rev Ecol Syst 18:293–320. https://doi.org/10.1146/annurev.ecolsys.18.1.293

Roundtree M, Roundtree H (2014) Diet in Early Virginia Indian Society. In: Encyclopediavirginia.org. https://www.encyclopediavirginia.org/Diet_in_Early_Virginia_Indian_Society#start_entry. Accessed 1 Jul 2019

Roundtree H, Clark W, Mountford K, Barber M (2008) John Smith's Chesapeake voyages. University of Virginia Press, Charlottesville, pp 1607–1609

Saksena M, Okrent A, Anekwe T (2019) America's eating habits: Food away from home. In: Ers.usda.gov. https://www.ers.usda.gov/webdocs/publications/90228/eib-196.pdf. Accessed 29 May 2019

Schoeninger M, DeNiro M, Tauber H (1983) Stable nitrogen isotope ratios of bone collagen reflect marine and terrestrial components of prehistoric human diet. Science 220:1381–1383. https://doi.org/10.1126/science.6344217

Smith J (1608) A true relation of such occurrences and accidents of noate as hath hapned in Virginia since the first planting of that collony. London: Printed for Iohn Tappe, and are to be sold at the Greyhound in Paules-Church-yard, by W.W. In Barbour PL (ed.) The complete works of Captain John Smith (1580–1631). (1986) 3 vols. 1:3–118. University of North Carolina Press, Chapel Hill

Smith J (1612) A map of Virginia. Oxford: Joseph Barnes, pp.1–3. In The complete works of Captain John Smith (1580–1631), edited by Philip L. Barbour, (1986) 3 vols., 1:119-1:190. Chapel Hill: University of North Carolina Press. Available in Capt. John Smith's complete works: http://www.virtualjamestown.org/exist/cocoon/jamestown/fha-js/SmiWorks1

USDA (2014) 2012 census of agriculture - Farm demographics highlights | USDA - National Agricultural Statistics Service. In: Nass.usda.gov. https://www.nass.usda.gov/Publications/Highlights/2014/Farm_Demographics/index.php#how_many. Accessed 6 May 2019

USDA (2018) USDA ERS—Feedgrains sector at a glance. In: Ers.usda.gov. https://www.ers.usda.gov/topics/crops/corn-and-other-feedgrains/feedgrains-sector-at-a-glance/. Accessed 1 Jul 2019

Vaughan A (1978) "Expulsion of the salvages": English Policy and the Virginia Massacre of 1622. The William and Mary Quarterly 35:57. https://doi.org/10.2307/1922571

A Fishing Trip: Exploiting and Managing the Commons of the Chesapeake Bay

Benjamin E. Cuker and Matthew Balazik

Give a person a fish, and they eat for a day.
Teach a person to fish, and they eat for the rest of their life.
Teach too many people to fish, and nobody eats.

Abstract

Captain John Smith and his band of English colonists arrived at the Chesapeake Bay in 1607, discovering an estuary little changed by human activity. This contrasted sharply with the European rivers of that era. The estuaries he knew from England and Europe all suffered from dam-building, overfishing, sedimentation, and pollution. The decedents of the original colonizers, and the many thousands that came subsequently reshaped the Bay by their exploitation of its food resources. Over harvesting, damming of waterways, and oyster reef destruction decimated oysters (*Crassostrea virginica*) and migratory species of finfish, including; shad (*Alosa sapidissima*), alewife (*Alosa pseudoharengus*) blueback herring (*Alosa aestivalis*) and striped bass (*Morone saxatilis*). In the twentieth century, application of scientific fisheries methods built around the concept of maximum sustainable yield (MSY) and habitat restoration has yielded only partial recovery of these and other species. Ecosystem based management (EBM) is discussed as a sustainable alternative to MSY that takes into account non-human consumers of the exploited resource. The chapter also reviews the various fishing techniques used on the Bay, including those invented by Native Americans and later methods that facilitated industrial scale harvesting.

Keywords

Over fishing · MSY · Ecosystem based management · Oysters · Shad · Striped bass · Menhaden

A Bay Full of Seafood The abundance and diversity of seafood in the Chesapeake Bay impressed Capt. John Smith. He wrote in his Map of Virginia, "Of fish we were best acquainted with Sturgeon, Grampus, Porpus, Seales, Stingraies, whose tailes are very dangerous. Brettes, mullets, white Salmonds, Trowts, Soles, Plaice, Herrings, Conyfish, Rockfish, Eeles, Lampreyes, Catfish, Shades, Pearch of 3 sorts, Crabs, Shrimps, Crevises, Oysters, Cocles and Muscles. But the most strange fish is a smal one so like the picture of St. George his Dragon, as possible can be, except his legs and wings, and the Todefish which will swell till it be like to brust, when it commeth into the aire. Fish. (Smith 1612)." While Smith and the other early English wealth-seeking colonists didn't come to the Chesapeake Bay for its abundance of seafood, they certainly did notice it and partake of the bounty. Indeed, for many of them it meant the difference between survival and starvation during their first few years of residence in the New World (Roberts 2007).

John White, artist and 13 month resident of the failed English colony at Roanoke provided a famous water color that depicted the wealth of seafood in the region that so

B. E. Cuker (✉)
Department of Marine and Environmental Science, Hampton University, Hampton, Virginia, USA
e-mail: Benjamin.cuker@hamptonu.edu

M. Balazik
US Armey Corps of Engineers, Research Development Center, Vicksburg, MS, USA
e-mail: balazikmt@vcu.edu

B. E. Cuker (ed.), *Diet for a Sustainable Ecosystem*, Estuaries of the World, https://doi.org/10.1007/978-3-030-45481-4_5

Fig. 1 Watercolor by John White depicting Algonquin fishing techniques. Fishers in the background use spears. Those in the dugout canoe appear to use a multi-pronged spear. The net in the stern of the vessel would help land the captured fish. A fish pound appears on the left. Netting strung on the stakes orient fish into the funnel that leads into the pound. The funnel prevents the confused fish from finding their way out of the trap. Fishers easily remove the captured fish from the pound with hand nets or spears. The fire in the canoe provides light, which attracts fish at night. Although White's depictions come from his 13 months spent at the failed colony of Roanoke Island, the Algonquin of the Chesapeake used the same techniques

enthralled Capt. John Smith. The oft-duplicated image portrays the various fishing techniques employed by the Algonquin Indians (Fig. 1).

Though the clear waters, expansive oyster (*Crassostrea virginica*) reefs, massive runs of shad (*Alosa sapidissima*), and abundant sturgeon (*Acipenser oxyrinchus*) that greeted Capt. John Smith are long gone, from early spring to late fall today's Chesapeake Bay still offers seafood and people eagerly in its pursuit. Wooden workboats haul baited crab pots or lift oysters with mechanical tongs. Anglers line the rails of larger party boats, sit in smaller private craft, or lean over the fishing piers, waiting for a bite on their baited hooks.

Rusty steal ships deploy purse seines to encircle schools of oily menhaden. Small boats use dip-nets to collect the catch of fish funneled into collection pounds by long nets strung between poles running perpendicular to the shore. If any single thing defines the relationship between people and the Chesapeake Bay, it is the pursuit of a free dinner. For many it's the dinner that counts most, for others its simply the pleasure of time spent on the water in hopes of getting it.

Of course the promise of the free meal isn't free of real or reel costs. Recreational and commercial fishers don't pay for the fish when they bring it ashore, as they would when checking out of a market. But they do pay significant upfront costs just to take the chance of catching something. The fishers must invest in gear, if it be a few dollars for a humble drop line and hook dangled from a pier, or thousands spent on fancy rods and reels, nets, traps, boats, and fish finding sonar. Add to that the cost of fishing line, bait, lures, trailers, boat slips, licenses, time, and fuel. The kilogram or so of free fish or crabs gathered by the recreational fisher, or the larger catches taken by commercial operations all comes at a cost. Fishery scientists may lump all of those costs into the term *fishing effort*, time and resources spent in pursuit of the catch.

The Chesapeake Bay is a *commons* for recreation, transportation, and the taking of seafood by private individuals and commercial enterprises. A commons is a resource shared by the entire population, and belonging to no single individual or entity. The concept of the commons is an early feature of the human food system, being the basis for nomadic people living off the land, going from one fertile common feeding ground to another as seasons changed. Settled civilizations retained aspects of the commons, demarking lands for the use of "commoners" to raise crops and livestock. Expanding human populations meant the potential for over use of the commons, reducing the fertility of the land and its yield of food. As such, successful cultures developed customs of local governance to regulate access to commons and thus sustain their usefulness to the food system (Shiva 2002). Terrestrial commons for crops and cattle offer the entire community a view of the state of the resource. It is easy to see evidence of overgrazing, or soils worn out from over cultivation and in need of a sabbatical. Aquatic commons may not be so transparent in revealing the state of the resources they offer. Today's cloudy waters of the Chesapeake Bay obscure from view most of its bounty of seafood. Only sustained episodes of reduced catches or results from systematic scientific surveys provide a metric for deducing the state of a fishery stock.

When Capt. John Smith and his party first entered the Chesapeake Bay in 1607, protecting the aquatic commons from overharvesting seemed not to be a problem. The populations of Algonquin and other tribes were small and so unlikely to significantly reduce the various populations of fish, mollusks, crabs, and other animals they pursued. Despite evidence of periodic localized reductions in oyster populations (Gallivan 2016), the abundance of other species probably suffered little from the indigenous fishers. This abundance of seafood and the clear waters of the Chesapeake Bay system that greeted Capt. John Smith stood in stark contrast to the situation in England and Europe that he knew. By the end of the first millennium, freshwater and estuarine stocks of fish in the Old World suffered from overfishing, the construction of dams, and the draining of wetlands. Sediments washed from agricultural fields, refuse and excrement dirtied the European Rivers, like the Thames in England (Roberts 2007). Medieval European rulers understood that overfished stocks of seafood needed protection. King Philip IV of France (1268–1314) banned various fishing methods and restricted some fishing to certain seasons. The King noted the reduction in; numbers of fish, their sizes (stunting), and market value (Roberts 2007). That Capt. John Smith began his inventory of Chesapeake Bay fish with sturgeon (*Acipenser oxyrinchus*) was no accident. For sturgeon were highly prized in England, and so diminished in number that by the thirteenth Century they were reserved for the royal family (Roberts 2007). Certainly the free access to fishing and hunting in the New World appealed to many of the immigrants from England and Europe that followed in Capt. John Smith's wake.

By the close of the seventeenth Century, the expanding population of European colonists forcibly extirpated most of the American Native residents of the Chesapeake Bay. This meant the colonists taking complete control of access to the fisheries of the Bay and having to grapple with growing pressure on the resource. In 1632 Charles I of England granted Lord Baltimore the charter of Maryland. That document specified, "Fishings of every kind of Fish as well as Whales, Sturgeons and other royal Fish in the Sea, Bays, Straits, or Rivers, within the premises and the Liberty of Fishing for Sea-Fish as well in the Sea, Bays, Straits and navigable rivers of the Province aforesaid (Casey 2018)." In 1680 the colonial government of Virginia passed the first law to regulate fishing in the Chesapeake Bay, forbidding the use of gigs and harping irons (harpoons) in several counties (VMRC 2018).

By the time of the adoption of the Virginia-Maryland Compact of 1785 that clarified the boundaries and trade relations between the two states, the question of access to fisheries on the Bay was important enough to include in the document. "The citizens of each state respectively shall have full property in the shores of Potowmack river. . . . the right of fishing in the river shall be common to, and equally enjoyed by the citizens of both states. Provided, That such common right be not exercised by the citizens of the one state, to the hindrance or disturbance of the fisheries on the shores of the other state; and that the citizens of neither state shall have a right to fish with nets or seines on the shores of the other (Virginia Compacts 2018)." Inclusion of fisheries in this compact signaled the growing importance of commercialization of the resource. Yet this agreement failed to prevent the nineteenth Century "oyster wars" on the Bay that involved armed conflict between fishers from both states, and between state authorities and pirate fishers (Wennersten 1981). Growing human populations, advances in fishing gear boat propulsion (engines), and expanding railroads combined to drive the industrialization of commercial fishing on the Bay in the nineteenth Century. Growing human populations meant more demand for seafood, improved gear and boats meant easier access to the resource, and railroads meant expanding the market to distant inland cities. The growing pressure on oysters and other valuable species of the Chesapeake Bay commons required new regulations to protect these resources from commercial, if not biological extinction. The oyster story is detailed in chapter "The Chesapeake Bay Oyster: Cobblestone to Keystone".

Protecting the Commons When extraction rates began to exceed the capacity of the commons to sustain commercially viable stocks of seafood, the jurisdictions governing the Chesapeake Bay did as the Europeans had done for many years before, placing regulations on fisheries. These laws aimed at maintaining populations of stocks took several forms; closures, size limits, catch limits, location restrictions, and gear restrictions. Violations of these laws today constitute a Class 2 misdemeanor in Virginia with penalties of up to $1000 and six months of imprisonment (VA Penalty 2018). Repeat offenders may lose their fishing licenses (VALaw 2018). Maryland uses a point system to levy fines on offenders and may suspend or revoke recreational and commercial licenses (MDDNR 2018a).

Full *Closures* of a fishery means prohibiting any taking of that species. The closures might be indefinite, lasting for as long as necessary for the species to recover population levels that could support renewed exploitation. But more often the closures are partial, prohibiting the taking during certain seasons, from specified areas, or restricting exploitation to certain life stages. Seasonal closures reduce overall pressure on the stock and may be designed to protect the species during critical stages of its life history, such as during mating and spawning season. The aim of a closure is to allow the stock to rebuild through natural reproduction to a population level and average body size that supports a sustainable fishery. Note that body size is as important as population size. A large population of stunted fish or small oysters offers little commercial value, small reward to the sport-fisher, and limited ecological value.

Catch limits restrict the number and size of a seafood item that can be taken over a designated time period. *Size limits* restrict the taking of generally smaller individuals. *Location restrictions* close certain areas to fishing or harvesting. Such areas may be closed temporarily to rebuild local populations or perpetually to provide a source of new recruits to the target population. *Gear restrictions* limit the kind of devices used to extract the resource. This includes regulating types of hooks, bait, nets, mesh size of nets, vessel propulsion, and bottom dredges.

Scientific Fisheries Management Scientific management of fisheries developed in the late nineteenth and early twentieth Century around the idea that, "... every fish population had the potential to produce a harvestable surplus and the largest surplus that could be harvested annually from that population (i.e., maximum sustainable yield) could be estimated by rigorous scientific analysis (i.e., stock assessment). The job of the fisheries manager at the time was to control fishing pressure, using various regulations, at a level such that sustainable catch levels could be achieved in perpetuity. However, fishing pressure, as always, was very difficult to control; many fisheries ended up being over harvested and yields eventually declined (Lackey 2006)." This narrow view of fisheries management discounted the ecological role of the managed species, focusing strictly on the amount of sustainable take for human use.

Scientific fisheries management is based on the assumption that knowing two things about a stock will allow it to be managed so as to produce a maximum sustainable yield (MSY) over time. The two things are the intrinsic rate of growth (*r*, similar to an interest rate for a savings account) for that population, and the maximum size of that population permitted by the natural resources and conditions available to support it in that environment (K, the *carrying capacity* of the environment). The theory of MSY is built on the work of two European luminaries of the early nineteenth Century who were trying to figure out the mathematics that best describe how human populations might expand over time. In 1798 the Rev. Thomas Robert Malthus of England articulated a mathematical understanding of the potential for rapid population growth, which he thought may result in human populations growing faster than the food supply needed to sustain them (Malthus 1817). Although his concern focused on the potential for overpopulation of humans in a quickly industrializing Europe, the mathematics worked as well for nonhuman species, such as fish.

From Malthus we received the *exponential* formula for population growth: $dN/dt = rN$, where N is the number of

Table 1 A comparison of number of fish for exponential and logistic growth for an example population of Atlantic menhaden with an intrinsic rate of growth (r) of 100%

Year	Exponential growth (population size)	Logistic growth (population size)
0	10	10
1	27	27
2	74	74
3	201	201
4	546	546
5	1484	1484
6	4034	4034
7	10,966	10,966
8	29,810	29,810
9	81,031	81,030
10	220,265	220,261
11	598,741	598,718
12	1,627,548	1,627,371
13	4,424,134	4,422,829
14	12,026,043	12,016,409
15	32,690,174	32,619,086
16	88,861,105	88,337,786
17	241,549,528	237,721,428
18	656,599,691	629,063,499
19	1,784,823,010	1,595,032,914
20	4,851,651,954	3,665,930,650
21	13,188,157,345	7,017,924,507
22	14,997,894,830	10,575,145,402
23	14,997,894,830	12,999,075,742
24	14,997,894,830	14,196,116,267
25	14,997,894,830	14,693,897,955
26	14,997,894,830	14,885,919,766
27	14,997,894,830	14,957,829,493
28	14,997,894,830	14,984,458,719
29	14,997,894,830	14,994,278,935
30	14,997,894,830	14,997,894,830

Note that little difference exists between the populations during the first 15 years, but the logistic population begins to slow its growth as it approaches the carrying capacity (K), which in this example is set at 15 billion fish

individuals in the population, dN is the instantaneous change in the number of individuals in a population, dt is the instantaneous change in time, and r is the intrinsic rate of growth per capita. This equation can be rewritten to determine the size of a population at some time in the future: $N_t = N_0e^{rt}$ where N_t is the size of the population in the future, N_0 is the beginning size of the population, e is Euler's constant (the natural logarithm), and t is the time in the future.

For example, consider a hypothetical population of Atlantic menhaden (*Brevoortia tyrannus*) consisting of ten individuals and with an intrinsic rate of growth (r) equal to 100% of the population, in other words the population doubles every year. This is a reasonable estimate of r, close to that observed in the early 1970s for Atlantic menhaden during a period of rapid increase (Fish Base 2018, personal communication, Dr. Genny Nesslage). In 4 years how many fish would there be in that population? $N_{4years} = 10e^{1*4} = 546$ fish. In 10 years, what would the population be? $N_{10\ years} = 10e^{1*10} = 220{,}260$ fish. And 26 years? $N_{26years} = 10e^{1*26} = 1{,}957{,}296{,}094{,}288$ fish! Going from just 10 fish to nearly 2 trillion fish in 26 years shows just how quickly populations can grow given the right conditions. Of course, the population of Atlantic menhaden or any other species can't grow beyond the environment's ability to support it (Table 1, Fig. 2).

Atlantic menhaden biomass for the East Coast, including the Chesapeake Bay was 2.3 million metric tons in 1958, more than twice that of estimates for recent years (ASMFC 2018b). Assuming the average menhaden weighs about 0.3 kg, that would put the 1958 population at 7.7 billion

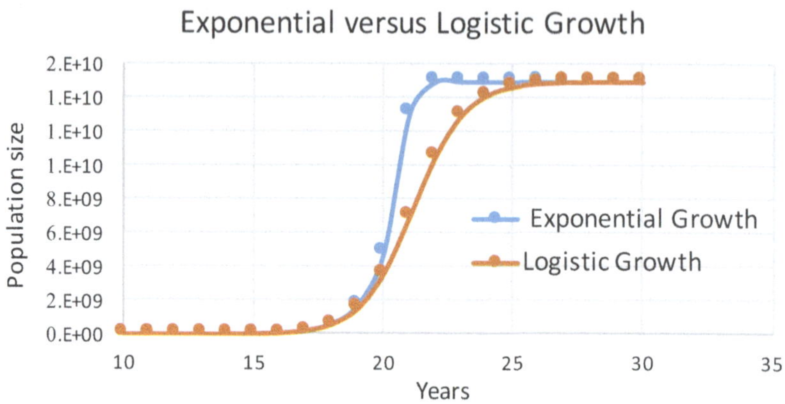

Fig. 2 The exponential and logistic models of population growth based upon the data for a fictitious population of Atlantic menhaden that starts with ten individuals and reaches a carrying capacity (K) of 15 billion fish. Note that the rate of population growth remains constant until it reaches K for the exponential model, while it begins to slow after exceeding the halfway point to K for the logistic formula

fish, much less than the 26 year population estimate of 2 trillion menhaden generated by the exponential equation in the example above. Would there really be habitat and food for the nearly 2 trillion menhaden in the Chesapeake Bay and coastal ocean from the example? Perhaps the true carrying capacity of an unfished stock of menhaden is closer to 15 billion, about twice the 1958 estimate of that fished population. Chapter "Menhaden, the Inedible Fish that Most Everyone Eats" explores the menhaden fishery of the Bay in more detail.

The exponential formula for population growth provided by Malthus failed to consider what happens to populations as they near the level of the environment's capacity to support them. In nature, a population's growth slows as it nears the carrying capacity of the environment. This is due in part to competition for limited resources such as food, water, space and shelter. Also, contagious diseases take a greater toll on overcrowded populations due to close proximity of individuals and compromised immune systems (due to stress and undernourishment). If the oversized population is a species that is preyed upon by another species, the second species is a predator. Large population sizes of prey may stimulate the growth of predator populations (by feeding them) and by the attraction of new of predators from elsewhere. All of these factors potentially slowing population increase, constitute *resistance to growth* evident as a population's size approaches the carrying capacity.

In the early nineteenth Century, the Belgian Pierre-Francois Verhulst developed the mathematical solution to the problem of characterizing the reduction of a population's growth rate as it nears carrying capacity (Cramer 2002). Malthus showed that populations were capable of exponential growth, and Verhulst noted that this rarely happens in nature due to various impeding factors. Verhulst coined the term *logistic* to describe a pattern of population growth that slows as it approaches its maximum level permitted by the environment. The equation for logistic growth is a modification of the exponential equation with the inclusion of a term for the carrying capacity of the environment (K). The logistic equation is: $dN/dt = rN(1-(N/K))$. Look what happens as the population size (N) approaches the carrying capacity (K). The larger N gets, the larger N/K becomes, making dN/dt (population growth rate) that much smaller. When N = K, the equation goes to zero, meaning no population growth once the carrying capacity is reached. The logistic equation can be rewritten to calculate a projected population's size, as was done with the exponential equation above: $N_t = K/(1 + [(K-N_0)/N_0]e^{-rt})$.

Suppose the carrying capacity (K) for Atlantic menhaden in the Bay and coastal Atlantic Ocean is 15 billion fish and we again start with an initial population of 10 fish. Using the Logistic equation, it would take about 25 years to reach K. $N_{25} = 15$ billion/$(1 + [(K-10)/10]\varepsilon^{-1*25}) = 14.99$ billion fish (Table 1). Absent the term for environmental resistance to growth, the exponential model predicts that the menhaden population will reach K several years sooner than the logistic equation. This is evident in the different shapes of the exponential and logistic growth curves (Fig. 2). The exponential curve shows that population growth stops abruptly when K is reached. The logistic curve shows slowing of growth starting just after the population hits the halfway mark to K.

The logistic growth curve forms the basis for scientific fisheries management with the goal of producing the maximum sustainable yield (MSY, Fig. 3). If fisheries managers permitted the harvest of 90% of a stock that was at carrying capacity, such as the one illustrated in Fig. 3, it might take 10 or 15 years for the population to return to commercially viability. If managers permitted the taking of only 10% of that same stock, then the adherents to MSY theory would say that the population was underexploited, leaving behind 40% of the possible catch. MSY theory suggests that optimum sustainable yield will result from regular fishing to keep the

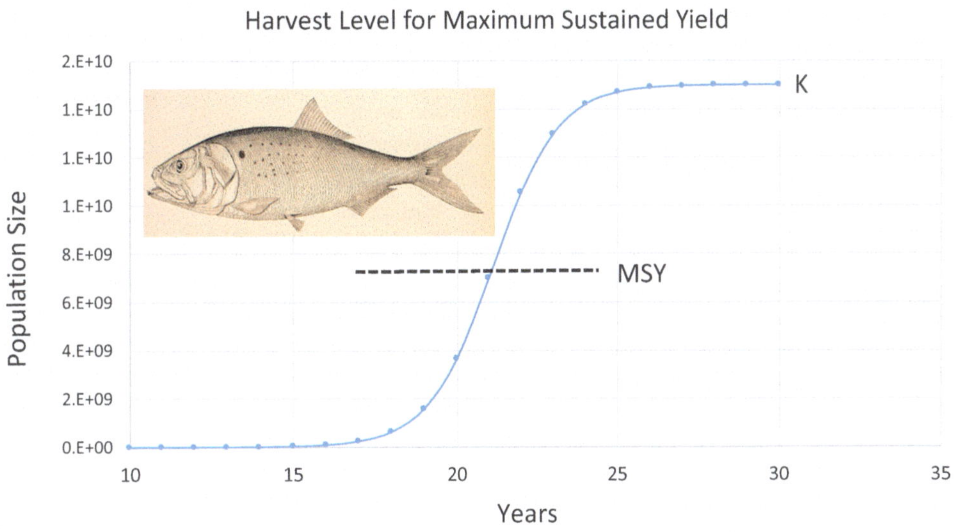

Fig. 3 The logistic model of population growth exhibits its most rapid growth about halfway to the carrying capacity. The slope of the line is steepest right before that inflection point. The theory of maximum sustained yield suggests harvesting the population down to that inflection point, as that will provide a large harvest and leave the population at its most productive density for recovering after the harvest. The data displayed comes from the fictitious Atlantic menhaden population illustrated in the text. Inset is an image of Atlantic menhaden (*Brevoortia tyrannus*) U.S. National Oceanic and Atmospheric Administration

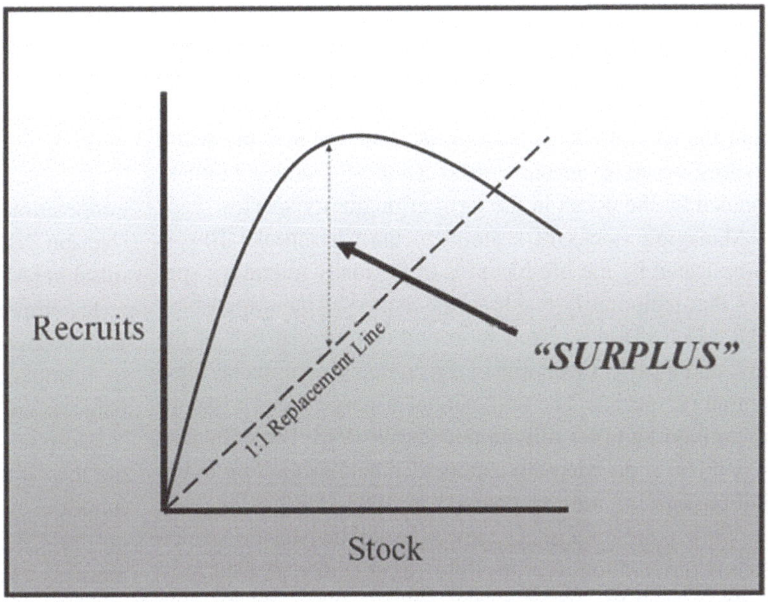

Fig. 4 The Ricker model of stock management emphasizes the optimization of recruitment of new individuals to the stock, noting that this occurs at about half the carrying capacity for that stock

stock at 50% of the environment's carrying capacity for the population. Any portion of the stock over the 50% of K level is viewed as excess production that should be harvested to maximize the yield of the fishery.

While MSY-driven fisheries science traces to the logistic population growth curve, managers take advantage of other related models to visualize the stocks they control. The Ricker curve addresses the production of *recruits* (new individuals produced by the stock population, Fig. 4). This model predicts that optimum recruitment occurs at half the stock carrying capacity. If stocks are overfished to low stock population levels, then recruitment is also low due to the relative absence of potential parents. If populations are instead unfished and allowed to reach K, recruitment is again low, but this time due to absence of resources, disease, and expanded predator populations.

MSY management suffers many problems. First, the manager must invest resources to produce a good estimate of the size of the stock every season. This may require various sampling techniques such as using tags for mark and recapture studies (Liljestrand et al. 2018), and analysis of recent landings data. The manager must also know the size of the population when at carrying capacity. Without a solid estimate of the current population size and its theoretical maximum, determining the MSY target is difficult. Second, seafood populations naturally vary from year to year, responding to the vagaries of the environment, such as shifts in temperature, salinity, densities of prey and predators, and diseases. Even without the intervention of humans, seafood populations might easily halve or double between one year and the next. Finally, MSY management ignores the important ecological role of all of that "excess" production escaping human consumption. Those "excess fish" may serve as food for other species such as striped bass (*Morone saxatilis*), brown pelicans (*Pelecanus occidentalis*), or bottlenose dolphins (*Tursiops truncates*). For example, Atlantic menhaden appear to suffer about 69% natural mortality, presumably most of this death is due to predators (Brainard 2017). The "under-harvested" production might also provide other important ecosystem services, such as reducing excess phytoplankton as is true for suspension feeding eastern oysters (*Crassostrea virginica*) and menhaden (*Brevoortia tyrannus*).

Striped Bass and Menhaden One of the reasons for the high biomass of available seafood in estuaries is that many species spend a portion of their lives in the adjacent ocean. While away from the estuary these species add mass by eating from the oceanic food web. When they migrate back to the estuary they bring with them the food energy and nutrients harvested from the sea. While naturally productive from the mix of nutrients from the land and sea, the warm shallow waters of estuaries also feature the added biomass donated by the ocean in the form of migratory species.

Managing stocks of seafood in the Chesapeake Bay is complicated by the life-histories of the many migratory species that temporarily reside in the estuary. The striped bass (aka rock fish) is a prized sport and commercial fish of the Bay, and a perfect example of the complexity of managing a migratory species. The commercial fishery landed a record 6.7 million kg (14.7 million pounds) in 1973. But collapsed over the next decade with commercial landings falling to 0.8 million kg (1.7 million pounds) in 1983 (CBP 2018a). In response to the collapse, Congress passed the Atlantic Striped Bass Conservation Act in 1984. Maryland and Delaware closed striped bass fishing from 1985 to 1989, and Virginia closed its fishery for 1-year in 1989. The closures worked well with the stock mostly recovered by 1995 (Fig. 5, Richards and Rago 1999; NOAA 2018a; CBP 2018a). This species is anadromous, being spawned in freshwater streams. The larvae feed on plankton in the rivers and are carried into the estuary where they remain and grow for the next 4 years or longer. Portions of the stock migrate into the coastal Atlantic Ocean, heading north in the summers and south in the winters. Females return to rivers to spawn beginning at age 4-years, and males at age 2-years (VIMS 2018). Striped bass may live to age 30-years and obtain sizes up to 150 cm (59 inches) and 35 kg (77 pounds) (NOAA 2018a). The closure of the striped bass fishery succeeded because it took into account the migratory nature of the fish and included the

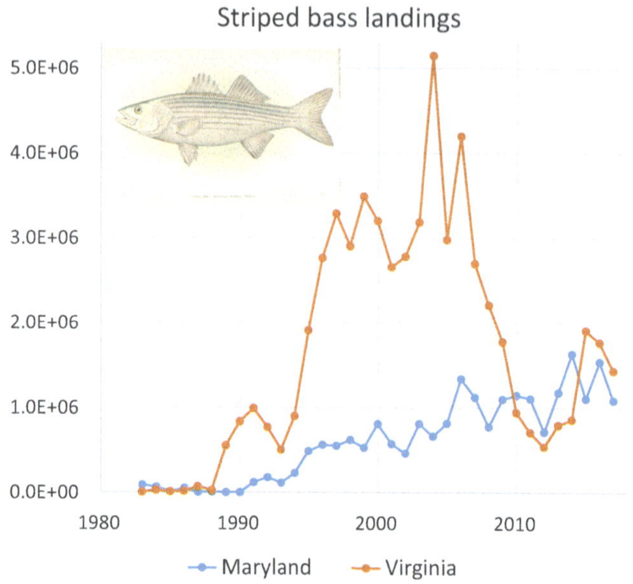

Fig. 5 Recreational landings (as number of fish) of striped bass since the fishery reopened for the Chesapeake Bay. Data from https://www.st.nmfs.noaa.gov/recreational-fisheries/data-and-documentation/queries/index. Inset image of the striped bass (*Morone saxatilis*) is from Smith, Hugh M. (1907) Fishes of North Carolina, North Carolina Geological and Economic Survey, vol.II, Raleigh, NC: E. M. Uzzell & Co

cooperation of coastal states all the way to New England (Nelson 2018). The Atlantic States Marine Fisheries Commission (ASMFC) managed the closure. ASMFC is an interstate compact approved by the US Congress in 1942. It works with NOAA Fisheries and the US Fish and Wildlife Service to formulate regulations aimed at maintaining sustainable fisheries (ASMFC 2018a).

Despite significant recovery in the population size from the low of the early 1980s, some fishers and scientists don't think the striped bass closure fully restored a healthy fishery in the Chesapeake Bay. They point to two troublesome markers. The high incidence of mycobacteriosis disease among striped bass, and their emaciated appearance (CBP 2018b). *Mycobacterium sp.* infects >50% of Chesapeake Bay striped bass, causing external ulcers and granulomatous inflammation of the viscera, including the liver, spleen and kidneys (Gauthier et al. 2008). The widespread infection is associated with poor water quality, and strongly linked to undernourishment, which appears to compromise the immune system (Jacobs et al. 2009). Why undernourished striped bass? The management strategy for striped bass emphasizes top-down trophic control by reducing mortality from fishing. This approach misses the bottom-up trophic relationship that potentially controls the striped bass population through reduced abundance of the proper food. The fate of striped bass may lay not only with how its fishery is managed, but that of one of its most important prey, the

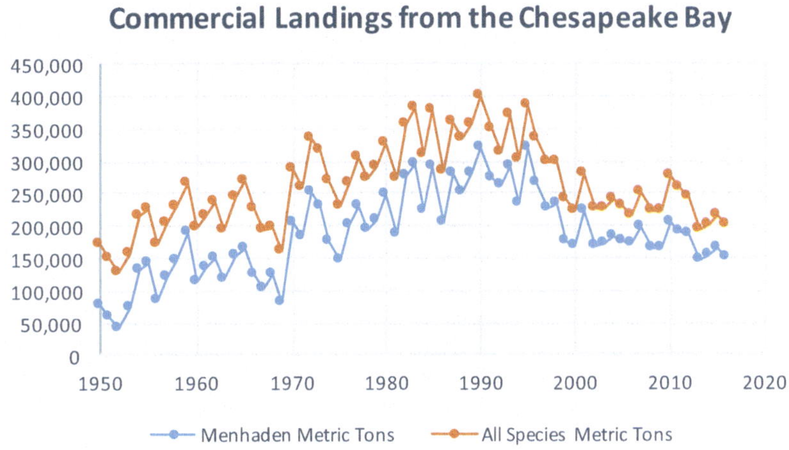

Fig. 6 Total annual commercial landings of fish from the Chesapeake Bay. The Atlantic menhaden accounted for about half of the annual commercial landings during the 1950s and 1960s, and became the overwhelming majority of the catch from the mid-1970s onward. The total catch between 1950 and 2016 was 12.5 million metric tons for menhaden, and 17.5 million metric tons for all species combined. Thus menhaden comprised 71% of the total commercial fishery during that 66-year period. Data from NOAA fisheries (NOAA 2018c)

Atlantic menhaden (*Brevoortia tyrannus*) (Uphoff 2003). The large populations of oily and energy-rich menhaden are the link between the very productive plankton of the Bay and adjacent coast and much of the rest of the food chain, including striped bass. The largest fishery in terms of catch on the Bay (and the entire East Coast of the US) is that for menhaden (NOAA 2018b), and as a consequence striped bass and a host of other species appear to suffer from reduced access to their favorite food. Full recovery of the striped bass fishery depends on changing management to include increased abundance of menhaden. Chapter "Menhaden, the Inedible Fish that Most Everyone Eats" addresses the role of the menhaden fishery in the Bay's ecosystem and food system.

The Rise and Fall of Fisheries The story is the same for almost all commercial fisheries, be they in the Chesapeake Bay, the neighboring Atlantic Ocean, the far-flung Southern Ocean, or anywhere else (Roberts 2007). A commercial fishery begins when fishers find an exploitable population and catch enough of it to sell to a market willing to pay a price that brings sufficient profits to continue the enterprise. There is a lot in that sentence. Simply being able to land lots of seafood doesn't mean it will be sold at a price that will make it profitable for the fisher. Consider the case of the Atlantic menhaden that today constitutes the largest fishery on the Chesapeake Bay (Figs. 6 and 7). Despite the historic enormity of the menhaden population, it wasn't until the fishery was industrialized after the Civil War that fishers on the Chesapeake paid much attention to the unpalatable fish. Considered inedible, its role in the food system was indirect. It was the best bait available, and good for fertilizing fields, but not much else. The development of the reduction industry that turned menhaden into commercial oils and animal-feed made it a more desirable catch. The expanding nineteenth century infrastructure of railroads and steamboats brought the processed catch to new markets. It was then worth the effort to exploit the Chesapeake Bay population. That enticed more *effort* to the fishing of menhaden. That *effort* took the form of more time, better gear, bigger boats, and more workers involved in the fishing. Soon the expanded effort reduced stocks to the point that there was less *catch per unit effort*. Seeing less return on investment of effort leaves the fishing industry with two choices. One is to get out of the business, and the other is to stay and further increase effort in pursuit of an increasingly scarcer stock. After all, the scarcity does increase market value of the limited catch that is landed. Left unchecked by regulators, persistent fishers will drive the stock to commercial if not biological extinction. Fortunately, government regulations arrived soon enough to save the menhaden from the fate of the Chesapeake Bay oyster industry (See chapter "The Chesapeake Bay Oyster: Cobblestone to Keystone" for details).

What would happen to the Chesapeake Bay's striped bass and Atlantic menhaden fisheries in the absence of effective regulation? They would share the fate of the Bay's oyster industry (Rothschild et al. 1994), and that of so many other failed fisheries around the world (Roberts 2007). When fisheries first develop, relatively little effort is needed to capture the fish (meaning finfish, oysters, muscle, crabs, etc.). This attracts increased effort as more people enter the fishery. There is also more incentive to invest in improved technology such as larger boats and more efficient gear. Soon the increased effort reduces the size of the stock and therefore harvest. This means less *catch per unit effort* (CPUE). The average size of the harvested fish also declines, as the increased effort differentially removes many of the older and larger individuals, particularly when regulations require doing such. The reduction in average individual size is also do to artificial selection for fish that breed earlier in life as a result of the fishing pressure on larger individuals (Zhou et al. 2010). The dwindling CPUE drives additional effort, as fishers invested in the enterprise attempt to maintain their

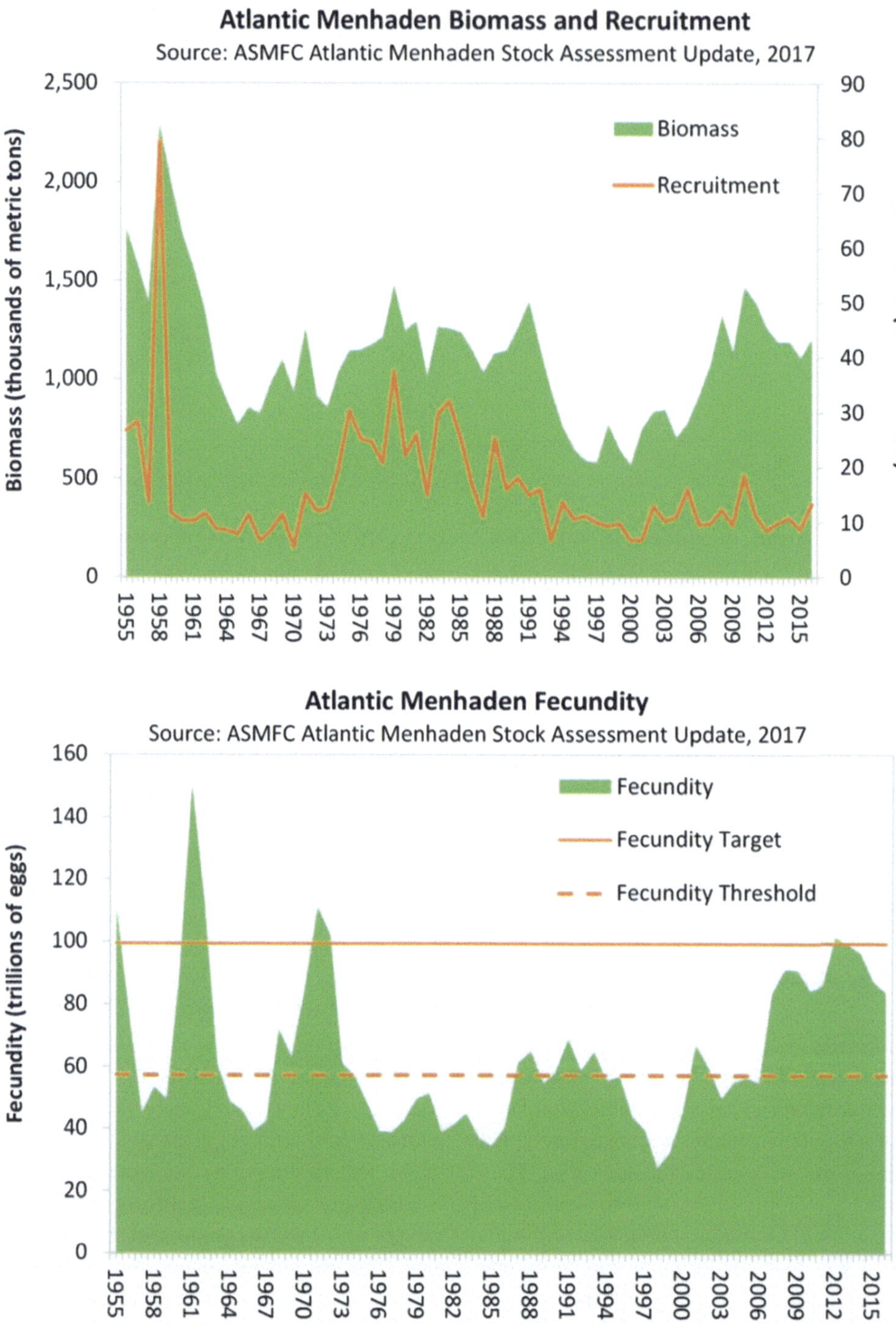

Fig. 7 Historic patterns of stock size, recruitment of new individuals to the population, and fecundity (production of fertilized eggs) for the Atlantic menhaden (ASMFC 2018c) http://www.asmfc.org/species/atlantic-menhaden

income flow. Unless checked by effective regulation the added effort may run the fishery to economic, if not biological extinction (Fig. 8). But, as long as some breeding stock remains, won't adding regulations recover the over-exploited fishery? Not necessarily.

The damage from overfishing a particular stock may so alter the ecosystem that recovery efforts may fail or be extremely difficult at best. Despite much investment over several decades, the habitat damage wrought by the oyster fishery in combination with shifts in agricultural practices and other factors means only a very partial recovery of about 3% for that once most important fishery in Bay (Rothschild et al. 1994; Mann and Powell 2007).

The New England fishery for Atlantic cod (*Gadus morhua*) provides a more recent example of the consequence of overfishing. Salted Atlantic cod fed colonists and enabled

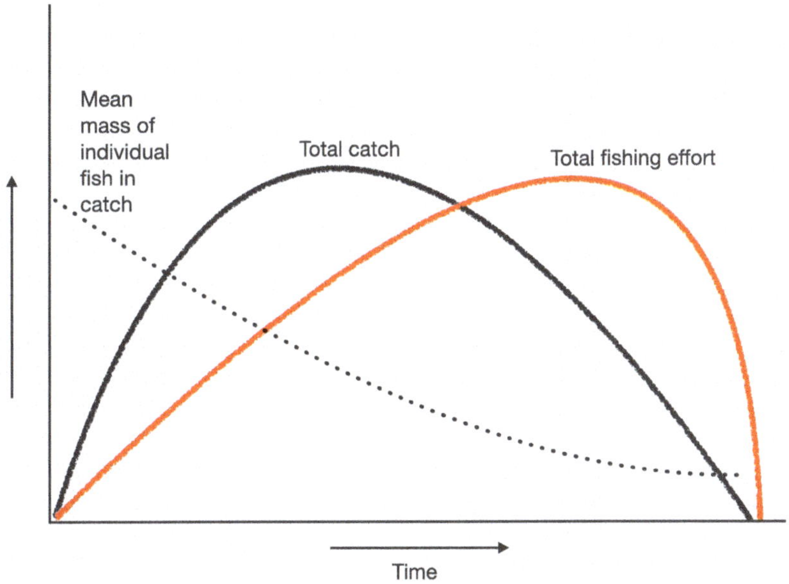

Fig. 8 The history of a poorly regulated fishery beginning with initial exploitation and ending with commercial extinction. Note that in the early year's, addition of fishing effort yields additional returns. As more effort is added, the stock declines, returning lesser yields, and less Catch per Unit Effort (CUE). As CUE declines, additional effort is applied to pursue the increasingly scarcer fish. Note also that average size of landed fish declines with higher fishing effort. Eventually the remaining stock of scarce and undersized fish bring little return per unit effort, and fishers leave the industry. Image modified from the FAO publication, Manual of Fisheries Science, Part 2 to show a fishery in collapse (FAO 2018)

the slave industry. This once super-abundant fishery employed over 40,000 Canadians and many others in New England. Expanded fishing effort in the Gulf of Maine in the 1970s led to huge catches in the 1980s. Unheeded fisheries scientists predicted the inevitable crash of the early 1990s. Despite nearly three decades of strict closures and regulations put in place in the 1990s, the fishery is essentially commercially extinct, with no sign of recovery. The intense fishing effort also meant habitat disruption by bottom trawls, reducing potential reproduction (Roberts 2007). At the same time a changing climate disrupted the growth of juvenile cod (Swain et al. 2016). The overfishing set up a regime shift, where cod now play a minor role in the ecosystem, being replaced by the American lobster (*Homarus americanus*). The once abundant cod used to prey intensely on juvenile lobsters. Now free of that mortality, the lobsters flourish (Guilford 2018).

Atlantic Sturgeon (*Acipenser oxyrinchus*) Of all the seafood that greeted Capt. John Smith and his band of colonists, none impressed him more than the hulking Atlantic sturgeon. Weighing up 363 kg (800 pounds) and reaching lengths of 5.5 m (18 feet), just one of these anadromous tasty fish and their eggs (caviar) could provide nourishment for many hungry colonists. And that is exactly what the sturgeon did upon their return trip to the area of Jamestown in the spring of 1609. Smith wrote, "We had more sturgeon than could be devoured by dog and man, of which the industrious by drying and pounding, mingled with caviar, sorrel, and other wholesome herbs, would make bread and good meat. Others would gather as much tockwhogh roots in a day as would make them bread a week, so that those wild fruits and what we caught we lived very well in regard of such a diet (Jamestown 2018)." The cruel winter of 1609 and lack of food meant the death of two thirds of the 214 Jamestown residents. Sturgeon and the other fruits of spring nourished the 70 or so starving colonists that survived that terrible winter.

Resupply from England and adaptation to their new surroundings enabled the remaining colonists and new arrivals to retain and expand their foothold in the new world. Sturgeon not only saved the colonists, but figured in their efforts to build local industries to produce goods to send back to England. They exported barrels of pickled sturgeon. Underscoring the importance of this species to the nascent colony, in 1612 Jamestown's Governor Thomas Dale required that all sturgeon caught be declared to him. Those failing to comply risked loosing their ears or being swung from the gallows (Roberts 2007). Perhaps the most severe penalties ever posted for violating fishing regulations on the Chesapeake Bay! Alas, preservation of the sturgeon meat proved unreliable and that enterprise faded away.

Atlantic sturgeon are toothless bottom feeders, using barbells on their snouts to detect mollusks, worms, and crustaceans in the sediments. The fish migrate out of the estuary at about the age of 2 years, feeding on the coastal benthic community. They return to the estuary for spawning, where historically they faced intensive fishing pressure.

During the eighteenth century sturgeon lost their appeal to residents of the Chesapeake and the rest of the developing country. The former delicacy once reserved for European royalty was reduced to the status of feed for hogs and bait for more desirable species. Racism may have contributed to this change in taste, as the plentiful sturgeon was considered a "... food of servants and negro slaves." Fishers of shad and other species also despised the sturgeon for damaging their nets (Tower 1908).

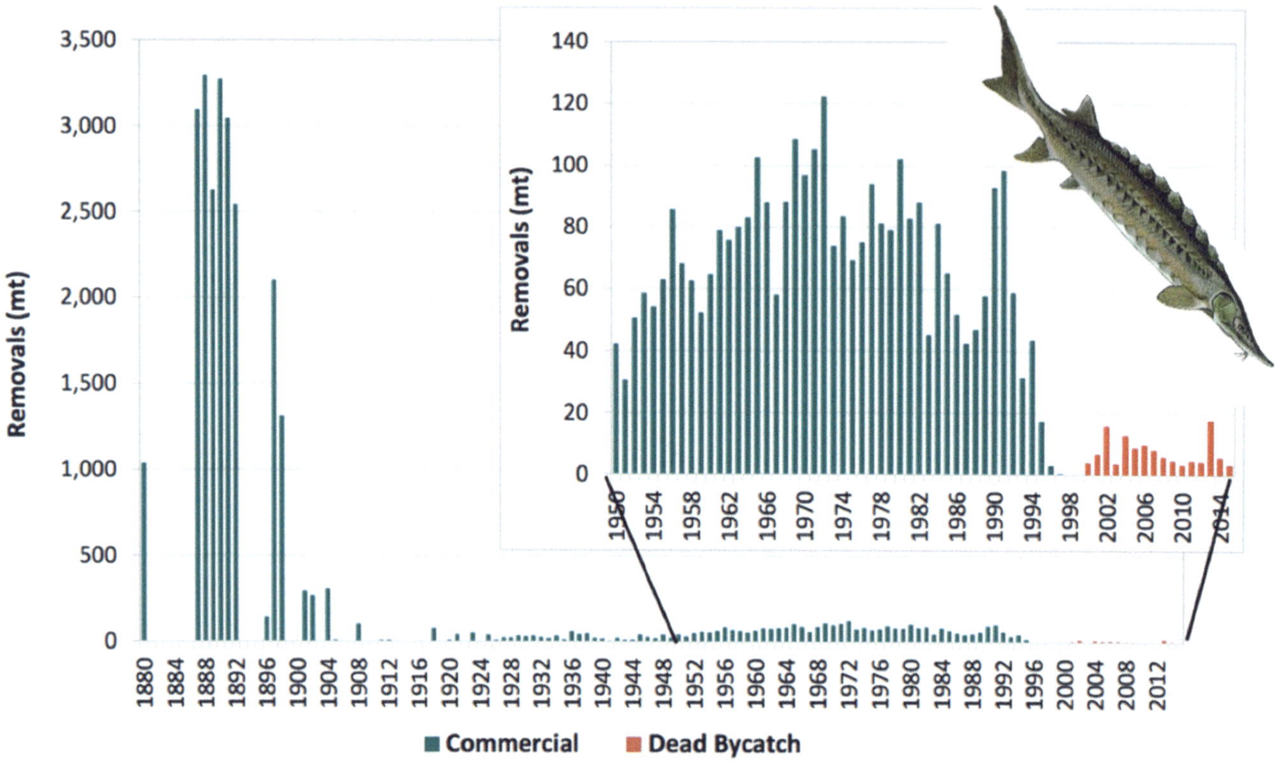

Fig. 9 Commercial landings (metric tons) and bycatch of Atlantic sturgeon for the entire US Atlantic coast. Landings are in metric tons. Note the near commercial extinction of the fishery in the early twentieth century, followed by an extended period of poor catch that motivated its closure. Figure from ASMFC (2018d)

Unfortunately for the fish, Atlantic sturgeon came back into culinary vogue in the late nineteenth century. Improved preservation techniques, smoking of the flesh, and European demand for caviar drove the expansion of the once small fishery. The fishery peaked in 1890 (329,000 kg, 726,000 pounds), and collapsed in 1905 (74,000 kg, 163,000 pounds). Despite the population collapse, fishers continued to pursue the increasingly scarce sturgeon (Tower 1908). By 1920 Chesapeake Bay fishers landed only10,400 kg (23,000 pounds) of Atlantic sturgeon. In 1924 Virginia set a length minimum of 1.2 m (4 feet) for legal taking of the fish. In 1974 Virginia closed the fishery for the scarce sturgeon (Fig. 9, Bushnoe and Musick 2018).

Overfishing, sedimentation from agricultural runoff suffocating gravel spawning grounds, removal of suitable spawning banks to facilitate navigation, and building of dams that blocked spawning runs, all came together to eliminate most of the breeding stock. Unintentional entanglement in fishing gear and vessel strikes also took their toll (NOAA 2018d). That some of the stock survived the onslaught of three centuries of fishing effort and habitat destruction is remarkable considering their life history characteristics and requirements for reproduction (Balazik et al. 2012; NOAA 2018e). Females of these long-lived fish (60 years) mature anywhere between their seventh and 27th year, and require 2–5 years between bouts of spawning (Smith 1985; Murdy et al. 1997). While they do produce many eggs, the success of the brood depends on stony substratum washed by clear oxygen-rich waters. Such habitat is in short supply in today's Chesapeake Bay.

In 2012 NOAA (National Oceanic and Atmospheric Administration) Fisheries Service listed the Atlantic sturgeon from the Chesapeake Bay, and other East Coast areas as *endangered* under the Endangered Species Act (ESA). This means protection of spawning habitat and closure of any fishing. Consequently, NOAA issued a final rule in 2017 designating 773 km (480 miles) of critical habitat for Atlantic sturgeon in the Chesapeake Bay. This pertains to stretches of the Potomac, Rappahannock, York, Pamunkey, Mattaponi, James, Nanticoke Rivers and Marshyhope Creek (Federal Register 2018). Critical habitat is defined under the ESA as "...specific geographic areas that contain features essential to the conservation of an endangered or threatened species and that may require special management and protection." Note that this includes presently unpopulated areas that exhibit potential for reintroduction of the species (USFWS 2018). The critical habitat designation may mean restricting activities in area, including various forms of fishing for other species. Critical habitat designation doesn't preclude all potentially harmful actions, nor does it require removal of developed lands and structures. It does require users of the habitat to justify activities and to minimize their harmful impacts.

In 1990 the Atlantic States Marine Fisheries Commission instituted a management plan with the goal of restoring populations of the Atlantic sturgeon. In practical terms this means removing dams to reopen stretches of rivers for spawning and to reconstitute the stony habitat needed to support the deposition of eggs (Fig. 10).

Atlantic Sturgeon Status and Restoration

Due to long generation times the species is slow to recover. Recent studies indicate that several tributaries of the Chesapeake Bay support Atlantic sturgeon populations, with spawning migrations in the Nanticoke, Potomac, Rappahannock, York, and the James River (the most robust). Since 2007, over 750 adult Atlantic sturgeon have been captured in the James River. Conservative mark-recapture estimates suggest more than 7000 adults in the James River, which may be the highest number for any river along the fish's entire range. The strong population in the James River traces to sturgeon fishery closures in the 1980s and 1990s, as well as improved water quality.

While adults are abundant in many river systems of the Chesapeake Bay, young juveniles seem to be scarce. Rivers outside the Bay system that have dedicated sampling to target these young fish show much more success compared to Chesapeake Bay tributaries. Between 2007 and 2017, only five young juveniles were captured in the James River, which is very worrying for managers. It is unknown whether low juvenile collection is due to sampling methods or recruitment failure. However, in 2018 over 250 age-0 sturgeon were captured during standard annual trawl sampling. This is very exciting considering that none were caught in the previous decade. The extremely wet spring and fall of that year may have been a factor for the strong year class. The heavy flow allowed easier access to high quality spawning habitat throughout the fall line that usually is not accessible during most years. The heavy flows may also have flushed sediment from spawning habitat downstream of the fall line as well. Every Atlantic sturgeon population is different and it might be that Chesapeake Bay tributaries have low levels of spawning success during most years and have bumper cohorts during extremely wet years. Considerable effort is planned to follow the development of this potential 2018 bumper crop. The hope is we will see a big increase in age 1–2 collections during the next 2 years. If we do not see these fish in our future sampling, it suggests that something may be hindering the survival of young Atlantic sturgeon and more drastic restoration efforts, such as hatchery work, might be required to maintain Chesapeake Bay populations.

Shad and River Herrings Although not as charismatic as the giant Atlantic sturgeon, another group of anadromous fish played an essential role in the historic food system of the Chesapeake Bay. These are the members of the family Clupeidae, or clupeids (Table 2, Fig. 11). The menhaden discussed above is a member of this group, but only an indirect player in the food system, being unpalatable. The other main clupeids of the Bay proved a favorite of both native peoples and the colonists that displaced them. These are the American shad (*Alosa sapidissima*) and the two species of so-called river herring, alewife (*Alosa pseudoharengus*) and blueback herring (*Alosa aestivalis*) (Kennedy 2018). Alewife and herring often school together, and fishers typically can't or don't distinguish between them, so most reports of catch simply list the combined numbers as river herrings. Other species of clupeids also frequent the Bay, but prove less important than American shad and the river herrings. All the clupeids play a critical role in the food chain, transforming plankton into fish flesh, which then becomes available to larger fish, birds, and mammals (including humans).

The American shad and river herrings occupy similar niches. All three return to the Chesapeake Bay to spawn in the spring, but at different times (Table 2). American shad are larger than the river herrings and were a more important part of the historic food system (Fig. 11). These anadromous fish should be understood as temporary residents of the Chesapeake Bay, and part of the greater coastal Atlantic Ecosystem. American shad from all over the Atlantic coast migrate to the Gulf of Maine for the summer and fall. Alewife migrates to deeper offshore waters in the winter.

As important as sturgeon proved to be for the starving residents of Jamestown, shad played an even larger role in the food system of the Chesapeake Bay. As discussed in chapter "The Algonquin Food System and how it Shaped the Ecosystem and Interactions with the English Colonists of the Chesapeake Bay", American shad and river herrings were essential to the Algonquin, providing welcome nourishment at the end of winters, and by use of preservation, for months later. The native fishers taught their techniques to the colonists, who also made shad and river herring central to their new food system.

Shad figured into the story of General and later President George Washington. Stores of salted shad saved Washington's troops from starvation while encamped at Valley Forge during the frigid winter of 1777–1778. Washington

Fig. 10 Albert Spells of the US Fish and Wildlife Service holding an Atlantic sturgeon caught on the Chesapeake Bay. Spells is working on restoring spawning habitat removing obstacles to migration in the James River (Gatenby 2018). Image USFWS with permission from A. Spells

Table 2 Comparison of the most important species of the family Culpeidae in the Chesapeake Bay; American shad (*Alosa sapidissima*), alewife (*Alosa pseudoharengus*) and blueback herring (*Alosa aestivalis*)

Species	Max adult length (SL) (cm)	Life span year	Adult mass (kg)	Diet	Time of spawning	Spawning location	Age of spawning year
American Shad	60	10	3.04	Zooplankton, small fish	Jan–Jun	Faster freshwater	4–6
Alewife	38	7	0.25	Diatoms, zooplankton	Mar–Apr	Shallow sluggish brackish water	3–5
Blueback herring	38	8	0.20	Zooplankton	Apr–May	Fast, deep freshwater rivers, streams	3–5
Atlantic menhaden	38	12	0.30	Phytoplankton, zooplankton, detritus	Mar–May, Sept–Oct	Off-shore shelf	1–3

Maximum adult length is standard length (mouth to base of tail). Table compiled from data in Murdy et al. (1997)

ran a large fishery for shad on shores of his Mount Vernon property located on the Potomac River. He ordered the preservation of each year's initial catch to assure food for his slaves. If the shad continued to run strong, as was usually the case, the subsequent catch was sent to market (FWS 2018; Kennedy 2018).

The American Shad and river herring dominated the Bay's fishery until those stocks collapsed circa 1970 (Fig. 12). The large catches of shad during the eighteenth and nineteenth centuries contributed in three ways to the food system of the era; mostly for direct human consumption, and secondarily the roe as feed for pigs, and when very abundant the fish, as fertilizer for farmer's fields (McPhee 2002; FWS 2018). The abundance of fish contributed to the growing human population of the Chesapeake Bay Watershed. This of course meant more fishing effort, and the construction of more dams to facilitate other human needs, that blocked fish migration. As early as 1740 colonial Virginians complained of fish scarcity, tying it to the proliferation of dams (Kennedy 2018).

Large seines set in the Susquehanna, Potomac and James Rivers, and in lesser flows around the Bay captured millions of migrating fish every spring. The spotty record keeping of the eighteenth and nineteenth centuries leaves the true number landed during that era a guess at best. However, a picture emerges of the abundance of those times from calculations by William Massmann in 1961. Using records from just one of

Fig. 11 The three main species of clupeid fish in the Chesapeake Bay food system, the American shad (*Alosa sapidissima*), alewife (*Alosa pseudoharengus*) and blueback herring (*Alosa aestivalis*)

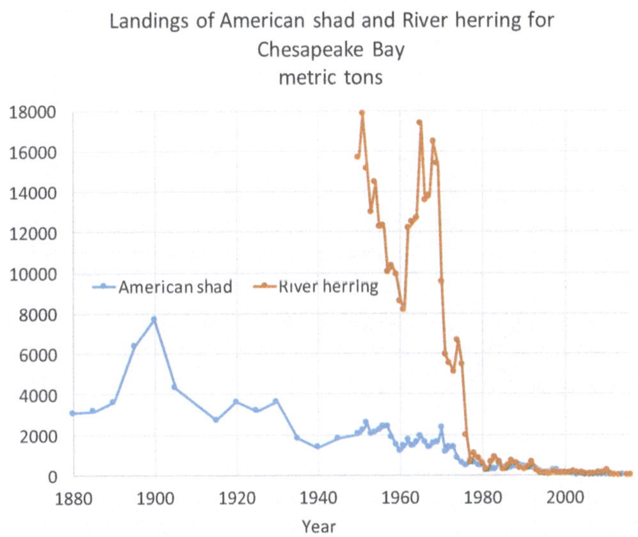

Fig. 12 The decline of the American shad (*Alosa sapidissima*) and river herring fishery of the Chesapeake Bay. Note that river herring refers to the combination of alewife (*Alosa pseudoharengus*) and blueback herring (*Alosa aestivalis*) as these species school together and most catch records don't distinguish between them. River herring data is not available before 1950. Data for American shad for 1880–1945 from NRC 2018. Data for 1950–2016 from NOAA 2018c

the many fishing locations on the Potomac River between 1814 and 1824, he estimated annual harvest ranging from 28,000 to 181,000 for shad, and 343,000 to 1,070,000 for river herring. That was equivalent to a third of the entire Potomac fishery for 1946–1956 (Kennedy 2018).

While all of the major shad and river herring runs of the Chesapeake suffered from the construction of dams, the Susquehanna river was the most impacted. The construction of canals in the 1820s and 1830s to allow navigation around the rapids above Harve de Grace, Maryland meant the damming of many important tributaries that formerly supported spawning grounds (Shank 2001). In 1928, the completion of the Conowingo Dam for hydroelectric power effectively ended the shad and river herring runs for most of the length of the Susquehanna river (Conowingo 2018).

Restoring the Shad and River Herring Restoring the fisheries for American shad and river herring requires two things; reestablishing access to suitable spawning grounds for these migratory species, and building populations of breeding fish. The first means the elimination of dams or installing fish ladders to provide a step-wise path around the structures. The second mandates closures to allow the stocks to rebuild naturally, and supplementing this with production of juveniles from fish hatcheries.

Dam Removal The Chesapeake Bay Program recorded the reopening of 5449 km (3386) miles of streams and rivers to passage of migratory fish between 1989 and 2017. Since 2011 52% of the restoration took place in Pennsylvania, 46% in Virginia, and 2% in Maryland. Dam removal is 124% ahead of the goal specified in the Chesapeake Bay Agreement (Fig. 13). Most of the nearly 50 dam removals took place in first and second order streams (headwaters) many km distant from the Chesapeake Bay (CBP 2018c). The important progress to date still leaves the vast majority of streams with impoundments that obstruct fish migration.

While removal of some obsolete dams will continue, the waterways won't return to the conditions experienced by Chief Powhatan and Capt. John Smith. Many dams will

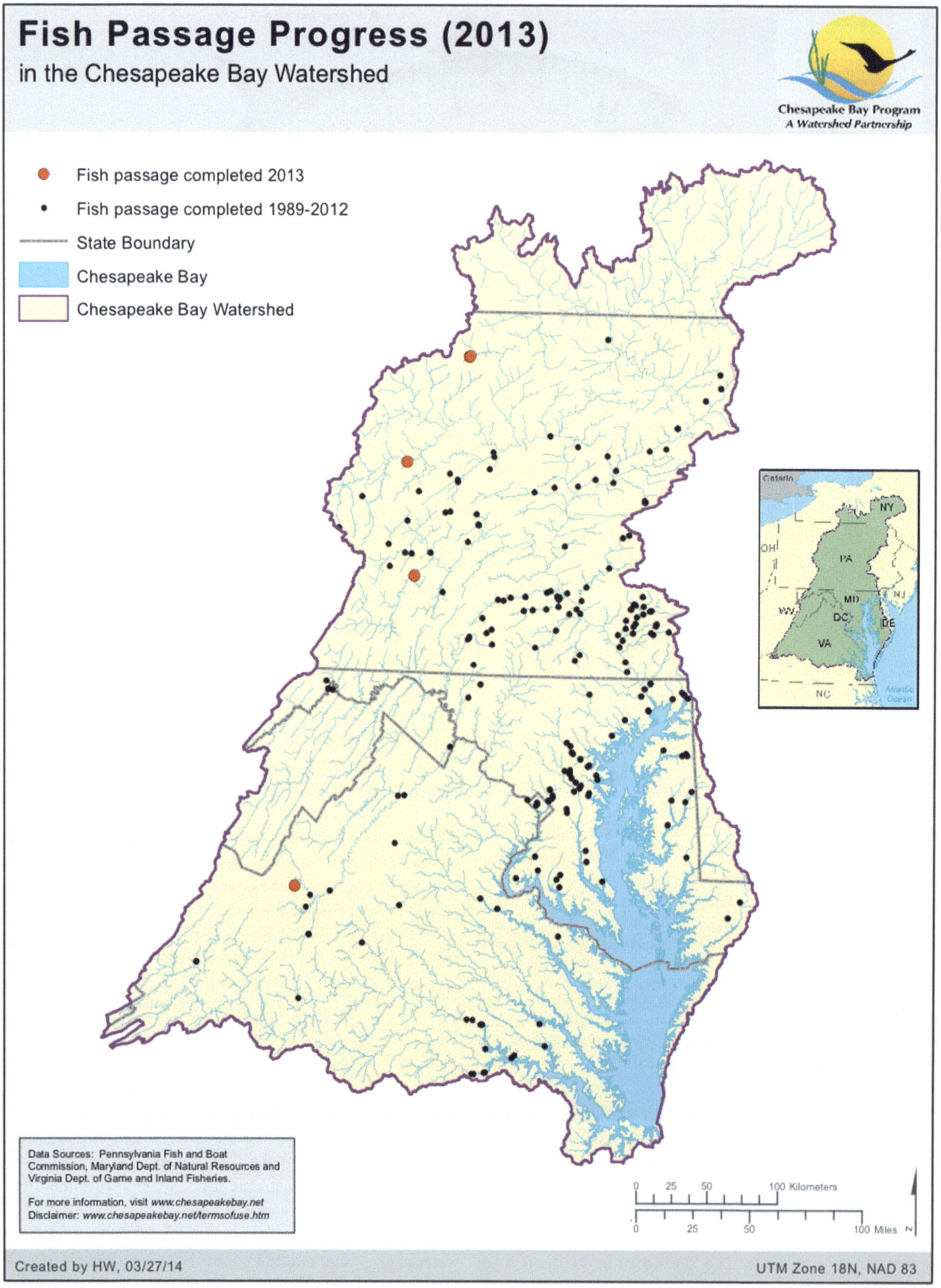

Fig. 13 Removal of dams in the Chesapeake Bay watershed to reestablish passages for migratory fish as of 2013. From CBP (2018b)

remain in place for several reasons. The Conowingo facility will continue to provide hydropower, and as a block against releasing nearly a century of deposited sediments into the Bay. Entire communities are built around the recreational opportunities offered by smaller impoundments across the watershed. Other dams provide water for drinking or irrigation. And some dams protect communities from floods. Thus, the social benefits garnered from maintaining many of the dams will outweigh those perceived from restoring waterways to conditions conducive to migratory fish runs.

Hatcheries The clupeids reproduce via open spawning, the females releasing their eggs into the water, with the males doing the same with their sperm. For the American shad and river herrings successful breeding requires the release of gametes in suitable habitat that facilitates both fertilization and survival of the developing fry. It takes fish to make fish. While dam removal reopens historic spawning grounds, it doesn't mean that there are sexually mature fish there to take advantage of the restored habitat. Given time, individuals from remaining stocks in the Bay may find their way to the newly liberated ancestral spawning grounds. To right the historic wrongs done to stocks, humans use fish hatcheries to supplement and hasten natural spawning.

Hatchery operators typically use nets to to capture ripe fish. After sorting the males and females, workers squeeze and stroke the fish beginning at the anterior end. This milking process releases the gametes through the posterior vents and into vessels of river water. The roe is extracted from the females first. Then the sperm containing milt is squeezed from a male. The extracted gametes are mixed, allowing the sperm to fertilize the eggs. The process usually takes place in the boat where the fish are landed so as to work with the freshest gametes as possible. Modern hatchery operators understand the importance of genetic diversity in sustaining viable populations, so obtain gametes from a variety of males and females.

The hatchery provides tanks that allow the fertilized eggs to develop into fry in an environment free of predators and other hazards. Hatchery workers expose the developing fish to oxytetracycline. These antibiotics combine with calcium in boney structures in the fish, leaving a detectable tag. This allows fishery scientists to ascertain the origin of returning adult fish (Reinert et al. 1998).

Decedents of two of the Algonquin tribes that greeted Capt. John Smith and his party, built hatcheries on their ancestral namesake tributaries, the Paumunkey and Mattaponi. These rivers drain into the York River. The tribes constructed the hatcheries in the early twentieth century, in keeping with the Powhatan tradition of giving back to nature. Both tribes maintain small populations in the vicinity of their ancestral lands, and the hatchery provides employment for some of their members (Graves 2018; Solomon 2018). The hatcheries are a link to their heritage, a culture tied closely to a food system built around shad, herrings, stripped bass, and other fish. The era of Jim Crow racism made the maintenance of food system heritage as much a necessity as a nicety for the two tribes. As did African Americans, Indians in the apartheid Virginia of that era suffered from government sanctioned segregation and racism (Mattaponi 2018). Survival of the tribes meant managing their fisheries and agricultural fields to assure food and income.

Today the Mattaponi counts about 450 in their ranks, about 75 of those residing on the reservation that houses their fish hatchery. The neighboring Pamunkey tribe includes 200 members, about 40 of whom live on their reservation, a portion of which works their hatchery (NPS 2018). Both tribes were recognized in treaties with the English colonial government (1646 and 1677), and only recently by the US Federal government (2016). To this date the Mattaponi still delivers its annual food system tribute of fish and wild game to the Governor of Virginia, fulfilling their end of the colonial era treaty.

The dam removals, closures and hatchery programs show some success in restoring the populations of shad and river herring in the Potomac and Rappahannock rivers, but much less so in the other main tributaries of the Bay (Fig. 14).

Ecosystem-Based Fishery Management (EBFM) The failure to properly include the importance of menhaden in the management of striped bass for the Chesapeake Bay is symptomatic of the historic reductionist approach to use science to manage fisheries. The frequent inability of single-species MSY fisheries management to protect the targeted stock and the rest of the ecosystem from which it is extracted has led to an alternative way of thinking about fisheries: Ecosystem-Based Fishery Management (EBFM). "The overall objective of EBFM is to sustain healthy marine ecosystems and the fisheries they support. In particular, EBFM should (1) avoid degradation of ecosystems, as measured by indicators of environmental quality and system status; (2) minimize the risk of irreversible change to natural assemblages of species and ecosystem processes; (3) obtain and maintain long-term socioeconomic benefits without compromising the ecosystem; and (4) generate knowledge of ecosystem processes sufficient to understand the likely consequences of human actions. Where knowledge is insufficient, robust and precautionary fishery management measures that favor the ecosystem should be adopted (Pikitch et al. 2004)."

Goals of EBFM include reducing bycatch (accidental capture/killing of non-target species), reducing the waste associated with discarded catch, protecting critical habitats, recovering ecosystem damaged by overharvesting, and

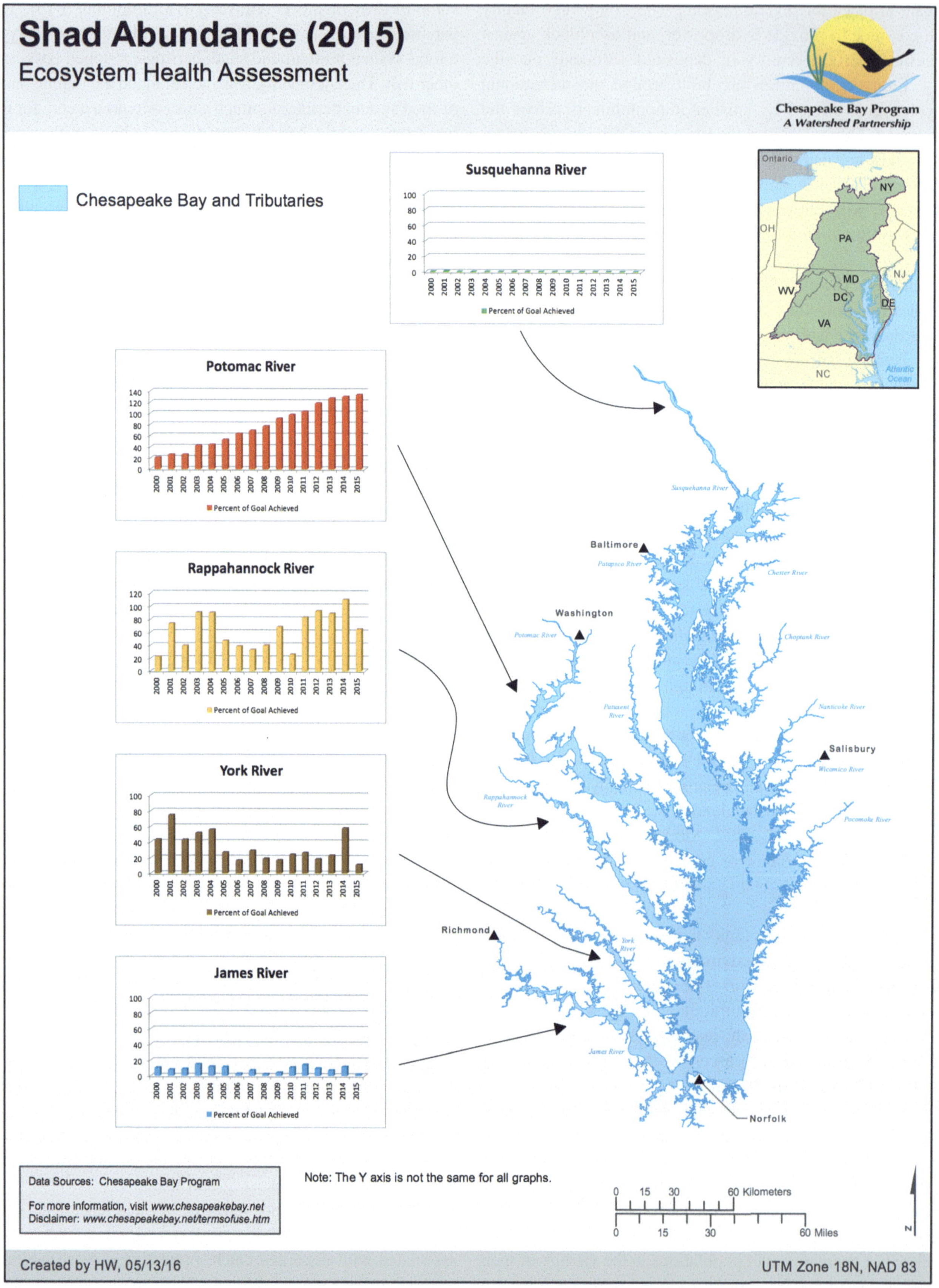

Fig. 14 Achievement of Chesapeake Bay Program goals for recovery of shad populations for the main tributaries. Note the success of shad recovery in the Potomac and Rappahannock Rivers. From: The Chesapeake Bay Program. https://www.chesapeakebay.net/images/maps/Shad_Abundance_2015.pdf

obeying the precautionary principle. "The precautionary principle. . . . has four central components: taking preventive action in the face of uncertainty; shifting the burden of proof to the proponents of an activity; exploring a wide range of alternatives to possibly harmful actions; and increasing public participation in decision making (Kriebel et al. 2001)."

EBFM is clearly at odds with the profit-motivated commercial fishing industry. They aren't in the business of maintaining ecosystem integrity. Share-holders don't get dividends from idle fishing gear and protected habitats. Commercial fishers make their money from catching and selling fish, and in many cases receive perverse government subsidies to do so even when the sustainability of the stock suffers as a consequence (Sumaila et al. 2016). In the US between 1996 and 2004 fisheries received about $6.9 billion from federal (79%) and state (21%) sources. The subsides came in the form of fuel (44%), research (40%), state sales tax exemptions (5%), disaster aid (4%), fishing access payments (2%), surplus fish purchases (2%), capitol construction fund (1%), seafood marking (1%) and vessel/permit buy backs (1%) (Sharp and Sumaila 2009).

Implementation of EBFM requires reversing the government subsidies that promote overfishing, and replacing them with funds to help transition the fishers to new occupations. The concept is akin to the soil-bank program used on the terrestrial side of the food system. In that case farmers receive subsidies to rest their lands, preserving soil, building fertility, and preventing depressed prices from excess production flooding the market.

EBFM may garner more support from recreational fishers than their commercial counterparts. Those that fish the Chesapeake Bay primarily motivated by a need to supplement their diets with free seafood might resent restrictions imposed by EBFM. Strictly recreational fishers may see the value in EBFM, as it will improve both the diversity and numeric availability of the catch. Fishers more interested in the sport of the hunt and the time spent enjoying the water than a potential meal should find EBFM an appealing proposition. It certainly fits well with catch-and-release fishers who focus entirely on the sporting aspect of the activity.

Other stakeholders may also support EBFM. The ecotourism industry benefits from restoration and preservation of aquatic systems. Neighbors of over-worked odiferous fish and crab processing plants will enjoy sweeter smelling air and increased property values. Boaters will be freed from excessive numbers of crab-pot buoys, and gillnets strewn across the navigable waters of the Chesapeake Bay.

In the end, EBFM may generate as much or more total income from the Chesapeake Bay as does traditional single-species MSY approaches. However, that income will be spread across a diverse set of industries, rather than concentrated in the hands of commercial extractors and the economic chain they feed. In the long run EBFM is good news for commercial fishers. Restricted catches mean less seafood on the market, creating scarcity and unsatisfied

Fig. 15 Seining for shad on the Potomac River. Illustration scanned from Hudson, C. B. 1894. From Guzik (2018)

demand. This raises the price consumers are willing to pay for their meal. Ultimately this may mean more profit for less effort exerted by the fisher.

Fishery of the Future? Blue catfish (*Ictalurus furcatus*) were introduced to the Chesapeake Bay system in the 1970s. They now represent about 75% of biomass in the tidal portions of most Bay tributaries. This generalist predator eats various species of concern including; shad, herring, blue crabs and striped bass. Blue catfish also compete with native species for space and food. Many commercial fishers switched to targeting blue catfish out of necessity, as it has largely replaced their former catch. In 2016 fishers took 1.5 million kg (3.4 million pounds) of blue catfish from the Chesapeake Bay watershed (NOAA-GIT 2017). Even with increased efforts using traditional gears such as hoop-nets, trotlines and gill nets, blue catfish biomass continues to expand. To increase harvest, commercial fishers and scientist from Virginaia Comonwealth University and the USFWS collaborated to test the efficacy of commercial low-frequency electrofishing in the James and York Rivers. The project was funded through a series Fisheries Resource Grants provided by VA Sea Grant.

Low-frequency electrofishing works by putting a low amount of direct current electricity into the water, stimulating swimming movements in certain fish species. Catfish, not having scales, are susceptible to the low electrical current. The involuntary swimming brings these bottom-dwellers to the surface, making for easy collection with dip nets. Catfish remain stunned by the electricity for several minutes, until they become desensitized to the current and swim away. This technique allows commercial fishers to be very selective, eliminating bycatch mortality. The project started in 2014 and ended in 2017. During the 4 years of sampling

Fig. 16 Purse seine fishing for menhaden. From: Image ID: figb0092, NOAA's Historic Fisheries Collection. Credit: NOAA National Marine Fisheries Service

Fig. 17 Men hauling bunted purse seine full of menhaden. From: Radcliffe, Lewis (1923) Fishery Industries of the United States: Report of the Division of Fisheries Industries for 1921, Report of the United States Commissioner of Fisheries, 1922, Washington, DC: Government Printing Office

221 trips were made and just over one million pounds of invasive blue catfish were harvested. Low-frequency electrofishing has thc potcntial to significantly reduce the abundance of invasive blue catfish in the Chesapeake Bay without any bycatch or mortality caused by lost gear such as ghost nets and pots. With proper management, low-frequency electrofishing can reduce and maintain blue catfish abundance at a reasonable level to better benefit the entire Chesapeake Bay watershed (NOAA-GIT 2017).

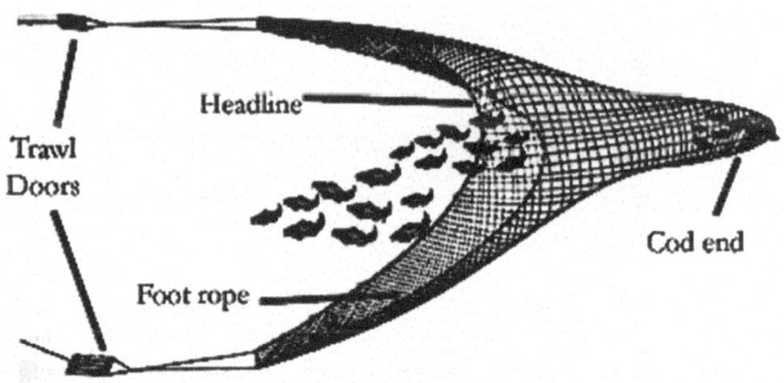

Fig. 18 A diagram of a trawl net as viewed from above. Note the otter boards (trawl doors) that provide the pressure to open the net. From: Anilocra, https://commons.wikimedia.org/wiki/File:Benthictrawl.jpg

Fishing Techniques

While the English brought with them their own notions of fishing, the early colonists benefited much from adopting the techniques taught to them by their native neighbors. This section reviews the various historic and contemporary methods used for collecting seafood from the Chesapeake Bay.

Spearing The vast abundance of fish in shoal waters encountered by Capt. Smith while exploring the Chesapeake Bay triggered a famous incident. On July 17, 1608 Capt. Smith used his sword to spear a cownose ray (*Rhinoptera bonasus*). The ray returned the favor, whipping its tail to land its barbed and poisonous sting deeply in John Smith's wrist. The Captain's ensuing pain, swelling, discoloration, and loss of motor function convinced Smith's crew of his imminent demise. Yet Smith survived his hard lesson on how one aspect of the Chesapeake food system works. The projection of land on the north side of the entrance to the Rappahannock River where Smith nearly met his death is named Stingray Point, reminding passing mariners and fishers of just one of the dangers they may encounter in the Bay (CBP 2018d).

Capt. John Smith noted how native fishers used bone-tipped spears and arrows while wadding or from dugout canoes. Fine lines attached to the arrows facilitated their retrieval, and hopefully with a skewered fish (Fig. 1, Smith 1612).

Hook and Line The native fishers used baited hooks of bone tied to lines and poles to angle for fish, as was the custom of the English (Smith 1612). Modern fishers keep the lines spooled on reels for easy control and storage of the string. Steel hooks take various forms, depending on what will hold best the intended prey. The bait may be alive, or cut from an animal, such as pieces of menhaden. Artificial lures shaped and colored to resemble a potential prey item often replace or supplement live bait. Fly-fishing is another version of angling. Fishers use artificial lures shaped like insects or other prey items that frequent the air-water interface. The fisher casts the lure and hook on to the surface, in hopes of enticing a bite.

Nets Various types of nets are used to capture seafood on the Bay. *Beach seines* are stout nets drawn through the water by boats or human power (Fig. 15). They entrap fish, swimming crabs, or other nekton that are too large to escape through their mesh. Typically, the nets are drawn onto a beach free of snags. Once landed, the catch is scooped-up by hand or with implements. In the absence of appropriate beaches, long wooden platforms may provide a smooth landing surface.

Fishers deploy *purse seines* from a pair of boats to encircle and trap large schools of fish, such as menhaden (Figs. 16 and 17). The name derives from an old fashion purse, the kind that was used in the era of Capt. John Smith. Once the two ends of the net are brought together by the boats maneuvering toward each other, the bottom of the net is closed by drawing tight a line passing through it. This is like closing an old fashion purse by drawing the strings tight. The bottom purse-line is attached to a heavy weight. Dropping the weight overboard pulls on the purse line, closing the bottom of the net. Then comes the hard work of pulling close the net full of fish. Until the invention of the hydraulic power block in 1950, that was all done by the work of many hands toiling for hours. In modern times large diameter hoses pump the fish aboard the mother vessel, replacing the dip nets of the era before mechanization.

Boats pull *trawl nets* through the water to collect seafood (Fig. 18). Some use otter boards placed on each side of the net opening. The otter boards are oriented so the pressure of the

Fig. 19 Setting a gill net for shad at night on the Susquehanna River Image ID: figb0164, NOAA's Historic Fisheries Collection. Credit: NOAA National Marine Fisheries Service

Fig. 20 A typical pound net configuration for the Chesapeake Bay. The pound (**a**) is where the fish collect, the heart (**b**) is funnel shaped to trap the fish and direct them toward the pound, the leader (**c**), which is longer in practice than illustrated intercepts the migrating fish and directs them to the heart. From MDDNR 2018b

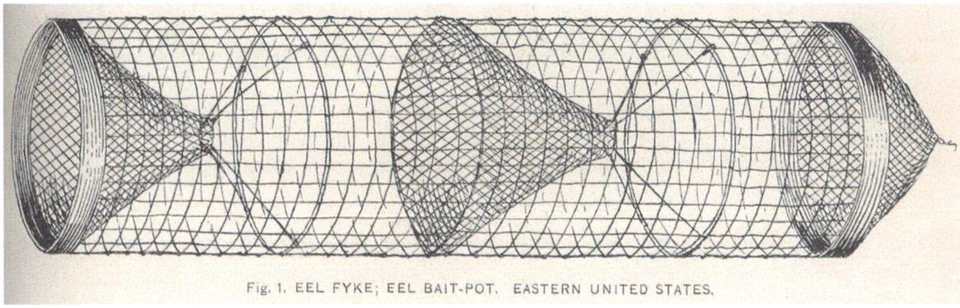

Fig. 21 Fyke net designed to catch eel. Designs vary depending on the targeted species. From: Smith, Hugh M. (1894) Fyke Nets and Fyke-Net Fisheries of the United States, with Notes on the Fyke Nets of Other Countries, Bulletin of the United States Fish Commission, vol. 12, 1892, Washington, DC: Government Printing Office

passing water pulls the ends of the net open. In lieu of otter boards, the net may be kept open with a long beam, hence a *beam trawl net*. The middle of the end of the trawl net is often elongated to form the cod-end, where the catch concentrates. The trawl net is hauled aboard the boat and its catch dumped on deck. Trawls may be rigged to pull near the surface, at some mid-water depth, or along the bottom. The weighted bottom trawls scrape the sediments, flattening the habitat and dislodging everything in their path including, sponges, oyster reefs, grass beds, and other critical features. Bottom trawling is highly destructive.

Fishers set *gill nets* to entrap fish passing through an area (Fig. 19). Gill nets are called such as an unsuspecting fish will swim into the net, not sensing the fine line from which it is made. Unable to pass through the mesh, the fish reverses course, but its protruding gill-covers catch on the line comprising the square in which it is stuck. Fishers typically set gill nets between two floats, and often deep enough to avoid ensnaring passing watercraft. Fishers retrieve the nets, dislodging the stuck fish into awaiting baskets. Gill nets differ from all other types in that the mesh of the net is what captures the fish. For all other nets, the mesh serves as a barrier to contain the captured fish, rather than to ensnare it.

Fishers use *cast nets* in quiet waters to catch schools of small fish, intended for bait. It takes some skill to toss out the net so it flares open and sets upon the school. The net closes around its catch upon retrieval.

John White's painting of Algonquin fishing practices features a *pound net* (Fig. 1). Pound nets are set perpendicular to shores or perpendicular to prevailing currents in off-shore flats (Fig. 20). The nets are stretched between a long series of poles driven into the bottom. The fine mesh of the *leader* nets prevents passage of the larger fish, forcing them to turn and follow the net until encountering an opening that directs them into the *pound*. The funnel shaped nets leading into the pound confuses the fish, preventing their escape. Fishers collect the catch from the pound by dip nets or lifting a net that comprises the bottom of the pound. The colonists readily adopted pound nets for fishing and to this day they are a regular feature of the shores of the Bay and its tributaries. They are popular for catching migrating shad and herring, but one can find almost any sort of pelagic species in the pound. These nets are also a favorite of fishing birds, such as double crested cormorant

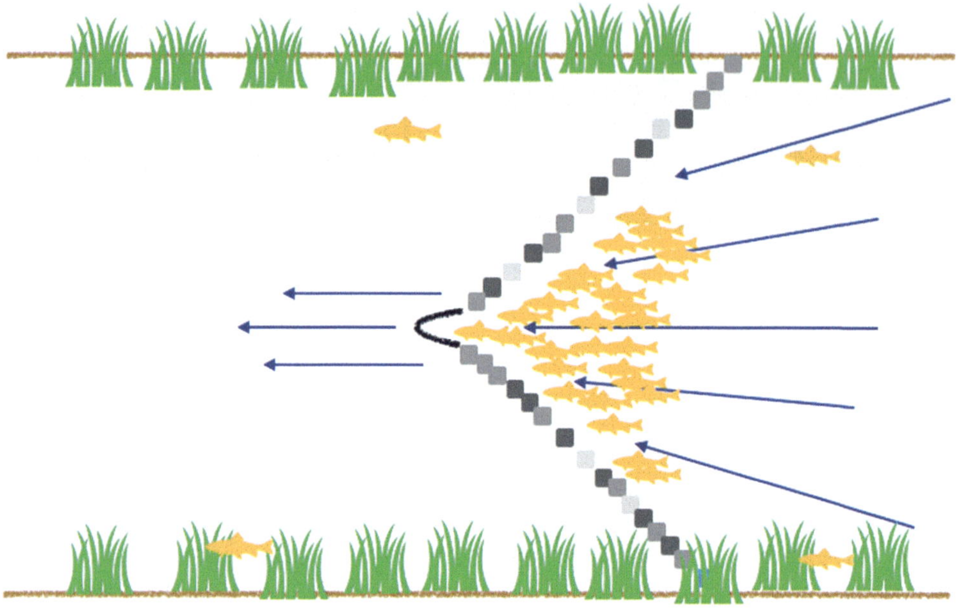

Fig. 22 Diagram of a fishing weir used by Native Americans on tributaries to the Chesapeake Bay. The weir is constructed of large river rocks rearranged in the shape of a "v." The arrows indicate the down stream flow from the ebbing tide. Fish are trapped behind rocks and focused into the tip of the v-shaped weir. There a wooden box or baskets receives the escaping fish

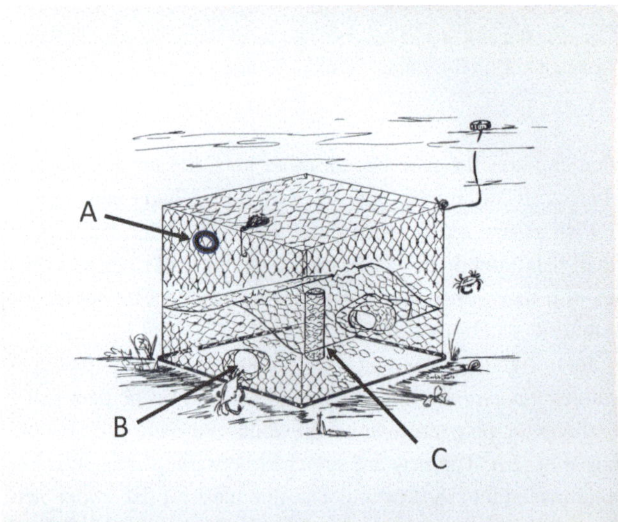

Fig. 23 The type of crab pot used to catch blue crabs (*Callinectes sapidus*) in the Chesapeake Bay. The cull ring (**a**) allows under-sized crabs to escape. Crabs enter through one of the funnel openings (**b**). Bait is placed in a tube (**c**) to attract the crabs. Modified from Image ID: fizh2058, NOAA's Historic Fisheries Collection. Credit: NOAA Central Library Historical Fisheries Collection

(*Phalacrocorax auritus*), brown pelican (*Pelecanus occidentalis*), and osprey (*Pandion haliaetus*) who appreciate perching upon the poles to await the arrival of a meal.

A *Fyke net* is typically a long cylinder or rectangular box with funnels formed by stiff rings or squares (Fig. 21). They are baited and staked to the bottom where they entrap fish or diamond back terrapins (turtles, *Malaclemys terrapin*) unable to navigate the systems of funnels to escape.

Dip nets allow fishers to collect their catch off of a spear or hook. With some skill they are direct weapons of capture for passing fish and swimming crabs.

Traps On the small scale Native fishers used reeds to weave cylindrical *eel pots* to catch individual or a few American eel (*Anguilla rostrata*). Variations of this design made of wire mesh are popular today for catching small fish for bait. On a much grander scale Native fishers built v-shaped fish weirs in the shallows of larger tidal waterways (Fig. 22). A wooden or woven box placed at the tip of the weir would concentrate fish trying to escape the ebbing waters of an impending low tide. Fishers could then use spears or dip nets to collect the entrapped fish. The weirs were built from materials at hand.

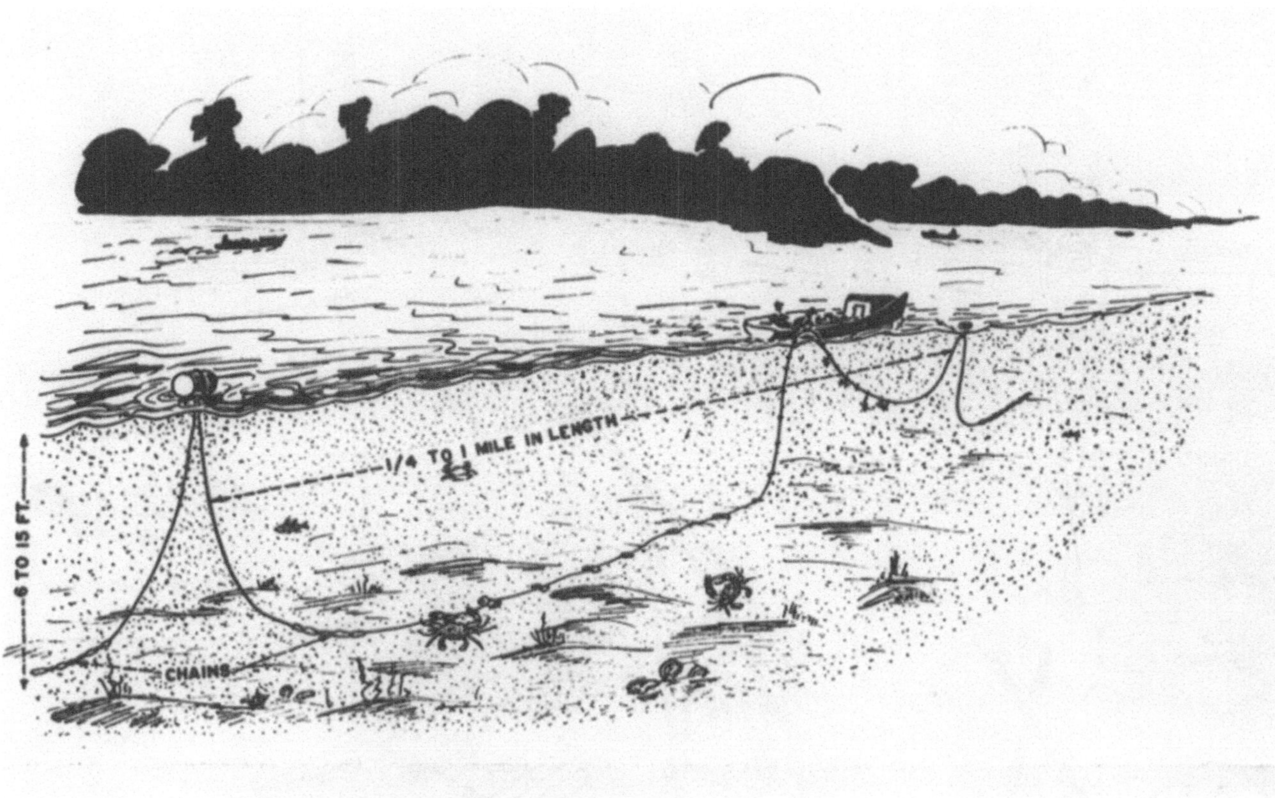

Fig. 24 Trotline being fished in Maryland waters of the Chesapeake Bay. The fishers retrieve the line over a roller on the side of their boat. They shake the crabs off the bait and into a dip net. From: Image ID: fizh2102, NOAA's Historic Fisheries Collection. Credit: NOAA Central Library Historical Fisheries Collection

In rocky rivers, large stones were shifted to create the weirs. Stakes with netting or woven mats provided more temporary structures (Cummins 2018).

The most common trap currently in use on the Bay is the *crab pot*, used for capturing blue crab (*Callinectes sapidus*). These are large wire mesh boxes measuring about 50 cm on a side (Fig. 23). Crabs are lured in to a funnel entrance by the odor of bait (fish heads or chunks of menhaden) placed in a wire tube. A small cull ring opening allow under-sized crabs to escape. The crab pot is tethered to a float which serves to mark the spot. The crabber uses a boathook to snag the line extending from the float and then hoists the pot into the boat. The pot is hinged on one side to allow the fisher to shake out the captured crabs into a basket. If needed, new bait is added to the pot before redeployment.

Trotlines The innate tenacity by which a crab clings to a piece of bait enables the efficacy of the *trot line*. Fishers set trotlines between two floats (Fig. 24). Bait is tied to the weighted trotline. The fishers need not wait long for any blue crab in the area to find and latch onto the bait. Owing to the determination to not loose the newly acquired meal, the blue crab holds tight, even after the line is pulled out of the water. The fishers give a few good shakes to encourage the crab to let go and fall into an awaiting dip net. Fishers now use trotlines to catch the invasive blue catfish (*Ictalurus furcatus*) in fresh water reaches of the Bay.

Tongs Native fishers wadded into shoal waters to collect oysters (*Crassostrea virginica*) by hand. Long handled tongs allow fishers to pull oysters from deeper waters on to a small boat. These inspired mechanized tongs deployed from a power winch to do the same thing in even deeper waters (Fig. 25).

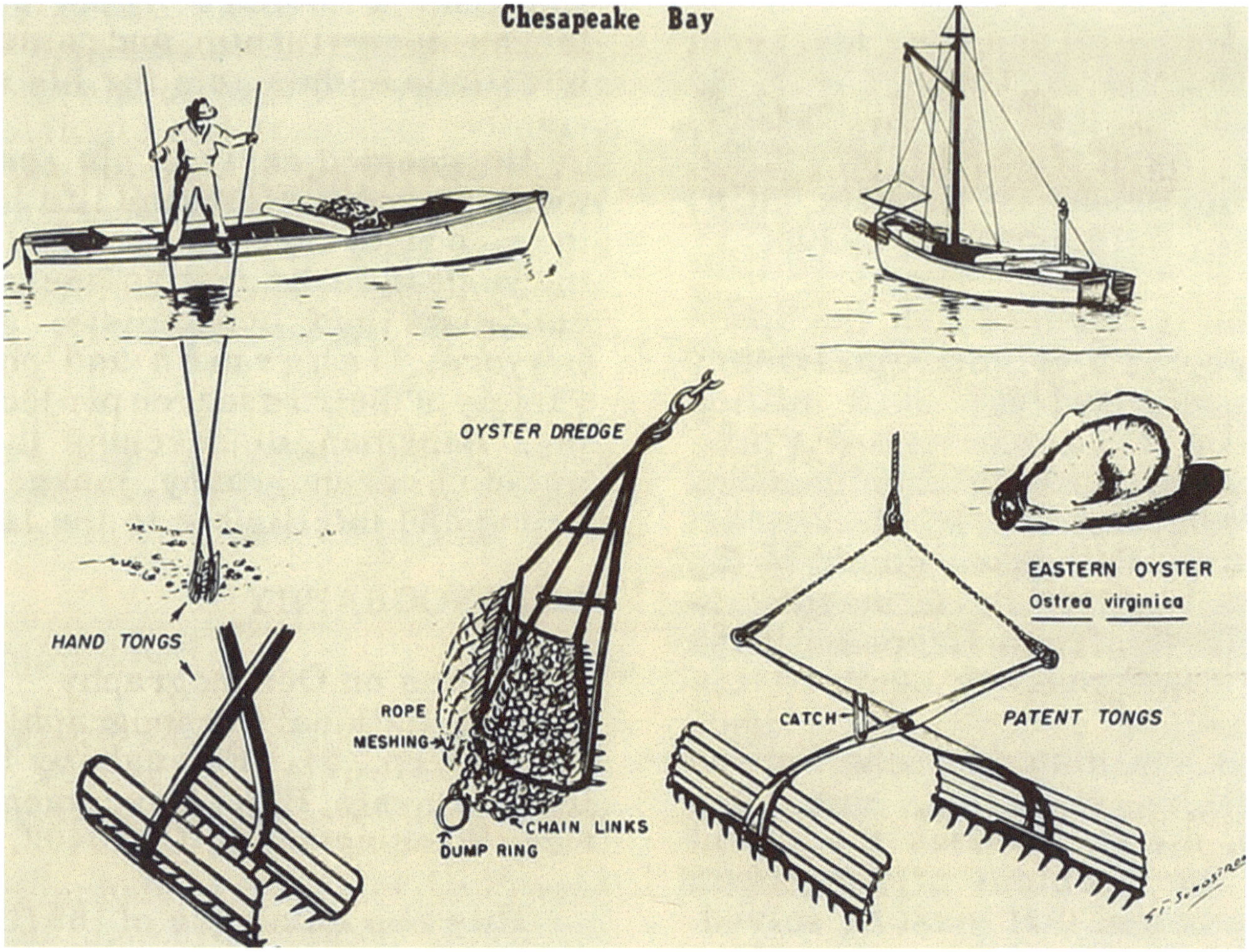

Fig. 25 Historic image illustrating gear used for fishing oysters. The patent tongs are used for oystering over reefs too deep to for hand tongs. From: https://www.flickr.com/photos/internetarchivebookimages/with/20046333613/

Scrapes and Dredges Boats pull stout iron or steel *scrapes or dredges* along the bottom to collect benthic animal living on or just in the surface of the sediments (Fig. 25). Oysters and overwintering blue crabs were favorite targets of this gear during the era of their abundance (circa pre 1950), but their use is now highly restricted.

References

ASMFC (2018a) About us—Atlantic states marine fisheries commission. In: Asmfc.org. http://www.asmfc.org/about-us/program-overview. Accessed 5 Nov 2018

ASMFC (2018b) In: Asmfc.org. https://www.asmfc.org/uploads/file/55089931S40_AtlMenhadenSAR_CombinedFINAL_1.15.2015-reduced.pdf. Accessed 9 Nov 2018

ASMFC (2018c) Species - Atlantic States Marine Fisheries Commission. In: Asmfc.org. http://www.asmfc.org/species/atlantic-menhaden. Accessed 12 Nov 2018

ASMFC (2018d) In: Asmfc.org. https://www.asmfc.org/uploads/file/59f8b1fdAtlanticSturgeonStockAssmtOverview_Oct2017.pdf. Accessed 18 Nov 2018

Balazik M, Garman G, Van Eenennaam J et al (2012) Empirical evidence of fall spawning by Atlantic sturgeon in the James River, Virginia. Trans Am Fish Soc 141:1465–1471. https://doi.org/10.1080/00028487.2012.703157

Brainard J (2017) Counting the fish in the sea: new evidence about menhaden could inform new approaches to managing them. Chesapeake Quarterly 16(2). http://www.chesapeakequarterly.net/V16N2/side2/ Accessed 11 Nov 2018

Bushnoe T, Musick J (2018) Essential spawning and nursery habitat of Atlantic sturgeon. (*Acipenser oxyrhynchus*) in Virginia. Unpublished manuscript. https://www.researchgate.net/profile/Arvind.../post/...rivers...spawn/.../3.pdf. Accessed 17 Nov 2018

Casey J (2018) In: Dnr.maryland.gov. https://dnr.maryland.gov/fisheries/Documents/history_of_comm_fishing.pdf. Accessed 3 Nov 2018

CBP (2018a) Striped Bass | Chesapeake Bay Program. In: Chesapeakebay.net. https://www.chesapeakebay.net/issues/striped_bass. Accessed 5 Nov 2018

CBP (2018b) Recent news: #dam removal | Chesapeake Bay Program. In: Chesapeakebay.net. https://www.chesapeakebay.net/news/blog/tag/dam_removal. Accessed 23 Nov 2018

CBP (2018c) Fish passage | Chesapeake Bay Program. In: Chesapeakebay.net. https://www.chesapeakebay.net/state/fish_passage. Accessed 23 Nov 2018

CBP (2018d) Cownose Ray | Chesapeake Bay Program. In: Chesapeakebay.net. https://www.chesapeakebay.net/S=0/fieldguide/critter/cownose_ray. Accessed 23 Nov 2018

Conowingo (2018) In: Conowingo Dam. Conowingodam.org. http://www.conowingodam.org/. Accessed 21 Nov 2018

Cramer J (2002) The origins of logistic regression. Papers.tinbergen.nl. https://papers.tinbergen.nl/02119.pdf. Accessed 5 Nov 2018

Cummins J (2018) In: Potomacriver.org. https://www.potomacriver.org/wp-content/uploads/2014/12/Return_of_American_Shad_Cummins.pdf. Accessed 26 Nov 2018

FAO (2018) Manual of fisheries science, Part 2—Methods of resource investigation and their application. In: Fao.org. http://www.fao.org/docrep/003/F0752E/F0752E01.htm. Accessed 12 Nov 2018

Fish Base (2018) Brevoortia tyrannus summary page. In: FishBase. https://www.fishbase.de/summary/Brevoortia-tyrannus.html. Accessed 9 Nov 2018

FWS (2018) American Shad. In: Fws.gov. https://www.fws.gov/chesapeakebay/shad.htm. Accessed 19 Nov 2018

Gallivan MD (2016) The Powhatan landscape: an archaeological history of the Algonquin Chesapeake. University Press of Florida, Gainesville

Gatenby C (2018) Virginia rivers opened for the first time in 100 years!. In: U.S. Fish and Wildlife Service Northeast Region. https://usfwsnortheast.wordpress.com/2016/12/13/virginia-rivers-opened-for-the-first-time-in-100-years/. Accessed 16 Nov 2018

Gauthier D, Latour R, Heisey D et al (2008) Mycobateriosis-associated mortality in wild striped bass (*Morone saxatalis*) from Chesapeake Bay, USA. Ecol Appl 18:1718–1727. https://doi.org/10.1890/07-2083.1

Graves L (2018) Mattaponi and Pamunkey tribes keep fish in river, food on table. In: Manataka.org. https://www.manataka.org/page1344.html. Accessed 22 Nov 2018

Guilford G (2018) The enigma behind America's freak, 20-year lobster boom. In: Quartz. https://qz.com/506376/lobsters-2/. Accessed 15 Nov 2018

Guzik S (2018) Blending art and science, Charles Bradford Hudson, division of fishes research highlight. In: Vertebrates.si.edu. http://vertebrates.si.edu/fishes/hudson_highlight/CBHudson_article.html. Accessed 26 Nov 2018

Jacobs J, Rhodes M, Baya A et al (2009) Influence of nutritional state on the progression and severity of mycobacteriosis in striped bass Morone saxatilis. Dis Aquat Org 87:183–197. https://doi.org/10.3354/dao02114

Jamestown (2018) Sturgeon | historic Jamestowne. In: Historicjamestowne.org. https://historicjamestowne.org/selected-artifacts/sturgeon-2/. Accessed 16 Nov 2018

Kennedy V (2018) Shifting baselines in the Chesapeake Bay. Johns Hopkins University Press, [S.1.]

Kriebel D, Tickner J, Epstein P et al (2001) The precautionary principle in environmental science. Environ Health Perspect 109:871. https://doi.org/10.2307/3454986

Lackey R (2006) Fisheries: history, science and management. In: Lehr JH, Keely J (eds) Water Encyclopedeia: surface and agricultural water. Wiely, NJ

Liljestrand E, Wilberg M, Schueller A (2018) Estimation of movement and mortality of Atlantic menhaden during 1966–1969 using a Bayesian multi-state mark-recovery model. Fish Res 210:204–213. https://doi.org/10.1016/j.fishres.2018.10.015

Malthus T (1817) Additions to the fourth and former edition of an essay on the principle of population. J. Murray, London

Mann R, Powell E (2007) Why oyster restoration goals in the Chesapeake Bay E are not and probably can not be achieved. J Shellfish Res 26:905–917. https://doi.org/10.2983/0730-8000(2007)26[905,worgit]2.0.co;2

Mattaponi (2018) Mattaponi reservation on the middle peninsula. In: Virginiaplaces.org. http://www.virginiaplaces.org/nativeamerican/resmatta.html. Accessed 23 Nov 2018

McPhee J (2002) The founding fish. Farrar, Straus and Giroux, NY

MDDNR (2018a) Penalty page. In: Dnr.maryland.gov. http://dnr.maryland.gov/fisheries/Pages/regulations/penalty.aspx. Accessed 5 Nov 2018

MDDNR (2018b) General information on pound nets. In: Dnr.maryland.gov. https://dnr.maryland.gov/fisheries/Pages/poundnets/general.aspx. Accessed 25 Nov 2018

Murdy E, Birdsong R, Musick J (1997) Fishes of Chesapeake Bay. Smithsonian Institution Press, Washington

National Park Service (2018) Meet the state-recognized Virginia Indian Tribes - Historic Jamestowne Part of Colonial National Historical Park (U.S. National Park Service). In: Home.nps.gov. https://home.nps.gov/jame/learn/historyculture/virginia-indian-tribes.htm. Accessed 23 Nov 2018

Nelson G (2018) Historical review of commercial fishery regulations for striped bass (Morone saxatilis Walbaum) in Massachusetts. Northeast Nat 25:143–160. https://doi.org/10.1656/045.025.0111

NOAA (2018a) Striped Bass—Fish Facts—chesapeakebay.noaa.gov. In: Chesapeakebay.noaa.gov. https://chesapeakebay.noaa.gov/fish-facts/striped-bass. Accessed 4 Nov 2018

NOAA (2018b) Menhaden—Fish Facts—chesapeakebay.noaa.gov. In: Chesapeakebay.noaa.gov. https://chesapeakebay.noaa.gov/fish-facts/menhaden. Accessed 5 Nov 2018

NOAA (2018c) Annual landings. In: St.nmfs.noaa.gov. https://www.st.nmfs.noaa.gov/commercial-fisheries/commercial-landings/annual-landings/index. Accessed 12 Nov 2018

NOAA (2018d) Atlantic sturgeon | NOAA fisheries. In: Fisheries.noaa.gov. https://www.fisheries.noaa.gov/species/atlantic-sturgeon. Accessed 16 Nov 2018

NOAA (2018e) Atlantic sturgeon population estimates | NOAA fisheries. In: Fisheries.noaa.gov. https://www.fisheries.noaa.gov/national/endangered-species-conservation/atlantic-sturgeon-population-estimates. Accessed 17 Nov 2018

NOAA-GIT (2017) Invasive catfish in the Chesapeake Bay. In: Chesapeakebay.net. https://www.chesapeakebay.net/documents/workshop_catfish_report_final_pdf.pdf. Accessed 7 Jan 2019

NRC (2018) In: NRC.gov. https://www.nrc.gov/docs/ML0823/ML082381111.pdf. Accessed 21 Nov 2018

Pikitch E, SAntora C, Babcock E, Bakun A, Bonfil R, Conover D, Dayton P, Doukakis P, Fluharty D, Heneman B, Houde E, Link J, Livingstgon P, Mangel M, McAllister M, Pope J, Sainsbury K (2004) ECOLOGY: ecosystem-based fishery management. Science 305:346–347. https://doi.org/10.1126/science.1098222

Register F (2018) Endangered and threatened species; designation of critical habitat for the endangered New York Bight, Chesapeake Bay, Carolina and South Atlantic Distinct Population Segments of Atlantic Sturgeon and the Threatened Gulf of Maine Distinct Population Segment of Atlantic Sturgeon. In: Federal Register. https://www.federalregister.gov/documents/2017/08/17/2017-17207/endangered-and-threatened-species-designation-of-critical-habitat-for-the-endangered-new-york-bight. Accessed 17 Nov 2018

Reinert T, Wallin J, Griffin M et al (1998) Long-term retention and detection of oxytetracycline marks applied to hatchery-reared larval striped bass, Morone saxatilis. Can J Fish Aquat Sci 55:539–543. https://doi.org/10.1139/f97-280

Richards R, Rago P (1999) A case history of effective fishery management: Chesapeake Bay striped Bass. N Am J Fish Manag 19:356–375. https://doi.org/10.1577/1548-8675(1999)019<0356:achoef>2.0.co;2

Roberts C (2007) The unnatural history of the sea. Island Press, Washington DC

Rothschild B, Ault J, Goulletquer P (1994) Decline of the Chesapeake Bay oyster population: a century of habitat destruction and overfishing. Mar Ecol Prog Ser 111:29–39. https://doi.org/10.3354/meps111029

Shank W (2001) The amazing Pennsylvania canals. American Canal and Transportation Center, York, PA

Sharp R, Sumaila U (2009) Quantification of U.S. Marine Fisheries Subsidies. N Am J Fish Manag 29:18–32. https://doi.org/10.1577/m07-170.1

Shiva V (2002) Water wars: Privatization, pollution and profit, 1st edn. South End Press, Cambridge, pp 23–26

Smith J (1612) A map of Virginia. Oxford: Joseph Barnes, pp.1-3. In: The complete works of captain John Smith (1580–1631), edited by Philip L. Barbour, (1986) 3 vols., 1. University of North Carolina Press, Chapel Hill, pp 119–190. Available in Capt. John Smith's complete works: http://www.virtualjamestown.org/exist/cocoon/jamestown/fha-js/SmiWorks1

Smith T (1985) The fishery, biology, and management of Atlantic sturgeon, Acipenser oxyrhynchus, in North America. Environ Biol Fish 14:61–72. https://doi.org/10.1007/bf00001577

Solomon P (2018) Virginia Indian archive | encounter: The Mattaponi River. In: Virginiaindianarchive.org. http://virginiaindianarchive.org/items/show/123. Accessed 22 Nov 2018

Sumaila U, Lam V, Le Manach F et al (2016) Global fisheries subsidies: an updated estimate. Mar Policy 69:189–193. https://doi.org/10.1016/j.marpol.2015.12.026

Swain D, Benoit H, Cox S, Cadigan N (2016) Comment on "Slow adaptation in the face of rapid warming leads to collapse of the Gulf of Maine cod fishery". Science 352:423–423. https://doi.org/10.1126/science.aad9346

Tower S (1908) The passing of the sturgeon: a case of the unparalleled extermination of a species. Popular Science Monthly 73:361–371

Uphoff J (2003) Predator-prey analysis of striped bass and Atlantic menhaden in upper Chesapeake Bay. Fish Manag Ecol 10:313–322. https://doi.org/10.1046/j.1365-2400.2003.00366.x

USFW (2018) Endangered species | what we do | listing and critical habitat | critical habitat | FAQ. In: Fws.gov. https://www.fws.gov/endangered/what-we-do/critical-habitats-faq.html. Accessed 17 Nov 2018

VALaw (2018) § 29.1–338. Revocation of license and privileges; penalties. In: Law.lis.virginia.gov. https://law.lis.virginia.gov/vacode/title29.1/chapter3/section29.1-338/. Accessed 5 Nov 2018

VAPenalty (2018) Taking game or fish during closed season or exceeding bag limit (§ 29.1–550)—Virginia Decoded - Virginia Decoded. In: Vacode.org. https://vacode.org/29.1-550/. Accessed 5 Nov 2018

VIMS (2018) Life history of striped bass | Virginia Institute of Marine Science. In: Vims.edu. http://www.vims.edu/research/departments/fisheries/programs/striped_bass_assessment_program/life_history/index.php. Accessed 4 Nov 2018

Virginia Compacts (2018) Maryland-Virginia compact of 1785. In: Law.lis.virginia.gov. https://law.lis.virginia.gov/compacts/maryland-virginia-compact-of-1785/. Accessed 3 Nov 2018

VMRC (2018) Virginia Marine Resources Commission historical highlights. In: Mrc.virginia.gov. http://mrc.virginia.gov/vmrchist.shtm. Accessed 3 Nov 2018

Wennersten JR (1981) The Oyster Wars of Chesapeake Bay. Tidewater Publishers, Centerville, MD

Zhou S, Smith A, Punt A et al (2010) Ecosystem-based fisheries management requires a change to the selective fishing philosophy. Proc Natl Acad Sci 107:9485–9489. https://doi.org/10.1073/pnas.0912771107

Menhaden, the Inedible Fish that Most Everyone Eats

Benjamin E. Cuker

Menhaden, the fish that nobody claims to eat,
finds its way into chicken and pig meat.
It's flesh once fertilized fields,
but now produces salmon and other aquaculture yields.
Its oil that once burned to create light,
is now taken by folks thinking it will stop heart disease and keep them in the fight.
The industry that catches this fish makes plenty of cash,
and if it keeps taking so many of them, the population might crash.
Those that chase striped bass already complain,
there's not menhaden enough for that sport fishery to sustain.
Menhaden eat the abundant plankton and turn that energy into fish,
for which so many animals it's their favorite dish.
Professional fisheries managers don't get much say,
since it's the politicians who make the rules for this fish at the end of the day.

Abstract

Atlantic menhaden *Brevoortia tyrannus* are an abundant member of the family Clupeidae. This fish is essential to the ecology of the Chesapeake Bay and coastal waters of the Atlantic Ocean, providing the trophic link between plankton and the menhaden's predators. These predators include the important sportfish striped bass (*Morone saxatilis*), other game fish, various piscivorous birds, and dolphins. Native Americans and early European colonists harvested the fish on a small scale to manure agricultural fields and feed pigs and chickens. New Englanders brought the menhaden reduction industry to the Chesapeake Bay in the 1870s. This led to the creation of the largest fishery on the East Coast of the US. The inedible menhaden were harvested en masse and processed to yield the commodities of fish meal and fish oil. In the early twentieth Century, fossil fuels took market share from both products, artificial N-fertilizer replacing fish meal, and kerosene replacing menhaden oil. The fish meal and oil became staples for the newly developed industry of concentrated animal feeding operations for chickens and pigs. The late twentieth and early twentyfirst centuries saw another shift in use of these products. Most of the fish meal and some of the oil went as feed to the lucrative and rapidly expanding global animal aquaculture industry. Most of the fish oil went to the new market of dietary supplements for the prevention of heart disease. Despite new evidence questioning the efficacy of fish oil for this medicinal application, the market remains strong. Mid-twentieth Century advances in mechanized harvesting of menhaden contributed to overharvesting that reduced the population well below historical levels. Political pressure from the industry and its allies restrains management measures. Sport fishers and conservationists advocate for reduced fishing pressure. The menhaden reduction industry provides important economic opportunities to a rural region and African American workers. A shift away from the animal product centered standard American diet may reduce the importance of this industry, returning menhaden to its prominent role in the natural food chain of the Chesapeake Bay.

Keywords

Brevoortia tyrannus · Reduction fishery · Striped bass · *Morone saxatilis* · Chesapeake Bay · Aquaculture feed · Fish oil supplements

B. E. Cuker (✉)
Department of Marine and Environmental Science, Hampton University, Hampton, Virginia, USA
e-mail: Benjamin.cuker@hamptonu.edu

B. E. Cuker (ed.), *Diet for a Sustainable Ecosystem*, Estuaries of the World, https://doi.org/10.1007/978-3-030-45481-4_6

Fig. 1 The smokestack is all that remains of an abandoned menhaden processing plant on the north shore of the Great Wicomico River near Reedville, VA. The relic greets mariners and menhaden ships as they make their way to Omega Protein, the last remaining fish reduction factory on the East Coast of the US. Image used by permission of the painter Jerry Spangler who created it in 2010 (https://jerryspangler.com/portfolio-viewer?collection=93790#lg=1&artworkId=530688)

The Birth of an Industrial Fishery for Fertilizer and Animal Feed The Great Wicomico River enters the Chesapeake from the west, some 10 km (6.2 miles) south of Smith Point, where the Potomac River joins the Bay. From the north, Cockrell Creek joins the Great Wicomico. On its shores is the small town of Reedville, VA. The banks of Cockrell Creek welcomes the entering mariner with a unique odor and the ruins of a dozen long-abandoned factories. The most imposing of these reminders of an earlier time is a tall brick smokestack built in 1902, rising 40 m (132 ft) from a now barren location on the northern shore of the river (Fig. 1). The smokestack is a monument to an industry that to this day annually produces more mass of fish than any other port on the East Coast of the US. The rotting factories were built in the late nineteenth and early twentieth centuries to process just one species of fish, the boney and oily *Brevoortia tyrannus* (common names include Atlantic menhaden, pogy, and bunker, Fig. 2).

Menhaden weren't processed for direct human consumption, as most people find the fish repulsive to the palate. Instead the factories cooked fish to a pulp and used presses to squeeze the plentiful oil from the flesh. This is called a reduction fishery, reducing the animal to the industrial products of protein-rich meal and useful oil (Fig. 3).

The reduction industry peaked in 1956, processing 712,100 metric tons of Atlantic menhaden in 20 different factories along the East Coast of the US. Overfishing and natural fluctuations in reproduction and recruitment reduced populations of the fish. In 2017, the industry processed 128,900 metric tons, which was below the total allowable catch for the species, a sign of a fishery in trouble. Despite lower catches compared to the 1950s, the annual take of menhaden still makes it the largest fishery on the East Coast of the US (ASMFC 2018).

The reduction work is now consolidated into one large industrial complex, less than one km beyond the towering smokestack on Cockrell Creek. That is the home to Omega Protein Corp. On days that it is rendering menhaden it produces a stench referred to by the locals as the smell of money. Although menhaden proves unpalatable for direct consumption by humans, it has played important roles in the food system as an animal feed, fertilizer and dietary supplement (Franklin 2007).

The menhaden reduction fishery created the small town of Reedville, Virginia that grew up on the shores of Cockrell Creek. Today it serves as the menhaden processing center for the Chesapeake Bay and much of the east coast of the US. In 1874,Capt. Elijah Reed brought the reduction fishery to Virginia, transferring his operation from Brooklyn, NY. The menhaden population along the New England coast thinned in response to a large and unregulated industry. So, Capt. Reed loaded his purse seins, kettles, and presses

Fig. 2 An adult (upper) and larval stage (lower) *Brevoortia tyrannus* (Atlantic menhaden). Adults typically range from 35 to 40 cm in length. Image from Chesapeake Bay Program (CBP 2019). https://www.chesapeakebay.net/S=0/fieldguide/critter/atlantic_menhaden

into two boats and sailed to Virginia where the menhaden population still flourished absent an industrial fishery (Haynie 1974).

Capt. Reed had good timing. Almost a decade after the Civil War ended, much of Northumberland County, Virginia still laid in ruins. Those were lean times for farmers and fishers living along the coast of the Chesapeake Bay. Former Confederates and emancipated African Americans suffered from a disrupted economy lacking the basic transportation and financial structure to support trade. The steamboats, that once connected Northumberland County to the rest of the Chesapeake community, had not run since the War. An absence of banks and post offices meant economic isolation. Capt. Reed built a factory to process the plentiful menhaden landed by his boats. This provided wages to workers of the area. Soon others followed in Reed's wake. Between 1880 and 1885, 15 menhaden factories sprung up on the banks of Cockrell's creek. The new town of Reedville arose on the river's shore, with expansive Victorian mansions for the factory owners, modest houses for the workers, a bank, post office, and stores. Schooners fishing the Bay and coastal waters kept a steady supply of menhaden going for the cooking tanks and oil presses in the factories. The intensity of fishing increased in 1894 as the first steam driven vessels displaced the schooners. The concentrated landing and processing of menhaden meant concentrated wealth. In 1911, Reedville had the highest income per capita of any town or city in the US (Haynie 1974).

What made the menhaden such a source of wealth if nobody wanted to eat it? The fish provided two important commodities needed to fuel the economy of late nineteenth and early twentieth century America: nitrogen rich fertilizer and oil. The name menhaden traces to the Algonquin Indian word "munnawhatteaug" which means fertilizer or manure. Pogy is another common name for the fish and it comes from the Abenaki Indian word, "paugagen," which also translates to fertilizer. Either name tells one exactly how the coastal Native Americans used the fish in their food system, not to eat but to add nutrients to their crops. The English colonists followed the native practice of using menhaden to fertilize their plantations of Indian corn (Franklin 2007).

Decades of tobacco and cotton agriculture exhausted the fields of the region. Farmers bought menhaden fish meal to return nitrogen and calcium to the soil. This menhaden fertilizer supplemented the supply of nitrogen-rich guano brought to the Bay in ships sailing from Peru.

Oil was the other item produced in the menhaden factories. Throughout most of history, humans relied upon plants and animals to provide oil for cooking, lighting, and lubrication. This changed in 1859 when Edwin Drake drilled the first commercial petroleum well in Titusville, PA, not far from the Chesapeake Bay Watershed. This began the

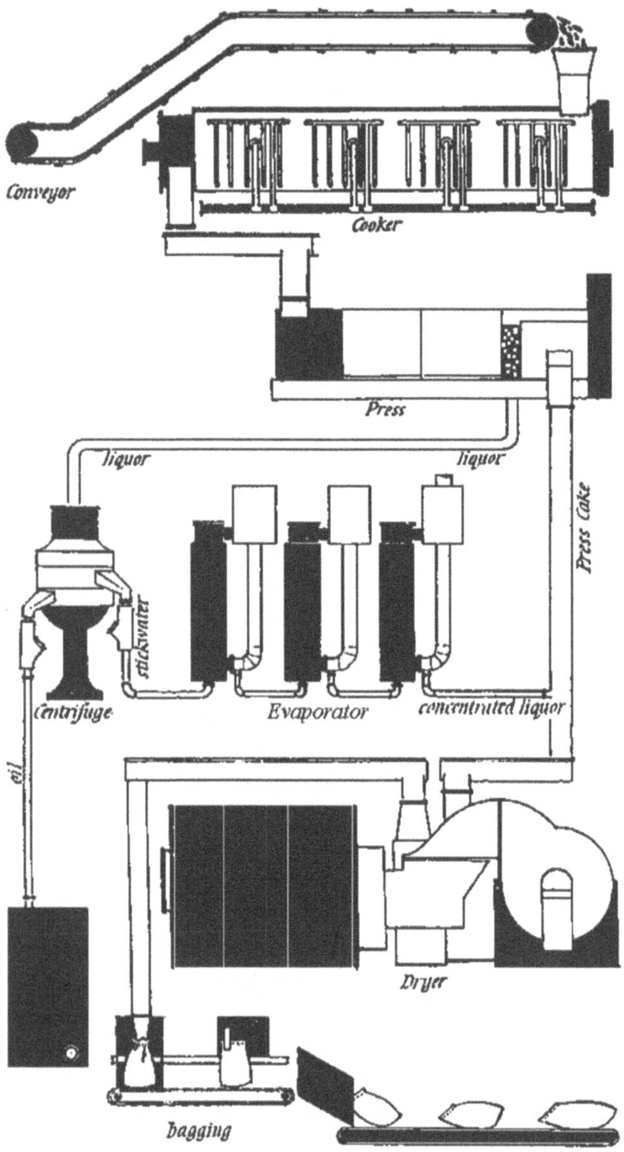

Fig. 3 The basic process for the production reduction fishery process for the production of meal and oil. Cooking and pressing separates the oil from the meal. Centrifugation clears the oil of water and particulates. Drying removes excess water from the meal. This is an energy intensive process typically using fossil fuels to create steam and electricity to run the machinery. Figure from the United Nations Food and Agriculture Organization (FAO 2019): http://www.fao.org/3/x5926e/x5926e01.htm

widespread replacement of biologically produced oils with petroleum. In 1863, John D. Rockefeller formed the Standard Oil Trust to monopolize the refining and distribution of petroleum. That corporation's production or kerosene for illumination and cooking stoves eventually took some of the profit from whale oil, perhaps helping to save several species of marine mammals from extinction (AOGHS 2019). But it took several decades for the world to embrace oil refined from petroleum. This kept up the demand for the oil pressed from menhaden well into the early twentieth century.

The development of the fossil fuel industry ultimately reduced the demand for *both*, menhaden oil and its meal as fertilizer. In the early twentieth century, German chemists, Fritz Haber and Carl Bosch developed a process to use fossil fuel to transform atmospheric nitrogen (N_2) into ammonia (NH_3) for making artificial fertilizer (See chapter "The journey from Peruvian Guano to Artificial Fertilizer ends with too Much Nitrogen in the Chesapeake Bay"). This new artificial fertilizer took market share away from menhaden meal.

From Feeding Chickens to Fueling Industrial Aquaculture The advent of plentiful petroleum redirected the use of menhaden meal and oil. No longer needed as much for fertilizer, the meal went to feed the developing animal agriculture industry. Concentrated Animal Feeding Operations (CAFOs) used the meal and oil to supplement the diets of chickens and pigs (See chapter "Livestock and Poultry: the Other Colonists who Changed the Food System of the Chesapeake Bay"). The rise of the broiler chicken industry on the Delmarva Peninsula in the 1920s meant a new and quickly expanding market for menhaden meal as animal feed (Edwards and May 1965; Stoner et al. 1990). This persisted into the 1980's when the worldwide expansion of aquaculture (the farming of aquatic plants or animals) opened a new and more lucrative market for fishmeal. As of 2019, Omega Protein sells about half of its fish meal to the aquaculture industry with the rest going to high end pet food, veterinary feeds, and specialized livestock supplements. This includes products for poultry, pigs, beef cattle, dairy cows, and horses. Omega Protein isn't completely out of the fertilizer market as it produces some refined liquid fertilizers for crop production (Omega Protein 2019a, Monty Diehl, Vice President of Operations, Omega Protein, personal communication).

Animal aquaculture is a practice distinct from fishing. Fishers use various techniques to collect their prey from the wild (see chapter "A Fishing Trip: Exploiting and managing the commons of the Chesapeake Bay"). Aquaculture instead grows animals at high densities in a confined space. Animal aquaculture can be split into two different general approaches.

The first practice uses suspension feeders such as oysters and mussels. These animals filter their food directly from the natural water surrounding them. The animals are placed in cages or tied to rafts where they feed on phytoplankton and other suspended particles. This is a highly sustainable practice in areas well flushed by tidal currents. The beds of suspension feeders generally improves the surrounding water quality as it reduces excess algae that is typical of eutrophic coastal systems. However, suspension feeders farmed in quieter waters may add excess nitrogen to the water and foul the sediments with oxygen demanding wastes (Gallardi 2014).

The second type of animal aquaculture is typically unsustainable yet accounts for 70% of the industry's production (Fry et al. 2018). It is a system of converting low market value animals or plants into the flesh of higher priced animals. The cultivated fish or shrimp are crowded into ponds, tanks, pens, or cages and supplied with feed. That feed is mostly generated from wild-caught species of low market value such as menhaden. These aquatic versions of CAFOs pack populations of fish or shrimp into unnaturally small spaces. And like their terrestrial counterparts destroy habitat, pollute the environment with excrement, promote the spread of disease, and cause stress to the confined animals. The cultured animals often escape from their pens and interbreed with wild populations causing genetic pollution (Martinez-Porchas and Martinez-Cordova 2012). The intensive reduction fisheries used to produce aquaculture feed also mean reduced populations of forage fish to support natural marine food chains.

Overharvesting of menhaden from the Chesapeake and coastal waters leaves less food for striped bass (*Morone saxatilis*), osprey (*Pandion cristatus*), dolphins (*Tursiops truncates*), and sandbar sharks (*Carcharhinus plumbeus*). Menhaden meal processed in Reedville may be exported to feed salmon crowded in pens along the coasts of Canada, Norway, or Chile leaving behind hungry predators in the Bay.

In 2016, the world food system produced 20 kg of animal seafood per capita. Half of that supply came from animal aquaculture. To appreciate the growth of aquaculture consider that in 1985 that industry produced about five million metric tons of fish worldwide in comparison to 82 million tons secured by the capture fishery. By 2014, each sector produced about 83 million metric tons of seafood. That means aquaculture increased over 16 fold, while the capture fishery remained static for the last four decades (FAO 2016). This vast expansion of fed aquaculture comes at the expense of overfishing of menhaden and other forage fish of low economic value (Kristofersson and Anderson 2006). This includes many species otherwise directly consumable by humans (Cashion et al. 2017).

Keep in mind that not all of the animal feed becomes tissue in the animal that eats it. Most fed-aquaculture operations report remarkable feed conversions ratios ranging between 1.2 and 2.5. For Atlantic salmon, this means that 1.3 kg of fish feed produces 1.0–0.5 kg of salmon. However, not all the nutrients (protein and calories) in the fish-feed become available to human consumers. For example, Atlantic salmon retain only about 28% of the protein and 24% of the calories from their feed (Fry et al. 2018). To obtain feed at lower cost, the aquaculture industry now includes plant-based alternatives to replace or supplement fish meal and oil in its feed (Anderson et al. 2016).

Although the Chesapeake Bay suffers the indirect effects of aquaculture from the harvesting and export of menhaden, it has been spared the direct environmental consequences of fed animal operations. Most aquaculture in the Bay focuses on sustainable farming of oysters, mussels, and clams. Oysters (*Crassostera virginica*) aquaculture yields 1.3–1.4 times more energy than invested and provides a higher net benefit to society than fed aquaculture (Williamson et al. 2015).

The Canadian based firm Cooke Inc. acquired Omega Protein for about $500 million in December of 2017. Cooke is a vertically integrated aquaculture firm, so is naturally interested in fishmeal and oil.

Heart Disease Creates a New Market for Menhaden Oil While petroleum displaced menhaden oil from lighting lamps and lubrication, it continued to be used as an animal feed supplement, as well as for waterproofing, cosmetics, paint, and other industrial applications throughout the twentieth Century. Fish oil took on new value in the 1980s as a source of omega-3 fatty acids. In the early 1970s, scientists discovered that Greenlandic Eskimos had lower cholesterol and triglyceride levels than predicted by their meat-rich diets (Bang et al. 1971). This work and other studies led to the belief that fish oils rich in omega-3 fatty acids could reduce the incidence of heart disease. When tested in epidemiological studies over time, fish oil consumption initially appeared to reduce the incidence of heart attacks (Burr 2001). This led to the American Heart Association recommending fish oil supplements to reduce chances of heart attacks. With this initial evidence for promoting heart health the lowly menhaden oil went from chicken feed and water proofing to the realm of a medically potent food. In 1997, fish oil received the status of Generally Recognized as Safe (GRAS) from the USDA (Omega Protein 2019b). This allowed the company to enter the lucrative dietary supplement market. The oil was also sold to manufacturers of margarine and other processed foods, who wanted to tout the presence of omega-3 fatty acids. Understanding the growing demand for fish oils rich in omega-3 fatty acids, the company (previously Haynie Snow Co. 2013, Zapata Haynie Co. 1972, Zapata Protein 1984) took on its present name, Omega Protein in 1997. The new name said it all in its two words, a company that makes protein rich fish meal and omega-3 rich fatty acid fish oil. In 2012, 7.8% of US adults consumed fish oil supplements daily (NIH 2019). By 2018, annual sales of dietary fish oil supplements reached between $15 and $30 billion (Greenberg 2018).

A series of recent studies refuted the heart healthy benefits attributed to fish oil. Some studies even found higher incidence of death associated with consumption of fish oil pills (Abdelhamid et al. 2018). Consequently, the American Heart Association backed away from supporting fish oil supplementation. Omega-3 fatty acids are important for a healthy diet and can be gotten from ground flaxseed and other plant sources, free of the contamination often found in fish oils (see chapter "Instead of Eating Fish: the Health Consequences of Eating Seafood from the Chesapeake Bay Compared to Other Choices"). Time will tell if dissemination of the current science on the topic will dampen the public's demand for fish oil supplements and Omega Protein's profitability. The fish oil companies spent twenty years and significant advertising dollars to convince the general public and medical industry of the efficacy of their product. There's no entity with the budget nor motivation to retell the story differently. So the public may continue to purchase fish oil supplements for some time to come, thinking they are improving their chances to beat heart disease.

When Elijah Reed founded the industrial menhaden fishery in 1874, it was only partially part of the food system, supplying fish meal for fertilizer. The oil it produced went for largely nonfood system uses. Today, the industry sees itself as firmly in the food business. Its 2016 annual report states that "Omega Protein Corporation is a nutritional products company that develops, produces, and delivers nutritious products throughout the world to improve the nutritional integrity of foods, dietary supplements, and animal feeds (OPAR 2019)."

Menhaden Ecology *Brevoortia tyrannus* (Atlantic menhaden) belong to the family Clupeidae, along with the Chesapeake Bay food system fish, American shad (*Alosa sapidissima*), alewife (*Alosa pseudoharengus*), and blueback herring (*Alosa aestivalis*). Clupeids eat low on the food chain and are often called forage fish, supporting larger piscivorous fish, piscivorous birds and marine mammals. The clupeids of the Chesapeake Bay undergo seasonal migrations from the estuary to the coastal ocean. Menhaden differ from the closely related shad and herring in several important ways. The shad and herring swim into the lower salinity waters of tributaries to spawn. Menhaden avoid migrating very far up-stream and instead spawn in saltier coastal waters. This life history characteristic saved the menhaden population from being excluded from spawning habitat by dams, which so impacted the shad and herring of the Bay. Menhaden also eat a unique diet of mostly phytoplankton, detritus, and small zooplankton. That diet gives their flesh a very pungent taste and odor that repels most humans, but so strongly attracts their natural predators that is the preferred bait for fishers.

The menhaden lifecycle begins with adults spawning eggs and sperm in the Atlantic Ocean. Ocean currents sweep the larvae and juveniles that develop form the fertilized eggs into the Chesapeake Bay. Rich concentrations of phytoplankton in the Bay fuel the development of the juveniles into young adults. After a year or so, the young adult fish migrate back into the Atlantic Ocean (Fig. 4).

Fig. 4 Two adult *Brevoortia tyrannus* (Atlantic menhaden) with mouths open for suspension feeding. Modified from image by Photo by Gene Helfman (NOAA 2019) https://chesapeakebay.noaa.gov/fish-facts/menhaden

Menhaden are suspension feeders like oysters and mussels. They are one of a very few species of fish able to effectively capture tiny particles in the size range of phytoplankton and bacteria (5–9 μm). This makes them well-adapted to the nutrient rich Chesapeake Bay and coastal waters that support large populations of phytoplankton. The fish swim with open mouths to collect the plankton (Fig. 5). As the water exits though the gill openings, it passes through closely spaced gill rakers that feature tiny branches (branchiospinule) and mucus used to trap the phytoplankton, detritus and zooplankton for ingestion. Crossflow filtration propels the captured particles to the esophagus for ingestion. Gill raker spacing increases with age, reflecting the larger particles encountered by larger juveniles and adults in their migration to the coastal ocean (Friedland 1985; Friedland et al. 2006). Menhaden swim in tightly packed schools as they seek dense patches of plankton. Schools of juveniles are often seen swimming fast in a tight circle for about a minute and then appear to go into a feeding mode. The circle swimming may effectively centrifuge the water, concentrating plankton and detritus particles in a ring of water for easier feeding.

Diatoms comprise much of the phytoplankton of the Chesapeake Bay and coastal waters (Marshall et al. 2003). These algae store most of the energy they have gotten from photosynthesis in lipids. Since lipids contain more than two times the energy of carbohydrates and proteins, diatoms make for a calorie-dense meal for the menhaden. The diatom's lipids are also the source of the omega-3 fatty acids concentrated in menhaden tissues (Obata et al. 2013; Budge et al. 2014).

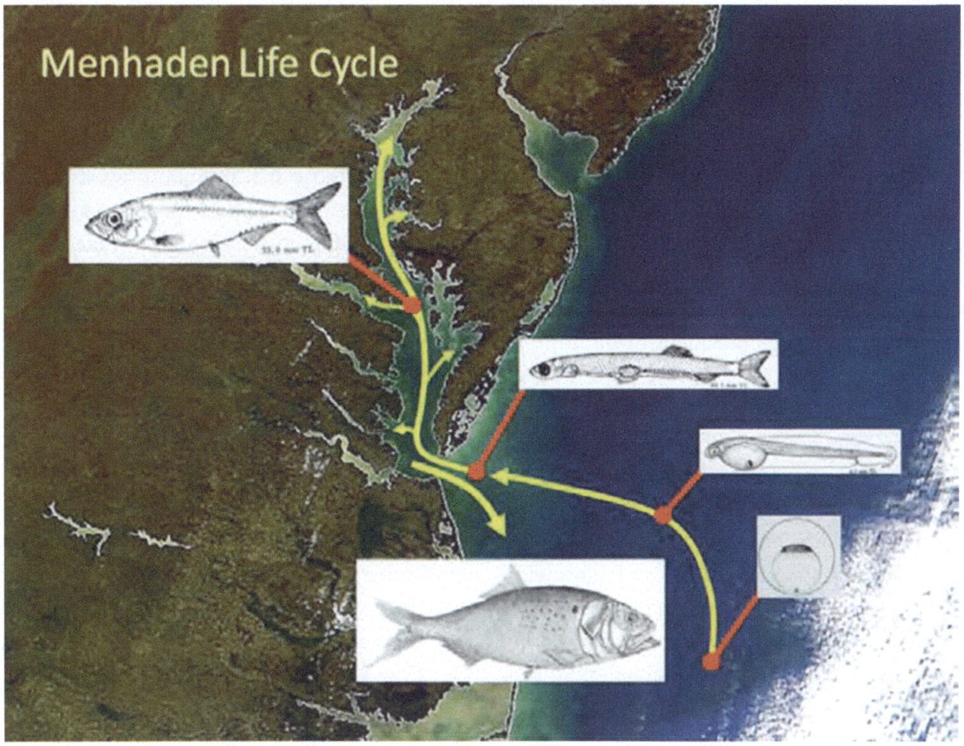

Fig. 5 Life cycle of *Brevoortia tyrannus* (Atlantic menhaden) in the vicinity of the Chesapeake Bay. Unlike the other clupeid fish of the Bay, menhaden spawn offshore. Ocean currents bring the larval fish into the Bay where they develop into juveniles and young adults. They migrate back into the Atlantic Ocean as they mature. Figure from Ed Houde and used by permission http://houdelab.cbl.umces.edu/Menhaden%20Target.html

In addition to the living phytoplankton and zooplankton, the menhaden diet includes detritus, primarily dead plant material from marshes and the watershed. Gulf menhaden (*Brevoortia patronus*), a species closely related to the Atlantic menhaden, detritus comprises about 30% of their diet. Most of the energy in such detritus is tied up in cellulose. Menhaden guts contain cellulase, the enzyme needed to digest cellulose and provide glucose to the fish (Deegan et al. 1990). The cellulase is probably produced by mutualistic microbes in the gut biome, as is typical of herbivores.

The importance of menhaden to the Bay ecosystem stems from their tight link to the base of the food chain and their sheer numbers. The Chesapeake Bay and surrounding coastal waters support about 15 billion of these migratory fish (ASMFC 2018). Adults may reach a length of 46 cm and live 10–12 years. Spawning begins in the first or second year of life. Adult Atlantic menhaden migrate into the Bay in spring and remain through early autumn. Juveniles may overwinter in the Bay (CBP 2019).

The vast number of menhaden in the Bay during the warmer months of the year suggest that their suspension feeding may help reduce the overabundance of phytoplankton that occurs at the same time (Friedland et al. 2006). The young of the year appear to have the greatest impact on algal removal and sequestration of nitrogen (the primary nutrient controlling phytoplankton in the Bay), yet the overall impact of menhaden on water quality appears to be small (Lynch et al. 2010). But what if the fishery for Atlantic Menhaden ended? Even after nearly a century of fishing, the Atlantic menhaden population featured 80 billion recruits in 1958, five times the current population (ASMFC 2018). A ban on fishing for the reduction industry could bring the population back to historic levels within a few years, given the very high reproductive capacity of menhaden and the excessive phytoplankton populations that plague the Bay. A return to super abundance of menhaden could help reduce algal concentrations as well as fuel the expansion of populations of the many species of fish, and birds, dependent on this oily fish. Virginia remains the only state on the East Coast to allow industrial purse-seining and the continuation of the reduction fishery for menhaden. The menhaden factories of New Jersey, Delaware and Maine long ago closed in response to the diminished fish populations and legal restrictions on odor. Sport fishers yearn for a return of abundant menhaden to help recover populations of striped bass and other game fish (Tolliver 2019).

Politics and Power of the Reduction Fishery Why is the menhaden reduction fishery still allowed to take so many fish? Being the largest fishery on the East Coast means that it produces income, jobs, taxes, funds to pay lobbyists, and money for donations to politicians. For 2016, Omega Protein reported $390,831,000 in revenues with $113,899,000 of gross profits (OPAR 2019). It's employees number about 250 people in Reedville (Fairbrother 2013). Local 400 of the United Food and Commercial Workers Union represents the workers on the ships used to catch the fish. These workers are among the highest paid in the Chesapeake Bay food system, earning about $50,000 for the eight monthlong fishing season. They also have a strong benefit package including good health insurance. The fishers get $600 per week of base pay. The rest depends on the weight of fish they catch (personal communication with vice president of the union). The workers at the processing facility aren't represented by the union.

Between 2008 and 2018, Omega Protein paid $2,617,000 for lobbying services, and donated $75,965 to candidates for federal offices (65% to Republicans, 35% to Democrats, CRP 2019). Over the same period, the company gave $466,299 to candidates for the Virginia's state legislature and state-wide offices (54% to Republicans, 45% to Democrats, VPAP 2019). The investment in lobbying and political donations makes sense when considering how the government regulates the menhaden industry.

Two state agencies regulate fishing in Virginia; the Virginia Department for Inland Fisheries for freshwater species, and the Virginia Marine Resources Commission (VMRC) for saltwater species. Atlantic Menhaden would naturally fall under the purview of fishery scientists at the VMRC, but it didn't until Democrats took control of the state's government in 2020. Prior to than the state legislature formulated regulations for menhaden. This was the only species with such special status. However, Atlantic menhaden are as much citizens of the other mid-Atlantic and New England states as they are of Virginia. Thus this highly migratory species also falls under the jurisdiction of the Atlantic States Marine Fisheries Commission (ASMFC), an entity set up under the US Congress in 1942 to promulgate effective regulations for the management of migratory fishery resources. Each state is represented by three commissioners, the director of the state's marine fisheries management agency, a member of the state's legislature, and a representative of the state's governor. ASMFC formulates recommendations for managing migratory species, but leaves it up to each state to implement plans. However, if ASMFC is unsatisfied with an individual state's action, it may recommend to the Secretary of Commerce for a closure of fishing for the species in question in that state's waters. That's what happened in 2019 when Omega Protein deliberatly exceeded that year's qutoa for menhaden harvested directly from the Bay by 30%. Doing such resulted in the Secretary's ordered shutdown as well as the legislative decision to transfer regulation of the indusry to the VMRC (Vogelsong 2019; Wheeler 2019).

Regulating menhaden catch is complicated by the fish's pattern of migration between the Bay and coastal waters. The younger fish tend to aggregate in the Bay, while the older ones move further away into coastal waters. Harvesting the younger fish within the Bay concerns conservationists the most, as its those fish that are preyed upon by striped bass, osprey, and a host of other species. These younger fish in the Bay are also the ones filtering the water of excess algae. And most importantly, the younger fish are the future of the stock and must be protected from excessive mortality to ensure recovery of the population.

In 2006, ASMFC capped the annual harvest from the Bay at 109,020 metric tons (the average of the previous 5 year's takes). In 2013, it dropped the limit to 87,216 metric tons, and in 2018 to 51,000 metric tons. However, responding to pressure from Omega Protein, the Virginia state legislature failed to adopt the lower limit. Omega Protein says it only took 35,000 metric tons out of the Bay in 2018, well below the ASMFC cap, but still lobbied against Virginia codifying the new limit, and simply ignored it in 2019 (Blankenship 2019). The Pew Charitable Trusts marine conservation group lists five reasons for reducing menhaden harvesting in the Bay: (1) The Chesapeake Bay is the Atlantic's most important nursery, accounting for 70% of the new recruits. (2) The predators and coastal communities relying on sportfishing will suffer with fewer menhaden. (3) Presently commercial menhaden fishing concentrates in and around the Bay, resulting in juveniles comprising 31–40% by weight of the catch, threatening the future of the stock. (4) The menhaden population in the Bay isn't coming back as it is in states further north, suggesting a critical localized depletion in the Bay that may roll into an unsustainably low population. (5) A restrictive cap prevents further harm and allows menhaden and their predators to recover (Gordon 2018).

Ecosystem Based Management of the Menhaden Fishery Chapter "A Fishing Trip: Exploiting and managing the commons of the Chesapeake Bay" provides an extensive discussion of the menhaden fishery and the difference between managing for maximum sustainable yield and a healthy ecosystem. The ASMFC intends to start managing the menhaden fishery based upon *ecosystem based reference points* (EBRFP) rather than the traditional approach of focusing on maximum sustainable yield over time. However, as of this writing, it has yet to settle on a set of EBRFP and continues managing the fishery much as it has since the 1940s. EBRFP for the Bay will undoubtedly include population estimates of the many species that prey upon menhaden. Most notable is striped bass, an important sport fish favored by anglers on the Bay and its tributaries. Computer models are used to project the impacts that different rates of menhaden harvest will have on the striped bass and other predators

Fig. 6 Crew of African American men working in coordinated effort to haul a purse seine loaded with Atlantic menhaden (*Brevoortia tyrannus*) in the era before mechanization of the process with the hydraulic power block. Image from the Smithsonian Institution: https://festival.si.edu/2004/water-ways/reedville/smithsonian

as well as competitors. One extensive and recent model projected that reducing the menhaden catch significantly increases the population and potential catch of striped bass, and to a lesser extent weakfish. Piscivorous birds and sharks would also benefit. On the other hand, increased fishing pressure on menhaden benefits medium sized pelagic fish, Atlantic cod (*Gadus morhua*), anchovies (*Anchoa mitchilli*) and alosines through indirect food web interactions (Buchheister et al. 2017). Most sport fishers would say the model is stating the obvious, claiming that present rates of menhaden take are leaving the striped bass undernourished and at reduced populations densities.

African American Labor and Menhaden Capt. Elijah Reed's picking Cockrell Creek for his menhaden reduction business was no accident. He understood that the future Reedville offered access to both the fish and the labor force to catch and process them. Menhaden are caught with purse seins. A mother ship sails up to a school of menhaden spotted from the crow's nest, and in more modern times by aircraft. The ship puts out two smaller boats to draw the net around the school. After the boats close the circle of netting, a heavy weight is dropped overboard. That weight is attached to a line that runs through the bottom of the net, and as it sinks it draws together the bottom of the net. It's like an old fashioned draw top purse but upside-down. The "set" might capture 250,000 fish or more. Getting the fish out of the net and loaded onto the mother ship is the hardest part of the enterprise. Up until the invention of the hydraulic power block in 1950, this was all done by the backs and hands of black men crowded into the two boats.

The men would lean over the side of the boats grasping handfuls of net. Over the course of several hours most of the net would be pulled to the surface and bunted, concentrating a mass of thousands of struggling menhaden in the small portion of the purse seine that remained in the water (Fig. 6). The men would then tie off the net and use pitchforks and shovels to cast the fish into the hold of the mother ship. The mother ship would sail or, in later times steam back to the factory in Reedville.

Pulling the nets required coordination of the crew. Everyone needed to hoist together. The workers used songs to organize their effort. Old timers in the industry say that the first person hired by the captain was the man to lead the singing to coordinate the work. To celebrate and preserve the history of menhaden work songs, the Reedville community helped form the Northern Neck Chanty Singers in 1991. The group started with men long retired from the industry

Fig. 7 The mechanization of gathering a menhaden purse seine and loading the fish aboard the mother ship. Two hydraulic power blocks are shown dangling from cranes fitted to the catch-boats. In this photo, the blocks have done their job of hoisting the seine. The large diameter hose is connected to a high capacity pump and is used to suck the menhaden aboard the ship. The power block and fish pump reduced the need for labor and hastened the work of completing a set of the seine. Historic image beyond copyright. http://omegaprotein.com/wp-content/uploads/2013/12/1955-Hydraulic-power-blocks.jpg

who in younger years sang the songs as they hoisted the nets (AAREG 1998; Anderson 2000). Working on a menhaden boat paid better than most jobs that African Americans could find during the Jim Crow era in Virginia. Many of the Northern Neck Chantey Singers used the wages they saved as young men to help pay for college or start a business.

The Power Block and Fish Pump The hauling of the menhaden purse seine net got easier with the introduction of the hydraulic block. Fishers passed the edge of the net over the shive in the block which was turned by hydraulic pressure and took most of the load of the work. This meant fewer hands needed to work the menhaden boats and a quicker recovery of the net. The industry also introduced large pumps connected to wide hoses to suck the netted fish into the hold of the mother ship (Fig. 7). By 1956, the power block and fish pump system helped to set a record in menhaden landings and set the stage for much lower stocks in the subsequent decades (ASMFC 2018). Despite the mechanization of the industry, African Americans continue to make up a large fraction of the reduction fishery labor force.

Menhaden Ships and War When the US entered World War I in 1917, it needed more ships and took possession of many of the menhaden steamers. Their wooden hulls made them ideal minesweepers, protecting them from the magnetic fuses used to detonate the anchored explosive devices. Some of the ships that survived the conflict returned to go back to war against menhaden. World War II yields a similar story but in reverse. After the end of that conflict, the military had a surplus of transport vessels. Many of these were converted to menhaden ships. Omega Protein continued to operate the seven and eight decade old ships through 2017.

The Future of the Menhaden Fishery on the Chesapeake Bay As long as residents of the Bay and beyond cling to the Standard American Diet (SAD) demand will remain for the products of the reduction fishery. The SAD is built around copious consumption of animal products and most of what Omega Protein sells goes to feed those animals. The SAD also means continuing the epidemic of coronary artery disease which in turn prompts the consumption of fish oil supplements by people who still believe the pills will save them from heart attacks, despite evidence otherwise. If people move toward a more sustainable and healthier whole plant based diet, the menhaden reduction industry would lose much of its market (Esselstyn 2001; Tuso et al. 2015).

Sport fishers and conservationists may also shape the future of the industry. While recreational fishers love to buy menhaden from the bait industry, most reject the reduction industry for its depauperating the fish needed to nourish striped bass, blue fish, and other game species (Olander 2019). Conservationists are of that same mind seeing the reduction industry as destroying food needed by the fish, birds, and marine mammals that they cherish. Yet, the reduction industry has strong political allies that will continue to protect it from extensive regulation.

Nature may have the final say. That Omega Protein wasn't able to catch its quota since about 2015 is not a good sign for the health of the menhaden stock associated with the Chesapeake Bay. Yet forage fish such as menhaden are famous for large swings in population size. Fluctuations in water temperature, the direction and strength of currents, water quality and the density of phytoplankton all add natural variability to the picture (Buchheister et al. 2016). On their own, fish populations will increase and decrease regularly. But add to that intensive fishing pressure and the downswing may go beyond the point of no return, tipping the system into a different stable state. This is what happened to New England's Cod fishery in the 1990s that still shows no sign of recovery (see chapter "A Fishing Trip: Exploiting and managing the commons of the Chesapeake Bay"). Omega Protein more than anyone else needs a sustainable menhaden population and should welcome management efforts to that end. But as is true for all fisheries, the investment in ships, gear, factories and labor makes it difficult to accept closures or reductions in fishing effort.

References

AAREG (1998) The Northern Neck Chantey Singers preserve African American work songs—African American Registry. In: African American Registry. https://aaregistry.org/story/the-northern-neck-chantey-singers-preserve-african-american-work-songs/. Accessed 7 Jun 2019

Abdelhamid A, Brown T, Brainard J et al (2018) Omega-3 fatty acids for the primary and secondary prevention of cardiovascular disease. Cochrane Database Syst Rev. https://doi.org/10.1002/14651858.cd003177.pub3

Anderson H (2000) Menhaden Chanteys: An African American Maritime Legacy. In: Maryland Sea Grant. Mdsg.umd.edu. https://www.mdsg.umd.edu/sites/default/files/files/MN18_1_MenhadenChanteys.PDF. Accessed 7 Jun 2019

Anderson A, Alam M, Watanabe W et al (2016) Full replacement of menhaden fish meal protein by low-gossypol cottonseed flour protein in the diet of juvenile black sea bass *Centropristis striata*. Aquaculture 464:618–628. https://doi.org/10.1016/j.aquaculture.2016.08.006

AOGHS (2019) First American Oil Well - American Oil & Gas Historical Society. In: American Oil & Gas Historical Society. https://aoghs.org/petroleum-pioneers/american-oil-history/. Accessed 4 Jun 2019

ASMFC (2018) Species—Atlantic States Marine Fisheries Commission. In: Asmfc.org. http://www.asmfc.org/species/atlantic-menhaden. Accessed 12 Nov 2018

Bang H, Dyerberg J, Nielsen A (1971) Plasma lipid and lipoprotein patten in Greenlandic west-cost Eskimos. Lancet 297:1143–1146. https://doi.org/10.1016/s0140-6736(71)91658-8

Blankenship K (2019) Bay Journal—Article: VA won't be penalized over menhaden regs if it stays under cap. In: Bayjournal.com, March 5, 2019. https://www.bayjournal.com/article/va_wont_be_penalized_over_menhaden_regs_if_it_stays_under_cap. Accessed 7 Jun 2019

Buchheister A, Miller T, Houde E et al (2016) Spatial and temporal dynamics of Atlantic menhaden (Brevoortia tyrannus) recruitment in the Northwest Atlantic Ocean. ICES Journal of Marine Science: Journal du Conseil 73:1147–1159. https://doi.org/10.1093/icesjms/fsv260

Buchheister A, Miller T, Houde E (2017) Evaluating ecosystem-based reference points for Atlantic Menhaden. Marine and Coastal Fisheries 9:457–478. https://doi.org/10.1080/19425120.2017.1360420

Budge S, Devred E, Forget M et al (2014) Estimating concentrations of essential omega-3 fatty acids in the ocean: supply and demand. ICES J Mar Sci 71:1885–1893. https://doi.org/10.1093/icesjms/fsu003

Burr M (2001) Reflections on the Diet and Reinfarction Trial (DART). European Heart Journal Supplements 3:D75–D78. https://doi.org/10.1016/s1520-765x(01)90124-5

Cashion T, Le Manach F, Zeller D et al (2017) Most fish destined for fishmeal production are food-grade fish. Fish Fish 18:837–844. https://doi.org/10.1111/faf.12209

CBP (2019) Atlantic Menhaden | Chesapeake Bay Program. In: Chesapeakebay.net. https://www.chesapeakebay.net/S=0/fieldguide/critter/atlantic_menhaden. Accessed 6 Jun 2019

CRP (2019) Center for responsive politics: Lobbying Spending Database - Omega Protein Corp, 2018 | OpenSecrets. In: Opensecrets.org. https://www.opensecrets.org/lobby/clientsum.php?id=D000028138&year=2018. Accessed 7 Jun 2019

Deegan L, Peterson B, Portier R (1990) Stable isotopes and cellulase activity as evidence for Detritus as a food source for juvenile Gulf Menhaden. Estuaries 13:14. https://doi.org/10.2307/1351427

Edwards H, May K (1965) Studies with Menhaden oil in practical-type broiler rations. Poult Sci 44:685–689. https://doi.org/10.3382/ps.0440685

Esselstyn C (2001) Resolving the coronary artery disease epidemic through plant-based nutrition. Prev Cardiol 4:171–177. https://doi.org/10.1111/j.1520-037x.2001.00538.x

Fairbrother A (2013) Bay Journal - Article: Omega Protein makes good on threat to cut jobs; but it doesn't have to. In: Bayjournal.com March 31, 2013. https://www.bayjournal.com/article/omega_protein_makes_good_on_threat_to_cut_jobs_but_it_doesnt_have_to. Accessed 7 Jun 2019

FAO (2016) The State of World Fisheries and Aquaculture 2016. Contributing to food security and nutrition for all. Rome. 200 pp. In: Fao.org. http://www.fao.org/3/a-i5555e.pdf. Accessed 4 Jun 2019

FAO (2019) Fish meal. In: Fao.org. http://www.fao.org/3/x5926e/x5926e01.htm. Accessed 9 Jun 2019

Franklin H (2007) The most important fish in the Sea. Island Press, Washington, DC

Friedland K (1985) Functional morphology of the branchial basket structures associated with feeding in the Atlantic Menhaden, Brevoortia tyrannus (Pisces: Clupeidae). Copeia 1985:1018. https://doi.org/10.2307/1445257

Friedland K, Ahrenholz D, Smith J et al (2006) Sieving functional morphology of the gill raker feeding apparatus of atlantic menhaden. J Exp Zool A Comp Exp Biol 305A:974–985. https://doi.org/10.1002/jez.a.348

Fry J, Mailloux N, Love D et al (2018) Feed conversion efficiency in aquaculture: do we measure it correctly? Environ Res Lett 13:024017. https://doi.org/10.1088/1748-9326/aaa273

Gallardi D (2014) Effects of bivalve aquaculture on the environment and their possible mitigation: a review. Fish Aquac J. https://doi.org/10.4172/2150-3508.1000105

Gordon J (2018) 5 reasons to keep the new cap on Menhaden fishing in the Chesapeake Bay. In.: Pewtrusts.org. https://www.pewtrusts.org/en/research-and-analysis/articles/2018/04/26/5-reasons-to-keep-the-new-cap-on-menhaden-fishing-in-the-chesapeake-bay. Accessed 7 Jun 2019

Greenberg P (2018) Omega principle. Penguin Books, New York

Haynie M (1974) Reedville, 1874–1974. Greater Reedville Association, Reedville, VA., Reedville, VA

Kristofersson D, Anderson J (2006) Is there a relationship between fisheries and farming? Interdependence of fisheries, animal production and aquaculture. Mar Policy 30:721–725. https://doi.org/10.1016/j.marpol.2005.11.004

Lynch P, Brush M, Condon E et al (2010) Net removal of nitrogen through ingestion of phytoplankton by Atlantic menhaden *Brevoortia tyrannus* in Chesapeake Bay. Mar Ecol Prog Ser 401:195–209. https://doi.org/10.3354/meps08389

Marshall H, Lane F, Nesius K (2003) Long-term phytoplankton trends and related water quality trends in the lower Chesapeake Bay, Virginia, U.S.A. Environ Monit Assess 81:349–360

Martinez-Porchas M, Martinez-Cordova L (2012) World aquaculture: environmental impacts and troubleshooting alternatives. Sci World J 2012:1–9. https://doi.org/10.1100/2012/389623

NIH (2019) 7.8% of U.S. adults (18.8 million) used Fish oil/Omega-3 fatty acids | NCCIH. In: NCCIH. https://nccih.nih.gov/research/statistics/NHIS/2012/natural-products/omega3. Accessed 5 Jun 2019

NOAA (2019) Menhaden - Fish Facts - chesapeakebay.noaa.gov. In: Chesapeakebay.noaa.gov. https://chesapeakebay.noaa.gov/fish-facts/menhaden. Accessed 6 Jun 2019

Obata T, Fernie A, Nunes-Nesi A (2013) The central carbon and energy metabolism of marine diatoms. Meta 3:325–346. https://doi.org/10.3390/metabo3020325

Olander D (2019) Commercial menhaden operation buys respectability. In: Sportfishingmag.com. https://www.sportfishingmag.com/commercial-menhaden-operation-buys-respectability/. Accessed 10 Jun 2019

Omega Protein (2019a) Omega nutrient. In: Omeganutrient.com. https://omeganutrient.com/. Accessed 5 Jun 2019

Omega Protein (2019b) Our history- Omega protein. In: Omega Protein. https://omegaprotein.com/who-we-are/our-history/. Accessed 9 Jun 2019

OPAR (2019) Omega Protein Annual Report 2016. In: Annualreports.com. http://www.annualreports.com/HostedData/AnnualReports/PDF/NYSE_OME_2016.pdf. Accessed 6 Jun 2019

Stoner G, Allee G, Nelssen J et al (1990) Effect of select menhaden fish meal in starter diets for pigs. J Anim Sci 68:2729. https://doi.org/10.2527/1990.6892729x

Tolliver L (2019) Menhaden: The little fish causing a big stink. In: The Virginia Pilot. April 22, 2019. https://pilotonline.com/distinction/outdoors/article_6eefdfca-5f9d-11e9-9b84-6b486d97a010.html. Accessed 6 Jun 2019

Tuso P, Stoll S, Li W (2015) A plant-based diet, atherogenesis, and coronary artery disease prevention. The Permanente Journal 62-67. doi: https://doi.org/10.7812/tpp/14-036

Vogelsong S (2019) Federal government orders Virginia menhaden fishery shut down. In: Virginia Mercury. https://www.virginiamercury.com/2019/12/19/federal-government-orders-virginia-menhaden-fishery-shut-down/. Accessed 18 May 2020

VPAP (2019) The Virginia Public Access Project. In: Vpap.org. https://www.vpap.org/donors/137690-omega-protein/?start_year=2008&end_year=2018. Accessed 7 Jun 2019

Wheeler T (2019) Virginia menhaden fishery threatened with moratorium. In: Bay Journal. https://www.bayjournal.com/news/fisheries/virginia-menhaden-fishery-threatened-with-moratorium/article_553501a2-33df-11ea-b8ce-4fb25e299e03.html. Accessed 18 May 2020

Williamson T, Tilley D, Campbell E (2015) Emergy analysis to evaluate the sustainability of two oyster aquaculture systems in the Chesapeake Bay. Ecological Eng 85:103–120. https://doi.org/10.1016/j.ecoleng.2015.09.052

Blue Crabs: Beautiful Savory Swimmers of the Chesapeake Bay

Rochelle D. Seitz

Tough shells and tender meat
Requires lots of work before one can eat

Abstract

Blue crabs have been an important economic and ecological species within Chesapeake Bay throughout history. They have been collected and consumed from the days of Captain John Smith in the 1600s to the present. They have a complex life cycle that includes many molts, a post-larval phase offshore of the Chesapeake Bay, and re-invasion of the Bay with post-larvae. Once thought of as "inexhaustible," blue crabs have been targeted by many fisheries including a peeler fishery, trot line fishery, crab pot fishery, soft crab fishery, and a recreational fishery. This species serves as both predator and prey in the Bay's food web. Predation and cannibalism are thought to account for most of the mortality of juvenile crabs. Blue crabs that survive an array of predators face their biggest challenge from humans, as fishing accounts for about 80% of mortality in large juvenile and adult crabs. The commercial fishery for blue crabs came into full swing in the mid nineteenth century, after the advent of ice for preserving the catch during transport. The growing crab industry of the nineteenth and early twentieth centuries provided skilled jobs for African Americans, and immigrants from Ireland, Italy, and other places. Extracting the meat from hard-shell blue crabs is hard and tedious work, and women did and still do most of the picking. Crab catches continued to rise through 1915, with a dip in 1920, and subsequently rose again until they declined during World War II. The crab harvest again increased from 1945 through 1995, but worrisome declines in the Chesapeake Bay blue crab population occurred in the mid-1990s through 2007. Desperate management actions averted a crisis in 2008 when the governors of Virginia and Maryland called for a 34% reduction in the harvest of female crabs in 2008 compared to 2004–2007. Major management efforts in 2008 triggered recovery of the population above the threshold for sustainable exploitation, and the current population is considered healthy.

Keywords

Callinectes sapidus · Soft-shell crab · Watermen · Economic role · Fisheries management · Seafood

Blue Crabs of the Chesapeake As Captain John Smith explored the Chesapeake Bay in the summer of 1607, he noted how the native people used various methods and devices to gather food from the Bay's waters. These included fishing from dug-out canoes, gathering oysters from the bottom by hand, and using traps to capture various species of fish and the blue crab (*Callinectes sapidus*), a favorite dish (Smith 1986; Rountree et al. 2008). In Greek, *Callinectes* means beautiful swimmer, and in Latin *sapidus* translates to savory (Brunkow 2019; Virginia Places 2018), and this beautiful swimmer from Chesapeake Bay has been consumed by humans for centuries (Fig. 1). The Powhatan Chief, Wahunsenacawh, served blue crabs to Capt. John Smith and other colonists during feasts of welcome (Farish 1943; Wharton 1957; Kennedy 2018). European settlers of the seventeenth century wrote of the consumption of blue crabs by the Algonquin speakers and other Native Americans. Gabriel Archer, a member of the first expedition to Jamestown, commented on the large Virginia crabs that

R. D. Seitz (✉)
Virginia Institute of Marine Science, William & Mary, Gloucester Point, VA, USA
e-mail: seitz@vims.edu

B. E. Cuker (ed.), *Diet for a Sustainable Ecosystem*, Estuaries of the World, https://doi.org/10.1007/978-3-030-45481-4_7

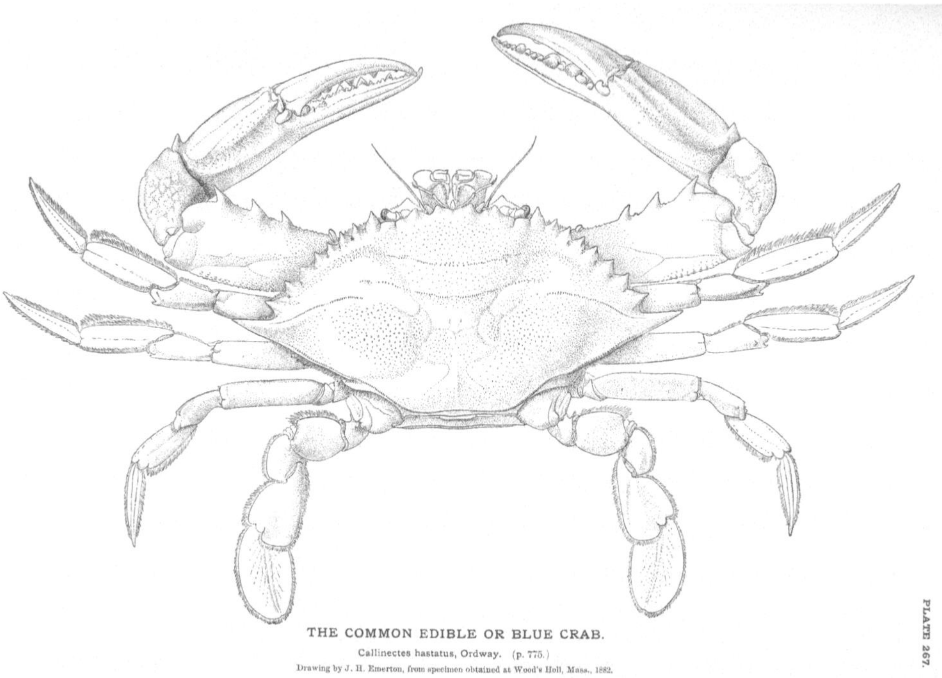

Fig. 1 Line drawing of the blue crab by J. H. Emerton, 1882. The blue crab species harvested in the early 1800s in Chesapeake Bay was thought to be *Callinectes hastatus* until the species was described by Mary Rathbun in 1896 and named *Callinectes sapidus* (Rathbun 1896). Wikimedia Commons. https://commons.wikimedia.org/wiki/File:FMIB_51203_Common_Edible_or_Blue_Crab.jpeg

settlers ate: "in the large sounds near the sea are multitudes of fish, banks of oysters, and many great crabs rather better in taste than ours, one able to suffice four men" (Archer 1607). Anyone who has eaten blue crabs would suspect this to be a bit of exaggeration, as it would be more likely that four crabs would suffice one man. But it may reflect the larger size of longer-lived blue crabs prior to intensive harvesting pressure. Blue crabs were just one of the many Chesapeake Bay seafood species chronicled by John Smith. Oysters, shad, sturgeon, and others seemed more important from his writings (Smith 1986). While John Smith enjoyed the blue crabs he ate, he had no way of foreseeing that this species would come to be the most economically important seafood produced by the Chesapeake Bay and directly consumed by humans, from the mid twentieth century onwards. This iconic species is synonymous with the Chesapeake Bay. "*C. sapidus* is of significant culinary and economic importance in the United States, particularly in the Chesapeake Bay" (Wikipedia 2019). It is the state crustacean in Maryland and currently remains as that state's largest commercial fishery (National Oceanic and Atmospheric Administration (NOAA) 2019a; Wikipedia 2019).

Before spawning, blue crabs are subjected to several fisheries throughout the Bay and throughout the latter stages of their lifecycle. Blue crabs have been harvested since colonial times (Churchill 1920), and present fisheries include a peeler fishery, trot line fishery, crab pot fishery, soft crab fishery, and a recreational fishery (Kennedy et al. 2007), which will be described in more detail below.

In the 1700s, crabs were caught and eaten locally (Wharton 1954; Kennedy 2018), and in the 1800s, blue crab abundances in the Chesapeake Bay were so great that H. N. Smith (US Fish Commission 1891) noted "...crabs... so great is the supply... that it seems almost inexhaustible." By the late 1800s, improvement in refrigeration and the expansion of railroads allowed marketing of seafood to the interior of the US (Van Engel 1958). Chesapeake Bay blue crabs became a delicacy enjoyed by diners across the country. Modern-day navigators in Chesapeake Bay see traces of that earlier time in the omnipresence of small buoys that float tethered to wire mesh crab pots resting on the bottom. These simple contraptions capture blue crabs efficiently. Over a million crab pot buoys dot the waters of the Bay and its tributaries from early spring to late fall (Miller et al. 2005), vexing mariners as they attempt to avoid snagging the pot lines. The scientific name and the armada of buoys proclaim the importance of the blue crab to the Chesapeake Bay food system. This chapter will explore the blue crab's biology, ecology, role in the food system, and efforts to restore the health and numbers of this iconic species.

Biology of the Blue Crab Mary Jane Rathbun (1896) first described the blue crab, *Callinectes sapidus*. A century of study illuminated its biology and complex life history.

Fig. 2 Blue crab molting (buster) from NOAA archives. Note that the old exoskeleton is being exited by the crab as it backs out of its shell and moves towards the lower right corner of the image. The newly molted crab is without a shell, soft, and is susceptible to predation and cannibalism for a few days before its new shell hardens. NOAA Photo Library. https://www.google.com/imgres?imgurl=http://www.photolib.noaa.gov/bigs/line0962.jpg&imgrefurl=http://www.photolib.noaa.gov/htmls/line0962.htm&h=1180&w=1820&tbnid=1UHC5fplWWi1qM&tbnh=181&tbnw=279&usg=AI4_-kQqDoXd8NK1lk-tbkpmoesPzkRjeA&vet=1&docid=WaNJdsSSTJeZ0M

Although an icon of the Chesapeake Bay, it ranges from Argentina to Nova Scotia (Williams 1974). Besides the accounts of John Smith from 1612 regarding the abundance of crabs (Hamor 1615), some of the earliest scientific knowledge of blue crab life history came from a Johns Hopkins scientist, William K. Brooks (1893). Though we now know the species as the "blue crab," in 1887, there were several vernacular names for it, including "blue crab," "edible crab," "common crab," "sea crab," or "channel crab," which was common along the middle and southern Atlantic coast, and "gulf crab" in the Gulf of Mexico (Rathbun 1887).

W. A. Roberts gave a colorful description of the blue crab's life cycle: "First, the 'hard crab,' or one in its natural condition; second, a 'snot,' or one that has just entered the shedding stage; third, a 'peeler,' when the old shell has begun to break; fourth, a 'buster,' when the new shell can be seen (Fig. 2); fifth, the 'soft crab;' sixth, a 'paper-shell,' or 'buckram,' when the new shell is beginning to harden" (Roberts 1905). It only takes a few hours after shedding for a soft crab to reach the 'paper-shell' stage, and about 2 days for the hardening to be complete. Among watermen of the time, crabs ready to molt were also called "shedders," "comers," "long-comers," or "short-comers" (the latter two terms referring to how soon before the crab was expected to molt [US Fish Commission 1891]). Current Chesapeake Bay vernacular primarily retains the terms "hard crab," "peeler," "shedder," and "buster."

Blue crabs are sexually dimorphic (the two sexes show different characteristics aside from their sexual organs), with the adult female having red-tipped claws and a wide abdomen on her bottom side (looks like the Washington D.C. Capitol building) under which she holds her brood of developing larvae (her 'sponge'), and with the adult male having a narrow apron (reminiscent of the Washington Monument). Following William Brooks, other researchers in Chesapeake Bay began adding to our understanding of blue crab biology, ecology, and fisheries in the first half of the twentieth century (e.g., Truitt 1934; Cronin 1947; Pyle and Cronin 1950; Van Engel 1958). Since then, many aspects of blue crab ecology, including life history, physiology, and recruitment, have been actively studied. As a result of all of these scientific efforts, the basic biology and ecology of the blue crab is relatively well understood, and books have expounded on this knowledge (e.g., Kennedy and Cronin 2007).

The blue crab is a decapod (10-legged) crustacean with a complex life cycle (Figs. 3 and 4). Crustaceans feature an external shell, or exoskeleton, that provides strength and protection for the animal. Growth and metamorphosis to a new life stage require shedding of the old shell and replacing it with a new one laid down beneath it, in the process of molting. The

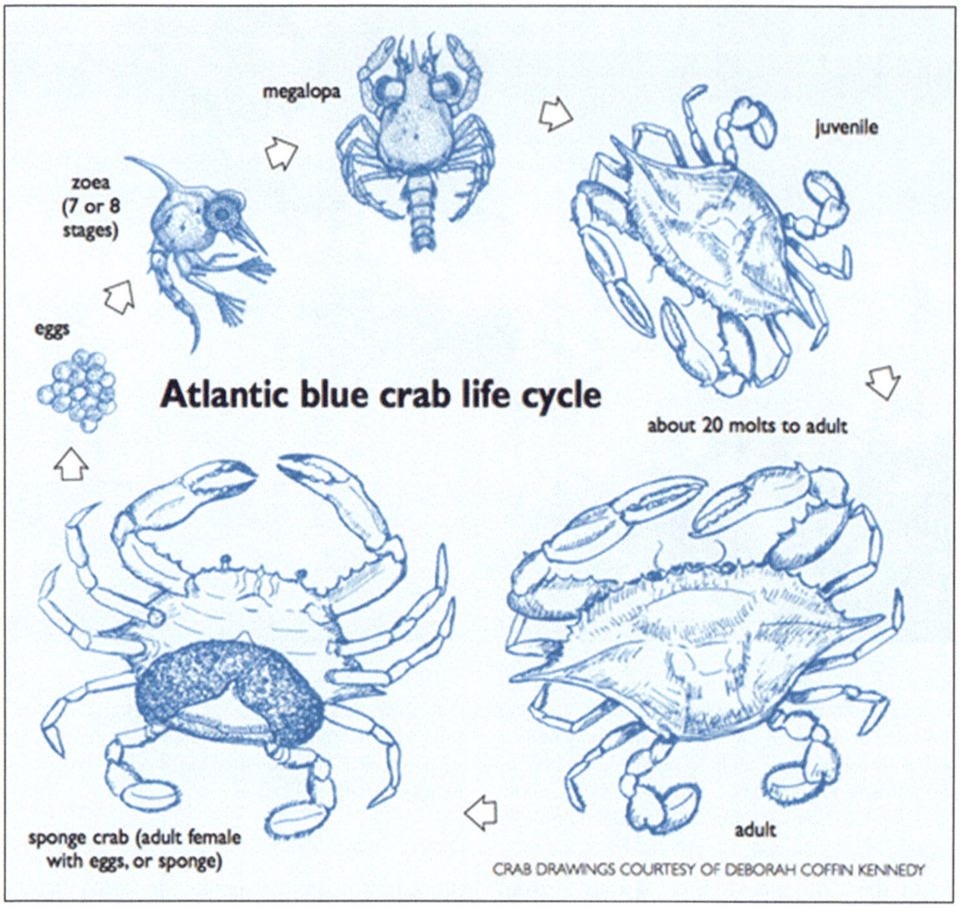

Fig. 3 Life cycle of the blue crab. Every new stage requires molting, the replacement of the old shell with a new one from underneath. Used by permission of the Chesapeake Bay Quarterly: https://www.chesapeakequarterly.net/V11N2/side2/ (Maryland Sea Grant 2012)

hours or days during which the new exoskeleton hardens are fraught with danger from predators and pathogens. Thus, the newly molted crabs seek refuge in structured habitats, like seagrass beds or woody debris, until the shell hardens.

Mating by Chesapeake Bay blue crabs typically occurs from May through October (Hines et al. 2003). A male crab will couple with a female crab immediately before her final molt to maturity (terminal molt). It is thought that females mature in the autumn of their second year, at 12–18 months of age, but they may delay maturity until 24–30 months of age (Miller et al. 2005). As soon as the female undergoes her terminal molt and is in the soft crab stage, the male crab will cradle her, mate with her, and protect her until her shell hardens (Van Engel 1958). Mating takes place in the lower-salinity waters of the Bay and its tributaries. This process takes several days. It begins with a courtship dance and release of chemical attractants (pheromones) by the male. A receptive female will allow the male to cradle her from atop in a precopulatory "doubler" posture. When the female begins her molt, she changes position to place her abdomen next to his. Then the male uses specialized appendages (gonopods) on his abdomen to deliver packets of sperm called spermathecae into her genital openings (gonopores) in her abdomen. She mates only once in her lifetime, relying on stored sperm to fertilize the eggs for each subsequent brood. In Chesapeake Bay, a female crab can produce up to nine million eggs each year in one to three broods (using only part of the sperm in her receptacles for each spawn; Van Engel 1958) and is thought to have six to seven broods per lifetime from a single mating (Hines et al. 2003). Soon after mating, the females migrate to the lower Bay where they extrude fertilized eggs that attach as a mass (sponge) to hair-like structures on the female abdomen. Females release their larvae into the full-salinity waters offshore of Chesapeake Bay, with peak spawning in July and August. Even though they can tolerate a wide range of salinities (Tan and van Engel 1966), because their evolutionary origins are as a tropical marine species, blue crabs require full strength, or marine, salinity for the development of their larvae. After larval development along the coastal shelf, currents transport the postlarvae back into Chesapeake Bay (Epifanio 2019). Within the Bay, the postlarvae progress through several molts

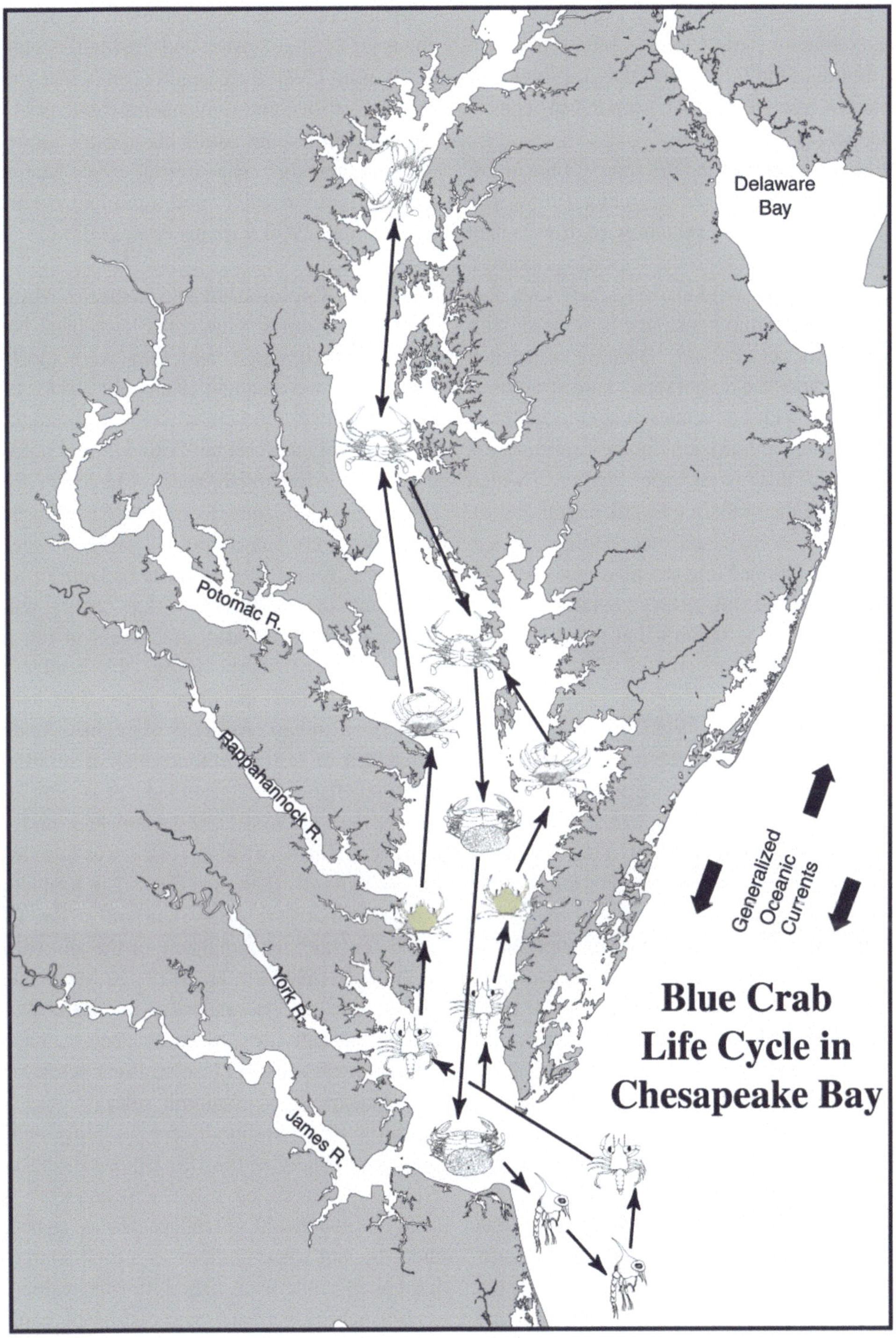

Fig. 4 Life cycle of the blue crab in Chesapeake Bay (modified from Seitz et al. 2001). The thick arrows in the offshore area are indicative of the dominant water flow in the summer. The directions of crab movement (thin arrows) are shown within the bay. Blue crab stages include zoeae (shown offshore side view), megalopa (shown offshore and entering lower Bay), first crab (shown mid-Bay, base of Rappahannock), juvenile (shown at base of Potomac and across the Bay near the mid-Bay islands), mature male (shown in upper Bay), and mature female sponge crab (shown in mid- and lower Bay). Note that the juveniles also move upriver into each of the tributaries where the life cycle is completed as in the mainstem

to the juvenile stage, disperse up-bay by riding favorable currents, and finally mature. Adult female blue crabs also overwinter in the mouth of the Bay, where they bury into the sediments and enter an inactive state of "torpor," or inactivity, to conserve energy.

Tagging studies have examined the life span of blue crabs, with one crab living to 5 years of age (Fischler 1965). The blue crab stock assessment (an evaluation of how many animals—the stock—are available in the region) validated a lifespan of 4 years at current exploitation levels and also estimated an upper limit for maximum age at 6 years with limited exploitation (Miller et al. 2005). Because crabs shed their external shell and therefore do not retain hard structures, it is difficult to age them with conventional ageing methods (i.e., counting growth rings in hard structures). Lipofuscin is a metabolic by-product that is an age pigment, which accumulates over time (in the brain or eyestalk neural tissue), and it has been used somewhat successfully to age crustaceans (Kilada et al. 2017). More recently, new methods for ageing, such as by measuring growth bands from the gastric mill ossicles (calcified structures that grind food) of lobsters, have been used (Gnanalingam et al. 2018) and may be applicable for blue crabs.

Ecology of the Blue Crab The blue crab is widely dispersed in the Western Atlantic Ocean, and throughout Chesapeake Bay it is very abundant (Norse 1977; Williams 1984; Hines et al. 1987; Lipcius and Van Engel 1990). This species is ecologically important, serving as both predator and prey in the Bay's food web (Baird and Ulanowicz 1989; Lipcius and Latour 2006). Predation and cannibalism are thought to account for most of the mortality of juvenile crabs. Predators of young juvenile crabs include striped bass (*Morone saxatilis*), Atlantic croaker (*Micropogonias undulatus*), mud crab (*Panopeus herbstii*), great blue heron (*Ardea herodias*), and other blue crabs. The blue crabs that survive this array of predators face their biggest challenge from humans. Fishing accounts for about 80% of mortality in large juvenile and adult crabs (Lipcius and Latour 2006).

Blue crabs shift their habitat as they grow. Juvenile blue crabs prefer structured habitats such as seagrass for nurseries (Lipcius et al. 2005). The nursery-habitat paradigm holds that juvenile crabs first inhabit seagrass and then, as they mature, they move to unstructured habitats, such as sand or mud. Crabs that are molting and soft crabs also prefer seagrass beds, which help to reduce mortality from predation while their usually protective shell is hardening (Ryer et al. 1997). After reaching >30 mm carapace width (CW), the larger juvenile crabs that have outgrown the protection of small refuges provided by seagrass beds (Lipcius et al. 2007) move to unstructured habitats that provide ample food sources for growth (Seitz et al. 2011).

Blue crabs are opportunistic feeders, and their diet changes as they grow and mature. Small juvenile crabs (<20 mm CW) feed mainly on small crustaceans, plant matter, detritus, and polychaetes (Seitz et al. 2011), whereas larger juveniles and adults incorporate more mollusks (up to 50% of their diet; Hines et al. 1990; Mansour and Lipcius 1991; Mansour 1992) and cannibalize smaller blue crabs (Hines et al. 1990; Lipcius et al. 2007).

History of Exploitation and Fisheries Management Since pre-colonial times, blue crabs have been harvested through much of their range, and they were mostly eaten nearby where they were caught (Kennedy 2018). In the early eighteenth century, blue crabs were so abundant that they were used not only as food but both soft and hard crabs were also used as bait for finfish, such as tautog (*Tautoga onitis*) and striped bass (*Morone saxatilis*) (Rathbun 1887). The commercial fishery came into full swing in the mid nineteenth century (Cronin 1998), after the advent of ice for preserving the catch during transport. Ice cutters would harvest the frozen surfaces of lakes and rivers in the northern Bay or more northern states during the winter and ship it to icehouses located in fishing ports. The fishery at that time comprised many different segments, and more recently supported "hundreds to thousands of small-scale, sometimes artisanal, fishermen" (Kennedy et al. 2007). The hard-crab fishery was the first to be established, and later, in 1880, a commercial fishery for soft crabs was founded in Chesapeake Bay (Churchill 1920) (see Fig. 8.2 in Kennedy 2018 for early landings [harvest] data). Currently, the Chesapeake Bay catch accounts for a majority of the annual blue crabs taken in the U.S. (Miller et al. 2005). In Maryland, blue crabs are responsible for a majority of seafood harvesting and processing activities. More on each segment of the fishery will be discussed later (see sections below on hard-shell and soft-shell fisheries' economic roles).

In the early twentieth century, Maryland had the largest blue crab fishery in the US. Upon visiting and interviewing crabbers (then called fishermen) and crab dealers, Roberts (1905) noted that Marylanders claimed to be the first to notice the taste and value of crabs as a food source. Both soft- and hard-shell crabs were fished, but the value of the soft-shell crab fishery was greater than that of the hard-crab fishery. In 1901, the Maryland crab fishery had 5399 men with 69 large vessels (55 used for crabbing and 14 used for transport of the catch), plus 4059 smaller boats. In that same year, the crab "catch was 12,910,746 soft crabs, valued at $202,563, and 29,474,379 hard crabs, valued at $85,884" (Roberts 1905). (Note that $288,447 worth of blue crabs caught in 1905 translates to $8,395,872 in 2019 dollars.) Crab catches continued to rise through 1915, with a dip in 1920, and subsequently rose again until they declined during World

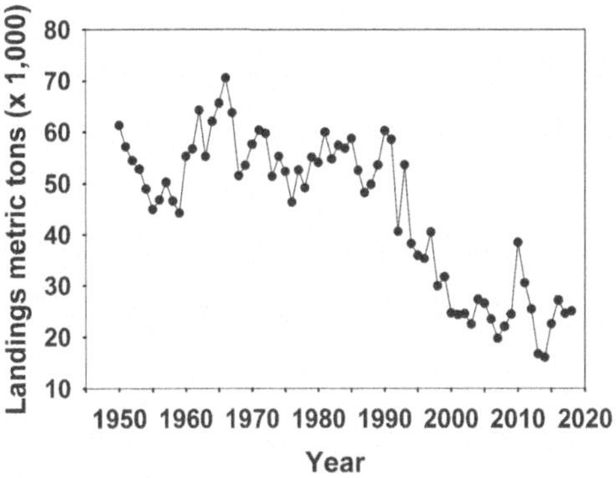

Fig. 5 Bay-wide blue crab landings (adjusted for reporting changes) from 1950 to 2018. Data compiled by R. Lipcius, VIMS, personal communication, from 2011 blue crab stock assessment and data for 1990 to 2018 from 2019 Chesapeake Bay Stock Assessment Committee report (CBSAC 2019)

War II. The crab harvest again increased from 1945 through the mid 1990s (Kennedy et al. 2007); information on abundance was inferred from landings (harvest) data (Fig. 5), which were likely to be underestimates.

Worrisome declines in the Chesapeake Bay blue crab population occurred in the mid-1990s through 2007 (Fig. 5), with record low landings of 21.2 million kg (46.8 million pounds) in 2007. Abundance declined by 81% and spawning stock by 84% from 1990 to 2001 (Lipcius and Stockhausen 2002). In addition, the average size of females declined by 8%. Reduction in catch per unit effort, diminishing size of individuals, and high exploitation rates signaled an overexploited fishery (see chapter "A Fishing Trip: Exploiting and managing the commons of the Chesapeake Bay" on fisheries). Destruction of coastal habitats joined fishing pressure in creating the crisis for crab populations. Global per capita fish supply was concurrently declining in 2000 (partially due to increasing world population growth), and management efforts were promoted to attain sustainable fisheries and aquaculture (Food and Agriculture Organization (FAO) 2002). State authorities adjust the number of commercial licenses issued and the length of the fishing season in response to the perceived population levels of crabs. In response to the declining catch of crabs in Chesapeake Bay, each Chesapeake Bay jurisdiction with a blue crab fishery (Maryland, Virginia, and District of Columbia) established their own regulations, and these included restrictions on numbers of licenses, catch limits, crab size limits, and time and area limits (including a Bay-wide spawning sanctuary designed to protect adult female crabs) (Seitz et al. 2001). Desperate management actions averted another crisis in 2008 when the governors of Virginia and Maryland called for a 34% reduction in the harvest of female crabs in 2008 compared to 2004–2007. Authorities closed the winter dredge fishery, established bushel limits on female crabs, restricted dates of female harvest, abolished the recreational crab license allowing five pots per license, required two additional crab escape rings (cull rings of 5.8 cm diameter) in each crab pot (VMRC 2019), and increased the minimum peeler size limit. These measures helped the fishery to recover from levels of around 52 million pounds in 2008 to over 92 million pounds in 2010 (Miller et al. 2011; R. Lipcius, personal communication, CBSAC 2019 winter dredge survey report).

Currently (2019), the US blue crab sustains one of the most valuable crab fisheries in the world and is a major fishery in several coastal states: Virginia, Maryland, North Carolina, Mississippi, Louisiana, and Texas. The harvest is worth over $24 million in Virginia and over $39 million in Maryland (NOAA 2019b), contributing to the livelihood of many thousands of full-time watermen whose jobs depend on the harvest of crabs each year (with over 1400 blue crab licensees in Virginia alone in 2018; Virginia Marine Resources Commission data, accessed August 2019). In 2012, it was estimated that the United States harvest of blue crabs was 77,000 metric tons, with an annual dockside value of $150 to $200 million (NMFS 2012).

The blue crab is fished at most of its later life stages, which puts heavy pressure on the crab population. On the bright side, blue crab densities in Chesapeake Bay were generally higher since enactment of the 2008–2009 regulations (CBSAC 2019 winter dredge survey report) and there is optimism for persistence of this iconic species, with a Bay-wide population level of 371 million in 2018 (Fig. 6; CBSAC 2019 winter dredge survey report), and the levels of young crabs estimated at 167 million, which was an increase of 34% over 2017.

The soft-shell crab fishery and its economic role Blue crabs have always been synonymous with the Chesapeake Bay region. As Kliman (2005) states: "…in the mid-Atlantic, we have our crab. More than any single foodstuff, the Chesapeake Bay blue crab has come to define this region." The harvest of this species has connected people to the Chesapeake Bay and supported many coastal communities (Kliman 2005). The industry features two different sectors; the soft-shell and the hard-shell. Let's first consider the soft-shell fishery. Anyone who has eaten whole, steamed, hard-shell blue crabs understands the amount of work and time required to get the meat out of the exoskeleton. This includes much handwork that involves tools for cracking the shells and picking out the hard-won meat. Newly molted soft-shell crabs present a much easier way to prepare a meal. Diners eat them whole, on sandwiches, or on salads.

While artisanal crabbers fed their families and friends in coastal communities, it was difficult to bring such perishable food to distant markets (Churchill 1920; Kennedy et al.

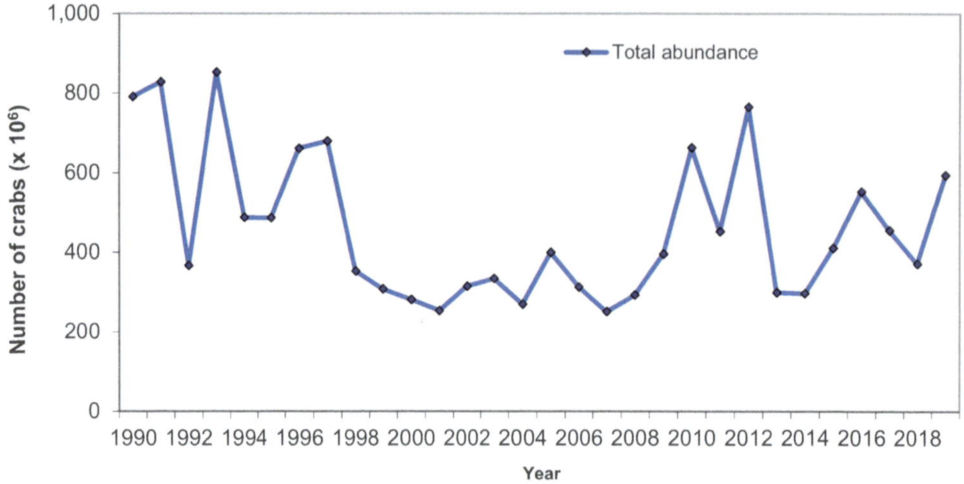

Fig. 6 Estimated Bay-wide blue crab total abundance (from the winter dredge survey) in millions of crabs from 1990 through 2019. This includes both juveniles and adults, and males and females. Modified from the 2019 Chesapeake Bay Stock Assessment Committee report (CBSAC 2019)

2007). Because crabs need to shed their shells (i.e., molt) to increase in size, crabbers took advantage of this by using this 'soft-shell' stage (between molts) as a food source. Watermen collected crabs nearing molting (peelers) and retained them in submerged traps so that they could be harvested as soon as the crabs shed their carapaces. The traps also protected the crabs from predators. Watermen monitored the crabs at least twice daily, removing the newly molted crabs immediately because "...soft crab left for any length of time with the hard crabs are liable to be injured by the latter" (Rathbun 1887). The first known crab shipper in Chesapeake Bay was Captain John H. Landon, from Crisfield, Maryland, who first transported crabs to markets in 1873 and 1874. He commented that he sold soft crabs to railroad employees who "...would take them to friends or sell them to game dealers in Philadelphia, which may account for the fact that Mr. Martin, to whom the first shipments were made, was familiar with their edible quality" (Roberts 1905). In those days, ice was not used in shipping, making it hard for them to get the crabs to market without spoilage (Roberts 1905).

Near Crisfield, Maryland, on the Eastern Shore, "the sounds, rivers, creeks, and marshes in the vicinity [...] may be said to teem with crabs," such that blue crabs served as an important soft-shell fishery (US Fish Commission 1891). Crabs were so plentiful and the soft-shell crab fishery was so economically viable, that they, "...during half the year, afford[ed] steady, profitable, and inarduous employment, to large numbers of persons (US Fish Commission 1891)." It was thought that crabs were so abundant that the supply was inexhaustible, as "only those [crabs] who have come up into comparatively shallow water are taken, while in deeper water they are undisturbed" (US Fish Commission 1891). And, "There seems to be no limit except the limit of natural food supply to restrict the future growth of this important industry" (Conservation Commission 1909). In 1909, crabs became such an important economic species, that they overtook oysters: "At Crisfield the oyster industry, formerly the town's chief support, has been to a certain extent superseded in importance by the crab trade. As many as 1000 boxes, each holding 15 dozen soft crabs, are sometimes shipped from this point by express in a single day" (Conservation Commission 1909).

Peelers and soft crabs were collected by several different means, and the soft crab industry was thriving in the late 1800s. Newly molted crabs seek refuge from predators in structured habitats such as seagrass beds. Crabbers understood this and collected most with scrapes (similar to oyster dredges but without teeth on the front bar and with a cotton, not chain, bag) (Fig. 7). Some crabs were caught by scoop nets (or dip nets), small seines, or incidentally on hard-crab trot lines (Roberts 1905). Similar to most decapod crustaceans, soft blue crabs do not feed until their claws and the rest of new shell hardens, so they would not likely be caught using bait. Hardened claws are necessary for handling their prey. The scrapes were deployed from sailing vessels, similar to the deployment of oyster dredges (Fig. 8). Usually, two light scrapes would be deployed from a sailing vessel with a crew of two men. The vessels travelled a few minutes (61–91 m or 75–100 yards) before the scrapes were hauled on board and their contents dumped and culled on the boat, using a broad, flat trough on the side of the vessel. Crabbers would attempt to catch crabs between "...daylight and ten o'clock in the forenoon, and between 3 o'clock in the afternoon and evening," believing that crabs would migrate to

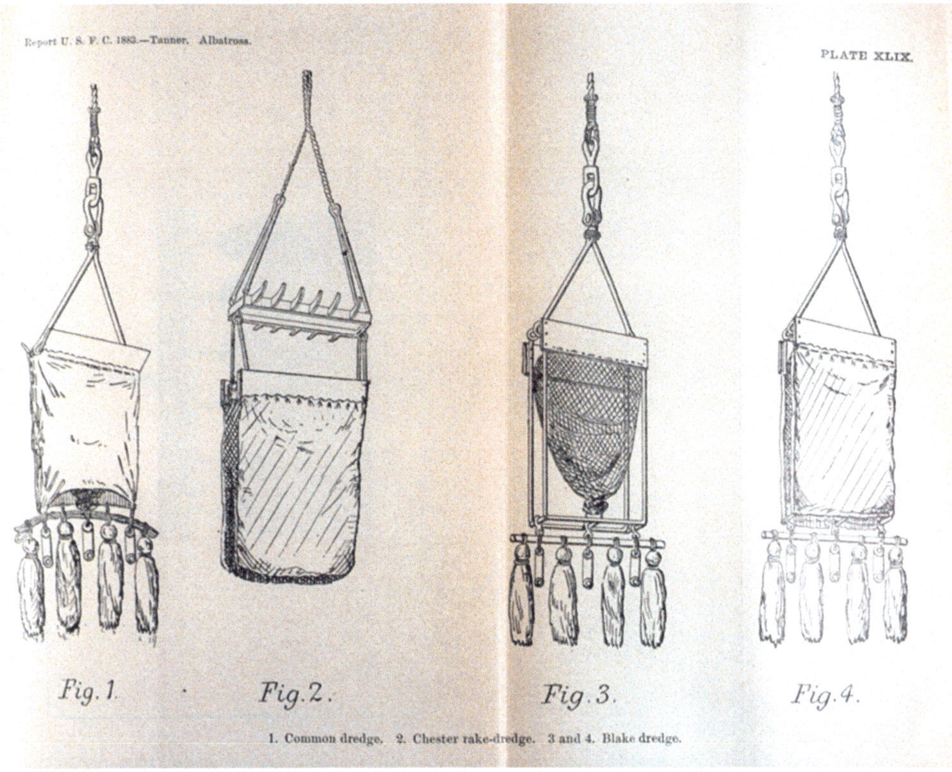

Fig. 7 Crab dredges (with teeth) and scrapes (without teeth), circa 1883. NOAA Photo Library. https://www.google.com/imgres?imgurl=http://www.photolib.noaa.gov/bigs/ship0423.jpg&imgrefurl=http://www.photolib.noaa.gov/htmls/ship0423.htm&h=1056&w=1356&tbnid=Mv7XExJCQfm3VM&tbnh=198&tbnw=254&usg=AI4_-kT2bwq26VvHK2Mz3WKiiJjdQc5tlg&vet=1&docid=MLhW0WWzzBfOzM

shelter with the bright sun in the late morning (Roberts 1905). While peelers can readily move, soft crabs tend not to. Watermen fished for soft-shell crabs in May through October, after which most turned to oyster tonging. June and July offered the best catches of peelers (Roberts 1905).

In the early years, most crabs were sold as 'peelers' or 'busters' (i.e., before shedding took place), with very few marketed as soft crabs. Watermen took crabs that were 'snots' to be shed in their floats (rectangular, 3 m × 1.2 m or 10 ft. × 4 ft. boxes with laths on either end and wooden planks on the bottom) until the soft crabs became 'peelers' (Fig. 9). Crabs were not fed in the floats. In the 1900s, crabs in floats suffered high mortality in the hot sun, so crabbers built canopies to shade the floats, ameliorating this problem. The added expense of the canopies was worth it, as crabs that died in the floats would "command a small price" if they were to be cooked and eaten within 2 or 3 h (i.e., dead crabs were sold to be eaten immediately) (Roberts 1905). Many of the dead crabs went to farms to feed hogs. Men working for the dealers would watch crabs in the floats and would remove crabs immediately after shedding so that their shells would not harden and they could be sold for the highest possible price. Employees would sort and pack soft crabs for market 3 or 4 times throughout the day to align with shipping them by rail or steamboat (Roberts 1905).

Careful packing of soft crabs on a layer of seaweed, close together, in a vertical, face-up position, and covered with another layer of seaweed, allowed them to stay alive for 60–70 h. This was enough to survive a trip from Crisfield, Maryland, to Canada. In the early 1900s, soft crabs would command, on average, a price of 1.5 cents each, with a range of 0.5–4 cents, depending on size (Roberts 1905). Four cents in 1905 is equivalent to $1.12 in 2019, a year when large females fetched $1.83, so prices remain relatively similar to those of the early twentieth century (Wiso's Crabs 2019). Initially, no ice was used in shipping soft crabs, and mortality was up to 50% (Churchill 1920). Crabbers would typically market their catch daily unless distance from a buying facility precluded sale. In that case, the crabbers would reside for months in shanties, and their catch would be sold to 'buy boats' nearby (Roberts 1905).

The start of commercial, industrial crabbing also brought on the use of floats for housing shedding crabs. The Maryland locations of Somerset County, near Pocomoke Sound, Tangier Sound, and the mouths of the Wicomico and Nanticoke Rivers, were considered the best soft-crabbing grounds due to

Fig. 8 Line drawing by C. B. Hudson of a Chesapeake Bay sailing vessel used for crab dredging. Wikimedia Commons. https://commons.wikimedia.org/wiki/File:FMIB_34249_Dredging_for_Crabs_Typical_Chesapeake_Canoe.jpeg

Fig. 9 Soft crab floats, Oxford, Maryland, circa 1885. Wikimedia Commons. https://commons.wikimedia.org/wiki/File:FMIB_38900_Floats_for_Soft_Crabs_at_Oxford,_Maryland.jpeg

MANY MILLIONS OF CRABS

Crisfield, on the Chesapeake Bay, Is Where They Are Caught.

SENT TO MARKETS IN THE NORTH

Recent Rapid Growth of an Industry Which Rivals the Oyster Business in Maryland Waters —Soft and Hard Shells.

CRISFIELD, Md., May 11.—Five million crabs! Think of this multitude of crustaceans in one immense, restless mass, and then wonder how it is that so many are sent to the markets of New-York and Philadelphia from this modest town. It is the annual total, and every year it is growing

Fig. 10 New York Times article published May 11, 1895, regarding the Crisfield, Maryland, blue crab industry (New York Times 1895)

the "grassy bottoms underlying the waters" where "young crabs congregate in vast numbers to shed" (Churchill 1920). Crisfield, Maryland, became the center of the soft crab industry in the late nineteenth century, taking advantage of the "large supply of crabs" in the region (New York Times 1895; US Fish Commission 1891, Fig. 10). Crabbers brought their catch to Crisfield and Deal Island (both in Somerset County, Maryland) for subsequent transport by steamboat and trains to markets across the US, making these two towns the largest suppliers of blue crabs in the nation (Roberts 1905). Dameron Marsh, Virginia, on the Western Shore of the Bay also produced many blue crabs, but lack of reliable transport to Crisfield across the Bay kept most of that catch from the national market.

Soft-shell crabs often command the highest price; thus, obtaining them at the soft-shell stage was and still is critical to watermen. In the summer of 2019, one vendor in Maryland marketed standard hard-shell crabs for $50 a half dozen, and the same number of soft-shell crabs for $65 (Cameron's Seafood 2019).

The Hard-Crab Industry and Its Economic Role In 1878 James McMenamin of Hampton, Virginia, initiated the hard-crab industry. He discovered a way "...for canning the meat of the hard crab" (Churchill 1920). Initially, it was difficult to find a market for hard crabs, as they "were even considered by many to be poisonous" (Churchill 1920). Crabs sent to market were typically 10–15 cm (4–6 in.) from tip to tip, with the largest ones spanning 18–25 cm (7–10 in.) in width (Rathbun 1887). In the 1880s, a new process came on the scene. The crabs were steamed, the shells cracked open, and the meat picked away from the exoskeleton. The meat could then be stored on ice or canned (Roberts 1905). This labor-intensive practice remains in place today.

Reflecting their abundance in the 1880s, crabs were caught mainly by scoop net or dip net. Fishers could easily see the crabs swimming in the clear waters alongside their boats and would capture them with long-handled nets. By 1920, they were mainly caught by dredge or scrape (similar to oysters). The growing demand for crabs spurred a new fishing technique in the early twentieth century. In winter, crabbers in Virginia used dredges to scoop the dormant crabs from the bottom sediments. Dredging led to increasing yields from 1901 to 1908, but the catch began to decline by 1916 (Churchill 1920).

Winter dredging exacts a large toll on the gravid females, reducing the next year's potential recruitment because the eggs that might have hatched were taken with the captured females.

In the early twentieth century, the towns of Oxford and Cambridge on Maryland's Eastern Shore landed the majority of hard-shell crabs. Cambridge shipped all of its catch live. Oxford sent only 1/3 of the catch to market live in July and August; the remaining catch was cooked, picked, and canned in "tin buckets." Live crabs were shipped in ventilated boxes or barrels under a layer of ice (Fig. 11). In 1905, crabbers received between 50 cents and $2 per barrel for hard crabs, depending on location of sale and whether the crabber marketed his own catch. Rising demand drove steadily increasing prices and the expansion of the industry to all corners of the Chesapeake Bay that had crabs for the taking (Churchill 1920).

Maryland fishers pursued crabs between June 1 and September 1, switching to tonging for oysters in the fall. Maryland fishers also used trotlines to catch crabs (Roberts 1905). This involved stringing bait along a line, dropping the anchored end to the Bay bottom and fastening the other end to a wooden float. Beef tripe, calf pelts, sting rays, hog chokers, and spoiled beef all served as bait. One man per 3.5–7.3 m (12–24 ft) boat would fish the trotlines 10–20 times per day. He pulled the trotline to the surface with the tenacious blue crabs loath to lose their grip on the bait, but

Fig. 11 Crabs in wooden crab barrels ready for ice and subsequent shipping. Coastal Ocean Program, NOAA. http://www.publicdomainfiles.com/show_file.php?id=13952301619058

shaking eventually did the job, dropping the catch into a prepositioned dipnet. Crabbers would typically reach the fishing grounds just after midnight or at least before 4 AM to lay the trot lines before the crabs began biting at sunrise. Because of the heat of the day and subsequent difficulty in keeping crabs alive, no crabbing would occur between 10 AM and 3 PM. Sometimes trotlines would attract male and female crabs mating, called "doublers" or "channeler and his wife" (Roberts 1905). The male crab was referred to as a "channeler," a "Jimmy," or a "Jim crab," whereas nowadays the term "Jimmy" is most commonly used.

Crab pots, another new technique of catching crabs, came on the scene in the 1930s. The pots are boxes of wire mesh with funnels to permit the entry of crabs responding to the odor of bait (often dead fish) secured in a compartment in the interior of the trap. A tether and float allow crabbers to fish the traps from boats (see chapter "A Fishing Trip: Exploiting and managing the commons of the Chesapeake Bay" on Fisheries for more details). Since the 1950s, crab pots have been responsible for the largest portion of reported crab landings (Kennedy et al. 2007; Miller et al. 2011). However, Maryland presently restricts their use in many areas, particularly Bay tributaries, keeping the practice of trot lining alive.

The growing crab industry of the nineteenth and early twentieth centuries provided skilled jobs for African Americans, and immigrants from Ireland, Italy, and other places. The work was tedious and fast-paced, but provided income to those shut out of other professions (Kraig 2013). The Great Depression reduced the demand for this luxury food item, closing many of the canneries. But after WWII, demand rose again, along with disposable incomes. By the 1970s, as many as 75 crab houses (picking plants) dotted the coast of the Bay (Warner 1976). These produced the valuable cleaned crab meat.

Technological developments to ensure a long shelf life of crab meat aided the industry. One development from 1951 was 'pasteurization' of crab meat, which was initiated in Crisfield and patented by Clifford Byrd of Byrd Seafood (Wharton 1954; Warner 1976). This entailed quickly steaming the crab meat to an internal temperature of 85 °C (185 °F) for 1 min, and then immersing it in an icy water bath. This improved the shelf life of refrigerated fresh crabmeat from 10 to 12 days to six months or a year, allowing the use of all crab meat from banner catches and other glut time periods (Warner 1976). When crab populations declined in the late 1990s, pasteurized crab meat from Asia and South America replaced that formerly produced locally. The Chesapeake Bay crab fishery has diminished (Fig. 5) as the marketing of crab cakes and imports of foreign crab meat have increased (Paolisso 2007). However, Chesapeake Bay blue crab still commands high prices, and several vendors have been prosecuted for fraud after being caught selling mislabeled imported crab meat as local crab product (Dujardin 2018).

Various fishing methods for blue crabs were important throughout the years, including the winter dredge, scrape, peeler pound and trap, trotline, and recreational fisheries. In 2000, the hard-crab and peeler-pot fisheries in Virginia took about 83% of the total catch, and the dredge fishery took about 15%, whereas in Maryland, the pot fishery took about 61–67% while the trot line fishery took up to 30–36% of the catch (Burreson et al. 2000). Only 1–2% of total blue crab landings were taken by the scrape and peeler-pound fisheries. The recreational harvest of blue crabs accounted for 15–25% of the catch in Maryland (Kennedy 2018) and about 15% of the catch in Virginia (Seitz et al. 2001). In 2008, in an effort to increase spawning, the winter dredge fishery, which targeted mature female crabs prior to spawning, was eliminated.

Preparing Crab for the Plate Cooks prepare blue crabs in several different ways. Soft-shell crabs may be steamed, sautéed, or fried and placed on salads or served in a sandwich. Hard-shell crabs require more work. If prepared at home, cooks typically steam or boil them in a large pot with a dose of spicy seasoning. The whole crabs are then traditionally served on a layer of newspaper, where the guests can crack and pick the meat from the exoskeleton. Bowls of melted butter are provided for dipping the picked meat before eating. Corn on the cob is often boiled or steamed with the crabs.

Commercial crab-picking operations do the tedious work of picking the crabmeat from the shell. The cleaned crab is sold fresh, canned, or frozen. Cooks use the meat in a variety of dishes, and the most famous is the crab cake. This is a delicacy that typically uses "back fin" meat (from the base of the crab's swimming leg) from a cooked hard crab, combined with something to bind the meat (such as an egg or mayonnaise), some seasonings, and possibly some filler like bread, bread crumbs, or crackers (Kraig 2013). The cake is typically fried in oil, but sometimes baked in an oven. It is unknown where the crab cake originated, but one theory holds that it emulated the New England cod cake, with crab meat substituted for the salted fish. Smaller versions of the cakes, called crab balls, often serve as appetizers.

Women, Children, and People of Color Provide Cheap Labor to Process Blue Crabs Extracting the meat from hard-shell blue crabs is hard and tedious work. After steaming the crabs in large vats for 20–40 min, the cooked crabs would be dumped in front of pickers "some of whom, with long experience, become very expert in extracting the meat" (Roberts 1905). They had to be good, as their pay depended on the amount of meat they picked. The superior meat that was white and firm was from the "hip" of the crab (i.e., base of the swimming leg) and was termed "flakes." In 1905, for about 0.9 kg (2 lbs) of crab meat (~1 quart), pickers received 4–5 cents for what amounted to about an hour's work. That's approximately $1.16 in 2019 dollars. Current pay for pickers is considerably better, $9.00 per kg (2.2 lbs) in Maryland, but still a low wage (see chapter "Ethics and Economics of Building a Food System to Recover the Health of the Chesapeake Bay and its People" on ethics).

Women did and still do most of the picking (Fig. 12). And they were often joined by children in the early years of the industry (Roberts 1905; Warner 1976). Labor laws and changing social norms eventually curtailed the exploitation of children in the crab houses. In the early years, pickers were drawn from the local population to work the seasonal job of processing blue crabs. Thus, they reflected the local demography. However, African American women pickers came to dominate the industry in the twentieth century. That started to change in the 1980s as itinerant seasonal workers from Mexico and Central America began to take their place. One crab house in Hampton, Virginia, employed African-American women almost exclusively in 1988. Thirty years later, the staff of pickers are all Latinos. By 2010, Latino workers on H2-B guest worker visas came to dominate the crab processing workforce on the Bay (The International Human Rights Law Clinic (IHRLC) 2010, see chapter "Ethics and Economics of Building a Food System to Recover the Health of the Chesapeake Bay and its People" on ethics).

Since crab picking is seasonal work, many women in the early days of the industry also engaged with other fishery activities. These included shucking oysters, gutting and salting fish, and pasting labels on cans of oysters (Kennedy 2018). The decline of the oyster and finfish industries around the Bay removed most of those employment opportunities.

Crab Packers Versus Crab Pickers In the 1930s, mostly African American women preformed the tedious task of picking the meat from crabs in the packing houses of Crisfield, Maryland. The Great Depression and recent closures of a garment factory in the area meant jobs were hard to find. In 1931 the owners of the packing houses reduced the pay to pickers from $0.35 per gallon (3.8 l) to $0.25 per gallon. This brought on several years of strikes, and vigilante attacks on labor organizers and Federal officials. After a 5-week-long strike in 1938, the workers prevailed and won back the higher rate of pay and official acceptance of their union, the Council for Industrial Organization's (CIO) Canary Union. The strike succeeded in part due to the coalescence of black and white workers. Although black women outnumbered their white co-workers by about eight to one, the remarkable cross-racial solidarity on the Eastern Shore of Maryland proved essential to winning the union victory. Understand that the Eastern Shore at that time was known for a history of lynching and other forms of intimidation aimed at its African American residents. The decline of the crab industry and other factors led to the loss of unionization of pickers in 1990 (Simpson 2019).

Fig. 12 Women picking crabs in a crab house (NOAA stock photo). NOAA Photo Library. https://www.google.com/imgres?imgurl=https://www.photolib.noaa.gov/bigs/fizh2129.jpg&imgrefurl=https://www.photolib.noaa.gov/htmls/fizh2129.htm&h=2059&w=2500&tbnid=IK79mASJ3IMEPM&tbnh=204&tbnw=247&usg=AI4_-kTKkFr8eFvtf9qZ9TkYgLGs1b79GA&vet=1&docid=7xkp150BhqXcaM

Blue Crab Byproducts The blue crab industry produces some very useful byproducts for other parts of the food system. The meat only comprises an estimated 10–29% of the picked crab, with the remainder as waste (Kennedy et al. 2007). In the 1800s, the dorsal carapace of the crab would be cleaned and used for serving deviled crab, and these were typically shipped with canned crab meat (Rathbun 1887). Up through the 1980s, solid crab waste was dried in commercial drying facilities and then converted into fertilizer or feed for chicken and other livestock (Hatem 1981; Kennedy et al. 2007). Everyone down-wind from the facility in Hampton, Virginia, knew exactly when crab processing was in progress. When meal plants began having trouble surviving in the early 1980s, crab waste was increasingly disposed of in landfills. With growing concerns about the potential for surface-water and ground-water pollution caused by the ammonia and nitrates in food-processing waste disposal, concerned stakeholders held a conference and discussed new ideas for the use of crab waste (Hatem 1981). Discarded shells and non-meat organs make for good fertilizer. The shells are rich in nitrogen, containing about 9% ammonia (Roberts 1905). Farmers sought this organic fertilizer to amend soils depleted by years of cultivation. More recently, waste from crabs has been developed into "an odorless, slow-release, garden fertilizer" by combining it with wood chips in a composting process (Kennedy et al. 2007).

Ghost Crab Pots Crabbers lose some of their crab pots over the course of the season. Pots are lost due to storms and whirling boat propellers that cut the line between the surface float and the trap that rests on the bottom. Such lost traps can continue to capture crabs and finfish for some time. To address this problem, the Virginia Marine Resources Commission and NOAA provided funding for watermen to search-out and collect the derelict traps. The project ran from 2008 to 2014, providing employment for as many as 70 crabbers during their off-season in winter. Side-scan sonar helped locate the traps that were grappled and pulled from the bottom. The crabbers collected 30,041 derelict traps. These contained 32,954 entrapped animals including, crabs, finfish, ducks, turtles, and whelks (Virginia Institute of Marine Science (VIMS) 2019a). Although ghost traps caused some blue crab mortality, the overall impacts on the blue crab population and fishery were minimal (CBSAC 2018).

Recent population levels and scientific surveys The Chesapeake Bay blue crab population has experienced "overfishing" (i.e., when the exploitation fraction is greater than the scientifically derived "threshold"; Miller et al. 2005) several times throughout history, such as in the early 1990s (Miller et al. 2011). The first stock assessment of blue crabs in Chesapeake Bay was not completed until 1997. It found that the stock was "moderately to fully exploited and at average levels of abundance" (Miller et al. 2005). Other stock assessments followed in 2003 and 2010 (Miller et al. 2011), and each time, management actions were amended to help improve population abundances for their economic and ecological contributions.

Traditionally, fisheries managers use commercial and recreational records of catch to assess the status of targeted populations. These provide a useful but limited understanding of population structure or composition since the catch excludes the earlier life stages and is dependent on reporting by industry and crabbers. To supplement the catch records, scientists conduct fisheries-independent surveys of blue crabs in Chesapeake Bay to assess population levels and to recommend management or conservation measures. Beginning in 1990, scientists implemented an annual Bay-wide winter dredge survey to estimate the adult blue crab population throughout Chesapeake Bay. The winter dredge survey samples 1500 sites annually (Sharov et al. 2003; VIMS 2019b) and is consistently cited as "a vital and indispensable tool for blue crab management" (Miller et al. 2011). For this survey, crabs are collected in the winter when they bury in the sediments and are not hiding in structured habitats or swimming freely in the water column. The Virginia Institute of Marine Science also conducts a trawl survey for crabs within the lower Chesapeake Bay tributaries (VIMS 2019c), and a survey for juvenile crabs in the shallow waters where the trawl and dredge boats cannot sample but where most young juvenile crabs reside (Ralph et al. 2013).

In 2001, fishery production of the blue crab in the Bay was the lowest on record, with stock biomass well below average at 21,539 metric tonnes (MT). Blue crab landings were on average 34,887 MT from 1945 to 2003 (Miller et al. 2005). In addition, the spawning stock suffered a significant and substantial reduction in the previous decade, where abundance and biomass declined by more than 80%, while larval abundance and recruitment deteriorated by nearly tenfold (Lipcius and Stockhausen 2002). In 2008, the blue crab fishery in Chesapeake Bay was declared a 'failure' by the U.S. Secretary of Commerce (Feltault 2009) and the blue crab population was in danger of collapse. Major management efforts were put into place in 2008, such as closure of the winter dredge fishery, restrictions on time of day for fishing, and restrictions on the numbers of crab pots deployed. These efforts triggered recovery of the population above the threshold for sustainable exploitation (Miller et al. 2011), and the current population is considered healthy (Fig. 6; CBSAC 2019; Maryland Department of Natural Resources (Maryland DNR) Maryland DNR 2018).

The most recent blue crab stock assessment concluded that the population was not overfished (Miller et al. 2011). The 2018 winter dredge survey also concluded that the population was healthy, with an estimated crab population of 371 million, even after a harsh winter with 35% mortality (VIMS 2019b). In fact, it had a 60% increase compared to the year before, showing the highest population numbers since 2012. However, because of concurrent low juvenile abundances, caution was warranted in relaxing any fishery regulations (VIMS 2019b). Nonetheless, managers changed regulations to allow modest increases in blue crab catches for the end of the crabbing season in 2019 (Wheeler 2019).

Fishery regulations and efforts to restore and improve blue crab habitat In colonial times, crab populations were thought to be inexhaustible. But industrial-scale harvesting in the early twentieth century resulted in decreased landings. Authorities established regulations to curtail the decline. The earliest known conservation attempts for Chesapeake Bay blue crabs based on relevant biological information were by E.P. Churchill in 1917, after he investigated blue crab packing houses (Churchill 1919; Pearson 1942). In 1917, Virginia enacted a law that prohibited the taking of an egg-bearing (sponge) crab during July and August, whereas in Maryland, the taking of sponge crabs was prohibited at any time of the year (Pearson 1942) (however, not many sponge crabs are found in Maryland). Another law was enacted by Maryland in 1917 making it illegal to possess a hard crab smaller than 5 in. (13 cm) from tip to tip, or a soft crab smaller than 3 in. (which was later changed to 3.5 in.), and Virginia enacted a similar law. Churchill (1919) also suggested that Virginia enacted a law to shorten the lower-Bay dredge season from November 1–May 1 to November 15–April 1. Low catches in 1920 prompted additional laws establishing a minimum size for peeler crabs (3 in., 7.6 cm), reducing the season for trot lining to April 1–November 30, and issuing a recommendation that "provisions be made for the collection of statistics and biological data continuously throughout future years in order that the effect of any new regulations may be known" (Pearson 1942). The final recommendation was not implemented, and circumstantial evidence was used to support the success of the management efforts when crab populations rose the following year.

More recently, additional laws were implemented to protect spawning females. Because spawning sanctuaries can potentially help protect a large percentage of the spawning stock, such sanctuaries can be useful for heavily fished populations. While in the lower Bay preparing to release their larvae, the female crabs have been protected from harvest by a spawning sanctuary, the Lower Bay Spawning Sanctuary (LBSS), enacted in 1942. The sanctuary consisted of 37,814 ha (93,440 acres). It was declared as a reaction to low harvests in 1940 and 1941, and it is effective from June 1–September 15 every year (Seitz et al. 2001). The sanctuary was expanded in 2000 to 172,235 ha, to protect not only spawning female crabs, but also those migrating to the lower Bay spawning grounds. The sanctuary roughly follows the 10.7 m depth contour throughout the lower Bay, up to the Maryland border (Lipcius et al. 2003). In 2002, the sanctuary was again expanded, this time to 240,092 ha, to roughly follow the 9 m depth contour, which protects ~50% of spawning females. This sanctuary has been effective in protecting female crabs from being captured by the fishery, as demonstrated in tagging studies (Lambert et al. 2006).

Chesapeake Bay fisheries were productive throughout the late nineteenth century, but many segments, including the native oyster, diamondback terrapin, shad, river herring, sturgeon, and others, later experienced declines by the beginning of the twentieth century. However, blue crabs remained resilient, showing less dramatic reductions over time, even though they were much more abundant in the nineteenth century than at present (Kennedy 2018). Their resilience is likely owing to their opportunistic feeding habits and ability to use multiple habitat types. In addition to overfishing, the influence of increasing nutrients in Chesapeake Bay, the influx of pollutants, and other environmental onslaughts have degraded the ecosystem to a new baseline. The Chesapeake Bay food web has suffered drastic changes due to anthropogenic impacts, such as overfishing (Jackson et al. 2001) and eutrophication (Boesch 2000), as well as natural disturbances (Glaspie et al. 2020), that may prevent restoration of habitats to their once-pristine state. The Bay now regularly features a dead zone of hypoxic bottom water that excludes blue crabs from such habitats (see chapter "Eutrophication: Obesity of the Bay and its People" on eutrophication). Understanding what has happened through time with various fisheries in the Bay can help to set relevant goals for their restoration (Kennedy 2018).

The importance of certain habitats for blue crab populations has been recognized by scientists and citizens alike. There are various habitats throughout Chesapeake Bay that are essential for blue crabs due to their value in providing food and refuge from predation. Bottom-up control (or control by lower food levels) is occurring for blue crabs, as crab abundance correlates with that of its principal prey, infaunal clams (Seitz et al. 2003). In addition, crab growth and survival of blue crab juveniles are high in structured habitats that provide refuge from predation (Hines and Ruiz 1995; Bromilow and Lipcius 2017), and high prey density (Seitz et al. 2003). Crab postlarvae require suitable settlement substrates and nursery habitats (Lipcius and Stockhausen 2002), and structured habitats such as seagrass are the most important (Perkins-Visser et al. 1996). Using field experiments, it was estimated that some upper-Bay subestuaries (e.g., the Rhode River) had 40–90% mortality for juvenile blue crabs in the summer in unstructured habitats at depths greater than 70 cm, mainly due to cannibalism (Hines and Ruiz 1995).

Mortality of juvenile blue crabs is important for determining adult crab populations in the subsequent year, and mortality can be more important than recruitment to subsequent adult populations (Hines and Ruiz 1995). Recruitment of crab postlarvae and young juveniles is limited to certain suitable settlement and nursery habitats (Lipcius et al. 2007). Telemetry studies of movement of juvenile blue crabs showed that they remain within subestuaries until they grow to maturity and mate (Hines and Ruiz 1995).

Unfortunately, the areal coverage of seagrass beds in Chesapeake Bay has been declining in the recent past, as is occurring worldwide (Orth et al. 2006, 2010). Increased warming of the Chesapeake Bay due to climate change could be detrimental for cold-water seagrass species, especially eelgrass, *Zostera marina* (Najjar et al. 2010; Lefcheck et al. 2017; Richardson et al. 2018). Eelgrass is a foundation species in the Bay that declined 29% from 1991 to 2017, partially due to climate warming (Lefcheck et al. 2017). Though reduction in nutrient pollution has led to increases for some seagrasses in Chesapeake Bay (notably those in the freshwater reaches; Lefcheck et al. 2018), continued coastal population increases and relaxing of environmental regulations since 2016 (U.S. Global Change Research Program 2018) could mean disaster for the lower Bay eelgrass that blue crabs so depend upon. Without reductions in emissions of greenhouse gases that affect our climate, coastal ecosystems are in danger of species losses (U.S. Global Change Research Program 2018), which could affect the communities that depend on coastal resources and fishing (see chapter "A New Food System for The Chesapeake Bay Region and a Changing Climate" on climate change).

There are many scientific efforts throughout Chesapeake Bay to maintain the blue crab population, which requires periodic stock assessments to determine whether overfishing is occurring (relative to reference points; Miller et al. 2011). Efforts are underway to restore and improve Bay habitats, which requires maintaining or improving water quality to promote nursery habitats such as seagrass beds (Kennedy and Cronin 2007; Lefcheck et al. 2018). All of these past and present scientific efforts are aimed at keeping the blue crab resource healthy for economic gain, and to preserve populations so that future generations can enjoy this beautiful, savory swimmer in the foreseeable future.

Acknowledgements I thank the many scientists and crabbers who came before me for their important contributions to our knowledge about blue crabs in Chesapeake Bay. This chapter was substantially improved by comments from Vic Kennedy, Rom Lipcius, and Pete Sheridan. Assistance was provided by Alison Smith of the VIMS Community Ecology lab. This is contribution number 3847 from the Virginia Institute of Marine Science.

References

Archer G (1607) The description of the now-discovered river and country of Virginia, with the likelihood of ensuing riches by Englands' aid and industry. In: Haile E (ed) (1998) Jamestown narratives: eyewitness accounts of the Virginia Colony. RoundHouse, Champlain, pp 118–121

Baird D, Ulanowicz R (1989) The seasonal dynamics of The Chesapeake Bay ecosystem. Ecol Monogr 59:329–364. https://doi.org/10.2307/1943071

Boesch D (2000) Measuring the health of the Chesapeake Bay: toward integration and prediction. Environ Res 82:134–142. https://doi.org/10.1006/enrs.1999.4010

Bromilow A, Lipcius R (2017) Mechanisms governing ontogenetic habitat shifts: role of trade-offs, predation, and cannibalism for the blue crab. Mar Ecol Prog Ser 584:145–159. https://doi.org/10.3354/meps12405

Brooks W (1893) Chapter VII. Fish and fisheries. In: Maryland. Its resources, industries, and institutions. Board of World's Fair Managers of Maryland, Baltimore, pp 239–263

Brunkow J (2019) Beautiful Swimmers | James River Association. In: Thejamesriver.org. https://thejamesriver.org/beautiful-swimmers/. Accessed 10 Sep 2019

Burreson E, Butler M, DuPaul W et al. (2000) The Virginia blue crab winter dredge fishery. Report of the Task Force on the Virginia Blue Crab Winter Dredge Fishery to the Governor and the General Assembly of Virginia. Senate Doc. 30

Cameron's Seafood (2019) Mail order blue crabs | Male Maryland Blue Crabs Online. In: Cameron's Seafood. https://www.cameronsseafood.com/buy-seafood-online/maryland-blue-crabs/2-male/#. Accessed 10 Sep 2019

CBSAC (2018) Chesapeake Bay Stock Assessment Committee. Impact of derelict trap removal on blue crab landings. In: Chesapeakebay.net. https://www.chesapeakebay.net/channel_files/27631/derelict_fishing_gear_removal_review_final.pdf. Accessed 18 Sep 2019

CBSAC (2019) Chesapeake Bay Stock Assessment Committee. 2019 Chesapeake Bay Blue Crab Advisory Report. In: Chesapeakebay.net. https://www.chesapeakebay.net/documents/CBSAC_2019_Blue_Crab_Advisory_Report_Final.pdf. Accessed 18 Sep 2019

Churchill E Jr (1919) Life-history of the blue crab, *Callinectes sapidus*, of the Chesapeake Bay. Bulletin, Bureau of Fisheries 36:39–128

Churchill E Jr (1920) Crab industry of Chesapeake Bay. US Commissioner of Fisheries, Washington DC, pp 1–25

Conservation Commission (1909) Report of the Conservation Commission of Maryland for 1908–1909. 155–179

Cronin L (1947) Anatomy and histology of the male reproductive system of *Callinectes sapidus* Rathbun. J Morphol 81:209–239. https://doi.org/10.1002/jmor.1050810205

Cronin L (1998) Early days of crabbing and a brief history of the Chesapeake Bay. J Shellfish Res 17:379–381

Dujardin P (2018) Federal prosecutors charge Casey's Seafood owner in blue crab case. In: Dailypress.com. https://www.dailypress.com/news/newport-news/dp-nws-casey-seafood-crab-charges-20180111-story.html. Accessed 14 Jul 2019

Epifanio C (2019) Early life history of the blue crab *Callinectes sapidus*: a review. J Shellfish Res 38(1):1–22

FAO (2002) Food and Agriculture Organization. The state of world fisheries and aquaculture. In: Fao.org. http://www.fao.org/3/a-y7300e.pdf. Accessed 14 Jul 2019

Farish H (1943, 1957) Journals and letters, 1773–1774. In: Wharton (ed) The bounty of the Chesapeake: Fishing in colonial Virginia. Colonial Williamsburg Incorporated, Williamsburg

Feltault K (2009) Standardizing the taste of tradition: Maryland crab cakes from Thailand. PhD Dissertation, American University

Fischler K (1965) The use of catch-effort, catch-sampling, and tagging data to estimate a population of blue crabs. Trans Am Fish Soc 94:287–310. https://doi.org/10.1577/1548-8659(1965)94[287,tuocca]2.0.co;2

Glaspie C, Seitz R, Lipcius R (2020) Are predator-prey model predictions supported by empirical data? Evidence for storm-driven shift to alternative stable state in crab-clam system. Accepted in Marine Ecology Progress Series. https://doi.org/10.1101/224097

Gnanalingam G, Butler M, Matthews T et al (2018) Directly ageing the Caribbean spiny lobster, Panulirus argus with validated band counts from gastric mill ossicles. ICES J Mar Sci 76:442–451. https://doi.org/10.1093/icesjms/fsy177

Hamor R (1615, 1998) A true discourse of the Present Estate of Virginia. In: Haile E (ed) Jamestown narratives: eyewitness accounts of the Virginia Colony. RoundHouse, Champlain, pp 792–856

Hatem M (1981) Crab byproducts and scrap. Maryland Sea Grant Program

Hines A, Ruiz G (1995) Temporal variation in juvenile blue crab morality: nearshore shallows and cannibalism in Chesapeake Bay. Bull Mar Sci 57:884–901

Hines A, Lipcius R, Haddon A (1987) Population dynamics and habitat partitioning by size, sex, and molt stage of blue crabs *Callinectes sapidus* in a sub-estuary of central Chesapeake Bay. Mar Ecol Prog Ser 36:55–64. https://doi.org/10.3354/meps036055

Hines A, Haddon A, Wiechert L (1990) Guild structure and foraging impact of blue crabs and epibenthic fish in a sub-estuary of Chesapeake Bay. Mar Ecol Prog Ser 67:105–126. https://doi.org/10.3354/meps067105

Hines A, Jivoff P, Bushmann P et al (2003) Evidence for sperm limitation in the blue crab, *Callinectes sapidus*. Bull Mar Sci 72:287–310

IHRLC (2010) The International Human Rights Law Clinic. Picked Apart: The hidden struggles of migrant worker women in the Maryland crab industry. In: Digitalcommons.wcl.american.edu. https://digitalcommons.wcl.american.edu/cgi/viewcontent.cgi?article=1001&context=fasch_rpt. Accessed 17 Sep 2019

Jackson J, Kirby M, Berger W et al (2001) Historical overfishing and the recent collapse of coastal ecosystems. Science 293:629–637. https://doi.org/10.1126/science.1059199

Kennedy V (2018) Shifting baselines in the Chesapeake Bay. Hopkins University Press, Baltimore

Kennedy V, Cronin L (eds) (2007) Biology and management of the blue crab. University of Maryland Press

Kennedy V, Oesterling M, Van Engel W et al (2007) History of blue crab fisheries on the U.S. Atlantic and Gulf Coasts. In: Kennedy V, Cronin L (eds) The blue crab *Callinectes sapidus*. Maryland Sea Grant College, College Park, pp 655–709

Kilada R, Reiss C, Kawaguchi S et al (2017) Validation of band counts in eyestalks for the determination of age of Antarctic krill, Euphausia superba. PLoS One 12:e0171773. https://doi.org/10.1371/journal.pone.0171773

Kliman T (2005) Crab imperialist. Washington City Paper

Kraig B (2013) The Oxford encyclopedia of food and drink in America, vol 1. Oxford University Press, Oxford

Lambert D, Lipcius R, Hoenig J (2006) Effectiveness of the Blue Crab Spawning Sanctuary in Chesapeake Bay. Virginia Institute of Marine Science, Gloucester Point

Lefcheck J, Wilcox D, Murphy R et al (2017) Multiple stressors threaten the imperiled coastal foundation species eelgrass (*Zostera marina*) in Chesapeake Bay, USA. Glob Chang Biol 23:3474–3483. https://doi.org/10.1111/gcb.13623

Lefcheck J, Orth R, Dennison W et al (2018) Long-term nutrient reductions lead to the unprecedented recovery of a temperate coastal region. Proc Natl Acad Sci 115:3658–3662. https://doi.org/10.1073/pnas.1715798115

Lipcius R, Latour R (2006) Food web interactions and modeling. American Fisheries Society

Lipcius R, Stockhausen W (2002) Concurrent decline of the spawning stock, recruitment, larval abundance, and size of the blue crab *Callinectes sapidus* in Chesapeake Bay. Mar Ecol Prog Ser 226:45–61. https://doi.org/10.3354/meps226045

Lipcius R, Van Engel W (1990) Blue crab population dynamics in Chesapeake Bay: variation in abundance (York River, 1972–1988) and stock-recruit functions. Bull Mar Sci 46:180–194

Lipcius R, Stockhausen W, Seitz R et al (2003) Spatial dynamics and value of a marine protected area and corridor for the blue crab spawning stock in Chesapeake Bay. Bull Mar Sci 72:453–469

Lipcius R, Seitz R, Seebo M et al (2005) Density, abundance and survival of the blue crab in seagrass and unstructured salt marsh nurseries of Chesapeake Bay. J Exp Mar Biol Ecol 319:69–80. https://doi.org/10.1016/j.jembe.2004.12.034

Lipcius R, Eggleston D, Heck K Jr et al (2007) Ecology of postlarval and young juvenile blue crabs. In: Kennedy V, Cronin L (eds) The blue crab *Callinectes sapidus*. Maryland Sea Grant College, College Park, pp 535–564

Mansour R (1992) Foraging ecology of the blue crab, *Callinectes sapidus* Rathbun, in lower Chesapeake Bay. PhD dissertation, The College of William and Mary

Mansour R, Lipcius R (1991) Density-dependent foraging and mutual interference in blue crabs preying upon infaunal clams. Mar Ecol Prog Ser 72:239–246. https://doi.org/10.3354/meps072239

Maryland DNR (2018) Maryland department of natural resources. Survey Shows Chesapeake Bay Blue Crab Population Healthy. In: News.maryland.gov. https://news.maryland.gov/dnr/2018/05/09/survey-shows-chesapeake-bay-blue-crab-population-healthy/. Accessed 15 Jun 2019

Maryland Sea Grant (2012) Chesapeake Quarterly Volume 11 Number 2: Atlantic blue crab life cycle. In: Chesapeakequarterly.net. https://www.chesapeakequarterly.net/V11N2/side2/. Accessed 13 Jul 2019

Miller T, Martell S, Bunnell D et al (2005) Stock assessment of the blue crab in Chesapeake Bay. University of Maryland Center for Environmental Science, Solomons

Miller T, Wilberg M, Colton A et al (2011) Stock assessment of the blue crab in Chesapeake Bay 2011. University of Maryland Center for Environmental Science, Solomons

Najjar R, Pyke C, Adams M et al (2010) Potential climate-change impacts on the Chesapeake Bay. Estuar Coast Shelf Sci 86:1–20. https://doi.org/10.1016/j.ecss.2009.09.026

New York Times (1895) Many millions of crabs. Crisfield, on the Chesapeake Bay, is where they are caught

NMFS (2012) National marine fisheries service. Fisheries Economics of the United States, NMFS

NOAA (2019a) National oceanic and atmospheric administration commercial fisheries statistics. In: National Oceanic and Atmospheric Administration. https://www.st.nmfs.noaa.gov/commercial-fisheries/commercial-landings/annual-landings/index. Accessed 30 Jun 2019

NOAA (2019b) National oceanic and atmospheric administration commercial fisheries statistics. In: National oceanic and atmospheric administration. https://foss.nmfs.noaa.gov/. Accessed 16 Sep 2019

Norse E (1977) Aspects of the zoogeographical distribution of *Callinectes sapidus* (Brachyura: Portunidae). Bull Mar Sci 27:440–447

Orth R, Carruthers T, Dennison W et al (2006) A global crisis for seagrass ecosystems. Bioscience 56:987–996. https://doi.org/10.1641/0006-3568(2006)56[987,agcfse]2.0.co;2

Orth R, Williams M, Marion S et al (2010) Long-term trends in submersed aquatic vegetation (SAV) in Chesapeake Bay, USA, related to water quality. Estuar Coasts 33:1144–1163. https://doi.org/10.1007/s12237-010-9311-4

Paolisso M (2007) Taste the traditions: crabs, crab cakes, and the Chesapeake Bay Blue crab fishery. Am Anthropol 109:654–665. https://doi.org/10.1525/aa.2007.109.4.654

Pearson JC (1942) Decline in abundance of the blue crab, (*Callinectes sapidus*), in Chesapeake Bay during 1940, and 1941, with suggested conservation measures. United States fish and wildlife service special scientific report No. 16., 27 pp

Perkins-Visser E, Wolcott T, Wolcott D (1996) Nursery role of seagrass beds: enhanced growth of juvenile blue crabs (*Callinectes sapidus* Rathbun). J Exp Mar Biol Ecol 198:155–173. https://doi.org/10.1016/0022-0981(96)00014-7

Pyle R, Cronin L (1950) The general anatomy of the blue crab, *Callinectes sapidus* Rathbun. Chesapeake Biological Laboratory

Ralph G, Seitz R, Orth R et al (2013) Broad-scale association between seagrass cover and juvenile blue crab density in Chesapeake Bay. Mar Ecol Prog Ser 488:51–63. https://doi.org/10.3354/meps10417

Rathbun R (1887) The crab, lobster, crayfish, rock lobster, and prawn fisheries. In: Goode G (ed) Fisheries and fishery industries of the United States. US Government Printing Office, Washington DC, pp 629–658

Rathbun M (1896) The genus *Callinectes*. In: Repository.si.edu. https://repository.si.edu/handle/10088/13419. Accessed 14 Jul 2019

Richardson J, Lefcheck J, Orth R (2018) Warming temperatures alter the relative abundance and distribution of two co-occurring foundational seagrasses in Chesapeake Bay, USA. Mar Ecol Prog Ser 599:65–74. https://doi.org/10.3354/meps12620

Roberts W (1905) The crab industry of Maryland. In Report of the Bureau of Fisheries 1904. Washington DC

Rountree H, Clark W, Mountford K et al (2008) John Smith's Chesapeake voyages, 1607–1609. University of Virginia Press, Charlottesville

Ryer C, van Montfrans J, Moody K (1997) Cannibalism, refugia and the molting blue crab. Mar Ecol Prog Ser 147:77–85. https://doi.org/10.3354/meps147077

Seitz R, Lipcius R, Stockhausen W et al (2001) Efficacy of blue crab spawning sanctuaries in Chesapeake Bay. In: Spatial processes and management of marine populations. University of Alaska Sea Grant, Fairbanks, pp 607–626

Seitz R, Lipcius R, Stockhausen W et al (2003) Potential bottom-up control of blue crab distribution at various spatial scales. Bull Mar Sci 72:471–490

Seitz R, Knick K, Westphal M (2011) Diet selectivity of juvenile blue crabs (*Callinectes sapidus*) in Chesapeake Bay. Integr Comp Biol 51:598–607. https://doi.org/10.1093/icb/icr098

Sharov A, Volstad J, Davis G et al (2003) Abundance and exploitation rate of the blue crab (*Callinectes sapidus*) in Chesapeake Bay. Bull Mar Sci 72:543–565

Simpson C (2019) 600 black women stand strong: The 1938 crab pickers strike. In: Washington Area Spark. https://washingtonareaspark.com/2012/12/05/600-black-women-stand-strong-the-1938-crab-pickers-strike/. Accessed 17 Sep 2019

Smith J (1986) A map of Virginia. The complete works of Captain John Smith (1580–1631). Available in Capt. John Smith's complete works: http://www.virtualjamestown.org/exist/cocoon/jamestown/fha-js/SmiWorks1

Tan E, van Engel W (1966) Osmoregulation in the adult blue crab, *Callinectes sapidus* Rathbun. Chesap Sci 7:30–35. https://doi.org/10.2307/1350986

Truitt R (1934) Preliminary report—blue crab investigations, 1932–1933. Division of purchase and printing, Richmond

US Fish Commission (1891) Notes on the crab fishery of Crisfield, MD. US Fish Commission

US Global Change Research Program (2018) The national climate assessment-impacts, risks, and adaptation in the United States. Melville House, Brooklyn

Van Engel W (1958) The blue crab and its fishery in Chesapeake Bay. Part 1: Reproduction, early development, growth, and migration. Commer Fish Rev 20:6–17

VIMS (2019a) Virginia Institute of Marine Science. Derelict Fishing Gear Removal Project | Virginia Institute of Marine Science. In: Vims.edu. https://www.vims.edu/ccrm/research/marine_debris/solutions/removal/index.php. Accessed 18 Sep 2019

VIMS (2019b) Virginia Institute of Marine Science. Winter Dredge Survey/Virginia Institute of Marine Science. In: Vims.edu. http://www.vims.edu/research/units/programs/bc_winter_dredge/index.php. Accessed 14 Jul 2019

VIMS (2019c) Virginia Institute of Marine Science. Juvenile Fish and Blue Crab Trawl Survey/Virginia Institute of Marine Science. In: Vims.edu. https://www.vims.edu/research/departments/fisheries/programs/juvenile_surveys/index.php. Accessed 14 Jul 2019

Virginia Places (2018) Crabs in Virginia. In: Virginiaplaces.org. http://www.virginiaplaces.org/natural/crabs.html. Accessed 14 Jul 2019

VMRC (2019) Virginia Marine Resources Commission, Chapter: pertaining to crab pots. In: Mrc.virginia.gov. https://www.mrc.virginia.gov/regulations/fr700.shtm. Accessed 18 Sep 2019

Warner W (1976) Beautiful swimmers: watermen, crabs and the Chesapeake Bay. Atlantic Little Brown, Boston

Wharton J (1954) The Chesapeake Bay crab industry. United States Department of the Interior, Fish and Wildlife Service

Wharton J (1957) The bounty of the Chesapeake: Fishing in colonial Virginia. University Press of Virginia, Charlottesville

Wheeler T (2019) Bay Journal—Article: Maryland, Virginia ease crabbing limits. In: Bayjournal.com. https://www.bayjournal.com/article/maryland_virginia_ease_crabbing_limits. Accessed 28 Jun 2019

Wikipedia (2019) *Callinectes sapidus*. In: En.wikipedia.org. https://en.wikipedia.org/wiki/Callinectes_sapidus#cite_note-Maryland-2. Accessed 20 Apr 2019

Williams A (1974) The swimming crabs of the genus *Callinectes* (Decapoda: Portunidae). Fish Bull 72:685–798

Williams A (1984) Shrimps, lobsters and crabs of the Atlantic coast of the Eastern United States, Maine to Florida. Smithsonian Institution Press, Washington DC

Wiso's Crabs (2019) Wiso's crabs: crab prices. In: Wisoscrabs.com. http://wisoscrabs.com/wordpress/crab-prices. Accessed 14 Jul 2019

The Chesapeake Bay Oyster: Cobblestone to Keystone

Kimberly S. Reece, Eugene M. Burreson, Deidre M. Gibson, Sierra S. Hildebrandt, and Ileana Fenwick

Oh, to be an oyster, slippery and slimy
Satisfying sailors, gritty and grimy.
Sitting at the bottom of the Chesapeake Bay
An eco-wonder of the past and today.
No two look alike, these oysters of the Bay
Each one attached to something hard and growing in its own way.
Rarely are they alone
Building big oyster cities to call home.
Too many oysters were taken for people to eat
And for the health of the Bay this meant defeat.
But with some work and luck,
Oysters are coming back to help clean up the muck.

Abstract

Throughout human history there are few foods that have been at the center of as much conflict, intrigue, murders, passionate emotions and poetry as the humble oyster. The history of the oyster in the Chesapeake Bay region includes a boom and bust period and is peppered with many colorful and fascinating anecdotes. It was a primary food source for Native Americans and early European settlers, and the targeted species of a thriving fishing industry in the 1800s during which over exploitation and competition resulted in "oyster wars" fought on several fronts. Today the oyster is the focus of a rapidly growing aquaculture industry and is recognized as a keystone species in the Bay providing important ecological functions such as water filtration and habitat through reef formation for a diverse community of Bay organisms including juvenile fish and crabs. This has motivated oyster restoration efforts led by both governmental agencies and non-profit organizations. Future challenges for the Chesapeake Bay oyster populations and aquaculture industry include multiple stressors; disease, changing climate, hypoxia, and harmful algal blooms.

Keywords

Oyster restoration · Oyster disease · *Crassostrea virginica* · *Crassostrea ariakensis* · Oyster aquaculture · Oyster wars

Distribution and Life Cycle Eastern oysters, *Crassostrea virginica*, are native to the US East and Gulf coasts and likely have been consumed since the earliest inhabitants arrived on the land along the banks of Chesapeake Bay and its tributaries (see chapter "The Algonquin Food System and how it Shaped the Ecosystem and Interactions with the English Colonists of the Chesapeake Bay"). Oysters are thought to have first colonized Chesapeake Bay 6500–8000 years ago as sea level rise from melting glaciers flooded the Susquehanna River Valley (Bratton et al. 2002).

Components of the oyster life cycle coupled with the shallow waters and tidal cycles of Chesapeake Bay are largely responsible for the success of this species in the region. Oysters are sequential hermaphrodites, beginning

K. S. Reece (✉) · E. M. Burreson
Department of Aquatic Health Sciences, Virginia Institute of Marine Science, Gloucester Point, VA, USA
e-mail: kreece@vims.edu; gene@vims.edu

D. M. Gibson · S. S. Hildebrandt · I. Fenwick
Department of Marine and Environmental Science, Hampton University, Hampton, VA, USA
e-mail: deidre.gibson@hamptonu.edu; sierra.hildebrandt@my.hamptonu.edu; ileanaf@email.unc.edu

B. E. Cuker (ed.), *Diet for a Sustainable Ecosystem*, Estuaries of the World, https://doi.org/10.1007/978-3-030-45481-4_8

Fig. 1 Watercolor depicting Native Americans eating oysters around a fire with a dugout canoe in the background that was the typical style of vessel used for fishing. Piles of discarded shells (i.e. middens) have been dated to as early as 2500 BC indicating that oysters were a staple food for thousands of years. (Image from VIMS archives and in the public domain due to age)

their adult lives as male, and switching to female as they grow and add sufficient energy reserves to support the production of eggs. These oysters are broadcast spawners. Females release millions of eggs into the water column where they are fertilized by sperm released from males. Spawning is cued by warming waters (≥20 °C) and by spawning of adjacent oysters. Fertilized eggs drift in the water column and the larvae develop cilia that allow them to move vertically in the water column. However, larvae are at the mercy of tidal currents and can be dispersed widely from the fertilization site. After about 2 weeks larvae congregate near the bottom and with an appendage called a foot search for a hard substrate on which to attach. That hard substrate usually is another oyster's shell, especially historically when there was no artificial hard substrate in the Bay such as dock pilings. The newly settled oysters are called spat. This is how oyster reefs develop vertically over time as more and more young oysters settle on them and grow. Vertical growth is limited by water level at high tide. The oysters can survive desiccation during low tide but require hours of submersion during high tide each day to feed and exchange gases with the surrounding water. As documented later in the chapter, this preference of larvae for settling on oyster shell also contributed to the decline of oysters, as the shells were relentlessly mined off by harvesters for more than a century and not returned to the Bay, thus severely limiting larval settlement sites (Hargis and Haven 1999).

Food Source for Early Americans The earliest evidence of oyster consumption in the Chesapeake region dates to about 4500 years ago (Haven et al. 1978; Hargis and Haven 1988). Many locations along the banks of the Bay feature ancient oyster middens, which are piles of discarded shells, indicating oyster consumption by the earliest Americans (Fig. 1). Some middens unburied by archeologists contained millions of shells, providing evidence that Native Americans consumed oysters in large quantities. Dating of shells from middens at White Oak Point on the south shore of the Potomac River reveal that Native Americans returned to the same place to gather oysters for over 3000 years (Haven et al. 1978; Rick et al. 2016).

Capt. John Smith and his crew of English Colonists arrived at the mouth of Chesapeake Bay in 1607. Carrying only limited supplies with them from England, Smith paid close attention to the food offerings of the New World. He observed Native Algonquin people gathering oysters from very large reefs. The colonists made oysters a staple in their diets. Other food items could be more difficult to acquire or to grow consistently, but oysters were abundant. One of Smith's crew members reported that oysters "lay as thick as stones" as they explored the Bay in 1608 (Hawks 1842; Wharton 1957; Alford 1973). Oyster reefs in the sixteenth and seventeenth centuries were so extensive in the bay that they were a menace to wooden ships and were deemed hazards to navigation by a visitor to the colonies in 1701 (Wharton 1957; Alford 1973). The oyster population of the Bay continued to thrive for two centuries until the people's love of this delicacy facilitated its decline.

Harvesting Native Americans and early colonists most likely collected oysters by hand from shallow water by wading or from dugout canoes or small boats. Around 1700 long-handled oyster tongs were developed that are essentially opposing rakes operated in a scissor-like manner (Fig. 2).

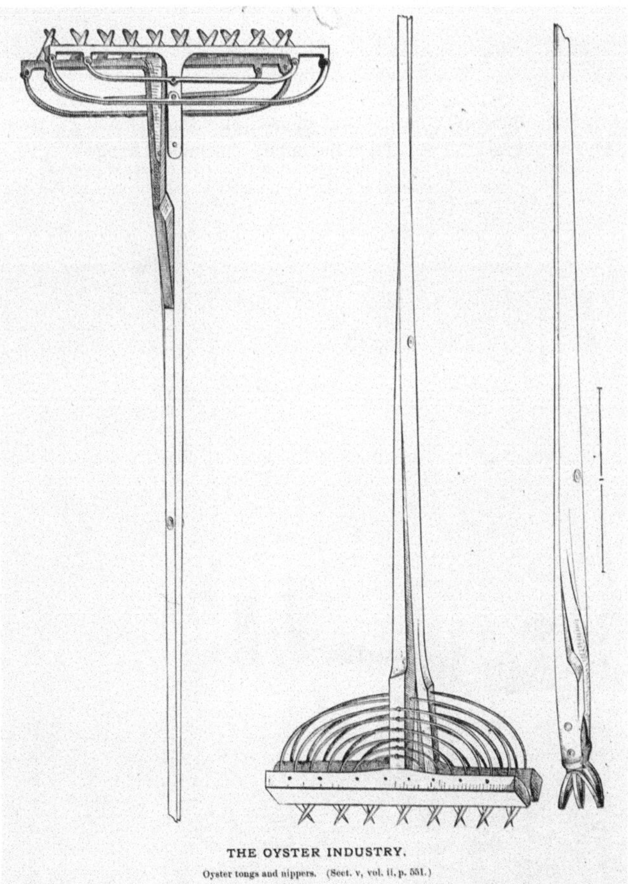

Fig. 2 Drawings of different types of oyster tongs (aka oyster rakes) used for harvesting. Drawing is from the book "The History and Present Condition of the Fishery Industries: The Oyster Industry", 1881 and in the public domain due to age

Patent tongs allowed harvest from slightly deeper waters, but they were an inefficient gear and there is little evidence of commercial activity at that time. Thus, the small human population around the Bay during the 1700s probably had little direct impact on oyster abundance. As the populations of Chesapeake Bay cities such as Norfolk, Richmond, Baltimore and Washington D.C. increased in the 1800s, so did the demand for seafood, including oysters. The advent of the railroads also opened up inland markets distant from the shores of the Chesapeake. In addition, oyster stocks were being depleted in New England waters, including those of Cape Cod Bay and Long Island Sound (Haven et al. 1978; Wennersten 1981). By the early 1800s the fishers from those areas started coming south to the Chesapeake looking for oysters (Ingersoll 1881). Some seafood brokers from New England set up business in Baltimore. Their use of dredges (Figs. 3 and 4) alarmed the locals. The dredges were efficient, but so damaged the reefs that they were banned or highly restricted by Virginia in 1810 and by Maryland in 1820. However, slack enforcement of the dredge laws meant the practice continued in protected waters, often under the cloak of darkness at night. From 1855 to 1860 about 80 schooners supplied New Haven, CT with 500,000–750,000 bushels of oysters each year from southern waters, either for direct shucking (75%) or for planting on leased ground (25%) (Wennersten 1981; MacKenzie 1996). Shucking is the process of removing the meat out of oyster and requires some skill with a knife. The oysters for planting went to replenish depleted local stocks.

From 1830 through the mid 1880s the oyster industry expanded significantly in Chesapeake Bay, especially in Maryland. Almost 117 million pounds (53 million kg) of oysters were harvested from the Bay in 1880, and there was the general belief that they were an unlimited resource (Alford 1973). Several factors contributed to the expanded harvest. Dredging was legalized again in 1865 in Maryland. Virginia subsequently also allowed dredging, but only from waters 6.1 m or deeper. Railroad lines to the Midwest from Baltimore were increasing and the development of the oyster canning industry in the 1840s made Baltimore the center of the Chesapeake Bay oyster trade (MacKenzie 1996). Numerous oyster packing houses were being built in Baltimore at the same time as the Baltimore and Ohio Railroad (Fig. 5). With the ability to can the oysters and then transport them across the country, there were no closed seasons and harvest was relentless. In the late 1860s, 9–10 million bushels were harvested in Maryland each year, two-thirds of which were shucked and canned.

In 1865 dredging yielded 3.6 million bushels of oysters in Maryland, and one million bushels in Virginia, while tongers harvested 1.2 million bushels in Maryland and almost one million bushels in Virginia for a total of 6.8 million bushels (one bushell = 35.2 L, Hargis and Haven 1988). Ten years

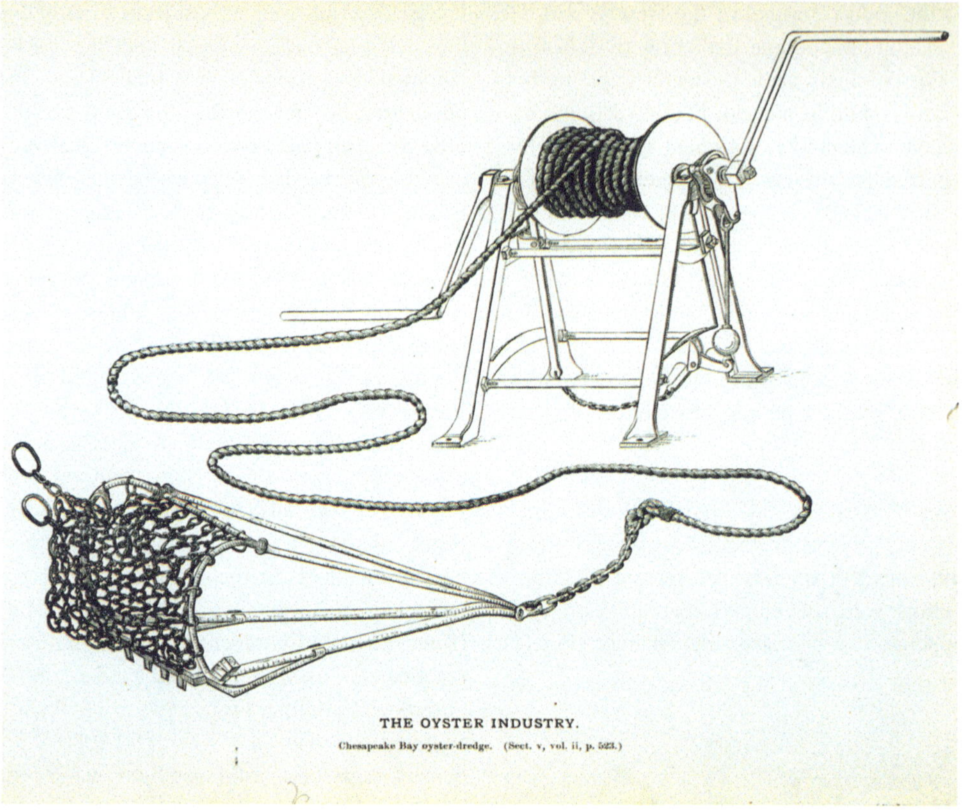

Fig. 3 Drawing of an oyster dredge that is towed along the bottom and over reefs to efficiently collect a large number of oysters. Drawing is from the book "The History and Present Condition of the Fishery Industries: The Oyster Industry", 1881 and in the public domain due to age

Fig. 4 Photo of a crew harvesting oysters with a dredge circa 1950. Manual dredges required several people to operate. Modern dredges are operated with electricity and some are hydraulic. (Photo: VIMS archives)

Fig. 5 Photo of a man in a Derby hat standing on the shell pile outside of the C. H. Pearson & Co. oyster canning company in Baltimore MD with the Aughinbaugh Canning Co. building in the background. A line of men with wheelbarrows full of oyster shells to deposit can be seen approaching (National Archives circa 1890. Photo is in the public domain due to age.)

later the total annual harvest in Chesapeake Bay more than doubled to 17 million bushels and continued to increase year after year until at least 1885 (Brooks 1891). This yield was almost entirely natural production as little oyster aquaculture or farming type production was done during this time period (Fig. 6 inset).

The state of Maryland set aside oyster beds for hand-tonging in the tributaries, with dredging allowed primarily on the deeper beds in the main stem of the bay. Dredgers, however, preferred harvesting in the quiet tributaries rather than the rough waters of the central Bay and poaching was common. Before 1865 oyster regulations were enforced by local sheriffs, but because of continued poaching by dredgers on tonging grounds, the state of Maryland established an oyster police force, the Oyster Navy, in 1868. Gunfire was often exchanged between hand tongers and dredgers, or the oyster police and dredgers. The conflict dubbed the "oyster wars" persisted for many years as a tumultuous sideshow to the oyster industry boom of the 1800s.

Oyster Wars Decades after the cannon and rifles of the US Civil War went silent, gunfire again erupted on the Chesapeake Bay. This time warriors clashed over access to lucrative oyster grounds. The oyster wars of Chesapeake Bay were fought on several fronts with many casualties on all sides. As mentioned above there were fierce conflicts between tongers and dredgers over harvest boundaries. Maryland's Oyster Navy and Virginia's Marine Police tried to patrol the tributaries to prevent illegal harvesting. They were severely understaffed, however, and lacked sufficient fire power and number of vessels to monitor all of the Bay tributaries needed to keep track of the unruly pirate dredge boats (Fig. 7) who would often work the reefs under the cover of darkness (Wennersten 1981). Enraged tongers would take matters into their own hands forming vigilante militias. They would keep 24-h watch on prime reefs from the shore and fire upon invading pirate dredge boats; cannons were even fired from shore in some places such as the Wicomico River. Of course, gunfire was returned, and battles ensued that involved several pirate vessels against most members of the local community with many men dying.

In addition, clashes between the Oyster Navy patrol boats and the pirate boats occurred frequently, as pirates were not intimidated by law enforcement and would stand their ground. In November 1888 pirate oyster schooners surrounded and shot at a police boat patrolling the Little Choptank River. It fled to Oxford Maryland, where the captain and crew disembarked and refused to go back out on patrol (Wennersten 1981). Wennersten also reports that after another encounter between a police boat and several pirate vessels on the Little Choptank, 14 dredgers boarded a docked police boat. The pirates forced the oyster police to take the

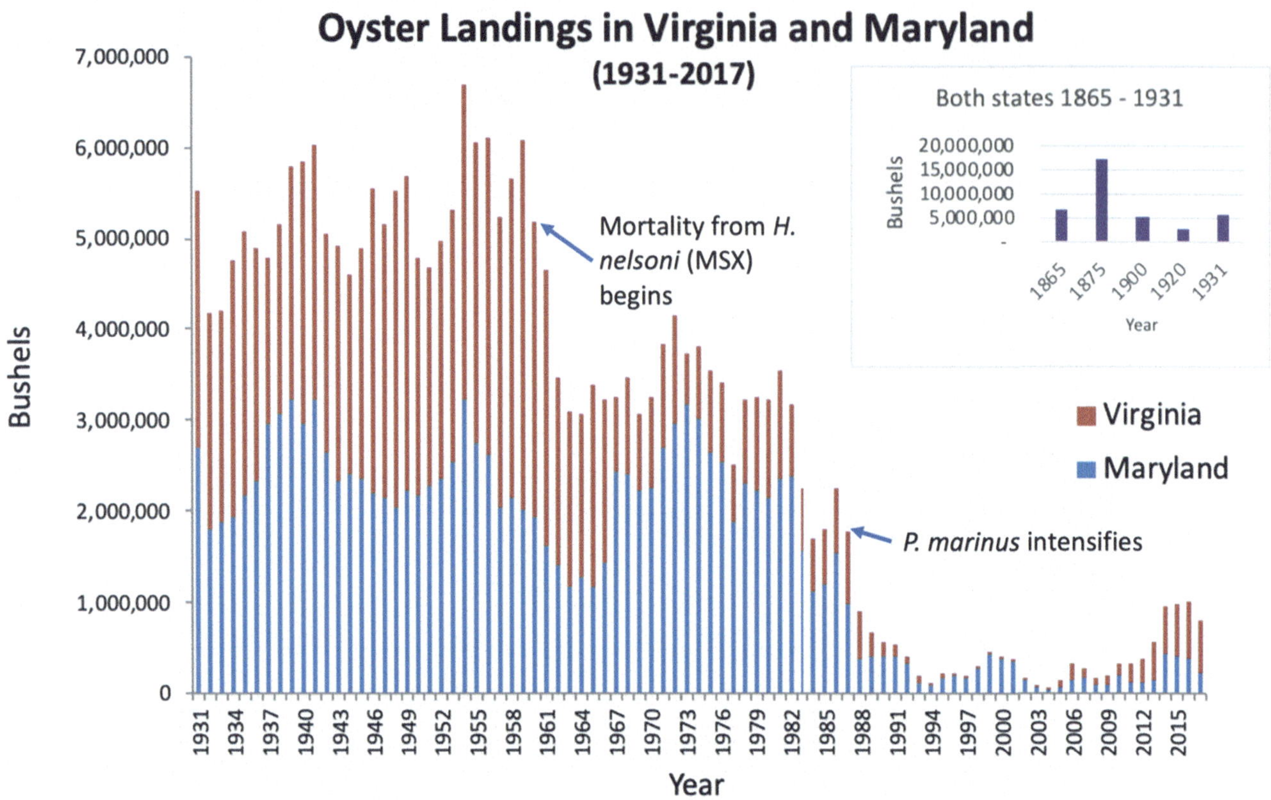

Fig. 6 Graph of oyster harvests (aka landings) from fishing and aquaculture from 1931 through 2017 in Maryland (blue portion) and Virginia (red portion). The times that MSX (*Haplosporidium nelsoni*) mortality began in the 1950s and *Perkinsus marinus* intensified in the late 1980s are indicated. A few data points of Chesapeake Bay harvests from 1865 to 1931 are shown on the inset graph. (Graph made by authors for this publication with data from the VA Marine Resources Commission)

boat back out on the water. The dredgers forced the captured police officers to illegally harvest oysters for several days. The pirates sent them ashore in a small rowboat. The large police boat was later found abandoned and stripped of its rigging many miles away.

In towns and villages along the Eastern Shore of Maryland and Virginia during the latter part of the nineteenth century, murders and other serious crimes including rape and arson were common and often not investigated. If they were, biased outcomes were not uncommon due to rampant corruption in law enforcement (Wennersten 1981). During the oyster boom period following the Civil War there was a high demand for workers who would be able and willing to do the grueling work required to harvest oysters. There was stiff competition among packers for skilled shuckers. Some unscrupulous dredge boat captains resorted to extreme measures to get their crew. During the oyster boom many Irish and German immigrants were arriving on US shores, often through Ellis Island. Shipping agents would recruit immigrants from New York and Philadelphia to sign documents for going to Baltimore to work on oyster dredge boats under extremely harsh conditions with scarce rations, and physical and psychological abuse. Unfortunately, mistreatment of workers was widespread, and some captains would even lock workers below deck at night to prevent escape. Unscrupulous recruiter/captain teams would resort to desperate measures including kidnapping of immigrants to bring them to Baltimore and load them onto ships essentially making them "oyster slaves." In 1884 after several days of torture, a dredge captain brutally murdered a German immigrant. After word reached lawyers and businessmen affiliated with the Maryland German Society in Baltimore, they worked to have the captain convicted of murder and get legislation passed to protect dredge boat workers. They were successful with the murder conviction; however, the legislators were reluctant to go up against the powerful oyster business. The German Society posted signs around New York, Philadelphia and Baltimore to warn the immigrants about the abusive situation on dredge boats. In 1886 a US marshal and deputies destroyed shacks in Dorchester County Maryland where shanghaied crewmen for dredge boats had been held. It was not until January of 1890, however, that a new maritime law was passed requiring harbor masters and seaport authorities to keep logs of dredge boat crews and their contracts for time and wages.

The oyster wars also involved conflicts between the watermen of Maryland and Virginia, as well as between the two states, which were involved in border disputes for centuries. There was disagreement about the boundary on

Fig. 7 Print of a wood engraving of oyster pirates dredging at night on Chesapeake Bay. (Harper's Weekly, March 1, 1884)

the Eastern shore in the Pocomoke Sound including the areas around Smith and Tangier Island's, as well as on the western side of the Chesapcakc along thc Potomac River. From the issuance of the Maryland Charter in 1634 that granted part of Virginia's land to the new state of Maryland, through to the time of a 1910 Supreme Court ruling that the low water mark of the Potomac River on the south bank was the Virginia Border with Maryland, there were several surveys and attempts at agreements to resolve to the satisfaction of legislators and citizens the state boundaries. The various border resolutions, however, were not usually accepted by whichever side felt that they had lost ground. Frequently watermen from the two states clashed over accusations of illegal removal of oysters each other's territory. Oyster poaching by the dredge boats from both the Maryland and Virginia sides continued to result in gun battles with law enforcement on the Potomac River as late as the 1950s.

Industry Impact on the Region The developing oyster industry had a huge impact on the Chesapeake Bay region. The economic opportunity created new wealth and led to the development of many small towns, especially on the Eastern Shore of Maryland. Crisfield, Oxford, and Cambridge became centers of oyster production with hundreds of shucking houses that provided employment for recently emancipated African Americans. In 1870 there were 563 vessels licensed in Maryland for dredging. The boats employed 2107 white and 1453 black workers. Smaller vessels and the land operations associated with oyster processing and transport employed 13,000 more workers in Maryland, primarily in Baltimore (Wennersten 1981). Although few women worked on the boats, many found employment in shucking and packing oysters (Fig. 8).

The increasing consumer demand for oysters led to the need for development of appropriate work boats that could carry the workers out to the reefs for harvesting. As mentioned above, traditionally Native Americans and early colonists had likely harvested oysters from dugout canoes or small boats (Fig. 1), possibly fitted with a small sail. The need for larger boats led to the development of the log canoe (Fig. 9). The hull of these vessels was made from three to nine shaped logs and was up to 35-feet (10.7 m) long. Fitted with sails they could cover a wider area than dugout canoes and could carry a much larger load of oysters; they were used primarily for tonging (Brewington 1963). Demand for oyster boats led to the development of shipyards throughout the region. Dredging for oysters, banned since 1820, was made legal again in 1865 in Maryland, but only by sail power. Log canoes were not powerful enough to pull a dredge, so larger log-hulled boats called bugeyes (Fig. 10) were developed. Bugeyes could pull a dredge but as logs suitable for their constructions became scarce and construction costs increased planked boats were developed (Brewington 1963). Eastern shore shipyards built the iconic skipjack (Fig. 11), which was produced by the thousands. Because of their huge sail area skipjacks provided the power to pull two oyster dredges at a

Fig. 8 Employees at a large oyster shucking house circa 1885. Note that a large number of the shuckers were women. (Photo: VIMS archives, in the public domain due to age)

Fig. 9 The Tilghman Island-style log canoe, Edmee S., built in the 1930s. (Photo from Wikimedia Commons, Author Acroterion, the file is licensed under the Creative Commons Attribution-Share Alike)

time (Vojtech 1993). The skipjack has become a symbol of Chesapeake Bay. The need for an engine-powered workboat for oyster tonging, fishing, and crabbing led to the development of the V-bottom workboat. The most popular was the deadrise workboat (Chowning 2007) (Fig. 12). Deadrise workboats have a shallow draft, a small forward cabin and a large open deck suitable for a variety of fishing efforts. They became extremely popular and are still the vessel of choice today for Chesapeake Bay watermen. Larger engine-powered buy boats were also developed to transport oysters

Fig. 10 The Bugeye, Edna E. Lockwood, built in 1889 for dredging oysters in Chesapeake Bay. (Photo from Wikimedia Commons, Author Acroterion, the file is licensed under the Creative Commons Attribution-Share Alike)

from tongers to leased beds or shucking houses (Chowning 2003).

Declining Harvest William K. Brooks, a Maryland oyster biologist, issued dire warnings of overfishing in Maryland in the 1890s as the oyster harvest declined drastically (Brooks 1891; Brooks and Paynter 1996). His warnings, and similar calls in Virginia, were largely ignored and relentless harvest continued, often with the reef-damaging dredges (Rothschild et al. 1994). By 1900 production fell to less than five million bushels a year in Maryland, prompting Brooks to issue his now famous 1905 statement: "We have wasted our inheritance by improvidence and mismanagement and blind confidence" (Brooks and Paynter 1996). Harvest continued to decline through the 1920s to about 2.5 million bushels per year in both states.

The depletion of the natural oyster reefs in Virginia in the late 1800s due to relentless harvesting caused concern. This prompted the Virginia Legislature to implement a survey in 1894. Lt. J. B. Baylor, who worked for the U. S. Coast and Geodetic Survey conducted a 2-year study of natural oyster beds, rocks, and shoals in Virginia. This massive survey located and mapped the ~250,000 acres (101,000 ha) of naturally productive oyster grounds throughout Virginia. These "Baylor" grounds have since been set aside for public shellfish harvesting and cannot be leased or used for other purposes. About 90,000 acres (36,000 ha) of Chesapeake Bay bottom in Virginia which are not part of the Baylor grounds can be leased from the State by private growers for shellfish harvesting/production.

Oyster harvest in Virginia was relatively stable in the early 1900s through 1958, varying between about 2.5 and 3.5 million bushels a year (Fig. 6). Harvest from the "public" beds, set aside for hand tong harvest after Baylor's survey, accounted for about 25% of the harvest. The remaining came from "private" grounds. Companies or individuals could lease unproductive bottom in the lower Bay from the Virginia Marine Resources Commission for a small fee. Small oysters were harvested from the "seed" beds in the upper James River by hand-tongers (Fig. 13). Because of the circulation gyre in the upper James River, oyster larvae are retained there, and settlement was high every year. However, low salinity hindered growth, so these oysters were hand-tonged, often by African American watermen, and transferred by buy boats (Fig. 14) to the leased grounds in the lower Bay, where salinity was higher and growth was rapid. Transplanted oysters were left on the leased grounds for 2 years to allow growth to market size. From the 1920s into the late 1950s about 700–800 boats were tonging seed oysters in the James River. As many as one million bushels of seed oysters a year were planted on leased grounds (MacKenzie 1996). This practice was highly successful and harvest from leased grounds accounted for about 80% of the production in Virginia by the late 1950s. Many of the leased grounds featured structures built on pilings, often a good distance from shore. These served as guardhouses to protect against poachers.

In 1925 there were 21 oyster-shucking houses in the Hampton Roads area. By the 1950s shucking was dominated by three large shucking operations. The J. H. Miles and Co., Inc. had 425 shuckers and other workers, the Ballard Fish and Oyster Co. had 275 and the J. S. Darling Co. had 75 workers, mostly African American men and women (MacKenzie 1996) (Fig. 15).

Disease Impact Everything changed in 1959 when high mortality of oysters occurred in the lower Bay from the protistan pathogen *Haplosporidian nelsoni*, commonly called MSX disease (Andrews 1966; Haskin et al. 1966). This pathogen had been reported 2 years earlier in Delaware Bay where it also caused high mortality (Haskin et al. 1966). The

Fig. 11 Fleet of oyster skipjacks in Baltimore Harbor circa 1890. (Photo: National Archives, in the public domain due to age)

Fig. 12 Photo of a deadrise workboat. These boats are still commonly used by oyster fishers and aquaculturists in Chesapeake Bay. (National Archives)

parasite *H. nelsoni* was introduced to the US from Asia (Burreson et al. 2000) with devasting effects during the 1960s and 1970s. Harvest plummeted, especially from leased grounds as they were in the same high salinity areas where *H. nelsoni* thrived. Mortality of oysters on leased grounds in the lower Bay averaged 80% over the 2-year period required for growth to market size and these grounds were not replanted after 1960 (Andrews 1984). Harvest declined from a peak of just over four million bushels a year in 1959 to around one million bushels a year by 1967 (Fig. 5). Oyster harvest in Maryland was not affected so dramatically because salinity is lower and the parasite was not common except in

Fig. 13 Oyster tongers in the James River, VA circa 1952. (Photo: VIMS archives)

Fig. 14 Harvesters transferring oysters to the buy boat in the James River, VA. (Photo: VIMS archives)

drought years and Maryland experienced excellent spat sets during the 1960s. Harvest in Virginia varied between 0.5 million bushels a year and one million bushels a year through 1988; harvest in Maryland actually increased during the period and varied between two million and three million bushels per year. The Virginia harvest during this period was split about evenly between oysters collected from private and public beds (Fig. 16).

Because of four consecutive drought years in the mid 1980s, another protistan pathogen, *Perkinsus marinus*, intensified beginning in 1988. This pathogen had been observed in Chesapeake Bay as early as the 1950s but caused

Fig. 15 Shucking house worker using a wheelbarrow to move oysters off the back of a truck and into the plant for processing. (Photo: VIMS archives)

minimal mortality until salinities increased because of the prolonged drought (Burreson and Andrews 1988). During the drought period *P. marinus* invaded all oyster beds in Maryland, while in earlier years the impact had been limited to more southern Maryland mainstem Bay reefs and those in the lower reaches of the branching tributaries. Because of high mortality from both *H. nelsoni* and *P. marinus*, oyster harvest in both Maryland and Virginia declined from 750,000 bushels per year in 1988 to less than 100,000 bushels per year by 1994. Maryland oysters did better than Virginia from 1995 to 2002, during which Virginia landings ranged between 17,500 and 55,500 bushels. Nonetheless, there were several drought years around the turn of the century and the oyster beds in Maryland were hard hit with the diseases. Maryland harvests were less than 100,000 bushels from 2003 to 2005 (Fig. 6). During drought years *P. marinus* also invaded the James River seed oyster beds in Virginia (Burreson and Andrews 1988), which were located in the upper portion of the river that had typically not been impacted by the parasitic diseases. Any oyster transplanted from the seed beds to leased grounds in higher salinity soon died. This realization essentially ended the practice of transplanting small James River oysters to leased grounds anywhere in Virginia waters. Therefore, harvest remained at or below 50,000 bushels per year in Virginia for over 10 years from 1994 to 2004.

Restoration: Importance of Reefs and Shell Although oyster reefs may have been a hazard to ship navigation in the colonial era, they also provided valuable ecological services to the Bay that were likely unrealized at that time. Oysters are filter feeders, removing particles from the water column, digesting phytoplankton and other food items, and depositing everything else on the bottom in the form of feces or pseudofeces. An adult oyster can filter over a gallon of water an hour (Newell 1988) so large oyster reefs are capable of cleaning large volumes of water. Some researchers have estimated that the oyster population in the early 1800s could filter the entire volume of Chesapeake Bay in about a week. With current oyster populations, it would take almost a year to filter the same volume of water (Newell 1988). Oyster reefs also provide habitat for a variety of other organisms important to the Bay ecosystem, particularly for larval fish and crabs (Peterson et al. 2003).

Up until the 1990s most shells from shucking houses were deposited in huge piles nearby (Fig. 17). Not only did harvesters remove the oysters, but they removed the most important settlement habitat as well. Oyster shells have long been used for a wide variety of purposes. Because oyster larvae prefer to settle on other oysters, this severely impacted the substrate available for settlement. Starting with the colonial days oyster shells were used to pave roads, as they provided a hard surface protecting against the formation of deep ruts in the mud formed by the pounding of horses' hooves and the grinding of wheels on the carts and carriages. In later years they were also used for railway embankments, ballast for vessels, a component of cement, chicken "grit", and for making lime. Much of the town of Crisfield, MD was built on wetlands filled with oyster shell. Oyster shells were

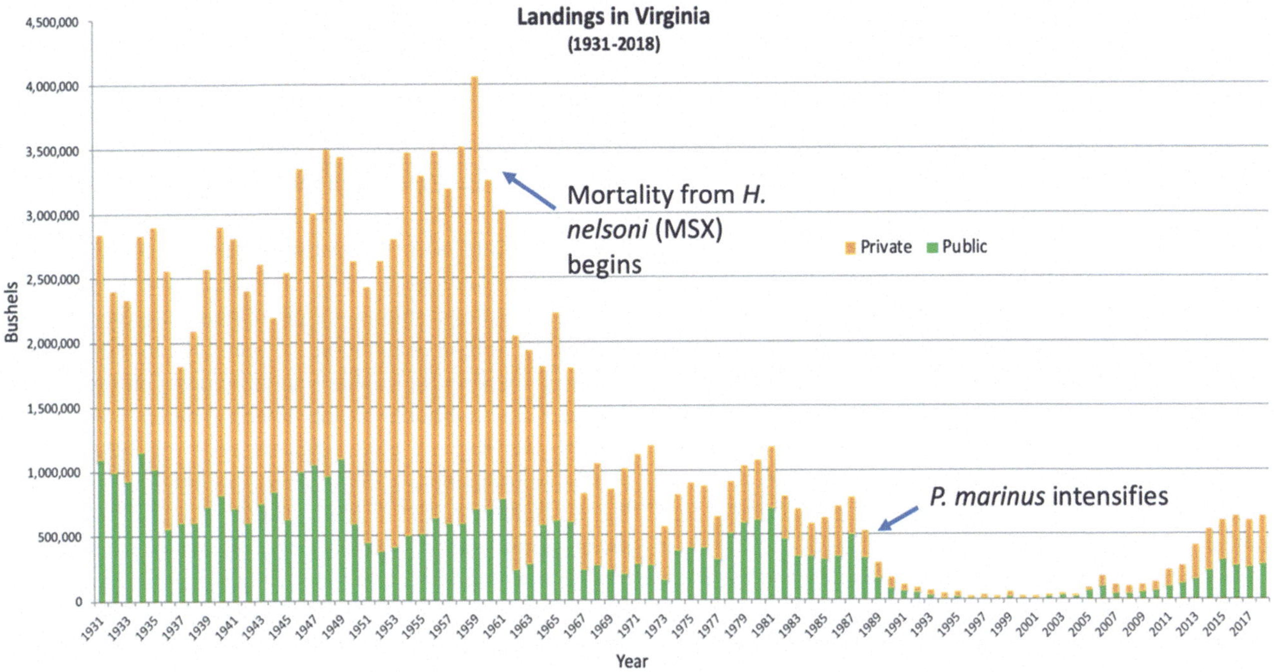

Fig. 16 Graph of Virginia harvest (aka landings) from 1931 to 2018 shown as the proportion from public grounds (green) and private leases (orange) each year. (Graph made by authors for this publication from VA Marine Resources Commission data)

Fig. 17 Large oyster shell pile outside of a shucking house in 1949. Most shells were not returned to the Bay but were used for other purposes such as making cement and lime, as chicken grit or vessel ballast and for railway embankments. (Photo: VIMS Archives)

burned to make lime, often without being shucked first, wasting the meats (MacKenzie 1996). Not only were shells from shucked oysters not returned to the Bay, the shell from long dead oysters that formed the base of reefs was also mined for these other uses, thereby destroying natural reefs that had developed over thousands of years (Hargis and Haven 1999). In 1921 there were 54 shell-processing plants nationwide; 29 (53.7%) were in the Chesapeake Bay region, 18 in Maryland and 11 in Virginia (U.S. Dept of Commerce Report for 1922). From 1920 to 1944 Virginia and Maryland combined produced over 2,777,000 tons of shell byproducts (Hargis and Haven 1999). Of them 1,555,000 tons were in the form of poultry grit, and 1,215,000 tons as ground or burnt lime. There are no adequate conversion programs to allow determination of the number of bushels of shell required for each ton of shell by-product, but it was clearly a very large number.

As discussed below, restoration rationale was originally directed primarily towards increasing oyster harvest. Later, rationale included a growing body of evidence that oyster reefs provide ecological services including water quality improvement by filtration and habitat provisions that support biodiversity (Luckenbach et al. 1999). These subsequent ecologically focused restoration efforts often involved programs to return shucked shell to oyster beds in recognition that substrate was important for their success.

Restoration: Non-native Oysters One of the proposals to restore oyster populations in Chesapeake Bay involved the introduction of non-native oysters. Two species were seriously considered, *Crassostrea gigas* and *C. ariakensis*, both native to Asia. *Crassostrea gigas*, commonly known as the Pacific oyster, is now widely cultured in Europe, along the west coast of the United States, and in Australia and New Zealand where it has been introduced successfully to all of these regions to replace depleted native stocks. *Crassostrea ariakensis,* called the Suminoe oyster, was also considered because it is an oyster that can tolerate moderate salinities that are more comparable to those found in Chesapeake Bay as opposed to the Pacific oyster, which is typically grown in higher salinity waters. In 1995 The Virginia Institute of Marine Science began research on both species under the Commonwealth of Virginia-mandated Rational Plan for Testing Application of Non-Indigenous Oyster Species, House Joint Resolution 450. The goal was to provide a science-based foundation for decisions concerning the introduction of non-native oyster species. Central to this research was field testing of non-native oysters in Chesapeake Bay, although these field studies were extremely controversial because non-native oysters were being placed into the waters. Genetically altered triploid individuals were used to greatly reduce the risk of a spawning event and natural colonization of the non-native species A strict biosecurity policy was put into place which included using non-reproductive triploid oysters to greatly reduce the risk of a spawning event and natural colonization of the non-native species (Allen and Burreson 2002). There was still concern, however, about the possibility of reproduction and unintentional introduction. Growth, survival and disease susceptibility of both non-native species in comparison to *C. virginica* were assessed in the Chesapeake Bay region. During these field studies *C. gigas* was relatively quickly eliminated from consideration for introduction. Although it performed well in high salinity environments on the seaside of the Eastern Shore, at the lower salinity sites within Chesapeake Bay *C. gigas* did not grow or survive well compared to the native *C. virginica.* Although *C. gigas* was less susceptible to the diseases (Calvo et al. 1999), perhaps the most important reason that it was eliminated from consideration was that the oyster industry members were unenthusiastic about the species. They felt that the taste was of lower quality and it had a poor shelf life compared to the native oyster, and its thin, brittle shell made it difficult to shuck. These issues, coupled with its poor growth performance at Bay salinities, led to *C. gigas* being eliminated from consideration for an introduction.

Similar field studies to those done with *C. gigas* were conducted with *C. ariakensis* side-by-side with the native *C. virginica* (Calvo et al. 2001; Kingsley-Smith et al. 2009). At the end of the Calvo et al. (2001) study, all *C. virginica* at the medium and high salinity sites had died from disease. Prevalence of *P. marinus* was low in *C. ariakensis* (0–28%) and intensity of infection was also low, while prevalence in *C. virginica* was 100% and many infections were moderate to high intensity. No *H. nelsoni* infections were found in *C. ariakensis* although they were prevalent in *C. virginica*. Growth and survival were higher in *C. ariakensis* at all salinities (84–87%) compared with *C. virginica* (0–19%). The Virginia Seafood Council conducted taste tests on *C. ariakensis* and found it tasted similar to *C. virginica*. At that time oyster industry members were extremely enthusiastic about *C. ariakensis* because it showed promise by growing fast and surviving well at Bay salinities and resisting local pathogens. In addition, it has a thick shell comparable to that of *C. virginica* making shucking similar to *C. virginica*.

A later field trial by Kingsley-Smith et al. (2009), however, was not quite as promising. The native oyster demonstrated significantly higher survival grown intertidally than the Suminoe oyster. In addition, although the disease levels were lower in the non-native and it did grow more quickly, there were no significant differences in survival between *C. ariakensis* and *C. virginica* at the lower salinity sites in Maryland within Chesapeake Bay. In addition, in subsequent field studies of triploid *C. ariakensis* in Bogue Sound, North Carolina the non-native oysters experienced high mortality and were found to be heavily infected with a previously unknown local parasite in the genus *Bonamia* (Burreson et al. 2004). There was fear of the spread of this pathogen into Chesapeake Bay if *C. ariakensis* was introduced. In addition, there were other concerns including the possible introduction of pathogens from Asia such as a herpes virus that the Suminoe oyster was found to carry (up to 40% prevalence) (Moss et al. 2007). Although the broodstock (i.e. reproductively capable) animals from Asia were held under strict quarantine conditions and were never to be introduced directly into local waters, there is the possibility of some aquatic pathogens, particularly viruses, to be vertically transmitted from parents to offspring (Brock and Bullis 2001; López Sanmartín et al. 2016; Bang Jensen et al. 2019). As mentioned above all non-natives introduced into the Chesapeake Bay region were to be triploid animals with three sets of chromosomes with low risk of reproducing. However, it was known that at a low frequency, triploids would undergo some reversion losing some of the third set of chromosomes becoming mosaics composed of both diploid and triploid cells. In addition, there was some rare gonadal and germ cell development in triploids raising some concerns about reproductive capabilities (Hallerman et al. 2001; Allen and Burreson 2002). Other studies such as one by Bishop and Peterson (2005) found that *C. ariakensis* was very susceptible to *Polydora* sp., mud worm infections that resulted in ugly shell blisters making the oysters very unappealing and unsuitable for the half-shell market. Risks associated with the pathogens along with the increasing focus on native aquaculture ultimately led to less interest from the growers. In their Environmental Impact Statement for oyster restoration in the Bay, The Army Corps of Engineers (ACOE) along with the states of Virginia and Maryland concluded that they would drop from consideration the introduction of the non-native oyster and would instead focus efforts on restoring the native oyster and developing aquaculture (US ACOE Norfolk 2009).

Restoration: Native Oysters In the Virginia waters of Chesapeake Bay, a form of oyster restoration was occurring as early as 1928 under the Oyster Repletion Act which was aimed at restoring "impoverished bottoms" in support of the fishery (Bryant and Pruitt 2007; Virginia Commission of Fisheries 1928). Shell was placed on public grounds to provide substrate for oysters to set and in many cases these grounds were planted with wild "seed" oysters moved from other locations; often from the James River. The Virginia Marine Resources Commission (VMRC) continues the practice of shell planting and seed transplantation to this day.

While early oyster restoration efforts were conducted entirely by the states for economic benefit, the more recent efforts are focused also on ecological benefits with the realization of the ecological importance of healthy oyster populations for improving water quality and habitat. As oyster harvest declined dramatically in the early 1990s, many attempts to limit harvest and initiate restoration projects aimed at preserving the oyster as an environmental resource were made in both Virginia and Maryland, but there was substantial industry and political opposition that greatly reduced their impact (Wesson et al. 1999). In 1990, Virginia Governor Wilder convened a Blue Ribbon Oyster Panel and a similar panel, the Maryland Oyster Roundtable, was established in Maryland in 1993 (Paynter 1999). The primary goal of both groups was to develop plans to not only restore the oyster resource, but also the oyster industry.

For years, researchers had stressed the importance of maintaining natural oyster reef height and cultch, but their advice was not heeded (Haven et al. 1978; Hargis and Haven 1988; Rothschild et al. 1994). Reef restoration was a major recommendation of the Blue Ribbon Oyster Panel (Bryant and Pruitt 2007) and in 1994 Governor Allen of Virginia signed the Chesapeake Bay Aquatic Reef Plan and the Oyster Fishery Management Plan, both of which called for the creation of 5000 acres of oyster reef habitat during the next 5 years. Two projects in the Piankatank River were implemented beginning in 1993 to investigate the value of reef structures for oyster survival and the methods by which reefs were constructed. Approximately 207,000 bushels of oyster and clam shells were deployed in the river as substrate (Luckenbach et al. 1999). The Blue Ribbon Oyster Panel also recommended establishing oyster sanctuaries in other tributaries (Bryant and Pruitt 2007). Subsequently, reefs also were constructed in the James and Lynnhaven Rivers. Following the panel's recommendation, one of the most important management decisions in Virginia was to close most areas to harvest and to restrict harvest in the remaining areas. A rotational harvest system (see below) with sanctuary areas was implemented, harvest seasons were shortened considerably and a 3″ cull limit was imposed.

Between 1999 and 2001 the state built 14 one-acre sanctuary reefs and stocked them with broodstock in the Rappahannock River under the Coastal Zone Management Program. Additionally, 500 acres of two-dimensional bars were constructed. Political pressure from oyster fishers to allow harvest on these reefs led to a management solution that balanced the ecological objectives and economic opportunity. The 500-acre harvest areas were divided into six sections and two sections were opened to harvest every 3 years, allowing oysters to grow to market size on the closed areas. In addition, the largest oysters harvested were required to be deposited on the sanctuary reefs to provide broodstock to support the harvest areas. This rotational harvest program is still in existence today. All reefs that are built in Virginia today are sanctuaries closed to harvest.

In Maryland the Oyster Roundtable Action Plan recommended seven areas for action: (1) Disease Management; (2) Habitat Restoration; (3) Hatchery Production Increase; (4) Aquaculture Enhancement; (5) Enhanced Research; (6) Oyster Recovery Areas; (7) Oyster Recovery Partnership (Paynter 1999). Oyster Recovery Areas were a critical part of the Action Plan and were designed to secure areas where aquaculture and rehabilitation pilot programs could be established. Three zones were established. In zone A harvest was prohibited as was planting of disease-infected seed oysters. In zone B harvest was allowed but planting of infected seed was prohibited. In zone C there were no restrictions beyond normal regulations. Maryland currently has 24 zone A sanctuaries including the entire Severn and Magothy Rivers.

As restoration efforts were shifting from the sole industry focus to increasing the oyster population for its ecological benefit, the general public became more aware and supportive of restoration. In turn, this attracted a wide variety of collaborators to oyster restoration projects including federal and state agencies such as the EPA, NOAA, the ACOE, the Maryland Department of Natural Resources, Maryland Department of the Environment, VMRC, and Virginia Department of Environmental Quality among others. There were tensions over the years, however, among the various entities involved regarding the emphasis on restoration priorities, that is economic or ecologic, and there was often pressure from the industry to open restored reefs to fishing (Luckenbach et al. 1999; Schulte 2017).

Scientific guidance has been provided over the years by scientists from the University of Maryland (UMD) and the Virginia Institute of Marine Science (VIMS). The UMD Horn Point oyster hatchery has provided billions of spat on shell for Maryland sanctuaries and spat for planting in Virginia has come from the Aquaculture Genetics and Breeding Technology Center at VIMS and from private hatcheries. Several non-profit organizations also have been involved in oyster restoration including the Chesapeake Bay Foundation and the MD Oyster Recovery Partnership, as well as many community groups focused on restoration of specific rivers, such as Lynnhaven River NOW.

In addition to the recommendations made by the Blue Ribbon Oyster Panel in the 1990s, the Virginia Marine Resources Commission established longer range goals to achieve no net loss of oyster standing stock over the next 5 years, and to achieve a doubling of the existing standing stock over the next 10 years. In reality, the depleted state of the oyster population in Virginia suggested that recovery would be extremely slow and limited to those areas where there were enough oysters for successful reproduction.

The ACOE got involved in oyster restoration in an effort to reduce pollution in the Bay and help meet standards of the Clean Water Act. Initial efforts were similar to Virginia's approach of "repletion", adding new shells and in some cases also seed to public oyster beds. By 2009 the ACOE concluded that Virginia's approach was not cost-effective because subsequent harvesting removed oysters and damaged beds as fast as they were 'replenished' (US ACOE Baltimore 2009). Essentially, it was a "put and take" fishery with little benefit to the ecosystem. They moved to promote restoration of only sanctuary areas and to approve alternative substrates for reef restoration in Maryland waters including clam shell, concrete, brick and cinderblock due to the shortage of oyster shell.

The conflict between the Virginia state and Federal restoration priorities of restoration for the fishery and ecosystem escalated in 2011 over repletion of reefs in the Great Wicomico River in Virginia. The ACOE and Virginia had cooperated since 2004 on creating the state's largest artificial reef. These sanctuary reefs were off-limits to harvest and in 2011 Virginia decided that it would not contribute its 25% cost share unless some acreage was open to harvest. Federal rules allowed restoration funding only for sanctuaries because of their ecological benefits, not for harvest areas. Thus, the ACOE spent only $600,000 of the planned $2.5 million for the Great Wicomico River that year. Virginia also rejected $200,000 for construction of new oyster reefs in the Lafayette River because that river is closed to harvesting due to bacterial pollution and reefs would provide no economic opportunity for watermen.

In the 2014 Chesapeake Bay Watershed Agreement, Virginia committed to restoring oyster reefs in the Lafayette, Lynnhaven, Piankatank, lower York, and Great Wicomico Rivers by 2025 (Chesapeake Bay Program 2014). As of 2018, 80 acres of restored reefs have been established in the Lafayette River and 91 acres were considered restored in the Lynnhaven River. In the Piankatank River there is a goal of having about 438 acres restored with 278 considered restored as of 2018. About 85 acres in the Great Wicomico River were still thought to have maintained sufficient reef height from the restoration efforts earlier in the twenty-first century, although additional surveys were planned. Only minimal oyster restoration work had occurred in the lower York River by 2018 (US ACOE Baltimore 2009).

Maryland has conducted an oyster repletion program since 1960 that involves transplanting small oysters from disease endemic areas to lower salinity where survival is higher. The drawback to this system is that diseases are moved into low salinity areas where they may not have previously existed (Paynter 1999). In the 2014 Chesapeake Bay Watershed Agreement, Maryland committed to restoring oyster reefs in Harris Creek, the Little Choptank River and the Tred Avon River. As of 2018 a Maryland Oyster Restoration Interagency Workgroup has been formed to address restoration efforts in

Maryland. This workgroup includes the US Army Corp of Engineers, Baltimore District, NOAA, the Oyster Recovery Partnership and the MD Dept of Natural Resources. By 2018, 350 acres were restored in Harris Creek, 339 acres in the Little Choptank River and 84 acres in the Tred Avon River.

The Chesapeake Bay Foundation has established Oyster Restoration Centers in Maryland and Virginia. They grow and plant seed oysters on man-made reefs. The Chesapeake Bay Foundation has been instrumental in developing the Save Oyster Shells program to acquire shucked oyster shell from restaurants in Virginia and Maryland. The shells are cleaned and cured for distribution on oyster reefs to promote spat settlement. At this time, most restored reefs in both Maryland and Virginia are sanctuaries where harvest is prohibited. Even with all the restoration efforts, however, commercial landings from the fishery have not dramatically increased in the last decade, as the wild market size oyster population does not seem to have increased substantially based on surveys from the Virginia portion of the Bay (VMRC, VOSARA data, http://cmap2.vims.edu/VOSARA/viewer/VOSARA.html). Much of the increase in harvest is resulting from the aquaculture sector.

Overall, the oyster restoration in Chesapeake Bay is progressing, whereas before 1990 very little had been accomplished (Luckenbach et al. 1999). Effective restoration, however, has proven to be somewhat more difficult, costly and contentious than anticipated (US ACOE Norfolk 2009; US ACOE Baltimore 2009; Schulte 2017) Nonetheless there is a commitment on the part of both the state and federal agencies (Chesapeake Bay Program 2014), as well as many NGO's (e.g. CBF, Lynnhaven NOW, Keith Campbell Foundation) to continue supporting oyster restoration in recognition of the essential role the species plays in the overall health of the Bay.

Aquaculture An early form of oyster "farming" or culture in Chesapeake Bay can be traced to when people realized that by returning shell to the water, they were more successfully capturing wild oyster spat on sites from which they would later harvest. This is still referred to as "traditional" aquaculture in the region. They also began moving seed oysters from more productive areas, such as public grounds in the James River, VA, in the late nineteenth century to private lease sites. These practices became even more common in the twentieth century (Hargis and Haven 1988). In the latter part of the twentieth century oyster culturing began to use genetically improved strains of oysters from hatcheries for large-scale planting on bottom (spat-on-shell) or in containerized gear.

Currently, most growers use disease-resistant triploid oysters in Chesapeake Bay. Historically selective breeding for oyster aquaculture in the Bay was focused on developing fast growing, disease resistant oysters that could tolerate a relatively wide range of salinities as is seen in the Bay region (NOAA 2019). Triploid oysters generally grow faster than diploids (Stanley et al. 1984; Baker and Mann 1991; Dégremont et al. 2012) and are generally more appealing for the profitable half shell market year-round in the US because they have little to no gonadal development. About 87% of oyster growers in Virginia reported planting triploids in 2017 (Hudson 2018).

After the initial devasting impacts on oyster populations by MSX in the late 1950s through the 1970's resistance to the parasite developing in wild populations (Haskin and Ford 1979). This was promising data and encouraged targeted selective breeding to develop oyster lines that could resist disease, particularly the introduced parasite, *H. nelsoni*. Scientists at Rutgers University developed a line in the 1970s that demonstrated increased survival over susceptible wild oysters and they found that mortality with MSX exposure decreased with each generation after three generations of selection (Haskin and Ford 1979). Selective breeding efforts were started in the 1990s at VIMS to develop lines with resistance to both parasitic diseases motivated by the intensification of *P. marinus* in the late 1980s and the fact that many areas of lower Chesapeake Bay were affected by both parasites (Ragone Calvo et al. 2003). After four generations of selection VIMS' DEBY oysters showed higher growth rate and survival along with lower prevalence of both diseases compared to the progeny of wild Chesapeake Bay broodstock. By the third summer in the field, the selected DEBYs suffered substantial mortalities, so that although they could tolerate the pathogens for a few years, they were not demonstrating true resistance (Ragone Calvo et al. 2003). While selection for MSX was relatively straight-forward (Haskin and Ford 1979; Ford and Haskin 1982), selecting for *P. marinus* tolerance or resistance had proven to be more complex (Dégremont et al. 2012; Frank-Lawale et al. 2014). These earliest triploid, disease resistant lines from Rutgers and VIMS would become the foundation for many more lines with more diverse characteristics such as more specific salinity range optimums (see below) and increased tolerance to the diseases.

The oyster aquaculture industry in Chesapeake Bay region was barely beginning in the late 1990s and has developed much more rapidly in the twenty-first century (Hudson 2018). One of the primary recommendations that came out of the decision to not introduce *C. ariakensis* to Chesapeake Bay, was for there to be federal and state support for development of aquaculture of the native oyster (US ACOE Norfolk 2009; Bosch et al. 2010). There are now state and federally supported research and extension programs for the growing shellfish aquaculture industry in both Maryland and Virginia. NOAA Sea Grant Programs in both states include promotion of shellfish aquaculture in their mission statements. The UMD Horn Point Hatchery provides oysters not only for restoration, but also produces eyed larvae and seed that aquaculturists can purchase. In 1997 the Virginia General Assembly passed an initiative that established the Aquaculture Genetics and Breeding Technology Center (ABC) at VIMS whose mission was to be "...to foster the development of breeding technologies to promote increased yields, improve disease resistance, and to make available genetically-improved broodstock for the aquaculture industry." The ABC program has developed a set of disease-resistant selected lines for different salinity regimes in the Bay that are available for hatcheries. These lines are demonstrating improved disease tolerance and provide more choices for the industry (Frank-Lawale et al. 2014; Proestou et al. 2016). Private hatcheries pay a nominal fee for diploid and tetraploid broodstock from lines of their choosing. They can then propagate either diploid or triploid eyed larvae, cultchless seed or spat-on-shell that are in turn sold to growers.

Aquaculturists in Chesapeake Bay use two main production methods, extensive and intensive, for growing oysters depending on the desired market. Extensive culture is also referred to as remote setting or spat-on-shell. Oyster eyed larvae are purchased from the hatchery and transported to setting sites, struck on containerized oyster shells and ultimately planted directly on the bottom. Because spat-on-shell cultivation produces oysters grown in clusters (similar to wild caught oysters), the primary product is predominantly for shucking, or meat market, rather than for half-shell consumption. Intensive culture is the process of growing single oysters in containers for predator protection. This is a more costly and labor-intensive method; however, the end result is a consistent and high-quality product that has the ability to obtain a higher price in the boxed, or half shell market. For containerized culture, oyster eyed larvae are settled in a downwelling system where they are provided a substrate of finely crushed oyster shell, or micro cultch, to allow individual settlement. Each larva ultimately outgrows the fine particle it settled on, producing a single oyster. When these single oysters, also called spat, are roughly 1.0 mm they are transferred out of the hatchery where they have received a controlled diet of algal food into a nursery system of upwellers (Fig. 18) where they are fed on phytoplankton from the local waters by the flow-through system. Once spat reach a size of about a quarter to one half of an inch (6–12 mm), they are typically transferred into mesh bags and placed by growers at their leased sites either on racks (Fig. 19), or in off-bottom cages (Fig. 20) or floats (Fig. 21). Rack and bag gear is often used for growing oysters intertidally where they receive periodic air exposure. This air exposure can reduce the amount of biofouling that the gear and oysters accumulate during grow-out therefore decreasing the amount of husbandry required. Bottom cages are the most common subtidal gear used by Chesapeake Bay growers and also are used at higher wave energy intertidal lease sites. Some growers use floating bag systems to grow oysters in surface waters and like intertidal growth this has the advantage of being able to easily dry oysters intermittently reducing biofouling.

Fig. 18 Upweller nursery system for shellfish. They are commonly held in these systems to receive flow-through natural water to grow after they leave the hatchery and before they are placed into gear and out into the waterways. (Photo: Gail Scott, VIMS)

The oyster aquaculture industry in Virginia is the largest on the US East coast and in 2017 the farm gate value of oysters was about $15.9 M USD (Hudson 2018). Although Virginia's oyster industry is larger than that of Maryland, the number of farmed raised oysters in Maryland grew from about 3300 in 2012 to over 74,000 in 2017 when it was worth about $6 M USD (MD DNR https://www.cbf.org/document-library/presentation-webinar-materials/2019-aquaculture-roscher.pdf). In Virginia oyster tourism is being promoted through "The Oyster Trail", which includes growers who provide tours of their farms for visitors. The trail is also promoted with Virginia restaurants serving oysters, as well as wineries, breweries, distilleries and local artists. Maryland is

Fig. 19 Rack and bag grow out gear, which is commonly used in Chesapeake Bay to grow oysters intertidally. Oysters are place in mesh bags that are tied to racks. (Photo: VIMS)

Fig. 20 Off-bottom cages are commonly used in Chesapeake Bay for growing oysters either intertidally or subtidally. Smaller oysters are often placed in bags inside cages as shown here. (Photo: Gail Scott, VIMS)

also promoting oysters for tourism along with their famous blue crabs through the "Maryland Crab and Oyster Trail".

The shellfish aquaculture industry is expected to continue to grow in the Bay region, although there have been some impediments to growth due to complaints from the public about visual pollution from the associated gear and obstruction to boating and other uses of the waterways (Luckenbach et al. 2008). In addition, some concerns have been raised about potential negative environmental impacts including effects on water quality and damage to submerged aquatic vegetation (SAV) areas by intensive aquaculture operations (Bricker et al. 2014; Smaal et al. 2018; Maryland Oyster Restoration Interagency Workgroup 2019). Effects to SAV are particularly disturbing because like oysters, SAV is targeted for restoration in the Bay due to its ecological importance. SAV beds are similar to oyster reefs in that they provides habitat for juvenile fish and crabs (Williamson et al. 2015) and oysters have been promoted as improving water quality, so it is somewhat ironic that the oyster aquaculture is thought to potentially threaten SAV and threaten water quality by altering nutrient cycling.

Acknowledgement of the concerns has prompted several research efforts (e.g. Murphy et al. 2016; Orth et al. 2017; Lunstrum et al. 2018) to determine if there are any negative

Fig. 21 Floating oyster aquaculture in Virginia waters. Oysters are grown in bags with floats hooked on lines strung between poles that are anchored to the bottom. (Photo: Clara Robison, VIMS)

impacts of intensive shellfish aquaculture and to identify mitigation strategies if necessary. In addition, the states of Maryland and Virginia, as well as federal agencies recognize the use conflicts and are taking steps to address them in an effort to keep the economic momentum going associated with the development of shellfish aquaculture and the recovering fishery (Luckenbach et al. 2008; Strickler 2018). Although the centuries of controversy that have surrounded the oyster continue to some extent today, they are muted by the recognition from all stakeholders that it is critically important for this species to not only survive, but to thrive. In the twenty-first century oyster populations and the aquaculture industry face the challenge of climate change and all the associated stressors in the Bay. Oysters are already being impacted by rising temperatures, diseases, hypoxia, eutrophication, and harmful algal blooms in Chesapeake Bay and may be increasingly affected by ocean acidification in the upcoming decades (Glibert et al. 2007, 2010; Bricker et al. 2008; Reece et al. 2016; Breitburg et al. 2018). Ultimately, for the health of the Chesapeake Bay ecosystem health to improve, and for the health and economy of the communities that surround it to be sustained, it is important for the residents of the Bay region to fully appreciate the value of this tasty little delicacy, *Crassostrea virginica*.

A Case Study on Oyster Restoration: Bringing Back Our Oysters in the Hampton River

Hampton, Virginia is a coastal community located at the southern end of the Chesapeake Bay that looks outward into the Hampton Roads Harbor. For centuries, residents of Hampton have relied on local waterways as a source of food, recreation, and economic well-being. However, after decades of pollution, the health of the waterways have declined. In 2012, many of the rivers in the area were listed in the Impaired Waters Report by the Virginia Department of Environmental Quality due to low dissolved oxygen levels, high levels of bacteria, and increased stormwater runoff.

The Hampton River, one of the rivers listed in the report, is a 3.2 mile (5.1 km) long tidal tributary located entirely within the city of Hampton. The river's biggest

(continued)

source of pollution is stormwater runoff. In 2016, the Chesapeake Bay Foundation (CBF) partnered with Hampton University (HU), a private Historically Black University situated on the banks of the river to help clean up the waterway. How did they plan to do this? OYSTERS! The goals of this work are to assess the temporal-spatio trends in Eastern oyster spat recruitments along the Hampton River, and to use that information to promote reestablishment of this bivalve's population.

Oysters serve as natural filtration systems. The more oysters, the more filtering. To better understand where oysters were naturally occurring and surviving in the Hampton river, a shoreline survey was conducted from November 2016 to January 2017 by members of the CBF and students from HU. From the survey, current oyster reefs were located and mapped using GIS, and the population size was determined to be more than 80,000 oysters.

In addition to the shoreline surveys, students at HU with the help of HU professors and CBF members, conducted spat recruitment surveys to determine where the oyster larvae were settling (Fig. 22) and to elucidate the ideal location for an artificial oyster reef. From 2016 to 2018, surveys were conducted from May to October at four study sites along the Hampton River (Fig. 23). At each site, three shell strings and oyster spat catcher cages were placed into the water column at the same depth and changed biweekly (Fig. 24). Shell strings are composed of ten pierced oyster shells strung along a metal wire (Fig. 25). During sampling events, a water quality sonde was used to measure a variety of environmental parameters including salinity, temperature, and dissolved oxygen (Fig. 26).

From the shell string data, and spat catcher data not shown, students determined that the ideal location for an artificial reef and oyster restoration studies, was at site 1, the mouth of the Hampton River near Hampton University. Figure 27 illustrates that the sites closer to the mouth had a higher recruitment of spat compared to sites further up river. Additionally, students determined that salinity played an important role in spat recruitment. Figure 28 illustrates the average salinity along the sites from 2017 to 2018, no physical parameters were measured during the first year of the project in 2016. However, the salinity in 2017 was higher than in 2018 due to a rainy summer season in 2018. The results of this research shows that the oyster spat recruitment, in this region, is heavily determined by salinity, and the results of more recent studies indicate that turbidity may also impact oyster larvae recruitment in the Hampton River.

Our research indicates that the number of oyster spat in 2018, at most sites, was lower than in 2017, mostly due to the lower salinity (Fig. 28).

This restoration project is the first of its kind in the Hampton River, and the data collected between 2016 and 2018 now serves as a baseline for additional research. Moving forward, undergraduate and graduate students are continuing to monitor sampling sites and adding layers to the research including: water quality impacts on recruitment, exploring new restoration techniques, and determining the impact of harmful algal blooms on oyster recruitment (Fig. 29).

This research is providing Hampton University students, from diverse backgrounds, an opportunity to engage in Chesapeake Bay oyster restoration while gaining hands-on experience, winning awards for their innovative oyster reef design to continue this work, and by giving back to the Hampton Roads community.

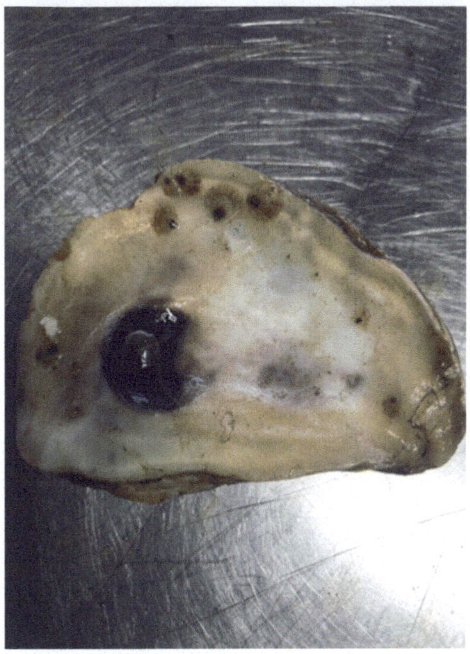

Fig. 22 Spat on shell after 3 weeks of growth

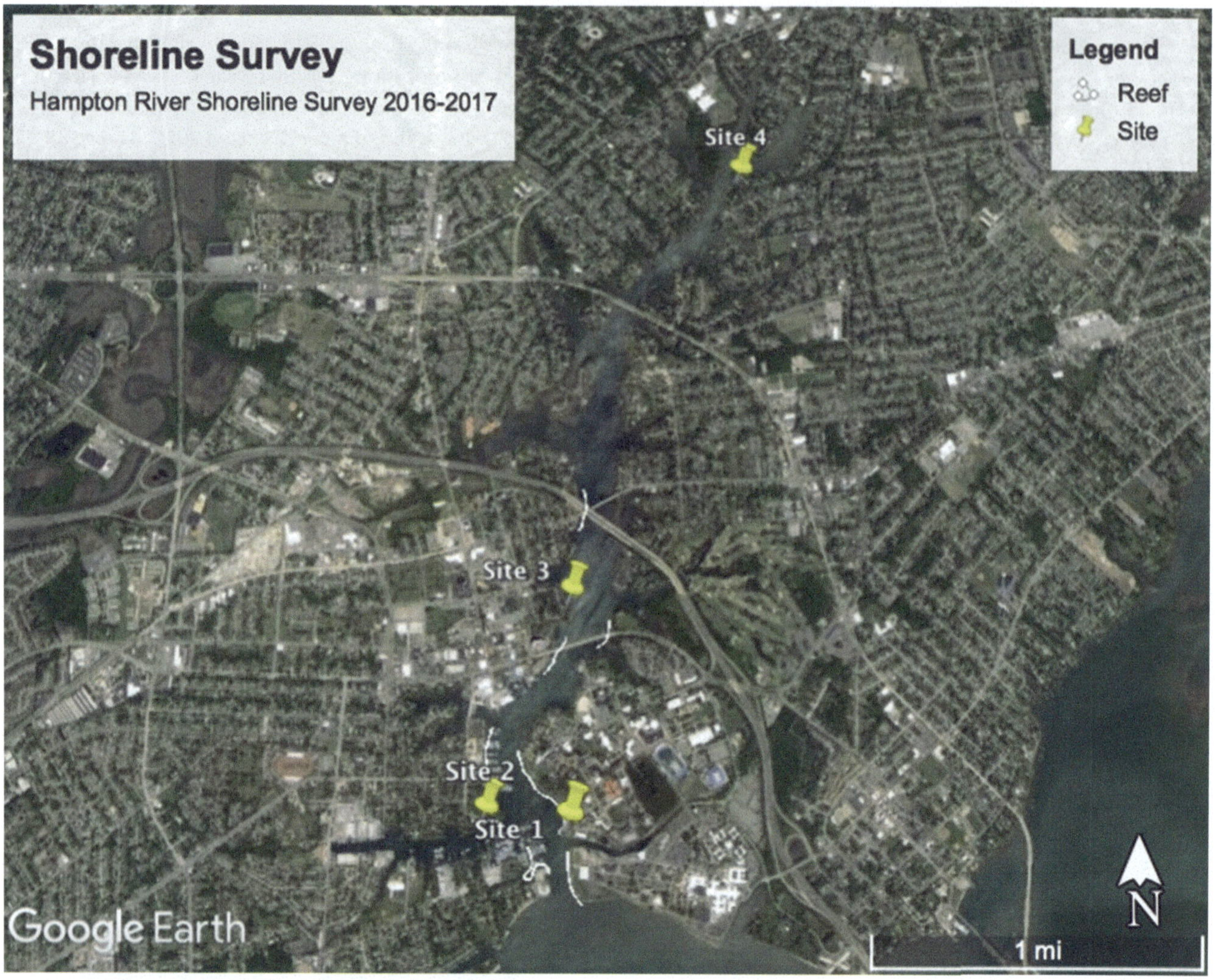

Fig. 23 Sampling Sites, yellow pins, and Oyster Shoreline Survey, white lines, and community oyster reef at site 4

Fig. 24 Hampton University undergraduate student, Ileana Fenwick, collecting oyster strings from Hampton River

Fig. 25 Oyster shell string after one month of colonization

Fig. 26 Hampton University graduate student, Sierra Hildebrandt, using a water quality sonde to collect environmental parameters such as dissolved oxygen, chlorophyll, turbidity, salinity and temperature

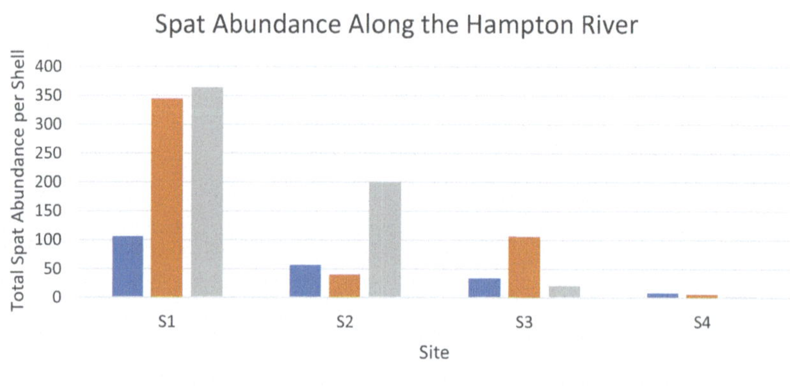

Fig. 27 Spat Abundance Along Hampton River

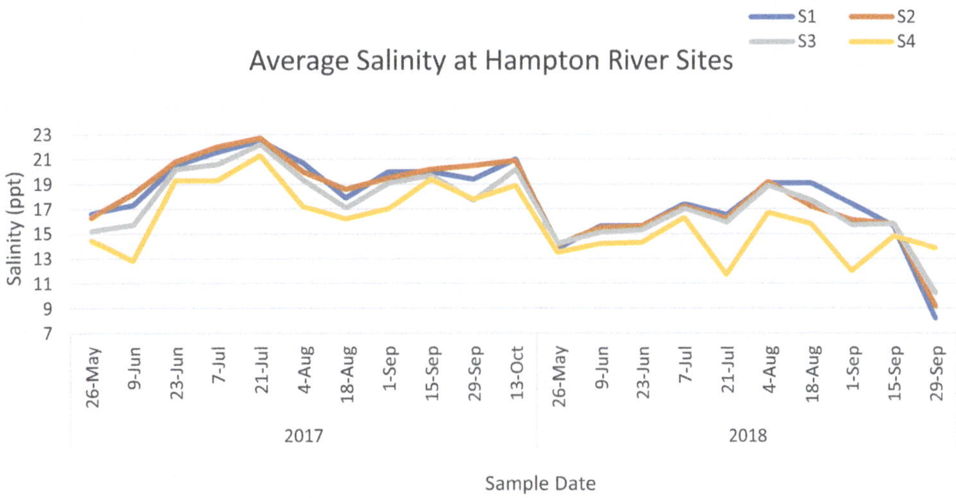

Fig. 28 2017–2018 Average Salinity along the Hampton River Sites

Fig. 29 Dr. Deidre Gibson and Ileana Fenwick assessing the benthic environment

References

Alford JJ (1973) The role of management in Chesapeake oyster production. Geogr Rev 63:44. https://doi.org/10.2307/213236

Allen SK, Burreson EM (2002) Standing policy for Virginia Institute of Marine Science non-native oyster research in Virginia. Virginia Institute of Marine Science/Aquaculture Genetics and Breeding Technology Center, Gloucester Point, VA

Andrews JD (1966) Oyster mortality studies in Virginia. V. Epizootiology of MSX, a protistan pathogen of oysters. Ecology 47:19–31

Andrews JD (1984) Epizootiology of diseases of oysters (*Crassostrea virginica*), and parasites of associated organisms in eastern North America. Helgolander Meeresunters 37:149–166. https://doi.org/10.1007/BF01989300

Baker BJ, Mann R (1991) Sterile triploid *Crassostrea virginica* (Gmelin, 1791) grow faster than diploids but are equally susceptible to *Perkinsus marinus*. J Shellfish Res 10:445–450

Bang Jensen B, Nylund S, Svendsen JC et al (2019) Indications for a vertical transmission pathway of piscine myocarditis virus in Atlantic salmon (*Salmo salar* L.). J Fish Dis 42:825–833. https://doi.org/10.1111/jfd.12990

Bishop MJ, Peterson CH (2005) Constraints to *Crassostrea ariakensis* aquaculture: season and method of culture strongly influence success of grow-out. J Shellfish Res 24:995–1006. https://doi.org/10.2983/0730-8000(2005)24[995,CTCAAS]2.0.CO;2

Bosch D, Kuminoff N, Stephenson K et al (2010) Evaluation of policy options for expanding oyster aquaculture in Virginia. Aquacult Econ Manag 14:145–163. https://doi.org/10.1080/13657301003776698

Bratton JF, Colman SM, Thieler ER, Seal RR (2002) Birth of the modern Chesapeake Bay estuary between 7.4 and 8.2 ka and implications for global sea-level rise. Geo-Mar Lett 22:188–197. https://doi.org/10.1007/s00367-002-0112-z

Breitburg D, Levin LA, Oschlies A et al (2018) Declining oxygen in the global ocean and coastal waters. Science 359:eaam7240. https://doi.org/10.1126/science.aam7240

Brewington MV (1963) Chesapeake Bay Log Canoes and Bugeyes. Tidewater Publishers, Centreville, MD

Bricker SB, Longstaff B, Dennison W et al (2008) Effects of nutrient enrichment in the nation's estuaries: a decade of change. Harmful Algae 8:21–32. https://doi.org/10.1016/j.hal.2008.08.028

Bricker SB, Rice KC, Bricker OP (2014) From headwaters to coast: influence of human activities on water quality of the Potomac River Estuary. Aquat Geochem 20:291–323. https://doi.org/10.1007/s10498-014-9226-y

Brock JA, Bullis R (2001) Disease prevention and control for gametes and embryos of fish and marine shrimp. Aquaculture 197:137–159

Brooks WK (1891) The oyster: a popular summary of a scientific study. The Johns Hopkins Press, Baltimore, MD

Brooks WK, Paynter KT (1996) The oyster, 3rd edn. Johns Hopkins University Press, Baltimore, MD

Bryant PL, Pruitt WA (2007) Report and recommendations of the Blue Ribbon Oyster Panel

Burreson EM, Andrews JD (1988) Unusual intensification of Chesapeake Bay oyster diseases during recent drought conditions. In: Proceedings of OCEANS '88. "A partnership of marine interests". pp 799–802

Burreson EM, Stokes NA, Carnegie RB, Bishop MJ (2004) *Bonamia* sp. (Haplosporidia) found in nonnative oysters *Crassostrea ariakensis* in Bogue Sound, North Carolina. J Aquat Anim Health 16:1–9. https://doi.org/10.1577/H03-008.1

Burreson EM, Stokes NA, Friedman CS (2000) Increased virulence in an introduced pathogen: *Haplosporidium nelsoni* (MSX) in the Eastern oyster *Crassostrea virginica*. J Aquat Anim Health 12:1–8. https://doi.org/10.1577/1548-8667(2000)012<0001:IVIAIP>2.0.CO;2

Calvo G, Luckenbach M, Allen S, Burreson E (1999) Comparative field study of *Crassostrea gigas* (Thunberg, 1793) and *Crassostrea virginica* (Gmelin, 1791) in relation to salinity In Virginia. J Shellfish Res 18:465–473

Calvo GW, Luckenbach MW, Allen SK, Burreson EM (2001) A comparative field study of *Crassostrea ariakensis* (Fujita 1913) and *Crassostrea virginica* (Gmelin 1791) in relation to salinity in Virginia. J Shellfish Res 20:221–229

Chesapeake Bay Program (2014) Chesapeake Bay Watershed Agreement

Chowning LS (2007) Deadrise and cross-planked. Tidewater Publishers, Centreville, MD

Chowning LS (2003) Chesapeake Bay Buyboats, 2nd edn. Tidewater Publishing

Dégremont L, Garcia C, Frank-Lawale A, Allen SK (2012) Triploid oysters in the Chesapeake Bay: comparison of diploid and triploid *Crassostrea virginica*. J Shellfish Res 31:21–31. https://doi.org/10.2983/035.031.0103

Ford SE, Haskin HH (1982) History and epizootiology of *Haplosporidium nelsoni* (MSX), an oyster pathogen in Delaware Bay, 1957–1980. J Invertebr Pathol 40:118–141. https://doi.org/10.1016/0022-2011(82)90043-X

Frank-Lawale A, Allen SK, Dégremont L (2014) Breeding and domestication of eastern oyster (*Crassostrea virginica*) lines for culture in the Mid-Atlantic, USA: line development and mass selection for disease resistance. J Shellfish Res 33:153–165. https://doi.org/10.2983/035.033.0115

Glibert PM, Alexander J, Meritt DW et al (2007) Harmful algae pose additional challenges for oyster restoration impacts of the harmful algae *Karlodinium veneficum* and *Prorcentrum minimum* on early life stages of *Crassostrea virginica* and *Crassostrea ariakensis*. J Shellfish Res 26:919–925. https://doi.org/10.2983/0730-8000(2007)26[919:HAPACF]2.0.CO;2

Glibert PM, Allen JI, Bouwman AF et al (2010) Modeling of HABs and eutrophication: status, advances, challenges. J Mar Syst 83:262–275. https://doi.org/10.1016/j.jmarsys.2010.05.004

Hallerman E, Leffler M, Mills S, Allen S (2001) Aquaculture of triploid *Crassostrea ariakensis* in Chesapeake Bay—a symposium report. Sea grant. College of William and Mary, Williamsburg, VA

Hargis WJ, Haven DS (1999) Chesapeake oyster reefs, their importance, destruction and guidelines for restoring them. In: Luckenbach MW, Mann R, Wesson J (eds) Oyster reef habitat restoration: a synopsis and synthesis of approaches. Virginia Institute of Marine Science Press, Gloucester Point, VA, pp 329–358

Hargis WJ, Haven DS (1988) Rehabilitation of the troubled oyster industry of the lower Chesapeake Bay. J Shellfish Res 7:271–299

Haskin HH, Ford SE (1979) Development of resistance to *Minchinia nelsoni* (MSX) mortality in laboratory-reared and native oyster stocks in Delaware Bay. Mar Fish Rev 41:54–63

Haskin HH, Stauber LA, Mackin JA (1966) *Minchinia nelsoni* n. sp. (Haplosporida, Haplosporidiidae): causative agent of the Delaware Bay oyster epizootic. Science 153:1414–1416

Haven DS, Hargis WJ, Kendall PC (1978) The oyster industry of Virginia: its status, problems and promise. A comprehensive study of the oyster industry in Virginia. Sea Grant, Gloucester Point, VA, Virginia Institute of Marine Science

Hawks FL (1842) The adventures of Captian John Smith, The founder of the colony of Virginia. D. Appleton and Co, New York

Hudson K (2018) Results of the 2017 Virginia shellfish aquaculture crop reporting survey. Virginia Shellfish Aquaculture Situation and Outlook Report No 2018-9:1–20

Ingersoll E (1881) A report on the oyster industry of the United States. The Government Printing Office, Washington, DC

Kingsley-Smith PR, Harwell HD, Kellogg ML et al (2009) Survival and growth of triploid *Crassostrea virginica* (Gmelin, 1791) and *C. ariakensis* (Fujita, 1913) in bottom environments of Chesapeake Bay: implications for an introduction. J Shellfish Res 28:169–184. https://doi.org/10.2983/035.028.0201

López Sanmartín M, Power DM, de la Herrán R et al (2016) Evidence of vertical transmission of ostreid herpesvirus 1 in the Portuguese oyster *Crassostrea angulata*. J Invertebr Pathol 140:39–41. https://doi.org/10.1016/j.jip.2016.08.012

Luckenbach MW, Lipton D, Webster D et al (2008) A framework for native oyster aquaculture development in Maryland. CRC, Edgewater, MD

Luckenbach MW, Mann R, Wesson JA (1999) Oyster reef habitat restoration: a synopsis and synthesis of approaches; proceedings from the symposium, Williamsburg, Virginia, April 1995. Virginia Institute of Marine Science, The College of William and Mary, Gloucester Point, VA

Lunstrum A, McGlathery K, Smyth A (2018) Oyster (*Crassostrea virginica*) aquaculture shifts sediment nitrogen processes toward mineralization over denitrification. Estuar Coasts 41:1130–1146. https://doi.org/10.1007/s12237-017-0327-x

MacKenzie CLJ (1996) History of oystering in the United States and Canada, featuring the eight greatest oyster estuaries. Mar Fish Rev 58:1–78

Maryland Oyster Restoration Interagency Workgroup (2019) 2018 Maryland Oyster Restoration Update: progress toward achieving the Chesapeake Bay Watershed Agreement Oyster Outcome in Harris Creek Little Choptank River, Tred Avon River, and Upper St. Mary's River. Chesapeake Bay Program's Sustainable Fisheries Goal Implementation Team, Annapolis, MD

Moss JA, Burreson EM, Cordes JF et al (2007) Pathogens in *Crassostrea ariakensis* and other Asian oyster species: implications for non-native oyster introduction to Chesapeake Bay. Dis Aquat Org 77:207–223. https://doi.org/10.3354/dao01829

Murphy AE, Emery KA, Anderson IC et al (2016) Quantifying the effects of commercial clam aquaculture on C and N cycling: an integrated ecosystem approach. Estuar Coasts 39:1746–1761. https://doi.org/10.1007/s12237-016-0106-0

Newell RIE (1988) Ecological changes in Chesapeake Bay: are they the result of overharvesting the American oyster *Crassostrea virginica*? Understanding the estuary: advances in Chesapeake Bay research. CRC, Solomons, MD, pp 534–546

NOAA CBO, NOAA RO (2019) 2018 Virginia oyster restoration update: progress in five estuaries. NOAA, Annapolis, MD

Orth RJ, Dennison WC, Lefcheck JS et al (2017) Submersed aquatic vegetation in Chesapeake Bay: Sentinel species in a changing world. Bioscience 67:698–712. https://doi.org/10.1093/biosci/bix058

Paynter KT (1999) Managing around oyster diseases in Maryland and Maryland oyster roundtable strategies. In: Luckenbach MW, Mann R, Wesson JA (eds) Oyster reef habitat restoration: A synopsis and synthesis of approaches. Virginia Institute of Marine Science Press, Gloucester Point, VA, p 12

Peterson C, Grabowski J, Powers S (2003) Estimated enhancement of fish production resulting from restoring oyster reef habitat: quantitative valuation. Mar Ecol Prog Ser 264:249–264. https://doi.org/10.3354/meps264249

Proestou DA, Vinyard BT, Corbett RJ et al (2016) Performance of selectively-bred lines of Eastern oyster, *Crassostrea virginica*, across Eastern US estuaries. Aquaculture 464:17–27. https://doi.org/10.1016/j.aquaculture.2016.06.012

Ragone Calvo LM, Calvo GW, Burreson EM (2003) Dual disease resistance in a selectively bred Eastern oyster, *Crassostrea virginica*, strain tested in Chesapeake Bay. Aquaculture 220:69–87. https://doi.org/10.1016/S0044-8486(02)00399-X

Reece KS, Vogelbein WK, Carnegie RB (2016) Strategies to minimize the impact of mid-Atlantic emerging harmful algal bloom (HAB) species on *Crassostrea virginica*. NOAA

Rick TC, Reeder-Myers LA, Hofman CA et al (2016) Millennial-scale sustainability of the Chesapeake Bay Native American oyster fishery. Proc Natl Acad Sci U S A 113:6568–6573. https://doi.org/10.1073/pnas.1600019113

Rothschild BJ, Ault JS, Goulletquer P, Héral M (1994) Decline of the Chesapeake Bay oyster population: a century of habitat destruction and overfishing. Mar Ecol Prog Ser 111:29–39

Schulte DM (2017) History of the Virginia Oyster Fishery, Chesapeake Bay, USA. Front Mar Sci 4:127. https://doi.org/10.3389/fmars.2017.00127

Smaal AC, Ferreira JG, Grant J et al (eds) (2018) Goods and services of marine bivalves. Springer, Berlin Heidelberg, New York, NY

Stanley JG, Hidu H, Allen SK (1984) Growth of American oysters increased by polyploidy induced by blocking meiosis I but not meiosis II. Aquaculture 37:147–155. https://doi.org/10.1016/0044-8486(84)90072-3

Strickler M (2018) Clam and oyster aquaculture work group. Office of the Secretary of Natural Resources, VA

US ACOE Baltimore (2009) Chesapeake Bay oyster restoration using alternative substrate; Maryland. US Army Corps of Engineers, Baltimore District, Baltimore, MD

US ACOE Norfolk (2009) Final programmatic environmental impact statement for oyster restoration in Chesapeake Bay including the use of a native and/or nonnative oyster. US Army Corps of Engineers, Norfolk District, Norfolk, VA

Virginia Commission of Fisheries (1928) Report of the commission of fisheries of Virginia. David Bottom, Superintendent of Public Printing, Richmond, VA

Vojtech P (1993) Chesapeake Bay Skipjacks. Tidewater Publishers, Centreville, MD

Wennersten JR (1981) The oyster wars of Chesapeake Bay, 1st edn. Tidewater Publishers, Centreville, MD

Wesson J, Mann R, Luckenbach M (1999) Oyster restoration efforts in Virginia. In: Oyster reef habitat restoration: a synopsis and synthesis of approaches; proceedings from the symposium, Williamsburg, Virginia, April 1995. Virginia Institute of Marine Science Press

Wharton J (1957) The bounty of the Chesapeake: fishing in Colonial Virginia, Virginia 350th Anniversary Celebration, Williamsburg, VA. The University Press of Virginia, Charlottesville

Williamson TR, Tilley DR, Campbell E (2015) Emergy analysis to evaluate the sustainability of two oyster aquaculture systems in the Chesapeake Bay. Ecol Eng 85:103–120. https://doi.org/10.1016/j.ecoleng.2015.09.052

Passenger Pigeon and Waterfowl: Flights to Extinction and Not

Benjamin E. Cuker

Flights of passenger pigeons once darkened the skies of the Chesapeake Bay.
Below them water fowl of all sorts would play.
Hunters with shotguns pursued these birds.
At first mostly to feed their own families, according to their words.
Then came the locomotives and steamships.
The hunters sent their slain birds on long trips.
To distant restaurants the fowl did go.
Enriching the market gunners, the bird's deadly foe.
Soon three billion pigeons turned to none.
Only new laws kept the ducks and geese from also being done.

Abstract

Hunting of fowl was a part of the Chesapeake Bay Region's food system from the earliest times. English colonists introduced the shotgun as an effective weapon against swimming, flying and roosting birds. Despite the new weaponry, populations of game-birds remained fairly abundant for the first century after colonization, since most hunting was for feeding families. The advent of the steamboat and railroad opened up new markets for fowl. These new forms of transportation brought ice for preserving the kill from northern states, and allowed hunters to send the chilled carcasses of slain birds to distant restaurants, hotels, and market centers. The railroads also spurred development of wildlands and the expansion of agriculture, reducing habitat for terrestrial and water birds. The combination of over-hunting and habitat loss drove the nineteenth Century extinction of the passenger pigeon (*Ectopistes migratorius*), a migratory species which once was the most populous bird in N. America. The loss of the passenger pigeon informed efforts to prevent the same fate befalling widely hunted waterfowl species. Many of the political elite hunted waterfowl and helped push legislation to protect their pastime. This led to the Migratory Bird Treaty Act of 1918, the first national environmental legislation for the US. This act, subsequent legislation, and efforts to recover lost habitat combined to rebuild the populations of most species of waterfowl that frequent the Chesapeake Bay. The Canada goose (*Branta canadensis*) has gone from near extinction in 1920 to the status of a nuisance species in many locations due to its ability to forage on industrial grain farms. The recovery of other species, such as the canvas back duck (*Aythya valisineria*) is much less complete. This points to the need to continue tight regulation of hunting and investment in habitat restoration.

Keywords

Over hunting · Market gunning · Chesapeake Bay birds

The industrialization of hunting leads to the passing of a pigeon About 7.8 nautical miles (14.5 km), as the bird flies, from where the Patapsco River joins the Chesapeake Bay, a red brick tower rises 234 ft (71.3 m) above the shores of Baltimore Harbor (Fig. 1). Erected in 1828, it was the tallest building in the US until the completion of the spire on Trinity Church in New York City in 1846 (NPS.gov 2018; Carrollmuseums.org 2018). The Phoenix Shot Tower (or Old Baltimore Shot Tower) stands today as monument to the industrialization of hunting in nineteenth Century

B. E. Cuker (✉)
Department of Marine and Environmental Science, Hampton University, Hampton, Virginia, USA
e-mail: Benjamin.cuker@hamptonu.edu

B. E. Cuker (ed.), *Diet for a Sustainable Ecosystem*, Estuaries of the World, https://doi.org/10.1007/978-3-030-45481-4_9

Fig. 1 The 215-ft (65.5 m) tall Phoenix Shot Tower commanding the nineteenth Century skyline of Baltimore, where it industrialized the production of shot for the hunting of birds and other wildlife. Erected in 1828, it was the tallest structure in the US until 1846. https://www.carrollmuseums.org/history/towerhistory.html

America. It was built for the manufacture of ammunition. Much of what the tower produced was small diameter (2–9 mm) balls of lead called shot. Hunters use shot in shotguns to kill birds in flight and other animals at close range. To work well the shot needs a smooth spherical shape. Using a process developed in England, shot makers poured molten lead through a sieve at the top of the tower. The drops of hot lead tumbled through the air, forming tiny globes, and ending their brief journey with a splash in a tank of water that cushioned their landing. For the small shot (2–6 mm), the next time airborne would probably be on exit from the muzzle of a gun pointed at a flock of birds. The investment in laying the 1.1 million bricks of Baltimore's shot tower speaks to the economic importance of hunting in that era. The annual output of 1 million bags of 25-pound shot paid off nicely for the investors.

That today one can often spot urban birds perched atop the Old Baltimore Shot Tower is a bit ironic. Since shot manufactured there, helped shoot to extinction the formerly most common bird in North America. Migrating flocks of the passenger pigeon (*Ectopistes migratorius*) once blackened the skies along the shores of the Chesapeake Bay and much of the eastern portion of the U.S. (Figs. 2 and 3). Passenger pigeons went from a population size of 3–5 billion in the early nineteenth Century to extinction by 1901 (Fuller 2014).

The bounty of the eastern deciduous forest supported the once huge biomass of passenger pigeons. Specialized anatomy allowed the birds to consume large, energy dense nuts and fruits. Acorns and chestnuts passed through beaks that unhinged and down flexible throats and crops. In return, pigeons dispersed the nuts they failed to digest, planting new trees to perpetuate the forest. Their massive droppings also fertilized the soil.

The huge flocks and colonial nesting made the bird easy prey for humans. Pre-Columbian people and later arrivals sustainably hunted passenger pigeons for centuries. Advanced technology, industrialization of hunting, and mass-marketing of their meat changed that. A single blast of a shotgun could kill a dozen or more birds. It was so easy to kill passenger pigeons that few considered it real sport. In addition to shotguns hunters deployed huge and often baited nets (Fig. 4). A captured and tethered pigeon was often placed on a stool to lure its brethren to the site. While the species is long gone, the term "stool pigeon" continues to inform our language. Hunters went after nestlings with burning sulfur, causing the birds to suffocate and fall to the ground. Some even cut down trees to get at the juvenile birds. In addition to the lucrative market in meat, passenger pigeons yielded a buttery oil. Fifty birds produced a pound of feathers useful in the production of bedding (Fuller 2014).

Errol Fuller published *The Passenger Pigeon* in 2015 (Fuller 2014). He discovered an anonymous newspaper story (Fon du Lac Commonwealth, May 20, 1871) that conveys the immensity of a passenger pigeon flock and the scale of the hunt:

> Imagine a thousand threshing machines running under full headway, accompanied by as many steamboats groaning off steam, with an equal number of...trains passing through covered bridges... and you possibly have a faint conception of the terrific roar following the monstrous black cloud of pigeons as they passed in rapid flight...a few feet before our faces... nearly on a level with the muzzles of our guns...the contents of a score of double barrels was poured into their dense midst. Hundreds, yes thousands dropped into the open fields below. Not infrequently a hunter would discharge his piece and load and fire the third and fourth time into the same floc. The slaughter was terrible beyond any description.

Although the lack of skill required to shoot passenger pigeons seemed to take the sport out of the hunt for some, others took advantage of the bird's abundance and ease of capture to create a new pastime. The new sport of "trap shooting" started in 1825, using the ubiquitous passenger pigeon as targets. Birds trapped in nets would be freed to face the fire of contestants (Fuller 2014). Eventually clay pigeons catapulted into the air replaced the real birds. Clay pigeons can't go extinct, so the sport thrives today around the world, despite the loss of the species for which it was created (USAshooting.org 2018).

Fig. 2 A male (upper) and female (lower) Passenger Pigeon by Louis Agassiz Fuertes. Public domain. https://upload.wikimedia.org/wikipedia/commons/a/a0/Mershon%27s_The_Passenger_Pigeon_%28frontispiece%29.jpeg

The nineteenth Century food system eliminated this once populous species from the face of earth. Foremost, the passenger pigeon was hunted relentlessly with shotguns, nets, and fire. Profiteering drove the intensity of the hunt. The new railroads and steamboats of the mid-nineteenth Century gave hunters expanded markets for birds shipped on ice (Schorger 1955). Hunting went from artisanal to industrial. In this regard, the story of the eastern oyster (*Crassostrea virginica*)

Fig. 3 Depiction of a shooting of passenger pigeons in northern Louisiana, Smith Bennett, 1875, Library of Congress. https://www.loc.gov/item/2004671874/

is the same, save for escaping extinction (See chapter "The Chesapeake Bay Oyster: From Cobblestone to Keystone."). Another aspect of the food system doomed the passenger pigeon. Farmers replaced the deciduous forest with crop lands. This robbed the passenger pigeons of their habitat for roosting, foraging, and nesting. It also turned the farmer into an adversary of the species. Killing the pigeons meant both more meat on the table and preserving their crops from the hungry flocks (Fig. 5).

That passenger pigeon numbers could go from several billion to zero in a few decades seems almost impossible. Their sheer numbers spoke against any suggestion for implementing conservation measures. Yet, three aspects of the bird's biology made it particularly vulnerable to the industrial hunting. First, the birds reproduced once or maybe twice a year, each female laying a single egg. At that rate each mated pair would take two years to replace themselves with the next generation. Loss to predators and parasites would easily extend the replacement time another year or two. Their reproductive potential proved no match for the industrial hunters. The pigeon's biology also dictated the need for living in large populations. According to Fuller (2014) "...the peculiar evolution of the birds meant they could survive—psychologically, emotionally and physically—only in vast numbers." Hunting and habitat loss apparently thinned populations below the threshold needed to sustain the species. Finally, recent analysis of the bird's genome made from preserved museum specimens suggest that the species was given to eruptive population cycles (Biello and Biello 2014). Perhaps the intense mortality imposed by hunters made a natural decline into a death-spiral disaster.

By the 1870s government authorities sensed the reduction in populations and in various places passed laws to regulate hunting. These measures proved fruitless, as by that time the birds failed to recover. Note that market forces didn't deter the overhunting as the populations began to drop. As with fisheries, market forces might prevent commercial extinction, but never the actual extinction of the species. The pigeon hunters devised more efficient technologies to pursue a

Fig. 4 A pigeon trap awaiting a flock, painting by James Pattison Cockburn in 1829. Such a net could capture several hundred passenger pigeons at one time.

declining resource. As they realized a diminished return for their investments in time, guns and nets, they simply increased their effort until the day all the birds were gone.

In many ways the passenger pigeon was the forage fish of the sky and land. Its huge flocks mirrored the tight schools of menhaden (*Brevoortia tyrannus*) and shad (*Alosa sapidissima*). It ate low on the food chain, and supported a host of predators such as minks, martens, raccoons, owls, hawks, and eagles. As with forage fish, many were captured with nets, and over-harvesting damaged the environment. The analogy works best today for menhaden. The flocks of passenger pigeons would darken the skies of the Chesapeake Bay while schools of menhaden did the same for the waters, each sporting populations in the many millions. Individuals in each of the species required the company of thousands of others in order to exist. Both species found a home in the environs of the Chesapeake, and well beyond its boundaries. The pigeons migrated throughout much of N. America. The menhaden populate waters from Maine to Mississippi. While pigeon flesh proved delectable to the nineteenth Century pallet, menhaden didn't, so their flesh went to animal feed, fertilizer and industrial oils. Perhaps this distinction saved menhaden from the fate of the passenger pigeon (see chapter "Menhaden, the Inedible Fish that Most Everyone Eats" for more about menhaden).

Waterfowl weather the storm A look at the navigation charts for the various sections of the Chesapeake Bay and its tributaries reveals the remains of a common feature of the nineteenth and early twentieth Century food system. Almost any expanse of shallow waters (circa 2 m, or about 6 ft deep) near a shoreline will sport the remnants of one and often many *duck blinds* (Fig. 6). Hunters drove wooden piles (poles) into the shallow sediments, upon which they built square platforms from boards. During hunting season the hunters festooned the platforms with small tree branches and dislodged shrubs (Fig. 7). The tangle of branches hid the waiting hunters from migrating waterfowl that would come to dine on the submerged plants growing from the shallow bottom. To entice the targets to swim within range

Fig. 5 A steam train near Alexandria Virginia during the Civil War. Railroads expanded the market for the passenger pigeon, ducks and oysters during the nineteenth Century. From the Library of Congress. https://www.loc.gov/pictures/item/2012649966/resource/

of their guns, the hunters floated tethered decoys of wood carved and painted to resemble their quarry of hope. When sufficient numbers of the birds arrived the gunners would blast the flock with shot, often killing dozens at a time (Sullivan 2003). The duck blind architects chose stout and rot-resistant timbers to support their platforms. Coast-wise mariners traversing these old hunting grounds in shoal-draft boats may come to grief when the hulls of their vessels smack one of the half gone antique piles hidden beneath the waves. Savvy anglers often cast their lines near the charted remains, as the piles sport miniature reefs of barnacles, mussels, oysters, seaweeds, crabs, and other life that attracts various species of fish.

Whether a hazard to navigation or a fish attracting reef, the ubiquity of these derelict and decaying structures speaks to an earlier time when flocks of migrating waterfowl filled the skies and waters of the Chesapeake. Capt. John Smith appreciated the diversity and abundance of birds offered by the Chesapeake, and understood their migratory nature when he wrote, "Of birds the Eagle is the greatest devourer. Hawkes there be of diverse sorts, as our Falconers called them, Sparowhawkes, Lanarets, Goshawkes, Falcons and Osperayes, but they all pray most upon fish. Partridges there are little bigger then our Quailes, wilde Turkies are as bigge as our tame. There are woosels or blackbirds with red shoulders, thrushes and diverse sorts of small birds, some red, some blew, scarce so bigge as a wrenne, but few in Sommer. In winter there are great plenty of Swans, Craynes, gray and white with blacke wings, Herons, Geese, Brants, Ducke, Wigeon, Dotterell, Oxeies, Parrats and Pigeons. Of all those sorts great abundance, and some other strange kinds to us unknowne by name. But in sommer not any or a very few to be seene. Birds (Smith 1612)."

Not fully understood by Capt. John Smith at that time, the Chesapeake Bay is a central feature of the Atlantic Flyway (Fig. 8). Birds migrating between summering grounds in northern Canada and wintering grounds as far south as the Caribbean, spend weeks to months feeding and breeding in

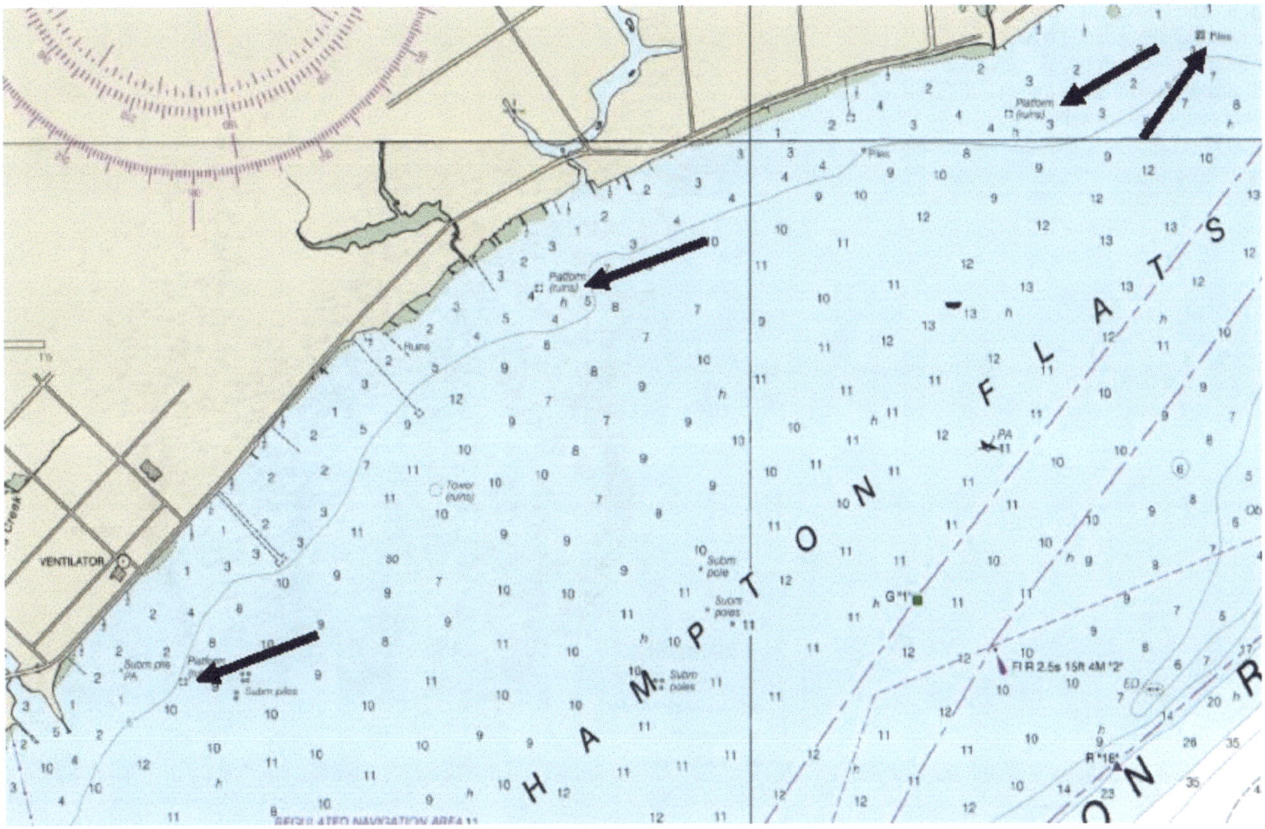

Fig. 6 Section of a navigation chart showing the wrecks of four duck blinds along a 2 mile (3.2 km) stretch of shore in Hampton Roads, the southern portion of the Chesapeake Bay (arrows point to charted remains)

Fig. 7 A contemporary duck blind in the Chesapeake Bay off the shore of Kent Island, Romancoke, MD, with the Eastern Shore in the background by Cyndy Sims Parr. https://commons.wikimedia.org/wiki/File:Chesapeakeduckblind.jpg

Fig. 8 The Chesapeake Bay is in the center of the Atlantic flyway for migratory waterfowl, as indicated by arrow (https://www.flickr.com/photos/usfwshq/29995934172)

the Chesapeake Bay. Some, such as contemporary sub-populations of the Canada goose establish year-round presence on the Bay. Beginning in colonial times this mostly seasonal wealth of water bird biomass attracted the attention of hunters from around the Bay, as well as wealthier sportsmen willing to travel some distance from such places as New York and Philadelphia (Sullivan 2003).

Probably all of the birds of the Chesapeake were hunted to some degree by natives, colonists, and their descendants. Many of these were sought not for food, but for their plumage. The Algonquin and other Native Americans valued feathers for dress and ceremony. Hunting pressure for plumage increased in the nineteenth Century as the women's fashion industry produced hats and accessories festooned in feathers (Doughty 1975). In contrast, the food system targeted species that provided the best eating rather than the best looks for the effort. On the Chesapeake Bay that meant waterfowl: duck, geese, and swans (family Anatidae). These are mostly large birds that offer a flesh suitable to the pallet. There is little evidence that other water birds, such as cormorant, pelicans, osprey, gulls, terns, herons, and egrets were killed for consumption in a large way. This may result from those birds eating higher on the food chain than waterfowl. The higher trophic level meant lower potential population densities, as so much energy is lost as it moves up the food chain. The higher trophic level feeding may have also rendered the flesh of those birds rather fishy-flavored and less palatable.

Two general types of ducks inhabit the waters of the Chesapeake (Fig. 9). Dabbling or marsh ducks (tribe Anatini) feed on seeds, grasses, sedges, other wetland plants and shallow beds of submerged vegetation. To gain access to food on the bottom of very shallow waters they tip their bodies forward, extending their necks as they forage. Diving or bay ducks (tribes Aythyini, Mergini, Oxyurini) submerge their entire bodies as they swim to the bottom to feed on vegetation and small animals. These two groups evolved contrasting body forms, each suitable to their way of feeding (Raikow 1973). The dabbling ducks feature longer legs placed at about mid-body and longer wings. They easily balance their bodies on their legs as they walk on land. Dabbler leg placement and large wings enables them to execute a rocket takeoff, springing quickly from the water to the air. Dabblers retain air in their feathers, adding to their effective volume, reducing their density, thus making them more buoyant.

The plumage of diving ducks permits air to escape when submersing. This reduces their buoyancy, and energy expenditure for diving. Diving ducks sport shorter legs placed back further on their body plan. They also have larger webbed feet and shorter wings. The arrangement makes for effective diving and swimming underwater, but means they are

Fig. 9 Contrasting body forms of dabbling duck (left) and diving duck

awkward on land and must work harder to gain flight. Diving ducks require an energetic running start called skittering. It combines quick swimming with the feet and numerous beats of the wings. Diving ducks are no more graceful when landing on water, coming down with an awkward splash and glide (Baldassarre 2014).

The primary species of dabbling ducks on the Chesapeake are the mallard (*Anas platyrhynchos*), American wigeon, or baldpate (*Anas americana*), American black duck (*Anas rubripes*), Northern Pintail (*Anas acuta*), Green-winged teal (*Anas crecca*), Blue-winged teal (*Anas discors*), and Wood duck (*Anas discors*). The wood duck is unique in building nests in trees, while the rest nest on the ground.

The most common species of diving ducks on the Bay include the Canvasback (*Anthya valisineria*), Redhead (Anthya americana), Lesser scaup (*Aytha affinis*), Hooded merganser (*Lophodytes cucullatus*), Common goldeneye (*Bucephala clangula*), Bufflehead (*Bucephala albeola*), and Ruddy duck (*Oxyura jamaicenis*). The canvasback was and is the favorite of hunters due to its massive size (as large as 1.6 kg or 3.5 lb) and taste (White 1989).

The American coot (*Fulica americana*) is a rail (family Rallidae) but exhibits convergent evolution with ducks, looking very much the part as it walks, swims, and dives. Its diet is also akin to ducks, feeding mostly on vegetation and some invertebrates. Yet it produces a flesh considered much less palatable than most ducks. The recipes for coot meat feature ingredients designed to disguise the flavor of flesh (Carlson 2013). Despite the dubious culinary reward, they still receive the attention of hunters.

The Tundra swan (*Cygnus columbianus*) is the largest waterfowl of the Chesapeake Bay with a wingspan of 203 cm (80 in.). It feeds mostly on clams it digs from the bottom, often sharing its meal with several different species of gulls that steal the catch from the swan's bill. Geese are the next largest waterfowl of the Bay with wing spans of 173 cm (68 in.) (Baldassarre 2014). The once threatened Canada goose (*Branta canadensis*) came back from the brink of extinction to be so common as to now enjoy the status of annoying pest for people with lawns abutting bodies of water. The geese feeds on grasses, seeds and berries. The Canada goose population on the Chesapeake now consists of seasonal migrants and year-round residents. The snow goose (*Chen caerulescens*) is somewhat smaller than the Canada, with wingspan of 150 cm (59 in.). It feeds primarily on berries, seeds, stems, tubers and roots. Swans and geese are enthusiastically hunted for their size and flavor (White 1989).

The majority of the aforementioned waterfowl have suffered from the food system, some nearing extinction before being rescued by legislation and law enforcement. The story of the canvasback duck illustrates how the food system impacted waterfowl on the Bay. The food system challenged the canvasback in four ways. First and most obvious was the mortality from legal (Geis 1959) and illegal (Walsh 1971, Haramis et al. 1993) hunting. Second, the food system polluted the Bay, destroying the beds of celery grass that fed the ducks (Bayley et al. 1978; Perry et al. 2007; Haramis et al. 2001). Third, the food system degraded the breeding grounds in Canada, draining wet lands and reducing essential habitat (Bartzen et al. 2010). Finally, pesticides used to protect crops along with other industrial chemicals became chronic contaminates of the birds (Perry 1974; Rattner and McGowan 2007). However, eating lower on the food chain, these contaminants proved less of a problem for canvas back duck (*Aythya valisineria*) than for predatory birds such as cormorant (White et al. 1979).

This chapter opened with the sad story of the terrestrial passenger pigeon, the tasty and once super abundant terrestrial bird driven to extinction by the industrializing food system. Perhaps the lessons learned from the passenger pigeon saved the also intensely hunted ducks, geese, and swan of the Chesapeake from suffering the same fate. The waterfowl also benefited from the remoteness and more difficult hunting provided by their habitat. Hunters of passenger pigeons could easily walk through the forests and fields to shoot or net their prey. Water fowling on the Chesapeake meant navigating perilous marshes, swamps, and waterways and often enduring wintery conditions. Shooting and trapping waterfowl required more sophisticated equipment and skills than needed by pigeon hunters. Despite these obstacles, by the early twentieth Century hunters of duck, geese and swan were walking and boating their way down the same path that led to the extinction of the passenger pigeon a few years before.

The skill and effort required to hunt waterfowl added to the desirability of their flesh for people of economic means. Hunting duck was "manly" work and although women sometimes participated, the sport was built on somewhat strict gender lines (Smalley 2005).

During the nineteenth Century wealthy city dwellers from New York, Baltimore, Philadelphia, Washington, DC, Richmond, VA, and other centers of privilege and power built elaborate hunting clubs on the shores of the Chesapeake Bay and its tributaries. A few hours ride on a train or steamboat brought them to a weekend of wintery water fowling. The hunting was considered great sport for those who could afford it. Considerer the cost they paid to shoot ducks; rail or steamboat tickets, room and board, guns, powder, shot, decoys, sometimes a retrieving dog, hiring a guide, membership in a hunt club, special hunting garb, and renting a boat. The seemingly "free" meat from the commons was anything but (Sullivan 2003).

As noted earlier, water fowling was a way of life for those that lived on the shores of the Bay during the seventeenth to nineteenth Centuries, and certainly not restricted to the wealthy. Yet the moneyed class exerted an outsized influence on both the killing of the birds and their eventual conservation. The wealthy hunters congregated mostly around the northern Chesapeake Bay along its western shore. Prime hunting locations stretched from the Gunpowder River (about 24 km, or 15 miles) north of Baltimore to the entrance of the Susquehanna River at the top of the Bay. The name Gunpowder River seems apt for one of the prime gunning locations on the Bay. The origin of the name may trace to deposits of an essential component of black powder, namely saltpeter (potassium nitrate), found along its banks by early colonists (Kingsville MD 2018). This section of the Bay hosted many exclusive hunt clubs, the first one known to have been established in 1819 as the "Maxwell's Point Gunning Club" (Sullivan 2003). In many cases the clubs built their houses on land rented from farmers, providing an important source of cash to that node in the food system. Along with the house came the right to use the property for access to the shore and hunting.

The money of wealthy industrialists and aristocrats built the gunning clubs and they often entertained other society elites. The San Domingo Farm Club hosted President Grover Cleveland in March of 1886. President Benjamin Harrison was a regular at the Bengies Ducking Club in the 1890s (Figs. 10, 11, and 12; Sullivan 2003).

In addition to the exclusive gunning clubs, much of the waterfowl hunting was staged out of Harve de Grace, Maryland, a town situated where the Susquehanna joins other tributaries and the Chesapeake Bay. The slowing of these rivers as they meet the Chesapeake created an extensive shallow underwater plain called the Susquehanna Flats, mostly to the east of the channel that cuts to the Bay. The flowing Susquehanna River provided a constant supply of inorganic nutrients to fertilize the wild celery (*Vallisneria americana*) and other submerged freshwater plants that thrived on the shallow and sun-lit sediments. This was the perfect feeding grounds for the migrating waterfowl that thrived on the wild celery. The hunters appreciated how the wild celery flavored the duck's flesh, as well as concentrating them for easy gunning. The large canvasback duck (*Aythya valisineria*) proved the favorite target as they provided a large and sumptuous mass of flesh (Fig. 13). Note the scientific name of the canvasback is informed by that of their favorite food, thanks to the father of American ornithology Alexander Wilson who first described the duck in 1814 (Baldassarre 2014). Hunters and diners considered the canvasback the "king of ducks" (Associated Press 2017).

Hunting intensity reduced the populations of the favored canvasback and the other waterfowl by the end of the nineteenth Century. This was reflected in the market prices for Canvasback, which sold for $1.25 per pair in the fall of 1873, jumping to $4.00 a pair in the fall of 1890 (Sullivan 2003). A classic example of reduced supply driving up prices as the duck population declined. The price on the table of a restaurant was of course higher. An 1896 menu form the Waldorf in New York City offered Canvasback for $4.00 (presumably for a single bird) and redhead for $3.25 (Menus 2018).

The prices above for canvasback also tells much about that duck's role in the food system of the time. Adjusting the 1890 canvasback price per pair for inflation yields $110.76 in 2017 dollars. Also consider that typical wages for the era ran around $0.15–$0.25 per hour, with only the most skilled workers earning $0.50 per hour (USBL 1905). Clearly an ordinary working person in 1890 couldn't easily afford a dinner of canvasback duck, as the entre priced at about a half a week's wages. The high price for canvasback meant that market gunners could make handsome money. A newspaper account from 1890 tells of one man killing 520 ducks in a single day of gunning out of Harve de Grace (Sullivan 2003). That works out to $1040 in 1890 dollars, or $28,796 in 2017 currency. That far exceeds a working person's annual wages of the era. While these calculations don't take account of capital costs for the gunners, and are based on a record daily take, they do demonstrate the financial incentives for market gunners of the era. Not surprisingly the market gunning industry persisted in an underground economy for decades after the practice was outlawed (Walsh 1971).

It wasn't just the gunning part of the food system that so reduced the canvasback population. By the mid-twentieth Century water quality declined on the Flats, eliminating almost all of the wild celery (Walsh 1960). Excess fertilizer and dislodged soil from poor farming practices were the culprit. The excess nutrients caused algal blooms that shaded

Fig. 10 A political cartoon depicting President Benjamin Harrison and superimposing issues of the day on to his pastimes of duck hunting and fishing. Library of Congress, image circa 1888. https://www.loc.gov/pictures/item/91482919/

the benthic plants. The suspended sediments eroded from the fields brought additional excess nutrients and shading. This drove the canvasbacks to alter their diet to include Baltic clams (*Macoma baltica*), other small animals, and corn from farms (*Zea mays*) (Lovvorn 1989; Haramis et al. 2001; Perry et al. 2007). A portion of the population also moved south to other bodies of water (Baldassarre 2014).

While being involved in the hunt added spice to the taste of the game, the wealthy didn't confine their consumption of wild waterfowl to that which they garnered themselves. Fancy hotels, restaurants, and private clubs featured wild duck, geese and swan as the centerpiece of gourmet meals. This created employment for the market gunners, crusty watermen of the Chesapeake who braved the elements to kill thousands of waterfowl each season to satisfy the orders from the establishments catering to the upper class (Walsh 1971).

Congress acts to save the birds By the early twentieth Century the deadly combination of subsistence, sport, and market gunning, coupled with habitat destruction threatened the once mighty flocks of waterfowl. Although the wealthy man's taste for the water birds help drive down their populations, it also may have helped save them from extinction. The wealthy hunters and consumers of waterfowl lived privileged lives. They occupied the chairs at the fancy duck-roast table as well as the seats of political power. This put them in position to enact and enforce legislation for the protection of their beloved waterfowl. This is exactly what they did when Congress adopted the Lacy Act in 1900 (outlawed transport of poached species across state lines), the 1913 Weeks-McLean Migratory Bird Act (ruled unconstitutional by the courts), the Federal Migratory Bird Law of 1913, and the Migratory Bird Treaty Act of 1918 (MBTA 2018). These Acts of Congress marked the beginning of the federal government taking legislative steps to protect the environment. Prior to these federal acts, it was up to the states to legislate protective measures for hunted species. Throughout the nineteenth Century, Maryland passed a variety of bills to regulate hunting in specific areas of the state.

To appreciate the significance of the Lacy Act and the federal laws that followed, consider the national context of the era. Despite the Union victory in the Civil War, the individual states still ruled supreme in governing the lives of citizens. It was the Jim Crow era when African Americans and other people of color were relegated to second class

Fig. 11 President Benjamin Harrison posing with his gun and a take of ducks from the Chesapeake Bay

status by restrictive state and municipal laws. Retention of Jim Crow required restraining the reach of a Congress that might legislate for its removal. Therefore, for those so concerned, any national level laws were unwelcome intrusions into states' rights. It seems that the strength of the coalition between the conservationists of the Audubon Society and the duck hunters, exceeded the fears of other elements that abhorred federal government overreach. The supporters of the states' rights argument went to the Supreme Court in 1920 to argue their case and lost (Greenspan 2018).

In subsequent years the Migratory Bird Treaty Act was modified to include the bald eagle (1940) and a host of other species (1972) (MBTA 2018). Congress passed the original act in order to reduce hunting pressure on vulnerable species of birds. However, by the 1970s authorities began to interpret and enforce the act to include the unintentional taking of protected species (Vazquez 2017). This motivated agricultural, energy and mining companies to institute best practices to reduce injury and death. Many birds die from pesticides, landing in oiled or toxic waters, flying into flared gas from oil wells, or colliding with wind turbines (Barrios and Rodríguez 2004, Mitra et al. 2011; Ramirez 2010; Ratner and McGowan 2007). However, in 2017 Congress rolled back enforcement of unintentional take to free offending agribusiness, energy companies, and other industries from potential fines. Rep. Liz Cheney (daughter of the former Vice President and energy company executive) championed the legislation to restrict enforcement of the act (Greenspan 2018). Interestingly, former VP Cheney is a duck hunter and resident of the Chesapeake Bay (St. Michaels, MD, Dean 2014; Kilborn 2006).

Under President Franklin Roosevelt the New Deal Congress of 1934 passed the Migratory Bird Hunting and Conservation Stamp Act (MBHCSA 2018). This required hunters to purchase federally issued stamps to allow them to hunt waterfowl or face a fine. The stamps provided a tool for law enforcement officers and a source of funds to restore the degraded wetlands required for successful breeding. Duck stamps have taken on a life of their own. Artists compete in an annual contest to win the honor of having their work featured on the stamp. Collectors buy and trade the stamps. Non-hunters also purchase them to support the habitat restoration efforts (Fig. 14).

Too few ducks and too many geese The legislative protections afforded to waterfowl by the Federal as well as various state and local governments certainly saved many species from extinction but failed to return the flock sizes of ducks to levels seen in the early eighteenth Century. Habitat destruction, environmental degradation, illegal hunting, new forms of human induced mortality (collisions with vehicles, buildings, poisonings, and others) all exacted their toll. Those species most evolved to take advantage of how humans reshaped the environment did the best.

Good data on waterfowl populations in North America exists beginning in 1955. All species showed varying levels of fluctuation in their population size since then. The combined estimated population of all North American ducks was 48 million in 2017. Over the life of the census mallards were consistently the most common duck, fluctuating between 6 and 12 million, showing at 11 million in 2017. Canvasback ducks remained at about the same numbers over the years and reached about 0.7 million in 2017. American widgeon also stayed close to the same, reporting at about 3 million in 2017. Several species of ducks showed strong increases since 1955. These include the blue winged teal, green winged teal, northern shoveler, and redhead. Only the northern pintail and scaup show a sustained decline since 1955. For 2017 Canada goose populations of the Atlantic flyway hover around 195,000, close to the average of the preceding decade. There are about 5 million Canada goose in N. America (USFWS 2018a).

Note that canvasback duck populations remained consistently low (0.6 million or less) since the start of the census in

Fig. 12 President Grover Cleveland gunning for ducks circa 1890. He frequently hunted waterfowl on the northern Chesapeake Bay

Fig. 13 Male (left) and female canvasback ducks https://www.flickr.com/photos/pazzani/4286214360/

1955. Yet their smaller relative, the redhead went from 0.5 million in 1955 to a 1. 3 million in 2017. Both species practice reciprocal nest parasitism, the hens laying some of their eggs in nests that belong to other breeding pairs. The unsuspecting adults incubate the introduced eggs as if they were their own. Redheads parasitize canvasbacks at a much higher rate than the converse (Péron and Koons 2012). Redheads may first lay eggs in the nests of canvasbacks before laying a brood in their own nest. Sometimes the redheads rely entirely upon parasitism for their annual clutch (Baldassarre 2014). This relationship may make canvasback population recovery that much more difficult (USFWS 2017).

In 1972 the US Fish and Wildlife Service (USFWS) closed all hunting of canvasback ducks. The nearly two million canvasbacks estimate in the early 1950s declined to about 0.5 million by the early 1970s. The USFWS noted that adult canvasbacks suffer as much as 50% mortality and that half of that is traced to hunting. Since the available breeding ground habitat of the prairie pothole region in Canada actually doubled during the period of population decline, removing pressure from hunting was the appropriate management tool (USFWS 1972). Hunting of canvasback returned in a

Fig. 14 A duck stamp issued by the US Department of the Interior that features a male (foreground) and female canvasback duck from the art of by Adam Grimm. https://www.fws.gov/birds/get-involved/duck-stamp/federal-duck-stamp-gallery-2014-2015.php

limited way, with short legal open seasons and small daily bag limits to prevent the excessive exploitation of earlier years. Data for the entire Eastern Flyway suggest an average of 0.6 million canvasback between 1955 and 2015, with as many as 0.7 million in 2016 (USFWS 2018a). This indicates some recent recovery, but nothing close to the millions of canvasbacks that once darkened the skies of the nineteenth Century Chesapeake Bay. The story for the other ducks of the Chesapeake mostly mirrors that for the canvasback, still too few.

In stark contrast to the course of canvasback duck, the geese of the Chesapeake, both the Canada goose and the snow goose have recovered well since teetering near extinction in the 1920s. Protection from hunting, active restoration of small breeding stocks, and changes in agriculture brought both species to super abundance by the 1990s (Ankney 1996). Much of the expansion of geese populations traces to an expanded farming system that leaves spilled grains (high in protein and calories) for the birds to eat in fall and winter (Fox et al. 2005). The geese thrive in the human built environment feasting on lawns and crops, and the dearth of predators releases the geese from the natural controls under which they originally evolved.

The overabundance of the geese makes them a nuisance for owners and managers of natural fields, cultivated lands, and turf environments situated near bodies of water. In many areas geese so well adapted to the human landscape that they established year round populations. The super abundant geese populations also serve as reservoirs of pathogens that sicken other animals, including humans. Canada geese distribute oocytes of the diarrhea causing intestinal parasite, *Cryptosporidiurn parvurn* in their feces, contaminating surface waters and feeding grounds (Graczyk et al. 2007). The oocytes survive chlorination, requiring filtration for elimination from drinking water supplies (Oram 2020). Canada geese transmit a variety of other diseases including; *Giardia* spp., *Toxoplasma gondii*, *Campylobacter jejuni*, *Chlamydiosis spp.*, pathogenic strains of *Escherichia coli*, *Listeria spp.*, *Pasturella multocida*, Avian Influenza, and West Nile virus disease (Hygnstrom 2010).

The geese that continue the habit of annual migrations bring their superabundance to remote natural settings. Their sheer numbers strain the carrying capacity of breeding and feeding grounds, degrading the habitat for other species. This is evident in the arctic feeding grounds around Hudson Bay. The migratory snow geese feed almost exclusively on underground plant parts in a foraging technique called grubbing. This results in the loss of both the targeted plant food and cohabiting species, advancing desertification in the coastal and inland marshes (Jano et al. 1998; Jefferies et al. 2006).

Weapons in the war on waterfowl Much of the language of the Migratory Bird Treaty Act of 1918 codified protection of waterfowl by the prohibition of the most effective methods of hunting. The lure of monetary reward sparked the industrialization of waterfowl hunting. The market gunners used their understanding of animal behavior and ecology to devise a variety of massively lethal techniques. This includes modifying age-old methods and inventing a variety on new ones.

Decoys The hunters understood that waterfowl preferred to land in places perceived as safe and providing good foraging. A duck or goose would look for the presence of others of their same species, or guild (a different species that ate similar food) as an indicator of a good place to visit. As noted above,

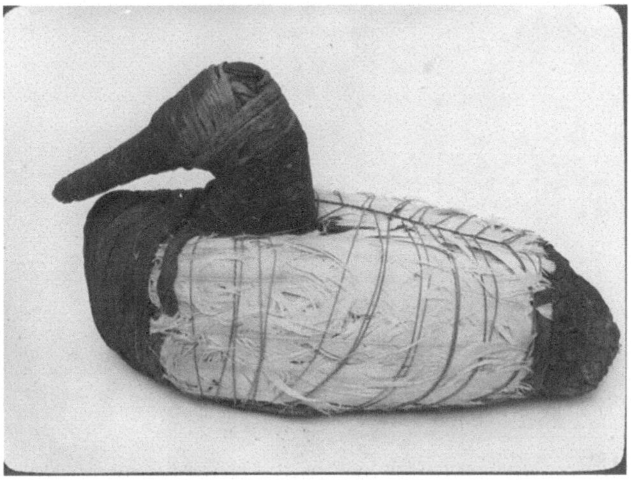

Fig. 15 A canvasback duck decoy discovered in the Lovelock Cave in Nevada, constructed by Native Americans about 2000 years B.P. Braun Research Library Collection, Autry National Center, Los Angeles. LS.3302. https://commons.wikimedia.org/wiki/File:Lovelock_Cave_decoy_Autry.jpg

passenger pigeons behaved similarly. Capitalizing on this behavior, hunters deployed decoys to entice the landing of their potential prey. Some 1600 years before the time of Capt. John Smith, Native Americans used decoys of feathers and reeds in the form and size of canvasback ducks (Fig. 15).

By the nineteenth Century the production of decoys became an important cottage industry everywhere hunters gunned for waterfowl on the Chesapeake Bay. Local artisans carved and painted wooden decoys to help hunters lure their intended prey. Today collectors actively trade and curate the surviving antique decoys. While some modern era waterfowl hunters still deploy the carved wooden versions, most use inexpensive imported plastic molded decoys, paying as little as $40 for a pack of 12. The art of decoy carving survives to serve the home decoration and folk art markets (Krikstan 2013). Expect to pay $50–$300 for decorative decoys. The Decoy Museum in Harve de Grace celebrates decoys and other artifacts of waterfowl hunting that once dominated the economy of this town at the top of the Chesapeake Bay (Decoy Museum 2018). The carved wooden duck decoy remains an icon of the Chesapeake Bay and its food system.

While manufactured decoys worked well in luring waterfowl, hunters discovered that the dead, and better yet living bodies of their intended prey proved more potent. In the nineteenth and early twentieth Century individual hunters and small industries captured and tamed ducklings and goslings to serve as live decoys. Typically, hunters clipped the wings of the tame decoys and tethered one of their legs to a weight to keep the bird swimming in the desired location (Walsh 1971).

Anti-waterfowl artillery Typical eighteenth Century guns for waterfowl shooting consisted of one or two barrels designed to fire loads of many small diameter balls of lead. The guns were loaded from the muzzle, requiring a fair amount of time between shots. By the end of the nineteenth century most hunters switched to breech loaded weapons, allowing for fast replacement of ammunition. In the 1870s automatic loading through pump-action appeared, allowing the hunter to rapidly discharge the weapon several times at a flock of birds. These and other advances in gun powder and weapon design improved the efficiency of the hunt. But market gunners sought to obtain more waterfowl than even the most advanced pump action shotgun could kill.

To increase the efficiency of the process the market gunners took two approaches, use a single bigger gun, or fire multiple guns at one time. Punt gunning followed the first approach. Hunters built oversized shot guns out of large diameter pipe. The guns were so large they couldn't be held by the hunter, but they were instead mounted on a small boat called a punt (Fig. 16). Essentially a small artillery piece, the recoil from the discharge of the punt gun would drive the small boat backwards. Battery gunning took the second approach, mounting from three to twelve somewhat smaller shotguns together in the bow of a boat (Fig. 17). A mechanism permitted simultaneous firing of the array. Market gunners could kill 50–100 birds at a time, and a night of hunting might yield five times that. Market gunning was a dangerous occupation. The normal operation of the artillery would damage hearing. Misfires and exploding gun barrels left some hunters blind, maimed, or dead (Walsh 1971).

Much market gunning took place at night with the aid of the directed beam of light from a lantern (Fig. 17). Although the Migratory Bird Act Treaty banned market gunning, the illegal practice persisted for decades. Night gunning proved effective in producing large yields and in eluding game wardens (Walsh 1971).

Lead kills twice As noted at the beginning of the chapter, hunting with shot meant firing a spray of small lead balls at the target. Lead was readily available, easily formed into the desired shape, and its high density transmitted the kinetic energy imparted from the exploding powder into the body of the target. Yet lead is also dangerous to waterfowl and other birds without having to pierce their flesh. Ingested lead shot is toxic to waterfowl and raptors. Waterfowl often consume spent lead pellets they encounter while foraging on bottom sediments and vegetation. The pellets are the size of gastrolites (grit) that the birds consume to providing grinding action in their gizzards. Wounded or uncollected killed waterfowl may be consumed by predators and scavengers which incidentally ingest the lead shot lodged in their meal (Scheuhammer and Norris 1996).

Fig. 16 A very large caliber punt gun designed for market hunting of waterfowl positioned in a punt boat, circa 1900–1912. https://upload.wikimedia.org/wikipedia/commons/d/d4/Sydney_Harold_Smith_boarding_gun_punt_YORYM-S234.jpg

Ingested lead causes a disease called plumbism in birds, leading to neurological, behavioral, and digestion disorders. It is often fatal after 3 weeks, killing about three million birds in the US annually. Plumbism is most common in the dabbling waterfowl associated with marshes, such as mallard ducks, teal, and Canada goose (De Francisco et al. 2003). Diners risk ingestion of lead from their meal which sometimes exceeds the daily recommended limit (6 µg/day) set by the US Food and Drug Administration (Scheuhammer et al. 1998; FDA 2018). Considering the environmental and human toxicity of lead, it was banned in the US for shooting waterfowl and coots (USGS 2018). The ban eliminates the issue of lead pellets in the flesh, but it doesn't protect the birds and their consumers from the centuries long legacy of lead contaminating aquatic sediments, marshes, and fields. Moreover, lead is still legal for shooting upland species, and some of that spent shot may find its way down slope. Alloys of steel, bismuth, copper, tungsten, iron, tin, and nickel are less toxic and approved for use (USFWS 2018b).

The Sinkbox By the early nineteenth Century, hunters discovered a potent weapon that permitted positioning themselves in the very midst of their prey. The sink box was a very small boat built to accommodate one or sometimes two hunters. Hunters would lay at anchor in the sink box, waiting for their decoys to attract the prey. The name of the boat came from the design which enabled the tinny vessel to float just above the surface. Ridding so low in the water apparently reduced the perception of threat on the part of the waterfowl. To help the boat sink low in the water the hunters placed special decoys of cast iron on the "wings" of the vessels (Figs. 18 and 19). The sink boxes often proved fatal to hunter as well as waterfowl, because unlucky hunters drowned when their vessels were swamped by a wave, or accidently shot through the hull by a mishandled gun (Sullivan 2003). Sink boxes were banned in the US by the Migratory Bird Act Treaty.

Baiting and trapping Hunters often baited an area with corn to lure waterfowl. They typically did so for about a week to habituate the birds to the site that often contained a blind that would eventually hide the shooter. Hunters placed the corn ashore or on a marsh to attract dabbling ducks and geese. For diving ducks, they poured corn into the water where it quickly sank to the sediments. The canvasback and other divers would quickly discover the bait, with many birds soon congregating for the feast. Hundreds of waterfowl might be feeding when the hunter finally fired his weapon.

Hunters also used bait to lure the waterfowl into traps (Fig. 20). The traps usually consisted of serval nested cones of netting set in marshy land. The birds would enter the trap

Fig. 17 The deadly combination of a lantern and a battery gun used to illegally hunt waterfowl at night by market gunners. Image by Tom Knapp. https://www.pinterest.com/pin/428897564494574186/

to eat the corn and have difficulty navigating their way back out. The nested cone traps were essentially the same as those set below the surface of the water to catch fish.

Illegal hunting Hunting out of season, killing more than the specified limit, baiting, trapping, market gunning, sink boxes, and night hunting have been illegal on the Chesapeake Bay for a century. Yet these practices remained in place in remote areas for many years after the ban went into effect. Reminiscent of illegal distillers of alcohol, the outlaw gunners would ship their contraband bounty to restaurants hidden under soft shelled crab or in secret compartments in trucks (Walsh 1971). Even in more recent years some hunters seem willing to risk an encounter with the game warden, fines and imprisonment (Phillips 1999; CNB 2016).

Habitat restoration Strong regulation of hunting has done much to restore the waterfowl populations of the Chesapeake Bay. However, the full recovery of duck populations requires restoration of degraded habitat on the Bay and in the remote reaches of their migration pathways. As noted above, the federal duck stamp system funds such efforts. State authorities, intergovernmental agencies, and private groups play major roles in planning, paying for, and implementing waterfowl habitat. Ducks Unlimited, is a private organization of mostly hunters, focused on habitat restoration (Yerkes 2018). This group conserved 14,500 ha (48,000 acres) of wetlands in Virginia and 4900 ha (19,060 acres) in Maryland (Ducks Unlimited 2018a, b).

Fig. 18 Sketch of a sink box and hunter from the book "American Game Bird Hunting" (1882) by John M. Murphy, Orange Judd, New York. https://commons.wikimedia.org/wiki/File:Sinkbox.jpg

Fig. 19 A sink box with cast iron decoys on its wings to help lower its waterline. Note the railroad bridge in the background, crossing the Susquehanna River

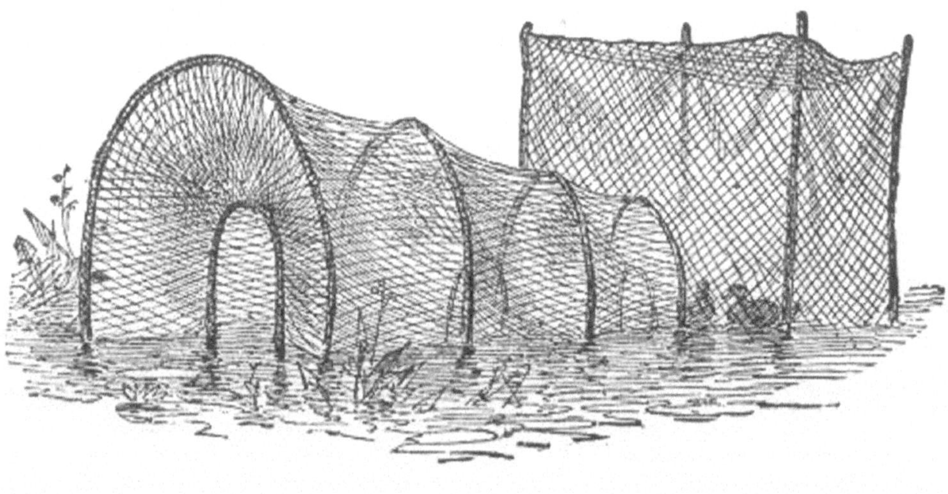

Fig. 20 A trap using a series of cones and typically baited with corn used to capture waterfowl. from The Project Gutenberg EBook of Camp Life in the Woods and the Tricks of Trapping and Trap Making, by William Hamilton Gibson, January 2, 2007. [EBook #17093] No copy right restrictions. http://www.gutenberg.org/files/17093/17093-h/17093-h.htm

Summary The food system caused the biological extinction of the passenger pigeon and commercial extinction of the ducks on the Chesapeake Bay. Industrialization of the production of shot, powder, and guns increased hunting pressure, but it was railroads and ice that enabled the hunters to ship their game to market. The effects of the food system on these migratory species went far beyond the boundaries of the watershed. While poor farming practice reduced habitat and forage within the Chesapeake ecosystem, agriculture also damaged seasonal breeding and feeding grounds used by these temporary residents all over North America. Informed by the demise of the passenger pigeon and the precipitous decline of waterfowl populations, in the early twentieth Century the US Congress passed landmark legislation to protect migratory birds. The measures saved the waterfowl of the Chesapeake from extinction, but populations of most species never returned to the levels of the early nineteenth Century. However, Canada and snow geese so well adapted to opportunities of the industrialized food system and built environment, that they display superabundance that degrades ecosystems and proves a nuisance for human residents of the Chesapeake Bay. Recovery efforts include strict controls on hunting and restoration of waterfowl habitat.

References

Ankney C (1996) An embarrassment of riches: too many geese. J Wildl Manag 60(2):217. https://doi.org/10.2307/3802219

Associated Press (2017) In Maryland, canvasback duck 'represents the hunting culture of the area'—Outdoor news. Retrieved from http://www.outdoornews.com/2017/02/06/maryland-canvasback-duck-represents-hunting-culture-area/

Baldassarre G (2014) Ducks, geese, and swans of North America, vols 1 & 2. Johns Hopkins University Press, Baltimore

Barrios L, Rodríguez A (2004) Behavioural and environmental correlates of soaring-bird mortality at on-shore wind turbines. J Appl Ecol 41(1):72–81. https://doi.org/10.1111/j.1365-2664.2004.00876.x

Bartzen B, Dufour K, Clark R et al (2010) Trends in agricultural impact and recovery of wetlands in prairie Canada. Ecol Appl 20 (2):525–538. https://doi.org/10.1890/08-1650.1

Bayley S, Stotts V, Springer P et al (1978) Changes in submerged aquatic macrophyte populations at the head of Chesapeake Bay, 1958-1975. Estuaries 1(3):171–182

Biello D, Biello D (2014) 3 billion to zero: what happened to the passenger pigeon? [online] Scientific American. Available at: https://www.scientificamerican.com/article/3-billion-to-zero-what-happened-to-the-passenger-pigeon/. Accessed 30 Mar 2018

Carlson J (2013) Hunting for dinner: making coot edible (and a recipe for Coot Cotechino) | simple, good and tasty. Retrieved from http://simplegoodandtasty.com/2013/12/10/hunting-for-dinner-making-coot-edible

Carrollmuseums.org (2018) Carroll Museums: making history yours. [online]. Available at https://www.carrollmuseums.org/history/towerhistory.html . Accessed 29 Mar 2018

CNB (2016) Maryland natural resources police arrest suspects for illegal goose hunting and oyster poaching. Retrieved from http://www.gloucestercitynews.net/clearysnotebook/2016/02/maryland-natural-resources-police-arrest-suspects-for-illegal-goose-hunting-and-oyster-poaching.html

Dean G (2014) Former VP Cheney hunts with Grand National Waterfowl Association. My Eastern Shore MD. [online] Available at. http://www.myeasternshoremd.com/news/dorchester_county/former-vp-cheney-hunts-with-grand-national-waterfowl-association/article_c793d975-acdd-5cdb-a191-0ab35b117189.html. Accessed 7 May 2018

Decoy Museum (2018) HOME. Retrieved from https://decoymuseum.com/

De Francisco N, Ruiz T, Agüera E (2003) Lead and lead toxicity in domestic and free living birds. Avian Pathol 32(1):3–13. https://doi.org/10.1080/0307945021000070660

Doughty R (1975) Feather fashions and bird preservation: a study in nature protection. University of California Press, Berkeley, CA, 183 p

Ducks Unlimited (2018a) Virginia conservation projects. Retrieved from http://www.ducks.org/Virginia/Virginia-Conservation-Projects

Ducks Unlimited (2018b) Maryland conservation projects. Retrieved from http://www.ducks.org/Maryland/Maryland-Conservation-Projects

FDA (2018) Questions and answers on lead in foods. Retrieved from https://www.fda.gov/Food/FoodborneIllnessContaminants/Metals/ucm557424.htm

Fox AD, Madsen J, Boyd H, Kuijken E, Norriss DW, Tombre IM, Stroud DA (2005) Effects of agricultural change on abundance, fitness components and distribution of two arctic-nesting goose populations. Glob Chang Biol 11(6):881–893

Fuller E (2014) The passenger pigeon. Princeton University Press, Princeton, NJ, pp 1–139

Geis A (1959) Annual and shooting mortality estimates for the canvasback. J Wildl Manag 23(3):253–261. https://doi.org/10.2307/3796883

Graczyk T, Cranfield M, Fayer R et al (2007) Infectivity of Cryptosporidium parvum oocysts is retained upon intestinal passage through a migratory water-fowl species (Canada goose, Branta canadensis). Trop Int Health 2(4):341–347. https://doi.org/10.1111/j.1365-3156.1997.tb00149.x

Greenspan J (2018) The history and evolution of the migratory bird treaty act. [online] Available at http://www.audubon.org/news/the-history-and-evolution-migratory-bird-treaty-act. Accessed 6 May 2018

Haramis G, Jorde D, Bunck C (1993) Survival of hatching-year female canvasbacks wintering on Chesapeake Bay. J Wildl Manag 57 (4):763–771. https://doi.org/10.2307/3809077

Haramis G, Jorde, D, Macko S et al (2001) Stable-isotope analysis of canvasback winter diet in upper Chesapeake Bay. Auk 118(4):1008. https://doi.org/10.1642/0004-8038(2001)118[1008:SIAOCW]2.0.CO;2

Hygnstrom S (2010) Canada geese (Branta canadensis) public health. Retrieved from http://icwdm.org/handbook/birds/CanadaGeese/HumanHealth.aspx

Jano A, Jefferies R, Rockwell R (1998) The detection of vegetational change by multitemporal analysis of LANDSAT data: the effects of goose foraging. J Ecol 86(1):93–99. https://doi.org/10.1046/j.1365-2745.1998.00232.x

Jefferies R, Jano A, Abraham K (2006) A biotic agent promotes large-scale catastrophic change in the coastal marshes of Hudson Bay. J Ecol 94(1):234–242. https://doi.org/10.1111/j.1365-2745.2005.01086.x

Kilborn P (2006) Weekends with the president's men. New York Times [online]. Available at https://www.nytimes.com/2006/06/30/travel/escapes/30michaels.html. Accessed 7 May 2018

Kingsville-md.us (2018) Kingsville, MD—history [online]. Available at http://www.kingsville-md.us/history/. Accessed 8 May 2018

Krikstan C (2013) Decoy carvers show Chesapeake Bay birds at Waterfowl Festival | Chesapeake Bay Program. Retrieved from https://www.chesapeakebay.net/news/blog/decoy_carvers_show_chesapeake_bay_birds_at_waterfowl_festival

Lovvorn J (1989) Distributional responses of canvasback ducks to weather and habitat change. J Appl Ecol 26(1):113. https://doi.org/10.2307/2403655

MBHCSA (2018) Migratory bird hunting and conservation stamp act. Retrieved from https://www.fws.gov/laws/lawsdigest/mighunt.html

MBTA (2018) Retrieved from https://legcounsel.house.gov/Comps/Migratory%20Bird%20Treaty%20Act.pdf

Menus (2018) Retrieved from http://menus.nypl.org/menu_pages/41245/explore

Mitra A, Chatterjee C, Mandal F (2011) Synthetic chemical pesticides and their effects on birds. Res J Environ Toxicol 5(2):81–96

NPS (2018) Baltimore travel itinerary—shot tower [online]. Available at https://www.nps.gov/nr/travel/baltimore/b29.htm. Accessed 29 Mar 2018

Oram B (2020) Report #1: Giardia, Cryptosporidium and Waterborne Disease. In: Water-research.net. https://water-research.net/index.php/about/15-reports. Accessed 18 May 2020

Péron G, Koons D (2012) Integrated modeling of communities: parasitism, competition, and demographic synchrony in sympatric ducks. Ecology 93(11):2456–2464. https://doi.org/10.1890/11-1881.1

Perry M (1974) Looking out for the canvasback, Part IV. Ducks Unlimited 38:21–27

Perry M, Wells-Berlin A, Kidwell D et al (2007) Temporal changes of populations and trophic relationships of wintering diving ducks in Chesapeake Bay. Waterbirds 30(sp1):4–16. https://doi.org/10.1675/1524-4695(2007)030[0004:TCOPAT]2.0.CO;2

Phillips A (1999) The raid of Hoopers Island: for Maryland, it's no game. Retrieved from https://www.washingtonpost.com/archive/sports/1999/03/14/the-raid-of-hoopers-island-for-maryland-its-no-game/689a4c4e-1956-42f5-a141-de6af12a6c09/?utm_term=.ec585cc244e5

Raikow R (1973) Locomotor mechanisms in North American ducks. Wilson Bull 85(3):295–307

Ramirez P (2010) Bird mortality in oil field wastewater disposal facilities. Environ Manag 46(5):820–826

Rattner B, McGowan P (2007) Potential hazards of environmental contaminants to avifauna residing in the Chesapeake Bay Estuary. Waterbirds 30(sp1):63–81. https://doi.org/10.1675/1524-4695(2007)030[0063:PHOECT]2.0.CO;2

Scheuhammer A, Norris S (1996) The ecotoxicology of lead shot and lead fishing weights. Ecotoxicology 5(5):279–295. https://doi.org/10.1007/bf00119051

Scheuhammer A, Perrault J, Routhier E et al (1998) Elevated lead concentrations in edible portions of game birds harvested with lead shot. Environ Pollut 102(2–3):251–257. https://doi.org/10.1016/s0269-7491(98)00083-9

Schorger A (1955) The passenger pigeon: its natural history and extinction. University of Wisconsin Press, Madison, WI

Smalley A (2005) 'I just like to kill things': women, men and the gender of sport hunting in the United States, 1940-1973. Gend Hist 17(1):183–209. https://doi.org/10.1111/j.0953-5233.2005.00377.x

Smith J (1612) A map of Virginia. Joseph Barnes, Oxford, pp 1–3. In The complete works of captain John Smith (1580-1631), edited by Philip L. Barbour, (1986) 3 vols., 1:119–190. University of North Carolina Press, Chapel Hill. Available in Capt. John Smith's complete works. http://www.virtualjamestown.org/exist/cocoon/jamestown/fha-js/SmiWorks1

Sullivan C (2003) Waterfowling on the Chesapeake, 1819-1936. Johns Hopkins University Press, Baltimore

Usashooting.org (2018) History of USAS | USA shooting [online]. Available at http://www.usashooting.org/about/history-of-usas. Accessed 30 Mar 2018

USBL (1905) US Bureau of Labor, wages and hours of labor in manufacturing industries, 1890 to 1904 | FRASER | St. Louis Fed. Retrieved from https://fraser.stlouisfed.org/content/?toc_id=498511&filepath=/files/docs/publications/bls/bls_v11_0059_1905.pdf&start_page=5

USFWS (1972) Fish and Wildlife Service [online]. Available at https://www.fws.gov/news/Historic/NewsReleases/1972/19720831.pdf. Accessed 6 May 2018

USFWS (2017) US Fish and Wildlife Service waterfowl population status, 2017. Retrieved from https://www.fws.gov/migratorybirds/pdf/surveys-and-data/Population-status/Waterfowl/WaterfowlPopulationStatusReport17.pdf

USFWS (2018a) Flyways.us. Retrieved from https://flyways.us/status-of-waterfowl/population-estimates/2016-canvasback-population-estimates

USFWS (2018b) Nontoxic shot, U.S. Fish & Wildlife Service—migratory bird program | conserving America's birds. Retrieved from https://www.fws.gov/birds/bird-enthusiasts/hunting/nontoxic.php

USGS (2018) USGS National Wildlife Health Center—lead poisoning. Retrieved from https://www.nwhc.usgs.gov/disease_information/lead_poisoning/

Vazquez M (2017) Clipping the wings of industry: uncertainty in interpretation and enforcement of the migratory bird treaty act. SSRN Electron J. https://doi.org/10.2139/ssrn.3020267

Walsh R (1960) Gunning the Chesapeake: duck and goose shooting on the Eastern Shore. Tidewater, Cambridge, MD, p 118

Walsh H (1971) The outlaw gunner. Tidewater, Centerville, MD, p 178

White C (1989) Chesapeake Bay: nature of the estuary—a field guide. Tidewater, Centerville, MD, pp 88–111

White D, Stedndell R, Mulhern B (1979) Relations of wintering canvasbacks to environmental pollutants-Chesapeake Bay, Maryland. Wilson Bull 91(2):279–287

Yerkes T (2018) Conserving the Chesapeake Bay [online]. Ducks.org. Available at http://www.ducks.org/conservation/where-ducks-unlimited-works/us-great-lakes-system/conserving-the-chesapeake-bay. Accessed 30 Mar 2018

Part III

Industrial-Chemical Agriculture Reshapes the Bay's Ecosystem

The Journey from Peruvian Guano to Artificial Fertilizer Ends with Too Much Nitrogen in the Chesapeake Bay

Benjamin E. Cuker and Jeffery Cornwell

The element nitrogen is essential to all life on earth.
But most of that what's in the air doesn't seem much worth.
The dinitrogen that dominates the atmosphere is joined together by a triple bond.
Of this plants are not very fond.
They need the help of bacteria to make that airy nitrogen reactive for them to use.
Or an artificial process to make synthetic nitrogen for people to abuse.
Reactive nitrogen from the chemist does require lots of fossil fuel to make.
And it changes form in the soil, leaving more air pollution in its wake.
When applied as a fertilizer it makes crops and weeds grow fast.
And turns the water so green with algae that it looks nothing like in the past.

Abstract

By the mid-nineteenth century, farmers in the Chesapeake Bay Watershed began importing fertilizers to restore productivity to depleted fields. This began with guano from remote Peruvian shores, which created a market for soil amendments that switched to artificial fertilizer in the twentieth century. The widespread use of artificial fertilizers and animal manures to supply reactive forms of nitrogen to crops resulted in increased productivity per unit area, but also many unintended consequences including; pollution of ground and surface waters, increases in weeds and agricultural pests, air pollution, reduced biodiversity and release of global warming gases. Nitrogen pollution of the Chesapeake Bay caused the onset of persistent eutrophication in the later half of the twentieth century. The USEPA is working with jurisdictions to reduce nitrogen pollution through a series of best management practices that facilitate preserving the current food system while reducing its impact on the environment. Moving the food system away from a focus on animal agriculture and toward producing crops for a whole-food-plant-based-diet will speed the recovery of the Chesapeake Bay ecosystem from the ravages of excessive reactive nitrogen.

Keywords

Nitrogen Pollution · Chesapeake Bay · Artificial fertilizer · Restoration

A ship from Peru A ship bearing cargo from Peru tied-up in the Baltimore Harbor in 1832. It delivered two casks of guano to John A. Skinner, editor of the *American Farmer*. Thus began a trade that bound the food system of the Chesapeake region and beyond, to a place half-a-world away (Fig. 1). Farmers saw their yields triple after application of guano to soils exhausted by years of tobacco, corn and cotton cultivation (Craven 1926). The farmers pressed Congress to ensure a steady supply of this imported manure. To this day, the US still lays claim to distant isles it first took under the Guano Islands Act of 1856 (NMAH 2018). In 1861, ships delivered 53,959 tons of guano to Baltimore, where factories prepared it for shipping to farms throughout the Chesapeake watershed and beyond (Lesher 2008; Guano 2018). The guano trade did more than fertilize depleted fields.

B. E. Cuker (✉)
Department of Marine and Environmental Science, Hampton University, Hampton, Virginia, USA
e-mail: Benjamin.cuker@hamptonu.edu

J. Cornwell
University of Maryland Center for Environmental Science, Cambridge, MD, USA
e-mail: cornwell@umces.edu

B. E. Cuker (ed.), *Diet for a Sustainable Ecosystem*, Estuaries of the World, https://doi.org/10.1007/978-3-030-45481-4_10

Fig. 1 The Olivbank, a ship built in 1892 for the guano trade. In this photograph the boat is anchored at the Seychelles Islands and loading guano from a smaller vessel. In the nineteenth and early twentieth Centuries ships like this brought guano from Peru and other locations to the port of Baltimore for use in the Chesapeake Bay watershed and beyond to meet the demand for nitrogen rich fertilizers. From: https://commons.wikimedia.org/wiki/File:Olivebank_(ship,_1892)_-_SLV_H99.220-4071.jpg

It also developed the market of dependency on imported fertilizers to subsidize US farming. This eventually led to the era of chemical-industrial agriculture that reshaped farming, the food system and the people of the Chesapeake region during the twentieth century.

The rocky shores of Peru and its islands provided almost 90% of the guano shipped to Baltimore (Lesher 2008). Seabirds along the Peruvian coast feast on the immense schools of fish that churn the waters. Prevailing winds from the south that work on surface waters create a pattern of circulation called an upwelling. The upwelling drives deep water to the surface. This newly arrived water is rich in inorganic nutrients including nitrogen and phosphorus. These nutrients stimulate blooms of phytoplankton that fuel one of the most productive food chains in the world; phytoplankton -> zooplankton -> small fish (anchovy) -> seabirds. While other animals partake in the bounty, it is the seabirds that connected this far-flung ecosystem to the Chesapeake region in the nineteenth century. Relatives of the Chesapeake's avian fauna, the Peruvian pelican (*Pelecanus thagus*) and the Guanay Cormorant (*Phalacrocorax bougainvillii*) spend their days fishing Peruvian anchovy (*Engraulis ringens*). The seabirds nest and roost on the rocky shore, covering it in a constant rain of droppings. All birds use a common cloacae to eliminate wastes from the intestines and kidneys. The resulting guano is a rich mixture of nitrogen, phosphorus, potassium, and other plant nutrients. Seabird guano is especially high in nitrogen as these birds get all of their calories from eating muscular and oily fish. Recall that nitrogen is a key constituent of muscle tissue (IUCN 2018).

The accumulated droppings of seabirds from thousands of years of habitation produced deposits over 60 m (200 ft) deep in many places. Mining depleted many of the best guano deposits by the end of the nineteenth century. This also destroyed considerable roosting habitat (IUCN 2018). Ultimately the invention of synthetic nitrogen fertilizer in the early twentieth century ended the era of massive guano exports for Peru. The Peruvians switched to harvesting the anchovy directly, building an economy based upon capturing live fish rather than what remained after passing through the guts of seabirds. Ever since, the Peruvian anchovies supported the largest fishery in the world, with more weight of that fish landed than any other species. Most of the catch becomes fish meal and animal feed (Iwamoto et al. 2010).

Why did the farmers of the Chesapeake region need to go so far away to get guano? After all, the Chesapeake has its own anchovy (bay anchovy, *Anchoa mitchilli*) and birds that eat it. While the Chesapeake Bay is productive, it can't match the output of the Peruvian coastal upwelling ecosystem in producing fish, fish-eating birds, and bird guano (Tarazona and Arntz 2001). And what guano is produced on the shores of the Chesapeake Bay won't accumulate as it does in Peru. The humidity of the Bay's climate and frequent rains prevent the formation of the type of thick dried deposits of guano that characterize the dry cliffs of the Peruvian shore.

Well before the era of imported guano, farmers understood the benefits of amending their fields with manures. This ancient practice informed both the native peoples of the Chesapeake and European colonists. Indeed, Capt. John Smith used the term "manuring" as synonymous with agriculture. For example, he wrote in his *Map of Virginia*, "... heaven and earth never agreed better to frame a place for man's habitation being of our constitutions, were it fully manured and inhabited by industrious people (Smith 1612)." The name menhaden for the very important bay and coastal fish (*Brevoortia tyrannus*) is a corruption of the Wampanoag Indian name (*munnawhatteaug*) for manure (Franklin 2007). Given a choice, most people would rather eat something other than the boney, oily, and smelly menhaden, hence its adoption as a fertilizer by coastal Native Americans and colonists (Goode 1880).

The basis of soil fertility This chapter is about how the food system altered the use and cycling of the key plant-nutrient, nitrogen in the Chesapeake Bay ecosystem. To understand the story of nitrogen requires an appreciation of some basic soil science. The fertility of agricultural soils depends upon numerous factors including mineral particle size, pH, temperature, ability to hold and exchange ions, organic content, microbial and small animal activity, aeration, water holding ability, drainage, slope, topsoil depth, and the availability of inorganic nutrients needed to support plant growth (Brady and Weil 2017).

The best agricultural soils are called loams, a roughly even mix of soil grain sizes called clay (<0.002 mm), silt (0.002–0.06 mm) and sand (0.06–2 mm). The small clay and silt particles provide a huge surface area to hold and exchange nutrient ions and water molecules. The spaces between the larger sand particles creates pores in the soil. Such soil porosity is critical for aeration. This allows for exchange of gases between the soil and the atmosphere, which brings oxygen to plant roots and releases carbon dioxide, methane and other volatiles to the atmosphere. Good soil porosity also allows for the penetration and drainage of water. This reduces ponding of water on the surface, erosion of soil by runoff, and perpetually water-logged soils absent of oxygen. Soils dominated by clays are dense, of low porosity, prone to ponding and runoff, difficult to cultivate, and resistant to deep root penetration. Soils dominated by sands are too porous and unable to store nutrient ions or water (Brady and Weil 2017).

Ideal loams don't exist in many areas of the Chesapeake Bay Watershed (CBW). Clays eroded from the Appalachian Mountains dominate some areas, while sands deposited by ancient seas characterize much of the coastal plain. So farmers must amend the soil with plant or animal manures to compensate for an imbalance of clay, silt and sand. Adding organic matter to both clay-dominated and sand-dominated soils improves their quality. In the case of clay-dominated soils, the decomposing organic matter and resistant plant fibers creates larger particles and spaces in the soil, improving porosity. For sandy soils, manures bind and release both nutrient ions and water molecules, making up for the lack of clay and silt. Manures may originate on-site by cutting cover crops and leaving them to decompose. Manures may also come from off-site, such as straw grown on other fields or dung collected from cow and chicken operations.

In addition to compensating for an imbalance of grain size distribution, the microbial decomposition of manures provides a source of key inorganic nutrients including ions of nitrogen, phosphorus, and potassium. Importantly, manures tend to release these nutrients over the entire growing season as decomposition progresses (Gale et al. 2006). The lack of nitrogen often limits crop growth, so farmers prize manures rich in this factor. Hence the popularity of guano discussed above.

Most crops prefer slightly acid to neutral soils in the range of pH 6.0–7.0. Amending the soil with lime ($CaCO_3$) or lye ($NaOH$) corrects the low-pH of clay dominated soils. Adjusting pH also increases the availability of nitrogen and other nutrient ions (Nicol et al. 2008).

Although the coastal-plain portion of the Chesapeake Bay Watershed shows little relief, most of the lands draining into the system are sloped. Thus farmers should employ contour plowing to create fields of mini-terraces that check the surface runoff of water and the attendant erosion of fertile soil. This practice protects both soil fertility and the Bay from sediment and nutrient pollution.

Agricultural soils reflect the history of crops grown on them over the years. The very nature of agriculture is to produce food or fiber to be used off-site. This means that a portion of the soil fertility is exported in plant tissues with each bushel of produce or bale of straw. The continuous weathering of soil particles and atmospheric deposition replaces some of the lost fertility, but over time the productivity drops as the fields become exhausted. Tobacco, corn and cotton famously took their toll on the soils of Virginia and Maryland (Craven 1926). There are five approaches to replenishing or preventing depleted soils: Poly-cultures of complementary plants, subsidies of imported animal and plant manures, cover cropping, crop rotation and inorganic fertilizers.

Native Americans practiced poly-culture with the three sisters of maize, squash and beans (Landon 2008). Planting these in close proximity (the beans would wrap around the stalks of maize) meant no one crop would unbalance the soil with its particular demand for nutrients. The beans would also bring nitrogen to the soils as discussed later. Poly-culture doesn't lend itself to mechanical sowing, weeding and harvesting.

Amending soil with manures generated offsite is quite effective in restoring soil fertility, as discussed above. But it requires access to either fields for growing plant manures, or concentrated stores of dung.

Cover-cropping involves planting a fast growing species during the gap between fall-harvest and spring-sowing of the intended food or fiber crop. The cover-crop is either mowed and left to decompose in place, or worked into the soil with a tilling device. Cover corps restore soil fertility in three ways. Those that can fix atmospheric nitrogen (legumes) return that nutrient to the soil. The roots of the cover crop hold the soil in place and protects against erosion. When the cover crop is mowed or tilled, it decomposes and enriches the soil with organic matter and inorganic nutrients.

Crop rotation takes advantage of the understanding of how different plants add to and subtract from the fertility of the soil. It is similar to poly-culture in that regard, but separates the crops by time. Typically, a nitrogen-fixing crop will help return fertility to the soil during the rotation.

Poly-culture, manures, cover-cropping and crop rotation all come out of traditional agriculture and lend themselves to the practice of organic farming as discussed in chapter "An Organic-Based Food System: A Voyage Back and Forward in Time". The fifth method, fertilizing with inorganic nutrients became part of the later colonial-era landscape. George Washington experimented with gypsum (calcium sulfate) (Craven 1926). But such soil amendments didn't gain wide use until the development and marketing of super-phosphate fertilizer in 1870 (Dixon 2018). This, of course, was only effective for crops limited by phosphorus. The Haber-Bosch process introduced N-based artificial fertilizer, ushering in the post WWII era of chemical (conventional) agriculture which currently dominates the landscape of the Chesapeake Bay and the US in general. This involves industrial processes to isolate specific nutrients from mineral sources for easy application to the soil in liquid or solid form. The big three inorganic nutrients are nitrogen (N), phosphorus (P) and potassium (K). They are typically marketed in a mix, so the farmer applies all three at once in a set ratio.

Nitrogen as the limiting factor To understand the relative importance of the various nutrients in the soil consider what elements make up the bulk of plant tissue. The carbohydrates, proteins, lipids, and nucleic acids of plants consist of

Fig. 2 Liebig's barrel used to illustrate the law of the minimum. The shortest stave determines the maximum height of water in the barrel. For soils, that required nutrient that is in the shortest relative supply will set the limit for plant growth, and called the *limiting factor*. Note that the relative and not the absolute amount of the nutrient present is what counts. So if a plant needs the nutrients carbon, nitrogen and phosphorus in the ratio of 106 (C):16 (N):1(P), and the availability in the soil (air for carbon) is 106 (C):10 (N):1(P), then N would be the limiting factor. For the illustrated barrel, the shortest stave would correspond to the availability of N from the soil. Image modified from: https://commons.wikimedia.org/wiki/File:Minimum-Tonne.svg

molecules built from atoms of carbon (C), oxygen (O), hydrogen (H), nitrogen (N), phosphorus (P) and sulfur (S). The first three of these elements, C, O and H, are the building blocks for all of the organic molecules of life. It is not surprising that these three elements make up 90% of the dry mass of most terrestrial plants. In every gram of dry mass of terrestrial plant tissue there is approximately 450 mg of C, 400 mg of O, and 50 mg of H (Knecht and Goransson 2004). The environment generously supplies these three elements from the atmosphere (CO_2 captured in photosynthesis) and water. Soil primarily provides the other three main macronutrients; N, P and S, along with a host of other atoms needed in lesser amounts (micronutrients) such as potassium, calcium, iron, magnesium, copper, etc. The addition of N, P and S is what transforms the carbohydrate skeleton into amino acids (proteins), fatty acids (lipids), and nucleic acids (genetic material and energy carrying molecules). So although N, P and S comprise only a small fraction of a plant's mass, the availability of these three elements from the soil largely determines the productivity of crops.

We owe much of our understanding of soil nutrients and crops to the nineteenth century German chemist, Justus von Liebig (1803–1873). He noted that plants required nutrients in particular ratios, and if a specific element was in short supply relative to others that it would prevent further growth. He called the nutrient in short supply relative to the others the *limiting factor*. Additions of the limiting nutrient to soil would satisfy the plant's requirements for growth (Fig. 2). This is why adding fertilizers to soils promotes the growth of crops. This practice provides more of which ever nutrient (s) that was (were) limiting growth. Liebig further understood that manures contained a mix of various plant nutrients, most of which were not limiting crop growth at any particular time. So he advocated for application of the specific limiting inorganic nutrients, particularly nitrogen (Liebig 1847). Indeed, nitrogen is the nutrient most often limiting agricultural production around the world (Rütting et al. 2018).

Liebig's work set the stage for two other German chemists to bring us into the age of chemical agriculture. Fritz Haber (1868–1934) won the Nobel Prize in Chemistry in 1918 for inventing a high-pressure and energy-intensive process to convert atmospheric nitrogen (N_2) in to ammonia (NH_3). Carl Bosch (1874–1940) won the Nobel Prize in Chemistry in 1932 for industrializing Haber's process. The Haber-Bosch process affected many spheres of industrial production that required reactive nitrogen, and two, in particular, the making of munitions for war and agriculture (Fig. 3).

The three Germans, Liebig, Haber and Bosch, took a reductionist approach to the issue of nitrogen limitation, setting up a solution to the problem using a synthetic process to isolate that element and to work around nature. Unwittingly, another German set the stage for a very different approach to the problem of nitrogen limitation for agriculture. German immigrant and Missouri famer, Moses Carver purchased a slave named Mary in 1855. Raiders stole Mary and infant son, George in 1862. Moses Carver regained the infant, but not George's mother. Moses Carver and his wife raised George as their son, George Washington Carver (circa 1860–1943).

George Washington Carver went on to earn a Ph.D. in Botany and lead an agricultural revolution to restore nitrogen-depleted farms across the US, including in the watershed of the Chesapeake Bay. Although stationed at the Tuskegee Institute in Alabama, he directed his work at African American famers across the US, using conferences, speeches, and publications to reach them. In one address at Hampton Institute (located at the southern end of the Chesapeake Bay) in 1899, he admonished the farmers to, "embrace mutual relationship of the animal, mineral and vegetable kingdoms, and how utterly impossible it is for one to exist in a highly organized state without the other (Hersey 2006)." In essence, he espoused an escape from rural poverty through adoption of ecological principles in agriculture. Although his work targeted African American farmers, his techniques certainly helped recover southern agriculture in general (Hersey 2006). George Washington Carver promoted crop rotation and the restorative ability of legumes to return nitrogen to the soil. Quite appropriately, his name sake, founding father George Washington, experimented with such legumes and crop rotation on his Chesapeake Bay estate a century earlier (Craven 1926). In contrast to Haber and Bosch, George Washington Carver used his holistic understanding of nature and agriculture to develop an ecological approach to solving the problem of nitrogen limitation, working with nature rather than around it (Kremer 2011). His work set the stage for the development of the modern organic farming movement that developed in the twentieth century, and is the subject of chapter "An Organic-Based Food System: A Voyage Back and Forward in Time".

Production of Artificial Nitrogen Fertilizer comes to the Chesapeake Bay In 1928, the Allied Signal Corporation opened the world's largest ammonia manufacturing plant in Hopewell, VA on the shores of the James River (Honeywell 2019). This plant supplied artificial fertilizer to

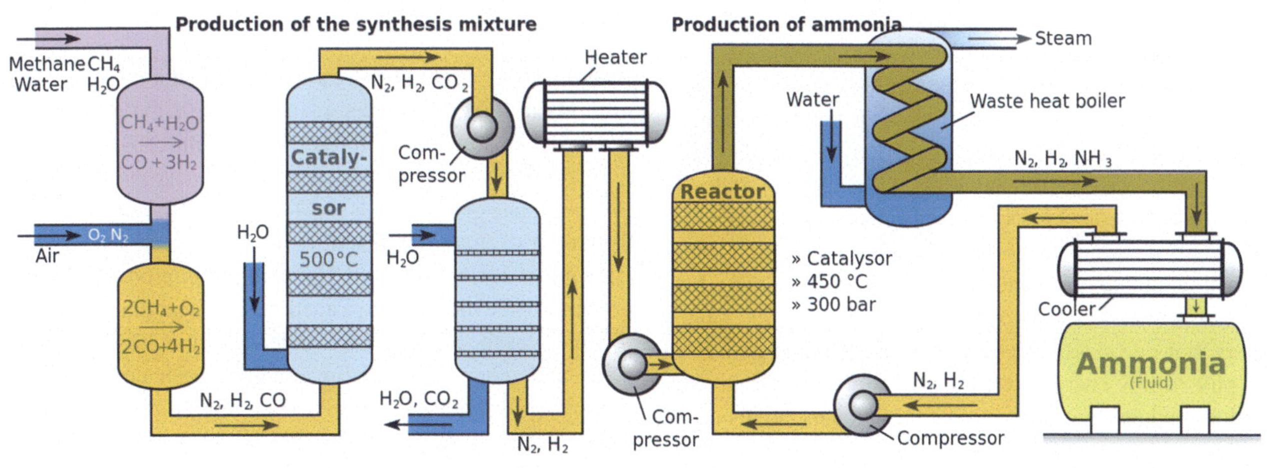

Fig. 3 The Haber-Bosch process to reduce atmospheric nitrogen (N_2) into ammonia (NH_3). This process uses energy and high pressure to break the triple bond that that joins the two nitrogen atoms together in N_2. https://commons.wikimedia.org/wiki/File:Haber-Bosch-En.svg

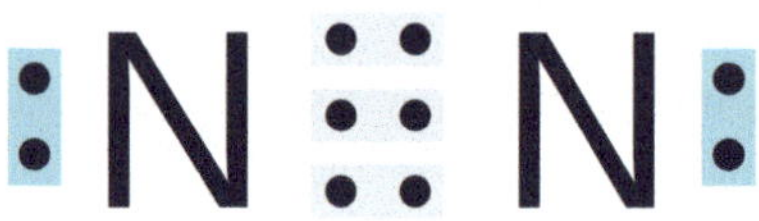

Fig. 4 Lewis dot structure of dinitrogen (N_2), illustrating the electron sharing in the very strong triple bond between the two atoms. The high energy needed to either reduce or oxidize N_2 restricts the availability of reactive nitrogen in the environment. https://commons.wikimedia.org/wiki/File:Electron-fomura_triple-bond_NN.svg

the entire Chesapeake Bay region, its strategic location facilitated transport of its products by rail, road and waterways.

The nitrogen cycling network To understand the role of nitrogen in the Chesapeake Bay food system and ecosystem let's explore the fundamental chemistry of this element and how it cycles in nature. The element nitrogen has an atomic number of seven, yielding three free-electrons in orbitals that readily participate in covalent bonding with other atoms. This includes bonding with another nitrogen atom to form the very stable and un-reactive molecule, dinitorgen (N_2). The stability and inertness of N_2 traces to the strength of the triple covalent bond holding the two atoms together (Fig. 4). That the lack of N most often limits agricultural production would suggest that the element is relatively rare in nature, yet this is far from the truth. Nitrogen comprises 78% of the atmosphere, being one of the most common elements on Earth. Its mass in the atmosphere, hydrosphere and biosphere exceeds that of the combination of C, P, O and S. However, atmospheric nitrogen is almost all in the unreactive form of N_2. It is ever present, yet unavailable to participate in the chemistry of life, *unless changed to one of the forms of reactive nitrogen* (Nr) (Bernhard 2010). For N to become reactive, it must become part of a compound that is either reduced (with electrons to donate) or oxidized (with the ability to accept electrons) (Galloway et al. 2003).

The nitrogen cycle network is as much about supplying energy to the microbial community as it is about providing a building block to plants and animals (Kuypers et al. 2018). As Nr cycles through the environment, it changes states of oxidation/reduction. The Nr provides plants and algae with a source of nitrogen for building amino acids and other nitrogen-containing organic compounds. For *nitrifying* bacteria, Nr provides energy to fix atmospheric CO_2 into simple sugars. For one class of *denitrifying* microbes, Nr acts as an acceptor of low energy electrons in the absence of oxygen, and for another Nr is a source of energy for metabolism.

Nitrogen fixation The process of *nitrogen fixation* transforms N_2 to Nr. It takes much energy to break the triple bond of N_2. The bond-dissociation energy for N_2 is 226 kcal/mol, more than twice that for the double bond of O_2 (119 kcal/mol) or a single C-C bond (84 kcal/mol, Blanksby and Ellison 2003). In nature, that energy to break the N_2 triple bond comes from either abiotic or biotic sources. Lightning is the primary abiotic source of natural N-fixation, accounting for 2.5% of the total annual natural planetary budget (Fowler et al. 2013). The high-energy electrical discharge converts N_2 to NOx (nitrous oxides), that dissolve in atmospheric water droplets. Rain brings the NOx to earth where it fertilizes soils.

Biology is the source of over 95% of the global natural N-fixation (Fowler et al. 2013). Unlike lightning which produces nitrous oxides, biological N-fixation yields the reduced compound, ammonia (NH_3). Various microbes (prokaryotes and Archaea) are *diazotrophs*, possessing the nitrogenase enzyme complex to catalyze the reaction that converts N_2 to NH_3. The summation reaction for this four-step process is: $N_2 + 8H + 8e^- + 16\ ATP = 2NH_3 + H_2 + 16\ ADP + 16\ Pi$. ATP is the high energy molecule adenosine triphosphate which cycles to the lower energy adenosine diphosphate and Pi (inorganic phosphate). Dizatrophs include free living forms common in soil such as some Cyanobacteria and *Azotobacter*, and others living in mutualistic symbiosis with plants, such as *Rhizobium* in the root nodules of legumes (Fig. 5). *Acetobacter diaszorophicus* is an N-fixing mutualistic bacteria living in the aerial portions of sweet potatoes (*Ipomoea batatas*, Yonebayashi et al. 2014). Diazotrophs include both aerobic and anaerobic species (ex. *Clostridium*, purple sulfur bacteria, purple non-sulfur bacteria, green sulfur bacteria). The natural solution to N-depleted soils promoted by George Washington Carver and the modern organic farming movement capitalizes on promoting biological N-fixation primarily through the planting of leguminous crops. The mutualistic *Rhizobium* associated with legumes, and the *Acetobacter* of sweet potato supplies the host plant with fixed-N in return for energy-rich photosynthetic compounds (Fig. 6).

Assimilation provides building blocks to plants and algae Fixed-N enters plants through the root hairs, primarily as ammonium ($NH_4^{\ +}$) from anaerobic soils and nitrate (NO_3^-) under aerobic conditions. Anaerobic soils tend to accumulate organic acids that lower the pH. This drives the formation of ammonium (NH_4) from ammonia (NH_3). The equation is $H^+ + NH_3 \rightarrow NH_4^+$. Nitrate is produced in the process of nitrification explained below. The plant or algae incorporates the ammonia or nitrate into amino acids, nucleic acids and other N-containing organic molecules and thus producing *organic nitrogen*. Increasing N-assimilation in crops means more production and is the point of applying N-rich fertilizers (Xu et al. 2012).

Ammonification feeds energy to decomposers The organic nitrogen produced in assimilation eventually returns

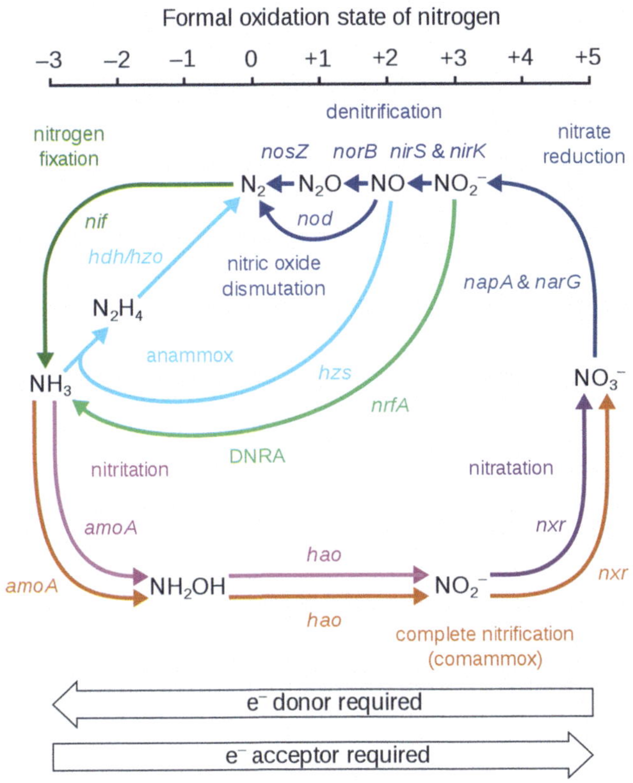

Fig. 5 Diagram of the biological component of the nitrogen cycle and the accompanying flow of energy, including the genes responsible for making the necessary enzymes. Genes and associated enzymes: nif (nitrogenase complex), hdh/hzo (amamox hydrazine dehydrogenase), amoA (ammonia monooxygenase), hao (hydroxylamine oxidoreductase), nxr (nitrite oxidoreductase) napA & narG (nitrite reductase), DNRA (Dissimilatory nitrate reduction to ammonium), nrfA (nitrite reductase), nirK & nirS (nitrite reductase), norB (nitric oxide reductase), nosZ (nitrous-oxide reductase). Image by Vinicius Waldow: https://en.wikipedia.org/wiki/Nitrogen_cycle#/media/File:Nitrogen_Cycle_-_Reactions_and_Enzymes.svg

to ammonium. Excreted waste and dead organisms supply decomposers (bacteria and fungi) with organic nitrogen. The decomposers obtain energy from the organic nitrogen as they mineralize it back to NH_4. Mineralization of organic-N is what happens as manures decompose. This makes Nr available to the crop. Since decomposition is a temperature dependent biological process, it fertilizes the soil over an extended period of time and particularly during the warm months of the growing season (Kaur et al. 2005).

Nitrification feeds energy to chemoautotrophs Ammonium is converted to nitrate, the form most available to plants, in the two-step process of *nitrification*. Bacteria such as *Nitrosomonas* convert NH_4^+ to nitrite (NO_2^-). *Nitrobacter* and other species oxidize nitrite to nitrate (NO_3^-). Formulas for the two steps are: $NH_3 + O_2 \rightarrow NO_2^- + 3H^+ + 2e^-$, followed by $NO_2^- + H_2O \rightarrow NO_3^- + 2H^+ + 2e^-$. Both steps produce ATP that these chemoautotrophic bacteria use to fix carbon from the atmosphere (CO_2) in to simple sugars. Nitrification may also result in the production of nitrous oxide (N_2O), an important greenhouse gas and air pollutant (Ji et al. 2015).

Denitrification enables anaerobic respiration, converting Nr to N_2 Anaerobic bacteria such as *Pseudomonas* and *Clostridium* breakdown organic molecules to obtain energy. Lacking available free oxygen in their environment they need an alternative acceptor for the low-energy electrons released in respiration. Nitrate serves this purpose. Accepting the low energy electrons transforms the nitrate into dinitrogen (N_2). Denitrification may be a multi stage process that takes advantage of the step-wise redox status of intermediary products (from best to poorest electron acceptor in the order is $NO_3^- > NO_2^- > NO > N_2O$). Denitrification ultimately results in the production of N_2. The overall equation is : $2\ NO_3^- + 10\ e^- + 12\ H^+ \rightarrow N_2 + 6\ H_2O$ (Seitzinger et al. 2006). Denitrification may also produce nitrous oxide.

Dissimilatory nitrate reduction to ammonium (DNRA) also enables anaerobic respiration and conserves Nr As shown above, some anaerobic bacteria in need of a low-energy electron acceptor convert nitrate to N_2 in the process of denitrification. DNRA provides an alternative pathway for the low-energy electron acceptor. In this case a three-step process that reduces NO_3^- to NO_2^- which is further reduced to $NH4^+$. DNRA is more efficient than denitrification in collecting low-energy electrons, as it takes more of these electrons to produce NH_4^+ than it does to yield N_2. So DNRA may predominate when nitrate concentrations are low. In contrast to denitrification, DNRA conserves Nr in the ecosystem (Lam and Kuypers 2011).

Anaerobic Ammonia Oxidation (anammox) provides energy to bacteria, converting Nr to N_2 Anammox bacteria use a combination of ammonium and nitrite ions as a source of energy, resulting in production of N_2. The general reaction is: $NH_4^+ + NO_2^- \rightarrow N_2 + 2H_2O$ + energy. This relatively recently discovered process may account for much of the loss of Nr to N_2 in aquatic systems (Kuypers et al. 2018).

Nitrous oxide production from nitrification, denitrification and combustion N_2O is produced by bacteria during both nitrification and denitrification. It is also a product of combustion of fossil fuels and other materials.

Artificial fertilizers, fossil fuels and agricultural biological nitrogen fixation adds reactive nitrogen to the environment Humans began managing the nitrogen cycle many centuries before the element was ever identified.

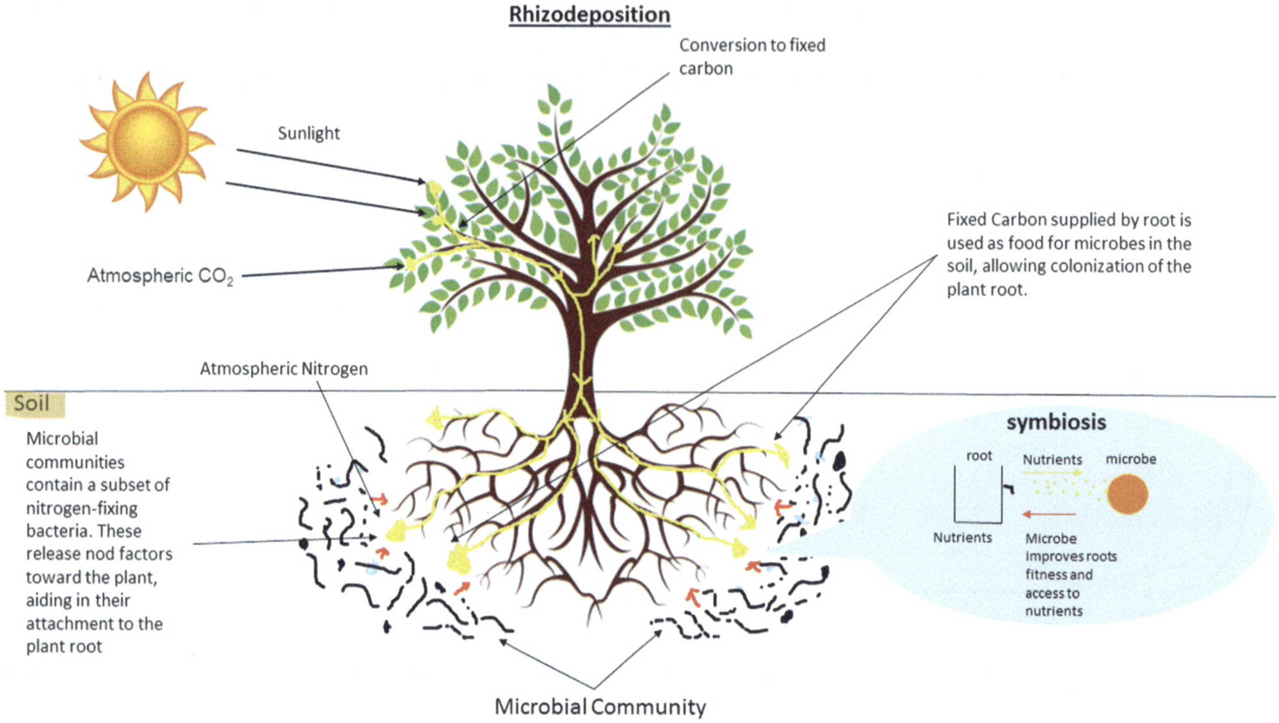

Fig. 6 Diagram illustrating the mutualistic symbiotic relationship between plants and their microbial nitrogen-fixing partners. The host plant provides photosynthates (sugars) to the bacteria that live in the special nodules on the roost that provide low-oxygen environments. The bacteria fix atmospheric nitrogen (N_2) in to the reduced molecule ammonia (NH_3), making reactive nitrogen available to the host plant. Author: Hallrob3. https://upload.wikimedia.org/wikipedia/commons/d/dc/Rhizodeposition.png

Harvesting crops such as cotton and tobacco removed nitrogen from the soil, sending it away in the fibers of clothing or in clouds of nicotine laced smoke. Planting N-fixing leguminous crops and fertilizing with nitrogen-rich manures added or returned this nutrient to the soil. The Haber-Bosch process freed the farmer from the necessity to work with nature as a partner, elevating the twentieth century agriculturist to the position of a lord of the land, or at least to make him think so. Humans contributed additional Nr to the ecosystem that they created by purposeful chemistry for agriculture and war. The growing dependence on fossil fuels also added Nr to the world's ecosystems. After all, fossil fuels come from time, pressure and heat working on the residue of life. And the molecules of life include nitrogen and sulfur, and various other atoms in addition to the hydrogen and carbon that creates the energy rich hydrocarbons of petroleum, natural gas, coal, and bitumen. For example, depending on the type of coal, it consists of 60–95% C and 1–3% N (Fossil Fuels 2018). So burning fossil fuels releases significant amounts of nitrogen to the atmosphere, primarily in the form of nitrous oxides (Fowler et al. 2013).

The annual budget for global N-fixation is about 413 Tg (Tera gram = 10^{12} g) of Nr, half of that coming from human activity. The natural portion of N-fixations is 203 Tg per year with; 58 Tg from biological nitrogen fixation (BNF) on land, 140 Tg BNF to the ocean and 5 Tg from lightning. The anthropogenic N-fixation is 210 Tg; with 120 Tg from Haber-Bosch fertilizer production, 60 Tg from agricultural BNF, and 30 Tg from fossil fuel combustion. So in terms of percent, the anthropogenic production of Nr is 57% from Haber-Bosch, 29% from agricultural BNF, and 14% from fossil fuels (Fowler et al. 2013).

Extra Nr from the food system changes the Chesapeake Bay Ecosystem As noted above, most of anthropogenic Nr comes in the reduced form from the Haber-Bosch process for making fertilizer. As intended, this changes the agro ecosystem by eliminating N-limitation of crops. However, using artificial N fertilizer results in several important unintended consequences that create problems on the farm and in the greater ecosystem.

Artificial N-fertilizers increase insect pest attack, the growth of weeds and the use of toxins used to control them An unintended consequences of applying Haber-Bosch Nr to crops is the increase in pest and weed populations. Inorganic N fertilizer increases the populations of various invertebrate pests including European red mite (*Panoychus ulmi*), two-spotted spider mite (*Tetranychus telarius*), clover mite (*Bryobia praetiosa*), greenhouse thrip (*Heliotrips haemorrhoidalis*), green peach aphid (*Myzus persicae*), greenbug (*Schizaphis graminum*), corn leaf aphid

(*Rhopalosiphum maidis*) and spotted alfalfa aphid (*Therioaphis maculate*). Between them, these pests attack apples, beans, peaches, tomatoes, Brussel sprouts, oats, rye, sorghum, and alfalfa. The Nr-fertilizer induced increase in levels of nitrate and amino acid levels in foliage appears to attract the pests. The high density of food and habitat provided for the pests by the fertilized crops attracts more of them and supports increases in their population sizes (Altieri and Nicholls 2003; Rashid et al. 2017; SARE 2018). Artificial fertilizer is also associated with intensive agricultural practices that purposefully eliminate biodiversity in favor of large monocultures, further adding to the irruption of pests (Matson 1997).

Application of inorganic Nr to fields stimulates the growth of weeds that are N-limited. Red root pigweed (*Amaranthus retroflexus*) is one of the most important weeds in the Chesapeake Bay region and grows prolifically when stimulated with Nr (Blackshaw and Brandt 2008). Inorganic N-fertilization stimulates many other weed species. Farmers are cautioned that, "Depending on weed species and density, additions of N fertilizer can increase the competitive ability of weeds more than that of the crop, with little or no increase in crop yield attained (Blackshaw et al. 2003)." Considering the stimulatory effect of artificial Nr on weeds, it's not surprising that commercial products called "weed and feed" combine herbicides with fertilizers. Using artificial N fertilizer to increase crop production means not only more weeds and pests, but also the application of more toxic pesticides and herbicides.

The pesticides and herbicides used to counter the consequences of inorganic N fertilizers lose effectiveness as artificial selection creates resistant strains of noxious invertebrates and plants (Heap 2014). Glyphosate (Roundup) is the most widely used herbicide word-wide, largely because it is marketed with genetically modified glyphosate-resistant corn and soybean seeds. Decades of use have produced 38 glyphosate-resistant species of weeds world-wide (Heap and Duke 2017). This includes glyphosate resistant pigweed (*Amaranthus palmeri*) in the heavily fertilized fields of Virginia and other parts of the Chesapeake Bay Watershed (VCE 2018). Many of the broad spectrum pesticides targeting pests also destroy their predators, worsening the situation after the pests evolve resistance to the chemical (Liu et al. 2016). These pesticides also impede many pollinators, even at sublethal doses (Desneux et al. 2007).

Rather than relying on spraying pesticides to protect corn from various pests, Monsanto Corporation (now owned by Beyer Corporation) patented a transgenic version of maize that incorporates genes from the bacterium *Bacillus thuringensis* (Bt) which produce the cry protein that is toxic to many insects. In use since the 1990s, Bt corn lost effectiveness against corn earworm (*Helicoverpa zea*) as the insect evolved resistance to the cry protein (Dively et al. 2016). See chapter "Pesticides Bring the War on Nature to the Chesapeake Bay" for detailed discussion of pesticide use in the Bay ecosystem.

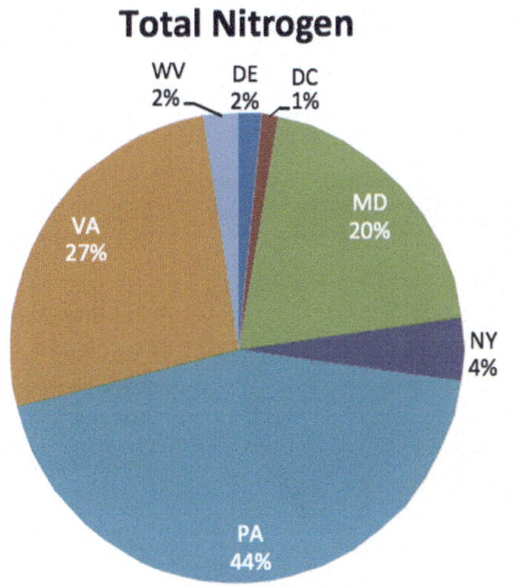

Fig. 7 Source of total nitrogen loading to the Chesapeake Bay by jurisdiction in the watershed based upon modeling for 2009. Note that half of the loading comes from Pennsylvania and New York, states that have no water frontage on the Bay proper. https://www.epa.gov/sites/production/files/2014-12/documents/cbay_final_tmdl_section_4_final_0.pdf (USEPA 2018b)

Inorganic N fertilizers escape from the farm to pollute groundwater, surface water, and the atmosphere On a global basis, about 71% of the Nr applied to crops is directly lost to the environment through runoff, infiltration to ground water, and denitrification. Leaving only 29% for *potential* consumption by humans *if* they ate a wholly plant based diet. However, since most humans now also eat agro-animals, the population gets only 12% of the added Nr in their diets. Meaning that 17% of applied Nr is lost as waste from animal feeding operations (Galloway et al. 2003). Not accounted for here, but a portion of Nr is applied to biofuel crops and represents more loss from the potential available for the food system (Smith and Searchinger 2012). The inefficiency in retention of applied Nr means that we are wasting most of the money and energy invested in artificial N fertilizers, and creating unintended pollution.

The area devoted to agriculture in the CBW makes it the second largest land-use behind forested or open wooded areas (69%) (USEPA 2010). The 87,000 farming operations cover 22% of the 166,000 km^2 (64,000 square miles) region. About 42% of the Nr pollution entering the Bay comes from agriculture, a large portion of that (17% of total entering the Bay) attributes to artificial fertilizers (Figs. 7 and 8; Parsons 2018).

Agriculture includes the production of non-food items such as fiber (cotton and flax), tobacco, and biofuels.

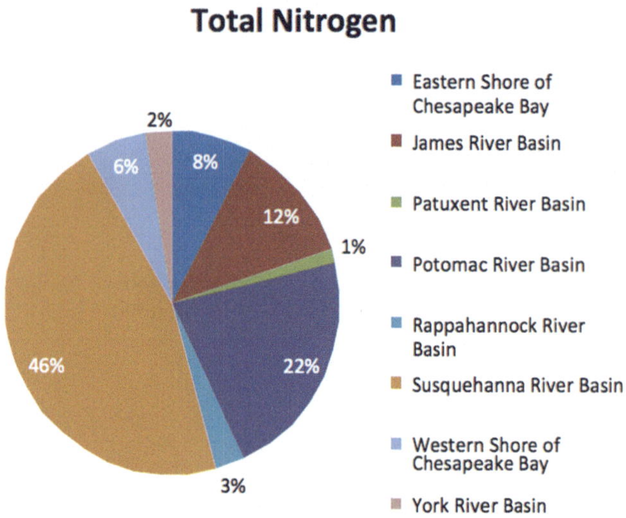

Fig. 8 Source of total nitrogen loading to the Chesapeake Bay by major tributary, based upon modeling of 2009 data. https://www.epa.gov/sites/production/files/2014-12/documents/cbay_final_tmdl_section_4_final_0.pdf (USEPA 2018b)

However, the food system dominates farming in the CBW, with pasture, hay, corn, wheat, soybeans, vegetables and fruits accounting for most of the production. For Virginia, Pennsylvania and Maryland the percent of harvested land devoted to food system crops in 2017 was 95.3%, 99.9% and 100% (USDA 2018a). Most of the land devoted to plant production goes to feedstock for agricultural animals. As such, animal operations make up 60% of the area's farm product sales. This includes dairy, beef, pigs, eggs, broilers (chicken), and turkey. Poultry production is concentrated on the Delmarva Peninsula. "Agriculture is the largest single source of nitrogen, phosphorus, and sediment loading to the Bay through applying fertilizers, tilling croplands, and applying animal manure (USEPA 2010)." The sum of these activities accounts for 44% of the N loads delivered to the Bay (USEPA 2010). Pennsylvania, followed by Virginia, and Maryland account for most of the N pollution entering the Bay (Table 1, Figs. 7 and 8). Despite its relatively small size, Maryland accounts for 27% of the point source N pollution, reflecting the large concentration of poultry CAFOs located on its portion of the Delmarva Peninsula (Ator and Denver 2015), a topic addressed in chapter "Livestock and Poultry: Other Colonists Who Changed the Food System of the Chesapeake Bay".

Nitrate contamination of groundwater is bad for the health of the Bay and the people who live there A portion of the highly soluble nitrate applied to crops infiltrates through the soil to contaminate groundwater. Much of the groundwater eventually emerges into streams that drain into the Bay. Nr is the primary nutrient (along with phosphorus) driving eutrophication of the Bay, as discussed in detail in chapter "Reversing the Eutrophication of the Chesapeake Bay and Its People". The legacy of extensive inorganic fertilizer application that began in the 1950s lives on in the aquifers of the watershed. This is augmented by the expansion of animal agriculture in the CBW, which produces manures that leach Nr into groundwater. This is most noted for the Delmarva Peninsula where centuries of cultivation and several decades of expanding poultry CAFOs produce very high nitrate concentrations of groundwater, and result in total N loads per unit area twice that of the average for the rest of the watershed (Ator and Denver 2015). Emerging groundwater accounts for 54% of the total annual flow of streams in the CBW. About half (48%) of the nitrate in the streams traces to groundwater (Philips et al. 2018). The median age of base flow from ground water into streams ranges from 40 to 60 years (Sanford and Pope 2013). In addition, sub soils may retain significant amounts of organic Nr for later release to ground water (Van Meter et al. 2016). This means that if fertilizer application ceased completely, it would take 4–6 decades before significant levels of excess Nr pollution would quit seeping into the Bay from aquifers in the watershed (Sanford and Pope 2013). This, of course, frustrates mangers and stakeholders that expect to see results from investments to reduce Nr pollution (Van Meter et al. 2016; Terziotti et al. 2018).

Contamination of groundwater with nitrate presents health problems when it is used for drinking. The USEPA and WHO (World Health Organization) set safe limits of nitrate in drinking water at 10 mg/l and 50 mg/l, respectively (WHO 2004; Puckett et al. 2011). Ingestion of excessive nitrate is most dangerous for infants. A portion of the ingested nitrate is reduced to nitrite, which combines with hemoglobin to form methemoglobin, reducing the ability of red blood cells to obtain and circulate oxygen. The resulting disease is called methemoglobinemia or blue baby syndrome. It may result in brain damage and death. However, it appears that other contaminants and health factors also play a role in the syndrome. Adults possess higher levels of the enzyme that reverse the effects of nitrites in formation of methemoglobin. Yet, adults as well as children may suffer other maladies from excessive nitrate consumption. Nitrate contaminated groundwater is linked to esophageal and gastric cancers (Ward et al. 2005). As noted above, ground water of the Delmarva Peninsula is highly contaminated with nitrates (Fig. 9), showing some of the highest values in the nation for agricultural areas, rendering many wells unsafe for drinking water (Ator and Denver 2015) (Fig. 10).

Nitrogen fertilizer reduces biodiversity Both the added Nr that stays on agricultural fields and the balance that escapes to the environment reduce biodiversity. As addressed above, artificially fertilizing fields leads to use of pesticides and herbicides that reduce biodiversity on site. As with the Nr, a portion of these toxins escapes to the environment,

Table 1 Percentage of total nitrogen delivered to the Chesapeake Bay from each jurisdiction in the watershed and by land use type

Jurisdiction	Agriculture (%)	Forest (%)	Stormwater (%)	Point Source (%)	Septic (%)	Atmospheric deposition (%)
Delaware	3	1	1	0	2	1
D.C.	0	0	1	5	0	0
Maryland	16	14	28	27	36	27
New York	4	7	3	3	5	5
Pennsylvania	55	46	33	25	30	42
Virginia	20	27	33	39	24	25
West Virginia	3	4	2	1	2	1
Total	100	100	100	100	100	100

This is based upon modeling of 2009 data. Point source refers to municipal wastewater facilities, industrial wastewater facilities, combined sewer outfalls (storm + sewage), permitted construction sites and permitted concentrated animal feeding operations. From USEPA (2010). https://www.epa.gov/sites/production/files/2014-12/documents/cbay_final_tmdl_section_4_final_0.pdf

threatening biodiversity in adjacent communities. The escaped Nr may also reduce biodiversity in adjacent areas. The Nr will favor certain species that respond to fertilization with faster growth over those that don't. In poorly buffered soils, the Nr may also reduce the pH, eliminating species unable to adapt to the new soil chemistry (Isbell et al. 2013).

Escaped Nr and phosphorus that typically accompanies it, also reduces biodiversity in streams, rivers, ponds, lakes, and the Chesapeake Bay (see chapter "Reversing the Eutrophication of the Chesapeake Bay and Its People"). In ponds and the Bay, the added nutrients stimulate the growth of a few species of phytoplankton that outcompete others and deny light to bottom dwelling plants and algae. In streams and rivers, fertilization leads to reduction in biodiversity of both producers and consumers (Brush 2009).

Nitrogen fertilizer promotes global warming and air pollution About 4% of Nr used in agriculture is converted to nitrous oxide (N_2O) as a consequence of both aerobic nitrification ($NH4^+ \rightarrow NH2OH \rightarrow NO_2^- \rightarrow NO_3^-$) and anaerobic denitrification ($NO_3^- \rightarrow NO_2^- \rightarrow NO \rightarrow N_2O \rightarrow N_2$). Analysis of N_2O trapped in ice cores revealed atmospheric concentrations of 270 ppb (by volume) in 1850, some 20% less than the current levels of more than 320 ppb. This traces to accelerated use of Nr fertilizers, particularly since 1960 (Smith et al. 2012).

Nitrous oxide is a potent greenhouse gas, with an effect 300 times that of carbon dioxide, and with an atmospheric residence time of 114 years. About 77% of NO_2 comes from agriculture, and this gas accounts for 6% of total greenhouse gas emissions (USEPA 2018b).

NO_x refers to the sum of oxides of nitrogen found in the atmosphere and consists primarily of nitrous oxide (NO_2) and nitric oxide (NO). NO_2 irritates human lungs and airways, exasperating asthma attacks. In the presence of sunlight, NO_x forms photochemical smog (USEPA 2018c). NO_x reacts with free oxygen and water to form nitric acid, an important component of acid precipitation. Acid precipitation damages crop production, natural plant communities, and poorly buffered aquatic systems (Havas and Hutchinson 1980; Schindler 1988). It is also a source of Nr that causes eutrophication in the Chesapeake Bay and other aquatic systems (USEPA 2010; Table 1).

Lawns as farms On a nation-wide basis, household and institutional lawns combine to constitute the fourth largest crop after corn, soy and wheat. Lawns comprise 1.9% of total continental US surface area and represent the single largest irrigated "crop" in the US, with a surface area about three times that for irrigated corn (Milesi et al. 2005). Although not a part of the food system, contemporary lawn cultivation methods derive from the industrial-chemical approach that dominates conventional agriculture. The ideal lawn is marketed as a lush, deep-green monoculture. This requires irrigation, Nr fertilizers, pesticides and herbicides. Scotts ™ is a large national firm that markets lawn care products. Their annual plan calls for several applications of different products; a crabgrass herbicide fertilizer combination in early spring, broadleaf herbicide and fertilizer combination in late spring, more fertilizer in July and September (Scotts 2018). As with food crops, a portion of the fertilizer applied to lawns in the CBW escapes to contaminate groundwater and surface waters, including the Chesapeake Bay. About 16% of the Nr entering the Bay comes from urban runoff (Table 2), much of this traces to escaped lawn fertilizers.

For the CBW, 1.54 million ha (35 million acres) is planted in lawns. Turf grass is the largest crop in Maryland (0.53 million ha), nearly equaling the sum of all other crops combined (0.61 million ha). Maryland lawns receive about 39 million kg (86 million pounds) of N fertilizer per year, close to that used in total for agriculture. Indeed, N fertilizer applied to lawns in Maryland is 26% greater than used by corn-growers (29 million kg) in that state where maize is the dominant crop (Dewar et al. 2011). Heavily fertilized lawns often produce so much growth that homeowners feel compelled to attach collection bags to their mowers, packaging and sending the clippings off as if it were a food or fiber crop, but instead destined for community landfills or possible

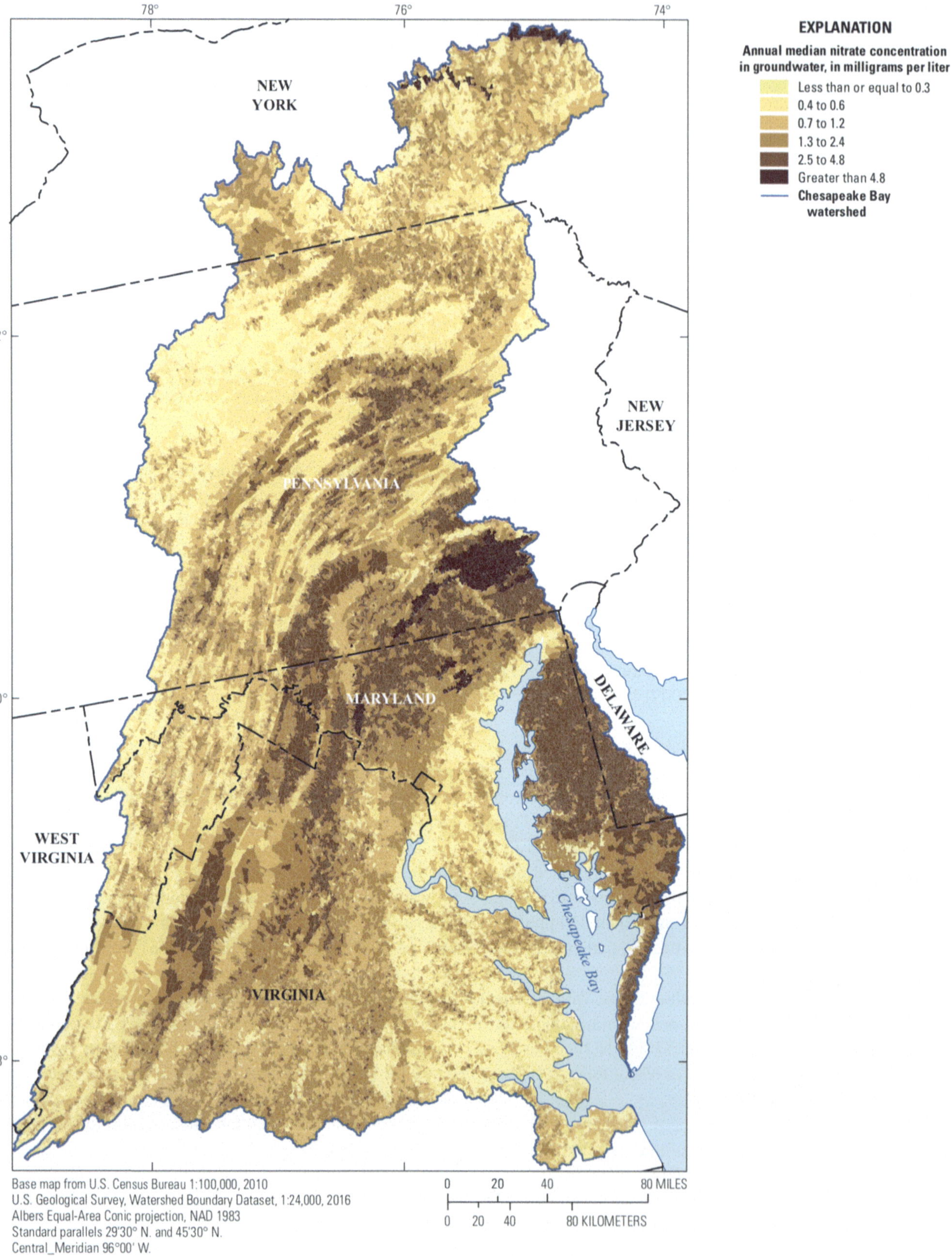

Fig. 9 The annual median nitrate concentration in ground water in the Chesapeake Bay Watershed. Figure from Terziotti et al. (2018). Note the high concentrations of nitrate in ground water associated with the agricultural areas of the Delmarva Peninsula and southeastern Pennsylvania

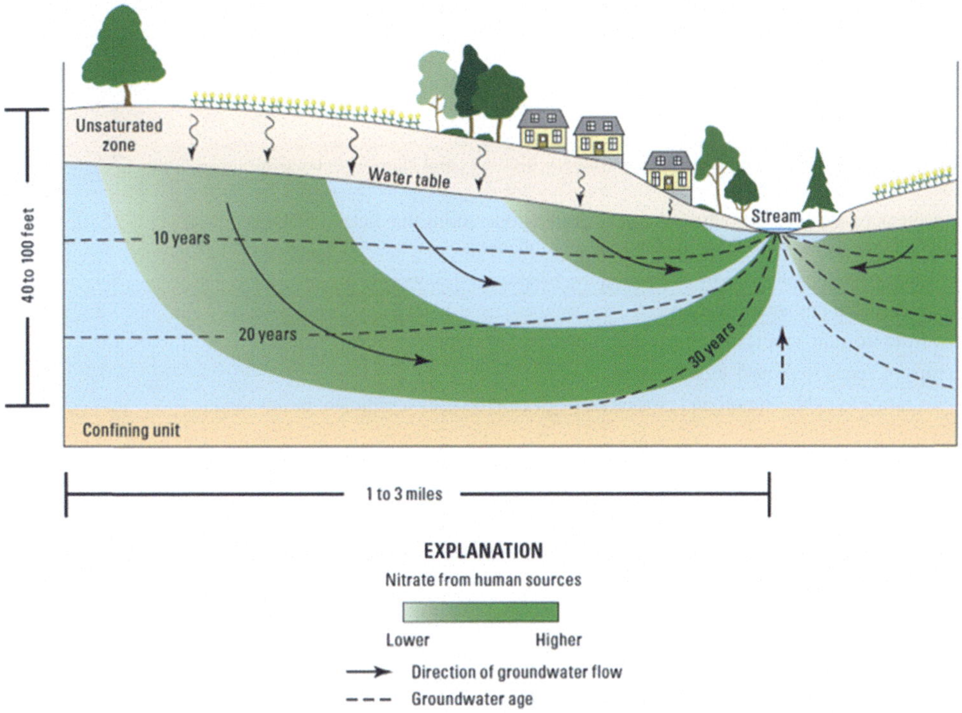

Fig. 10 Diagram showing the flow of nitrates in groundwater and illustrating the longer time of residence in deeper portions of the aquifer. Note that 40–100 ft = 12.2–30.5 m, and 1–3 miles = 1.6–4.8 km. Image from Terziotti et al. (2018)

composting sites. Since exporting clippings means removal of nitrogen, this helps motivate continued purchases of fertilizers.

Maryland passed a law to regulate the sale and application of lawn fertilizers in 2011 (MDA 2018). It sets limits on the amount N that may be applied per unit area, and heavily restricts phosphorus. The law also requires certification of lawn care professionals, and levies fines for non-compliance.

Pretending that lawns are crops and using chemicals as weapons to control nature, with the unintended consequence of polluting the Chesapeake Bay is a perfect model of the conventional food system that dominates the CBW. The alternative is to let nature take its course and provide highly diverse and self-maintaining lawns that require only mowing. Absent the fertilizer, pesticide and herbicide applications, the property manager would save money at the expense of the chemical corporations and at the benefit of the Bay (Marshall et al. 2015).

Putting the Chesapeake Bay on a Diet: Total Maximum Daily Loads (TDMLs) By the mid 1980s, the jurisdictions bordering the Chesapeake Bay came under sufficient pressure from residents to commence action to reverse the eutrophication of the system (see chapter "What Nature, Politics and Policy Demand of the Chesapeake Bay and Its Food System"). This primarily meant reducing nutrient pollution from agriculture and wastewater treatment. Those initial efforts proved ineffective in restoring the health of the Bay, leaving the USEPA to establish TMDLs for the primary pollutants under the Clean Water Act. On December 29, 2010, the USEPA promulgated its largest ever TMDL, encompassing all of the Chesapeake Bay Watershed (CBW) jurisdictions. This became popularly known as a "pollution diet," *a most appropriate nickname considering that most of the newly regulated pollution derived directly (agriculture) or indirectly (sewage) from the food system.* The TMDLs called for a 25% reduction in nitrogen, 24% reduction in phosphorus, and 20% reduction in sediment washed from the land. This corresponds to annual limits of 185.9 million pounds (84.3 million kg) of N, 12.5 million pounds (5.7 million kg) of P, and 6.45 billion pounds (2.9 billion kg) of sediment. The pollution diet centered on the main meals of surface water, ground water, and wastewater, but also included a snack of atmospheric inputs, setting the goal of reducing airborne N entering the Bay from 17.9 to 15.7 million pounds (8.1–7.1 million kg) per year. Responsibility for achieving the TMDLs was spread across all CBW jurisdictions. TDMLs were to be realized by 2025, with a preliminary target of 60% reduction by 2017 (USEPA 2018a). The Chesapeake Bay Program established a list of best management practices (BMPs) to address and reduce the pollution of the Bay (Table 2).

Table 2 The best management practices (BMPs) for reducing Nr and other pollution entering the Chesapeake Bay and their projected annual cost

Best management practice	Cost in $ millions	Percent of Bay total
Alternative systems for watering of livestock to keep them out of streams where they cause erosion and add pollutants from their wastes *	100.7	11.2
Ammonium emissions reduction using alum (aluminum sulfate) or biofilters and lagoons in concentrated animal feeding operations *	7.5	0.8
Animal waste management systems for concentrated animal feeding operations that keep pollutants out of ground water and the Chesapeake Bay while storing and handling manures*	255.7	28.3
Barnyard runoff control to prevent pollutants from entering the Chesapeake Bay*	5.4	0.6
Capture and reuse of water from rain and irrigation	7.8	0.9
Carbon sequestration by converting cropland to hayland to build carbon content of soil	1.5	0.2
Commodity cover crops of cereal to protect soils in winter	20.2	2.2
Conservation plans for farms to reduce runoff and erosion through crop rotations, sediment basins, and conservation tillage	6.5	0.7
Conservation tillage by reducing disturbance to soil with plows and cultivators, prior to planting, and during wedding	0	0
Cover crops (non-harvested) to reduce erosion and nutrient leaching	45.9	5.1
Cropland irrigation management to match the application of water and fertilizer to maximize productivity and uptake of nutrients by the crop	0	0
Decision agriculture based upon site specific data including soils, crops, pests, nutrients, moisture, sustainability and crop yields	32.3	3.6
Enhanced nutrient management that optimizes use of fertilizers	17.4	1.9
Forested buffers planted along waterways and intended to check runoff and erosion while absorbing nutrients	32.4	3.6
Grass buffers planted along waterways and intended to check runoff and erosion while absorbing nutrients	15.9	1.8
Horse pasture management that uses fencing to protect waterways and rotational grazing to reduce nutrient runoff	0.8	0.1
Land retirement by planting permanent cover crop over cropland	12.5	1.4
Liquid poultry manure injection into root zone with soil closure at time of application	22.3	2.5*
Loafing lot management to stabilize surfaces of areas used intensively by people, animals, or vehicles	1.1	0.1
Manure transporting operations modified to reduce loss to the environment from animal wastes*	12.4	1.4
Mortality composters to treat animal carcasses and minimize nutrient loss to the environment*	25.7	2.8
Nutrient management plan to minimize fertilizer loss while maintaining yield and includes testing of soils, plants and manures, timing of application, and location of application	1.5	0.2
Poultry and swine phytase enzyme to maximize uptake of phosphorus	0	0
Precision intensive rotational grazing to optimize forage quality and damage to soils by overgrazing and compaction	26.7	3
Prescribed grazing to optimize forage quality and damage to soils by overgrazing and compaction*	11.5	1.3
Stream access control with fencing to exclude livestock from riparian zones along waterways*	186.6	20.7
Steam restoration including legacy sediment removal and reestablishing habitat and flow patterns	4.3	0.5
Tree planting in non-riparian areas to convert portions of cropland	15.1	1.7
Water control structures using boards to manage surface drainage ditches	0.5	0.1
Wetland restoration to re-establish natural hydraulic conditions prior installation of subsurface or surface drainage systems	32.7	3.6
Totals	**$903.2**	**100%**
Total due directly to animal agriculture	**$627.7**	**69.6%**

BMPs that pertain directly to animal agriculture are indicated by *. Note that 69.6% of the annual BMP cost attributes to animal agriculture. From Kaufman et al. (2014)

The TMDLs appear to be producing some success with reductions in loading of; N (11%), P (21%) and sediment (10%). The reductions in P and sediments exceed the 2017 goals, but that for N falls short by 9%. The Chesapeake Bay Program attributes the pollution reduction to upgrades of wastewater treatment plants and expanding best management practices in agriculture, such as cover-crop planting (CBP 2018, Table 3).

Denitrification augments the TDML diet The great irony of Nr in the Chesapeake Bay ecosystem and for artificially fertilized food systems world-wide is that societies using industrial chemical agriculture get more than they wish for. Societies using food systems built on artificial fertilizers wish to produce more food and are willing to pay the upfront price in money, but rarely account for the attendant damage to their ecosystem as a consequence of this attempt to control nature.

Table 3 Reduction in total nitrogen pollution entering the Chesapeake Bay between 2009 and 2017 and load targets for 2017 and 2025

Sector	2009 Load kg × 10^6	2017 Load kg × 10^6	2017 Target Load kg × 10^6	2025 Target Load kg × 10^6
Agriculture	51.7	48.2[a]	40.2	32.6
Urban runoff	18.0	18.8[a]	15.0	13.0
Waste water & Combined Sewer Overflow	23.7	15.2	19.8	17.2
Septic	3.8	4.0[a]	3.2	2.9
Forest & Non-tidal water deposition	20.9	20.7	21.2	21.6
Atmospheric deposition to watershed	1.4	0.4	0.6	0
Atmospheric deposition to water	8.8	7.3[a]	0.8	6.9
Total	128.3	114.6	100.8	94.2

Data modified from CBP (2018)
[a]Missing 2017 target

They want more Nr and get it, but the bargain includes substantial escapes to the environment that wreck havoc on the rest of the ecosystem. This is exactly what happened to the Chesapeake Bay. The TDML diet imposed by the USEPA is designed to reduce the application of Nr and to prevent its transfer from crops to the general ecosystem. In the simplest terms, when an obese human goes on a "diet" the intent is to lose weight by reducing calories. The TDML diet for the Chesapeake Bay seeks to reduce "fattening" Nr by applying less to the land, keeping it there, and thus lowering the amount getting into the water. In the long run, this may work but wouldn't getting rid of the excess Nr speed the process? Adding exercise to a weight-loss diet speeds the rate of trimming down, since the physical activity burns calories and raises basal metabolic rates. Denitrification is to the Chesapeake Bay what vigorous exercise is to the overweight dieter. Exercise alone is usually insufficient to obtain desired weight loss, but it does augment an effective change in eating pattern. And denitrification alone can't solve the problem of excessive Nr in the Chesapeake Bay, but it can augment the reduction of inputs.

Resorting oysters can help remove excess nitrogen from the Bay Over harvesting, habitat destruction, and pathogens have left the Bay with less than 1% of the oyster population that greeted Capt. John Smith in 1607 (see chapter "The Chesapeake Bay Oyster: Cobblestone to Keystone"). The removal of excess nitrogen in the Bay is one of the ecosystem services lost with the demise of this suspension feeder. Transferring the nitrogen in plankton to the shells of oysters helps reduce the amount of this element circulating in the Bay. In addition, oyster feeding coupled with microbial activity on reefs created by this mollusk leads to increased rates of denitrification. Kellogg et al. (2013) examined the nitrogen budget of a restored oyster reef and an unrestored control area. Both were located in the Choptank River, a sub-estuary of the Chesapeake Bay, located on the Eastern Shore of Maryland. Denitrification at the recreated reef removed 556 kg N ha^{-1} $year^{-1}$ more than at the control site. The authors estimate that restoring oyster reefs to all the suitable areas of the Choptank River would remove about 48% of the annual external loading of total N to that ecosystem. Expanding the current Bay-wide efforts to restore historic oyster reefs would help meet TMDL goals.

The Brush narrative considers both the addition of Nr and denitrification The same food system that pumped excess Nr in to the Chesapeake Bay also reduced the capacity for the CBW to eliminate some of that pollution through denitrification. To understand this, consider the narrative on how the ecological history of the Bay developed by Dr. Grace Brush (Fig. 11). She used paleoecological evidence and chemistry from cores of Bay sediments to reconstruct the evolving Chesapeake Bay ecosystem over the past 14,000 years. By studying the nutrient content and micro fossils of pollen, algae and protozoans buried in bottom sediments, she illuminated how people and their changing food systems reorganized the ecosystem (Brush 2009). The rest of this section summarizes the Brush narrative. Before Captain John Smith's arrival in 1608, forests and wetlands dominated the landscape. These included nitrogen fixing species such as Alder (*Alnus* sp.), the leguminous tree black locust (*Robinia pseudo-acacia*), and various herbaceous legumes. The sediment cores she studied contained little nitrogen from pre-colonial times, and showed evidence of a Bay with clear water that allowed the penetration of light and thus supported beds of submerged macrophytes and benthic diatoms (microscopic algae).

Agriculture in the early colonial period (1610–1740) focused on a mix of tobacco and food crops, using a system of rotation that allowed the land to lay fallow for some years so as to recover fertility. The expanding population cleared forests and drained wetlands to put about 40% of the land under cultivation. In this process humans replaced beavers (*Castor Canadensis*) as the primary ecosystem engineers of the CBW. Colonists trapped the dam-building beaver

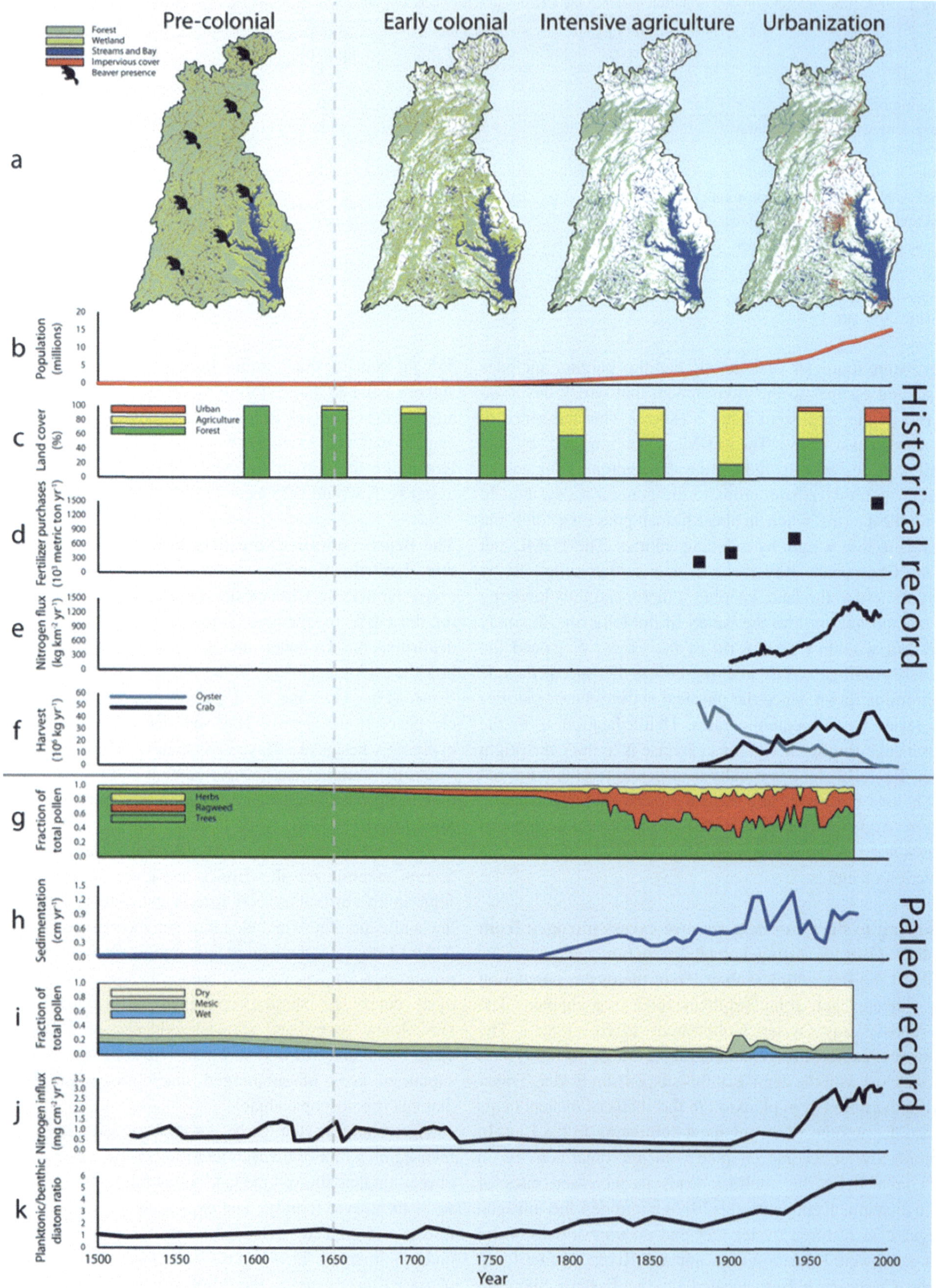

Fig. 11 (a) The Chesapeake watershed at different historical periods: precolonial up to 1650 showing a forested landscape with a large beaver population; early colonial when about 40% of the land was deforested for agriculture (in white); intensive agriculture when up to 80% of the land was under cultivation; and urbanization when about half of the watershed is in forest and urban and suburban centers characterize some

intensely, ultimately driving it to localized extinction in most of the CBW. Many of the beaver dams remained functional for years after the animals were extirpated, and colonists built their own impoundments to power mills for grinding grain and supplying fresh water. Consequently, much of the CBW remained wet in the early colonial period, providing the anaerobic conditions that favored denitrification. This meant relatively little of the Nr that escaped field and forest was making its way to the Bay.

Human population growth in the CBW over the next century was mostly rural, providing the labor to clear and farm the land. The timber produced by clearing forests supported the construction of ships used to export surplus grain. By 1850, nearly half the CBW was under cultivation. In another 50 years, the area devoted to farming reached its maximum at 80% (circa 1900). The extensive cultivation left its mark in the Bay as eroded soils settled to the bottom of river beds and shipping ports, hindering navigation and motivating dredging operations.

A period of urbanization and intensive farming followed at the beginning of the twentieth century, reducing area cultivated to 40% of the watershed by 1950 and 22% by 2000. Farms grew larger and relied on an industrial approach built on artificial fertilizer, pesticides, herbicides and mechanization. The growing urban areas now cover about 20% of the CBW, similar to the land area under cultivation. Afforestation of abandoned farmland returned about 60% of the landscape to forest by 2000.

The annual use of fertilizers in the CBW expanded rapidly during the twentieth century, going from about 300,000 tons circa 1900, to 1,500,000 tons by 2000. Most of that increase took place in the second half of the century, during the post WWII period. This is matched by increasing fluxes of total N deposited in the Bay sediments, going from 1 mg N cm^{-2} $year^{-1}$ in 1950 to thrice that rate by 2000.

The conversion of the CBW from dominance by natural forests with beavers to human-managed and well drained grasslands (grain crops) meant a drier landscape absent of most of the perpetually wet soils that previously characterized the system. Perpetually wet soils provide the low-oxygen environments required by denitrifying bacteria. Thus the massive expansion of agriculture in the nineteenth century disabled most of the denitrifying capacity of the ecosystem. So when farming embraced subsidies of Nr from artificial fertilizers in the twentieth century, the reduced capacity of denitrification couldn't balance the massive unintentional loss of fertilizer from farms to the environment. This led to the onset of eutrophication in the Chesapeake Bay during the second half of the twentieth century. The eutrophication process produces extensive zones of anoxic water and sediments during the warmer months of the year. These seasonal environments provide habitat for denitrification of some of the Nr, but only after it has already done the work of eutrophication. Yet, it is still important in reducing the return of Nr to the surface waters during mixing events.

Brush (2009) closes her narrative of Nr in the Chesapeake Bay ecosystem with suggestions for intercepting and removal of Nr escaped from farms. This includes surrounding agricultural fields with vegetated buffers, aquatic environments with anoxic habitats, and wetlands. The vegetated buffers will capture the Nr in their own biomass, while the aquatic systems and wetlands will restore habitat for denitrification.

Changing the food system to end Nr pollution of the Chesapeake Bay also means curtailing the production and consumption of animals The dependence of conventional agriculture on synthetic Nr mirrors the relationship between that of the US population and their food supply. In 2014, the US population spent \$1.46 trillion for food (USDA 2018b), and \$64 billion for weight loss products (Statista 2018). The weight loss expense estimate includes diet foods and food replacements, diet medications, surgery, commercial weight loss programs, and more. That means that for every dollar spent on food, the population spends \$0.44 on attempting to reverse one of the main effects of consuming it! In 2000, farmers in the CWB purchased 1.5 million metric tons of fertilizer. Assuming a cost of \$160 per ton in 2000 (DAP 2018), yields \$240 million spent on fertilizer that year. Implementing the TMDL best management practices will cost \$903.2 million per year (Kaufman et al. 2014). The USEPA budgeted \$77 million for the Chesapeake Bay Program in 2018, most of which goes to support the TMDL program to reverse the effects of nutrient pollution (Blankenship 2018). The balance will be absorbed by farmers

Fig. 11 (Continued) parts of the watershed. (**b**) The increase in the human population. The data were derived from spatially allocating historical county census data from the US Census Bureau to inside the Bay watershed boundary. The spatial allocation was done using weighting factors derived from the 2000 Census Block Group database distributed to 30-m residential road density cells. (**c**) The historical record of land use. (**d**) Sales of fertilizer in the Chesapeake watershed. (**e**) Historical record of nitrogen fluxes into the Potomac River. (**f**) The historical record of the crab and oyster harvest. (**g**) Pollen representation of land use in the upper estuary. (**h**) Paleoecological record of sedimentation rates in the upper estuary. (**i**) Pollen profile showing a gradual increase in dry taxa after colonization. (**j**) Nitrogen influx into the Chesapeake Bay. (**k**) The change in the ratio of planktonic to benthic diatom species in the central mesohaline section of the estuary. Image and caption text taken directly from Brush (2009). See that paper for specifics on data sources

Table 4 Area in cultivation for the three main agricultural states of the Chesapeake Bay Watershed, showing the primary crops used in animal feed, and the reduction in needed land if the food system quit producing fodder and feed for animals

Crop	Pennsylvania	Virginia	Maryland
Total area in cultivation of all crops (thousands of ha)	3997	1627	724
Total Corn (thousands of ha)	543	192	190
Total Soy (thousands of ha)	0.237	158	200
Total Hay (thousands of ha)	735	488	0
Corn for animal feed (thousands of ha)	250	88	88
Soy for animal feed (thousands of ha)	0.166	111	140
Total area for animal feed (thousands of ha)	985	687	228
% Reduction of total area for crops if food system switches to plant-based	75.3	56.7	68.6

The last row in the table shows the amount of land that could be taken out of agricultural production if the food system no longer supported the production of animal based foods. Crop data for 2017, from USDA (2017a). Estimation for portion of corn and soy crops used for animal feed (40% and 70% respectively) from NCGA (2018)

and jurisdictions. That means that for every dollar spent on Nr rich fertilizers per year, the people of the CBW will spend $3.76 to reverse detrimental effects to the Bay due to its application and other Nr polluting activities. We are wasting money by eating the wrong diet for our health, and we are wasting more money by amending the soil in the wrong way to support that diet. The health of the Bay ecosystem suffers from an agricultural system that supports the diet that sickens us (see chapters "Reversing the Eutrophication of the Chesapeake Bay and Its People" and "Ethics and Economics of Building a Food System to Recover the Health of the Chesapeake Bay and Its People").

Full recovery of ecosystem function will require changes in the food system. This includes revising agricultural practices to promote persistent wet areas for denitrification, the reintroduction of crop rotation and expansion of cover cropping to promote biological N-fixation, and the disuse of artificial N fertilizers. Adopting organic farming techniques, as addressed in chapter "An Organic-Based Food System: A Voyage Back and Forward in Time", will move the food system away from dependence on chemical fertilizers.

While the TMDL program is often described as putting the Bay on a nutrient diet, we can only hope that it's more successful than dieting regimes adopted by people trying to lose weight. Most such diets fail, and dieters often eventually gain additional weight (Mann et al. 2007). However, individuals that don't go on a regime that just modifies their Standard American Diet, but instead change to eating a Whole-Food-Plant-Based-Diet (WFPBD) do succeed in sustained weight reduction (Rizzo et al. 2011; Wright et al. 2017).

The BMPs designed to achieve the TMDLs will help the current food system to operate with less loss of Nr to the environment (Table 2). As important as that is, more pollution reduction could be achieved by changing the food system, moving from production for an animal-based human diet, to agriculture for a WFPBD. Shifting the food system away from animal agriculture and toward producing for a WFPBD will greatly reduce the amount of land needed to support the population. In the US, 46 % of the corn crop is used for animal feed, 10% for direct human consumption within the country, and the balance goes to biofuel (30%) and exports (13.5%) (NCGA 2018). Also in the US, 70% of the soybean crop goes to animal feed (46% for export), 15% to direct human consumption after processing, and 5% for biofuel (USDA 2018c; USDAERSA 2019). How much could we reduce the land area devoted to cultivation, if we quit producing animal feed in the CBW? By removing the portion of corn and soybean going to animal feed and the land used to grow hay for fodder, the percent of agricultural land could be reduced by 56% in Virginia, 75% in Pennsylvania, and 70% in Maryland (calculations based upon current cropland in those states per USDA 2018a). These three states comprise most of the CBW agriculture and are responsible for 91% of the Nr load to the Bay (Fig. 7). Reducing the land under cultivation by just 25% would cut the cost of recovering the health of the Bay by 60% (Kaufman et al. 2014). This goal is easily achieved by eliminating cultivation of crops used to feed animals (Table 4). This will further reduce N loading from agricultural lands, as well as from concentrated animal feeding operations (Crews and Peoples 2004). Eliminating animal agriculture in CBW would cut the cost of BMPs to reach the TMDL by nearly 70% (Table 2). Shifting away from animal-agriculture would save money, the environment, and the health of the people eating from the food system.

References

Altieri M, Nicholls C (2003) Soil fertility management and insect pests: harmonizing soil and plant health in agroecosystems. Soil Tillage Res 72:203–211. https://doi.org/10.1016/s0167-1987(03)00089-8

Ator S, Denver J (2015) Understanding nutrients in the Chesapeake Bay watershed and implications for management and restoration—the Eastern Shore (ver. 1.2, June 2015): U.S. Geological Survey Circular 1406, 72 p. https://doi.org/10.3133/cir1406

Bernhard A (2010) The nitrogen cycle: processes, players, and human impact. Nat Educ Knowl 3(10):25

Blackshaw R, Brandt R (2008) Nitrogen fertilizer rate effects on weed competitiveness is species dependent. Weed Sci 56:743–747. https://doi.org/10.1614/ws-08-065.1

Blackshaw R, Brandt R, Janzen H et al (2003) Differential response of weed species to added nitrogen. Weed Sci 51:532–539. https://doi.org/10.1614/0043-1745(2003)051[0532:drowst]2.0.co;2

Blankenship K (2018) Bay journal—article: new federal budget keeps funds for Bay Program at steady level. Bayjournal.com. https://www.bayjournal.com/article/new_federal_budget_keeps_funds_for_bay_program_at_steady_level. Accessed 16 Sept 2018

Blanksby S, Ellison G (2003) Bond dissociation energies of organic molecules. ChemInform. https://doi.org/10.1002/chin.200324299

Brady N, Weil R (2017) The nature and properties of soils, 15th edn. Pearson, Upper Saddle River NJ

Brush G (2009) Historical land use, nitrogen, and coastal eutrophication: a paleoecological perspective. Estuaries Coasts 32:18–28. https://doi.org/10.1007/s12237-008-9106-z

CBP (Chesapeake Bay Program) (2018) Chesapeake progress—2017 and 2025 Watershed Implementation Plans (WIPs). Chesapeakeprogress.com. https://www.chesapeakeprogress.com/clean-water/watershed-implementation-plans. Accessed 9 Sept 2018

Craven A (1926) Soil exhaustion as a factor in the agricultural history of–Virginia and Maryland, 1606-1860. University of Illinois, Urbana, IL

Crews T, Peoples M (2004) Legume versus fertilizer sources of nitrogen: ecological tradeoffs and human needs. Agric Ecosyst Environ 102:279–297. https://doi.org/10.1016/j.agee.2003.09.018

DAP (2018) DAP fertilizer—monthly price—commodity prices—price charts, data, and news—IndexMundi. Indexmundi.com. https://www.indexmundi.com/commodities/?commodity=dap-fertilizer&months=240. Accessed 16 Sept 2018

Desneux N, Decourtye A, Delpuech J (2007) The sublethal effects of pesticides on beneficial arthropods. Annu Rev Entomol 52:81–106. https://doi.org/10.1146/annurev.ento.52.110405.091440

Dewar H, Cronin M, Landers T (2011) Urban fertilizers & the Chesapeake Bay: an opportunity for major pollution reduction. Environmentmaryland.org. https://environmentmaryland.org/sites/environment/files/reports/Environment-Maryland%2D%2D-Urban-Fertilizers%2D%2Dthe-Chesapeake-Bay-WEB.pdf. Accessed 14 Sept 2018

Dively G, Venugopal P, Finkenbinder C (2016) Field-evolved resistance in corn earworm to cry proteins expressed by transgenic sweet corn. PLOS One 11:e0169115. https://doi.org/10.1371/journal.pone.0169115

Dixon M (2018) Chemical fertilizer in transformations in world agriculture and the state system, 1870 to interwar period. J Agrarian Change. https://doi.org/10.1111/joac.12259

Fossil Fuels (2018) Fossil fuels—chemistry encyclopedia—water, gas, coal, coke, petroleum, natural gas. Chemistryexplained.com. http://www.chemistryexplained.com/Fe-Ge/Fossil-Fuels.html. Accessed 5 Sept 2018

Fowler D, Coyle M, Skiba U et al (2013) The global nitrogen cycle in the twenty-first century. Philos Trans R Soc B Biol Sci 368 (1621):20130164. https://doi.org/10.1098/rstb.2013.0164

Franklin H (2007) Most important fish in the sea, 1st edn. Island Press, Washington DC, p 14

Gale E, Sullivan D, Cogger C et al (2006) Estimating plant-available nitrogen release from manures, composts, and specialty products. J Environ Qual 35:2321. https://doi.org/10.2134/jeq2006.0062

Galloway J, Aber J, Erisman J et al (2003) The nitrogen cascade. BioScience 53:341. https://doi.org/10.1641/0006-3568(2003)053[0341:tnc]2.0.co;2

Goode G (1880) The use of agricultural fertilizers by the American Indians and the Early English Colonists. Am Nat 14:473–479. https://doi.org/10.1086/272593

Guano (2018) The Guano trade. National Museum of American History. http://americanhistory.si.edu/norie-atlas/guano-trade. Accessed 10 Aug 2018

Havas M, Hutchinson T (1980) Effects of acid precipitation on terrestrial ecosystems. In: NATO conference series I:V4 Ecology

Heap I (2014) Herbicide resistant weeds. In: Pimentel D, Peshin R (eds) Integrated pest management. Springer, Dordrecht

Heap I, Duke S (2017) Overview of glyphosate-resistant weeds worldwide. Pest Manag Sci 74:1040–1049. https://doi.org/10.1002/ps.4760

Hersey M (2006) Hints and suggestions to farmers: George Washington Carver and rural conservation in the South. Environ Hist 11:239–268. https://doi.org/10.1093/envhis/11.2.239

Honeywell (2019) Honeywell history. Honeywell. https://www.honeywell.com/who-we-are/our-history. Accessed 13 Jan 2019

Isbell F, Reich P, Tilman D et al (2013) Nutrient enrichment, biodiversity loss, and consequent declines in ecosystem productivity. Proc Natl Acad Sci USA 110(29):11911–11916

IUCN (2018) A bird droppings biodiversity paradise—the Guano Islands and Capes National Reserve System, Peru. IUCN. https://www.iucn.org/content/bird-droppings-biodiversity-paradise-–-guano-islands-and-capes-national-reserve-system-peru. Accessed 10 Aug 2018

Iwamoto T, Eschmeyer W, Alvarado J (2010) *Engraulis ringens*. The IUCN red list of threatened species 2010: e.T183775A8174811. https://doi.org/10.2305/IUCN.UK.2010-3.RLTS.T183775A8174811.en

Ji Q, Babbin A, Jayakumar A, Oleynik S, Ward B (2015) Nitrous oxide production by nitrification and denitrification in the Eastern Tropical South Pacific oxygen minimum zone. Geophys Res Lett 42:10755–10764. https://doi.org/10.1002/2015gl066853

Kaufman Z, Abler D, Shortle J et al (2014) Agricultural costs of the Chesapeake Bay total maximum daily load. Environ Sci Technol 48:14131–14138. https://doi.org/10.1021/es502696t

Kaur K, Kapoor K, Gupta A (2005) Impact of organic manures with and without mineral fertilizers on soil chemical and biological properties under tropical conditions. J Plant Nutr Soil Sci 168:117–122. https://doi.org/10.1002/jpln.200421442

Kellogg M, Cornwell J, Owens M, Paynter K (2013) Denitrification and nutrient assimilation on a restored oyster reef. Mar Ecol Prog Ser 480:1–19. https://doi.org/10.3354/meps10331

Knecht M, Goransson A (2004) Terrestrial plants require nutrients in similar proportions. Tree Physiol 24:447–460. https://doi.org/10.1093/treephys/24.4.447

Kremer G (2011) George Washington Carver. Greenwood, Santa Barbara, CA

Kuypers M, Marchant H, Kartal B (2018) The microbial nitrogen-cycling network. Nat Rev Microbiol 16:263–276. https://doi.org/10.1038/nrmicro.2018.9

Lam P, Kuypers M (2011) Microbial nitrogen cycling processes in oxygen minimum zones. Annu Rev Mar Sci 3:317–345. https://doi.org/10.1146/annurev-marine-120709-142814

Landon A (2008) The "how" of the three sisters: the origins of agriculture in Mesoamerica and the human niche. Digitalcommons.unl.edu. https://digitalcommons.unl.edu/cgi/viewcontent.cgi?article=1039&context=nebanthro. Accessed 19 Aug 2018

Lesher P (2008) A load of Guano: Baltimore and the fertilizer trade in the nineteenth century. Northern Mariner/le marin du nord XVIII (3–4):121–128. https://www.cnrs-scrn.org/northern_mariner/vol18/tnm_18_3-4_121-128.pdf

Liebig J (1847) Chemistry in its application to agriculture and physiology. Edited by Playfair L T.B. Peterson ... Taylor & Francis, London

Liu Y, Li X, Zhou C, Liu F, Mu W (2016) Toxicity of nine insecticides on four natural enemies of Spodoptera exigua. Sci Rep. https://doi.org/10.1038/srep39060

Mann T, Tomiyama A, Westling E et al (2007) Medicare's search for effective obesity treatments: diets are not the answer. Am Psychol 62:220–233. https://doi.org/10.1037/0003-066x.62.3.220

Marshall A, Orr D, Bradley L, Moorman C (2015) Review of organic lawn care practices and policies in North America and the implications of lawn plant diversity and insect pest management. HortTechnology 25(4):437–446

Matson P (1997) Agricultural intensification and ecosystem properties. Science 277:504–509. https://doi.org/10.1126/science.277.5325.504

MDA (2018) Maryland's Lawn Fertilizer Law. Mda.maryland.gov. https://mda.maryland.gov/resource_conservation/Documents/fertilizerwebpage.pdf. Accessed 14 Sept 2018

Milesi C, Elvidge C, Dietz J et al (2005) A strategy for mapping and modeling the ecological effects of US lawns. In: ISPRS proceedings XXXV 8-W27. http://www.isprs.org/proceedings/XXXVI/8-W27/milesi.pdf

NCGA (2018) World of corn 2018. Worldofcorn.com. National Corn Growers Association. http://www.worldofcorn.com/#corn-usage-by-segment. Accessed 17 Sept 2018

Nicol G, Leininger S, Schleper C, Prosser J (2008) The influence of soil pH on the diversity, abundance and transcriptional activity of ammonia oxidizing archaea and bacteria. Environ Microbiol 10:2966–2978. https://doi.org/10.1111/j.1462-2920.2008.01701.x

NMAH (2018) The Guano Islands Act. National Museum of American History. http://americanhistory.si.edu/norie-atlas/guano-islands-act. Accessed 10 Aug 2018

Parsons W (2018) Agriculture | Chesapeake Bay Program. Chesapeakebay.net. https://www.chesapeakebay.net/issues/agriculture. Accessed 7 Sept 2018

Philips S, Focazio J, Bachman M (2018) Discharge, nitrate load, and residence time of ground water in the Chesapeake Bay watershed by Scott W. Phillips, Michael J. Focazio, and L. Joseph Bachman. Pubs.usgs.gov. https://pubs.usgs.gov/fs/fs15099/. Accessed 8 Sept 2018

Puckett L, Tesoriero A, Dubrovsky N (2011) Nitrogen contamination of surficial aquifers—a growing legacy. Environ Sci Technol 45:839–844. https://doi.org/10.1021/es1038358

Rashid M, Ahmed N, Jahan M et al (2017) Higher fertilizer inputs increase fitness traits of brown planthopper in rice. Sci Rep. https://doi.org/10.1038/s41598-017-05023-7

Rizzo N, Sabate J, Jaceldo-Siegl K et al (2011) Vegetarian dietary patterns are associated with a lower risk of metabolic syndrome: the Adventist health study 2. Diab Care 34:1225–1227. https://doi.org/10.2337/dc10-1221

Rütting T, Aronsson H, Delin S (2018) Efficient use of nitrogen in agriculture. Nutr Cycling Agroecosyst 110:1–5. https://doi.org/10.1007/s10705-017-9900-8

Sanford W, Pope J (2013) Quantifying groundwater's role in delaying improvements to Chesapeake Bay Water quality. Environ Sci Technol 47:13330–13338. https://doi.org/10.1021/es401334k

SARE (2018) Impacts of fertilizers on insect pests. Sare.org. https://www.sare.org/Learning-Center/Books/Manage-Insects-on-Your-Farm/Text-Version/Managing-Soils-to-Minimize-Crop-Pests/Impacts-of-Fertilizers-on-Insect-Pests. Accessed 6 Sept 2018

Schindler D (1988) Effects of acid rain on freshwater ecosystems. Science 239:149–157. https://doi.org/10.1126/science.239.4836.149

Scotts (2018) How to Start the Scotts® 4 Step® Program. Scotts.com. https://www.scotts.com/en-us/library/lawn-food/how-start-scotts-4-step-program. Accessed 14 Sept 2018

Seitzinger S, Harrison J, Böhlke J et al (2006) Denitrification across landscapes and waterscapes: a synthesis. Ecol Appl 16:2064–2090. https://doi.org/10.1890/1051-0761(2006)016[2064:dalawa]2.0.co;2

Smith J (1612) A map of Virginia. Joseph Barnes, Oxford, pp 1–3. In The complete works of Captain John Smith (1580-1631), edited by Philip L. Barbour, (1986) 3 vols., 1:119–190. University of North Carolina Press, Chapel Hill. Available in Capt. John Smith's complete works http://www.virtualjamestown.org/exist/cocoon/jamestown/fha-js/SmiWorks1

Smith K, Searchinger T (2012) Crop-based biofuels and associated environmental concerns. GCB Bioenergy 4:479-484. https://doi.org/10.1111/j.1757-1707.2012.01182.x

Smith K, Mosier A, Crutzen P, Winiwarter W (2012) The role of N2O derived from crop-based biofuels, and from agriculture in general, in Earth's climate. Philos Trans R Soc B Biol Sci 367:1169–1174. https://doi.org/10.1098/rstb.2011.0313

Statista (2018) Total weight loss U.S. market size 2013-2015 | Statistic. Statista. https://www.statista.com/statistics/625673/total-weight-loss-market-size-in-us/. Accessed 16 Sept 2018

Tarazona J, Arntz W (2001) The Peruvian coastal upwelling system. In: Seeliger U, Kjerfve B (eds) Coastal marine ecosystems of Latin America. Ecological studies (analysis and synthesis), vol 144. Springer, Berlin

Terziotti S, Capel P, Tesoriero A, Hopple J, Kronholm S (2018) Estimates of nitrate loads and yields from groundwater to streams in the Chesapeake Bay watershed based on land use and geology: U.S. Geological Survey Scientific Investigations Report 2017–5160, 20 p. https://doi.org/10.3133/sir20175160

USDA (2018a) USDA—National Agricultural Statistics Service Homepage. Nass.usda.gov. https://www.nass.usda.gov. Accessed 9 Sept 2018

USDA (2018b) USDA ERS—Food prices and spending. Ers.usda.gov. https://www.ers.usda.gov/data-products/ag-and-food-statistics-charting-the-essentials/food-prices-and-spending/. Accessed 16 Sept 2018

USDA (2018c) USDA Factsheet Soybeans. Usda.gov. USDA Factsheet Soybeans. US Department of Agriculture. https://www.usda.gov/sites/default/files/documents/coexistence-soybeans-factsheet.pdf. Accessed 17 Sept 2018

USDAERSA (2019) Oil crops outlook: April 2019. Ers.usda.gov. https://www.ers.usda.gov/publications/pub-details/?pubid=92830. Accessed 13 May 2019

USEPA (2010) Sources of nitrogen, phosphorus, and sediment to the Chesapeake Bay. Epa.gov. https://www.epa.gov/sites/production/files/2014-12/documents/cbay_final_tmdl_section_4_final_0.pdf. Accessed 8 Sept 2018

USEPA (2018a) Chesapeake Bay TMDL Fact Sheet | US EPA. US EPA. https://www.epa.gov/chesapeake-bay-tmdl/chesapeake-bay-tmdl-fact-sheet. Accessed 8 Sept 2018

USEPA (2018b) Overview of greenhouse gases | US EPA. US EPA. https://www.epa.gov/ghgemissions/overview-greenhouse-gases. Accessed 14 Sept 2018

USEPA (2018c) Basic information about NO2 | US EPA. US EPA. https://www.epa.gov/no2-pollution/basic-information-about-no2#Effects. Accessed 14 Sept 2018

Van Meter K, Basu N, Veenstra J et al (2016) The nitrogen legacy: emerging evidence of nitrogen accumulation in anthropogenic landscapes. Environ Res Lett 11:035014. https://doi.org/10.1088/1748-9326/11/3/035014

VCE (2018) Prevention and control of Palmer Amaranth in soybean. Pubs.ext.vt.edu. http://pubs.ext.vt.edu/content/dam/pubs_ext_vt_edu/2808/2808-1006/2808-1006_pdf.pdf. Accessed 6 Sept 2018

Ward M, deKok T, Levallois P et al (2005) Workgroup report: drinking-water nitrate and health—recent findings and research needs. Environ Health Perspect 113:1607–1614. https://doi.org/10.1289/ehp.8043

WHO (2004) Guidelines for drinking water quality. 3rd edn, Vol. 1, Recommendations. World Health Organization, Geneva

Wright N, Wilson L, Smith M et al (2017) The BROAD study: a randomised controlled trial using a whole food plant-based diet in the community for obesity, ischaemic heart disease or diabetes. Nutr Diab 7:e256–e256. https://doi.org/10.1038/nutd.2017.3

Xu G, Fan X, Miller A (2012) Plant nitrogen assimilation and use efficiency. Annu Rev Plant Biol 63:153–182. https://doi.org/10.1146/annurev-arplant-042811-105532

Yonebayashi K, Katsumi N, Nishi T et al (2014) Activation of nitrogen-fixing endophytes is associated with the tuber growth of sweet potato. Mass Spectrom 3:A0032–A0032. https://doi.org/10.5702/massspectrometry.a0032

Pesticides Bring the War on Nature to the Chesapeake Bay

Benjamin E. Cuker, Indu Sharma, Kendra Dorsey, Olivera Stojilovic, Nefertiti Smith, and Andrew Justice

All good things must come to an end
The gnat opened a bottle that he thought was gin
He stepped to the bar for a drinking spree
What do you know, it was DDT!
. . .
A verse from the 1951 song, Insect Ball, *by Big Jay McNeely, commenting on the most famous pesticide of that era*
. . . .
Hey farmer farmer
Put away the DDT
I don't care about spots on my apples
Leave me the birds and the bees
Please!
. . . .
A verse from the 1970 song, Big Yellow Taxi, *by Joni Mitchell, which bemoaned the war on nature.*

Abstract

The end of World War II brought DDT and other organic pesticides to Chesapeake Bay agriculture. These new insecticides and herbicides supported the industrialization of farming that primarily produced feedstock for the manufacture of animal-based foods. The widespread use of these new chemicals constituted a war on nature with the intent of poisoning non-human animals that ate crops and weeds that competed for land, space, light, nutrients, and water. The pesticides tended to fail as the artificial selection they imposed on their targets promoted the evolution of resistant pests. In the 1960s, herbicides replaced insecticides as the dominant pesticide. Genetic engineered pesticide resistant corn and soybean facilitated the marketing of the generalist herbicides glyphosate, and later dicamba. The genetically engineered crops and pesticide dominate the animal feedstock agriculture in the region. Years of pesticide use have contaminated water, soils, fish and wildlife. The pesticides threaten both human health and that of the environment. Adopting sustainable organic agricultural practices can end the war on nature and help recover the health of the Chesapeake Bay ecosystem and that of its people.

Keywords

DDT · Glyphosate · Organic farming · Herbicides · Insecticides · Sustainable agriculture

The War on Nature Comes to the Chesapeake Bay The city squares and roadsides of the Chesapeake Bay Watershed bristle with monuments to the wars fought on its shores long ago. These include conflicts between tribes of Native Americans, the genocide practiced on Indians by Europeans, Bacon's Rebellion, Nat Turner's Rebellion, the American Revolutionary War, The War of 1812, and the American Civil War. Long silent cannon and tarnished bronze markers pay tribute to battles and heroes of those

B. E. Cuker (✉) · K. Dorsey · O. Stojilovic · N. Smith · A. Justice
Department of Marine and Environmental Science, Hampton University, Hampton, Virginia, USA
e-mail: Benjamin.cuker@hamptonu.edu

I. Sharma
Department of Biological Sciences, Hampton University, Hampton, VA, USA

B. E. Cuker (ed.), *Diet for a Sustainable Ecosystem*, Estuaries of the World, https://doi.org/10.1007/978-3-030-45481-4_11

momentous events from the past. Yet, one war in the Chesapeake Bay region that started in 1945 is still contested on many fronts today. It's the war on nature, fought in the name of the food system with toxic chemicals and genetically engineered crops.

Although different from all the other wars fought on the Chesapeake's soil that preceded it, this quiet conflict even has its own bronze markers to commemorate one of its heroines. Historical signs celebrate the birthplace of Rachel Carson in Springdale, PA and her home in Silver Spring, MD. A bronze disk along the Points of Light walkway in Washington, DC also acclaims her accomplishments. That disk includes a quote in which Carson addresses the war on nature, "...challenged as mankind has never been challenged before to prove our maturity and mastery, not of nature, but of ourselves." Carson advocated for ending the war on nature and replacing it with "...respectful coexistence of mankind and the environment (Fig. 1)." Rachel Carson fought on the side of nature. Her 1962 book, "Silent Spring," gave a powerful voice to birds and other casualties of the war on nature who couldn't speak for themselves. This heroic scientist and author brought to the attention of the world the consequences of spreading poisons across the landscape to sustain a food system that produced an excessive surplus.

The end of WWII marks the formal start of the war on nature. Demobilization meant converting munitions factories to chemical plants to produce three general classes of chemicals for agriculture: (1) Pesticides designed to enhance crop production by killing organisms that ate the crops. These include insects, other invertebrates, small vertebrates, and fungi. (2) Herbicides designed to kill plants (weeds) that competed with crops for space, water, and nutrients. (3) Artificial fertilizers designed to boost the availability of nutrients for enhanced crop production (See chapter "The Journey from Peruvian Guano to Artificial Fertilizer Ends with Too Much Nitrogen in the Chesapeake Bay" for discussion of fertilizers). These new chemical weapons reshaped and industrialized the agricultural landscape and the food system that it supported. The term pesticide is comprehensive and refers to any agent designed to work by killing organisms of any sort (Environmental Protection Agency 2018).

General Classification of Pesticides Three different systems prevail for classifying pesticides:

1. By target organism. Examples are insecticides for insects, herbicides for plants, fungicides for fungi, nematicides for nematodes, etc.
2. By molecular structure. Examples are metallic compounds, organophosphates, organochlorines, carbamates, etc.
3. By the mode of action (MOA). These refer to the physiological system disrupted by the chemical and include growth regulators, neurotoxins, biosynthesis inhibitors, etc.

Additionally, pesticides may be classified by their solubility (hydrophobic vs. hydrophilic), persistence (half-life), and side effects (carcinogens, teratogens, endocrine disrupters, etc.).

Why Poison Nature to Produce Food? To answer this question, we must first understand something of human cultural evolution. Ancient humans discovered the basics of plant lifecycles allowing for the development of agriculture. They found that they could regenerate food sources by sowing seeds from the edible plants. With this discovery, people went from gatherers of nature's bounty to engineers of agricultural ecosystems. The planting of seeds and harvesting the resulting crops allowed for human cultural evolution. They no longer relied solely upon foraging, hunting or fishing. Agriculture provided economic and social stability.

People of the Fertile Crescent of Mesopotamia domesticated wheat, barley, peas, lentils, chickpeas, bitter vetch, and flax. Residents of China and India grew rice and millet. The Sahel of Africa offered rice and sorghum. Civilizations in the Americas grew up around corn, squash, potatoes, beans, and sunflowers. Mastering cultivation meant the ability to support larger human populations centered in villages, towns, and cities. It also meant extensive plantings that proved attractive to non-human consumers. Insects led the way in exploiting this new concentrated source of food. Granivorous, frugivorous and herbivorous birds and mammals joined in the feast. Nematode worms, mites, viruses, fungi, and various protozoa found their seats at the dinner table set by this new marvel called agriculture (Cohen 2009).

The ancient farmers became proficient at producing large fields planted in monocultures, to the approval of the pests that would swarm in to enjoy the feast. On the other hand, some food systems developed around the use of polycultures, with intercropping multiple different species. Such was the case for the three sisters plantings typical of Native American agriculture along the shores of the Chesapeake Bay (Smith 1612). The planting of corn, beans, and squash immediately next to each other diluted the potential food sources for pests that specialized on any one of them, as well as making better use of soil nutrients (Lewandowski 1987).

Monocropping and its attendant pests dominated the agricultural scene of the Old World. Devoting an entire field to a single crop simplified the processes of planting, weeding, and harvesting. It also set the stage for mechanization. Farmers deployed tractors and other farm machines to address that one

Fig. 1 Image of a bronze plaque celebrating the life of Chesapeake Bay resident, scientist, and author. Her book, *Silent Spring*, drew the nation's attention to the environmental crisis created by the widespread use of pesticides. Image used by permission from Allen C. Brown

crop. But the benefits of monocropping came at the expense of promoting pests. Farmers looked for ways to reduce the percentage of their crop lost to animal pests and microbial diseases.

It wasn't just the animal, bacterial, and fungal competitors for their crops that frustrated ancient farmers. They also contended with uninvited plants whose seeds, stolons and rhizomes found their way into the agricultural fields. These weeds competed with the crops for space, sunlight, nutrients, and water. Famers quickly learned that their fields required regular weeding to produce a good crop. And the situation worsened as farmers learned how to fertilize their fields. The added nutrients stimulated the growth of both crops and weeds (Dyck and Liebman 1994).

The First Inorganic Pesticides Inventive planters used a variety of inorganic compounds to deter pests from consuming seeds and crops. These included aluminum sulfate (alum), antimony, arsenic, calcium and magnesium salts, calcium carbonate, cobalt, copper sulfate, calcium hydroxide (lime), iron salts, mercury, potassium nitrate (salt peter), sodium carbonate, potassium carbonate (potash), calcium oxide (quick lime), ammonium chloride, sodium chloride and sulfuric acid (Smith and Secoy 1976; Davis 2013). Arsenic, lead, and mercury based pesticides proved dangerous to farmworkers and consumers, accounting for many deaths and disabilities to those suffering high levels of exposure (Carson 1962). Some apple orchards in Chesapeake Bay Watershed presented dangerous levels of arsenic in their soils decades after cessation of using pesticides with that metal (Robinson et al. 2007).

The First Organic Pesticides Ancient farmers also used the mix of organic and inorganic compounds found in smoke against insects and fungi. They used burnt straw, dried fish, animal dung, animal horns, and other foul-smelling fuels to produce smoke used to fumigate the standing crop. About 4000 BCE, the Persians pioneered the use of an organic pesticide, obtaining pyrethrum from daisies (*Anacetum cinerariaefolium*). Pyrethrum remains in use today as an insecticide (MGK 2019; Unsworth 2010).

One War Begins as Another One Finishes The first battles fought in the war on nature in the Chesapeake Bay Watershed took place years before the formal start of the all-out war on nature. That's not unusual for human conflicts. Abolitionist John Brown and his men raided the armory at Harper's Ferry, VA in 1859, two years before South Carolina fired the official first shot of the Civil War. The early battles in the chemical war on nature were fought with inorganic pesticides including Paris green (copper acetorarsenite) and lead arsenate in the nineteenth Century to kill pests on crops. Occasionally, consumers poisoned by pesticide residues on food became collateral damage in the war on nature. The Pure Food and Drug Act of 1906 helped protect consumers from arsenic on fruit and vegetables. The Federal Insecticide Act of 1910 insured consumers and farmers of the effectiveness of their chemical agents, but was silent on the effects of these substances on human health or the environment (Davis 2013). The USEPA administers the Federal Insecticide, Fungicide, and Rodenticide Act (FIFRA 1947, 1972) which is the updated version of 1910 Law.

DDT and the Atomic Bomb The atomic bombing of the Japanese cities Hiroshima and Nagasaki in the summer of 1945 marked the end of WWII. This freed the industrial war effort to target a new enemy, nature. Technology, industrial production, and innovative new weapons saved the day in the fight against fascism. The public was ready for this winning trio to do the same to solve basic domestic problems, such as ensuring an adequate food supply. In 1945, the public understood that the war caused death and despair by starvation as much as by bombs and bullets. And that didn't end with the signing of peace treaties. The disruptions of WWII left a legacy of impaired agricultural production and distribution, creating food insecurity for millions of people (Shaw 2007). DDT (dichloro diphenyl trichloroethane, Fig. 2) became the atomic bomb of agriculture. Austrian chemist, Othmar Zeidler, synthesized this chlorinated hydrocarbon in 1874, but it was the Swiss chemist Paul Muller who discovered its deadly effect on insects. It works by disrupting the sodium channels of the neurons in insects (Coats 1990). The US military used DDT to delouse troops and civilians living in the filth of war (Fig. 3), ending epidemics of typhus transmitted by the lice. DDT was also used to kill mosquitos carrying malaria. Paul Muller won a Nobel prize in 1948 in recognition of DDT's success stopping typhus.

After the war, the public embraced DDT to kill insects everywhere. Unlike its arsenic and lead predecessors, DDT displayed no acute effects on human health. It took years before the chronic conditions associated with the pesticide emerged. These include cancers of the liver, pancreas, and breast. DDT exposure is also associated with diabetes, spontaneous abortion, reduced duration of lactation, birth defects, early onset of puberty in girls, reduced semen quality in males, neurodevelopment disorders, poor psychomotor development, depression of thyroid activity, and compromised immune systems (Eskenazi et al. 2009).

Despite widespread use around the home and in urban areas to control mosquitos, most DDT went to agriculture. Between 1945 and 1974, about 612 million kg (1.35 billion pounds) of DDT was applied to the US landscape. Most went on cotton fields (80%) and the remainder entered the food chain through the soybean and peanut crops. Usage peaked at 36 million kg (80 million pounds) in 1959, dropping to 5.4

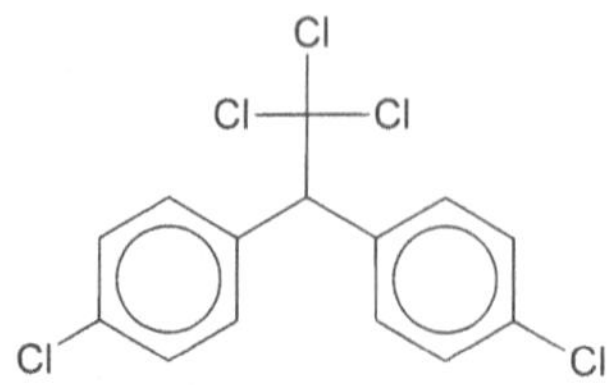

Fig. 2 The molecular structure of DDT. Note the presence of chloride atoms. DDT was the first of the chlorinated hydrocarbons to be used as a pesticide. https://commons.wikimedia.org/wiki/File:DDT.svg

Fig. 3 DDT powder being applied as a delousing agent during World War II. Image from the US Department of Health. Public domain. http://phil.cdc.gov/phil/details.asp?pid=2620

million kg (12 million pounds) in the early 1970's when it was banned (USEPA 1975).

DDT Legacy in the Chesapeake Bay *Silent Spring,* by Rachel Carson, documents the consequences to the environment wrought by less than two decades of exposure to DDT. DDT and its derivatives destroyed beneficial insects (pollinators and predators of pest insects), killed trout and other fish, and devastated bird populations by weakening the shells of their eggs. Effects were most pronounced at higher trophic levels, as DDT accumulated in fatty tissues and biologically magnified up the food chain. Absent a mechanism for excreting or metabolizing DDT, the pesticide stays resident in an animal until it is eaten by a predator. The predator then assimilates the DDT in its own tissues. And so it goes with each step up the food chain (USEPA 2012). Consider some specific examples from the Chesapeake Bay.

Barn owl (*Tyto alba*) nests sampled in 1972 and 1973 along the Potomac River yielded eggs with high concentrations of DDE (a breakdown product of DDT) and other organochlorines. Egg shell thickness significantly declined with increasing concentrations of the containments. Contamination reduced reproductive success below the level needed for replacement (Klaas et al. 1978).

By the 1960s, bald eagles (*Haliaeetus leucocephalus*), osprey (*Pandion haliaetus*), double crested cormorant (*Phalacrocorax auratus*), and brown pelicans (*Pelecanus occidentalis*) nearly disappeared from the Chesapeake Bay owing to the effects of DDT. Populations of these top predators only began to recover decades after DDT was banned in 1974. Proving Carson right, eliminating DDT eventually led to near complete recovery of bald eagle populations on the Chesapeake Bay (Watts et al. 2008). And the number of mated pairs of osprey on the Chesapeake went from a perilous 1500 in 1972 to 10,000 by 2013 (Learn 2016). Despite this success, the ghost of DDT still lingers in the Chesapeake Bay food chain. DDE continues its presence in fish tissue and a fivefold biological magnification in osprey eggs. However, concentrations of DDE appear low enough to have little effect on osprey reproduction (Rattner et al. 2004; Lazarus et al. 2016).

Insecticides After DDT Despite the heavy dependence upon DDT after WWII, the industry moved quickly to introduce other organic pesticides to agriculture. Notwithstanding the initial effectiveness of DDT in the war on nature, it was a general insecticide and incompatible with crops requiring living insects for pollination. Muller spent a good portion of his Nobel Prize lecture discussing these fast developing alternatives to DDT (Muller 1948). These synthetic organic insecticides fall into five different chemical families: organophosphates, carbamates, organochlorines, pyrethroids, and thiocarbamates (Fig. 4).

By the time Silent Spring debuted in 1962, DDT was no longer the pesticide of choice. The rapid lifecycles and high fecundity of insects meant that heritable traits conferring resistance would quickly come to dominate the genomes of populations under the artificial selection of DDT. For example, apple growers in Virginia found DDT no longer effective against codling moth (*Cydia pomonella*) and turned to organophosphates instead (Croft 1982). Yet, insects soon evolved resistance to the DDT replacements. On the Eastern Shore of Virginia, the Colorado potato beetle, *Leptinotarsa decemlineata* showed rise in resistance with more frequent dosing of fields (Tisler and Zehnder 1990).

Another insect nerve disruptor was introduced a year after DDT for the war on nature. Organophosphate pesticides came from WWII where the German war machine weaponized these molecules into a nerve gas, sarin. Organophosphates destroy cholinesterase, the enzyme that breaks down the neurotransmitter acetylcholine (Everts 2016). These organophosphate pesticides proved not only to be deadly and debilitating to insects but to other species with nervous systems including humans. DDT gained favor in part for lacking apparent acute effects on humans. In contrast, organophospahtes proved harmful and even lethal at low doses.

The chlorinated organophosphate insecticide, Chlorpyrifos, is of special concern for the Chesapeake Bay. It is a neurotoxin (cholinesterase inhibitor) that is found throughout the Bay. Its linked to learning disabilities, developmental delay, attention deficit syndrome, and other human health problems (CBF 2019). As of this writing, a Federal court has ordered the USEPA to ban its use, despite pressure from the chemical agriculture industry to the contrary (Chase-Lubitz 2019).

Cross Resistance Insects that develop resistance to one pesticide often develop resistance to other pesticides with the similar modes of action. To be effective, the pesticide inhibits or binds an enzyme or neural transmitter. A mutation in the target that prevents the binding of the pesticide will result in resistance. These resistant gene may confer cross resistance to an array of pesticides with the same mode of action (MOA). This phenomenon of cross resistance was seen for the codling moth mentioned above (Dunley and Welter 2000).

The Kepone Battle in the War on Nature The James River winds its way through the City of Richmond making passage to Hampton Roads, where it flows into the Chesapeake Bay. Some 37 km (23 miles) southeast of Richmond and 63 km

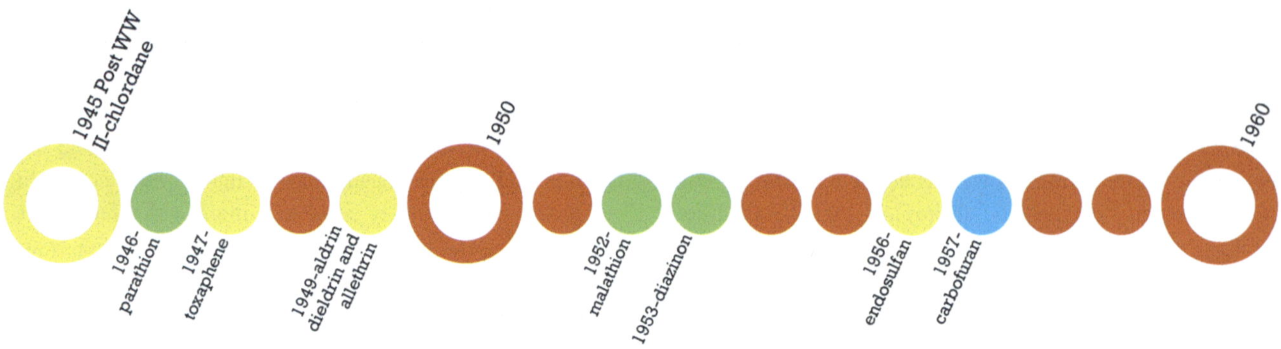

Fig. 4 Chronology of the insecticides introduced after DDT in the post WWII years, the yellow circles state that the insecticide belongs to the chlorinated hydrocarbons, the green to organo-phosphates, blue to methyl carbamate. Note that allethrin belongs to pyrethroid group. Figure by: Olivera Stojilovic

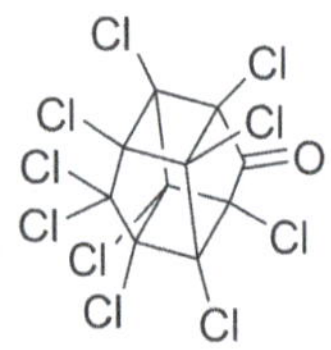

Fig. 5 The chemical structure of Kepone (chlordecone), a chlorinated pesticide that contaminated the James River during the 1960s and 1970s. Image by Edgar 181, https://commons.wikimedia.org/wiki/File:Chlordecone.png

(39 miles) west-north west of Capt. John Smiths' Jamestown Island, the Appomattox River joins the James. Here lies the small city of Hopewell, VA (population 23,000) situated on the banks of the James river where it intersects with the Appomattox river before it empties into the Chesapeake Bay. Towers, tanks and buildings of the Honeywell Corporation (formerly AlliedSignal Corporation) industrial park reach skyward from the banks of the western shore of the James River. This chemical plant opened in 1928 as the world's largest manufacturer of reactive nitrogen for artificial fertilizer. It has since shifted to produce the chemical precursor of Nylon, and toxic chemicals to kill agricultural pests. Between 1966 and 1975 it produced the pesticide called Kepone (chlordecone, Fig. 5).

Kepone is a chlorinated hydrocarbon comparable to DDT (Dichlorodiphenyltrichloroethane) in structure, purpose and longevity in the environment. AlliedSignal sold it as pesticide for killing agricultural and household insects. Like DDT, it worked by disabling the nervous system of its targets and wreaking havoc with the physiology of non-targeted animal species that it encountered.

AlliedSignal and a contractor (Life Sciences Corp.) located on the same site, produced 1.4 million kg (three million pounds) of Kepone over their decade of operation. Mishandling of the pesticide meant that 7% (90,000 kg) of the powdery Kepone escaped to the environment, mostly being washed into the James River. Plant workers labored without respirators or any protective clothing in a building with several cm of the pesticide dust covering the floor and even lunch tables. Kepone caught the attention of state and federal authorities in 1975 when 29 of those workers displayed profound neurological damage and other maladies from handling the toxin. This opened the way to investigating Kepone accumulation in the sediments and fish of the James River. The pesticide also poisoned the microbial community in the municipal wastewater treatment plant. AlliedSignal tore down the plant in 1975 and paid a fine of $13.2 million for poisoning its workers and the environment. The Life Sciences Corporation escaped paying its $four million fine through declaring bankruptcy. As a consequence of the decade-long Kepone release, the James River remained closed to fishing between 1975 and 1988. Yet for years after, fish from the river still tested positive for trace amounts of Kepone. Migratory fish and birds distributed the Kepone packaged in their tissues throughout the Chesapeake Bay. The Kepone disaster informed regional news for many years after it first came to light. Yet, pesticides remain central to the system of conventional chemical industrial farming that still dominates agriculture in the region (Foster 2005).

Weeds as the Second Front in the War on Nature The first big battle in the war on nature began with a DDT assault on crop eating insects in 1945. Other chemicals reinforced the effort in subsequent years as farmers perceived insects and other invertebrates as their primary enemies. Although obscured by the big push against insects, the chemical companies had launched an assault on a second front in the war on nature, weeds. This started with another pesticide developed during WWII, the herbicide 2,4-D (2,4-Dichlorophenoxyacetic acid).

Earlier in the twentieth century, botanists identified auxin, a plant hormone that controlled the growth of plants. Experiments showed that excessive auxin applied to plants would cause them to grow prolifically for a very short time, and then die from the disruption of their metabolism. Synthetic versions of auxin proved particularly effective at killing plants. This caught the attention of the Department of War which was working on biological weapons. Secret research began in the Chesapeake Bay Watershed at Camp Detrick in Frederick, MD. The military hoped to find a chemical it could use to kill crops in Japan and Germany, starving them into surrender. The conflict ended before herbicide based biological warfare could be tested. While never used for such purposes during WWII, the new organic herbicide, 2,4-D, proved effective in killing the broadleaved weeds that infested fields of grain crops. And most importantly, 2,4-D didn't harm the corn, wheat, or other grain producing grasses. It was specific for broad-leaved plants. This allowed grain production without the labor needed to clear the fields of weeds (Peterson 1967).

After 17 years of deployment in the war on nature on the home front, 2,4-D became a biological weapon against a foreign power. Beginning in 1961, the US military began spraying the countryside of Vietnam with 2,4-D as a defoliant. The idea was to kill the forests that obscured and harbored enemy forces. The 2,4-D was mixed with another herbicide, 2,4,5-T in a solution called Agent Orange. Agent Orange also contained dioxin (2,3,7,8-tetrachlorodibenzo para dioxin) a persistent toxicant that causes cancer, birth defects, endocrine disruption, immune system suppression, and neurological disorders (Stellman et al. 2003). The spraying killed or maimed 400,000 Vietnamese, caused half a million birth defects, and two million people to suffer from cancer or other diseases. Over two million US personnel were exposed to Agent Orange. Thousands later died from illness directly linked to that exposure. The US Department of Veteran Affairs administers funds to compensate Agent Orange casualties (USDVA 2019).

It wasn't just the battle ground of Vietnam that got flooded with herbicides in the 1960's and early 1970's. Plant-killing pesticides rapidly gained favor as a weapon against weeds in the US food system. In 1960, US agriculture used about 900 million kg of pesticides, 58% being insecticides and only 18% herbicides. By 1980, total pesticide use tripled, going to about 280 million kg. The increase came from a 15-fold increase in herbicides, while insecticide application remained virtually unchanged (Fig. 6). By 1968, the herbicide use had already tripled. The USDA presents a comparison of the types of pesticides used that year with the assortment applied forty years hence (Fig. 7). In 2008, the herbicide glyphosate (which didn't exist in 1968) dominated pesticide usage, accounting for 38% of the total applied.

In 2008 corn and soy combined to receive 61.2% of pesticides used that year (Fig. 8). Most remarkably, only a small fraction of each of those crops goes directly for human consumption. Most of the soy and half of the corn crop is for animal feed. The other half of the corn is fermented to produce ethanol for mixing with gasoline.

Big Corporations and Genetic Engineering Open a New Front in the War on Nature The transition from gathering to growing foods brought the beginning of agriculture. Farmers chose their crops from the biodiversity of nature. Through artificial selection, agriculturalists transformed the decedents of ancestral wildtype plants into specialized varieties that displayed desirable traits. Controlled breeding and grafting yielded thousands of varieties of plants best suited for local conditions. *This meant that saving seed for the next year's crop was central to farming.*

The industrialization of farming brought specialization to the production of seeds. Some farms began to grow crops not for food or fiber, but to produce seeds for sale to other farmers. These seed farms sped up the process of artificial selection for the growing farmer who was willing to pay for an improved variety. However, the growing farmer only needed to purchase the new seed once, as she could save seed from the harvest at the season's end. To ensure a steady customer base, seed producers developed hybrid varieties with highly desirable traits, such as fast growth and larger size of fruit. Because they were hybrids of two different breeds these plants didn't produce seeds that displayed all the desired traits. That meant that growing farmers were required to purchase the hybrid seed every year. This still left the growing farmer with choices. She could buy the hybrid seeds annually or do the additional labor of saving and storing seed from the non-hybrid seed harvest, just as farmers have always done.

In the 1970s, a new technology emerged called genetic engineering. Its application to agriculture would radically change the role of the seed seller. Genetic engineering uses techniques to change DNA composition, thus producing genetically modified organisms (GMOs). Genetic engineers either alter existing genes, or introduce new genes from a different species (transgenic engineering) (Byrne 2014, Nongmoproject 2019). The proponents of GMOs promised crops with a variety of highly desirable traits, such as faster growth, resistance to pests, resistance to frost, higher nutritional value, and resistance to herbicides (Key et al. 2008, Mogilna and Magufwa 2009).

Genetic engineering also meant producing seeds that would be protected under the Plant Patent Act of 1930 and the Plant Variety Protection Act of 1970. This gave the seed seller new control over their market. A farmer who had not bought the GMO seed but caught growing a crop that contained the patented trait found only in that GMO seed, would be sued by the seed seller. For example, the Monsanto Corporation introduced its GMO products in the 1990's and by 2003 had won legal actions against 147 farmers, winning on average $412,000 per case (CFS 2005). Ironically, this includes settlements against organic farmers with crops

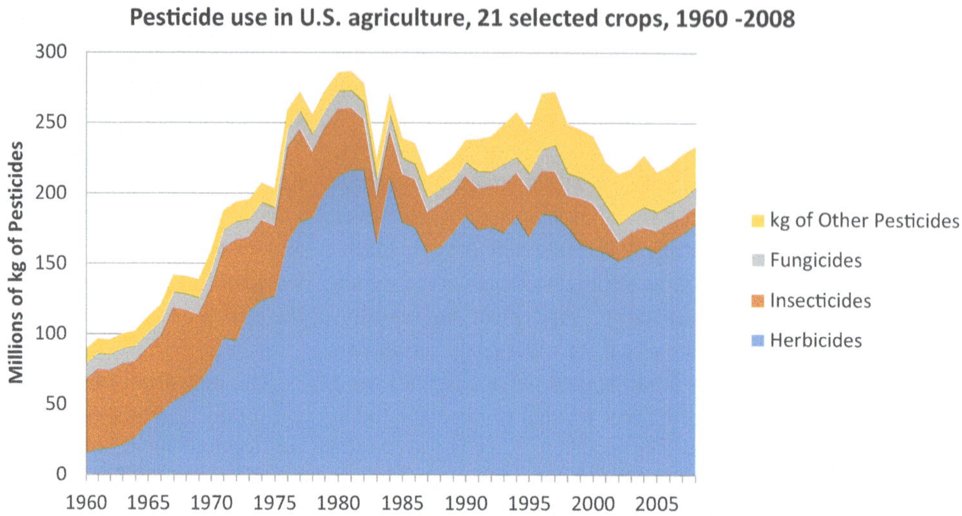

Fig. 6 The changing pattern of pesticide use in the US between 1960 and 2008 illustrates reduced application of insecticides and increased use of herbicides. In 1960, insecticides accounted for 58% of pesticides applied. This dropped to 17% in 1980. In the same interval herbicide application went from 18% to 74%. By 2008, insecticides went to 6% of pesticides applied, compared to 76% for herbicides. Data from USDA-ERS 2019

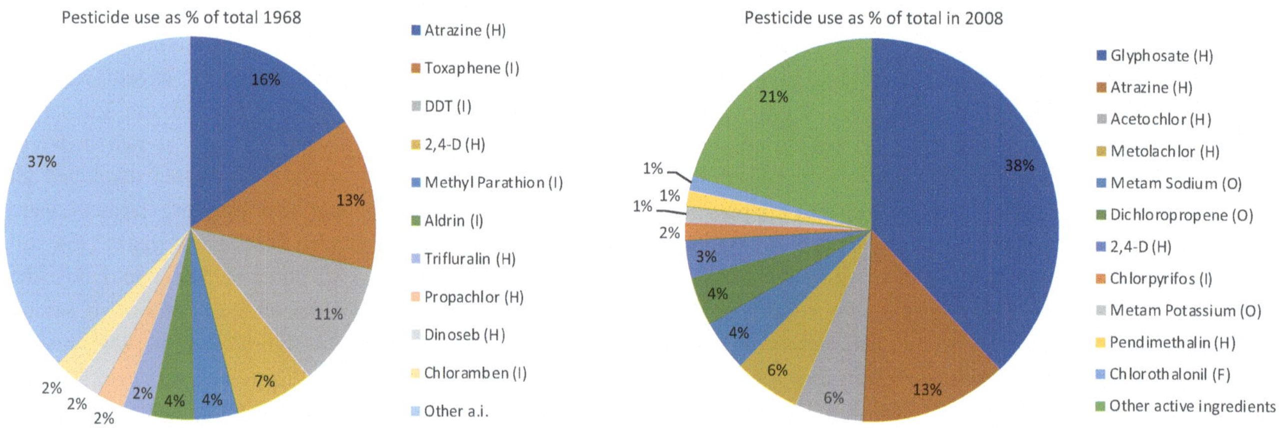

Fig. 7 Change in the kind of pesticides used in the US in 1968 and 2008. Use is noted by letter, H = herbicide, I = insecticide, O = other. By 2008, the herbicides glyphosate and atrazine accounted for half the pesticide usage. Data from USDA-ERS 2019

contaminated by pollen from nearby fields of Monsanto GMO plants.

Bt on and in Crops *Bacillus thuringiensis* (Bt) is a bacterium widely distributed in soils. It produces a crystal protein (Cry) and other polypeptides that are toxic to a variety of arthropods. The Cry proteins appear to create holes in the gut walls of susceptible insects. In the 1920s, farmers in Europe began spraying preparations containing Bt on their crops to kill pests. Bt was certified as a *topical pesticide* in the US in 1961, and remains a popular organic alternative to chemical products (Sanahuja et al. 2011). Beginning in 1996, the Monsanto Corporation used genetic engineering to transfer the genes for making Cry proteins into corn, soy, cotton and a variety of vegetable crops. This turned Bt into a *systemic pesticide* that can't be washed off of fruits or vegetables. While traditional topical Bt application proved safe to the consumer and environment, uncertainty remains about the Cry proteins in GMO food crops. Unlike topical application, the GMO Cry proteins are ingested by the consumer. The proteins are also edited by the genetic engineers and further altered by the GMO host, meaning that assumptions about the safety of the GMO Bt shouldn't be based on the topical version (Latham et al. 2017). Alas, pests are developing resistance to Bt GMO plants, limiting the future of these crops (Taylor 2017).

Glyphosate Monsanto was a large agricultural chemical corporation long prior to the advent of genetic engineering. Its laboratories produced many toxic products that left their mark on the environment, including PCBs (polychlorinated biphenyls), DDT, and Agent Orange.

In 1974, Monsanto introduced a broad spectrum herbicide called glyphosate (Fig. 9), and marketed it as Roundup™. The story of glyphosate begins with a 1961 patent awarded to the Stauffer Chemical Co. It was a chemical developed to descale pipes and chelate metals. In 1970, the Monsanto Corporation's, John Franz discovered that this pipe-cleaning

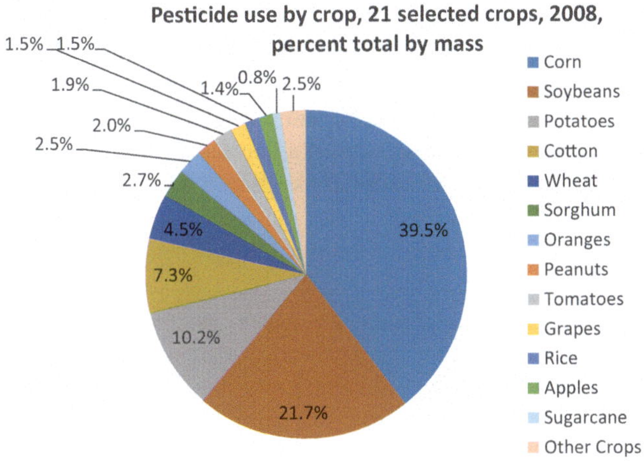

Fig. 8 Pesticide use by type of crop. Corn and soy accounted for most of the pesticide use in the food system in 2008. Data from USDA-ERS 2019

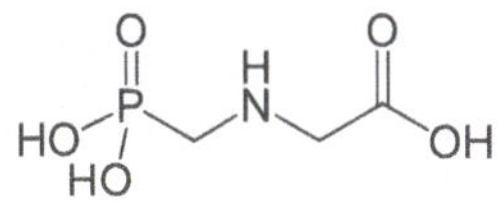

Fig. 9 Structure of glyphosate, the most commonly used herbicide in the Chesapeake Bay Watershed and the world between 1998 and 2019 (last year or record). Its primary use is on animal feed crops of corn and soybeans. By Yikrazuul, open source. SVG development W3C icon.svg https://commons.wikimedia.org/wiki/File:Glyphosate.svg

agent would also kill plants. Glyphosate works by inhibiting an enzyme called 5-enolpyruvylshikimate-3-phosphate (EPSP) synthase, which is central to the formation of the aromatic amino acids (phenylalanine, tyrosine, and tryptophan), hormones, vitamins, and other essential metabolites. It is a systemic poison, translocating to all parts of the plant including the roots. This was useful for killing all plants in an area, such as those growing through cracks in a sidewalk or along a roadside (Dill 2005). But how could it be useful in agriculture as it would kill crop and weed alike?

Enter genetic engineering (GE). Monsanto inserted a glyphosate-insensitive gene into canola plants (*Brassica napus*) in 1995, soy bean (*Glycine max*) in 1996, Corn (*Zea mays*) in 1998, alfalfa (*Medicago sativa*) in 2011, and sugar beets (*Beta vulgaris*) in 2012. This meant that Monsanto could market is GE Roundup Ready ™ seeds in tandum with the herbicide. Paring the herbicide with GE crops meant glyphosate use in the US went from 4.5 million kg in 1992 to 123 million kg in 2014. Of this, about 120 million kg went on soybeans and 80 million kg went on corn (Benbrook 2016; Duke 2017).

The three most planted food system crops in the US are corn, soy and wheat, in that order. Corn and soy enter the food system primarily as animal feed. Only 10% of corn and 15% of soy goes for direct human consumption (NCGA 2018, USDAERSA 2019). Wheat is produced for direct human consumption, with 50% of the US crop destined for export. In 1996, Monsanto introduced *Roundup Ready* soybeans. The soybeans were engineered to resist glyphosate by inserting genes from the *Agrobacteria* that expressed a modified EPSP enzyme which was insensitive to Roundup (Funke et al. 2006). Roundup Ready corn followed in 1998. Although Monsanto created glyphosate resistant wheat, they pulled it from certification in response to intense political pressure from Europe and Asian markets for the traded grain (Person 2009). Monsanto added cotton and other plants to the Roundup Ready corral.

Roundup Ready crops provided Monsanto with control of much of the world food supply while making money on selling both the seeds and the pesticide. One source estimates that 93% of soybeans, 82% of corn, and 85% of cotton in the US is planted in glyphosate resistant GMO crops (BP 2019).

Nature as Terminator of GMO Technology In order to protect their patented technology from free use by farmers, Monsanto and other bioengineering firms often packaged their inventions in seeds bred to produce infertile plants in the second generation, so called terminator seeds. This meant forcing farmers to buy new seeds each year. Yet nature provides its own terminator scheme in response to Bt and glyphosate GMO plants. Artificial selection provided by decades of assaulting plant-eating animals with Bt and weeds with glyphosate has created new resistant versions of each type of pest. Increasing numbers of pest insects now show resistance to Bt GMO plants (Tabashnik et al. 2013). Super weeds that resist glyphosate have farmers resorting to new chemical herbicides (Wilkerson and Chow 2015). As was the case with DDT and all other pesticides, within species genetic diversity combined with intensive artificial selection will eventually render useless GMO toxins. As of 2017, 37 weeds have evolved resistance to the herbicide. This marks the beginning of the end of the world's best-selling pesticide. Bayer Corporation, which bought Monsanto in 2018, is contending with thousands of pending lawsuits.

Toxicity of Glyphosate Based Herbicide for Humans Early proponents of Bt GMO crops pointed to the benefit of saving the environment and people from a soaking with chemical pesticides. Yet the very company that first patented Bt GMO plants, Monsanto, went on to produce Roundup Ready crops that *required* the extensive use of a chemical pesticide.

In 2018, a California jury awarded Dewayne Johnson, a grounds keeper, $289 million in a lawsuit against Monsanto.

The jury found that exposure to Roundup caused Mr. Johnson to contract non-Hodgkin's lymphoma. A judge reduced the award to $78.5 million. Monsanto went on to lose similar cases (Wamsley 2018; Tyko 2019). Despite jury decisions consistent with the World Health Organization's 2015 designation of glyphosate as a probable human carcinogen, many of the world's government regulatory bodies fail to agree. Glyphosate is the most widely used pesticide in the world. "Worldwide 9.4 million tons of the chemical has been sprayed on fields—enough to spray nearly half a pound of Roundup on every cultivated acre of land in the world" (Malkan 2019). Any finding that it is toxic to humans engenders enormous economic consequences to the corporations. Thus, Monsanto engages in the sponsoring of ghost written articles in scientific journals and the popular press to influence the peer review process and agency decisions (McHenry 2018). Beyond cancer, glyphosate exposure is also linked to reduced gestation period, higher rates of spontaneous abortion, congenital abnormalities, non-alcoholic fatty liver disease, alteration in the gut microbiome, anxiety, depression, toxicity to honeybees, and reduced monarch butterfly populations (Malkan 2019).

It wasn't so much the 2,4 D in agent orange that poisoned so many people in Vietnam, but it was the dioxin in the mix that proved so damaging. For Roundup™ the story may be similar, as other ingredients show toxic effects (Benachour and Séralini 2009). Farmworkers may absorb the toxins through contact and breathing.

Glyphosate Produces Antibiotic Resistance Monsanto patented glyphosate as an antibiotic, since many bacteria share the shikimate pathway that the herbicide blocks. Both Roundup™ and Kamba (Monsanto herbicide with dicamba) increase the rate that pathogenic strains of bacteria (*Salmonella typhimurium* and *Escherichia coli*) acquire cross resistance to ciprofloxacin by a factor of 1000–1000,000 (Kurenbach et al. 2018). These herbicides add to the crisis of antibiotic resistance caused by the food system (see chapter "Livestock and Poultry: The Other Colonists Who Changed the Food System of the Chesapeake Bay").

Glyphosate and the Chesapeake Bay Some of the original testing of glyphosate resistant GMO soy took place in the Chesapeake Bay Watershed near Queenstown, MD (Zilberfarb 2017). Glyphosate GMO crops won quick acceptance in the watershed, as they became part of the no-till methods used to reduce soil erosion. Rather than removing weeds with a cultivator, the no-till methods used herbicide. By 2001, glyphosate largely replaced Atrazine and other earlier generation pesticides. This reduced the total mass of herbicides applied, but due to increased potency, not the overall toxicity (Hartwell 2011). Glyphosate is highly water soluble, and readily binds to soil. It has a half-life of 35–96 days in soil, and its decomposition product, AMPA (Aminomethylphosphonic Acid), persists longer with a half-life of 76–240 days (Schueler and Youngk 2015).

Roundup™ may also be contributing to the eutrophication and disruption of the Chesapeake Bay food chain. A field experiment found it shifted the community structure of phytoplankton, increasing nano-plankton (cyanobacteria) by 40 fold while doubling productivity. Apparently, the P liberated from glyphosate stimulated nano-algal growth, while toxicity from the herbicide depressed the micro-plankton (green algae, diatoms, and dinoflagellates). This means less food for the zooplankton that nourish the rest of the food chain (Pérez et al. 2007).

Dicamba With glyphosate suffering from resistance in weeds and resistance from litigious humans who think it is giving them cancer, Monsanto turned to another herbicide to market with its GMO crops. They chose Dicamba, a pesticide in use since 1962. It is similar to 2,4 D in structure and mode of action, being an auxin mimic, and specific for broadleaved plants. As such, it was used with corn, a grass, and therefore not susceptible to the agent (Hartzler 2017). Dicamba wasn't appropriate for the broadleaved soybean crop, until 2015 when Monsanto received approval to use it with a GMO soy plant that could tolerate the herbicide. Monsanto added the gene to soybeans already possessing the added gene for glyphosate tolerance. They named the new GMO plant, Roundup Ready 2 Extend™ and marketed it with dicamba. Dicamba readily volatilizes and often drifts on to neighbors fields, killing their non-GMO crops. Monsanto's GMO dicamba resistant soy accounted for 40% of the US crop in 2018 (Abbott 2018). Figure 10 presents a chronology of post WWII herbicides.

Pesticide Contamination of the Chesapeake Bay Ecosystem Decades of pesticide application left its mark on the Chesapeake Bay ecosystem (Pimental 2009; USEPA 2012; USGS 2012). A 1985 study detected 20 different pesticides in ground water with the herbicide Atrazine being the most common (Mostaghimi et al. 1993). A sampling of Virginia's coastal plain in 1992–1993 found pesticides (atrazine, cyanazine, alachlor, metolachlor, and carbofuran) absent from surface waters of the Bay and offshore sediments, but present in upland soils, groundwater and in groundwater emerging into the estuary (Gallagher et al. 1996). Streams and ground water of the Potomac River basin sampled in 1992–1996 had measurable amounts of 12 different pesticides (Ator et al. 1998). The Susquehanna River drains highly productive agricultural land in Pennsylvania and New York. In 1997–1998, the river provided large annual loads of pesticides to the Chesapeake Bay, including

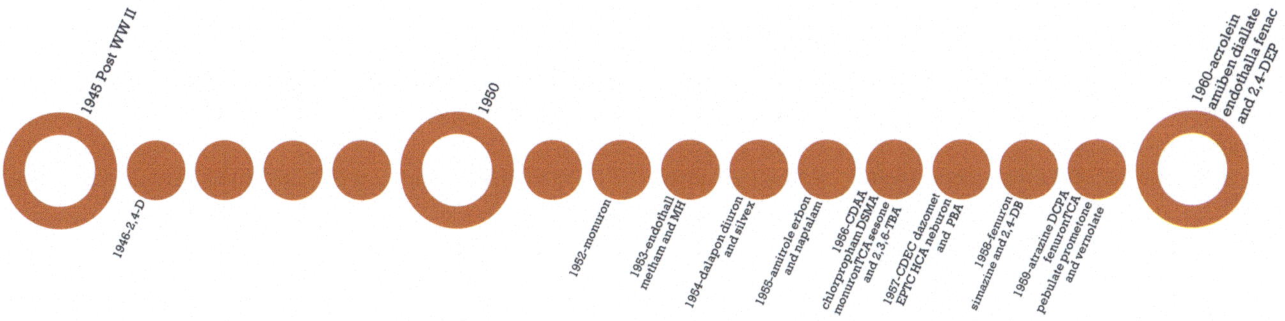

Fig. 10 Chronology of the herbicides introduced in the post WWII era. The names in the figure represent the common names of the herbicides. Figure by Olivera Stojilovic

5523 kg of herbicides and 98 kg of insecticides. Atrazine and its degradant, 6-amino-2-chloro-4-(isopropylamino)-s-triazine (CIAT) were the most common herbicides. Diazinon, B-endosulfan and chlorpyrifos combined for over half of the pesticide load. Some 23 years after DDT was banned, DDE (degraded DDT) had an annual loading rate to the Bay of 7 kg/y from the Susquehanna. In the 1990s, other tributary rivers (Potomac, James, Patuxent, Choptank) presented measurable levels of various pesticides dissolved in the water (Liu et al. 2002). Herbicides, Atrazine, Deethylatrazine, Metolachlor, Simazine appeared in all 12 drinking water sources sampled along the Potomac River (Kingsbury et al. 2008).

In 2000 and 2004, the largest study to date explored pesticide concentrations in the main stem of the Bay and its tributaries. Herbicides and their degradation products dominated the identified chemicals, with concentrations well above those for organochlorine insecticides. Most of the Bay's contamination originates in tributaries draining agricultural lands, much of which is planted in corn, soybeans and small grains. These typically receive doses of glyphosate, chlorothalonil, atrazine and other herbicides. In 2000, Atrazine and metolachlor appeared in all 18 stations on the main stem of the Bay. Atrazine, CIAT and other herbicides were almost always most concentrated in the tributaries draining croplands, particularly those of the Delmarva (McConnell et al. 2007).

Pesticides contaminate sediments in all parts of the Bay, with higher concentrations in the Susquehanna Flats and other sections draining agricultural areas. DDT (and its metabolites), cyclodiene (dieldrin, etc.) and chlorinated benzenes (fungicides) were found in sediments throughout the Bay (Hartwell and Hameedi 2007).

Pesticide Transport The problem of pesticides contaminating the environment starts with their very use. If farmers refused to spread the poison on their fields, it would not be an issue. However, pesticides are now part of conventional industrial agriculture which is promoted by companies selling the chemicals, and agricultural college curricula in Chesapeake Bay region. Often the pesticides show up in places other than the intended targets. The runoff from rainstorms and irrigation transports a portion of the pesticides in to drains and streams. Some pesticides leach through soils. Improper disposal of excess chemicals and accidental spills contaminate soils and waterways. Sprayed pesticides often drift away from targeted crops to adjacent land and water. Pesticides survive the wastewater treatment process, contaminating both the sludge (biosolids) and effluent from the facilities. The contaminated sludge and effluent may be applied to agricultural lands as fertilizer and irrigation (PMCBW 2009).

Pesticides in the Food Chain Although farmworkers may absorb pesticides through contact and breathing, it's the food system that exposes most of the population to these chemicals. Persistent pesticide residue contaminate fruits, vegetables, and grains. In 2016, pesticides contaminated 47% of domestically produced food in the US. Washing fruits and vegetables often doesn't remove all of the pesticide. Systemic insecticides reside in the tissues of plant, so are immune from any manner of cleaning. Neonicotinoids are systemic and presently the most widely used insecticide worldwide (Hyland et al. 2019).

The tendency for hydorphobic organic pesticides to accumulate in fatty tissue and biologically magnify with each step up the food chain often leads to contamination of animal-based foods. Thirteen different agricultural pesticides contaminated Chesapeake Bay oysters sampled in the 1990's (Lehotay et al. 1998). In 1990s, Chlordane and DDT concentrated in Asiatic clams (*Corbicula fluminea*) and yellow bullhead catfish (*Ameiurus natalis*) (Zappia 1996). Chlordane contaminates fish in the Anacostia and Potomac Rivers (Pinkney 2009). Pesticide induced stress inhibit immune system function in small mouth bass (*Micropterus dolomieu*) and promote skin lesions (Blazer et al. 2010).

Atrazine interferes with metamorphosis in freshwater fish and amphibians (Rohr and McCoy 2010). Atrazine concentrations in the Potomac basin correlate with development of testicular oocytes (intersex fish) and in small mouth bass (Kolpin et al. 2013).

Carbofuran caused a dozen bird die-off event in Virginia between 1985–1990 involving primarily 10 different coastal species (Stinson et al. 1994). The presence of DDT and DDE in osprey and eagles were discussed above.

Total pesticide mass used in the in Maryland declined between1985 and 2004, yet the greater toxicity of the newer chemicals phased in during this period meant that the total toxic load remained the same or increased. This is probably true for the entire Bay watershed (Hartwell 2011).

Pesticide in Our Food The widespread use of pesticides means that those chemicals contaminate much of the food purchased in the Chesapeake Bay region and the rest of the US. One study that included stores in Maryland and Virginia found that: glyphosate contaminated 100% of oat-based breakfast cereals (360 ppb avg.), and 100% of pinto beans (509 ppb avg.); neonicotinoids contaminated 73% of apple sauces (0.037 nmol/g avg.), 80% of spinach (0.065 nmol/g); organophosphates contaminated 100% of apple sauces (0.76 nmol/g avg.), 61% of whole apples (0.32 nmol/g avg.) and 32% of spinach (0.388 nmol/g avg.) (Klein 2019).

The USDA, periodically, tests conventionally grown produce for the presence of pesticides. This is after washing and peeling as would be done by the consumer. The Environmental Working Group (EWG) released a compilation of that data in 2019. Pesticides appeared on nearly 70% of all US produce. This included 225 different chemicals (pesticides and their breakdown products). EWG lists the twelve most contaminated items as; strawberries, spinach, kale, nectarines, apples, grapes, peaches, cherries, pears, tomatoes, celery, and potatoes. Most items on the list contained two or more pesticides. Kale, known as a healthy food, contained up to 18 different pesticides. Nearly 60% of the kale tested positive for Dacthal (DCPA), a pesticide banned in Europe and listed as suspected carcinogen by the USEPA (EWG 2019).

Pesticides and Human Health Some older organophosphate pesticides caused acute and often fatal effects in humans. Newer ones might lack the acute effects of their predecessors, but many prove dangerous and even lethal over time. According to the Maryland Pesticide Network, "Long-term, chronic exposure to low-level concentrations of pesticides may be a chronic disease health risk for residents of the Chesapeake Bay watershed". A growing body of epidemiological research suggests an association between pesticide exposures and chronic diseases such as certain cancers, as well as reproductive, neurological, respiratory and developmental disorders (MPN 2009).

A systematic review of the literature revealed pesticide exposure associated with various malignancies, including non-Hodgkin lymphoma, leukemia, and cancers of the brain, breast kidneys, lungs, ovaries, pancreas, prostate, and stomach (Basil et al. 2007).

Consider some of the pesticides commonly used in the Chesapeake Bay Watershed and their human side effects. Glyphosate is associated with non-Hodgkin lymphoma. Exposure to the pesticide increases the chances of contracting that cancer by 41% (Zhang et al. 2019). Acute exposure to Atrazine causes diseases of the heart, lungs, and kidneys; low blood pressure, muscle spasms, weight loss and damage to adrenal glands. Chronic exposure damages the heart and retina; causes muscle degeneration, and cancer (CDC 2019b). Chlorpyrifos is associated with reduced IQ of children, a 60% increase in autism spectrum disorder, reduced birth size, endocrine disruption, lung and prostate cancer PAN 2019. See Table 1 for other pesticides used in the region and their human health consequences.

The USEPA requires manufacturers to conduct toxicity testing prior to releasing a new pesticide to the market. The companies typically contract this work to various laboratories. However, conclusions drawn from those industry-funded studies frequently don't match the outcomes from other independent work. Poor study design and unsupported conclusions plague the industry bought studies (Mie et al. 2018). So, it's not surprising that some pesticides that initially win approval eventually lose their sanction and must be removed from the market.

Why Most Pesticides Ultimately Fail Despite agribusiness' chemical war on nature, pests and diseases continue to reduce harvests of global crops by 30% (FAO 2017). Farmers strike a Faustian bargain when they spread poisons on crops. They reason that some level of toxicity in the food system is compensated by the benefits of a better harvest. Sadly, the improved harvest is often short-lived. The targeted pests typically have short life cycles and the capacity for rapid reproduction. The combination of these traits means that genes conferring resistance to a pesticide will quickly spread to subsequent generations, rendering that pesticide useless. Consider a population of insects in which 5% of the individuals display genetic resistance to a pesticide. The first use of that chemical will kill 95% of the pest insect, leaving the farmer quite happy. The initially elated farmer will continue to spray the toxin over the next few seasons,

Table 1 List of common pesticides that have been used in the Chesapeake Bay watershed

Name	Chemical description	Classification	Period of use	Target	Application	Half-life	Known human health effects
DDT (dichloro diphenyl trichloroethane)	Colorless, tasteless, and almost odorless crystalline chemical compound. $C_{14}H_9Cl_5$, chlorinated hydrocarbon	Insecticide	1945–1972	Nervous system	Applied directly on crops	2–15 years, DDE produced in breakdown also toxic.	Neonatal hypotonia, carcinogen
Rotenone	Colorless and odorless, $C_{23}H_{22}O_6$, organic heteropentacyclic compound, from plants	Insecticide/ Piscicide	1947–2005, 2010–Present	Mitochondrial dysfunction	Applied on crops and directly into bodies of water	3.5 days	Parkinson's disease
Imidacloprid (N-nitroimidazolidin-2-imine)	Colorless, with slight odor, $C_9H_{10}ClN_5O_2$, imidazolidine	Insecticide (nicotine mimic)	1994–Present	Disrupts nervous system function of insects	Applied directly on crops	26–229 days	Skin irritation, numbing and tingling on fingers and lips, facial numbness and swelling, lethargy and nausea, autism specturm
Kepone (Chlordecone)	White solid powder, $C_{10}Cl_{10}O$, a chlorinated polycyclic hydrocarbon	Insecticide	1958–1975	Nervous system dysfunction	Applied directly to crops	3.8–46 years	Nervous system dysfunction (tremors, ataxia), reproductive issues, liver toxin
Chlorpyrifos	Colorless to white powder, $C_9H_{11}C_{13}NO_3PS$ chlorinated organophosphate	Insecticide	1965-pesent	Nervous system	Applied directly to crops	60 days–1 year	Nervous system, convulsions, psychiatric symptoms, partial paralysis
1,3-Dichloropropene	Colorless liquid, sweet smell, $C_3H_4Cl_2$, derived from a hydride of propene	Insecticide, Nematicide	1950s–Present	Metabolic dysfunction	Applied directly in soil and on crops	2–54 days	Fulminant hepatitis, probable carcinogen
2,4-D (2,4-Dichlorophenoxyacetic acid)	Odorless, white/tan. $C_0H_6Cl_2O_3$, chlorophenoxyacetic acid that is phenoxyacetic acid and has a of ring hydrogens at positions 2 and 4 are substituted by chlorines	Herbicide	1940–Present Day	Mimic of plant hormone that causes uncontrolled cell growth in broadleaf	Post-Emergent, applied directly on weeds	15–186 days	non-Hodgkin's lymphoma
Paraquat (1,1′-Dimethyl-4,4′-bipyridinium dichloride)	Bright green liquid with pungent smell, $C_{12}H_{14}Cl_2N_2$, organic cation derived from hydride	Herbicide, broad spectrum, killing all green plants	1961–Present	Inhibits photosynthesis	Applied directly on crops	1.5–13 years	Parkinson's disease, organ failure
Atrazine (1-Chloro-3-ethylamino-5-isopropylamino-2,4,6-triazine)	Odorless, solid and liquid form, $C_8H_{14}ClN_5$	Selective herbicide, used for corn, sorghum, other corps	1959–Present	Photosynthesis inhibitor	Applied directly to crops	13–261 days	Endocrine disrupter

(continued)

Table 1 (continued)

Name	Chemical description	Classification	Period of use	Target	Application	Half-life	Known human health effects
Glyphosate (N-(phosphonomethyl) glycine)	Colorless, odorless, $C_3H_8NO_5P$, phosphonic acid	Herbicide	1974–Present	Inhibits production of plant growth proteins	Direct application to crops	2–197 days	non-Hodgkin's lymphoma

Only two pesticides, DDT and Kepone have been banned in the US, although many of these pesticides have negative effects on human and environmental health. For example: 2,4-D has been associated with Non-Hodgkin's Lymphoma; Rotenone has been associated with Parkinson's disease; Paraquat has been associated with Parkinson's disease and organ failure; 1,3-D has been associated with Fulminant hepatitis; and Atrazine (banned by European Union in 2004) is a known endocrine disrupter (Gleason et al. 1957; Longnecker et al. 1997; NPIC 1999; Jervais et al. 2008; Gervais et al. 2008; Tanner et al. 2011; Bunch et al. 2012; Saravu et al. 2013; Waśkiewicz et al. 2014; CDC 2019a; Cimino et al. 2016; EPA 2019; Greenbook 2019; Kim et al. 2019). Chlorpyrifos was on the verge of being banned by the USEPA as of this writing. Note that half-life estimates are often variable and dependent upon the location of the pesticide: soil, water, or air

only to see a constantly diminishing effect of the pesticide. The farmer might increase the dose, hoping once again to beat back the pest. Yet soon the pest population becomes nearly 100% resistant to the pesticide, making the farmer a master of artificial selection.

A second reason for pesticide failure lies with the ignorant notion that underlies its use. The idea that spreading a toxin in the agricultural ecosystem will harm only the targeted pest, with minimum damage to the rest of the biotic community. Often the pesticide does more damage to the species that prey upon the pest, than it does to the pest population. The toxin vanquishes predatory insects, birds, bats and other top-down controls provided by nature. This leaves the farmer with nothing more than soon-to-be obsolete chemicals to defend against hordes of resistant insects intent on devouring the crop. Yet, the chemical industry stands at the ready to step into the breach with the next generation of pesticides, continuing the cycle for another turn. The Chesapeake Bay Watershed offers an example of the pesticide unintentionally killing a natural predator of the targeted pest. Many farmers in the region are practicing conservation tillage (not plowing the land before planting) to reduce soil erosion. The lack of plowing promotes elevated populations of introduced invasive slugs (*Deroceras reticulatum*). These mollusks cause extensive damage to seedlings of crops in the region including soybean (*Glycine max*) (Douglas and Tooker 2012). The use of neonicotinoids to control insect pest of soybeans makes the situation worse, as that pesticide moves up the food chain and kills the beneficial ground beetle (*Chlaenius tricolor*) that eats the slugs, resulting in reduced crop yield (Douglas et al. 2014).

Alternatives to Chemical Pesticides Agriculture fed the multitudes for millennia without artificial pesticides. However, the post WWII food system demanded surplus production for export and the feeding of animals at an ever-increasing rate. This meant a premium was placed on yield, the amount of food (primarily corn and soy) that could be produced per unit area. High yield, of course, invited pests to eat the crops and weeds to grow in the fertilized soil.

The successive failures of pesticides and growing concern about damage they caused to human and environmental health, prompted some crop scientists of the 1950's to propose an ecological approach to the problem. This gave birth to the field of Integrated Pest Management (IPM) (IPM 2019). The UN Food and Agriculture Organization (FAO) and various government agencies created programs to promote IPM in order to reduce the use of pesticides. This includes university extension services to the Chesapeake Bay Watershed. For example, the University of Maryland

Extension Service explains IPM as a, "… research-based holistic approach to pest management that emphasizes biological (e.g., attracting natural enemies), cultural (e.g., planting disease resistant varieties), and physical (e.g., hand removal of insect pests) approaches to prevent problems and control pests and diseases at acceptable levels. Monitoring and using organic or other low-risk pesticides only when pest or disease levels are unacceptable are also part of this management approach (UMExtension 2019)."

When practiced, IPM lowers frequency of detectible pesticide residue in food. In one study, about 73% of conventionally produced foods tested positive for pesticide residue, compared to 47% for IPM items. But organic produce is twice as good with detection frequency reduced to 23%. Presumably, overspray or processing events contaminated the organic food (Baker et al. 2002).

Despite the laudable goals and some success stories, less than 20% of farmers of the Chesapeake Bay region use IPM (Hellkamp et al. 1998; Malone et al. 2004). IPM requires expert monitoring of pests and investment of time (Matyjaszczyk 2019). The continued growth of pesticide use since the 1970's (when IPM was introduced) demonstrates the lack of enthusiasm for IPM (Ehler 2006).

Some IPM Practices The biological, cultural and physical control techniques used in IPM are the same as those employed by organic farmers (Flint 2012). Biological controls include promoting natural enemies of pests, such as predatory insects and parasitoids (typically wasps that lay their eggs in pest insects), and pathogens (bacteria and fungi) that attack pests. Pheromones placed in traps attract pest insects, preventing them from mating. Releasing sterile male insects can reduce reproductive success of pest populations, as done in Maryland and Virginia with potato beetles (*Leptinotarsa decemlineata*) (Baker et al. 2008). Trap plants attract pest insects away from the crop. In Virginia, mustard (*Brassica kbar*) and rape (*B. naups*) are used to attract harlequin bugs (*Murgantia histrionica*) away from broccoli (B. oleracea) (Ludwig and Kok 1998).

Cultural controls involve modifying crop and landscape management practices to decrease pest establishment, reproduction, and dispersal. This includes rotating crops and changing times of planting. Physical controls include hand or mechanical removal of pests (weeding), steam pasteurization of soil, exposing soil to the sun by cultivation, pest barriers such as screens, netting and fencing, mechanical traps for rodents and insects, and sticky traps for insects.

Unquestionably, adoption of IPM by all conventional farmers would reduce artificial pesticide use. However, is any level of these chemicals really required in the food system to provide the populace with good nutrition?

In the end, crops produced under IPM may or may not have been grown with pesticides. The uncertainty robs IPM crops of the market share provided by consumers interested in buying food that's healthier for their families and the environment. Enter USDA Organic Certification.

Organic Agriculture to End the War on Nature The organic farming movement in the US started in 1940 when J.I. Rodale bought a farm in Emmaus, PA, on the edge of the Chesapeake Bay Watershed (see chapter "An Organic-Based Food System: A Voyage Back and Forward in Time" on organic agriculture). The movement developed and promoted ecological techniques of farming that eschewed chemical fertilizers and pesticides. In 2002, the USDA adopted a certification plan for labeling organic agricultural products. The official USDA seal of approval informs customers that their food was produced without artificial chemicals or genetically modified organisms. This appeals to consumers for two reasons. First, they can eat food free of toxic pesticides. Second, they perceive organically produced food as less harmful to the environment than conventional crops raised with artificial chemicals.

Organic farming eliminates the use of pesticides and provides additional environmental benefits including less fossil fuel input, higher soil water, organic and nitrogen content, and yields similar to conventional methods (Pimentel et al. 2005). Organic foods reduce exposure to pesticides. One interventional study revealed that switching families from diets of conventionally produced to organically grown foods significantly reduced the presence of metabolites of pesticide in urine for 13 different chemicals, including a 83% reduction of clothianidin, 95% reduction of malathion, and 61% reduction of chlorpyrifos (Hyland et al. 2019). A population study demonstrated an inverse relationship between metabolites of organophosphate pesticide metabolites in urine and the consumption of organically produced food (Curl et al. 2015). Organic food reduced exposure to pesticides and contained higher concentrations of health-promoting antioxidant phytonutrients such as phenolic acids, flavanones, stilbenes, flavones, flavonols and anthocyanins (Barański et al. 2014).

Reducing exposure to pesticides improves health outcomes. A large study in France showed that choosing organic foods reduced the odds of getting cancer by about 25% (Baudry et al. 2018). Higher intake of pesticide contaminated fruits and vegetables correlates with reduced rates of conception and a lower probability of live births in humans (Chiu et al. 2018).

Conclusion Farmers and consumers in the Chesapeake Bay region can effectively end the war on nature by opting for a food system based upon organic agriculture. Conventionally produced food is generally less expensive than organic versions of those products. Yet, the savings doesn't include the cost to the health of the ecosystem and the consumer.

"Prices of organic foods include not only the cost of the food production itself, but also a range of other factors that are not captured in the price of conventional food, such as: 1. Environmental enhancement and protection (and avoidance of future expenses to mitigate pollution). For example higher prices of organic crops compensate for low financial returns of rotational periods which are necessary to build soil fertility; 2. Higher standards for animal welfare; Avoidance of health risks to farmers due to inappropriate handling of pesticides (and avoidance of future medical expenses); 3. Rural development by generating additional farm employment and assuring a fair and sufficient income to producers (FAO 2019)."

And the price of organic foods is falling. Consumers paid 9% more for organic foods in comparison to conventional offerings in 2014. That's down to 7.5% in 2019 (AP 2019). The amount of land farmed organically in the US increased over fourfold, from 935,000 acres (378,381 ha) in 1992 to four million acers (1,618,746 ha) in 2016. Sales of organic food went from $13.8 billion in 2005 to $43.3 billion in 2015. As of 2019, about 5% of the food consumed in the US is certified organic (SARE 2019). This leaves a long way to go in ending the war on nature, but we are moving in the right direction.

Much of the resistance to making peace with nature will come from the pesticide industry, which reported $9 billion in sales in 2012 (EPA 2019). This doesn't include the profits from selling genetically modified pesticide-resistant corn and soybean seeds to go with the chemicals. Farmers will also have to relearn the art and science of working with nature to produce their crops, rather than trying to fight it.

References

Abbott C (2018) Monsanto expects its dicamba-tolerant seeds on 40% of U.S. Soy Acres. In: Successful farming. https://www.agriculture.com/crops/soybeans/monsanto-expects-its-dicamba-tolerant-seeds-on-40-of-us-soy-acres. Accessed 18 Jun 2019

AP (2019) Associated Press Good news if you buy organic food — it's getting cheaper. In: MarketWatch. https://www.marketwatch.com/story/heres-why-prices-of-organic-food-are-dropping-2019-01-24. Accessed 25 Jun 2019

Ator S, Blomquist J, Brakebill J (1998) USGS CIRC 1166: water quality in the Potomac River Basin, Maryland, Pennsylvania, Virginia, West Virginia, and the District of Columbia, 1992–96. In: Pubs.usgs.gov. https://pubs.usgs.gov/circ/circ1166/index.html. Accessed 19 Jun 2019

Baker B, Benbrook C, Groth E III, Benbrook K (2002) Pesticide residues in conventional, integrated pest management (IPM)-grown and organic foods: insights from three US data sets. Food Addit Contam 19:427–446. https://doi.org/10.1080/02652030110113799

Baker M, Dastur S, Jaffe B, Wong T (2008) Mating competition in colorado potato beetles (coleoptera: chrysomelidae) does not show a cost of insecticide resistance. Ann Entomol Soc Am 101:371–377. https://doi.org/10.1603/0013-8746(2008)101[371:mcicpb]2.0.co;2

Barański M, Średnicka-Tober D, Volakakis N et al (2014) Higher antioxidant and lower cadmium concentrations and lower incidence of pesticide residues in organically grown crops: a systematic literature review and meta-analyses. Br J Nutr 112:794–811. https://doi.org/10.1017/s0007114514001366

Basil K, Vakil C, Sanborn M et al (2007) Cancer health effects of pesticides: systematic review. Can Fam Physician 53:1704–1711

Baudry J, Assmann K, Touvier M et al (2018) Association of frequency of organic food consumption with cancer risk. JAMA Intern Med 178:1597. https://doi.org/10.1001/jamainternmed.2018.4357

Benachour N, Séralini G (2009) Glyphosate formulations induce apoptosis and necrosis in human umbilical, embryonic, and placental cells. Chem Res Toxicol 22:97–105. https://doi.org/10.1021/tx800218n

Benbrook C (2016) Trends in glyphosate herbicide use in the United States and globally. Environ Sci Eur. https://doi.org/10.1186/s12302-016-0070-0. Accessed 15 Feb 2019. https://www.ncbi.nlm.nih.gov/pmc/articles/PMC5044953/

Blazer V, Iwanowicz L, Starliper C et al (2010) Mortality of centrarchid fishes in the potomac drainage: survey results and overview of potential contributing factors. J Aquat Anim Health 22:190–218. https://doi.org/10.1577/h10-002.1

BP (2019) Herbicide tolerant crops. In: Beyond pesticides. https://www.beyondpesticides.org/programs/genetic-engineering/herbicide-tolerance. Accessed 17 Mar 2019

Bunch T, Gervais J, Buhl K, et al (2012) *Dicamba General Fact Sheet*; National Pesticide Information Center, Oregon State University Extension Services. http://npic.orst.edu/factsheets/dicamba_gen.html

Byrne P (2014) *Genetically Modified (GM) crops: techniques and applications – 0.710 – extension.* [online] Colorado State University Extension. Available at: https://extension.colostate.edu/topic-areas/agriculture/genetically-modified-gm-crops-techniques-and-applications-0-710/ (Accessed 15 Mar 2019). https://royalsociety.org/topics-policy/projects/gm-plants/what-is-gm-and-how-is-it-done/

Carson R (1962) Silent spring. Houghton Mifflin Company, Boston

CBF (2019) Insecticides: chlorpyrifos. In: Chesapeake bay foundation Cbf.org. https://www.cbf.org/issues/chemical-contamination/insecticides-chlorpyrifos.html. Accessed 20 Jun 2019

CDC (2019a) CDC | Facts about Paraquat. In: Emergency.cdc.gov. https://emergency.cdc.gov/agent/paraquat/basics/facts.asp. Accessed 1 Apr 2019

CDC (2019b) Environments well water atrazine and your health – CDC tracking network. In: Ephtracking.cdc.gov. https://ephtracking.cdc.gov/showAtrazineHealth.action. Accessed 25 Jun 2019

CFS (2005) Monsanto vs. U.S. Farmers. In: Centerforfoodsafety.org. https://www.centerforfoodsafety.org/files/cfsmonsantovsfarmerreport11305.pdf. Accessed 16 Mar 2019

Chase-Lubitz J (2019) Appeals court orders EPA to act on demands to ban chlorpyrifos. In: POLITICO. https://www.politico.com/story/2019/04/19/epa-chlorpyrifos-court-3115794. Accessed 20 Jun 2019

Chiu Y, Williams P, Gillman M et al (2018) Association between pesticide residue intake from consumption of fruits and vegetables and pregnancy outcomes among women undergoing infertility treatment with assisted reproductive technology. JAMA Intern Med 178:17. https://doi.org/10.1001/jamainternmed.2017.5038

Cimino A, Boyles A, Thayer K, Perry M (2016) Effects of neonicotinoid pesticide exposure on human health: a systematic review. Accessed 17 Apr 2019

Coats J (1990) Mechanisms of toxic action and structure-activity relationships for organochlorine and synthetic pyrethroid insecticides. Environ Health Perspect 87:255. https://doi.org/10.2307/3431033

Cohen M (2009) Introduction: rethinking the origins of agriculture. Curr Anthropol 50:591–595. https://doi.org/10.1086/603548

Croft B (1982) Arthropod resistance to insecticides: a key to pest control failures and success in North American apple orchards. Entomol Exp Appl 31:88–110. https://doi.org/10.1111/j.1570-7458.1982.tb03121.x

Curl C, Beresford S, Fenske R et al (2015) Estimating pesticide exposure from dietary intake and organic food choices: the multi-ethnic study of atherosclerosis (MESA). Environ Health Perspect 123:475–483. https://doi.org/10.1289/ehp.1408197

Davis F (2013) Banned: a history of pesticides and the science of toxicology. Yale University Press, New Haven

Dill G (2005) Glyphosate-resistant crops: history, status and future. Pest Manag Sci 61:219–224. https://doi.org/10.1002/ps.1008

Douglas M, Tooker J (2012) Slug (Mollusca: Agriolimacidae, Arionidae) ecology and management in no-till field crops, with an emphasis on the mid-Atlantic region. J Integr Pest Manag 3:C1–C9. https://doi.org/10.1603/ipm11023

Douglas M, Rohr J, Tooker J (2014) Neonicotinoid insecticide travels through a soil food chain, disrupting biological control of non-target pests and decreasing soya bean yield. J Appl Ecol 52:250–260. https://doi.org/10.1111/1365-2664.12372

Duke S (2017) The history and current status of glyphosate. Pest Manag Sci 74:1027–1034. https://doi.org/10.1002/ps.4652

Dunley JE, Welter SC (2000) Correlated insecticide cross-resistance in azinphosmethyl resistant codling moth (Lepidoptera: Tortricidae). J Econ Entomol 93(3):955–962

Dyck E, Liebman M (1994) Soil fertility management as a factor in weed control: the effect of crimson clover residue, synthetic nitrogen fertilizer, and their interaction on emergence and early growth of lambsquarters and sweet corn. Plant Soil 167:227–237. https://doi.org/10.1007/bf00007949

Ehler L (2006) Integrated pest management (IPM): definition, historical development and implementation, and the other IPM. Pest Manag Sci 62:787–789. https://doi.org/10.1002/ps.1247

Environmental Protection Agency (2018) What is a pesticide? | US EPA. In: US EPA. https://www.epa.gov/minimum-risk-pesticides/what-pesticide. Accessed 09 Feb 2019

EPA (2019) Pesticides industry sales and usage: 2008–2012 estimates. In: Epa.gov. https://www.epa.gov/sites/production/files/2017-01/documents/pesticides-industry-sales-usage-2016_0.pdf.. Accessed 25 Jun 2019

Eskenazi B, Chevrier J, Rosas L et al (2009) The pine river statement: human health consequences of DDT use. Environ Health Perspect 117:1359–1367. https://doi.org/10.1289/ehp.11748

Everts S (2016) The Nazi origins of deadly nerve gases | October 17, 2016 Issue – Vol. 94 Issue 41 | Chemical & Engineering News. In: Cen.acs.org. https://cen.acs.org/articles/94/i41/Nazi-origins-deadly-nerve-gases.html. Accessed 18 Jun 2019

EWG (2019) EWG's 2019 Shopper's guide to pesticides in produce™. In: Ewg.org. https://www.ewg.org/foodnews/summary.php. Accessed 25 Jun 2019

FAO (2017) FAO – News article: concerted action needed to stop diseases and pests from ravaging the food chain. In: Fao.org. http://www.fao.org/news/story/en/item/469269/icode/. Accessed 18 Jan 2019

FAO (2019) Organic agriculture: why is organic food more expensive than conventional food?. In: Fao.org. http://www.fao.org/organicag/oa-faq/oa-faq5/en/. Accessed 25 Jun 2019

Flint M (2012) IPM in practice. University of California Agriculture and Natural Resources, Oakland, CA

Foster R (2005) Kepone: The 'Flour' factory. In: richmondmagazine.com. https://richmondmagazine.com/news/kepone-disaster-pesticide/. Accessed 14 Jan 2019

Funke T, Han H, Healy-Fried M et al (2006) Molecular basis for the herbicide resistance of Roundup Ready crops. Proc Natl Acad Sci 103:13010–13015. https://doi.org/10.1073/pnas.0603638103

Gallagher D, Dietrich A, Reay W et al (1996) Ground water discharge of agricultural pesticides and nutrients to estuarine surface water. Groundwater Monitor Remed 16:118–129. https://doi.org/10.1111/j.1745-6592.1996.tb00579.x

Gervais G, Luukinen B, Buhl K et al (2008) 2,4-D general fact sheet. In: Npic.orst.edu. http://npic.orst.edu/factsheets/24Dgen.html

Gleason M, Gosselin R, Hodge H (1957) Clinical toxicology of commercial products. Williams and Williams, Baltimore, p 1,160

Greenbook (2019) Telone II – Dow AgroSciences – | Greenbook. In: Greenbook.net. https://www.greenbook.net/dow-agrosciences/telone-ii. Accessed 17 Apr 2019

Hartwell S (2011) Chesapeake Bay watershed pesticide use declines but toxicity increases. Environ Toxicol Chem 30:1223–1231. https://doi.org/10.1002/etc.491

Hartwell S, Hameedi J (2007) Magnitude and extent of contaminated sediment and toxicity in Chesapeake bay – NCCOS coastal science website. In: NCCOS coastal science website. https://coastalscience.noaa.gov/data_reports/magnitude-and-extent-of-contaminated-sediment-and-toxicity-in-chesapeake-bay/. Accessed 24 Jun 2019

Hartzler B (2017) Dicamba: past, present, and future | integrated crop management. In: Crops.extension.iastate.edu. https://crops.extension.iastate.edu/blog/bob-hartzler/dicamba-past-present-and-future. Accessed 18 Jun 2019

Hellkamp AS, Shafer SR, Campbell CL, Bay JM, Fiscus DA, Hess GR, McQuaid BF, Munster MJ, Olson GL, Peck SL, Easterling KN, Sidik K, Tooley MB (1998) Assessment of the condition of agricultural lands in five Mid-Atlantic states. Environ Monit Assess 51 (1/2):317–324

Hyland C, Bradman A, Gerona R et al (2019) Organic diet intervention significantly reduces urinary pesticide levels in U.S. children and adults. Environ Res 171:568–575. https://doi.org/10.1016/j.envres.2019.01.024

IPM (2019) Integrated Pest Management Program (IPM) | National Institute of Food and Agriculture. In: Nifa.usda.gov. https://nifa.usda.gov/program/integrated-pest-management-program-ipm. Accessed 13 Jun 2019

Key S, Ma J, Drake P (2008) Genetically modified plants and human health. J R Soc Med, 101(6), 290–298. https://www.ncbi.nlm.nih.gov/pmc/articles/PMC2408621/. Accessed 05 Mar 2019

Kim S, Chen J, Cheng T, et al (2019) PubChem 2019 update: improved access to chemical data. Nucleic Acids Res. 2019 Jan 8; 47(D1): D1102–D1109. https://doi.org/10.1093/nar/gky1033. [PubMed PMID: 30371825]

Kingsbury J, Delzer G, Hopple J (2008) Anthropogenic organic compounds in source water of nine community water systems that withdraw from streams, 2002–05: U.S. Geological Survey Scientific Investigations Report 2008–5208, 66 pages. https://pubs.usgs.gov/sir/2008/5208/pdf/sir2008-5208.pdf. Accessed 20 Jun 2019

Klaas E, Wiemeyer S, Ohlendorf H et al (1978) Organochlorine residues, eggshell thickness, and nest success in barn owls from the Chesapeake Bay. Estuaries 1:46. https://doi.org/10.2307/1351649

Klein K (2019) Pesticides uncovered in store brand cereal, applesauce, beans and produce. https://1bps6437gg8c169i0y1drtgz-wpengine.netdna-ssl.com/wp-content/uploads/2019/02/FoodTestingReport_final.pdf. Accessed 11 Apr 2019

Kolpin D, Blazer V, Gray J et al (2013) Chemical contaminants in water and sediment near fish nesting sites in the Potomac River basin: Determining potential exposures to smallmouth bass (*Micropterus dolomieu*). Sci Total Environ 443:700–716. https://doi.org/10.1016/j.scitotenv.2012.09.063

Kurenbach B, Hill A, Godsoe W et al (2018) Agrichemicals and antibiotics in combination increase antibiotic resistance evolution. Peer J 6:e5801. https://doi.org/10.7717/peerj.5801

Latham J, Love M, Hilbeck A (2017) The distinct properties of natural and GM cry insecticidal proteins. Biotechnol Genet Eng Rev 33:62–96. https://doi.org/10.1080/02648725.2017.1357295

Lazarus R, Rattner B, McGowan P et al (2016) Chesapeake Bay fish-osprey (*Pandion haliaetus*) food chain: Evaluation of contaminant exposure and genetic damage. Environ Toxicol Chem 35:1560–1575. https://doi.org/10.1002/etc.3386

Learn J (2016) Ospreys recover in Chesapeake despite pollutants | The Wildlife Society. In: Wildlife.org. https://wildlife.org/ospreys-recover-in-chesapeake-despite-pollutants/. Accessed 12 Jun 2019

Lehotay SJ, Harman-Fetcho JA, McConnell LL (1998) Agricultural pesticide residues in oysters and water from two Chesapeake Bay tributaries. Mar Pollut Bull 37(1–2):32–44

Lewandowski S (1987) Diohe'ko, the three sisters in Seneca life: implications for a native agriculture in the finger lakes region of New York State. Agric Hum Values 4:76–93. https://doi.org/10.1007/bf01530644

Liu B, McConnell L, Torrents A (2002) Herbicide and insecticide loadings from the Susquehanna River to the Northern Chesapeake Bay. J Agric Food Chem 50:4385–4392. https://doi.org/10.1021/jf010133j

Longnecker M, Rogan W, Lucier G (1997) The human health effects of DDT (dichlorodiphenyltrichloroethane) and PCBS (polychlorinated biphenyls) and an overview of organochlorines in public health. Annu Rev Public Health 18:211–244. https://doi.org/10.1146/annurev.publhealth.18.1.211

Ludwig S, Kok L (1998) Evaluation of trap crops to manage harlequin bugs, *Murgantia histrionica* (Hahn) (Hemiptera: Pentatomidae) on broccoli. Crop Prot 17:123–128. https://doi.org/10.1016/s0261-2194(97)00107-5

Malkan S (2019) Glyphosate: health concerns about the most widely used pesticide – U.S. right to know. In: U.S. right to know. https://usrtk.org/pesticides/glyphosate-health-concerns/. Accessed 17 Mar 2019

Malone S, Herbert D, Pheasant S (2004) Determining adoption of integrated pest management practices by Grains Farmers in Virginia. In: Joe.org. https://www.joe.org/joe/2004august/rb6.php. Accessed 21 Jun 2019

Matyjaszczyk E (2019) Problems of implementing compulsory integrated pest management. Pest Manag Sci 75:2063–2067. https://doi.org/10.1002/ps.5357

McConnell L, Rice C, Hapeman C et al (2007) Agricultural pesticides and selected degradation products in five tidal regions and the main stem of Chesapeake Bay, USA. Environ Toxicol Chem 26:2567. https://doi.org/10.1897/06-655.1

McHenry L (2018) The Monsanto Papers: poisoning the scientific well. Int J Risk Safety Med 29:193–205. https://doi.org/10.3233/jrs-180028

MGK (2019) Pyrethrum Nature's Insecticide – about Pyrethrum. In: Pyrethrum.com. https://pyrethrum.com/About_Pyrethrum.html. Accessed 18 Jan 2019

Mie A, Rudén C, Grandjean P (2018) Safety of safety evaluation of pesticides: developmental neurotoxicity of chlorpyrifos and chlorpyrifos-methyl. Environ Health 17:77. https://doi.org/10.1186/s12940-018-0421-y

Mogilna N, Magufwa A (2009) Genetic modified foods: advantages and disadvantages. In: Essuir.sumdu.edu.ua. https://essuir.sumdu.edu.ua/bitstream/123456789/7916/1/14.pdf. [Accessed 15 Mar 2019]

Mostaghimi S, McClellan P, Cooke R (1993) Pesticide contamination of groundwater in Virginia: BMP impact assessment. Water Sci Technol 28:379–387. https://doi.org/10.2166/wst.1993.0440

MPN (2009) Maryland pesticide network pesticides and the Maryland Chesapeake Bay Watershed. In: Mdpestnet.org. http://www.mdpestnet.org/wp-content/uploads/2012/11/MPN-2009WhitePaper.pdf. [Accessed 7 Mar 2019]

Muller P (1948) Dichloro-diphenyl-trichloroethane and newer insecticides. In: Nobelprize.org. https://www.nobelprize.org/uploads/2018/06/muller-lecture.pdf. Accessed 12 Jun 2019

NCGA (2018) World of Corn 2018. In: Worldofcorn.com. National Corn Growers Association http://www.worldofcorn.com/#corn-usage-by-segment. Accessed 17 Sep 2018

Nongmoproject.org (2019). What is a GMO? – The Non-GMO Project. [online] Available at: https://www.nongmoproject.org/gmo-facts/what-is-gmo/. Accessed 15 Mar 2019

NPIC (1999) National Pesticide Information Center In: Npic.orst.edu. http://npic.orst.edu/factsheets/ddtgen.pdf. Accessed 1 Apr 2019

PAN (2019) Pesticide action network pesticides database: chemical active ingredient search. In: Pesticideinfo.org. http://www.pesticideinfo.org/Search_Chemicals.jsp. Accessed 1 Apr 2019

Pérez G, Torremorell A, Mugni H et al (2007) Effects of the herbicide roundup on freshwater microbial communities: a mesocosm study. Ecol Appl 17:2310–2322. https://doi.org/10.1890/07-0499.1

Person D (2009) WestBred Sale Could Change Wheat Industry – Food Manufacturing. In: Web.archive.org. https://web.archive.org/web/20111006070940/http://www.foodmanufacturing.com/scripts/ShowPR~RID~12434.asp. Accessed 17 Mar 2019

Peterson G (1967) The discovery and development of 2,4-D. Agric Hist 41:243–254

Pimental A (2009) Pesticide runoff can pose risk to humans, wildlife in Chesapeake Bay Region | Chesapeake Bay Program. In: Chesapeakebay.net. https://www.chesapeakebay.net/news/blog/pesticide_runoff_can_pose_risk_to_humans_wildlife_in_chesapeake_bay_region. [Accessed 1 Mar 2019]

Pimentel D, Hepperly P, Hanson J et al (2005) Environmental, energetic, and economic comparisons of organic and conventional farming systems. BioScience 55:573. doi: https://doi.org/10.1641/0006-3568(2005)055[0573,eeaeco]2.0.co;2

Pinkney A (2009) Analysis of contamination concentrations in fish tissue collected from the waters of the District of Columbia. U.S. Fish and Wildlife Service, Chesapeake Bay Field Office, Annapolis, MD. CBFO-C08–03

PMCBW (2009) Pesticides and the Maryland Chesapeake Bay Watershed. In: Maryland Pesticide Network & Maryland Pesticide Education Network. http://www.mdpestnet.org/projects/pesticides-and-the-chesapeake-bay-watershed/. Accessed 20 Jun 2019

Rattner B, McGowan P, Golden N et al (2004) Contaminant exposure and reproductive success of Ospreys (*Pandion haliaetus*) nesting in Chesapeake Bay Regions of concern. Arch Environ Contam Toxicol 47:126–140. https://doi.org/10.1007/s00244-003-3160-0

Robinson G, Larkins P, Boughton C et al (2007) Assessment of contamination from arsenical pesticide use on orchards in the Great Valley region, Virginia and West Virginia, USA. J Environ Qual 36:654. https://doi.org/10.2134/jeq2006.0413

Rohr J, McCoy K (2010) A qualitative meta-analysis reveals consistent effects of Atrazine on freshwater fish and Amphibians. Environ Health Perspect 118:20–32. https://doi.org/10.1289/ehp.0901164

Sanahuja G, Banakar R, Twyman R et al (2011) *Bacillus thuringiensis*: a century of research, development and commercial applications. Plant Biotechnol J 9:283–300. https://doi.org/10.1111/j.1467-7652.2011.00595.x

Saravu K, Sekhar S, Pai A et al (2013) Paraquat – a deadly poison: report of a case and review. Indian J Crit Care Med 17:182–184. https://doi.org/10.4103/0972-5229.117074

SARE (2019) Organic farming and ranching resources. Sustainable agriculture and research education. In: Sare.org. https://www.sare.org/Learning-Center/Topic-Rooms/Organic-Production. Accessed 25 Jun 2019

Schueler T, Youngk A (2015) Potential benefits of nutrient and sediment practices to reduce toxic contaminants in the Chesapeake Bay

Watershed Report 2: removal of toxic contaminants from the agriculture and wastewater sectors. In: Chesapeakebay.net. https://www.chesapeakebay.net/channel_files/23275/part_2_review_draft_122915.pdf. Accessed 17 Mar 2019

Shaw D (2007) FAO's origins. In: World Food Security. Palgrave Macmillan, London

Smith J (1612). A map of Virginia. Oxford: Joseph Barnes, pp. 1–3. In The Complete Works of Captain John Smith (1580–1631), edited by Philip L. Barbour, (1986) 3 vol. 1:119–190. Chapel Hill: University of North Carolina Press. Available in Capt. John Smith's complete works.: http://www.virtualjamestown.org/exist/cocoon/jamestown/fha-js/SmiWorks1

Smith A, Secoy D (1976) A compendium of inorganic substances used in European pest control before 1850. J Agric Food Chem 24:1180–1186. https://doi.org/10.1021/jf60208a048

Stellman J, Stellman S, Christian R et al (2003) The extent and patterns of usage of Agent Orange and other herbicides in Vietnam. Nature 422:681–687. https://doi.org/10.1038/nature01537

Stinson E, Hayes L, Bush P et al (1994) Carbofuran affects wildlife on Virginia corn fields. Wildl Soc Bull 22:566–575

Tabashnik B, Brévault T, Carrière Y (2013) Insect resistance to Bt crops: lessons from the first billion acres. Nat Biotechnol 31:510–521. https://doi.org/10.1038/nbt.2597

Tanner C, Kamel F, Ross G et al (2011) Rotenone, Paraquat, and Parkinson's disease. Environ Health Perspect 119:866–872. https://doi.org/10.1289/ehp.1002839

Taylor A (2017) Insects are increasingly evolving resistance to genetically modified crops. In: The Scientist Magazine®. https://www.the-scientist.com/the-nutshell/insects-are-increasingly-evolving-resistance-to-genetically-modified-crops-30750. Accessed 17 Mar 2019

Tisler A, Zehnder G (1990) Insecticide resistance in the Colorado Potato Beetle (Coleoptera: Chrysomelidae) on the Eastern Shore of Virginia. J Econ Entomol 83:666–671. https://doi.org/10.1093/jee/83.3.666

Tyko K (2019) Monsanto Roundup cancer lawsuit: California man awarded $80 million. In: usatoday. https://www.usatoday.com/story/money/2019/03/27/monsanto-roundup-cancer-lawsuit-california-man-awarded-80-million/3293824002/. Accessed 18 Jun 2019

USGS (2012) Toxic contaminants in the Chesapeake Bay and its Watershed: extent and severity of occurrence and potential biological effects. In: Chesapeakebay.net. https://www.chesapeakebay.net/channel_files/22127/ii.c._toxic_contaminants_summary_report_2012.pdf. Accessed 18 Feb 2019

UMExtension (2019) How to ID, prevent, and manage plant problems | University of Maryland Extension. In: Extension.umd.edu. https://extension.umd.edu/hgic/topics/how-id-prevent-and-manage-plant-problems. Accessed 21 Jun 2019

Unsworth J (2010) History of pesticide use. In: Agrochemicals.iupac.org. https://agrochemicals.iupac.org/index.php?option=com_sobi2&sobi2Task=sobi2Details&catid=3&sobi2Id=31. Accessed 16 Jan 2019

USDA-ERS (2019) Pesticide Use in U.S. Agriculture: 21 Selected Crops, 1960–2008. In: Ers.usda.gov. https://www.ers.usda.gov/publications/pub-details/?pubid=43855. Accessed 14 Jun 2019

USDVA (2019) Agent Orange Settlement Fund – Compensation. In: Benefits.va.gov. https://www.benefits.va.gov/compensation/claims-postservice-agent_orange-settlement-settlementFund.asp. Accessed 14 Jun 2019

USEPA (1975) DDT regulatory history: a brief survey (to 1975) | About EPA | US EPA. In: Archive.epa.gov. https://archive.epa.gov/epa/aboutepa/ddt-regulatory-history-brief-survey-1975.html. Accessed 11 Jun 2019

USEPA (2012) U.S. environmental protection agency, U.S. geological survey, and U.S. Fish and Wildlife Service, 2012, Toxic contaminants in the Chesapeake Bay and its watershed – Extent and severity of occurrence and potential biological effects – Technical Report – December, 2012: Annapolis, Md., U.S. Environmental Protection Agency Chesapeake Bay Program Office, 175 p

Wamsley L (2018) California Judge Cuts award to $78.5 million in Monsanto weedkiller case. In: Npr.org. https://www.npr.org/2018/10/23/659848853/california-judge-cuts-award-to-78-5-million-in-monsanto-weedkiller-case. Accessed 17 Mar 2019

Waśkiewicz A, Gładysz O, Szentner K et al (2014) Role of glutathione in abiotic stress tolerance. Oxidative Damage Plants:149–181. https://doi.org/10.1016/b978-0-12-799963-0.00005-8

Watts B, Therres G, Byrd M (2008) Recovery of the Chesapeake Bay Bald Eagle nesting population. J Wildl Manag 72:152–158. https://doi.org/10.2193/2005-616

Wilkerson J, Chow B (2015) Why roundup ready crops have lost their Allure – science in the news. In: Science in the News. http://sitn.hms.harvard.edu/flash/2015/roundup-ready-crops/. Accessed 17 Mar 2019

Zappia H (1996). Chlordane, DDT, PCB's, and other selected organic compounds in Asiatic Clams and Yellow Bullhead in the Potomac River Basin, 1992. Water-resources investigations report 96–4210 US Department of the Interior, US Geological Survey. [Accessed 11 Mar 2019]

Zhang L, Rana I, Shaffer R et al. (2019) Exposure to glyphosate-based herbicides and risk for non-Hodgkin Lymphoma: a meta-analysis and supporting evidence. Mutation research/reviews in mutation research. https://doi.org/10.1016/j.mrrev.2019.02.001

Zilberfarb S (2017) The surprising link between GMOs and the Chesapeake Bay. In: U.S. Soy. https://ussoy.org/the-surprising-link-between-gmos-and-the-chesapeake-bay/. Accessed 17 Mar 2019

Livestock and Poultry: Other Colonists Who Changed the Food System of the Chesapeake Bay

Benjamin E. Cuker

Lace up your boots and we'll broom on down
To a knocked out shack on the edge of town
There's an eight-beat combo that just won't quit
Keep walkin' 'til you see a blue light lit
Fall in there and we'll see some sights
At the house of blue lights
There's fryers and broilers and Detroit barbecue ribs . . .
From the House of Blue Lights*, a 1946 song by Don Raye and Freddie Slack (emphasis added). The song celebrates two of the food-animals brought to Jamestown by the English and changes wrought by World War II that inform and altered the food system of the Chesapeake Bay.*

Abstract

The introduction of domesticated animals into the Chesapeake Bay regions' food system had profound consequences for the health of the ecosystem and its people. Draft animals provided the labor to clear and cultivate the majority of the Chesapeake Bay Watershed by 1900. Pigs brought by English colonists became important sources of food for both the settlers and native peoples. Initially, introduced chickens played a persistent but minor role in the food system. This changed with the development of the chicken for meat (broiler) industry on the Bay's Delmarva Peninsula in the 1920s. The broiler industry pioneered the use of concentrated animal feeding operations (CAFOs) and vertical integration of all aspects of production, a model now dominating the entire livestock enterprise. This propelled a 50% rise in per capita consumption of animal-based foods and attendant rise in human diseases, particularly coronary artery disease. Industrialization of chicken also increased its share of the animal products market sevenfold. The broiler industry pioneered the use of antibiotics as growth promoters, resulting in development of antibiotic-resistant pathogens. Poultry workers face various health problems, including increased rates of respiratory disease and certain cancers. The poultry industry is the major source of *Salmonella* poisoning. Broiler chickens suffer crowded and polluted conditions in CAFOs. These CAFOs dominate the landscape of the Delmarva, producing air and water pollution. Nitrogen as a waste product from the industry reaches the Bay and is a major source of eutrophication.

Keywords

CAFOs · Broilers · Animal agriculture · Antibiotics · Chicken industry · Pollution

Domesticated Animals Restructure the Food System The English men and women who colonized the Chesapeake Bay Region weren't the only exotic animals to arrive at Jamestown in 1607–1608. Chickens and pigs came on the first English ships and, by 1611, the list of newly introduced domesticated animals included cattle and goats. The site of these creatures apparently startled the resident Algonquin tribespeople, as they kept no domesticated animals other than dogs (Roundtree 2014).

The domesticated animals brought by the English reflected the food system of the Old World. They raised chickens, pigs, cattle and goats for food (eggs, dairy and meat). The uneatable portions of the food animals provided

B. E. Cuker (✉)
Department of Marine and Environmental Science, Hampton University, Hampton, Virginia, USA
e-mail: Benjamin.cuker@hamptonu.edu

B. E. Cuker (ed.), *Diet for a Sustainable Ecosystem*, Estuaries of the World, https://doi.org/10.1007/978-3-030-45481-4_12

a second source of resources, goods made from animal parts. These included leather from dried skins and feathers used in bedding. Oxen, mules, and horses played essential roles in the food system by providing the labor to clear and cultivate the land, and to transport the harvest to market. English settlers used animals to muscle wooden plows through the soil. Whether pulling a wagon along a bumpy road or tugging a barge in a canal, animals also provided the labor to move the harvest to market. Like their Native American neighbors, the English used dogs to help in the hunting of game and birds.

The food system implemented by the English colonists treated domesticated animals as whole beings of nature. The ability to domesticate particular species hinged on their level of sociality. Horses, donkeys, cattle, pigs, sheep, goats, ducks, geese, and chickens are all highly social beings with strong structural relationships. Humans inserted themselves into the social orders of these animals, assuming the alpha, or dominant position. Domesticated animals also learned to associate humans with rewards of food. In a sense, the domesticated animals and their keepers struck a cultural evolutionary bargain. The humans provided shelter, protection, and access to food in exchange for the animal's labor or the certainty of being slaughtered at some point in its lifecycle. Such was the way of life on a family farm for generations in the Chesapeake Bay region. The working animals were always given names and sometimes others raised as future meals were afforded the same respect. Some might say the farm animals were treated as part of the family, but more accurately, farmers and their domesticated animals belonged to a common interspecies social structure that was controlled by the farmer.

Ethicists might argue that the traditional use of animals for labor, food, and other products was inherently exploitive. But, certainly, those animals came much closer to living a life for which their ancestors had evolved rather than that experienced by creatures in the modern food system. Full industrialization of animal agriculture meant a reductionist approach steeped in the masculine view of the necessity of man's dominance over nature to maximize profit. Modern animal agriculture disaggregates life histories to obtain the maximal growth at the least cost. Conception, gestation or incubation, lactation or nurturing, and growth to maturity are now just stops along an assembly line. For some species, controlled encounters replace evolved mate finding and courtship rituals. For cattle, chickens, pigs, and sheep, life now usually begins with artificial insemination. The animal industry obtains the semen by inducing ejaculation in males fooled by decoys or fingering of genitals. This facilitates complete human control of conception, and allows for the breeding of animals so oversized and disproportioned as to preclude safe copulation (Ax et al. 2016). Mechanical incubators replace the warmth and egg turning attention of brooding chickens. The hog industry invented lactation crates in which the suckling piglets feed by inserting their heads through steel bars. After weaning or hatching, animals are "finished" in concentrated animal feeding operations (CAFOs) where the creatures live in densely populated factory buildings. They stand in their own manure and breath ammonia contaminated air as they imbibe machine delivered feed and water. The only time these animals see daylight or smell fresh air is in the back of trucks bringing them to slaughter. Chickens are no longer chickens, but "broilers" to produce breasts and thighs, or "layers" to deliver eggs (Davis 2009). Cattle are milk, cheese, yogurt, hamburger, steak, roast loin, and hot dogs. Pigs are barbecue, ham, bacon, and sausage. Industrialized agriculture makes its money by removing any sense of natural being from the creatures it uses to produce market commodities.

CAFOs and the complete industrialization of animal agriculture profoundly affects the ecology of the Chesapeake Bay and the health of its human residents. Often, the vast distances between where soy and corn crops are grown, and where that feed is turned into animal flesh, urine, and feces means the breaking of nutrient cycles. The nitrogen, phosphorus, potassium, and other nutrients released by the animals rarely makes it back to the soy and corn fields. Instead it pollutes the Bay and ground water, while growers of feeds use artificial chemicals to fertilize their fields. The CAFOs pollute the air with ammonia, methane, nitrous oxides and carbon dioxide. The CAFO system engenders the use of growth hormones and antibiotics. These additives contaminate both the Bay and its animal products. CAFOs are also breeding grounds for diseases that infect people, domesticated animals and wildlife.

Animal Labor Reshapes the Watershed The importance of animal labor to shaping the food system and its impact on the Chesapeake Bay can't be understated. Agricultural experimenters, including Chesapeake Bay resident Thomas Jefferson, redesigned the plow (Jefferson 1953). First wooden, then iron, and finally steal plows were drawn across fields by oxen, mules, or horses. Various designs accommodated the needs of local conditions. Most plows used the moldboard design to dig deep furrows with adjacent linear mounds of over-turned sod. This method buried weeds and loosened the soil to accept plantings. Such plowing also promoted soil erosion, particularly on sloped fields (Craven 1926). Runoff from rainstorms carried vast quantities of soil into the tributaries of the Chesapeake Bay. When the soil dried out between rains, winds eroded the exposed earth carrying away clay and other fine particles. Much of that dust also found its way into the Bay.

Fig. 1 A mule pulling a plow in July 1937. Despite the arrival of engine powered tractors at the end of the nineteenth century, many farmers couldn't afford the device and relied on draft animals for cultivating and other farm work into the 1960s. The team of animal, farmer, and plow placed the majority of land in the Chesapeake Bay Watershed under cultivation by 1900. Image from the Library of Congress (http://cdn.loc.gov/service/pnp/fsa/8b15000/8b15300/8b15388v.jpg)

The animal powered food system transformed the Chesapeake Bay and its watershed. Draft animals helped clear timber from forests, making the way for putting 40% of the land under cultivation by the mid eighteenth century. Application of fertilizers opened otherwise marginal land to cultivation, moving the watershed to 50% farmed by 1850, and 70% by 1890 (Fig. 1). Ultimately mechanization expanded the portion of the watershed under cultivation to a peak 80% in 1930. That meant that animal labor did most of the work to transform the watershed prior to the introduction of steam and internal combustion engines in the late nineteenth century (Brush 2008).

General William T. Sherman's Field Order No. 15 of January 16, 1865 allocated 40 acres (16.2 ha) of confiscated land *and a mule* to each emancipated slave family in several southern states (Myers 2005). While this Civil War era order never came to fruition, Sherman's inclusion of the mule in the promise makes the point of the necessity of a draft animal for a functioning farm.

Chesapeake Bay sediments record the transformation of the watershed by animal power. The plowed watershed released soil to the Bay six times faster than pre-colonial times. The onslaught of sediments filled previously navigable channels and suffocated oyster reefs. Submerged grass beds succumbed to both burial and the lack of sunlight transmitted by sediment laden waters (Brush 2008).

Steam and internal combustion power virtually eliminated the working animal from agriculture by the middle of the twentieth century. Steam- and diesel-powered trains brought farm products to markets well beyond those reachable by draft animals and carts. First steam, and then internal combustion powered tractors took over the duties of mules, oxen, and horses.

Animals to Eat The combination of mechanization and fossil fuel energy subsides of agriculture meant an end to significant animal labor by the end of the twentieth century. While draft animals lost their prized position as unpaid farm hands, other species moved up in importance in agriculture, not as co-workers, but as sources of food. It's not that the English settler's and those that followed didn't eat domestic animals, as they certainly did. However, in the long run, the work of a draft animal meant the difference between sustenance farming and bringing in enough harvest to make money. In their day, draft animals exceeded the importance of chickens, pigs, or cattle. The tractor, truck and fossil fuel

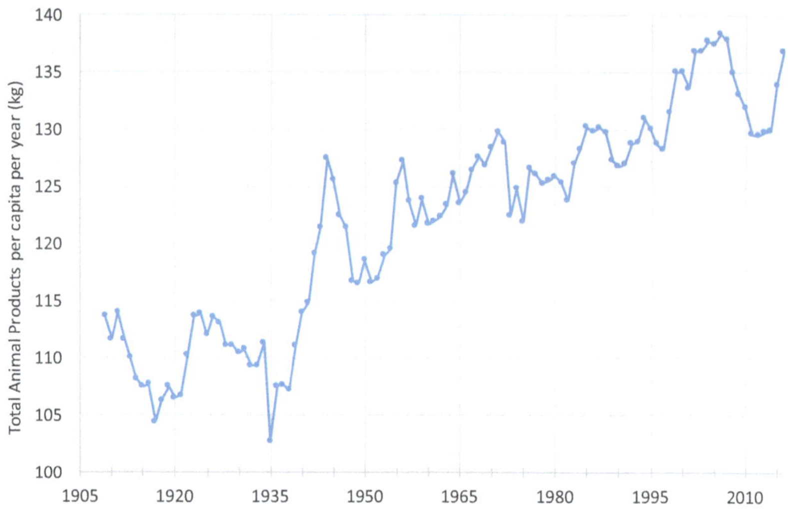

Fig. 2 The disappearance from the market and presumed per capita consumption of all animal foods over time in the US. The line represents data from the USDA (USDA 2019a). This data overestimates actual food consumption by about 20%, as it isn't corrected for losses due to spoilage and other sources. The foods included are; dairy, beef, pork, poultry, eggs, lamb, veal, fish, and shellfish. Note that consumption of animal foods increased sharply in the post-World War II era after 1945

freed the farmer to focus on raising animals for eating. The growing prosperity after World War II meant that consumers could afford to buy more meat and they did. Per capita consumption of animal products went from about 110 kg per year in 1910 to 137 kg per year in 2016, a 25% increase (Fig. 2). Growing per capita demand for animal products coupled with an expanding population provided a lucrative market for the producers and sellers of flesh and dairy.

The second half of the twentieth century saw not only an increase in the consumption of animal products, but also a shift in the proportions of which species people ate. In the early twentieth century, dairy comprised 35–40% of the animal mass in the US diet. Beef, pork, and eggs, each accounted for 15–20% of the animal mass consumed, followed by poultry and fish/shellfish, each at 5%, with veal and lamb each contributing 2–3%. Note that veal is juvenile beef.

The pattern changed considerably after World War II. Beef consumption doubled, peaking at 32% of total animal food intake in 1975, then declined back to 18% by 2016. Dairy consumption declined after the war to circa 25% between 1965 and 2016. Eggs also lost favor, comprising 11% of the animal product consumption in 2016. Pork remained fairly steady at around 15%. Fish/shellfish changed little, still accounting for only 5% of animal food consumed in 2016. Poultry consumption displayed the most dramatic shift after the war, going from only 5% in the early twentieth century to 25% of animal food consumed by 2016. Poultry is now the number one source of animal food in the US diet (Fig. 3). The five-fold increase in poultry consumption over the course of five decades has profound implications for the ecology of the Chesapeake Bay, since the chicken industry dominates the landscape of its Eastern Shore.

The two terrestrial animals used for food that most informs the Chesapeake Bay food system are pigs and chicken. These pioneer domesticated animals explain much about how the food system influenced the health of the Bay and its people. Although pigs were initially more important than poultry in feeding Bay area residents, it was adaptations of innovations in the breeding, growing, processing, and selling of chickens that revolutionized animal agriculture around the world. And that story began along the shores of the Chesapeake Bay. But let's first look at Pigs.

Pigs The English colonists brought pigs (*Sus domesticus*) with them to Jamestown. Those European domesticated pigs descended from the wild boar of Western Asia (Giuffra et al. 2000). The colonists left a country where all land belonged to somebody; the Crown, private landlords, or the community in the form of a common. As such, fences and pens restrained pigs from wandering the English countryside. America presented a different story. The colonists encountered a system where land was more controlled by a particular native tribe than it was owned. Wars and genocide pushed the Native Americans from their home territories. Eventually, the English royalty divvied up the stolen land in a system of ownership with grants awarded to favored individuals. But in the years before such formal transfer of land from the native residents to the new English lords, the colonists let their pigs run wild. This freed the colonists from having to feed the pigs, as was the practice in England. Instead, the free-ranging omnivorous hogs could feast on acorns and other offerings of the Virginia landscape. When desired, the colonists would kill a pig for its flesh and skin. The Algonquians were accomplished hunters and added this new invasive species to their diet. To quell disputes over ownership of the free ranging pigs, in the 1690s, the Virginia government instituted a system of earmarks (incisions in the floppy ears of the pigs) to indicate ownership (Roundtree 2014).

Pigs are highly social animals with the capacity for fast population growth. Both sexes reach sexual maturity between

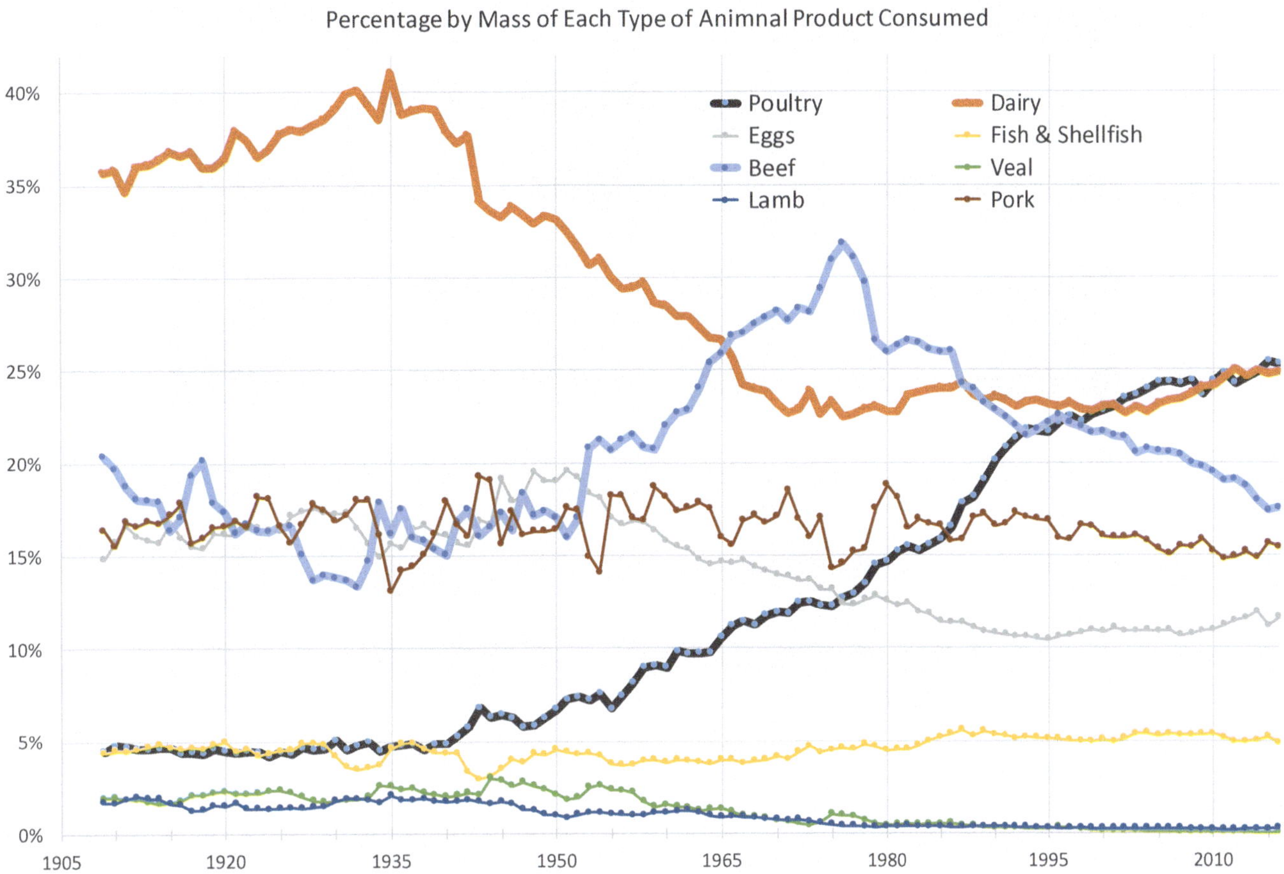

Fig. 3 Changes in the proportions of various animal species comprising the US diet between 1909 and 2016. The percentages are calculated based on wet mass, with all dairy reported as milk equivalents (the amount of whole milk needed to make the dairy product). The relative proportion of types of animal products in the diet remained fairly constant until the end of World War II, when consumption of dairy declined people began to consume more poultry. Beef consumption also increased after WWII, reaching a peak circa 1975. Subsequently the fraction of the diet attributed to beef declined to pre-war levels by 2016. Data from the USDA (USDA 2019a)

5 and 8 months of age. The estrus cycle is rapid, with females becoming receptive to males every 19–23 days (Anderson 2009). Litter size is variable, but typically 10 piglets survive weaning (McGloughlin 1976). The piglets are about 1 kg (2.2 pounds) at birth, and 6.3 kg (12 pounds) when weaned three weeks later. Modern commercial feeding operations provide a ration of soy and corn. They are fed about 1.1 kg per day for 7 weeks, bringing the pigs to a mass of 25 kg (55 pounds). After that the ration is increased to 3.6 kg (8 pounds) per day for 16 weeks. The pigs reach market size in just 6 months after their birth. The typical mass of a marketed pig is 128 kg (282 pounds) and yields a 96 kg (211 pound) carcass that is dissected into at least twelve different food products. Selective breeding, altered feeds, industrialized barns, and antibiotics means that modern pigs are 1.6 times the size of those produced in 1959 (National Pork Board 2019).

Pigs figured prominently in the early colonial history of the Chesapeake Bay. Eleven km (7 miles) south of Jamestown, a peninsula called Hog Island juts out from the southern shore of the James River. In about 1609, the English introduced pigs to this isolated area in order to protect them from hunting by the native residents. The 60 colonists that survived the "starving time" (winter of 1609–1610) fled Jamestown in a ship on June 7, 1610, anchoring the night at Hog Island. Lord De La Warr arrived the next day on a resupply ship from England and persuaded the settlers to return to Jamestown and rebuild their crumbling colony. Hog Island remained an important producer of salted pork for many years and gave rise to the famed Smithfield Ham (Rouse 1994; Jamestown 2019). The Surry Nuclear Power Plant and a surrounding wildlife refuge replaced the pig colony in 1968.

Industrialization of the hog growing industry largely moved the pig production out of the Chesapeake Bay Watershed. Few pigs raised to make Smithfield hams are grown near Smithfield, Virginia, anymore. Smithfield Foods owns only 23 company farms in Virginia (Smithfield Foods 2019a). Instead, most pigs destined for slaughter and packaging in Smithfield arrive by truck from North Carolina and other states. The coastal plain of North Carolina houses

Fig. 4 Pigs in a Concentrated Animal Feeding Operation (CAFO). Relatively few such pig CAFOs exist in the Chesapeake Bay Watershed. Most pigs entering the Bay food system come from North Carolina and other states. Image from USEPA website. https://upload.wikimedia.org/wikipedia/commons/7/72/Hog_confinement_barn_interior.jpg

thousands of pig CAFOs (Fig. 4, 225 owned by Smithfield Foods), primarily situated in rural areas with high African American populations. The disproportionate effect of pollution (air, water, noise) and pathogens produced by these CAFOs on the poor and people of color is an environmental justice issue (Nicole 2013). In one location, neighbors of Smithfield Foods CAFOs sued the company and were awarded in $473.5 million in damages in 2018. That decision went for appeal and the awards are likely to be reduced (Brown 2018).

The shift of pig CAFOs to North Carolina spared the Chesapeake Bay from the industry's factory farm pollution. Yet, the large slaughterhouse and packaging facilities in Smithfield proved capable of supplying plenty of pollution to the Pagan River, a winding tributary to the James River. In 1999, the fourth US Circuit Court of Appeals upheld a lower court's judgment of $12.6 million for polluting the Pagan River. At the time, it was the largest fine ever imposed under the Clean Water Act (USEPA 1999). Subsequently, Smithfield Foods purchased the right to direct its pre-treated effluent to the Hampton Roads Sanitation District, the municipal wastewater treatment system for the area.

Among states in the US, North Carolina's nine million head of pigs ranks only second to Iowa's 23 million head. The states of the Chesapeake Bay watershed combine to account for only 2.3% of the US pig population (Table 1). Yet, the slaughtering and packaging of pigs imported from other states remains an important part of the Chesapeake Bay food system. Smithfield Foods is headquartered in Smithfield Virginia, just 27 km (17 miles) from Hog Island, home to the first large pig colony on the Chesapeake. It is the world's largest pork producer, doing $15 billion in sales in 2017 and employing 54,000 people. It has operations in 27 states across the US, Mexico, England, Poland and Romania. Founded by Joseph Luter in 1936, it was bought by WH Group of China for $7.1 billion in 2013. It was the largest stock acquisition of an American company by a Chinese firm in history. However, the company points out that WH Group isn't owned by the government of China, nor does Smithfield Foods sell any pork from China in the US (Smithfield Foods 2019a).

The facilities located in the town of Smithfield slaughter and process 10,400 pigs per day when running at full capacity (Fig. 5). This complex employs 2410 people. In addition, Smithfield Foods has 87 workers producing pet food, 52 involved in distribution, 495 in headquarters' offices, and 191 at Virginia farms (personal communication, Ari Durall, Communications Manager, Smithfield Foods).

But it's not all animals for Smithfield Foods, as it employs 36 workers in its Gourmet Nut Products Division located in Toano, VA. The area around Smithfield supports many large peanut farms, making the legumes a natural, if minor secondary protein rich food for the pork-producing giant.

Table 1 The pig populations as of December 2018 in Iowa, North Carolina, and the states of the Chesapeake Bay Watershed

Jurisdiction	Rank Among States	Population of Pigs	% of Total US
Iowa	1	23,330,000	31.29
North Carolina	2	9,000,000	12.07
Pennsylvania	12	1,310,000	1.76
Virginia	18	345,000	0.46
New York	31	46,000	0.06
Maryland	33	17,000	0.02
Delaware	40	6500	0.01
West Virginia	44	4000	0.01
US	–	74,550,200	1

Note that Chesapeake Bay region account for only 2.3% of the US pig population. Data from the USDA (2019b)

Fig. 5 View of the large Smithfield Foods meat processing complex in Smithfield, VA taken September 30, 2017. The plant is adjacent to the Pagan River, a tributary to the lower James River. Image from: Nyttend, https://commons.wikimedia.org/wiki/File:Smithfield_Foods_processing_plant_from_Ivy_Hill_Cemetery.jpg

The Chesapeake Bay workforce of Smithfield Foods reflects the shifting demographics of food system employees for the entire area. African Americans from surrounding rural communities traditionally provided much of the labor at the facilities, as they did for seafood and chicken processing plants around the Bay. By the end of the twentieth century, many of those positions went to Hispanic workers coming from Mexico and more recently Central America. However, one aspect of the food system employment picture at Smithfield Foods remains fairly static, as whites dominate management positions (92% in 2017), while people of color make up 68% of the blue collar staff that does the slaughtering, packing, shipping, and farm work. On a company-wide basis, the break down for non-management positions is: 30% Hispanics, 29% black or African American, 7% Asian, and 1% Native Americans. Males account for 90%

of the management and 64% of the overall positions in the company. The corporation acknowledges the disparities of race and gender and claims to have programs in place to address the situation (Smithfield Foods 2019b).

In the 1990s, Joe Luter (the son of Joseph) embraced a business model of vertical integration pioneered by the poultry industry. Vertical integration meant Smithfield Foods controlled every aspect of the industry from conception to feed production, growing, slaughtering, packaging, and marketing. However, they claim no ownership of the manure produced in their CAFOs nor the risk taken by their contract pig farmers.

Despite the relatively few hogs still raised in the Chesapeake region, its food system remains strongly connected to pork production through the oily clupeid fish called menhaden (*Brevoortia tyrannus*). Discussed in detail in chapter "Menhaden, the Inedible Fish that Most Everyone Eats", the menhaden fishery is the largest on the Bay. The fish are reduced to oil and meal, both of which end up in commercial hog and chicken feed (Carson 1945; Auchterlonie 2017).

Chickens and Chicken People The chickens that arrived with Capt. John Smith and company at Jamestown traveled more than the distance from England to Virginia to get there. These exotic birds journeyed halfway around the world and through 5 millennia on their way to the Chesapeake Bay. The origins of domesticated chickens trace to the red jungle foul (*Gallus gallus*) of Southeast Asia (Fig. 6). Archeology and genetics indicate the first of multiple domestications about 3400 years BCE (Storey et al. 2012). Evolution restricted red jungle fowl to Southeast Asia until humans transported their domesticated decedents around the world (Lawler 2016). Along the way, chicken keepers applied artificial selection to produce a wide variety of breeds for various purposes including production of eggs, production of meat, sporting (fighting and attendant betting), and show.

Most Chesapeake Bay area farmers and many townsfolk kept a small flock of chickens for a regular supply of eggs, and a limited supply of meat. The arrival of the first African slaves at Jamestown in 1619 shaped the trajectory of agriculture and the food system for the next four centuries. Chattel slaves formed the backbone of the plantation system that produced the cash crops of tobacco and cotton, in addition to much of the grain and produce that fed the region. As slaves weren't paid for laboring for their masters, they needed some other way to obtain cash or barter items to participate in the economy. Many slaveholders allowed their human property to raise and sell chickens. These stolen Africans brought with them their own traditions of raising chickens, and in America earned the reputation as masters of the art (Lawler 2016).

Despite the status of being one of the first domesticated animals to arrive from England, chickens played a limited, yet persistent role in the early colonial food system in regard to flesh. Wild birds, pigs, sheep, goats, and cattle, as well as, shad, sturgeon, and shellfish dominated the animal foods of that time (Lawler 2016). Alarmed that some slaves bought their freedom with proceeds from the selling of larger animals, the Virginia Assembly put a stop to this in

Fig. 6 Red jungle foul (*Gallus gallus*), the ancestor of the domesticated chicken. The male is pictured on the left and the female on the right. Both images taken in India. Male image by C K Subramanya https://commons.wikimedia.org/wiki/File:Red_jungle_fowl.png and female by Lip Kee https://commons.wikimedia.org/wiki/File:Gallus_gallus_female_Kaziranga_3.jpg

1692 by outlawing such further transactions involving horses, cattle, and pigs. However, the prohibition didn't include chickens. "Chicken bones account for a full third of all bones found among the African American quarters of Maryland and Virginia plantations, and plantation records show frequent cash payments to slaves for poultry (Lawler 2016)."

Perhaps to stimulate poultry sales to the wealthy white class of slaveholders and merchants, slaves and freedmen introduced their customers to the African version of fried chicken. Chicken meat dredged in flour and fried in lard became the signature dish of the southern cuisine. While the Scotts and Italians evolved similar traditions of cooking chicken, it would have been the West African version that slaves brought to their master's and client's tables (Graber and Twilley 2016).

While meat from chickens was a part of the Chesapeake Bay food system since the birds made landfall in Jamestown, most people raised them primarily for their eggs. Killing and eating the chicken was reserved for special occasions. Up until World War II, people consumed three times more mass of eggs than they did poultry. The War changed that. Poultry escaped the rationing placed on other animal flesh, creating greater demand for chicken to replace beef and pork (Sundin 2018). The wartime taste for chicken led to a continuous growth in the industry. By 1975, chicken flesh tied eggs in consumption at about 12.5% of animal food for each. Chicken meat consumption doubled that 40 years later in 2016, checking in at 25% of animal consumption. Chicken eggs continue to hover around 12% of total animal flesh consumed (Fig. 3).

The advance of the chicken-flesh market created two different types of breeds, chickens for producing eggs ("layers") and chickens for producing meat ("broilers"). Neither of these new specialized classes of breeds much resembled the various breeds of all-purpose chickens that populated the barnyards of the Chesapeake Bay region until the mid-twentieth century.

Chicken Mostly for Eggs The pre-industrial breeds of hens that roamed the barnyards of the Chesapeake Bay region reached sexual maturity at about 6 months and would lay about 3–4 eggs per week over their decade long lifespan. The annual egg production of about 150 eggs by traditional captive breeds was tenfold that of their red jungle fowl ancestor (Table 2). After their tenures as egg producers expired, the spent hens would end up in a pot of soup. Farmers also used some of the young males for meat before they matured and could engage in territorial fights with the other roosters kept for breeding. The system of traditional small chicken flocks presented a sustainable source of animal food for family farms. While the farmers might supplement the chicken's diet with grain and kitchen waste, the free roaming birds ate insects, seeds, and whatever else the barnyard offered. This channeled otherwise unavailable nutrition and energy to the human food system. The traditional barn yard chickens lived a life different from their ancestors, yet it retained all of the basic features experienced by the red jungle fowl. Roosters protected and mated with their hens. The hens incubated and rotated their eggs. The newly hatched chicks found protection and instruction from their brooding mothers. The chickens formed social structures, built nests, scratched the earth and pecked the ground in long-evolved foraging behaviors. They lived an altered, but full life.

Industrialized Egg Production In 1800, about 94% of the US and the Chesapeake Bay Region population resided in rural areas. This declined to 40% by 1900, and 19% by 2010 (USCensus 2019). Modern agriculture is so thoroughly mechanized, industrialized, and subsidized by fossil fuel that by 2016, only 1.5% of the US population worked in this sector (USBLS 2019). In 1910, 87.7% of farms reported having chickens. That dropped to 5.6% by 1992 (USDA 2019c). This means that scarcely 1% of chicken eggs now come from barnyards or backyards. Instead, huge factory farms now produce almost all eggs for eating. About 89% of US chicken egg production keeps the hens in cages for their entire lives. The newly laid eggs roll through holes in the cages and onto conveyor belts that carry them to machinery for washing, size-sorting, and boxing (UEP 2019, Fig. 7).

Chicken for Meat Although African American slaves and freedmen proved central in developing the early chicken business of the Chesapeake Bay Region, it was a white women in Delaware who inadvertently pioneered the industrialization of producing chicken exclusively for meat. In 1923 Cecilia Steele, the wife of a Coast Guardsman ordered 50 chicks from a hatchery with the intention of raising a flock of laying hens. By accident the hatchery delivered 500 chicks. She succeeded in bringing 387 to a 0.9 kg (2 pound) market

Table 2 A comparison red jungle fowl (*Gallus gallus*) and derived industrial commercial breed types

Breed type	Longevity (months)	Number of Eggs Produced per Hen per Year	Mass at Maturity or Market Size (kg)
Red Jungle Fowl	180–240	10–15	0.9
Broiler	1.5–3.0	0	2.3
Layer	12–24	260–300	2.9

Females are used for both the meat producing broilers and the egg producing layers. Both broilers and layers reach sizes two to three times the ancestral stock and experience much shorter lives (UEP 2019)

Fig. 7 A typical facility used for the production of chicken eggs. The laying chickens live their entire lives in cages, reaching their heads through the bars to access the feed from troughs and water from nipples. The eggs roll through the floors for collection in a trough as pictured, or on to a conveyor belt. Image courtesy of People for the Ethical Treatment of Animals

weight, selling them for 67 cents a pound, or fivefold the price fetched for spent chickens from the egg industry (Spinrad 2015, McKenna 2017a, b). The Steele's used the profits to buy another thousand chicks to raise for meat production. It wasn't locals who bought Ms. Steele's chickens, as most were shipped to New York City to be sold in markets servicing the growing Jewish population interested in observing traditional religious dietary requirements. The Delmarva hens were slaughtered in view of the purchasers using kosher techniques. Ms. Steele created what is now called the *broiler* industry, the raising of chickens in concentrated feeding facilities, and shipping them off to slaughter to service remote markets. In 1925 the Delmarva Peninsula produced 50,000 chickens. The adoption of Ms. Steele's broiler model meant seven million chickens brought to market from the region in 1935. The modern era of raising chicken for meat was underway (Lawler 2016). Ms. Steele's original 24 m^2 (256 ft^2) chicken shed is a broiler industry shrine, preserved at the Delaware Agricultural Museum (Del. Ag. Museum 2019).

World War II, Delmarva Chickens and Antibiotics While the broiler business grew during the 1920s and 1930s, it was WWII that accelerated the expansion of the industry on the Delmarva Peninsula. As noted above, wartime rationing of beef and pork brought attention to chicken as an alternative source of animal protein. Much of that attention came from the Department of War, which contracted for essentially all of the Delmarva broiler production for feeding the troops. The civilian market denied Delmarva hens, turned instead to Georgia, Arkansas, and elsewhere for its chicken. However, WWII drained many of the broiler farms and slaughterhouses of their laborers. Some 33,000 Delmarva workers left to be soldiers and sailors. To alleviate the labor shortage, the government put German prisoners of war to work in the broiler industry. One contingent of 550 prisoners of war worked the Armour Company chicken processing plant in Harrington, DE (Brown 2017).

The end of WWII left the broiler industry with overcapacity. Lifting of wartime rationing meant consumers returning to eating beef and pork. The oversupply of broilers depressed the price of chicken, driving producers to search for ways to cut the cost of production. They did so using four different approaches; (1) adding growth promoters to feed, (2) developing a new class of breed exclusively for meat production, (3) restructuring the industry along the lines of vertical integration, (4) making it easier for consumers to buy chicken by bringing most of it to market as cut-up pieces or highly processed and battered un-chicken-like shapes.

Antibiotics as Growth Promoters In 1928, Dr. Alexander Fleming discovered that a chemical produced by the mold *Penicillium notatum* would kill pathogenic bacteria. During

WWII, the US government invested in the production of this new antibiotic drug dubbed penicillin. Its use during the war saved thousands of lives by curing infected battlefield injuries. After the war, penicillin went on to save thousands of civilians from death by infection. This promoted the search for other such antibiotics produced by microorganisms.

The Lederle Corporation used the Actinobacteria *Streptomyces aureofaciens* to make a new antibiotic, Aureomycin (a tetracycline). Dr. Thomas Jukes was a nutrition scientist interested in chickens, and employed by Lederle. Acting on a tip from a scientist at Merck Laboratories (a rival business) in 1948, he performed an experiment to see if adding an antibiotic to chicken feed would improve the growth of the birds. Jukes didn't use the expensive refined drug, but instead fed his chickens with the brew-mash of bacteria from which the Aureomycin was extracted. The point was to find an inexpensive supplement, not to dose the chickens with costly drugs. It worked. Hens eating the antibiotic containing bacterial mash gained two and one-half times the mass of those on the control diet. Most remarkably, only 60 g of mash produced an extra 167 g of chicken mass! Jukes had discovered a *growth promoter* that reshaped the broiler industry and animal agriculture in general (Stokstad and Jukes 1950; McKenna 2017a, b). Low doses of antibiotics and the high growth rates they provided helped return the post-WWII broiler industry to profitability.

Throughout his life, Jukes (he died in 1999) remained a steadfast proponent of his innovation of using antibiotics as growth promoters. He also championed all other aspects of chemical industrial agriculture. He attacked Rachel Carson for publishing *Silent Spring*. He saw DDT as a miracle pesticide rather than as a persistent environmental toxin that nearly drove to extinction raptors and many other species of birds (Stoll 2012, see chapter "Pesticides Bring the War on Nature to the Chesapeake Bay").

Why do antibiotics promote growth in chickens (and other farmed animals)? Chickens and other livestock packed densely into CAFOs present ideal conditions for the spread of bacterial infections. Close proximity means easy transmission of contagious disease and a stressful environment that reduces immune response. The low-dose of antibiotics in the feed might help thwart infection and thus allow the animal to direct more of its nutrition into growth. However, another mechanism seems at play. The antibiotics alter the animal's microbiome, promoting the growth of species of bacteria and fungi in the intestine that enhance the uptake of nutrients, particularly energy-rich fats (Lin et al. 2013; Lin 2014).

After their introduction to the broiler business in the 1940s, antibiotic growth promoters became a central feature of industrial animal agriculture in the US and Europe. In 2015, about 70% by mass of all medically relevant antibiotics produced in the US went to promote the growth of chicken and other livestock (Pew 2016). In addition, between 1955 and 1965, the industry used antibiotics as a meat preservative. Spraying the drugs on the dead animal parts meant they could be shipped and stored refrigerated (not frozen) for a month before reaching the consumer (McKenna 2017a, b).

The extensive use of antibiotics in the industrial animal food system created a health care system crisis. The use of low doses of antibiotics to promote growth and preserve meat in industrial animal agriculture rendered those life-saving drugs nearly useless for much of the human population of the US. The pathogenic bacteria present in the guts of chickens and other livestock are normally held in check by the animal's immune systems. However, exposure to the antibiotics exerted artificial selection on those small populations of pathogens, initially killing most of them. The surviving pathogens were the few individuals with genetics that conferred resistance to the antibiotic. Over time the few became the many. Making matters worse, antibiotic resistance is transferred between different species of bacteria via R-plasmids that contain the information in segments of DNA (Levy and Marshall 2004; Tazzyman and Bonhoeffer 2014). Consequently, humans sickened with antibiotic resistant bacteria cultured in the guts of industrial animals have one less tool to fight infections (Fig. 8, CDC 2017a).

Each year in the US, 400,000 people contract antibiotic resistant food borne diseases (CDC 2018). The list of multiple-antibiotic resistant pathogenic bacteria is growing with *Salmonella spp.* and *Campylobacter spp* causing substantial increases in morbidity and mortality (Economou and Gousia 2015). Food sourced Salmonella infects one million Americans annually, causing 23,000 hospitalizations and 450 deaths (CDC 2019a). *Campylobacter*, which traces almost exclusively to poultry, infects 1.2 million people annually but is less deadly than *Salmonella* (CDC 2017b). Other antibiotic resistant pathogens from animal agriculture include, *Listeria*, enterococci, and some strains of *E. coli* (Levy and Marshall 2004).

In 1945, Alexander Fleming warned of the potential for development of antibiotic resistant bacteria with the overuse of his penicillin. The livestock industry proved him right about an array of antibiotics (Rosenblatt-Farrell 2009). By 1977, it was clear that the use of antibiotics to promote growth in farm animals was causing antibiotic-resistance in important human pathogens. Thus, Donald Kennedy, the then newly appointed head of the Food and Drug Administration, moved to ban their use. However, opposition from powerful members of Congress with links to the animal agriculture industry derailed Kennedy's efforts (McKenna 2017a, b). As of 2020, the use of antibiotic growth promoters remains virtually unregulated in the US. This contrasts with the European Union, which began banning antibiotic growth

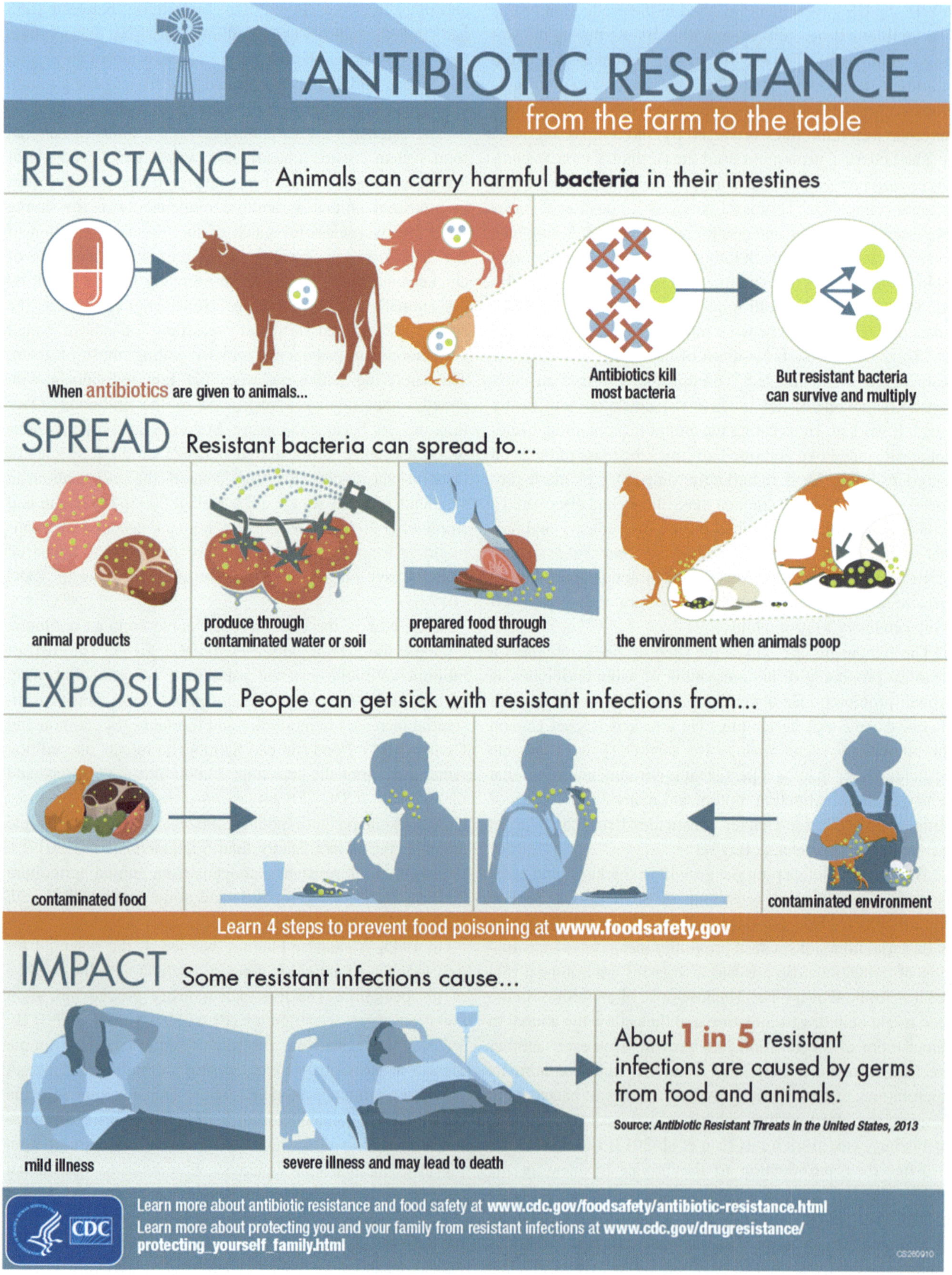

Fig. 8 US Center for Disease Control poster explaining how antibiotic use in animal agriculture leads to the development and spread of antibiotic-resistant pathogenic bacteria that sicken humans. The poster neglects to mention that most antibiotics used on farms is for promoting growth. Image from https://www.cdc.gov/foodsafety/challenges/from-farm-to-table.html (CDC 2017a)

promoters for livestock in the 1990s, and prohibited them for use in chickens in 2003 (Castanon 2007).

While the official line of the CDC is to discourage the use of antibiotics for growth promotion, the agency does support their use to; "Treat disease in animals that are sick, control disease for a group of animals when some of the animals are sick, and **prevent disease in animals that are at risk for becoming sick** (emphasis added) (CDC 2018)." Administering antibiotics to prevent disease is a way around the growing public dislike of using the drugs for growth promotion. Note that such preemptive use of antibiotics is discouraged in human medicine and reserved for certain conditions and surgeries where postoperative infection is likely (Enzler et al. 2011).

Although the animal food industry continues to escape meaningful government regulation of antibiotic abuse, recently consumers have forced the issue through the marketplace. Both Perdue and Tyson, the two major broiler companies on the Delmarva Peninsula, now advertise the absence of antibiotic growth promoters in the diets of their chickens (McKenna 2017a, b).

Finding a New Breed of Chicken for Meat The new broiler industry called for a new breed of chicken. The lineage of the modern broiler chicken traces to the winners of a contest held in the Chesapeake Bay's University of Delaware's Agricultural Experiment Station, in 1948. After WWII, USDA partnered with the pioneering A&P (The Great Atlantic and Pacific Tea Company) grocery chain to find the "chicken of tomorrow." To understand this, we need to take a voyage back in time and explore how the A&P changed the food system in the US.

Founded in 1859 as a tea and coffee shop, the A&P evolved to a chain of nearly 200 stores in 1900 that sold dry groceries (grains, beans, flour, etc.). By 1930, the A&P became the first supermarket, adding dairy, meat and fresh produce to their offerings. By then, it was the world's largest retail company, with 16,000 stores in the US and Canada. The A&P was revolutionary. Prior to its invention of the supermarket, shopping for groceries meant visiting several stores to get the fixings for a meal. Milk and cheese came from a dairy shop; meat from a butcher; flour, oils, canned goods and cereals from the dry grocer; fruit and vegetables from the green grocer; breads and cakes from the bakery. The A&P Supermarket put this all under one roof. A&P also created self-serve shopping, eliminating the person behind the counter that gathered the list of desired goods for the shopper. Although A&P succumbed to more innovative competitors and ceased operations in 2015, it forever changed how consumers interfaced with the food system (Funding Universe 2019).

Part of the A&P legacy was the ability of supermarkets to deliver uniform low-cost food. The Chicken of Tomorrow Contest invited farmers across the nation to offer-up a breed that would best fill the new demand for a tasty chicken with a large amount of meat on its bones. The winner of the contest wasn't one of the hundreds of distinct breeds that farmers developed to fit best to their local situations. Instead, it was a hybrid between two of those breeds. "The winner was Charles Vantress from California, who had crafted a red-feathered hybrid out of the New Hampshire, the most popular meat bird among East Coast growers, and a California strain of Cornish (McKenna 2018)." That the winner was a hybrid and not a true breeding line meant that fertile eggs supplied by Charles Vantress's company were destined to dominate the fledgling broiler industry. Only the company possessing the two different breeds had the capacity to keep the appropriate mating going to produce the desired hybrid. Chicken growers wouldn't get the same results if they let the hybrids copulate. This is exactly what seed companies began doing years earlier, selling high-yielding hybrid varieties that required the farmers to buy new seed each year. Seed saved from the hybrid corn or wheat wouldn't be like the parent generation.

The emerging broiler industry promoted the 1948 Chicken of Tomorrow competition with a parade and celebration in Georgetown, Delaware. The Delmarva chicken industry turned this into an annual event featuring the world's largest frying pan capable of cooking 800 chicken quarters at a time, beauty queens, games, art, and food. The industry ended the festival in 2014, redirecting funds to lobbying (Old Line Plate 2015).

The hybridization industry continued to "improve" the broiler chicken, going for faster growth, larger size, and greater efficiency of converting feed to chicken mass. The modern (2005 strain) broiler reaches a mass of 4202 g in 56 days. That's fourfold the 905 g size of a common strain from 1957 given the same feeding regime and time to grow (Fig. 9). The 2005 strain also converts feed to mass of chicken 50% more efficiently (Zuidhof et al. 2014). In general, larger homeotherms (warm blooded birds and mammals) feature lower surface area to volume ratios, meaning less loss of heat to the environment and therefore less food energy diverted from growth to keeping the animal warm.

The rapid growth, giant size and more efficient food conversion of the modern broiler hen is a boon to the industry but comes at a cost to the chickens. They suffer from cardiovascular disease, pulmonary hypertension, deformities, bone defects, and lameness (Julian 1998). This creates significant pathology for a bird over its extremely abbreviated 7–9 week life span. The industry

Fig. 9 The increasing size of commercial broiler chickens with genetic selection over time. These are images of actual chickens produced with the same simultaneous feeding regime for strains used in the broiler industry in 1957, 1977 and 2005. From Zuidhof et al. (2014)

sends the birds to market prior to their reaching sexual maturity for two reasons. First, pre-sexual chickens don't present the antagonistic behaviors that would make coexistence impossible in crowded CAFO's. Second, none of the energy and nutrition of the feed is diverted away from growth to sexual reproduction. Ironically, the selective breeding means that most of the juvenile hens sent to slaughter suffer from aliments otherwise restricted to old-age.

And what of the males? Skilled workers determine the gender of chicks, and dispose of the males in machinery that turns them into animal feed.

Vertical Integration of the Broiler Industry In the post WWII era, the broiler industry pioneered a new model for animal agriculture, called vertical integration. This means that a single company owns or controls all aspects of production and marketing (McKenna 2017a, b). The pork industry discussed above saw the success of the poultry business and adopted the same approach in the 1970s.

The broiler companies contract farmers to raise the hens. Farmers find the capitol to build and maintain the chicken houses, provide the labor, pay for utilities, and are responsible for disposal of dead birds and the huge quantity of manure produced in the operation. The broiler corporations (called integrators) provide the farmers with newly hatched chicks, feed, pharmaceuticals, veterinary services, technical guidance, and transport of the market-weight hens to slaughter/processing plants. The corporation pays the farmers based on the amount of broilers they produce and how that performance ranks in comparison to other contractors in the area. It is called the tournament system, with highest rewards going to the best producers. "Fees are determined in the following way. The integrator measures the average cost of the inputs that the integrator provided to growers for chickens delivered to the processing plant in a week—the total value of feed, chicks, and veterinary services provided to growers divided by the total weight of chickens delivered that week. The company develops this calculation for each grower. Each grower is then paid a base fee, and those growers whose costs are lower than the average for all growers receive a premium over the base fee; those whose costs exceed the average for all growers receive a deduction from the base. The amount of the premium or deduction reflects the size of the cost difference (USDA 2014)."

High mortality or poor growth means little compensation for the farmer. Despite the risk of failed flocks, the median family income for contract broiler producers in 2011 was $68,455 compared to $57,050 for all farm households, and $50,504 for all US households. Yet, this obscures the fact that 20% of contract poultry growers earn less than the federal poverty limit (USDA 2014).

Access to capital for building and maintaining the CAFOs helps determine the demography of contract broiler growers. They are overwhelmingly white males (85%) with an average age of 55 years. Only 16% completed college. In contrast, much of the workforce in the slaughter houses consists of African Americans, Latinx (circa 50%) and women. About 25% of the slaughter house workers are undocumented immigrants (NCFH 2014).

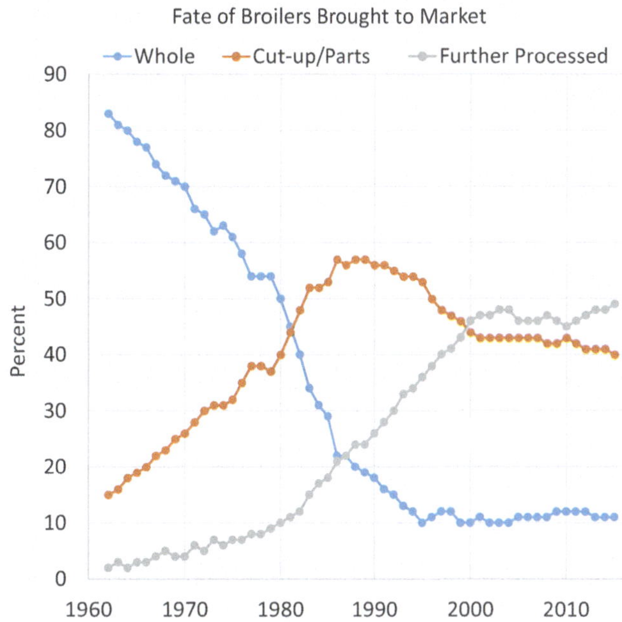

Fig. 10 Fate of chickens brought to market in the US since the 1960s. Prior to 1965 over 80% of the broiler chickens were sold as whole carcasses. In 2015, that fell to 11%, with most sold as cut-up parts (40%) and processed foods (49%) such as batter fired nuggets (NCC 2019)

Making Chicken Easier to Eat Residents of the Chesapeake Bay Region, like the rest of US, buy chicken in a very different way than their ancestors did two generations ago. In 1962, consumers bought 83% of their broilers as whole birds. By 2015, that dropped to 11% of the market, with cut-up parts (40%) and highly processed foods (49%) accounting for the vast majority of the chicken consumed (Fig. 10). This shift in the way people eat chicken accounts for much of its growing popularity illustrated in Fig. 3. Buying a whole chicken means the consumer either roasts the bird whole, or must butcher it prior to cooking (frying, baking, using in other dishes). That labor is saved by buying chicken parts. Easier still is purchasing preformed processed masses of chicken flesh already battered, fried and frozen, as invented in 1963 by Dr. Robert Baker of Cornell University. Baker didn't patent his "chicken sticks" (McKenna 2017a, b). The McDonalds Corporation brought its version of the chicken stick to market as "Chicken McNuggets" in 1983. The popularity of the battered bits of processed tissue quickly made that company the second largest seller of chicken meat, behind KFC (formerly Kentucky Fried Chicken) which specialized in more identifiable chicken parts (Rude 2016). Chicken McNuggets and similar products sold frozen at the grocery, or in fast food restaurants, provides the consumer with chicken flesh absent of any of the trappings of real animals. There is no identifiable skin, bones, tendons, blood vessel, or other tissue to remind one they are eating the remains of a bird. This is quite unlike the experience of ancestors who would decapitate a spent laying chicken in the backyard to produce the Sunday dinner. Thus, the fried and battered piece of processed chicken is very easy to eat, both physically and mentally.

Broiler CAFOs Dominate the Food System Landscape of the Delmarva Peninsula Several thousand imposing steel chicken houses dot the landscape of the Delmarva Peninsula. They stand as the legacy of Cecilia Steele's invention of the broiler industry in the 1920s, and the subsequent new-found hunger for the consumption of chicken meat. On a clear day, a peak out of the window of an airliner reveals a never ending string of broiler CAFOs stretching nearly the full length of the Delmarva Peninsula (Figs. 11 and 12).

Modern broiler CAFOs on the Delmarva Peninsula average about 2000 m^2 in floor space and house 20,000 to 30,000 hens (Fig. 13). This means a density of about 12 birds per m^2. The flock arrives as new hatchlings and is taken to slaughter in about 45 days. They spend the entire time in the chicken house, standing in a mix of their own manure and sawdust, eating feed from troughs and drinking water from nipples attached to pipes (Fig. 14).

Chicken Dominates the Terrestrial Animal Food Scene of the Bay The advent of the broiler industry changed the nation's eating habits and the local food system. As addressed above, the production of pigs for regional consumption has been mostly farmed out to North Carolina and other states, but the processing of pork is still a big industry in the lower Bay. Dairy and beef cattle production remains present throughout the watershed, but those industries concentrate in the states of New York and Pennsylvania (Table 3). Neither of those jurisdictions physically touches the Bay, yet effluents from their dairy and cattle producers drain to the Chesapeake via the Susquehanna River. However, its poultry that's left its mark on the Bay ecosystem more than any other of the terrestrial animals exploited for food.

Working Conditions and Health Hazards for Broiler Industry Workers During 2015 and 2016 Poultry workers rank 12th among professions in severe injury reports, with Tyson Foods (one of the two dominant broiler companies on the Delmarva Peninsula) ranking fourth for all employers. In 2016, Tyson received \$700,000 in OSHA (Occupational Safety and Health Administration) fines. Injuries included cuts, amputations, bone fractures, contusions, and chemical burns (Berkowitz and Hedayati 2017). Fast production line speeds, 12–14 hour shifts, dangerous equipment and blood slickened floors make for hazardous working conditions in slaughter and packaging houses. Repetitive injuries also afflict these workers (SPLC 2013).

Injures are also a regular feature of one specialized class of employment, chicken catchers. These mostly Latinx and African American workers must physically collect each of the 25,000 or so chickens in a CAFO house and place the armloads (seven at a time) of unwilling hens in cages on trailers for transfer to the slaughter house. Chicken catchers suffer from back, hand, and respiratory injuries (SPLC 2013).

In addition to physical injuries, broiler industry workers suffer health problems from exposure to polluted air and pathogens. The high density of chickens in CAFOs and no system for removing dung means the buildup of high levels of ammonia, hydrogen sulfide, volatile organic compounds, and particulate matter in the air. Yet CAFOs are agricultural facilities and therefore exempt from regulation by the USEPA and most state laws other than direct discharge into waterways (USEPA 2017).

Chickens eliminate excess nitrogenous waste as urea. Bacteria in soil or on the floor of a chicken house convert it to ammonia. The gas irritates the mucus membranes of the hens, increasing susceptibility to respiratory infections, conjunctivitis of the eyes, and lesions of the cornea (Aziz 2010). It also causes chemical burns on the feet and legs of the chickens. Ammonia exposure causes acute and chronic respiratory pathologies in chicken catchers and poultry house workers. Poultry workers exhibit significantly more chronic occurrences of burning eyes, sneezing, stuffy noses, running noses, cough, bronchitis, phlegm and chest tightness than other laborers. Reducing the protein and increasing the fiber content of the feed helps reduce ammonia production (Naseem and King 2018). But that would mean slower growth rates and reduced profits.

Slaughterhouse poultry workers suffer significantly higher rates of various cancers including; "...buccal and nasal cavities and pharynx (base of the tongue, palate and other unspecified mouth, tonsil and oropharynx, nasal cavity/middle ear/accessory sinus), esophagus, recto-sigmoid/rectum/anus, liver and intrabiliary system, myelofibrosis, lymphoid leukemia and multiple myeloma... (Johnson et al. 2010a)." Poultry workers also display an 8.6-fold increase in the chances of death from penial cancer. Poultry slaughterhouse workers show an 8.9 fold increase in chances of contracting pancreatic cancer and 9.1 fold increase for liver cancers (Felini et al. 2011). A possible cause of the high cancer rates is oncogenic retroviruses that induce various tumor types in chickens (Johnson et al. 2010b). Poultry plants became centers of outbreak for the Covid-19 virus epidemic of 2020. The close proximity of workers on the production line makes for rapid spread of disease, similar to the situation for chickens in CAFOs. Purdue and Tyson plants reported hundreds of cases of infected workers, including fatalities in each of their facilities across the entire Delmarva Peninsula on the Cheasapeake Bays's Eastern Shore (Ash 2020; Sternberg 2020; Vogelsong and Bravender 2020).

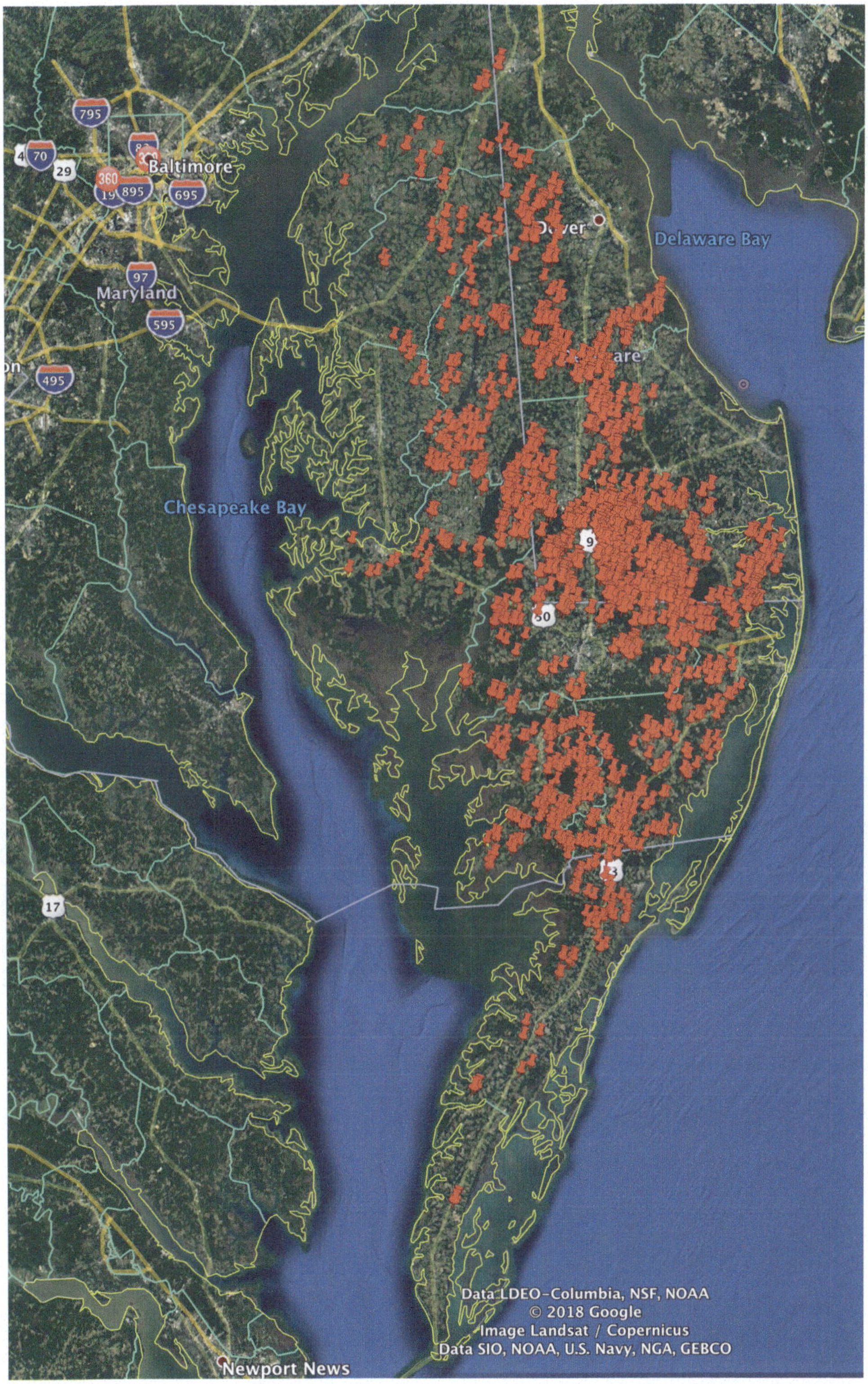

Fig. 11 The distribution of broiler house Concentrated Animal Feeding Operations (CAFOs) on the Delmarva Peninsula. Image is from Google Earth based upon satellite images taken in 2016. Each pin represents one CAFO. Each CAFO contains 1–14 broiler houses, with an average of 3.3 houses per each of the 1116 pined location. This yields a count of 3687 broiler houses on the Delmarva, with a calculated density of 0.26 units per km^2. Image Landsat/Copernicus, Data SIO, NOAA, US Navy, NGA, GEBCO. The estimated number of houses presented here may be low, as another source reports 5091 units for the Delmarva (Gustin 2018)

Fig. 12 Pins indicate broiler chicken CAFOs in the vicinity of Laurel, Delaware. Image from Google Earth. Image Landsat/Copernicus, Data SIO, NOAA, US Navy, NGA, GEBCO

Health Hazards for Neighbors of Broiler CAFOs CAFO operators use large ventilation fans to control concentrations of ammonia in the houses. However, this pollutes the air surrounding the facilities, exposing neighbors of the facility to high levels of irritating ammonia (Naseem and King 2018). CAFOs with four or more chicken houses regularly surpass the 100 lbs. (45.4 kg) per day limit for ammonia that USEPA places on stationary facilities (Pescatore et al. 2005). The occurrence of exacerbated asthma attacks is associated with proximity to dairy and pig CAFOs (Rasmussen et al. 2017). Is this also true for broiler CAFOs? In 2017, residents of the Maryland portion of the Delmarva Peninsula pushed for legislation (Community Healthy Air Act) to assess air pollution and its health effects emanating from chicken CAFOs, and to determine compliance with state and federal laws. The measure failed in the state legislator, illustrating the political grip of the broiler industry in the region (Gustin 2018). However, in 2019, the Delmarva Poultry Industry funded a small scale pilot study to compare upwind and downwind air quality at two CAFOs. Critics worry about the validity of any results from a study funded by the broiler industry (Miller 2019).

Health Hazards for Consumers of Poultry and Eggs Consumers of poultry products face two types of hazards, pathogens and dietary components that cause non-contagious diseases.

Poultry and Pathogens As discussed above, poultry subjected to antibiotics for growth promotion present consumers with drug-resistant diseases. This is a serious problem, as poultry is the most common source of food-borne disease outbreaks in the US. One study found that between 1998 and 2012…"poultry was associated with

Fig. 13 A typical broiler chicken CAFO with three houses. Hoppers for supplying feed are located south of the midpoint of each house. Note the very green pond to the left of the houses. This suggests that nutrient runoff from the CAFO is causing eutrophication of the pond. We used a sample of 15 CAFOS to estimate mean sizes of the buildings. The houses averaged 141.3 m (SD = 30.9) long and 14.6 m (SD = 3.2) wide. Thus the houses averaged 2072 m^2 each. Flock sizes vary between 20,000 and 30,000 per house (Chicken Check In 2019). That means about 12 hens per m^2 of floor space

279 (25%) of cases, accounting for the highest number of outbreaks, illnesses, and hospitalizations, and the second highest number of deaths. Of the 149 poultry associated outbreaks caused by a confirmed pathogen, *Salmonella enterica* (43%) and *Clostridium perfringens* (26%) were the most common pathogens (CHAI et al. 2016)." Restaurants (37%), private homes (25%) and catering facilities (13%) were all associated with outbreaks. Food-handling errors accounted for 64% of the outbreaks. Yet, practicing safe handling of poultry appears beyond the resources and abilities of most consumers. Only 17.5% store raw poultry properly, and only 11% thaw frozen birds safely (Kosa et al. 2015).

The shift from selling whole chickens to chicken parts as discussed above, increases exposure to dietary pathogens. A comprehensive study done in 2012 found *Salmonella* and *Campylobacter* contamination on chicken parts at the rate of 26.3% and 21.4% respectively. This compares to 5.9% for *Salmonella* and 10.6% for *Campylobacter* for whole young chickens (USDA 2019d). The USDA sets a maximum acceptable percent positive standard for pathogens in poultry products (Table 4). The findings from the 2012 study well exceed those standards. Depending on the product, contamination rates for Salmonella of 9.8–25.0% is permissible. This means that unsafe handling or undercooking of the products is likely to result in significant rates of exposure for consumers (between one in ten and one in four), even if the poultry companies complied with the standards. *Raw poultry as sold in the US is a perfectly legal biohazard for consumers*. According to a 1974 court ruling, *Salmonella* is a natural part of poultry, destroyed by proper cooking, and therefore not subject to regulation (Legal 2019). A second ruling in 2001 prevents the USDA from closing down processing plants for producing *Salmonella* contaminated products (FindLaw 2019). Thus poultry plants producing *Salmonella* contaminated products don't face fines and can't be shut down by regulators. In contrast to the US, Sweden doesn't allow the sale of *Salmonella* contaminated poultry, eliminating the pathogen from broilers by 1972, and layers 20 years later. That nation instituted hygienic practices throughout the production process, including heating the feed to eliminate pathogens (Wierup et al. 1995).

As with the meat, chicken eggs also transmit *Salmonella* to consumers and account for 53% of infections. Pasteurization or complete cooking is recommended to reduce infection, but just handling the eggs may transmit the pathogen to consumers (Whiley and Ross 2015).

In addition to *Salmonella* and *Campylobacter*, poultry is a source of pathogenic strains of *Escherichia coli.* Extraintestinal Pathogenic *E. coli* (ExPEC) survive outside of the gut, as the name indicates. ExPEC causes 70–90% of urinary tract infections (UTIs) in humans (Singer 2015). Poultry appears to be the main source pool for UTI causing ExPEC (Manges 2016).

Fig. 14 Interior of a broiler chicken house illustrating the density of the hens, the system of pipes and troughs for watering and feeding the birds, and massive ventilation fans. A mix of saw dust or similar organic material and feces covers the floor. That manure remains in place during the 42 day-long residency of the flock. The manure removal and disposal takes place after the flock is sent to slaughter. Image from the USDA. It was taken Aug. 23, 2013 by Bob Nichols. http://www.publicdomainfiles.com/show_file.php?id=13986676816105

Table 3 The annual production of key animal products from the jurisdictions with drainages that reach the Chesapeake Bay

Jurisdiction	Annual Egg Production (millions)	Annual Broiler Production (millions)	Annual Dairy Production (million kg)	Number of Cattle
Delaware	0	259.8	45	16,000
Maryland	833.7	306.7	446	182,000
New York	1694.10	0.59	6396	1,450,000
Pennsylvania	8212.80	185.2	4901	1,620,000
West Virginia	273.1	86.1	64	380,000
Virginia	691.4	277.4	803	1,530,000
Total Bay States	11,705.1	1115.79	12,655	5,178,000
Total US	105,688.70	8831.20	94,648	93,700,000

The percentage of the US production accounted for by the products are 12.6% for broilers, 11.1% for chicken eggs, 13.4% for dairy, and 5.5% for cattle. Note that 89% of the dairy production takes place in New York and Pennsylvania, states with no land contacting the Bay. Data from USDA 2018 and USDAERS 2020

Poultry, Eggs and Non-communicable Disease Dietary cholesterol and saturated fats cause cardiovascular disease (CVD), the number one cause of death in the US and the world (Ference and Mahajan 2013). "Among 29,615 adults pooled from 6 prospective cohort studies in the United States with a median follow-up of 17.5 years, each additional 300 mg of dietary cholesterol consumed per day was significantly associated with higher risk of incident CVD (adjusted hazard ratio [HR], 1.17; adjusted absolute risk difference [ARD], 3.24%) and all-cause mortality (adjusted HR, 1.18; adjusted ARD, 4.43%) and each additional half an egg consumed per day was significantly associated with higher risk of incident CVD (adjusted HR, 1.06; adjusted ARD, 1.11%) and all-cause mortality (adjusted HR, 1.08; adjusted ARD, 1.93%) (Zhong et al. 2019)".

Table 4 The USDA maximum allowable rates (%) of contamination for Salmonella and Campylobacter on chicken parts

Product	Salmonella	Campylobacter
Broiler carcass	9.8	15.7
Broiler parts	15.4	7.7
Comminuted chicken	25.0	1.9

Comminuted chicken refers to the process of crushing the tissue to form a shaped product, which is usually sold breaded in the form of a nugget. The USDA considers facilities with contamination rates above these percentages to be in violation of the standard (USDA 2016, 2019e)

Most consumers consider chicken to be the healthiest form of meat, thinking it contains less cholesterol and saturated fat than beef or pork (Malone and Lusk 2017). Yet, a metanalysis of 124 randomized control studies with various diets found no significant difference in blood cholesterol and lipids for beef, chicken or fish consumption (Maki et al. 2012). This is consistent with the modern broiler bred to contain up to 10 times the fat per kg of chickens from 1896 (Wang et al. 2009). Modern broilers differ little in their cholesterol and saturated fat content from beef, pork, and seafood. And chicken eggs contain nearly six times the cholesterol found in beef tenderloin steak (Table 5).

About 12% of adults older than 20 years of age in the US have total cholesterol levels higher than 240 mg/dl, while any total concentrations above 150 mg/dl (and low density lipid cholesterol above 70 mg/dl) means increased risk of heart attack (O'Keefe et al. 2004; CDC 2019b). In addition to heart disease, dietary cholesterol and oxidized cholesterol are associated with a host of other maladies. These include; dementia, erectile dysfunction, inflammatory bowel disease, and macular degeneration (Schwartz and Kloner 2009; Poli et al. 2013; Testa et al. 2016).

After decades of decline, egg consumption in the US is on the rise (Fig. 3), which may attribute to intensive marketing that influences the perception of the healthfulness of the product and misleading egg-industry funded studies (AEB 2019; Fernandez 2012). The Freedom of Information Act allowed Dr. Michael Greger (2018) to obtain communication between the USDA and one company that sells chicken eggs. The USDA officer denied the companies' attempt to use any of the following terms for eggs in promotional literature; "nutritional powerhouse that aids in weight loss," "healthy snacking," "healthy," "nutritious," "relatively low in calories," "healthful," and "high in nutritional content." As the officer pointed out, the high level of cholesterol and saturated fats in eggs disqualifies them from labeling as healthy. However, the official suggested the statement, "can reduce hunger" as an alternative to words that refer to health. This interaction highlights the internal contradiction created by the mixed mandate of the USDA. It is supposed to both promote US agricultural products and to recommend healthy diets.

Consumers wishing to eliminate or reduce dietary cholesterol won't fulfill that desire by continuing to consume any sort of animal product. They should instead look to beans and nuts as protein-rich replacements for meat, since plants don't produce cholesterol (Table 5).

The High Cost of Cheap Chicken Meat for the Chesapeake Bay Ecosystem The system of producing the modern giant boiler chicken comes with oversized environmental consequences. Industrialized broiler production completely compartmentalizes the chicken's life cycle in time and space. Corn, soy, and amendments such as fish meal and leftovers from food processing plants form the feed. These products may hail from the Chesapeake Bay watershed, but feeds typically includ items trucked in from remote farms. About 48% of the feed turns into chicken tissue, as the modern broiler exhibits a remarkable Food Conversion Efficiency (FCE) of 1.9 (Zuidhof et al. 2014). So if half of the chicken feed becomes chicken parts, where goes the rest? Modern delivery systems mean only a small portion of the feed never finds its way into a hen's beak. The excreted metabolites from the ingested food yields manure rich in nitrogen and phosphorus. Noxious ammonia released from the manure pollutes the air of the chicken house and the surrounding communities. Carbon dioxide and nitrous oxides from the birds and their litter contribute to the stock of greenhouse gasses. Soluble fractions of the manure may runoff into the Bay immediately, or leach into groundwater where they are detained before eventually seeping into the waters of the estuary. Ideally, the industry would return all of the manure to fields that produced the chicken feed in the first place, as what used to happen in the days of small family farms. The nitrogen and phosphorus in the manure are critical plant nutrients. Manure from poultry should be prized for its particularly high concentrations of these elements (Table 6). Yet handling and transport costs keeps the bulk of the manure close to the CAFOs where it is more likely to contaminate the Bay. The current economics of broiler production means it pays to ship chicken feed from farms, but not to send the resulting chicken manure back to those fields. Instead, corn and soy farmers prefer the low price of artificial fertilizes to sustain crop production (see chapter "The Journey from Peruvian Guano to Artificial Fertilizer Ends with Too Much Nitrogen in the Chesapeake Bay").

The years of plentiful and cheap chicken litter resulted in manure-saturated fields throughout the Delmarva Peninsula. Adding excess manure is prohibited by state laws designed to reduce nutrient runoff and migration into ground water. To promote export of manure out of the Delmarva, the state of Maryland provides a $ one million per year ($18 per ton) subsidy to the giant Perdue Farms poultry company to process chicken manure into fertilizer at its AgriRecycle facility

Table 5 The nutrition content per 100 g (wet mass) of various food substances found in the Chesapeake Bay food system

Item	Energy (kcal)	Protein (g)	Total Saturated Fat (g)	Cholesterol (mg)	Total fat (g)	Protein (g/Kcal)
Atlantic croaker	104	17.8	1.09	61	3.17	0.17
Striped bass	97	17.7	0.51	80	2.33	0.18
Flounder	86	16.2	0.54	56	2.37	0.19
American shad	197	16.9	3.12	75	13.8	0.09
Oyster	107	12.5	1.79	62	4.46	0.12
Blue crab meat	119	23.81	0.20	107	1.19	0.20
Chicken breast with skin	172	20.9	2.66	64	9.25	0.12
Chicken breast no skin	107	23.2	0	58	0.89	0.22
Chicken egg	155	12.6	3.27	373	10.60	0.08
Ground beef 70% lean	332	14.4	11.75	78	30.00	0.04
Beef tenderloin steak	129	22.35	1.76	65	5.29	0.17
Cured Smithfield honey ham	125	17.9	0	54	3.60	0.14
Bacon Smithfield hickory smoked	471	29.4	14.71	118	41.80	0.06
Whole milk	62	3.33	1.88	10	3.33	0.00
Soy bean, organic	92	8.46	0.77	0	4.62	0.09
Navy beans	85	9.38	0	0	0.77	0.11
Peanuts boiled	278	13.89	4.17	0	22.22	0.05
Raw walnuts	700	16.67	1.5	0	66.67	0.02

Data is from the USDA (2019f). The scientific names of the foods are: Atlantic croaker, *Micropogonias undulatus*, Blue fish, *Pomatomus saltatrix*, Striped bass, *Morone saxatilis*, Flounder, *Pleuronectes americanus*, American shad, *Alosa sapidissima*, Oyster, *Crassostrea virginica*, Blue crab, *Callinectes sapidus*, Chicken, *Gallus gallus domesticus*, Cattle (beef), *Bos Taurus*, Pig (ham and bacon), *Sus scrofa domesticus*, Soybean, *Glycine max*, Navy beans, *Phaseolus vulgaris*, Peanuts boiled, *Arachis hypogaea*, Walnuts, *Juglans sp.*

Table 6 The production of manure, nitrogen and phosphorus per 1000 kg of animal mass per day of different livestock types

Production in kg per 1000 kg of animal mass			
Live stock	Manure	Nitrogen	Phosphorus
Beef	26.81	0.10	0.05
Dairy	36.29	0.20	0.03
Pigs	28.62	0.19	0.07
Chickens (layers)	27.44	0.38	0.14
Chickens (broilers)	36.29	0.50	0.15
Turkeys	19.78	0.34	0.13

Note that poultry manure is significantly richer in nutrients than that from cattle or pigs. Data from NRCS 1995

in Delaware. The plant processes 50,000 tons of Delmarva chicken manure annually. Its product goes not to the farmer's fields that produce the feed crops, but to retailers of fertilizer for golf courses and home gardens (Dance 2017). This short-circuits nutrient recycling, contributing to the environmental unsustainability of the broiler industry.

What's in the industrial chicken manure may also deter farmers from using it as fertilizer. The enteric pathogens discussed above (*Salmonella*, *Campylobacter*, and *E. coli*) and residues of antibiotics render manure less useful for growing crops, as farmers seek to avoid contamination of their harvest. The broiler industry also poisoned the manure with arsenic. The crowding of hens into CAFOs created the ideal situation for diseases to rapidly spread through the flocks. Crowding meant suppression of the immune system and easy transmission of pathogens. Outbreaks of the intestinal parasite coccidiosis (*Eimeria* sp.) plagued the emerging broiler industry. In the 1940s, they began adding the arsenic-containing-compound Roxarsone (3-nitro-4-hydroxyphenylarsonic acid) to chicken feed to control the disease and promote faster growth (Reid 1990). However, public response to arsenic contamination of poultry meat, soils, and surface waters eventually ended the practice with bans enacted in the European Union in 1999, Maryland in 2012 and the entire US in 2013 (Garbarino et al. 2003; Fisher et al. 2015). Arsenic is a carcinogen and a general toxin, interfering with over 400 enzymatic pathways in the human body (Ratnaike 2003). Although no longer permitted for use in poultry, the legacy of arsenic contamination continues to pollute soils, ground water and the Chesapeake Bay food chain.

While antibiotics, arsenic and other pollutants from chicken manure threaten the Chesapeake Bay and our food system, it's the excess nitrogen that causes the most visible effects. Chapter "The Journey from Peruvian Guano to Artificial Fertilizer Ends with Too Much Nitrogen in the Chesapeake Bay" and "Eutrophication: Obesity of the Bay and Its People" discusses how nitrogen from the chicken industry is helping to drive eutrophication of the Bay. The Clean Water Act of 1972 (discussed in chapter "What Nature, Politics and Policy Demand of the Chesapeake Bay and Its Food System") authorizes the USEPA to regulate pollution from point sources (end of pipe) entering waterways.

Table 7 Key pollutants from livestock operations and animal manure

Pollutant	Description of pollutant	Pathways to the environment	Potential impacts
Nitrogen	Organic forms (e.g., urea) and inorganic forms (e.g., ammonium and nitrate) in manure may be assimilated by plants and algae	• Overland discharge • Leachate into ground water • Atmospheric deposition as ammonia	• Eutrophication and harmful algal blooms (HABs) • Ammonia toxicity to aquatic life • Nitrate linked to methemoglobinemia
Phosphorus	As manure ages, phosphorus mineralizes to inorganic phosphate compounds that may be assimilated by plants	• Overland discharge • Leachate into ground water (water soluble forms)	• Eutrophication and HABs
Potassium	Most potassium in manure is in an inorganic form available for plant assimilation; it can also be stored in soil for future plant uptake	• Overland discharge • Leachate into ground water	• Increased salinity in surface water and ground water
Organic Compounds	Carbon-based compounds decomposed by micro-organisms. Creates biochemical oxygen demand because decomposition consumes dissolved oxygen in the water	• Overland discharge	• Eutrophication and HABs • Dissolved oxygen depletion, and potentially anoxia • Decreased aquatic biodiversity
Solids	Includes manure, feed, bedding, hair, feathers, and dead livestock	• Overland discharge • Atmospheric deposition	• Turbidity • Siltation
Salts	Includes cations (sodium, potassium, calcium, magnesium) and anions (chloride, sulfate, bicarbonate, carbonate, nitrate)	• Overland discharge • Leachate into ground water	• Reduction in aquatic life • Increased soil salinity • Increased drinking water treatment costs
Trace Elements	Includes feed additives (arsenic, copper, selenium, zinc, cadmium), trace metals (molybdenum, nickel, lead, iron, manganese, aluminum), and pesticide ingredients (boron)	• Overland discharge • Leachate into ground water	• Aquatic toxicity at elevated concentrations
Volatile Compounds Including Greenhouse Gasses	Includes carbon dioxide, methane, nitrous oxide, hydrogen sulfide, and ammonia gases generated during manure decomposition.	• Inhalation • Atmospheric deposition of ammonia	• Eutrophication • Human health effects • Climate change
Pathogens	Includes a range of disease-causing organisms, including bacteria, viruses, protozoa, fungi, prions and helminths	• Overland discharge • Potential growth in receiving waters	• Animal, human health effects
Antimicrobials	Includes antibiotics and vaccines used for therapeutic and growth promotion purposes	• Overland discharge • Leachate into ground water • Atmospheric deposition	• Facilitates the growth of antimicrobial-resistance • Unknown human health and aquatic life effects
Hormones	Includes natural and synthetic hormones used to promote animal growth and control reproductive cycles	• Overland discharge • Leachate into ground water	• Endocrine disruption in fish • Unknown human health effects
Other Pollutants	Includes pesticides, soaps, and disinfectants	• Overland discharge • Leachate into ground water	• Unknown human health and ecological effects • Potential endocrine disruption in aquatic organisms

From USEPA 2013

However, it exempts the non-point source pollution released from agricultural operations (Janasie 2018). Table 7 presents the various pollutants emanating from livestock manure, their pathways of movement through the environment and their potential impacts on ecosystem health. These remain largely unregulated.

References

AEB (2019) Home – American Egg Board. In: Aeb.org. https://www.aeb.org/. Accessed 30 Mar 2019

Anderson L (2009) Reproductive biology of Pigs. In: Lib.dr.iastate.edu. https://lib.dr.iastate.edu/cgi/viewcontent.cgi?article=1483&context=ans_air. Accessed 10 Feb 2019

Ash S (2020) Hogan: Wicomico Co. has fourth-highest COVID-19 case rate in Md., update on Delmarva poultry plant outbreaks - 47abc. In: 47abc. https://www.wmdt.com/2020/05/hogan-wicomico-co-has-fourth-highest-covid-19-case-rate-in-md-update-on-delmarva-poultry-plant-outbreaks/. Accessed 15 May 2020

Auchterlonie N (2017) The benefits of Fishmeal and Fish Oil in Swine and Poultry Diets | IFFO – The Marine Ingredients Organisation. In: Iffo.net, reprinted from Milling and Grain magazine, February 2017. http://www.iffo.net/node/1056. Accessed 10 Feb 2019

Ax R, Dally M, Didion B et al (2016) Artificial insemination. In: Hafez B, Hafez E (eds) Reproduction in farm animals. Wiley, New York. https://doi.org/10.1002/9781119265306.ch26

Aziz T (2010) Harmful effects of ammonia on birds. [online] PoultryWorld. https://www.poultryworld.net/Breeders/Health/2010/10/Harmful-effects-of-ammonia-on-birds-WP008071W/ [Accessed 10 Mar. 2019]

Berkowitz D, Hedayati H (2017) [online] Nelp.org. https://www.nelp.org/wp-content/uploads/OSHA-Severe-Injury-Data-2015-2016.pdf. Accessed 10 Mar 2019

Brown J (2017) German POWs a common sight in Delaware during World War II. In: Dover post. https://www.doverpost.com/news/20170529/german-pows-common-sight-in-delaware-during-world-war-ii. Accessed 14 Mar 2019

Brown H (2018) North Carolina jury awards $473.5m in hog waste suit against Smithfield. In: New food economy. https://newfoodeconomy.org/north-carolina-jury-fines-smithfield-foods-nuisance-lawsuit-hog-farm-manure/. Accessed 10 Feb 2019

Brush G (2008) Historical land use, nitrogen, and coastal eutrophication: a paleoecological perspective. Estuar Coasts 32:18–28. https://doi.org/10.1007/s12237-008-9106-z

Carson R (1945) Fish and Shellfish of the Middle Atlantic Coast. In: Digitalcommons.unl.edu. http://digitalcommons.unl.edu/cgi/viewcontent.cgi?article=1002&context=usfwspubs. Accessed 10 Feb 2019

Castanon J (2007) History of the use of antibiotic as growth promoters in European poultry feeds. Poult Sci 86:2466–2471. https://doi.org/10.3382/ps.2007-00249

CDC (2017a) Antibiotic Resistance from the Farm to the Table | Food Safety | CDC. In: Cdc.gov. https://www.cdc.gov/foodsafety/challenges/from-farm-to-table.html. Accessed 21 Mar 2019

CDC (2017b) Campylobacter (Campylobacteriosis) | Campylobacter | CDC. In: Cdc.gov. https://www.cdc.gov/campylobacter/index.html. Accessed 22 Mar 2019

CDC (2018) Antibiotic resistance, food, and food-producing animals. In: Centers for disease control and prevention. https://www.cdc.gov/features/antibiotic-resistance-food/index.html. Accessed 22 Mar 2019

CDC (2019a) Salmonella Homepage | CDC. In: Cdc.gov. https://www.cdc.gov/salmonella/index.html. Accessed 22 Mar 2019

CDC (2019b) High cholesterol facts. In: Centers for disease control and prevention. https://www.cdc.gov/cholesterol/facts.htm. Accessed 29 Mar 2019

Chai S, Cole D, Nisler A et al (2016) Poultry: the most common food in outbreaks with known pathogens, United States, 1998–2012. Epidemiol Infect 145:316–325. https://doi.org/10.1017/s0950268816002375

Chicken Check In (2019). Space & housing: are chickens crammed in houses? Do chickens have enough space to move?. [online] https://www.chickencheck.in/faq/space-chicken-houses/. Accessed 9 Mar 2019

Craven A (1926) Soil exhaustion as a factor in the agricultural history of Virginia and Maryland, 1606–1860. University of Illinois, Urbana, IL

Dance S (2017) Maryland spends $1M a year to transport chicken litter, to the benefit of the Chesapeake — and poultry companies. The Baltimore Sun. [online] https://www.baltimoresun.com/features/green/blog/bs-md-manure-export-20170115-story.html. Accessed 7 Mar 2019

Davis K (2009) Prisoned chickens, poisoned eggs: an inside look at the modern poultry industry. Book Pub. Co., Summertown, TN

Delaware Ag. Museum (2019) Delaware Agricultural Museum and Village – Delmarva Poultry. In: Agriculturalmuseum.org. http://www.agriculturalmuseum.org/exhibits-features/great-hall-exhibits/delmarva-poultry/. Accessed 20 Feb 2019

Economou V, Gousia P (2015) Agriculture and food animals as a source of antimicrobial-resistant bacteria. Infect Drug Resist 8:49. https://doi.org/10.2147/idr.s55778

Enzler M, Berbari E, Osmon D (2011) Antimicrobial prophylaxis in adults. Mayo Clin Proc 86:686–701. https://doi.org/10.4065/mcp.2011.0012

Felini M, Johnson E, Preacely N et al (2011) A pilot case-cohort study of liver and pancreatic cancers in poultry workers. Ann Epidemiol 21:755–766. https://doi.org/10.1016/j.annepidem.2011.07.001

Ference B, Mahajan N (2013) The role of early LDL lowering to prevent the onset of atherosclerotic disease. Curr Atheroscler Rep 15(4):312. https://doi.org/10.1007/s11883-013-0312-1

Fernandez M (2012) Rethinking dietary cholesterol. Curr Opin Clin Nutr Metab Care 15:117–121. https://doi.org/10.1097/mco.0b013e32834d2259

FindLaw (2019) Supreme beef processors vs. the United States. In: FindLaw's United States Fifth Circuit case and opinions. https://caselaw.findlaw.com/us-5th-circuit/1429779.html. Accessed 28 Mar 2019

Fisher D, Yonkos L, Staver K (2015) Environmental concerns of Roxarsone in broiler poultry feed and litter in Maryland, USA. Environ Sci Technol 49(4):1999–2012

FundingUniverse (2019) History of the Great Atlantic & Pacific Tea Company, Inc. – FundingUniverse. In: Fundinguniverse.com. http://www.fundinguniverse.com/company-histories/the-great-atlantic-pacific-tea-company-inc-history/. Accessed 16 Feb 2019

Garbarino J, Bednar A, Rutherford D et al (2003) Environmental fate of Roxarsone in poultry litter. I. Degradation of Roxarsone during composting. Environ Sci Technol 37(8):1509–1514

Giuffra E, Kijas J, Amarger V, et al (2000) The origin of the domestic Pig: independent domestication and subsequent introgression. Genetics 154:1785–1791

Graber C Twilley N (2016) As American as fried chicken. In: The Atlantic. https://www.theatlantic.com/science/archive/2016/07/finger-lickin-good/491108/. Accessed 18 Feb 2019

Greger M (2018) Flashback Friday: who says eggs aren't healthy or safe? | NutritionFacts.org. In: Nutritionfacts.org. https://nutritionfacts.org/video/flashback-friday-who-says-eggs-arent-healthy-or-safe/. Accessed 27 Mar 2019

Gustin, G. (2018). Giant Chicken Houses Overrun Delmarva, and Neighbors Fear It's Making Them Sick. [online] InsideClimate News. https://insideclimatenews.org/news/23042018/poultry-industry-epa-asthma-respiratory-illness-chicken-houses-delmarva-legislature-health-study-delaware-maryland-virginia. Accessed 9 Mar 2019

Jamestown (2019) History Timeline | Historic Jamestowne. In: Historicjamestowne.org. https://historicjamestowne.org/history/history-timeline/. Accessed 10 Feb 2019

Janasie C (2018) The management of nonpoint source pollution under the Clean Water Act. In: Nsglc.olemiss.edu. http://nsglc.olemiss.edu/projects/ag-food-law/files/mgt-nonpoint-source-pollution-under-cwa.pdf. Accessed 29 Mar 2019

Jefferson T (1953) In: Betts E (ed) Thomas Jefferson's Farm book. Princeton University Press, Princeton

Johnson E, Zhou Y, Lillian Yau C et al (2010a) Mortality from malignant diseases—update of the Baltimore union poultry cohort. Cancer Causes Control 21:215–221. https://doi.org/10.1007/s10552-009-9452-6

Johnson E, Ndetan H, Lo K (2010b) Cancer mortality in poultry slaughtering/processing plant workers belonging to a union pension fund. Environ Res 110:588–594. https://doi.org/10.1016/j.envres.2010.05.010

Julian R (1998) Rapid growth problems: ascites and skeletal deformities in broilers. Poult Sci 77:1773–1780. https://doi.org/10.1093/ps/77.12.1773

Kosa K, Cates S, Bradley S et al (2015) Consumer-reported handling of raw poultry products at home: results from a national survey. J Food Prot 78:180–186. https://doi.org/10.4315/0362-028x.jfp-14-231

Lawler A (2016) Why did the chicken cross the world? Atria Books, New York, pp 192–195

Legal (2019) American Public Health Association | 511 F.2d 331 (1974) | 1f2d3311775 | Leagle.com. In: American Public Health Association v. Butz. https://www.leagle.com/decision/1974842511f2d3311775. Accessed 28 Mar 2019

Levy S, Marshall B (2004) Antibacterial resistance worldwide: causes, challenges and responses. Nat Med 10:S122–S129. https://doi.org/10.1038/nm1145

Lin J (2014) Antibiotic growth promoters enhance animal production by targeting intestinal bile salt hydrolase and its producers. Front Microbiol 5:33. https://doi.org/10.3389/fmicb.2014.00033

Lin J, Hunkapiller A, Layton A et al (2013) Response of intestinal microbiota to antibiotic growth promoters in chickens. Foodborne Pathog Dis 10:331–337. https://doi.org/10.1089/fpd.2012.1348

Maki K, Van Elswyk M, Alexander D et al (2012) A meta-analysis of randomized controlled trials that compare the lipid effects of beef versus poultry and/or fish consumption. J Clin Lipidol 6:352–361. https://doi.org/10.1016/j.jacl.2012.01.001

Malone T, Lusk J (2017) Taste Trumps health and safety: incorporating consumer perceptions into a discrete choice experiment for meat. J Agric Appl Econ 49:139–157. https://doi.org/10.1017/aae.2016.33

Manges A (2016) *Escherichia coli* and urinary tract infections: the role of poultry-meat. Clin Microbiol Infect 22:122–129. https://doi.org/10.1016/j.cmi.2015.11.010

McGloughlin P (1976) Some factors affecting litter size in Pigs. Ir J Agric Res 15:141–145

McKenna M (2017a) Big Chicken: the incredible story of how antibiotics created modern agriculture and changed the way the world eats. 40–285

McKenna M (2017b) Big Chicken. National Geographic Society, Washington, DC

McKenna M (2018) The surprising origin of chicken as a dietary staple. In: Nationalgeographic.com. https://www.nationalgeographic.com/environment/future-of-food/poultry-food-production-agriculture-mckenna/. Accessed 16 Feb 2019

Miller J (2019) Maryland air quality monitoring plan reignites poultry battle. In: Delmarvanow.com. https://www.delmarvanow.com/story/news/local/maryland/2019/01/30/maryland-air-quality-monitoring-plan-reignites-poultry-battle/2712135002/. Accessed 11 Mar 2019

Myers B (2005) Sherman's Field Order No. 15. In: New Georgia Encyclopedia. https://www.georgiaencyclopedia.org/articles/history-archaeology/shermans-field-order-no-15. Accessed 6 Feb 2019

Naseem S, King A (2018) Ammonia production in poultry houses can affect health of humans, birds, and the environment—techniques for its reduction during poultry production. Environ Sci Pollut Res 25(16):15269–15293

National Pork Board (2019) Typical Market Pig Today – Pork Checkoff. In: Pork Checkoff. https://www.pork.org/facts/stats/consumption-and-expenditures/typical-market-pig-today/. Accessed 10 Feb 2019

NCC (2019) How broilers are marketed. The National Chicken Council. https://www.nationalchickencouncil.org/about-the-industry/statistics/how-broilers-are-marketed/. Accessed 20 Mar 2019

NCFH (2014) Poultry workers. [online] Ncfh.org. http://www.ncfh.org/uploads/3/8/6/8/38685499/fs-poultryworkers.pdf. Accessed 10 Mar 2019

Nicole W (2013) In: Ehp.niehs.nih.gov. https://ehp.niehs.nih.gov/doi/full/10.1289/ehp.121-a182. Accessed 10 Feb 2019

NRCS (1995) Animal manure management | NRCS. [online] Nrcs.usda.gov. https://www.nrcs.usda.gov/wps/portal/nrcs/detail/national/technical/nra/?cid=nrcs143_014211. Accessed 7 Mar 2019

O'Keefe J, Cordain L, Harris W et al (2004) Optimal low-density lipoprotein is 50 to 70 mg/dl. J Am Coll Cardiol 43:2142–2146. https://doi.org/10.1016/j.jacc.2004.03.046

Old Line Plate (2015) The Delmarva Chicken Festival. In: Old line plate. https://oldlineplate.com/the-delmarva-chicken-festival/. Accessed 20 Feb 2019

Pescatore A, Casey K, Gates R (2005) Ammonia emissions from broiler houses. J Appl Poult Res 14:635–637. https://doi.org/10.1093/japr/14.3.635

Pew (2016) Antibiotics and animal agriculture: a primer. In: Pewtrusts.org. https://www.pewtrusts.org/en/research-and-analysis/fact-sheets/2016/12/antibiotics-and-animal-agriculture-a-primer. Accessed 21 Mar 2019

Poli G, Biasi F, Leonarduzzi G (2013) Oxysterols in the pathogenesis of major chronic diseases. Redox Biol 1:125–130. https://doi.org/10.1016/j.redox.2012.12.001

Rasmussen S, Casey J, Bandeen-Roche K et al (2017) Proximity to industrial food animal production and asthma exacerbations in Pennsylvania, 2005–2012. Int J Environ Res Public Health 14:362. https://doi.org/10.3390/ijerph14040362

Ratnaike R (2003) Acute and chronic arsenic toxicity. Postgrad Med J 79(933):391–396

Reid W (1990) History of Avian medicine in the United States. X. Control of Coccidiosis. Avian Dis 34(3):509

Rosenblatt-Farrell N (2009) The landscape of antibiotic resistance. Environ Health Perspect 117(6):A244–A250. https://doi.org/10.1289/ehp.117-a244

Roundtree H (2014) Domesticated animals by early Virginia Indians, uses of. In: Encyclopediavirginia.org. https://www.encyclopediavirginia.org/Domesticated_Animals_During_the_Pre-Colonial_Era. Accessed 3 Feb 2019

Rouse P (1994) When Porkers Ruled Hog Island. In: Dailypress.com Feb. 6. https://www.dailypress.com/news/dp-xpm-19940206-1994-02-06-9402030334-story.html. Accessed 10 Feb 2019

Rude E (2016) Secrets of the chicken nugget: a surprising history. In: Time. http://time.com/4431334/history-chicken-nuggets/. Accessed 20 Mar 2019

Schwartz B, Kloner R (2009) How to save a life during a clinic visit for erectile dysfunction by modifying cardiovascular risk factors. Int J Impot Res 21:327–335. https://doi.org/10.1038/ijir.2009.38

Singer R (2015) Urinary tract infections attributed to diverse ExPEC strains in food animals: evidence and data gaps. Front Microbiol 6:28. https://doi.org/10.3389/fmicb.2015.00028

Smithfield Foods (2019a) Smithfield Foods, Inc. Meat Products | Smithfield Foods Company USA. In: Smithfieldfoods.com. https://www.smithfieldfoods.com/about-smithfield. Accessed 10 Feb 2019

Smithfield Foods (2019b) Smithfield Foods, Inc. Meat Products | Smithfield Foods Company USA. In: Smithfieldfoods.com. https://www.smithfieldfoods.com/integrated-report/2017/people/diversity-inclusion. Accessed 17 Feb 2019

Sternberg S (2020) Chicken plants – and the food supply – take center stage in Delaware's COVID-19 fight. In: US Word and News

Reports. https://www.usnews.com/news/healthiest-communities/articles/2020-05-05/chicken-plants-take-center-stage-in-delaware-coronavirus-fight. Accessed 15 May 2020

Spinrad P (2015) Super-sizing the chicken 1923-present – 19 February 2015. In: Upc-online.org. http://www.upc-online.org/industry/150219_super-sizing_the_chicken.html. Accessed 18 Feb 2019

SPLC (2013) Unsafe at these speeds. [online] Splcenter.org. https://www.splcenter.org/sites/default/files/Unsafe_at_These_Speeds_web.pdf Accessed 10 Mar 2019

Stokstad E, Jukes T (1950) Further observations on the "animal protein factor". Exp Biol Med 73:523–528. https://doi.org/10.3181/00379727-73-17731

Stoll M (2012) Industrial and agricultural interests fight back | Environment & Society Portal. In: Environmentandsociety.org. http://www.environmentandsociety.org/exhibitions/silent-spring/industrial-and-agricultural-interests-fight-back. Accessed 22 Mar 2019

Storey A, Athens J, Bryant D et al (2012) Investigating the global dispersal of chickens in prehistory using ancient mitochondrial DNA signatures. PLoS One 7:e39171. https://doi.org/10.1371/journal.pone.0039171

Sundin S (2018) Make it do - meat and cheese rationing in World War II. In: Sarahsundin.com. http://www.sarahsundin.com/make-it-do-meat-and-cheese-rationing-in-world-war-ii-2/. Accessed 17 Feb 2019

Tazzyman S, Bonhoeffer S (2014) Plasmids and evolutionary rescue by drug resistance. Evolution 68:2066–2078. https://doi.org/10.1111/evo.12423

Testa G, Staurenghi E, Zerbinati C et al (2016) Changes in brain oxysterols at different stages of Alzheimer's disease: their involvement in neuroinflammation. Redox Biol 10:24–33. https://doi.org/10.1016/j.redox.2016.09.001

UEP (2019) Facts & Stats - United Egg Producers. [online] United Egg Producers. https://unitedegg.com/facts-stats/. Accessed 8 Mar 2019

USBLS (2019) Employment by major industry sector: U.S. Bureau of Labor Statistics. [online] Bls.gov. https://www.bls.gov/emp/tables/employment-by-major-industry-sector.htm. Accessed 8 Mar 2019

USCensus (2019) New census data show differences between urban and rural populations. [online] The United States Census Bureau. https://www.census.gov/newsroom/press-releases/2016/cb16-210.html. Accessed 8 Mar 2019

USDA (2014) USDA ERS – financial risks and incomes in contract broiler production. [online] Ers.usda.gov. https://www.ers.usda.gov/amber-waves/2014/august/financial-risks-and-incomes-in-contract-broiler-production/. Accessed 9 Mar 2019

USDA (2016) USDA finalizes new food safety measures to reduce Salmonella and Campylobacter in Poultry. In: Usda.gov. https://www.usda.gov/media/press-releases/2016/02/04/usda-finalizes-new-food-safety-measures-reduce-salmonella-and. Accessed 27 Mar 2019

USDA (2018). Chickens and Eggs 2017 Summary. [online] Nass.usda.gov. https://www.nass.usda.gov/Publications/Todays_Reports/reports/ckegan18.pdf. Accessed 8 Mar 2019

USDA (2019a) National Agricultural Library | United States Department of Agriculture. In: Data.nal.usda.gov. https://data.nal.usda.gov/dataset/food-availability-capita-data-system/resource/6195a96b-c4c6-4593-b18e-0a5d6869601f. Accessed 4 Feb 2019

USDA (2019b) Hogs and Pigs. In: USDA Economics, Statistics and Market Information System. https://usda.library.cornell.edu/concern/publications/rj430453j?locale=en. Accessed 10 Feb 2019

USDA (2019c). Poultry Production in the United States. [online] Ers.usda.gov. https://www.ers.usda.gov/webdocs/publications/42203/13403_aib748b_1_.pdf?v=41055. Accessed 8 Mar 2019

USDA (2019d) The nationwide microbiological baseline data collection program: raw chicken parts survey. In: Fsis.usda.gov. https://www.fsis.usda.gov/wps/wcm/connect/a9837fc8-0109-4041-bd0c-729924a79201/Baseline_Data_Raw_Chicken_Parts.pdf?MOD=AJPERES. Accessed 27 Mar 2019

USDA (2019e) Pathogen reduction – Salmonella and campylobacter performance standards verification testing. In: Fsis.usda.gov. https://www.fsis.usda.gov/wps/wcm/connect/b0790997-2e74-48bf-9799-85814bac9ceb/28_IM_PR_Sal_Campy.pdf?MOD=AJPERES. Accessed 27 Mar 2019

USDA (2019f) Food composition databases show foods list. In: Ndb.nal.usda.gov. https://ndb.nal.usda.gov/ndb/search/list. Accessed 29 Mar 2019

USDAERS (2020) State fact sheets. In: Ers.usda.gov. https://www.ers.usda.gov/data-products/state-fact-sheets/. Accessed 12 May 2020

USEPA (1999) 09/15/99: Appeals Court Upholds Ruling Against Smithfield Foods for Polluting Virginia River. In: Archive.epa.gov. https://archive.epa.gov/epapages/newsroom_archive/newsreleases/c7a68726816ff7b3852567ef0053e790.html. Accessed 11 Feb 2019

USEPA (2013) Literature review of contaminants in livestock and poultry manure and implications for water quality. Office of Water (4304T) EPA 820-R-13-002Washington, DC

USEPA (2017) Laws and regulations that apply to your agricultural operation by farm activity | US EPA. [online] US EPA. https://www.epa.gov/agriculture/laws-and-regulations-apply-your-agricultural-operation-farm-activity#LivestockPoultryAquaculture. Accessed 10 Mar 2019

Vogelsong S, Bravender R (2020) COVID-19 cases keep climbing at Virginia poultry plants; some members of Congress seek better protections. In: https://www.nbc12.com. https://www.nbc12.com/2020/05/05/covid-cases-keep-climbing-virginia-poultry-plants-some-members-congress-seek-better-protections/. Accessed 15 May 2020

Wang Y, Lehane C, Ghebremeskel K, Crawford M (2009) Modern organic and broiler chickens sold for human consumption provide more energy from fat than protein. Public Health Nutr 13:400. https://doi.org/10.1017/s1368980009991157

Whiley H, Ross K (2015) Salmonella and eggs: from production to plate. Int J Environ Res Public Health 12:2543–2556. https://doi.org/10.3390/ijerph120302543

Wierup M, Engström B, Engvall A et al (1995) Control of Salmonella enteritidis in Sweden. Int J Food Microbiol 25:219–226. https://doi.org/10.1016/0168-1605(94)00090-s

Zhong V, Van Horn L, Cornelis M et al (2019) Associations of dietary cholesterol or egg consumption with incident cardiovascular disease and mortality. JAMA 321:1081. https://doi.org/10.1001/jama.2019.1572

Zuidhof M, Schneider B, Carney V et al (2014) Growth, efficiency, and yield of commercial broilers from 1957, 1978, and 2005. Poult Sci 93(12):2970–2982. https://doi.org/10.3382/ps.2014-04291

Part IV

Consequences of and Alternatives to the Standard American Diet: Human and Ecosystem Health

Instead of Eating Fish: The Health Consequences of Eating Seafood from the Chesapeake Bay Compared to Other Choices

Benjamin E. Cuker

If you live near the Bay,
people may think that you fish all day.

Eating what you catch seems just fine,
and goes so well with white wine.

But do you really need to eat that fish
or should you choose a different dish?

Seafoods are often contaminated with toxins like mercury and PCBs,
would it be better to choose food from plants and trees?

You don't really need all that extra animal protein,
so would it be better to get some fiber from eating bean?

Fish, crabs, oysters and mussels deep fried in batter,
eating them will just make you fatter.

Even boiled or broiled the seafood is still animal flesh
and eating it may not with a healthy lifestyle mesh.

Better to choose the flax, grains, nuts, vegetables and fruits for your next meal,
in terms of health it will be the best deal.

Abstract

Seafood is promoted as an important part of a healthy diet. But is it? The science shows that animal seafood consumption is associated with poorer health-outcomes, including diabetes, heart disease and cancer. Seafood from the Chesapeake Bay is also a source of dietary toxins including heavy metals and organic pollutants, as well as cholesterol. Seafood is praised as an excellent source of protein, yet this nutrient is a false idol of nutrition, as most Americans eat twice the quantity they need in their diets. Consumption of animal protein is associated with several adverse health outcomes including cancer and kidney disease. Seafood is rich in omega-3 fatty acids that are promoted as preventative of coronary artery disease, yet findings from extensive studies don't support the efficacy of taking fish oil supplements. Seafood satiates eaters while providing no dietary fiber, a nutrient critical for health and most often missing in the Standard American Diet. The health benefits of choosing whole plant based foods that are widely available in the Chesapeake Bay region as an alternative to seafood consumption is discussed.

Keywords

Seafood · Health · Protein · Omega-3 fatty acids · Heart disease · Cancer

Seafood Mariners navigating the Chesapeake Bay may encounter someone fishing for some sort of seafood every day of the calendar year, as there is always at least one species in legal season for taking. Even in the actual absence of the fisher, their trace still informs most of the Bay's waters.

B. E. Cuker (✉)
Department of Marine and Environmental Science, Hampton University, Hampton, Virginia, USA
e-mail: Benjamin.cuker@hamptonu.edu

B. E. Cuker (ed.), *Diet for a Sustainable Ecosystem*, Estuaries of the World, https://doi.org/10.1007/978-3-030-45481-4_13

Buoys bob at the surface, tethered to traps for crabs. Flags mark the ends of gill nets. Long rows of wooden poles emerge perpendicular from the shoreline, forming the leaders of pound nets used to trap migrating fish. Considering the commercial landings between 1950 and 2016 one can appreciate the enormity of the fishing industry. In that interval commercial fishing on the Chesapeake yielded 17.5 billion metric tons with a value of $8.2 billion (NOAA 2018). This number excludes blue crabs and recreational fishing, so far underestimates the actual landings and their economic values.

This chapter considers the dietary choices offered by the Chesapeake Bay region from the perspective of comparisons with animal seafood. Best known for its crabs, oysters and fin fish, the region offers many other potentially healthier choices for the consumer to consider.

Promoting Seafood as Health Food Seafood from the Chesapeake brings nutrition, recreation, employment, and profit to many. But is the consumption of that seafood good for the health of those that eat it? This question may seem surprising since seafood that is free of contamination or pathogens is often considered a part of a healthy diet. That notion is promulgated by the Virginia and Maryland agencies charged with promoting the products of their respective seafood industries. In one publication the Virginia Cooperative Extension Service extols the health benefits of seafood; "Seafood tastes good, is low in saturated fat, is an excellent source of protein. . . ." That publication goes on to say, "Seafood, as part of a healthy diet, is associated with a lower risk of heart disease, obesity, and some types of cancer. It has also been shown to improve pregnancy and birth outcomes. These health benefits are largely due to the omega-3 fatty acids found in fish (Villalba et al. 2018)." The extension service promotes a minimum of two, 99 g (3.5 oz) servings of seafood per week.

Note that the Virginia Cooperative Extension Service's claim of seafood consumption being associated with protection from heart disease, obesity and cancer places it in the category of a *functional food*. Health benefits derived from consumption of a dietary item that go *beyond its basic nutritional content* allows its classification as a functional food. Typically, functional foods are associated with the prevention and treatment of chronic disease (Hasler 2002).

Maryland's promotional webpage for seafood spends about twice as much space on health concerns (chemical containments and pathogens) as it does on potential health benefits. However, they state, "It is agreed upon by the medical community that eating seafood 2–3 times a week has many unique health benefits. The seafood that comes from the Chesapeake Bay and other Maryland waterways contains many vitamins, minerals, and nutrients that are essential for human health." That page lists specific nutritional benefits associated with blue crabs, oysters and crabs (high quality protein, various vitamins, minerals and omega −3 fatty acids) (Maryland Seafood 2018).

The USDA (U.S. Department of Agriculture) and USDHHS (U.S. Department of Health and Human Services) recommends a minimum consumption of 227 g of seafood per week for all individuals two years of age or older (for a diet of 2000 calories per day, Health.gov 2018). Despite that 80% of Americans responding to a survey reported to having eaten seafood in the preceding month, depending upon age and sex, 82–92% failed to meet the minimum rate of recommended consumption set by the USDA and USHHS (Jahns et al. 2014). According to the USDA, most Americans eat only half the recommended amount of animal seafood (USDA 2018b). Understand that the USDA has conflicting missions. It is supposed to promote nutrition and food safety and simultaneously help market the full array of agricultural goods (including from fisheries) produced in the US (USDA 2018c). Despite universal agreement that tobacco use is unhealthy, the USDA actively promotes marketing of this dangerous agricultural product (USDA 2018d). So the USDA's promotion of seafood consumption is informed by its marketing mission perhaps as much as by nutrition science.

The promotion of seafood as healthy seems successful. In a national survey of 1062 people, 78% of respondents agreed with the statement, "I think seafood is good for your health," 19% were unsure, and only 3% disagreed. Yet when asked if pregnant women should eat seafood, only 21% agreed, 41% were unsure and 38% disagreed (Hicks et al. 2008). This suggests that the 57% of respondents who thought seafood is healthy only for non-pregnant individuals aren't troubled by consuming food unacceptable for a developing fetus. That group probably sees the health benefits of seafood as overriding potential detriments for all but the fetus. The two most often cited benefits of seafood consumption are the high contents of *high quality* protein and healthful fatty acids. Let's first consider the issue of protein.

Protein from Seafood The lean human body consists of about 62% water, 16% lipids, 16% protein, and 6% minerals (Weininger et al. 2018). The protein portion is assembled from 20 different types of amino acids (Fig. 1). Humans can synthesize 11 of these amino acids. The remaining nine are called *essential* amino acids, as they must be obtained from the diet. Foods with amino acid profiles most similar to that of humans are called *high quality* sources of protein. Animal seafood (fish and shellfish) provides high quality protein since it contains all of the essential amino acids and a general amino acid profile close to that of humans. That means that a

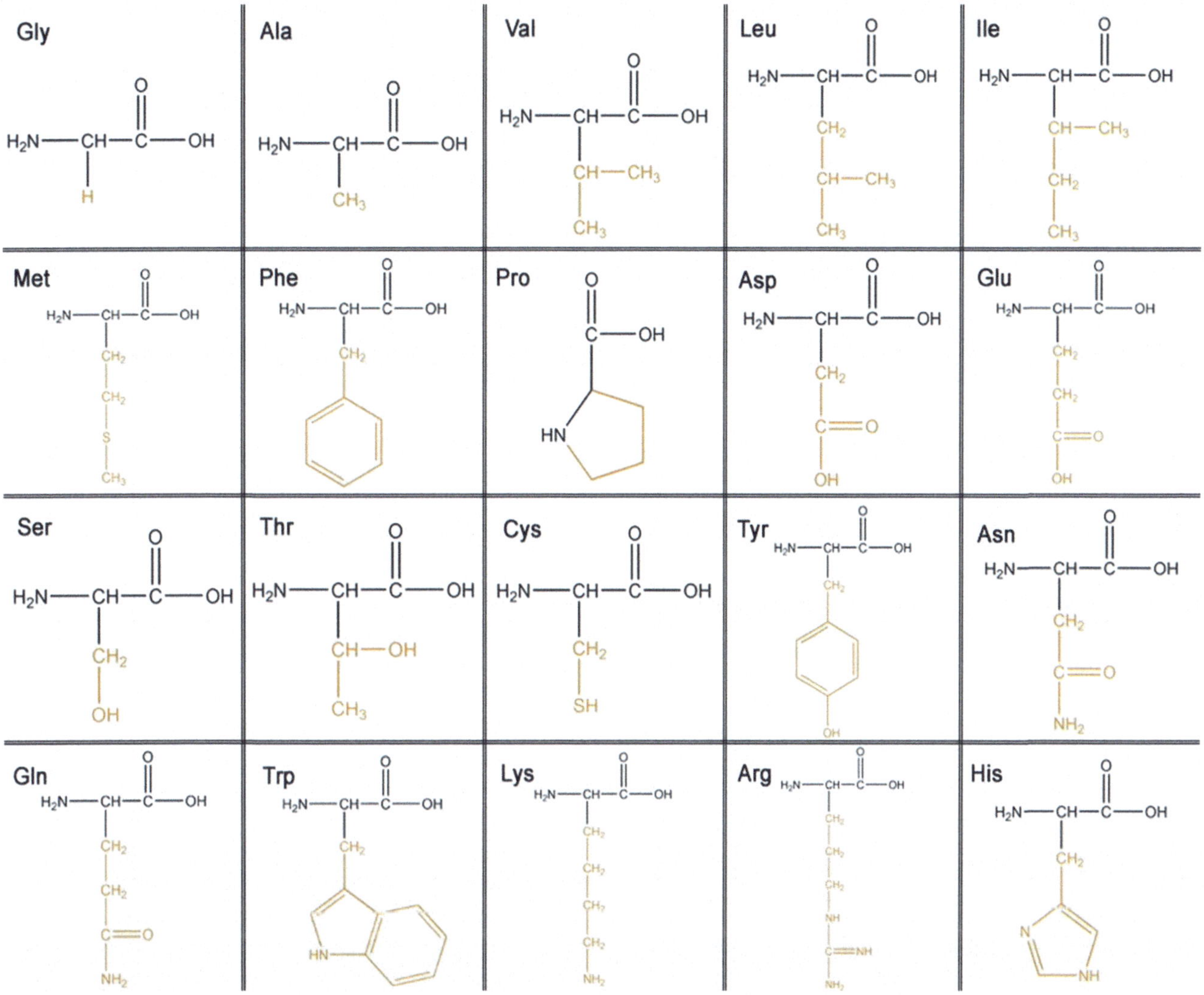

Fig. 1 The twenty amino acids found in human protein. Created by Dalibor Bosits. https://commons.wikimedia.org/wiki/File:Tablica_aminokiselina.jpg. The essential amino acids that can't be synthesized by humans are histidine (His), Isoleucine (Iso), Leucine (Leu), Lysine (Lys), Methionine (Met), Phenylalanine (Phe), Threonine (Thr), Tryptophan (Try), and Valine (Val). The others are Arginine (Arg), Cysteine (Cys), Glutamine (Gln), Glycine (Gly), Proline (Pro), Tyrosine (TYR), Alanine (Ala), Aspartic acid (Asp), Asparagine (Asn), Glutamic acid (Glu), and Serine (Ser)

diet rich in fish will assure sufficient quantities of the amino acids needed for human health. *But is lack of high quality protein really a concern for human health?*

The enthusiasm for protein as the most important macronutrient traces to the nineteenth century, when it was discovered by Dutch chemist, Gerhard Mulder (Mulder 1849). Mulder took the name from the Greek word *proteios*, meaning of prime importance. In 1890 the USDA recommended 110 g of dietary protein per day for a "working man," twice the current suggested level (Carpenter 1986). In 1932 Dr. Cicely Williams described the disease of infant and child malnutrition she saw in the Gold Coast of Africa, called Kwashiorkor. She attributed this to lack of dietary protein, but spent the last 20 years of her life refuting her own early conclusion (Waterlow 1984). However, her initial erroneous assumption spurred the agencies of the United Nations to try to close a perceived "protein gap" for the developing world, establishing the UN Protein Advisory Group in 1955 (Semba 2016). The US dairy industry gladly directed its excess production to the effort of developing special protein rich dietary supplements to alleviate the supposed "protein crisis" (Carpenter 1986). However, McLaren (1974, 2000) exposed what he called "The Great Protein Fiasco," demonstrating the fallacy of the dogma of protein deficiency. Essentially, the supposed protein gap disappeared when the UN Food and Agricultural Organization (FAO) revised its protein recommendations down to the level of traditional diets. That said, certain populations in the developing world that suffer calorie insufficiency may benefit from increased protein consumption that would come with eating more food (Semba 2016).

The human diaspora from Africa led to a remarkable dispersal to all corners of the globe, save for Antarctica. This meant the ability to adapt to a wide range of climates and therefore diets. Despite disparity in location, all humans begin life with essentially the same diet, breast milk (at least until the invention of baby formula in the nineteenth century). The protein content of breast milk is about 8.5 g/l or less than 1% by mass, the lowest of any mammal tested. Contrast this to the concentrations of protein found in milk of baboon (11.5 g/l), cow (33.6 g/l), elephant (37.1 g/l), sheep (54.1 g/l), cat (75.7 g/l) and rat (86.9 g/l) (Davis et al. 1994). Millions of years of evolution means that breast milk is the perfect food for a fast growing infant. While protein needs increase as the infant develops into a child and an adult, the low amino acid content of breast milk suggests that humans don't require particularly high levels of this macronutrient.

The very large Adventist Study 2 (n = 71,757) examined the diets of omnivores (Standard American Diet) and vegans (strict vegetarians). *Both groups consume about twice the recommended level of protein (42 g/d), 78 g/d for omnivores and 75 g/d for vegans* (Rizzo et al. 2013). While total protein intake differed little between the two groups, vegans showed better health outcomes. Omnivores presented an obesity rate of 33.3% and an average Body Mass Index (BMI) of 28.6, compared to vegans with 9.4% obesity rate and a BMI of 24.1. A BMI above 25 suggests overweight (Rizzo et al. 2013).

It was in 1971, the era of the "protein gap" that Francis Moore Lappé wrote her first edition of "Diet for a Small Planet," which shattered the myth that insufficient agricultural production drove widespread food shortages. She showed that the world food system could easily feed the human population if we fed crops directly to people, rather than first passing then through the guts of farm animals. The energy saved by eliminating one trophic level in the human diet would more than make-up any gap between food production and need. However, in the course of destroying the myth of insufficient food production, she inadvertently created a new myth, that of plant protein inferiority due to a dearth of essential amino acids. In that first edition she devoted much space to the concept of complementary proteins (Bressani 1988). This involved combining various plant and sometimes dairy based foods to achieve an amino acid profile similar to that of a chicken egg, a high quality protein food which she selected as an arbitrary standard. Lappé proposed combinations of legumes (beans, lentils, or peas) with whole grains to provide the necessary amino acid profile. Subsequently she revised her approach, understanding that a diet of whole plant foods that met calorie sufficiency also provided amino acid sufficiency (Lappé 2011; Young and Pellett 1994). Humans contain a large pool of circulating essential amino acids that compensates for transitory dietary insufficiency (Munro 1970). Yet the myth of the need for protein complementation persists, appearing in textbooks and in scientific journals (McDougall 2002). To be clear, while all plant proteins include the essential amino acids, they do differ from animal protein sources in their amino acid profiles, with plants containing less of the sulfur-amino acids (methionine and cysteine).

Since most Americans get twice the recommended level of protein in their diets (Rizzo et al. 2013), it is difficult to argue for the efficacy of eating seafood on the account of this macronutrient. Eating excessive protein from seafood is at best neutral to health, and possibly detrimental. This is for five reasons. *The excess protein: 1. may create direct health problems, 2. leaves less room for other nutrients that are lacking in the diet, 3. is accompanied by natural substances that harm health, 4. is accompanied by toxic contaminants. 5. is often cooked with the addition of unhealthy ingredients and with methods that produce toxins.* Let's consider each of these five scenarios.

1. The direct detriments of excess protein from animal seafood Proponents of animal seafood consumption often deconstruct this dietary choice to focus on its high protein content. Addressing the efficacy of consuming that protein means attempting to understand the effects of protein independent of everything else that comes packaged with it, a difficult task at best. After all, consumers sit down to a dinner of the whole foods of fish, oysters and crabs, not pills, powders or potions of isolated specific nutrients extracted from the seafood. We may use reductionism to try to understand what we are eating, but ultimately we dine holistically (except for those consuming fish oil or similar supplements). So the following discussion includes some factors attendant to the protein in seafood.

Excess animal seafood protein means excess calories. The calories come from the protein itself (4 kcal/g) and associated fatty tissue (9 kcal/g). Since 71% of Americans are overweight or obese, most don't need the added calories of seafood in their diets. This is especially true in the states around the Chesapeake Bay where obesity rates have gone from 12–13% in 1990 to 28–37% in 2016 (CDC 2017). The calories offered by animal seafood from the Chesapeake Bay range from 86 kcal/100 g for flounder to 197 kcal/100 g for American shad. This makes seafood a better choice than either 70% lean ground beef (332 kcal/100 g) or bacon (431 kcal/100 g), and marginally better than chicken (107–172 kcal/100 g) or ham (125 kcal/100 g). However, the 85–92 kcal/100 g offered by beans is about 65% lower than that gotten from the same mass of animal seafood (Table 1).

A ratio of protein to calories is a measure of efficiency of delivery of that macronutrient. For the animal seafood of the Chesapeake Bay that ranges from 0.09 for the oily American

Table 1 The nutrition content per 100 g (wet mass) of various food substances found in the Chesapeake Bay food system

Item	Energy (kcal)	Protein g	Total Sat. Fat (g)	Choleste. (mg)	Total fat (g)	Fiber (g)	Carbohydrate (g)	Protein/ Kcal	Fiber/ kcal
Atlantic croaker	104	17.8	1.09	61	3.17	0	0	0.17	0.00
Blue fish	124	20.0	0.92	59	4.24	0	0	0.16	0.00
Striped bass	97	17.7	0.51	80	2.33	0	0	0.18	0.00
Flounder	86	16.2	0.54	56	2.37	0	0	0.19	0.00
American shad	197	16.9	3.12	75	13.8	0	0	0.09	0.00
Oyster	107	12.5	1.79	62	4.46	3.6	3.6	0.12	0.03
Hard shell clam meat	153	25.88	0	65	1.76	0	4.71	0.17	0.03
Blue crab meat	119	23.81	0.20	107	1.19	0	0	0.20	0.00
Chicken breast with skin	172	20.9	2.66	64	9.25	0	0	0.12	0.00
Chicken breast no skin	107	23.2	0	58	0.89	0	0	0.22	0.00
Ground beef 70% lean	332	14.4	11.75	78	30	0	0	0.04	0.00
Cured Smithfield honey ham	125	17.9	0	54	3.6	0	7.14	0.14	0.06
Bacon Smithfield hickory smoked	471	29.4	14.71	118	41.8	0	0	0.06	0.00
Chicken egg	155	12.58	3.27	373	10.6	0	0.1	0.08	0.00
Whole milk	62	3.33	1.88	10	3.33	0	5.0	0.00	0.00
Corn, whole kernel	111	3.33	0	0	1.11	1.1	23.3	0.03	0.21
Soy bean, organic	92	8.46	0.77	0	4.62	2.3	6.9	0.09	0.08
Navy beans	85	9.38	0	0	0.77	5.4	15.4	0.11	0.18
Peanuts boiled	278	13.89	4.17	0	22.22	8.3	8.33	0.05	0.03
Whole wheat flour	333	13.33	0	0	1.67	13.3	70.00	0.04	0.21
Millet, raw	378	11.02	0.72	0	4.22	8.5	72.85	0.03	0.19
Oats, organic steel cut	375	12.5	0.5	0	6.25	10.0	67.5	0.03	0.18
Ground flax seed, organic	429	15.38	1.43	0	10.71	28.6	50.0	0.04	0.12
Yellow squash	16	1.0	0	0	0	1	3	0.06	0.19
Raw kale	35	2.92	0.18	0	1.49	4.1	4.42	0.08	0.13
Raw broccoli	34	2.82	0.11	0	0.37	2.6	6.64	0.08	0.20
Raw spinach	23	2.86	0.06	0	0.39	2.2	3.63	0.12	0.16
Raw sweet potato	86	1.57	0.18	0	0.05	3.0	21.2	0.02	0.25
Raw beets	43	1.61	0.03	0	0.17	2.8	9.56	0.04	0.22
Raw apple with skin	52	0.26	0.03	0	0.17	2.4	13.81	0.01	0.27
Raw strawberries	32	0.67	0.02	0	0.30	2.0	7.68	0.02	0.24
Raw walnuts	700	16.67	1.5	0	66.67	10.0	10.0	0.02	0.01

The last two columns of the table show the ratio of protein and fiber to calories for each food. Choleste. abbreviates total cholesterol. Data is from the USDA (2018a). The scientific names of the foods are: Atlantic croaker, *Micropogonias undulatus*, Blue fish, *Pomatomus saltatrix*, Striped bass, *Morone saxatilis*' Flounder, *Pleuronectes americanus*, American shad, *Alosa sapidissima*, Oyster, *Crassostrea virginica*, Hard shell clam, *Mercenaria mercenaria*, Blue crab, *Callinectes sapidus*, Chicken, *Gallus gallus domesticus*, Cattle (beef), *Bos Taurus*, Pig (ham and bacon), *Sus scrofa domesticus*, Corn, *Zea mays*, Soybean, *Glycine max*, Navy beans, *Phaseolus vulgaris*, Peanuts boiled, *Arachis hypogaea*, Wheat, *Triticum sp.*, Millet, *Pennisetum sp.*, *Panicum miliaceum*, Oat, *Avena sativa*, Flax, *Linum usitatissimum*, Yellow squash, *Cucurbita pepo*, Raw kale, *Brassica oleracea, var. acephala*, Broccoli, *Brassica oleracea, var. italic*, Spinach, *Spinacia oleracea*, Sweet potato, *Ipomoea batatas*, Beet, *Beta vulgaris*, Apple, *Malus pumila*, Strawberries, *Fragaria sp.*, Walnuts, *Juglans sp.*

shad to 0.19 for the lean winter flounder. The ratios for soybeans (0.09) and navy beans (0.11) are comparable to the American shad (Table 1). This ratio is important for populations needing more protein in their diets, which isn't the case around the Chesapeake Bay nor the nation at large.

About 8–10% of the US population develops kidney stones, and that rate is increasing (Scales et al. 2012). The chance of developing kidney stones is correlated to the consumption of animal protein, including fish (Robertson et al. 1979; Tracy et al. 2014; Turney et al. 2014).

Consumption of protein from animals, but not plants is linked to kidney disease. Diseased kidneys challenged with fish protein showed excessive glomerular filtration rate (GFR), with no effect from the same levels of protein from beans (Nakamura et al. 1989). Often switching patients to a vegan diet restores kidney function (Barsotti et al. 1990). Animal protein tends to lower the pH of the body, which promotes the progression of chronic kidney disease (CKD). Moving to a plant-based diet reverses acidosis and checks the progression of CKD (Uribarri and Oh 2012). The efficacy of the vegan diet on kidney function may go beyond the particular effects of animal protein and include benefits related to increased fiber intake, isoflavones from soy that reduce vasodilation, reduced sodium that lowers blood pressure, and elimination of animal fats (Moorthi et al. 2017). The preceding sentence points to the difficulty of applying the reductionist approach to understanding the health consequences of consuming protein from seafood, as that macronutrient is never consumed in isolation.

Eating excess animal protein increases incidence of human cancers. Insulin Like Growth Factor 1 (IGF1) is a growth hormone produced primarily by the liver. It promotes cell proliferation, a critical process for growth of infants and children, and for the regular replacement of aging cells. However, excess IGF1 is linked to tumor growth and the spread of cancer to new tissues (Dewell et al. 2007; Yang et al. 2011). Animal protein sources, including seafood, stimulates the liver to produce excess IGF1. Serum concentrations of IGF1 are 13% lower in vegan women than meat-eaters (Allen et al. 2002).

Animal protein is implicated in the first two stages of cancer development; *initiation* and *promotion*. Cancers originate from changes in the genetic code of a cell. Typically, an agent such as a chemical or virus *initiates* the mutation in DNA. If that mutation is transmitted in cell division, and if it causes subsequent unregulated production of daughter cells, cancer is established. Within the cell, enzymatic systems may repair the damaged DNA. Additionally, the immune system can destroy the cancerous cells and prevent further development of the disease. If either defense system fails, the cancerous cells multiply and form visible tumors in the process of *promotion*. Promotion (the second stage of cancer development) depends largely on the balance of dietary factors called promoters and anti-promoters. Promoters feed the developing tumor while anti-promoters check its growth. If the balance shifts in favor of anti-promoters, promotion is reversed. Animal protein appears to act as a cancer promoter. *Progression* is the third and final state of cancer development. The original tumor sheds cancerous cells that subsequently invade new tissues to establish new centers of cancer. This *metastasis* of the cancer makes treatment more difficult. Progression is often fatal unless countered with chemotherapy or other treatments that kill the pockets of cancer distributed in the body. Consumption of animal protein (casein from dairy) in excess of 10% of total calories is strongly linked to increased rates of cancer in animal studies (Campbell 2017, extensive review of topic in Campbell and Campbell 2016).

Some research links fish and fish protein to certain cancers. Breast cancer of the estrogen-receptor positive type is positively correlated to consumption of fish for post-menopausal women. That study eliminated cooking method and fat content as the causative agent, leaving protein as the possible driver (Stripp et al. 2003). Consumption of salted fish seems to cause nasopharynx and possibly stomach cancer (Secretan et al. 2009). A diet rich in seafood is linked to increase rates of thyroid cancer (Glattre et al. 1993). Yet other work finds seafood consumption protective against a variety of cancers, seemingly due to the presence of omega-3 fatty acids (Kaizer et al. 1989; Fernandez et al. 1999), discussed below.

Consumption of protein from fish (as well as from poultry, eggs and red meat) increases the risk for developing type-2 diabetes. Replacement of 5% of energy intake from animal proteins with plant proteins decreases risk of developing type-2 diabetes by 20–25% (Comerford and Pasin 2016). However, the type-2 diabetes promoting effects of the protein from the fish are difficult to separate from those also caused by persistent organic pollutants (Ruzzin 2012).

2. Eating animal seafood leaves less room in the diet for consumption of other healthier items. As we eat, our stomachs fill, causing the process of satiation which leads to the feeling of fullness called satiety. Satiation tells us to stop eating and satiety suppresses further consumption. Satiation and satiety are complex physiological processes influenced by sensory, cognitive, cultural and psychological factors. The brain, stomach, intestines, pancreas, adipose tissue, blood sugar concentration, and various hormones all play a role in satiation and satiety (Benelam 2018). These processes clearly evolved to optimize food consumption in a world very different from today. For most of our history we lived with periods of food scarcity. So our system of satiation probably evolved to match the rate of gut passage to maximize our ability to extract nutrition from our food. An environment of food scarcity wouldn't provide the evolutionary drive for prevention of obesity. Our system of satiation is no match for the modern era of super abundance and the calorie rich Standard American Diet (Benelam 2018).

The three most important factors determining the satiety of a food item are its contents of water, protein, and fiber (Holt et al. 1995). The high protein content of fish makes it one of the most satiating foods, outranking beef (Borzoei et al. 2006). Highly satiating foods reduce the likelihood of consuming other subsequent items in a meal. This may be a useful strategy for weight reduction and control if the satiating food is low in caloric density. However the caloric density of fish is three to four times that of most fresh vegetables (Table 1), which also have high satiety scores (Holt et al. 1995). Achieving satiation with seafood leaves little room in the meal for foods that offer a wide variety of nutrients beyond the protein, fat, and minerals found in seafood. As noted above, most people in the US and the Chesapeake Bay ecosystem consume twice the protein they need. *Yet the vast majority of Americans are deficient in fiber intake.* Only 3% of Americans meet the USDA recommended daily intake of fiber (25–38 g/d, depending on age and gender), most consuming about 16 g/d. African Americans consume significantly less than the average at 13 g/d, and Mexican-Americans significantly more at 18 g/d (Clemens et al. 2012; King et al. 2012; Hoy and Goldman 2014). Obese people report the lowest daily fiber intake (King et al. 2012). *There is a fiber gap, not a protein gap in the US and the Chesapeake Region.*

The American Association of Cereal Chemists (AACC 2001) defines dietary fiber as .... "the remnants of the edible part of plants and analogous carbohydrates that are resistant to digestion and absorption in the human small intestine with complete or partial fermentation in the human large intestine. It includes polysaccharides, oligosaccharides, lignin and associated plant substances. Dietary fiber exhibits one or more of either laxation (fecal bulking and softening; increased frequency; and/or regularity), blood cholesterol attenuation, and/or blood glucose attenuation."

Plants and fungi (mushrooms and nutritional yeast) produce fiber to form their cell walls. This provides structural integrity and is the reason trees like the tulip poplar (*Liriodendron tulipifera*) can reach heights of 46 m (150 feet) above the shores of the Chesapeake Bay. Cellulose is a major constituent of dietary fiber, and the primary component of wood. Cellulose and starch are polymers of the same simple sugar, glucose. The two polymers differ in how the starch monomers link to each other. The alpha bond of starch aligns the glucose molecules in the same direction. The beta bonds of cellulose cause the monomers to be rotated 180 degrees at each link. This provides the structural strength lacking in starch. However, breaking those beta bonds requires the enzyme cellulase. This is produced only by microbes that live freely in natural waters, sediments and soil, and those that reside in the guts of herbivores and detritivores like menhaden, termites, deer, and cows (Deegan et al. 1990). The cellulase from these microbes enable herbivores to access the energy stored in cellulose. Since humans lack the mutualistic bacteria in their guts that produce cellulase, they are unable to digest the cellulose and release the monomers of glucose for direct uptake in the small intestine. Yet, cellulose and the other constituents of dietary fiber do nourish the microbiome that lives in the lower gut, producing an acid environment in the colon that is protective from cancer, and is active in reducing inflammation and providing support for the immune system (Sonnenburg and Sonnenburg 2014).

The dearth of fiber in the standard American diet links to numerous diseases. Sufficient fiber (15 g/1000 kcal intake) is associated with reduced rates of; colon, breast cancer and prostate cancer, diabetes, coronary heart disease, stroke, obesity, diverticulitis, constipation, and Alzheimer's disease (Dewell et al. 2007; Martins and Fernando 2014; Greger and Stone 2015). The early thinking on fiber and diet emphasized its role as a bulking agent in formation of feces. More recently fiber is recognized for its role in sequestering cholesterol and containments, and production of protective factors by the microbiome in the lower gut (Burkitt 1979; Ito et al. 1987; Possemiers et al. 2011; Mudgil and Barak 2013; Sonnenburg and Sonnenburg 2014).

As with protein, fiber comes packaged in food that contains other important nutrients, so parsing the effect of fiber is difficult. The studies demonstrating the importance of dietary fiber in protecting against and treating the diseases of affluence really show the efficacy of eating whole plant foods that contain the fiber and all of the associated nutrients, rather than just that of the fiber in isolation (Eastwood and Kritchevsky 2005).

Consider the ratio of fiber to caloric content for typical items in the Chesapeake Bay food system. Most animal seafoods lack fiber altogether. It appears only in small amounts in oysters and hard-shell clams (0.03 g fiber per kcal, Table 1). The small amount of fiber in the oysters and clams is due to the algae and detritus retained in the guts of these mollusks. The highest efficiency of fiber deliver with regard to calories comes from whole fruits and vegetables, with ratios typically of 0.20 and higher. While whole grains, legumes and nuts offer similar levels of fiber per 100 g to that provided by fruits and vegetables, the ratio of fiber to calories is lower (0.03–0.21). This is due to the substantial caloric content of starch and or oils found in the whole grains, legumes and nuts (Table 1).

Fruits, Vegetables, Whole Grains, Legumes, and Nuts as Functional Foods Fruits, vegetables, whole grains, legumes and nuts that contain dietary fiber offer a host of other phytonutrients, mostly absent from animal seafood. These include vitamins, some minerals, antioxidants, and anti-cancer agents. As such, these groups are replete with functional foods that offer protection from chronic disease (Hasler 2002). To illustrate the health benefits of these whole plant foods, consider outcomes associated with some representatives from each group.

Fruits and Vegetables Fruit and vegetable consumption reduces rates of mortality and morbidity. Increased consumption of fruits and vegetables is correlated with longevity in a dose-response fashion. In a large study (N = 71,706) those with diets absent of fruits and vegetables lived 3 years shorter and had 53% higher mortality rates than others that consumed five servings a day. Considering the separate effects of each category, those who never consumed fruit lived 19 months shorter than those that ate just one fruit/day, and those who never consumed vegetables lived 32 months shorter than others who ate three servings/day (Bellavia et al. 2013). Higher expenditures for fruit and vegetables also predicts lower mortality rates for the elderly (age 65 and older, Lo et al. 2012).

Table 2 Consumption of fruits and vegetables in the US and the jurisdictions that comprise at least a portion of the Chesapeake Bay watershed

Location	Times Fruit Eaten Daily	Times Vegetables Eaten Daily	% Meeting Recomm. for Fruit	% Meeting Recomm. for Vegetables	% Population with Asthma
US	1.0	1.7	13.1	8.9	8.3
Delaware	1.0	1.5	12.8	7.5	9.2
District of Columbia	1.1	1.8	15.2	9.2	10.6
Maryland	1.1	1.7	13.2	8.4	8.8
New York	1.1	1.7	15.5	8.8	9.9
Pennsylvania	1.1	1.6	12.7	7.5	10.2
Virginia	1.1	1.7	13.4	8.8	7.9
West Virginia	1.0	1.6	7.7	6.6	10.8

Data from Moore and Thompson (2016) and CDC Asthma (2018)

On a global basis, correcting the low consumption rate of fruits and vegetables would reduce heart disease by 31%, stroke by 19%, and cancers of the stomach by 19%, esophagus by 20%, lung by 12% and colorectal region by 2% (Locke et al. 2005). Fruits and vegetables are rich in fiber (benefits discussed above), antioxidants and other phytonutrients, all of which appear to protect against various types of disease. Irrespective of the underlying mechanisms, consumption of fruits and vegetables is linked to better health outcomes (Bellavia et al. 2013).

Consider the role of fruits and vegetables in regulating asthma, an important disease afflicting residents of the US and the Chesapeake Bay area. Currently 8.3% of US residents suffer from this disease that restricts the ability to breath (CDC 2017). The rates are higher than the national average in all of the jurisdictions comprising the Chesapeake Bay watershed (Table 2). While asthma is linked to poor air quality, a diet rich in fruits and vegetables proves beneficial in reducing symptoms. Bronchial asthma patients placed on vegan diets showed strong reduction in symptoms that allowed withdrawal of medication in almost all cases (Lindahl et al. 1985). Diet manipulations demonstrated that adult patients eating five servings of vegetables and two servings of fruit per day were 2.26 times less likely to exacerbate (have an asthma attack) as those eating less than two serving of vegetables and one serving of fruit per day. Although researchers point to the antioxidants provided by fruits and vegetables as central to the mechanism of protection, treatment with isolated antioxidant supplements showed no response, indicating the efficacy of the whole food (Wood et al. 2012). Lung function in asthmatic children is also improved by consumption of fruits and vegetables (Chatzi et al. 2007; Cook et al. 1997).

The role of seafood consumption in asthma is mixed. Some studies show oily fish consumption to reduce incidence of asthma, linking this to anti-inflammatory properties of omega-3 fatty acids (Yang et al. 2013). Others indicate exposure to seafood is linked to causing asthma symptoms for a portion of the population. About 2.3% of the US population reports physician diagnosed allergies to seafood (Sicherer et al. 2004). Asthma is common in the complex of symptoms displayed by those with fish allergies (ACAAI 2018). Airborne allergens from seafood also present an occupational hazard, causing asthma for some workers (Lopata and Jeebhay 2013).

Despite the demonstrated benefits of consuming fruits and vegetables, most Americans fail to meet recommended daily requirements, 76% are deficient in fruit intake and 87% miss the mark for vegetables. This is also true for the residents of the Chesapeake Bay area (Moore and Thompson 2016, Table 2).

Legumes According to a study that scored a variety of whole plant foods, legumes provide the least expensive way to get the benefits of phytonutrients. The dry seeds of legumes, called pulses have the highest density of nutrients. Beans, peas and lentils are examples of pulses. Dried beans yielded the highest nutrient density (nutrient per calorie) based on six nutrients: dietary fiber; potassium; magnesium; and vitamins A, C, and K (Drewnowski and Rehm 2013). In addition to high nutrient content, legumes are functional foods, protecting against chronic disease. Consider just a few of the demonstrated benefits of eating pulses. Consumption of legumes is associated with increased longevity and decreased morbidity due to cardio-vascular disease (Nagura et al. 2009). A minimum of four servings of beans per week reduces coronary heart disease 22% (Bazzano et al. 2002). A single serving of legumes per day, significantly reduces blood pressure in both hypertensive and non-hypertensive individuals (Jayalath et al. 2013; Leiter et al. 2013), significantly lowers the LDL cholesterol in comparison to controls (Ha et al. 2014), reduces chances of a myocardial infarction (MI, heart attack) by 38% (Kabagambe et al. 2005) and in eight weeks reverses vascular impairment (atherosclerosis) due to peripheral artery disease (Zahradka et al. 2013). Dietary therapy with legumes proves effective in the treatment of type 2 diabetes without the side effects of drugs (Singhal et al. 2013).

Whole Grains As with fruits, vegetables, and legumes, whole grains are rich in dietary fiber (Table 1). Whole grains are functional foods that deliver phytonutrients that may supplement or facilitate the action of fiber in protection from disease. Consider the case of flaxseed (*Linum usitatissimum*), an early part of colonial agriculture in the Chesapeake region and a highly functional food. Its Latin species epitaph means "very useful" and foretells the importance of this crop. Note that this plant goes by a second common name, linseed which is usually associated with non-human-food commercial products and animal-feed.

Upon the instruction of the Virginia Company of London, Sir George Yeardley introduced flax to the Jamestown colony in 1619. The Virginia Company wanted the colonists to develop an industry of fiber production for linen cloth and rope (NPS 2018). Although flax never achieved the commercial success in the Chesapeake region that tobacco did, colonists planted small plots of it to supply their own needs. In addition to its important non-dietary fiber (that used for textiles, paper and rope), and industrial oil content (waterproofing, paints and linoleum floor covering), the plant produces numerous small, hard seeds rich in dietary fiber, short chain omega-3 fatty acids and a host of other phytonutrients. Grinding the seeds produces a course flour easily incorporated into baked goods, soups, and numerous other dishes. Ground flaxseeds mixed with water makes a binding substance with qualities similar to that of chicken eggs, and is used as an egg-substitute in baked goods.

Flaxseed is a functional food with an extensive list of documented beneficial health outcomes (Goyal et al. 2014 provides a thorough review). The appreciation of the functional aspect of flax as food goes back to at least 1200 BCE with its benefits in prevention and curing disease noted in Ayurvedic (ancient Indian) medicine and the writings of Hippocrates circa 650 BCE (NCCIH 2018).

Modern studies of flax indicate strong efficacy in the prevention and treatment of; cardiovascular disease, atherosclerosis, diabetes, cancer, arthritis, osteoporosis, kidney disease, obesity, autoimmune, and neurological disorders (Goyal et al. 2014). In addition to the diseases reviewed by Goyal et al. (2014), recent work shows efficacy for cancer and hypertension. Flaxseed is very rich in lignans (800 times higher than any other food), phytoestrogens that appear to be anti-cancer agents (Singh et al. 2011). Flaxseed fed to women in controlled clinical trials reduces the risk of breast cancer (Fabian et al. 2010) and reduces tumor growth in patients diagnosed with the disease (Thompson 2005). Flaxseed consumption also significantly reduces tumor proliferation in prostate cancer patients (Demark-Wahnefried et al. 2008). A double-blind controlled study demonstrated that consumption of flaxseed reduces blood pressures in hypertensive patients by 10–14 points (systolic) and 7 points (diastolic), reducing the risk of stroke by 46% and that of coronary heart disease by 27%. The efficacy of flaxseed in reducing hypertension proved better or comparable to most drug treatments (Rodriguez-Leyva et al. 2013).

3. Non-protein natural health-harming substances in seafood. Animal seafood is a source of dietary cholesterol and saturated fats. The cholesterol content of the animal seafood items representative of the Chesapeake Bay presented in Table 1 range 56–107 mg/100 g. This is comparable to the cholesterol content of terrestrial animal foods such as chicken, ground beef and ham. Only chicken eggs at 373 mg Cholesterol/100 g significantly exceeds that of seafood (Table 1). Note that plants don't produce cholesterol. The saturated fat content of representative Chesapeake Bay seafood ranges from 0 for hard shell clams to 3.12 g/100 g for American shad, and is similar to most other animal items. Saturated fat content of most plant-based foods tend to be one or two order of magnitudes less than that derived from animals, with the highest levels typically found in peanuts and tree nuts (Table 1). Both cholesterol and saturated fats prove detrimental to human health, as explained below.

Cholesterol Cholesterol is a sterol lipid molecule synthesized by all animal cells. It provides structural integrity to animal cell membranes and is functional in intercellular transport, cell signaling, nerve conduction, the synthesis of steroid based hormones, and the production of bile salts. While it is synthesized by all animal cells, its main center of production in humans is in the liver. Animal life is impossible without cholesterol, yet it is not a required nutrient since humans produce all they need (Anonymous 2018).

For most of the course of the 20 million years of human evolution our diets were similar to those of the other great apes, consisting of plant-based food that contained no cholesterol. The high fiber diet of early humans would tend to extract some cholesterol (a component of bile) in the intestines, preventing reuptake by the body, thus there was evolutionary pressure to retain this important lipid. The human diet expanded during the Paleolithic period (2.6 million to 12,000 years ago) to include significant additions of animal-based food and therefore cholesterol. The health problems associated with excess cholesterol arise mostly in later life, in the post reproductive period (Milton 2000; Nestle 2000; Jenkins et al. 2003). So there's no selective pressure for traits conferring protection from diseases associated with cholesterol in old-age to be passed on to the next generation.

Dietary cholesterol and saturated fats drive coronary artery disease, the leading cause of death in the US (CDC Mortality 2018, Table 3), and several other maladies. Coronary heart disease refers to the obstruction of the vessels that supply oxygen rich blood to heart muscle. Cholesterol appears in two major forms in the blood stream, HDL

Table 3 Heart disease and stroke mortality in jurisdictions of the Chesapeake Bay Watershed for 2016 and 2014 respectively (from CDC Heart Disease 2018; CDC Stroke 2018)

Jurisdiction	Heart Disease Deaths per 100,000	Stroke Deaths per 100,00
U.S.	165.5	36.6
Delaware	163.2	38.8
District of Columbia	211.7	35.5
Maryland	164.3	38.0
New York	177.8	26.1
Pennsylvania	176.2	36.7
Virginia	150.7	37.0
West Virginia	191.0	45.3

(high-density lipoprotein) and LDL (low-density lipoprotein). LDL cholesterol causes the formation of the atherosclerotic plaques on walls of blood vessels that lead to coronary artery disease and inhibit circulation in general. These plaques reduce blood flow by obstructing the lumen (inside space) of the vessel and interfering with vessel elasticity. Although the beginnings of atherosclerotic plaques appear in children who eat animal-based diets, the symptoms of cardiovascular disease tend to occur decades later in adulthood (Steinberg and Grundy 2012; Ference and Mahajan 2013). It is long known that populations that eat plant-based diets, and therefore consume no dietary cholesterol, show virtually no cardiovascular or coronary heart disease (Higginson and Pepler 1954; Thomas et al. 1960).

At levels between 41 and 60 mg/dl HDL Cholesterol protects the vascular system from the actions of excess LDL. HDL removes oxidized LDL from arterial walls, prevents its oxidation of free LDL, reverses the progression of atherosclerosis, and limits vascular inflammation (Barter 2005). However, recent work finds elevated levels of HDL predictive of cardiovascular disease and heart attack (Singh and Rohatgi 2018). This rejects the previous dogma that unlike LDL, high concentrations of HDL are protective.

The American College of Cardiology provides recommendations for serum levels of LDL to avoid the consequences of cardiovascular disease (CVD). These range from <55 mg/dl for those of extreme risk (displaying progressive CVD) to <130 mg/dl for those with no risk factors. However, evidence suggests that LDL levels of 50–70 mg/dl are required for complete protection from CVD. That is the range of LDL typical of plant-based populations and about half the 108 mg/dl which is average for adults in the U.S. (Clendenning 2005).

Although LDL presents risk and moderate HDL provides benefit, most of the public health literature focuses on total cholesterol, comprised of LDL and HDL. High total cholesterol is considered >200 mg/dl, and very high >240 mg/dl. Patients with total cholesterol >200 mg/dl are recommended to take cholesterol reducing drugs, such as statins that lower the cholesterol produced by the liver. Some healthcare providers also recommend the more effective treatment of eliminating dietary sources of cholesterol and adding plant-based foods with phytosterols known to reduce cholesterol concentration (Carr 2010). But few physicians have training in nutrition and many think that their patients won't be willing to make the required lifestyle change of adopting a plant-based diet (Hyman 2009).

Nearly 37% of adults in the US have high cholesterol, but only 55% of those take the recommended medication. About 7% of children ages 6–19 display high cholesterol based upon a target level of less than 200 mg/dl (CDC Cholesterol 2018). The percentage of those displaying high cholesterol would certainly be even higher if public health officials adopted the more protective target level of 150 mg/dl (Clendenning 2005). It is remarkable that the leading cause of death in the US is coronary artery disease and that it is essentially completely preventable by eating a plant-based diet. This of course means a diet absent of terrestrial and seafood animal products.

High cholesterol is implicated in conditions other than cardiovascular and heart disease. After coronary heart disease, stroke is the second leading cause of death worldwide, and fourth leading cause of death in the US (CDC Mortality 2018). Survivors often suffer severe debilitation. High total cholesterol increases chances of stoke (Tirschwell et al. 2004; Zhang et al. 2012). Ischemic stroke (particles stopping blood flow in the brain) accounts for 78% of all types of stokes in the US and mostly originates from a thrombosis (platelets and fibrin forming a clot) resulting from disturbance of arterial plaques.

The cholesterol-linked sources of mortality from coronary heart disease and stroke differ by race/ethnicity in the U.S. African Americans are 30% more likely to die from heart disease than non-Hispanic whites and also twice as likely to have a stroke (Graham 2015). Despite the claims of seafood being protective from heart disease (Jahns et al. 2014), non-Hispanic blacks consume nearly twice as much animal seafood (22.5 g/d/person) as non-Hispanic whites (10.2 g/d/person, Wang et al. 2010). Blacks (281.1 g/d/person) do consume slightly more meat per person than whites (254.9 g/d/person). This corresponds to blacks getting about 8% of their animal flesh from seafood compared to 4% for non-Hispanic whites. Other work confirms higher rates of seafood consumption by non-Hispanic blacks (USEPA 2018a).

Breast cancer proliferation is associated with high levels of cholesterol, and as the disease progresses, the rate of uptake by the growing tumors drastically drops blood concentrations. In essence, cholesterol feeds the growth of breast cancer (Danilo and Frank 2012). High total cholesterol also predicts cancers of the prostate and colon, but not others such as liver (Kitahara et al. 2011). Perhaps the sequestering of cholesterol by fast growing tumors reduces the blood levels (Buchwald 1992).

Cholesterol levels of over 250 mg/dl triples the chances of developing Alzheimer's disease (the leading cause of dementia, Notkola et al. 1998). Cholesterol appears to promote Alzheimer pathology through several different mechanisms related to circulation (Kalaria et al. 2012).

Saturated Fats Saturated fatty acids consist of chains of carbon atoms held together with single bonds, the remaining outer orbital electrons used to bond with hydrogen atoms. Consumption of diets high in saturated fats is linked to increased production of cholesterol and heart disease. This is well established in clinical metabolic ward trials with tight control of diet (Hooper 2001; Sacks and Katan 2002; Aranceta and Pérez-Rodrigo 2012; Chiu et al. 2017). However, animal-industry funded authors of recent meta-analyses of low-power observational cross-sectional studies showed no relationship between saturated fat and heart disease (Siri-Tarino et al. 2010; Chowdhury et al. 2014). Since the statistically more powerful controlled clinical trials clearly demonstrated the adverse effects of saturated fats on human health, there should be no doubt about the need to avoid those lipids in seafood or any other dietary item. In addition to heart disease, dietary saturated fats raise blood pressure (Rasmussen et al. 2006). It is also the basis for causing insulin insensitivity and type-2 diabetes (Santomauro et al. 1999, See chapter "Sugar Twice Enslaves: Consequences for the People of the Chesapeake Bay" for detailed discussion of type-2 diabetes). This is consistent with the association between seafood consumption and prevalence of type-2 diabetes for participants in the Women's Health Study (Djoussé et al. 2010).

4. Toxic and Pathogenic Contaminates of Seafood. A variety of chemicals contaminate the Chesapeake Bay ecosystem (Fig. 2). They derive from industrial, agricultural, municipal and pharmaceutical sources. The list includes; polychlorinated biphenyls (PCBs), dioxins and furans, petroleum hydrocarbons, pesticides, pharmaceuticals, household and personal care products, polybrominated diphenyl ether (PDEs), biogenic hormones and heavy metals. In 2010, the Chesapeake Bay Program (CBP) reported that 72% of the Chesapeake Bay's tidal waters are fully or partially impaired by these toxins (CBP 2018).

These toxins generally have limited solubility in water and quickly attach to sediment particles or incorporate into the tissues of living organisms. Living cells and tissue offer a mix of polar and non-polar environments that readily uptake the contaminants. A sample of filtered water from the Chesapeake Bay is unlikely to show significant concentrations of these contaminants, even when levels are evident in the plankton, sediment caught on the filter, bottom sediments, benthic organisms and fish. Many of these toxins remain in living tissues for extended periods, as the cells or organ systems of contaminated plants and animals often lack effective mechanics to decompose or excrete these chemicals. Once incorporated into bacteria or algae, they quickly move up the food chain as unintended portions of meals taken by worms, crabs, fish, etc. The concentrations of containments increase with each subsequent trophic level, the top predators displaying the largest body-loads, and therefore often demonstrate the most devastating effects of the toxin. This is the process of biological magnification. The location of the toxin within the human body depends on its chemistry. Metals tend to accumulate in muscle tissue, while organic compounds concentrate in fat cells.

Most of the chemical contaminants of seafood in the Chesapeake Bay cause chronic rather than acute toxicity. That is, a human eating a contaminated fish is unlikely to die or even notice any immediate effect other than a sense of satisfaction, if it was prepared well. However, continued consumption of contaminated seafood will lead to accumulation of the toxin(s) in the person's body. This body-burden of toxins may cause various symptoms including; disruption of the endocrine system, reduced immune system function, reduced fertility, mutations of the genetic material in sperm and eggs, birth defects and cancers. Let's consider the two major contaminates of seafood found in the Chesapeake Bay, PCBs and mercury.

Polychlorinated Biphenyls PCBs are a diverse group (209 congeners) of synthetic organic compounds consisting of ring-structures made of carbon, hydrogen and attached chlorine atoms. They were manufactured in the US from 1929 until 1979 when they were banned. PCB's found many industrial applications including adhesives, caulking, plastics, floor finishes, copy paper, use in a variety of electrical devices, paint, ink, lubricants, thermal insulation and electrical insulation. They were prized for their chemical stability, resistance to breaking down from heat and microbes, and non-flammability. These very properties of stability and resistances account for their persistence as an environmental contaminant of air, soil, water and living tissues for decades after production stopped. Many of the products containing PCBs are still in use. Leakage and

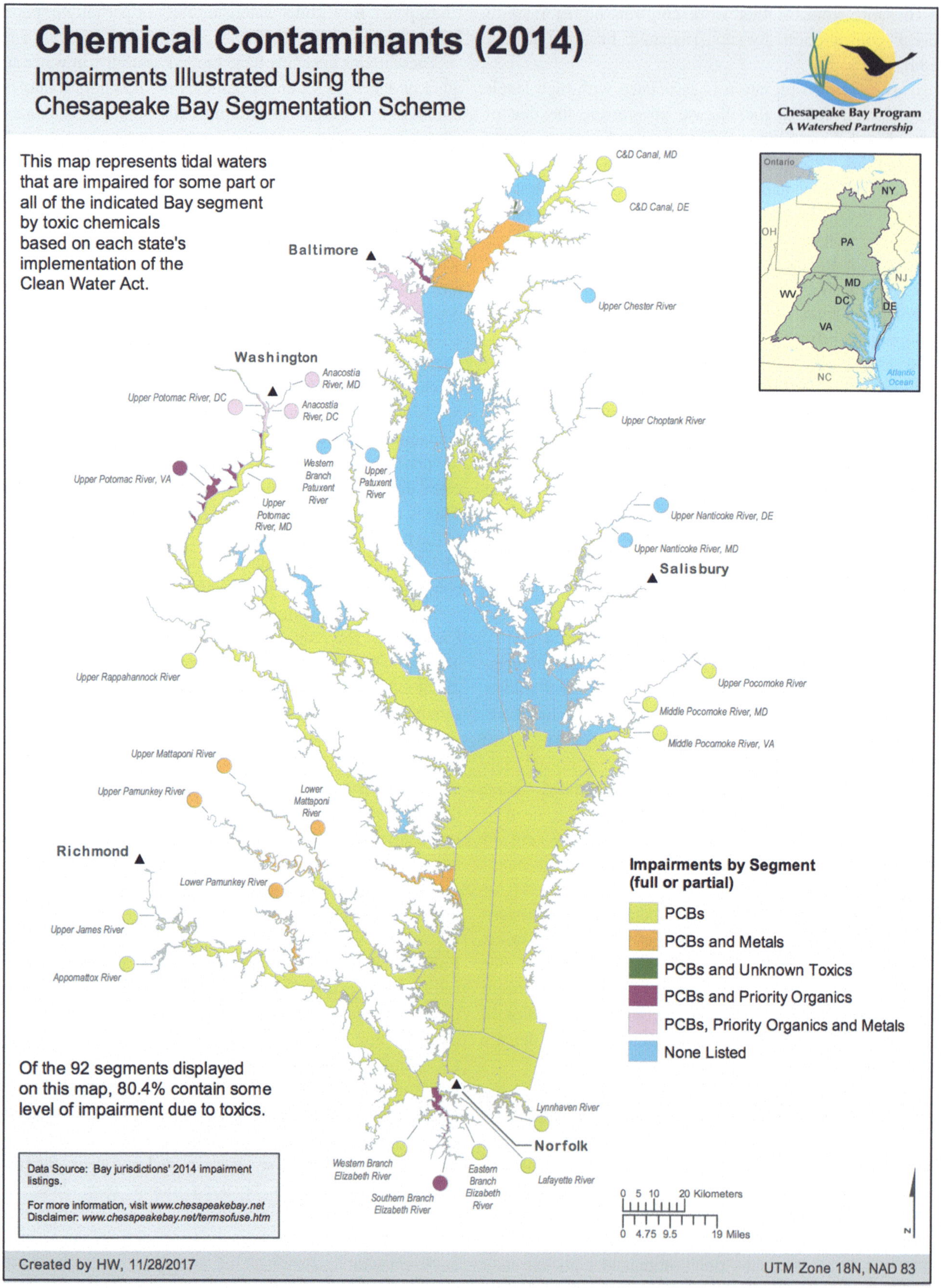

Fig. 2 The distribution of chemical contaminants in the Chesapeake Bay. From the Chesapeake Bay Program (CBP Maps 2018). Created by Howard Weinberg. https://www.chesapeakebay.net/what/maps/chemical_contaminants_2014

improper disposal of these items means a continuing stream of new contamination.

PCBs are carcinogenic and associated with cancers of the liver, skin and non-Hodgkin's lymphoma. Despite the celebrated chemical stability of PCBs, they do undergo some changes in the environment, and the forms most likely found attached to sediments or in the tissues of seafood are the most potent carcinogens (ATSDR 2020).

These toxins cause other health problems. PCBs suppress immune system function. This includes shrinking of the thymus, decreased resistance to Epstein-Barr virus and other infections. PCBs affect reproduction, reducing fertility, birth weight and gestational age. These effects on reproduction appear in populations exposed through the workplace and fishing activities. Both animal and human studies show that PCB contamination causes deficits in neurological development, including visual recognition, short-term memory and learning. PCBs act as endocrine disrupters, reducing thyroid hormone production, potentially influencing normal growth patterns. Other effects of PCB contamination include dermatitis, liver toxicity, and elevated blood pressure, serum cholesterol and increased risk of stroke in women (USEPA 2018b; Bergkvist et al. 2014). Thus, eating a PCB contaminated item of seafood may raise cholesterol levels via two mechanisms, the dietary component and disruption of body regulation by PCB toxicity. The half-life of PCBs in human tissue is 10–15 years, meaning that it will take that long for the body to rid itself of about 50% of its current load of the toxin (Ritter et al. 2010).

The FDA (Food and Drug Administration) sets an allowable limit 0.2–3.0 ppm of PCBs for all foods and 2.0 ppm for fish. The World Health Organization (WHO) sets the allowable daily intake limit at 6.0 μg/kg/d (ATSDR 2020). All of the responsible regulating jurisdictions in the Chesapeake Bay (six states and Washington, DC) have issued advisories for limiting consumption of seafood based upon PCB contamination. The most contaminated sites in the Bay are the Anacostia and Potomac Rivers in Washington, DC. The DC authorities banned the consumption of American eel (*Anguilla rostrate*), carp (*Cyprinus carpio*), Channel catfish (*Ictalurus punctatus*) and blue catfish (*I. furcatus*) (CBP 2018). Maryland published advisories for 23 species of fish and blue crab, corresponding to 87 areas of water, all based upon PCB contamination (MDE 2018). Seafood advisories for the Virginia portion of the Chesapeake Bay watershed are summarized in Table 4. Note that most of the advisories in Virginia are driven by PCBs, with relatively few issued on the basis of mercury contamination. PCBs are also the principal contaminant of the Chesapeake and Delaware Canal, driving fish advisories from Delaware (CBP 2018). The PCBs accumulate in the adipose tissues of seafood and the authorities issuing advisories suggest removal of as much fat as possible in the cooking process.

Mercury Mercury (Hg) is the most important metal contaminant of the Chesapeake Bay food system. Combustion of coal and municipal solid waste moves elemental mercury into the atmosphere. Dry and wet precipitation from the atmosphere brings elemental mercury into the water and watershed of the Bay. The Chesapeake Bay serves as a trap for mercury pollution from the atmosphere, exporting only about 5% of that which falls on the watershed to the ocean (Mason et al. 1999). This suggests that mercury may continue to contaminate seafood in the Bay for many years after society shifts away from burning coal. The action of bacteria transforms the elemental mercury into methylmercury (MeHg) in the organic and anoxic sediments of the Bay and in the soils of adjacent wetlands. MeHg is the most toxic and biologically available form of this metal. It is taken up by benthic animals from the sediments and by fish from particles in the water column (Chen et al. 2014). MeHg is biologically concentrated in living tissue and biologically magnified up the food chain ten-fold each step, realizing its highest tissue concentrations in top predators (Xu et al. 2013).

MeHg poisoning from seafood came to the attention of the world in the 1950's after a massive occurrence in Minamata, Japan. Industrial dumping of inorganic mercury waste into Minamata Bay resulted in the formation of MeHg, which them moved up the food chain to poison thousands of people who ate fish from the system. The symptoms of poisoning include severe tremors, loss of motor control, visual and auditory sensory loss, numbness of extremities, speech impairment, cerebral palsy in infants, seizures, growth and developmental problems, microcephaly (small head) in infants and mental retardation. Mercury is a general toxin, often exacerbating various diseases (Hachiya 2006). The half-life of mercury in humans is about 2 months (Yaginuma-Sakurai et al. 2012).

The USEPA considers 300 μg of total mercury (THg)/g of wet mass fish to illicit concern for human consumption (screening value). The USEPA and FDA set an oral reference dose (RfD) for MeHg of 0.1 μg/kg body mass of the consumer (EPA-FDA Fish 2018). This is the maximum daily dose that the agencies believe will engender no ill health effects over the long term. However, it only applies to children and women of child-bearing age. In practice, this means that a 60-kg (132-pound) women's safe dose is 42 μg MeHg per week (note that 1 ppm = 1 μg/g). The FDA reports mercury concentrations in common Chesapeake Bay species, bluefish (*Pomatomus saltatrix*) of 0.368 ppm (FDA 2018). A typical serving of fish is about 170 grams, so a meal of bluefish would yield 62.56 μg MeHg. That is in excess of the 42 μg MeHg per week calculated from the RfD.

Table 4 Species of seafood in the Virginia portion of Chesapeake Bay ecosystem with sufficient levels of contamination for official advisories on consumption (VDH 2018)

Species	Contaminants	Locations	Advisory
Largemouth bass *Micropterus salmoides*	Mercury	Hampton Roads, Harrison Lake, Potomac R, Mattaponi R., Mattaponi R.	Two meals/month
Smallmouth bass	PCBs	James R., Potomac R.	
Carp *Cyprinus carpio*	Mercury, PCBs	Hampton Roads, Maury R., Bull Run, Rappahannock R.	Two meals/month
Carp	PCBs	James R., Potomac R.	Do not eat
Spot *Leiostomus xanthurus*	PCBs	Hampton Roads, York R.	Two meals/month
Croaker *Micropogonias undulatus*	PCBs	Hampton Roads, Rappahannock R., York R.	Two meals/month
Gizzard shad *Dorosoma cepedianum*	PCBs	Hampton Roads, Mobjack Bay, Back R., James R., Mill Creek, Rappahannock R., York R., Paumunkey R.	Two meals/month
Gizzard shad	PCBs	James R., Lake Anna, Mattaponi R.	Do not eat
Striped bass *Morone saxatilis*	PCBs, Mercury	Chesapeake Bay & tributaries, Piankatank R., Potomac R., Rappahannock R.	Two meals/month
Redbreast sunfish *Lepomis auritus*	PCBs	Maury R.	Two meals/month
Rock bass *Ambloplites rupestris*	PCBs	Maury R.	Two meals/month
Yellow bullhead catfish *Ameiurus natalis*	PCBs	Maury R.	Two meals/month
American eel *Anguilla rostrata*	PCBs	James R., Potomac R., Rappahannock R.	Two meals/month
American eel	PCBs	Potomac R.	Do not eat
Bullheaded catfish *Ameiurus nebulosus*	PCBs	Potomac R.	Two meals/month
Flathead catfish *Pylodictis olivaris*	PCBs	James R.	Two meals/month
Flathead catfish >81 cm	PCBs	James R.	Do not eat
Channel catfish <46 cm *Ictalurus punctatus*	PCBs	Potomac R., Bull Run, Rappahannock R.	Two meals/month
Channel catfish >46 cm	PCBs	Potomac R.	Do not eat
Quilback carpsucker *Carpiodes cyprinus*	PCBs	James R.	Two meals/month
Blue catfish >81 cm *Ictalurus furcatus*	PCBs	James R.	Do not eat
Blue catfish	PCBs, Mercury	Rappahannock R., Paumunkey R.	Two meals/month
White catfish *Ameirus catus*	PCBs	Potomac R.	Two meals/month
Redear sunfish *Lepomis microlophus*	Mercury	Harrison Lake, Chickahominy Lake	Two meals/month
Chain pickerel *Esox niger*	Mercury	Harrison Lake, Chickahominy Lake	Two meals/month
Bowfin *Amia calva*	Mercury	Harrison Lake, Chickahominy Lake	Two meals/month
White perch *Morone americana*	PCBs	Potomac R., Mattaponi R.	Two meals/month
Yellow perch *Perca flavescens*	PCBs	Potomac R.	Two meals/month
Creek chub *Semotilus atromaculatus*	PCBs	Indian Run R.	Two meals/month

A sampling of 348 fin-fish from the lower Chesapeake Bay in 2009–2010 found MeHg in all of the species collected (white perch, striped bass, white catfish, Atlantic croaker, channel catfish, summer flounder (*Paralichthys dentatus*), weakfish (*Cynoscion regalis*), blue catfish, American eel and spot) ranging from 24.4–111.0 μg THg/kg wet mass. The value for striped bass, perhaps the most popular sport fish on the Bay, was 109.0 μg THg/kg wet mass. None of the species investigated reached the USEPA screening level (Xu et al. 2013). However, there are places on the Bay where mercury concentrations are sufficient to warrant fish consumption advisories (Table 4).

Pathogens and Natural Toxins in Seafood from the Bay The eastern oyster (*Crassostrea virginica*), Atlantic ribbed mussel (*Geukensia demissa*), cherrystone clam (*Mercenaria mercenaria*) and bay scallop (*Argopecten irradians*) are commonly harvested in the Bay. Contamination of these shell fish by enteric pathogens (such as strains of *Escherichia coli*) causes gastrointestinal illness in people eating these items (Sobsey et al. 1980). Local and state departments of health recognized the problem on the Chesapeake Bay in the early twentieth century and instituted regulations to protect the public, including closures of shell fishing areas (Schulte 2017, Fig. 3). Today the Bay jurisdictions belong to the National Shellfish Sanitation Program (NSSP 2018). The need to reduce contamination of shellfish helped drive the installation of early wastewater treatment plants around the Bay during the mid-twentieth century (see chapter "Finishing the Journey: Wastewater as a Misplaced Resource"). Despite the proliferation and improvement of wastewater treatment facilities, portions of the Bay continue to experience regular exposure to human enteric pathogens. Leaky pipes, spills, overflows, faulty septic systems, illegal discharges of human waste by ships and boats are all potential sources of pathogens.

In addition to human enteric bacteria, suspension and deposit feeding shellfish can accumulate naturally occurring bacteria that challenge health. *Vibrio parahaemolyticus* results in diarrhea and vomiting. *Vibrio vulnificus* is very dangerous and causes infections of the blood stream, skin lesions and may result in limb amputations. It kills one third of its victims. Vibrio infections come from eating raw or undercooked shellfish or through open wounds in the skin (CDC oysters 2018).

Reduction of water quality in the Bay during the twentieth century brought another danger to the consumption of molluscan shellfish. Intermittent blooms of toxic algae (HABs, harmful algal blooms), principally of certain dinoflagellates, such as *Dinophyssis sp.*, lead to accumulation of toxins they produce in the tissues of suspension feeding (oysters, mussels, scallops) and deposit feeding mollusks (clams) (Lloyd et al. 2013).

Fig. 3 Sign placed in the East River, of Mobjack Bay on the Western Shore of the Chesapeake Bay by the Virginia Department of Health indicating closure of an area for the harvesting of molluscan shellfish for consumption due to contaminants. Such signs are common in both rural and urban areas of the Chesapeake Bay. Photo by the author

Despite closures of shellfish grounds to commercial harvest, one may see boats flying the yellow quarantine flag working such restricted areas. They must move their harvest to a clean area of the bay to allow the shellfish to rid themselves of pathogens in the new environment. This is the process of *depuration* and takes several weeks. Cooking kills most pathogens and reduces the chances of getting sick from shellfish with pathogens. However, many people enjoy eating oysters raw.

5. Unhealthy methods of preparing seafood. As detailed above, seafood presents several challenges to the health of consumers. Methods of preparation and cooking may create additional problems for human health. Not cooking, as in the case of raw oysters on the half-shell, increases the chances of illness from pathogenic bacteria. However, most seafood from the Bay is traditionally cooked in some fashion before consumption.

Deep-Frying and Pan-Frying Deep-frying involves submerging the seafood in a vat of cooking oil heated to 177–191 °C (350–375 °F) for several minutes. Often the item is first coated with a batter of white flour or corn meal, typically made with milk, egg white, beer or water. Deep frying poses several risks. It adds additional calories and fat to the seafood. Deep-frying raises the caloric value of a raw fillet of flounder from 106 to 153 calories per 100 g and lipid content from 6 g to 7 g (USDA 2018a). Deep-frying increased fat content 120–180% for sea bream (*Sparus aurata,* Mnari Bhouri et al. 2010).

The high temperature cooking environment creates three classes of harmful organic compounds, HAs (Heteocyclic Amines), PAHs (Polycyclic Aromatic Hydrocarbons), and AGEs (Advanced Glycation End products). HAs are mutagenic and carcinogenic (Sugimura et al. 2004). PAHs are also carcinogenic (Boström et al. 2002). For both HAs and PAHs, exposure comes through eating the fried seafood and through breathing the fumes of the cooking process. AGEs form naturally in the body and during high-temperature cooking. They are produced in non-enzymatic reactions between reducing sugars, free amino acid groups of proteins and lipids. AGEs cause oxidative stress and inflammation, promote arthrosclerosis, high blood pressure, shortened-life spans, kidney failure and Alzheimer's disease. They also oxidize LDL cholesterol, making it potent in producing atherosclerosis and stoke (Uribarri et al. 2010). Consumption of deep-fried fish more than once per week is linked to increased risk of prostate cancer (Odds Ratio = 1.32, a ratio of 1 means linkage between the two factors, Stott-Miller et al. 2013).

Deep-frying changes the lipid composition in seafood, rendering it less healthy. During most of their evolution, humans consumed a diet with a ratio of omega-6 to omega-3 fatty acids of about 1 to 1, while Western diets (Standard American Diet) contain ratios of 15–16.7. The high omega-6 to omega-3 ratio is associated with cardiovascular disease, cancer, inflammatory and autoimmune disease, while low ratios are protective for breast cancer and rheumatoid arthritis (Simopoulos 2002a). Deep-frying alters the very low omega-6 to omega-3 ratios found in raw fish of 0.01–0.12, to the much higher values of 1.07–5.98, thus reducing the beneficial aspects of consuming fish (Candela et al. 1998). Pan-frying seafood uses much less oil than deep-frying and as such is probably healthier, but it still produces a high-temperature environment that may alter lipid ratios and produce harmful compounds.

Grilling, Broiling, and Baking Grilling, broiling and baking reduce the lipid content of seafood as the hot oils tend to volatilize or drip away. However, recipes that call for adding oils or butter reduces the fat-rendering effect. The high temperatures of these cooking methods create oxidation products as with deep-frying. Poaching, steaming and microwaving use boiling water or water vapor to cook seafood. Those methods provide the healthiest option, the lower cooking temperature preserves healthier omega-3 fatty acids and produces fewer unhealthy organic compounds (Moradi et al. 2011). However, for most seafood, poaching, steaming and microwaving don't provide the same appeal for diners as do the less healthy methods of cooking. One important exception is the blue crab which is classically steamed or boiled in its shell and seasoned with a mix of spices.

Breading and Fillers Cooks often add other food items in preparing seafood, yielding dishes richer in calories and fat than the base meat. Crab cakes are popular around the Chesapeake Bay. A typical recipe calls for pan-frying in oil 0.45 kg (1 pound) crab meat, 0.12 liter (0.5 cup) breadcrumbs, 2 eggs, 0.6 liter (0.25 cup) mayonnaise and various seasonings. The crab meat per 100 g contains: 88 kcal, 1.77 g total lipids, and 107 mg cholesterol. A 100 g serving of crab cake contains: 155 kcal, 7.52 g total lipids, and 150 mg cholesterol. The additives nearly doubled the calories, quadrupled the lipid content and increased cholesterol by 50% (USDA 2018a).

Omega-3 Fatty Polyunsaturated Fatty Acids (PUFAs) As noted at the beginning of the chapter, animal seafood receives praise as a source of omega-3 fatty acids (Greenberg 2019). The omega refers to the location of the first double valence bond from the beginning of the chain. These fatty acids come from the algae at the base of the aquatic food chain (Fig. 4). Fatty acids are the structural units of lipids (fats, oils and waxes). As with amino acids and proteins, the human body can synthesize most of the needed fatty acids, and requires a dietary source for others, deemed *essential fatty* acids. Linoleic acid (LNA) and alpha-linolenic acids (ALA) are the two essential fatty acids required in the human diet. Metabolic pathways in the body (mostly liver) convert these two 18-C fatty acids into the omega-3 fatty acids eicosatetraenoic acid (EPA) and docosahexaenoic acid (DHA, Fig. 5). Dietary sources of DHA are important supplements to that synthesized by the human body from LNA, as the liver

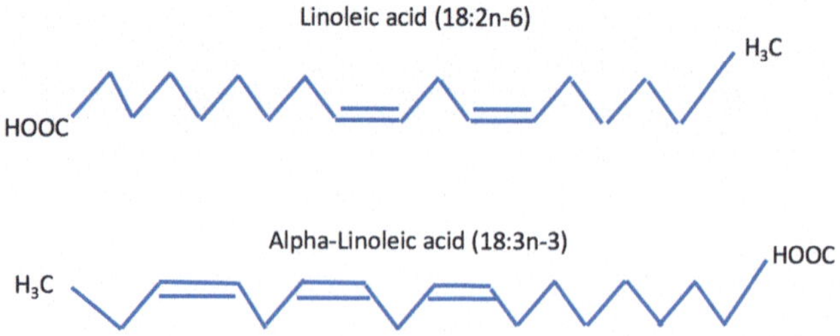

Fig. 4 Structures of the two essential fatty acids, linoleic and alpha-linolenic. Linoleic acid is converted to the omega-3 fatty acid, Alpha-linolenic acid (ALA) by human metabolism

ALA
EPA
DHA

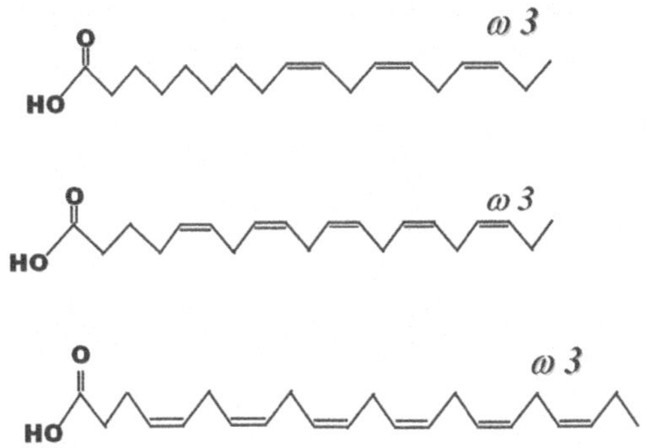

Fig. 5 Structures of Alpha-linolenic acid (ALA), eicosatetraenoic acid (EPA) and docosahexaenoic acid (DHA). In the human body ALA is converted to EPA, which is then converted to DHA. ALA is present in many vegetables, beans, nuts, seeds and fruits. The location of the omega bond is indicated for each molecule. Flax seeds and walnuts and various kinds of animal seafood offer high concentrations of ALA

is only about 15% efficient in the process (Burdge and Calder 2005). In the body, the omega-3 fatty acids may also be converted to omega-6 fatty acids. Algae, yeast, flaxseed, chia seed, walnuts and other plants are rich in omega-3 fatty acids (NIH 2018). Animal seafood contains both omega-3 fatty acids and synthesized omega-6 fatty acids.

Omega-3 fatty acids play an essential role in formation of the phospholipid matrix of cell membranes, particularly in the retina, brain and sperm. They and Omega-6 fatty acids form eicosanoids, signaling molecules that regulate inflammation, vasoconstriction and platelet aggregation. The omega-6 derived eicosanoids more effectively suppress inflammation than those from omega-3 s (Gabbs et al. 2015). Adequate consumption of omega-3 s has been associated with protection from a variety of maladies including depression and anxiety (Ginty and Conklin 2015), eye and brain functional deficiencies (Singh 2005), cardiovascular disease (Rangel-Huerta and Gil 2018), inflammation (Li et al. 2014), autoimmune diseases (Simopoulos 2002b) and breast cancer (Calado et al. 2018).

Despite much research demonstrating the health benefits of omega-3 consumption, a large meta-analysis and systematic review showed that consuming these fatty acids by *eating fish*, or via a dietary supplement of fish oil, proved ineffective in lowering the risk of all-cause mortality, cardiac death, sudden death or stroke (Rizos et al. 2012). A meta-analysis of controlled studies that tested omega-3 supplementation found no significant effect on the secondary prevention (treating people that already had heart attacks) of cardiovascular disease (Kwak et al. 2012). Another large study (n = 3314) of men with angina (heart induced chest pain) found that advice to eat oily fish or take fish oil supplements *increased* their risk of cardiac death (Burr et al. 2003). A meta-analysis of ten trials involving 77,917 individuals confirmed no effect of fish oil supplementation on reducing cardiovascular disease (Aung et al. 2018). Possibly the pervasive contamination of seafood by PCB's and other toxins mitigates the expected benefits of omega-3 s for cardiovascular health (Bergkvist et al. 2014). In 2002, the American Heart Association (AHA) recommended eating a minimum of two servings of oily fish per week. In 2017, it revised that recommendation to state that supplementation with fish oil for patients with coronary heart disease is only "reasonable." Currently, the AHA rejects the use of fish oil supplements for diabetic patients with heart disease, and for prevention of stroke (Jackson 2018). A recent press-release reaffirming the 2002 recommendation of two servings of oily fish per week links the reader to a page of four featured recipes; three with eggs and one with cheese, both foods high in cholesterol and saturated fats (AHA 2018).

Omega-3 s from seafood are often recommended for brain health, yet in one study cognitive performance in older adults was inversely associated with fish consumption (Danthiir et al. 2013). This is consistent with a study showing higher levels of omega-3 s to predict cognitive impairment and dementia (Laurin et al. 2003). This may attribute to contamination of the fish by MeHG, which even at low levels of dietary exposure from fish results in reduced cognitive function in adults (Yokoo et al. 2003). Yet higher levels of EPA and DHA are associated with slower brain shrinkage and aging in adults (Pottala et al. 2014). Indeed, vegans and omnivores that eat little fish, present reduced plasma omega-3 s. Supplementation with low-dose (250 mg) omega-3 s derived from algae brought vegans to levels of EPA and DHA associated with preservation of brain health (Sarter et al. 2015). Thus, in order to achieve the benefits of marine derived omega-3 s without contamination, diners in the Chesapeake region should consider a supplement of clean-sourced algae along with plant-based foods containing these fatty acids such as flaxseed and walnuts.

Despite the mounting evidence otherwise, omega-3 fish oil is heavily marketed to health-conscious consumers and physicians interested in reducing coronary artery disease. The Chesapeake Bay is truly at the heart of this industry as its largest fishery is for menhaden (*Brevoortia tyrannus*) which is used to produce fish-oil supplements (See chapter "Menhaden, the Inedible Fish that Most Everyone Eats" on menhaden).

Seafood or Not This chapter challenges the reader to reconsider the popular notion that consuming seafood promotes health. In making their dietary choices, people of the Chesapeake Bay and beyond should consider the presence of contaminants, pathogens, excess protein, and cholesterol in their food. They should also think about the health benefits conferred by a diet built around a variety of whole plant foods; fruits, vegetables, legumes and whole grains (Dewell et al. 2008).

References

AACC (2001) American association of cereal chemists. The definition of dietary fiber Cereal Foods World 46:112–129

ACAAI (2018) Fish allergy. In: American College of Allergy, Asthma and Immunology Public Website. https://acaai.org/allergies/types/food-allergies/types-food-allergy/fish-allergy. Accessed 20 Jul 2018

AHA (2018) Healthy for good. In: www.heart.org. https://www.heart.org/en/healthy-living. Accessed 30 Aug 2018

Allen N, Appleby P, Davey G et al (2002) The associations of diet with serum insulin-like growth factor I and its main binding proteins in 292 women meat-eaters, vegetarians, and vegans. Cancer Epidimiol Biomarkers Prevent 11:1441–1448

Anonymous (2018) Cholesterol: synthesis, metabolism, regulation. In: Themedicalbiochemistrypage.org. https://themedicalbiochemistrypage.org/cholesterol.php. Accessed 26 Jul 2018

Aranceta J, Pérez-Rodrigo C (2012) Recommended dietary reference intakes, nutritional goals and dietary guidelines for fat and fatty acids: a systematic review. Br J Nutr 107:S8–S22. https://doi.org/10.1017/s0007114512001444

ATSDR (2020) Polychlorinated Biphenyls (PCBs) 2014: What Standards and Regulations Exist for PCB Exposure? | ATSDR - Environmental Medicine & Environmental Health Education - CSEM. In: Atsdr.cdc.gov. https://www.atsdr.cdc.gov/csem/csem.asp?csem=30&po=8. Accessed 12 May 2020

Aung T, Halsey J, Kromhout D et al (2018) Associations of Omega-3 fatty acid supplement use with cardiovascular disease risks. JAMA Cardiol 3:225. https://doi.org/10.1001/jamacardio.2017.5205

Barsotti G, Cupisti A, Morelli E et al (1990) Vegan supplemented diet in nephrotic syndrome. Nephrol Dial Transplant 5(Suppl 1):75–77

Barter P (2005) The role of HDL-cholesterol in preventing atherosclerotic disease. Eur Heart J Suppl 7:F4–F8. https://doi.org/10.1093/eurheartj/sui036

Bazzano L, He J, Ogden L et al (2002) Legume consumption and risk of coronary heart disease in US men and women. NHANES I epidemiologic follow-up study. ACC Curr J Rev 11(2):31–32

Bellavia A, Larsson S, Bottai M et al (2013) Fruit and vegetable consumption and all-cause mortality: a dose-response analysis. Am J Clin Nutr 98:454–459. https://doi.org/10.3945/ajcn.112.056119

Benelam B (2018) Satiation, satiety and their effects on eating behaviour. [online] Onlinelibrary.wiley.com. https://onlinelibrary.wiley.com/doi/pdf/10.1111/j.1467-3010.2009.01753.x. Accessed 14 Jun 2018

Bergkvist C, Kippler M, Larsson S et al (2014) Dietary exposure to polychlorinated biphenyls is associated with increased risk of stroke in women. J Intern Med 276:248–259. https://doi.org/10.1111/joim.12194

Borzoei S, Neovius M, Barkeling B et al (2006) A comparison of effects of fish and beef protein on satiety in normal weight men. Eur J Clin Nutr 60(7):897–902

Boström C, Gerde P, Hanberg A et al (2002) Cancer risk assessment, indicators, and guidelines for polycyclic aromatic hydrocarbons in the ambient air. Environ Health Perspect 110:451–489. https://doi.org/10.1289/ehp.02110s3451

Bressani R (1988) Protein complementation of foods. In: Karmas E, Harris RS (eds) Nutritional evaluation of food processing. Springer, Dordrecht

Buchwald H (1992) Cholesterol inhibition, cancer, and chemotherapy. Lancet 339:1154–1156. https://doi.org/10.1016/0140-6736(92)90744-n

Burdge G, Calder P (2005) Conversion of α-linolenic acid to longer-chain polyunsaturated fatty acids in human adults. Reprod Nutr Dev 45:581–597. https://doi.org/10.1051/rnd:2005047

Burkitt D (1979) The protective value of plant fibre against many modern western diseases. Qualitas Plantarum. Plant Foods Hum Nutr 29(1-2):39–48

Burr M, Ashfield-Watt P, Dunstan F et al (2003) Lack of benefit of dietary advice to men with angina: results of a controlled trial. Eur J Clin Nutr 57:193–200. https://doi.org/10.1038/sj.ejcn.1601539

Calado A, Neves P, Santos T et al (2018) The effect of flaxseed in breast cancer: a literature review. Front Nutr 5:4. https://doi.org/10.3389/fnut.2018.00004

Campbell TC (2017) Cancer prevention and treatment by wholistic nutrition. J Nat Sci 3(10):e448

Campbell T, Campbell T (2016) The China study. BenBella Books, Dallas

Candela M, Astiasarán I, Bello J (1998) Deep-fat frying modifies high-fat fish lipid fraction. J Agric Food Chem 46:2793–2796. https://doi.org/10.1021/jf9709616

Carpenter K (1986) The history of enthusiasm for protein. J Nutr 116(7):1364–1370

Carr (2010) Cholesterol-lowering phytosterols: factors affecting their use and efficacy. Nutr Diet Suppl 59. https://doi.org/10.2147/nds.s10974

CBP (2018) Toxic contaminants in the Chesapeake Bay and its Watershed: extent and severity of occurrence and potential biological effects. Technical Report Dec. 2012. In: Federalleadership.chesapeakebay.net. https://federalleadership.chesapeakebay.net/ChesBayToxics_finaldraft_11513b.pdf. Accessed 31 Jul 2018

CBP Maps (2018) Maps | Chesapeake Bay Program. In: Chesapeakebay.net. https://www.chesapeakebay.net/what/maps/chemical_contaminants_2014. Accessed 2 Aug 2018

CDC (2017) *FastStats*. National Center for Health Statistics. Centers for Disease Control and Prevention, Obesity and Overweight [online] https://www.cdc.gov/nchs/fastats/obesity-overweight.htm. Accessed 27 Mar 2018

CDC Asthma (2018) CDC – Asthma – Most Recent Asthma State Data. In: Cdc.gov. https://www.cdc.gov/asthma/most_recent_data_states.htm. Accessed 20 Jul 2018

CDC Cholesterol (2018) High cholesterol facts. In: Centers for Disease Control and Prevention. https://www.cdc.gov/cholesterol/facts.htm. Accessed 27 Jul 2018

CDC Heart Disease (2018) Stats of the States – Heart Disease Mortality. In: Cdc.gov. https://www.cdc.gov/nchs/pressroom/sosmap/heart_disease_mortality/heart_disease.htm. Accessed 29 Jul 2018

CDC Mortality (2018) FastStats. Center for Disease Control. In: Cdc.gov. https://www.cdc.gov/nchs/fastats/deaths.htm. Accessed 27 Jul 2018

CDC oysters (2018) Oysters and vibriosis. In: Centers for Disease Control and Prevention. https://www.cdc.gov/features/vibrio-raw-oysters/index.html. Accessed 3 Aug 2018

CDC Stroke (2018) Stats of the States – Stroke Mortality. In: Cdc.gov. https://www.cdc.gov/nchs/pressroom/sosmap/stroke.htm. Accessed 31 Jul 2018

Chatzi L, Apostolaki G, Bibakis I et al (2007) Protective effect of fruits, vegetables and the Mediterranean diet on asthma and allergies among children in Crete. Thorax 62:677–683. https://doi.org/10.1136/thx.2006.069419

Chen C, Borsuk M, Bugge D et al (2014) Benthic and pelagic pathways of methylmercury bioaccumulation in estuarine food webs of the Northeast United States. PLoS One 9:e89305. https://doi.org/10.1371/journal.pone.0089305

Chiu S, Williams P, Krauss R (2017) Effects of a very high saturated fat diet on LDL particles in adults with atherogenic dyslipidemia: a randomized controlled trial. PLoS One 12:e0170664. https://doi.org/10.1371/journal.pone.0170664

Chowdhury R, Warnakula S, Kunutsor S et al (2014) Association of dietary, circulating, and supplement fatty acids with coronary risk. Ann Intern Med 160:398. https://doi.org/10.7326/m13-1788

Clemens R, Kranz S, Mobley A et al (2012) Filling America's fiber intake gap: summary of a roundtable to probe realistic solutions with a focus on grain-based foods. J Nutr 142(7):1390S–1401S

Clendenning R (2005) The optimal low-density lipoprotein is 50 to 70 mg/dl. J Am Coll Cardiol 45:1732. https://doi.org/10.1016/j.jacc.2005.02.051

Comerford K, Pasin G (2016) Emerging evidence for the importance of dietary protein source on glucoregulatory markers and type 2 diabetes: different effects of dairy, meat, fish, egg, and plant protein foods. Nutrients 8:446. https://doi.org/10.3390/nu8080446

Cook D, Carey I, Whincup P et al (1997) Effect of fresh fruit consumption on lung function and wheeze in children. Thorax 52:628–633. https://doi.org/10.1136/thx.52.7.628

Danilo C, Frank P (2012) Cholesterol and breast cancer development. Curr Opin Pharmacol 12:677–682. https://doi.org/10.1016/j.coph.2012.07.009

Danthiir V, Hosking D, Burns N et al (2013) Cognitive performance in older adults is inversely associated with fish consumption but not erythrocyte membrane n–3 fatty acids. J Nutr 144:311–320. https://doi.org/10.3945/jn.113.175695

Davis T, Nguyen H, Garcia-Bravo R et al (1994) Amino acid composition of human milk is not unique. J Nutr 124(7):1126–1132

Deegan L, Peterson B, Portier R (1990) Stable isotopes and cellulase activity as evidence for detritus as a food source for Juvenile Gulf Menhaden. Estuaries 13(1):14–19

Demark-Wahnefried W, Polascik T, George S et al (2008) Flaxseed supplementation (not dietary fat restriction) reduces prostate cancer proliferation rates in men presurgery. Cancer Epidemiol Biomark Prev 17:3577–3587. https://doi.org/10.1158/1055-9965.epi-08-0008

Dewell A, Weidner G, Sumner M et al (2007) Relationship of dietary protein and soy isoflavones to serum IGF-1 and IGF binding proteins in the prostate cancer lifestyle trial. Nutr Cancer 58:35–42. https://doi.org/10.1080/01635580701308034

Dewell A, Weidner G, Sumner M et al (2008) A very-low-fat vegan diet increases intake of protective dietary factors and decreases intake of pathogenic dietary factors. J Am Diet Assoc 108(2):347–356

Djoussé L, Gaziano J, Buring J et al (2010) Dietary omega-3 fatty acids and fish consumption and risk of type 2 diabetes. Am J Clin Nutr 93:143–150. https://doi.org/10.3945/ajcn.110.005603

Drewnowski A, Rehm C (2013) Vegetable cost metrics show that potatoes and beans provide most nutrients per penny. PLoS One 8 (5):e63277

Eastwood M, Kritchevsky D (2005) Dietary fiber: how did we get where we are? Annu Rev Nutr 25(1):1–8

EPA-FDA Fish (2018) EPA-FDA fish advice: technical information | US EPA. In: US EPA. https://www.epa.gov/fish-tech/epa-fda-fish-advice-technical-information. Accessed 3 Aug 2018

Fabian C, Kimler B, Zalles et al (2010) Reduction in Ki-67 in Benign Breast tissue of high-risk women with the Lignan secoisolariciresinol diglycoside. Cancer Prev Res 3:1342–1350. https://doi.org/10.1158/1940-6207.capr-10-0022

FDA (2018) Mercury Levels in Commercial Fish and Shellfish (1990–2012). In: Fda.gov. https://www.fda.gov/food/foodborneillnesscontaminants/metals/ucm115644.htm. Accessed 3 Aug 2018

Ference B, Mahajan N (2013) The role of early LDL lowering to prevent the onset of atherosclerotic disease. Curr Atheroscler Rep 15:312. https://doi.org/10.1007/s11883-013-0312-1

Fernandez E, Chatenoud L, La Vecchia C (1999) Fish consumption and cancer risk. Am J Clin Nutr 70:85–90. https://doi.org/10.1093/ajcn/70.1.85

Gabbs M, Leng S, Devassy J et al (2015) Advances in our understanding of oxylipins derived from dietary PUFAs. Adv Nutr 6:513–540. https://doi.org/10.3945/an.114.007732

Ginty A, Conklin S (2015) Short-term supplementation of acute long-chain omega-3 polyunsaturated fatty acids may alter depression status and decrease symptomology among young adults with depression: a preliminary randomized and placebo controlled trial. Psychiatry Res 229:485–489. https://doi.org/10.1016/j.psychres.2015.05.072

Glattre E, Haldorsen T, Berg J et al (1993) Norwegian case-control study testing the hypothesis that seafood increases the risk of thyroid cancer. Cancer Causes Control 4:11–16. https://doi.org/10.1007/bf00051708

Goyal A, Sharma V, Upadhyay N et al (2014) Flax and flaxseed oil: an ancient medicine & modern functional food. J Food Sci Technol 51:1633–1653. https://doi.org/10.1007/s13197-013-1247-9

Graham G (2015) Disparities in cardiovascular disease risk in the United States. Curr Cardiol Rev 11:238–245. https://doi.org/10.2174/1573403x11666141122220003

Greenberg P (2019) The omega principle. Penguin Books, New York

Greger M, Stone G (2015) How not to die, 1st ed. Flatiron books, New York

Ha V, Sievenpiper J, de Souza RJ et al (2014) Effect of dietary pulse intake on established therapeutic lipid targets for cardiovascular risk reduction: a systematic review and meta-analysis of randomized controlled trials. Can Med Assoc J 186(8):E252–E262

Hachiya N (2006) The history and the present of minamata disease – entering the second half century. JMAJ 49(3):112–118

Hasler C (2002) Functional foods: benefits, concerns and challenges—a position paper from the American council on science and health. J Nutr 132:3772–3781. https://doi.org/10.1093/jn/132.12.3772

Health.gov (2018) 2010 Dietary guidelines – health.gov. In: Health.gov. https://health.gov/dietaryguidelines/2010/. Accessed 31 Jul 2018

Hicks D, Pivarnik L, McDermott R (2008) Consumer perceptions about seafood – an Internet survey. J Foodserv 19(4):213–226

Higginson J, Pepler W (1954) Fat intake, serum cholesterol concentration, and atherosclerosis in the South African Bantu. Part II Atherosclerosis and coronary artery disease. J Clin Invest 33:1366–1371. https://doi.org/10.1172/jci103013

Holt S, Miller J, Petocz P et al (1995) A satiety index of common foods. Eur J Clin Nutr 49(9):675–690

Hooper L (2001) Dietary fat intake and prevention of cardiovascular disease: systematic review. BMJ 322:757–763. https://doi.org/10.1136/bmj.322.7289.757

Hoy M, Goldman J (2014) Fiber intake of the U.S. population What We Eat in America, NHANES 2009–2010. [online] Ars.usda.gov. https://www.ars.usda.gov/ARSUserFiles/80400530/pdf/DBrief/12_fiber_intake_0910.pdf. Accessed 15 Jun 2018

Hyman M (2009) The ecology of eating: the power of the fork. Altern Ther Health Med 15(4):14–15

Ito Y, Niiya Y, Otani M et al (1987) Effect of food intake on blood lead concentration in workers occupationally exposed to lead. Toxicol Lett 37(2):105–114

Jackson E (2018) Omega-3 Supplementation and CVD Prevention – American College of Cardiology. In: American College of Cardiology. https://www.acc.org/latest-in-cardiology/ten-points-to-remember/2017/03/14/14/42/omega-3-polyunsaturated-fatty-acid-fish-oil. Accessed 7 Aug 2018

Jahns L, Raatz S, Johnson L et al (2014) Intake of seafood in the US varies by age, income, and education level but not by race-ethnicity. Nutrients 6:6060–6075. https://doi.org/10.3390/nu6126060

Jayalath V, de Souza R, Sievenpiper J et al (2013) Effect of dietary pulses on blood pressure: a systematic review and meta-analysis of

controlled feeding trials. Am J Hypertens 27:56–64. https://doi.org/10.1093/ajh/hpt155
Jenkins D, Kendall C, Marchie A et al (2003) The Garden of Eden—plant based diets, the genetic drive to conserve cholesterol and its implications for heart disease in the 21st century. Comp Biochem Physiol A Mol Integr Physiol 136:141–151. https://doi.org/10.1016/s1095-6433(02)00345-8
Kabagambe E, Baylin A, Ruiz-Narvarez E et al (2005) Decreased consumption of dried mature beans is positively associated with urbanization and nonfatal acute myocardial infarction. J Nutr 135 (7):1770–1775
Kaizer L, Boyd N, Kriukov V et al (1989) Fish consumption and breast cancer risk: an ecological study. Nutr Cancer 12:61–68. https://doi.org/10.1080/01635588909514002
Kalaria R, Akinyemi R, Ihara M (2012) Does vascular pathology contribute to Alzheimer changes? J Neurol Sci 322:141–147. https://doi.org/10.1016/j.jns.2012.07.032
King D, Mainous A, Lambourne C (2012) Trends in dietary fiber intake in the United States, 1999–2008. J Acad Nutr Diet 112(5):642–648
Kitahara C, Berrington de González A, Freedman N et al (2011) Total cholesterol and cancer risk in a large prospective study in Korea. J Clin Oncol 29:1592–1598. https://doi.org/10.1200/jco.2010.31.5200
Kwak S, Myumg S, Lee Y et al (2012) Efficacy of omega-3 fatty acid supplements (eicosapentaenoic acid and docosahexaenoic acid) in the secondary prevention of cardiovascular disease. Arch Intern Med 172:686–694. https://doi.org/10.1001/archinternmed.2012.262
Lappé F (2011) Diet for a small planet. [S.l.]: Random House Publishing Group
Laurin D, Verreault R, Lindsay J, Dewailly É, Holub B (2003) Omega-3 fatty acids and risk of cognitive impairment and dementia. J Alzheimers Dis 5:315–322. https://doi.org/10.3233/jad-2003-5407
Leiter L, Kris-Etherton P, Vuksan V et al (2013) Effect of dietary pulses on blood pressure: a systematic review and meta-analysis of controlled feeding trials. Am J Hypertens 27(1):56–64
Li K, Huang T, Zheng J et al (2014) Effect of marine-derived n-3 polyunsaturated fatty acids on C-reactive protein, interleukin 6 and tumor necrosis factor α: a meta-analysis. PLoS One 9:e88103. https://doi.org/10.1371/journal.pone.0088103
Lindahl O, Lindwall L, Spångberg A et al (1985) Vegan regimen with reduced medication in the treatment of bronchial asthma. J Asthma 22:45–55. https://doi.org/10.3109/02770908509079883
Lloyd J, Duchin J, Borchert J et al (2013) Diarrhetic shellfish poisoning, Washington, USA, 2011. Emerg Infect Dis 19:1314–1316. https://doi.org/10.3201/eid1908.121824
Lo Y, Chang Y, Wahlqvist M et al (2012) Spending on vegetable and fruit consumption could reduce all-cause mortality among older adults. Nutr J 11:113. https://doi.org/10.1186/1475-2891-11-1133
Locke K, Pomerleau J, Causer L et al (2005) WHO | The global burden of disease attributable to low consumption of fruit and vegetables: implications for the global strategy on diet.Bul. World Health Organ 83(2):100–108
Lopata A, Jeebhay M (2013) Airborne seafood allergens as a cause of occupational allergy and asthma. Curr Allergy Asthma Rep 13:288–297. https://doi.org/10.1007/s11882-013-0347-y
Martins I, Fernando W (2014) High fibre diets and Alzheimer's disease. Food Nutr Sci 05(04):410–424
Maryland Seafood (2018) Seafood in Your Diet. [online] Seafood.maryland.gov. http://seafood.maryland.gov/seafood-in-your-diet/. Accessed 7 Jun 2018
Mason R, Lawson N, Lawrence A (1999) Mercury in the Chesapeake Bay. Mar Chem 65:77–96. https://doi.org/10.1016/s0304-4203(99)00012-2
McDougall J (2002) Plant foods have a complete amino acid composition. Circulation 25:05(25):e197; author reply e197
McLaren D (1974) The great protein fiasco. Lancet 13:93–96
McLaren D (2000) The Great Protein Fiasco revisited. Nutrition 16:464–465
MDE (2018) New interactive map to provide fish consumption advisories. In: News.maryland.gov. http://news.maryland.gov/dnr/2018/04/11/new-interactive-map-to-provide-fish-consumption-advisories/. Accessed 2 Aug 2018
Milton K (2000) Back to basics: why foods of wild primates have relevance for modern human health. Nutrition 16:480–483. https://doi.org/10.1016/s0899-9007(00)00293-8
Mnari Bhouri A, Jrah Harzallah H, Dhibi M et al (2010) Nutritional fatty acid quality of raw and cooked farmed and Wild Sea Bream (*Sparus aurata*). J Agric Food Chem 58:507–512. https://doi.org/10.1021/jf902096w
Moore L, Thompson F (2016) Adults meeting fruit and vegetable intake recommendations — United States, 2013. In: Cdc.gov. https://www.cdc.gov/mmwr/preview/mmwrhtml/mm6426a1.htm. Accessed 19 Jul 2018
Moorthi R, Vorland C, Gallant K (2017) Diet and diabetic kidney disease: plant versus animal protein. Curr Diab Rep 17:15. https://doi.org/10.1007/s11892-017-0843-x
Moradi Y, Bakar J, Motalebi A et al (2011) A review on fish lipid: composition and changes during cooking methods. J Aquatic Food Product Technol 20:379–390. https://doi.org/10.1080/10498850.2011.576449
Mudgil D, Barak S (2013) Composition, properties and health benefits of indigestible carbohydrate polymers as dietary fiber: a review. Int J Biol Macromol 61:1–6. https://doi.org/10.1016/j.ijbiomac.2013.06.044
Mulder G (1849) Chemistry of vegetable and animal physiology. Edinburgh
Munro H (1970) Free amino acid pools and their role in regulation. Mammalian protein metabolism, Vol. IV. Academic Press. pp 299–386
Nagura J, Iso H, Watanabe Y et al (2009) Fruit, vegetable and bean intake and mortality from cardiovascular disease among Japanese men and women: the JACC Study. Br J Nutr 102:285–292. https://doi.org/10.1017/s0007114508143586
Nakamura H, Takasawa M, Kashara S et al (1989) Effects of acute protein loads of different sources on renal function of patients with diabetic nephropathy. Tohoku J Exp Med 159:153–162
NCCIH 2018 (2018) Flaxseed and Flaxseed Oil. In: NCCIH (National Center for Complementary and Integrative Health). https://nccih.nih.gov/health/flaxseed/ataglance.htm. Accessed 24 Jul 2018
Nestle M (2000) Paleolithic diets: a sceptical view. Nutr Bull 25:43–47. https://doi.org/10.1046/j.1467-3010.2000.00019.x
NIH (2018) Office of Dietary Supplements – Omega-3 Fatty Acids. In: Ods.od.nih.gov. https://ods.od.nih.gov/factsheets/Omega3FattyAcids-HealthProfessional/. Accessed 6 Aug 2018
NOAA (2018). Annual landings. Retrieved from https://www.st.nmfs.noaa.gov/st1/commercial/landings/annual_landings.html
Notkola I, Sulkava R, Pekkanen J et al (1998) Serum total cholesterol, apolipoprotein E {FC12}e4 Allele, and Alzheimer's disease. Neuroepidemiology 17:14–20. https://doi.org/10.1159/000026149
NPS (2018) Flax Production in the Seventeenth Century – Historic Jamestowne Part of Colonial National Historical Park (U.S. National Park Service). In: Nps.gov. https://www.nps.gov/jame/learn/historyculture/flax-production-in-the-seventeenth-century.htm. Accessed 22 Jul 2018
NSSP (2018) National Shellfish Sanitation Program. In: Fda.gov. https://www.fda.gov/downloads/Food/GuidanceRegulation/FederalStateFoodPrograms/UCM505093.pdf. Accessed 3 Aug 2018
Possemiers S, Bolca S, Verstraete W et al (2011) The intestinal microbiome: a separate organ inside the body with the metabolic potential to influence the bioactivity of botanicals. Fitoterapia 82 (1):53–66

Pottala J, Yaffe K, Robinson J et al (2014) Higher RBC EPA + DHA corresponds with larger total brain and hippocampal volumes: WHIMS-MRI Study. Neurology 82:435–442. https://doi.org/10.1212/wnl.0000000000000080

Rangel-Huerta O, Gil A (2018) Omega 3 fatty acids in cardiovascular disease risk factors: an updated systematic review of randomised clinical trials. Clin Nutr 37:72–77. https://doi.org/10.1016/j.clnu.2017.05.015

Rasmussen B, Vessby B, Uusitupa M et al (2006) Effects of dietary saturated, monounsaturated, and n−3 fatty acids on blood pressure in healthy subjects. Am J Clin Nutr 83:221–226. https://doi.org/10.1093/ajcn/83.2.221

Ritter R, Scheringer M, MacLeod M et al (2010) Intrinsic human elimination half-lives of polychlorinated biphenyls derived from the temporal evolution of cross-sectional biomonitoring data from the United Kingdom. Environ Health Perspect 119:225–231. https://doi.org/10.1289/ehp.1002211

Rizos E, Ntzani E, Bika E et al (2012) Association between omega-3 fatty acid supplementation and risk of major cardiovascular disease events. JAMA 308:1024. https://doi.org/10.1001/2012.jama.11374

Rizzo N, Jaceldo-Siegl K, Sabate J et al (2013) Nutrient profiles of vegetarian and nonvegetarian dietary patterns. J Acad Nutr Diet 113:1610–1619

Robertson W, Peacock M, Hodgkinson A (1979) Dietary changes and the incidence of urinary calculi in the U.K. between 1958 and 1976. J Chronic Dis 32(6):469–476

Rodriguez-Leyva D, Weighell W, Edel A et al (2013) Potent antihypertensive action of dietary flaxseed in hypertensive patients. Hypertension 62:1081–1089. https://doi.org/10.1161/hypertensionaha.113.02094

Ruzzin J (2012) Public health concern behind the exposure to persistent organic pollutants and the risk of metabolic diseases. BMC Public Health 12:298. https://doi.org/10.1186/1471-2458-12-298

Sacks FM, Katan M (2002) Randomized clinical trials on the effects of dietary fat and carbohydrate on plasma lipoproteins and cardiovascular disease. Am J Med 113(9):13–24

Santomauro AT, Boden G, Silva ME, Rocha DM, Santos RF, Ursich MJ, Strassmann PG, Wajchenberg BL (1999) Overnight lowering of free fatty acids with Acipimox improves insulin resistance and glucose tolerance in obese diabetic and nondiabetic subjects. Diabetes 48(9):1836–1841

Sarter B, Kelsey K, Schwartz T et al (2015) Blood docosahexaenoic acid and eicosapentaenoic acid in vegans: associations with age and gender and effects of an algal-derived omega-3 fatty acid supplement. Clin Nutr 34:212–218. https://doi.org/10.1016/j.clnu.2014.03.003

Scales C, Smith A, Hanley J et al (2012) Prevalence of Kidney Stones in the United States. Eur Urol 62(1):160–165

Schulte D (2017) History of the Virginia Oyster Fishery, Chesapeake Bay, USA. Front Marine Sci. https://doi.org/10.3389/fmars.2017.00127

Secretan B, Straif K, Baan R et al (2009) A review of human carcinogens—Part E: tobacco, areca nut, alcohol, coal smoke, and salted fish. Lancet Oncol 10:1033–1034. https://doi.org/10.1016/s1470-2045(09)70326-2

Semba R (2016) The rise and fall of protein malnutrition in global health. Ann Nutr Metab 69(2):79–88

Sicherer S, Muñoz-Furlong A, Sampson H (2004) Prevalence of seafood allergy in the United States determined by a random telephone survey. J Allergy Clin Immunol 114:159–165. https://doi.org/10.1016/j.jaci.2004.04.018

Simopoulos A (2002a) The importance of the ratio of omega-6/omega-3 essential fatty acids. Biomed Pharmacother 56:365–379. https://doi.org/10.1016/s0753-3322(02)00253-6

Simopoulos A (2002b) Omega-3 fatty acids in inflammation and autoimmune diseases. J Am Coll Nutr 21:495–505. https://doi.org/10.1080/07315724.2002.10719248

Singh M (2005) Essential fatty acids, DHA and human brain. Indian J Pediat 72:239–242. https://doi.org/10.1007/bf02859265

Singh K, Rohatgi A (2018) Examining the paradox of high high-density lipoprotein and elevated cardiovascular risk. J Thoracic Dis 10:109–112. https://doi.org/10.21037/jtd.2017.12.97

Singh K, Mridula D, Rehal J, Barnwal P (2011) Flaxseed: a potential source of food, feed and fiber. Crit Rev Food Sci Nutr 51:210–222. https://doi.org/10.1080/10408390903537241

Singhal P, Kaushik G, Mathur P (2013) Antidiabetic potential of commonly consumed legumes: a review. Crit Rev Food Sci Nutr 54 (5):655–672

Siri-Tarino P, Sun Q, Hu F et al (2010) Meta-analysis of prospective cohort studies evaluating the association of saturated fat with cardiovascular disease. Am J Clin Nutr 91:535–546. https://doi.org/10.3945/ajcn.2009.27725

Sobsey M, Hackney C, Carrick R et al (1980) Occurrence of enteric bacteria and viruses in oysters. J Food Prot 43:111–113. https://doi.org/10.4315/0362-028x-43.2.111

Sonnenburg E, Sonnenburg J (2014) Starving our microbial self: the deleterious consequences of a diet deficient in microbiota-accessible carbohydrates. Cell Metab 20(5):779–786

Steinberg D, Grundy S (2012) The case for treating hypercholesterolemia at an earlier age. J Am Coll Cardiol 60:2640–2642. https://doi.org/10.1016/j.jacc.2012.09.016

Stott-Miller M, Neuhouser ML, Stanford JL (2013) Consumption of deep-fried foods and risk of prostate cancer. Prostate 73:960–969. https://doi.org/10.1002/pros.22643

Stripp C, Overvad K, Christensen J et al (2003) Fish intake is positively associated with breast cancer incidence rate. J Nutr 133:3664–3669. https://doi.org/10.1093/jn/133.11.3664

Sugimura T, Wakabayashi K, Nakagama H (2004) Heterocyclic amines: mutagens/carcinogens produced during cooking of meat and fish. Cancer Sci 95:290–299. https://doi.org/10.1111/j.1349-7006.2004.tb03205.x

Thomas W, Davies J, O'Neal R et al (1960) Incidence of myocardial infarction correlated with venous and pulmonary thrombosis and embolism. Am J Cardiol 5:41–47. https://doi.org/10.1016/0002-9149(60)90007-2

Thompson L (2005) Dietary flaxseed alters tumor biological markers in postmenopausal breast cancer. Clin Cancer Res 11:3828–3835. https://doi.org/10.1158/1078-0432.ccr-04-2326

Tirschwell D, Smith N, Heckbert S et al (2004) Association of cholesterol with stroke risk varies in stroke subtypes and patient subgroups. Neurology 63:1868–1875. https://doi.org/10.1212/01.wnl.0000144282.42222.da

Tracy C, Best S, Bagrodia A et al (2014) Animal protein and the risk of kidney stones: a comparative metabolic study of animal protein sources. J Urol 192(1):137–141

Turney B, Appleby P, Reynard J et al (2014) Diet and risk of kidney stones in the Oxford cohort of the European Prospective Investigation into Cancer and Nutrition (EPIC). Eur J Epidemiol 29 (5):363–369

Uribarri J, Oh M (2012) The key to halting progression of CKD might be in the produce market, not in the pharmacy. Kidney Int 81(1):7–9

Uribarri J, Woodruff S, Goodman S et al (2010) Advanced glycation end products in foods and a practical guide to their reduction in the diet. J Am Diet Assoc 110:911–916.e12. https://doi.org/10.1016/j.jada.2010.03.018

USDA (2018a). Food composition databases show foods list. [online] Ndb.nal.usda.gov. https://ndb.nal.usda.gov/ndb/search/list. Accessed 12 Jun 2018

USDA (2018b) A closer look at current intakes and recommended shifts – 2015–2020 dietary guidelines – health.gov. In: Health.gov.

https://health.gov/dietaryguidelines/2015/guidelines/chapter-2/a-closer-look-at-current-intakes-and-recommended-shifts/#food-groups. Accessed 31 Jul 2018

USDA (2018c) Mission Areas | USDA. In: Usda.gov. https://www.usda.gov/our-agency/about-usda/mission-areas. Accessed 31 Jul 2018

USDA (2018d) Cotton & Tobacco | Agricultural Marketing Service. In: Ams.usda.gov. https://www.ams.usda.gov/about-ams/programs-offices/cotton-tobacco. Accessed 31 Jul 2018

USEPA (2018a) Estimated Fish Consumption Rates for the U.S. Population and Selected Subpopulations. In: Epa.gov. https://www.epa.gov/sites/production/files/2015-01/documents/fish-consumption-rates-2014.pdf. Accessed 1 Aug 2018

USEPA (2018b) Learn about Polychlorinated Biphenyls (PCBs) | US EPA. In: US EPA. https://www.epa.gov/pcbs/learn-about-polychlorinated-biphenyls-pcbs. Accessed 1 Aug 2018

VDH (2018) Virginia Department of Health. Fish Consumption Advisories – Environmental Epidemiology. In: Vdh.virginia.gov. http://www.vdh.virginia.gov/environmental-epidemiology/public-health-toxicology/fish-consumption-advisories/. Accessed 1 Aug 2018

Villalba A, Jahncke M, Kuhn D et al (2018) Safe and Nutritious Seafood in Virginia. [online] Virginia Cooperative Extension, Publication AREC-156P. https://pubs.ext.vt.edu/AREC/AREC-156/AREC-156.html. Accessed 7 Jun. 2018

Wang Y, Beydoun M, Caballero B et al (2010) Trends and correlates in meat consumption patterns in the US adult population. Public Health Nutr 13:1333–1345. https://doi.org/10.1017/s1368980010000224

Waterlow J (1984) Kwashiorkor revisited: the pathogenesis of oedema in kwashiorkor and its significance. Trans R Soc Trop Med Hyg 78 (4):436–441

Weininger J, Truswell A, Carpenter K et al (2018) Human nutrition. [online] Encyclopedia Britannica. https://www.britannica.com/science/human-nutrition. Accessed 11 Jun 2018

Wood L, Garg M, Smart J (2012) Manipulating antioxidant intake in asthma: a randomized controlled trial. Am J Clin Nutr 96:534–543. https://doi.org/10.3945/ajcn.111.032623

Xu X, Newman M, Fabrizio M et al (2013) An ecologically framed mercury survey of finfish of the lower Chesapeake Bay. Arch Environ Contam Toxicol 65:510–520. https://doi.org/10.1007/s00244-013-9917-1

Yaginuma-Sakurai K, Murata K, Iwai-Shimada M et al (2012) Hair-to-blood ratio and biological half-life of mercury: experimental study of methylmercury exposure through fish consumption in humans. J Toxicol Sci 37:123–130. https://doi.org/10.2131/jts.37.123

Yang S, Miah A, Pabari A et al (2011) Growth Factors and their receptors in cancer metastases. Front Biosci 16:531. https://doi.org/10.2741/3703

Yang H, Xun P, He K (2013) Fish and fish oil intake in relation to risk of asthma: a systematic review and meta-analysis. PLoS One 8:e80048. https://doi.org/10.1371/journal.pone.0080048

Yokoo E, Valente J, Grattan L et al (2003) Low level methylmercury exposure affects neuropsychological function in adults. Environ Health. https://doi.org/10.1186/1476-069x-2-8

Young V, Pellett P (1994) Plant proteins in relation to human protein and amino acid nutrition. Am J Clin Nutr 59:1203S–1212S

Zahradka P, Wright B, Weighell W et al (2013) Daily non-soy legume consumption reverses vascular impairment due to peripheral artery disease. Atherosclerosis 230(2):310–314

Zhang Y, Tuomilehto J, Jousilahti P et al (2012) Total and high-density lipoprotein cholesterol and stroke risk. Stroke 43:1768–1774. https://doi.org/10.1161/strokeaha.111.646778

Sugar Twice Enslaves: Consequences for the People of the Chesapeake Bay

Benjamin E. Cuker and Michelle Penn-Marshall

Its sweetness guided us to consume more fruit
Providing carbs, antioxidants, and fiber to boot
We got greedy and separated the sweetness from the food
Because eating more sugar improved our mood
The sugar was extremely hard to make
Because labor from slaves was what it did take
When slave produced sugar came to the Chesapeake
Its consumption made us morally and physically weak
Today we enslave ourselves to sugar addiction
Making healthy eating a fiction
Every year the people get sicker and fatter
But we eat added sugar and meat no matter
When eating whole plant-based foods instead
Would lift us from our own sickbed
Returning to none but sugar in the fruit
Would help make obesity and diabetes moot

Abstract

The Chesapeake Bay region was highly involved in the colonial era triangle trade that connected the sugar colonies of the Caribbean to Western Europe and Africa. This brought both sugar and slavery to the Chesapeake. While chattel slavery has passed, virtual slavery to sugar persists as this addictive substance is widely consumed by African Americans and only somewhat less so by others in the Chesapeake region today. The excess calories from added sugars combines with increased consumption of animal products to set up widespread nonalcoholic fatty liver disease (NAFLD) that leads to diabetes and obesity. Per capita annual sugar consumption in the US increased 20-fold between 1820 (3 kg) and 2016 (60 kg). Sweetened beverages, candies and pastries, provide the bulk of added dietary sugars, but 22% is hidden in a large variety of processed foods including soups, meats, prepared meals and condiments. Non-caloric artificial sweeteners also promote weight gain. Retraining the pallets of consumers to accept shifting to a whole-food plant-based diet promises to free the population from type-2 diabetes and other obesity related diseases.

Keywords

Added sugars · Obesity · Diabetes · US diet · Processed foods

Baltimore the Sugar Port The huge, neon-lit Domino Sugar sign greets mariners on their port side as they make their way to Baltimore's Inner Harbor (Fig. 1). Nearing the Domino Sugar processing plant, a pungent sweetness wafts over them. A bucket dangling from an industrial crane digs deep in the hull of a sugar ship, transferring 7–10 tons of granules ashore with each mechanical bite (Mountford 2003). A sugar ship arrives about every two weeks and stays for 7 days to offload

B. E. Cuker (✉)
Department of Marine and Environmental Science, Hampton University, Hampton, Virginia, USA
e-mail: Benjamin.cuker@hamptonu.edu

M. Penn-Marshall
The Office of the Vice President for Research, Associate Provost and Graduate College, Hampton University, Hampton, VA, USA
e-mail: Michelle.penn-marshall@hamptonu.edu

B. E. Cuker (ed.), *Diet for a Sustainable Ecosystem*, Estuaries of the World, https://doi.org/10.1007/978-3-030-45481-4_14

Fig. 1 The Domino sugar processing plant located on the shores of Baltimore's inner harbor. https://commons.wikimedia.org/wiki/File:Domino_Sugars_plant_building,_Baltimore,_Maryland.JPG

32–44 million kg (70–98 million pounds) of its sweet cargo. The factory processes 3 million kg (6.5 million pounds) of sugar a day, or about 20% of the US demand (Samford 2015; Statistica 2018). The plant, built in 1922, continues the tradition of importing sugar to the Chesapeake that dates to the 1700's.

Large-scale sugar imports into Baltimore began during the Revolutionary War. By 1825, Baltimore featured 11 sugar refineries. Additional sugar-works on the Chesapeake operated in Alexandria, Virginia and Washington, D.C., along the shores of the Potomac River (Gutman et al. 2012; Samford 2015).

Today, Baltimore is the largest import center for sugar in the US. Initially the sugar islands of the Caribbean provided most of the import, but now shipments also come from distant ports in Africa, South America and elsewhere (Dinsmore 2017). The Domino factory is the last of the great manufacturing facilities that once lined the shores of Baltimore's Inner Harbor (Gutman et al. 2012). The persistence of this century-old factory says much about the sustained demand for sugar. It also speaks to the usefulness of the Chesapeake Bay as watery highway for delivery of raw material, and the effectiveness of the sugar lobby in producing favorable legislation for the business. Sugar, along with other processed and animal-based foods, defines the current Chesapeake Bay food system and the Standard American Diet.

The What and Why of Sugar Table sugar is sucrose, a disaccharide consisting of one glucose covalently bound to one fructose molecule (Fig. 2). Fructose is a pentose, or 5-carbon sugar. Glucose is a hexose, or 6-carbon sugar. Metabolism releases the energy stored in these sugars for use by the cells of bacteria, fungi, plants and animals. While other energy-rich molecules (lipids and amino acids) also fuel metabolism, it is glucose that plays the central role in delivering energy to cells, tissues, organs and organisms.

Most plants store sugar in a long-chain polymer called starch. Starchy plant tissues provide energy for future growth and reproduction. Tubers provide stored energy for the next season's vegetative growth for plants such as potatoes. Plants package starch and oils in seeds to fuel sprouting embryos. Eating starchy parts of plants such as potatoes, grains and beans provides a rich source of sugars for human metabolism. Amylases are digestive enzymes that break the bonds between the individual monomers in starch, making the sugars available for uptake and use by animals.

Not all plant-sugar is stored as starch. Plants evolved the capacity to concentrate sugars in specialized tissues to invite animal consumption. Floral nectars provide a sugar-rich drink to pollinating insects (bees) and birds (hummingbirds) that enable the plant's sexual reproduction. Many flowering plants produce another organ rich in sugars called fruit. Fruit envelops the plant's seeds. This invites animals to consume the fruit and seed package. Most of the hard-coated seeds survive passage through the animal's gut, emerging fully competent in the feces. The fruit-eating animal provides the plant with dispersal to a new location for its

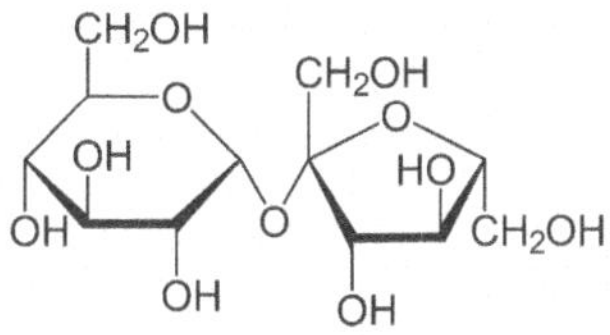

Fig. 2 Sucrose, consists of glucose (left) and fructose (right) joined by a covalent bond. That bond is broken in the first metabolic step to allow each sugar to be metabolized by cells. Image from: https://en.wikipedia.org/wiki/Sucrose#/media/File:Saccharose2.svg, and created by NEUROtiker

next generation and nutrient-rich feces to nourish the emerging seedling in its early days of life.

Human evolution traces to the great ape (Hominidae) family of primates, including gorillas, chimpanzees and orangutans. The extant non-human great apes are almost exclusively plant eaters, with chimpanzees and orangutans adding occasional insects or other small animals to their diets. Fruits play a large role in the diets of great apes (Dunn 2012). The human predilection to consuming sweet foods seems tied to the importance of fruit in our ancestral diets. Apparently, fruit consumption by humans remains a critical dietary component, as the massive Global Burden of Disease study finds the lack of dietary fruit to be the number one driver of mortality, and responsible for 4.9 million deaths annually around the world (Lozano et al. 2012). It's not just the calories delivered by the fruit, but a host of other critical nutrients including fiber and antioxidants that make fruit an essential part of a healthy human diet that prevents disease (de Oliveira et al. 2008; Chiba et al. 2011; Wood et al. 2012; Ananthakrishnan et al. 2013).

Other than the occasional dollops of honey gathered from beehives, the vast majority of human evolution took place absent a source of sugar separated from the plant that produced it. The successful human search for sweetness always ended with consuming a piece of ripe fruit until ingenuity and industry enabled the extraction of sugar from plants. The momentous invention changed the history of the world and the health of the people that populate it.

History of Produced Sugar Following is an account condensed from the late Sidney Mintz (Mintz 1985), a former professor at Johns Hopkins University and resident of the Chesapeak Bay. Sugar cane (*Saccharum officinarum*) is a grass that was probably first domesticated in New Guinea and subsequently brought to the Philippines, India and Indonesia about 8000 BCE. This large tropical to subtropical grass produces a stem of up to 5 cm (2 inches) in diameter, and 4.6 m (15 feet) in height (Fig. 3). Although it grows quickly in its first 6 weeks of life, it requires 9–18 months to "ripen," that is to have sufficient sucrose stored in its stem for harvest.

Ancients could release the sucrose from the cane by chewing and sucking out the sweetness. They discovered that crushing the fibers in the cane with instruments also released the sweet juice. Heating that juice in a vessel over a fire would evaporate most of the water, leaving behind a golden or brown sugar saturated liquid (molasses) and a portion that upon separation and further heating would produce crystals of sucrose (granular sugar). The Hindu religious document, the *Buddhagosa*, provides perhaps the first written evidence of sugar making around 500 AD. Clearly, sugar making was a long and labor-intensive process. As such, cooks used it sparingly as a spice or condiment to be sprinkled on a dish to provide some sweetness, in a similar manner as one might add salt or ground pepper seeds from a shaker. Sugar was also used as a medicinal to address a variety of ailments. However, the work required to produce meaningful quantities of sugar meant it was expensive and reserved for the wealthy (Mintz 1985).

Expanding networks of trade brought sugar cane to China, the Arab world, North Africa and other places around the Mediterranean. The Moors of North Africa carried Islam and sugar with them as they invaded Europe, conquering the Iberian Peninsula (Spain) in 711 AD. Moorish influence extended to other reaches of Europe. Assimilation and expulsion returned Spain to firm European control circa 1492, the same year of Columbus's first voyage of discovery to the New World. The Europeans returned the favor of invasion, with the Great Crusades (1095–1291) assaulting the Middle East. The Crusades further exposed Europeans to sugar and put them in charge of its production in places such as the kingdom of Jerusalem (1099–1187). The expulsion of the Crusaders in 1291 ended the European control of sugar production in the Middle East. Europeans retained access to sugar through the development of industries in the eastern Mediterranean (Crete, Cyprus, Morocco, Sicily, Spain) possessing a climate that supported its cultivation. Some of these plantations used slave labor. The center of sugar production shifted to the Portuguese and Spanish controlled Atlantic Islands (Madera Islands, Canary Islands, Azores and Cape Verde Islands) circa 1450. These plantations expanded the use of slaves. While profitable, sugar produced during this era was still expensive and a luxury.

Columbus brought sugar cane from the Canary Islands to Santo Domingo on his second voyage to the New World. The labor of enslaved Africans produced the first shipment of sugar from the New World to Europe in 1516. The sugar industry spread throughout the Caribbean and into Brazil. Producers built mills powered by animals, water and then wind to crush the cane. The ruins of the sugar windmills cap many a peak throughout the Caribbean. The Portuguese in Brazil, and the French, English and Dutch in the Caribbean took over the sugar industry and produced enough to move the product from expensive luxury to the status of common

Fig. 3 Sugar cane field in Madeira, image by Hannes Grobe. https://upload.wikimedia.org/wikipedia/commons/d/d6/Sugar_cane_madeira_hg.jpg

commodity in Europe. In 1660, England imported 1000 hogsheads (238,000 liters), in 1700, 50,000 hogsheads and by 1730 twice that.

Europeans primarily used sugar to sweeten the two new beverages of colonial origin, tea and coffee. These sweetened drinks became central to European food-culture during the industrial revolution (starting ca 1760). European factory bosses gladly gave the workers tea or coffee breaks, as the caffeine and sugar combination appeared to renew the energy of their employees. Soon, confectioners used sugar to produce candies, and bakers used it to sweeten their products.

The role of sugar in the industrialization of Europe went beyond fueling the labors of the new working class. Indeed, the sugar plantation system pioneered in the New World became the model of industrialization to produce all manner of manufactured goods. The only real difference being the reliance upon wage laborers instead of slaves.

The production of granulated sugar is associated with the byproduct, molasses. This syrup was and is also used as a food-sweetener. Importantly, it is also the basis for production of the alcoholic beverage called rum. The European industrial worker would drink the stimulatory sugary tea or coffee during the day and rum at night to relax.

Produced Sugar and Slavery Comes to the Chesapeake Region Ships from the Caribbean brought sugar and rum to the colonies of the Chesapeake Bay. The same vessels delivered Virginia's and Maryland's tobacco and cotton to European ports in the triangle trade. So, by the eighteenth century, sugar became as much of a staple in the Chesapeake Bay food system as it had for Europe.

It was not just the sugar and rum that entered the Chesapeake Bay food system from the Caribbean. The concept of using African slave labor to produce those commodities soon became central to agriculture of the Bay region. In 1619, John Rolfe (the father of the Virginia tobacco export industry) reported that "20 and odd" Africans were brought to the Chesapeake colonies by the English privateer, White Lion. The English ship took them from a

Portuguese slaver, which had gotten its human cargo from Angola. Before being taken by the crew of the White Lion, the Africans were destined for the West Indian port of Vera Cruz and a life of slavery on a sugar plantation. The sale of these Africans in the Chesapeake was the first step in bringing the slave-system developed for sugar production to the nascent tobacco and food system industries of North America (NPS 2018; Walsh 2003).

Other Sweeteners The tropical to sub-tropical climate required by sugar cane restricted where it could be grown. Native Americans from New England made their own sugar extracts from the sap of maple (*Acer* sp.) trees and less often birch (*Betula* sp.) trees. The European colonists adopted their techniques to produce maple syrup and sugar (Ciesla 2002). A group of abolitionists, including Thomas Jefferson (who paradoxically owned over 100 slaves), understood the link between cane sugar and slavery. They attempted to start a maple sugar industry in the Chesapeake Bay region to replace the imported sucrose tainted by enslavement. Only small mountainous regions of Virginia, such as Highland County, offered the maple trees a climate conducive to sugar production. The small Virginia maple sugar industry could never rival the imported sucrose from cane, yet still exists today and has been celebrated with an annual festival since 1958 (Theobald 2012; Maple Festival 2018).

Europeans also sought an alternative to cane sugar. They looked for a homegrown source of sucrose to supply their needs when war and lesser disputes interrupted trade from the tropical sugar producers. This led to the cultivation of a sucrose rich variety of beet (*Beta vulgaris*), called the sugar beet (Fig. 4). The sugar beet industry started in mid-eighteenth century Prussia and subsequently flourished in France due to support from Emperor Napoleon Bonaparte (Austin 1928). Inspired by the European success, American abolitionists founded the "Beet Sugar Society of Philadelphia" in 1836 seeking sucrose untainted by slavery, as was the goal of the preceding maple sugar project (Pedder 1836). It was not until the time of the Civil War that the US sugar beet industry took root, but it required government subsidies and tariffs on imported cane sugar to blossom in the early twentieth century. As of 2018, the US domestic production of sucrose is nearly evenly split between sugar beets and sugar cane (USDA ERS 2018). Presently, all US sugar production takes place in states outside of the Chesapeake Bay Watershed. Cane sugar production is concentrated in Florida, Louisiana and Florida, with sugar beets grown in the west and Midwest (ASA 2018).

In addition to sucrose from cane and beets, the food industry uses other sweeteners. Foremost is high fructose corn syrup (HFCS), discussed below. Other caloric sweeteners include all types of syrups: corn sweetener, dextrose, fructose, glucose, honey, invert sugar, lactose, malt syrup, maltose, molasses, trehalose; turbinado sugar and fruit juice concentrates (Bowman 2017).

High Fructose Corn Syrup (HFCS) In the 1970s, food processors sought alternatives to federally price-supported domestic cane and sugar beet sucrose, as well as the imported sucrose made more expensive by protective tariffs. They found the cheaper sweetener of their desire in the form of a new synthetic sugar, high fructose corn syrup (HFCS). Cane, beet, fruit and maple sugars were all produced by concentrating the sucrose naturally occurring in those plants. Instead, HFCS is produced by using one enzyme to produce corn syrup from cornstarch, and another enzyme to convert a portion of the glucose into fructose. Why convert the hexose to the pentose? Fructose is the sweetest sugar known (Bray and Popkin 2014), and to compete with sucrose from cane or beets, there must be sufficient amounts of the pentose. HFCS42 (42% fructose) is used to sweeten processed foods, while the HFCS52 (52% fructose) sweetens soft drinks (StarchEU 2013).

Added Sugar Is a Major Part of the Standard American Diet Between 1820 and 2000, people in the US increased their intake of added sweeteners from 3 kg per person per year to 68 kg per person per year, a 22-fold increase (Fig. 5). Sugar supplies 3.75 kcal per g (USDA 2018a). So, on average, a person in 1820 ingested about 11,250 extra calories (kcals) per year from sweeteners, and their counterpart in the year 2000 consumed approximately 255,000 calories from those added sugars. For perspective, consider that the recommended daily caloric consumption for moderately active adults in their mid-30's is 2000 calories for women and 2600 calories for men (USDA 2018b). Exceeding those daily caloric targets will likely result in overweight or obesity. The 255,000 added calories per year works out to about 699 calories per day from sweeteners. That is 35% of the recommended caloric consumption for women and 27% of that for men. This may over-estimate sugar consumption by women, as the per capita data from USDA doesn't account for gender. Note that the US Center for Disease Control (CDC) estimates that women and men presently obtain about 13% of their calories from added sugar (CDC 2018a). The discrepancy comes from most Americans consuming well above the recommended number of calories per day.

Sugar consumption varies by ethnicity in the US. The percent of calories in the diet of adults from added sugar is highest for Non-Hispanic African Americans (14.5% males, 15.2% females), followed by Mexican Americans (12.9% males,

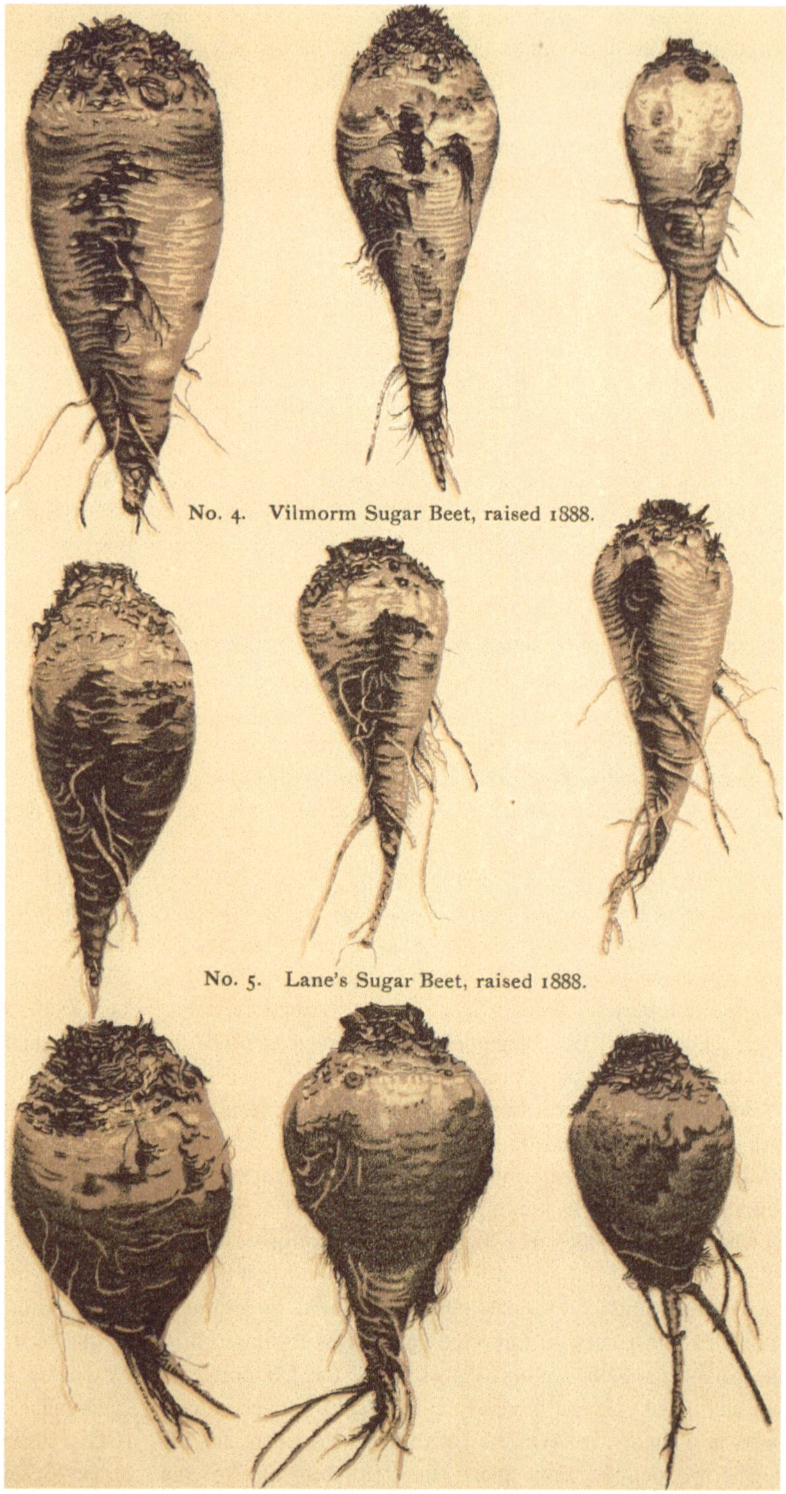

Fig. 4 Drawings of different varieties of sugar beet roots from: Annual report of the Secretary of the Board of Agriculture, Massachusetts. State Board of Agriculture, 1854–1915. https://upload.wikimedia.org/wikipedia/commons/4/42/Annual_report_of_the_Secretary_of_the_Board_of_Agriculture%22_%281854-1915%29_%2818744700114%29.jpg

12.6% females) and non-Hispanic whites (12.8% males, 12.6% females) (CDC 2018a).

The nature of caloric sweeteners changed with the invention of HFCS. This cheaper product entered the market ca. 1970 and by 2000 was consumed at the same rate as traditional refined sucrose (Fig. 6). However, public health concerns (discussed below) around HFCS caused a 38% decline in consumption of this synthetic sweetener between 2000 and 2017.

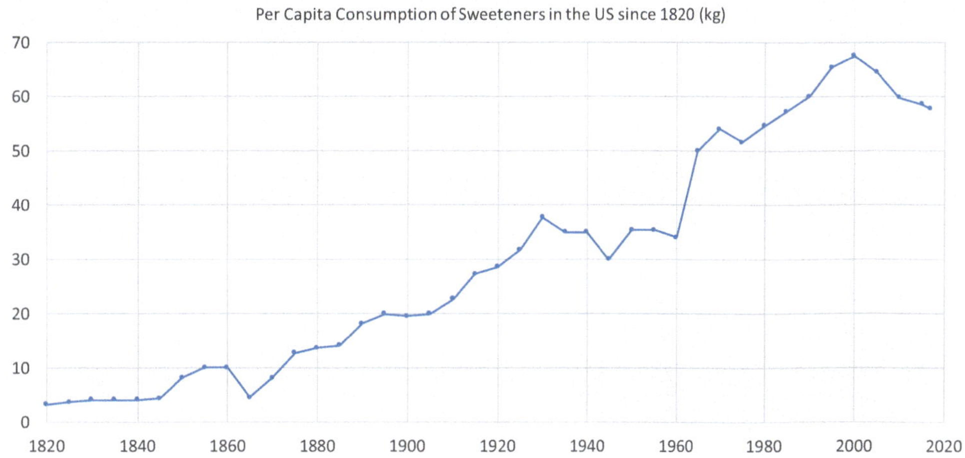

Fig. 5 Estimated per capita consumption of sweeteners. This includes all types of calorific sweeteners. Data from 1820–1965 is drawn from Guyenet (2012). Data from 1965–2018 is from USDA ERS (2018)

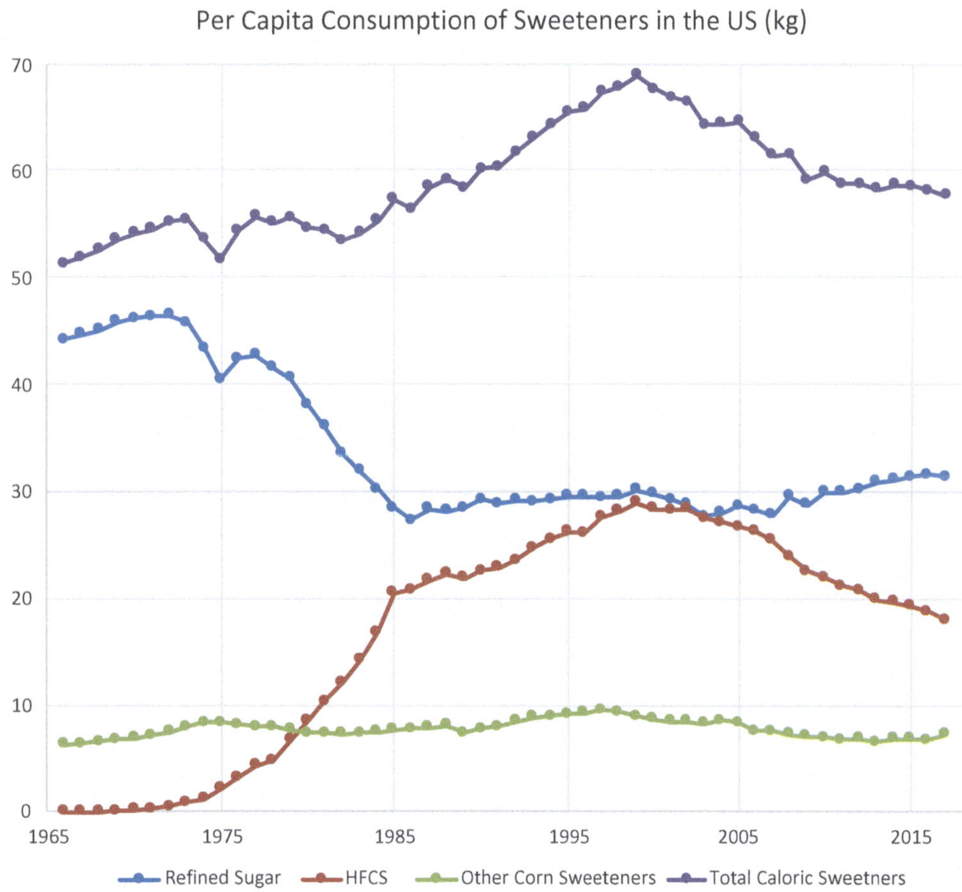

Fig. 6 Per capita consumption of sweeteners in the US since 1966. Note the advent of HFCS (High Fructose Corn Syrup) meant a reduction in the consumption of traditional refined sugar. Data from USDA ERS 2018

Sugar as Preservative The availability of food produced by the Chesapeake Bay Ecosystem varies by season. In the era before canning and refrigeration, it was important for pre-industrial residents to preserve food to get them through the lean months of winter and early spring. The Algonquin used smoking to cure (preserve by drying) fish. The low heat from the fire dried the fish and the smoke included anti-microbial compounds. European colonists followed this practice and introduced the desiccation (drying) of meat with either salt or sugar. The added salt or sugar provided high osmotic potential, drawing water from the meat. Depending on the food involved, sugar also promoted the growth of a microbial community that produced lactic acid or ethyl alcohol, which inhibited bacteria that caused spoilage (Anon 2018). In the Chesapeake region, sugar also gained favor for curing their primary export crop, tobacco. This also added to the desirability of the product (Northrup 2016).

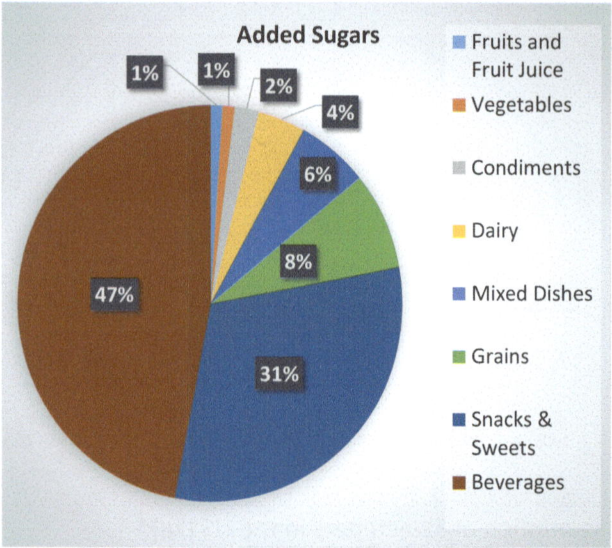

Fig. 7 The distribution of added caloric sweeteners in the US diet. Dairy is mostly sweetened milks and yogurt. Mixed dishes refer to prepared foods such as lasagna and frozen dinners. Grains include breakfast cereals and breads. Snacks and sweets mean candies and baked goods. Beverages refer to sweetened drinks other than milk and fruit juices, but includes soft drinks, fruit drinks, coffee & tea, sport & energy drinks, and alcoholic beverages. Data from the Sugar Association (2018)

Sources of Added Sugars Only a small portion of added sugar arrives in the Standard American Diet by the conscious effort of the consumer using it as a condiment to sweeten a dish or beverage. The vast majority comes packaged in pre sweetened beverages and processed foods. Nearly 75% of all foods and beverages in the Standard American Diet contain added sugar (Bray and Popkin 2014). The Sugar Association is the trade group for caloric sweeteners. They provide a breakdown that shows 78% of added sugar comes from beverages and manufactured snacks and sweets. Most of the other 22% appears in breads, prepared meals and condiments (Fig. 7, Sugar Association 2018).

Unless consumers read the nutrition facts label required on all processed foods, they unknowingly consume significant amounts of added sugar. For example, barbecue sauce is a popular condiment in the Chesapeake Bay region and is used to baste and flavor meats. Consumer admiration for this condiment explains the 180 different brands sold in the US. Most offer essentially very similar nutritional profiles per 38 grams (2 tablespoon) serving; about 60 calories from a sweetener and some tomato puree, with about 350 mg of sodium (Myfooddiary 2018). A typical ingredient list is; high fructose corn syrup, vinegar, water, tomato paste, modified cornstarch, molasses, salt, natural hickory smoke flavor, dried garlic, mustard flour, potassium sorbate, spice, dried onions, caramel color, paprika (Kraft Barbecue Sauce label information). As evident from Fig. 6, consumption of HFCS has dropped some 30% in recent years, as compared to its peak usage around the year 2000. While most barbecue sauces on grocer's shelves contain HFCS, some brands boast their use of sugar instead. No matter what the sweetener used, barbecue sauce adds significant numbers of calories to the dish.

Added Sugar for Breakfast Most brands of prepared breakfast cereals contain added sugars. Consider these popular choices. For each 237 ml (1 cup) serving the grams of added sugar are: *Cheerios* (1 g), *Kellogg's Corn Flakes* (3 g) *Honey Nut Cheerios* (9 g), *Kellogg's Rice Krispies* (9 g), *Kellogg's Froot Loops* (10 g), *Kellogg's Froot Loops with Marshmallows* (14 g), *French Archer Farms Vanilla Almond Granola* (32 g). What this means is that a bowl of sugary cereal such as *Froot Loops with Marshmallows* is 54% sugar. Breakfast pastries also contain much added sugar. *Frosted Strawberry Pop-Tarts* offer 16 g of added sugar per serving, but still less than the 20 g in the *Cookies and Cream* version of the pastry. The later item is 40% sugar by mass.

Added Sugar for Lunch Consider a typical lunch of sandwich and soup. Most manufactured whole wheat breads offer about 3 g (1–6 g) of added sugar per slice. Put some *Healthy Ones Honey Ham* on the bread to make a sandwich and gain an additional 3 g of added sugars. Add a slice of *Kraft American Cheese* and get another 1 g of sweetener. Top it off with 3 *Vlasic Bread and Butter Pickle Chips* to get another 5 g of sugar, and 15 g of *Bolthouse Farms Honey Mustard Yogurt Dressing* for 3 g of extra sugar. A cup (237 ml) of Progresso Garden Vegetable Soup to go with the sandwich provides 4 g of added sugar (and only 3 g of dietary fiber and 3 g of protein). The seemingly unsugary lunch of sandwich and soup yields 22 g of added sugar, the same as a half of a 355 ml (12 ounce) can of cola. Replacing the ham and cheese with 30 ml (2 tbsp) of *Jif Peanut Butter* (3 g sugar) and 15 ml (1 tbsp) of *Smucker's Natural Strawberry Jam* (10 g sugar) would yield about the same added sugar.

Added Sugar for Dinner Sugar sweetens many prepared meals. *P F Chang's Frozen Orange Chicken* meal contains 33 g of added sugar; more mass than the 20 g of protein that comes from the meat in the dish. Even prepared meals marketed to the health-conscious, spike their offerings with sugar. *Lean Cuisine's Lasagna with Meat Sauce* offers 9 g of added sugar per serving, half the mass of the reported protein and over twice the content of dietary fiber in the dish. Is it better to make one's own pasta sauce? Prepared tomato sauces for use on pasta provide about 10 g (4–24 g) of added sugar per 237 ml (1 cup).

Added Sweeteners Damage the Public's Health by Reducing Room for More Nutritious Foods By their very nature, extracted sugars detract from the healthful aspects of human diets. The act of extraction separates the sweetness from the nutrition offered by the plant. As noted above, humans attract to naturally sweet fruits. These offer high *nutritional density*. That is, fruit provides important quantities of nutrients essential for human health in addition to the simple carbohydrates used to fuel metabolism. Sucrose and HFCS offer very low nutritional density, providing calories absent any fiber, vitamins, antioxidants, protein or anything else (Table 1). This means that every calorie given over to a highly refined sweetener results in less consumption of a food with higher nutritional density. It is possible for people consuming 30% or more of their calories from added sugars to maintain a steady and seemingly healthy weight as long as they keep their total energy budget neutral (calories consumed = calories used in metabolism). Yet such a diet will result in potential nutrient deficiencies.

It is not that a sweetener offers nothing to its consumer. The starving colonists of Jamestown would certainly have benefited from adding sugar to their meager diets during the winter famine of 1609. Additional calories from any nontoxic source would have helped them survive. Yet the vast majority of current residents of the Chesapeake, and the US in general, don't suffer from a lack of calories in their diets. Quite the opposite is true, as many are sickened by a diet featuring excess calories. However, one sweetener stands out as providing meaningful nutrition in addition to calories. Molasses differs from refined cane and beet sugars, as well as HFCS, in offering significant amounts of calcium, iron and antioxidants (Table 1).

Added Sweeteners Help Drive Obesity Adding refined sugars to the diet means extra calories without the benefit of satiating fiber or protein. The additional calories fuel the production of fat storage in adipose tissue. It is certainly possible to eat a diet free of added sugars and achieve obesity by eating too many calories from almost any combination of carbohydrates, lipids and proteins. Yet, the epidemiology points to added sugars as a major driver of obesity (Bray and Popkin 2014).

The fructose in HFCS and sucrose is associated with disease beyond its contribution of excess calories (Lustig 2013). Non-Alcoholic Fatty Liver Disease (NAFLD) is the most common pathology of the liver and afflicts 30% of the US population, particularly obese individuals (Le et al. 2017). The accumulation of lipids in the liver reduces the functioning of this organ, and may lead to steatohepatitis, and eventually cirrhosis, complete failure or hepatic cancer (Mayo Clinic 2018). Fructose from added sweeteners, along with dietary cholesterol is closely associated with NAFLD. Fructose changes metabolism in the liver, alters the microbiome in the gut, and causes leakiness of the gut wall. These all contribute to NAFLD (Jensen et al. 2018). What about the fructose in fruit? Not deconstructing the whole food and leaving the fructose in fruit protects against liver pathology. The fiber and other phytonutrients in fruit apparently mitigate against NAFLD (Kitson et al. 2014).

Although not the only dietary factor, added sugars contribute to dental caries (tooth decay, Gupta et al. 2013). Skeletal remains of the Powhatan show the arrival of corn agriculture in the Chesapeake region with a dramatic increase in dental carries. The mashing and cooking of the corn to make a coarse flour made for breads and other dishes contributed free sugars to the diet (Gallivan 2016).

Regulation of Sugar in the Human Body and Its Failure Glucose is the primary fuel of metabolism and its regulation is critical to human health. In a healthy human, the concentration of glucose circulating in the bloodstream ranges from 60 to 90 mg/dl prior to a meal and less than 140 mg/dl 2 h after dining. Persistent blood sugar concentrations outside of this range cause serious pathologies, often resulting in death.

The beta cells of the islets of Langerhans in the pancreas produce a protein called insulin, in response to elevated levels of blood sugar. This occurs after eating. The insulin circulates throughout the body and latches on receptor cites on cell membranes. This sets into motion a cascade of enzymatic reactions that enable the uptake of glucose by cells, reducing its concentration in the blood. The glucose transported into the cells is used to fuel metabolism, stored temporarily as glycogen, or converted to fat. When blood sugar levels drop too low, the alpha cells from the islets of Langerhans release a peptide called glucagon. This hormone causes the liver to convert stored glycogen into glucose, replenishing its supply in the blood stream.

Diabetes mellitus is the failure of long-term blood sugar regulation. It appears in three general forms. *Type 1* diabetes is caused by insufficient production of insulin. *Type 2* diabetes is insulin insensitivity on the part of the muscle and other cells of body. *Gestational* diabetes appears in pregnant women with no prior history of diabetes. Type 2 accounts for about 95% of all diabetes cases (CDC 2017).

Type 1 diabetes generally appears in the first few years of life. It is considered to be an idiopathic (not well understood) autoimmune (body attacks itself) disease. The T-cells of the immune system attack the beta cells that produce insulin (Kahaly and Hansen 2016). Type 2 accounts for about 5% of people suffering from diabetes (CDC 2018b). Even if well treated with insulin, the diagnosis reduces life expectancy by a decade. Despite the designation of idiopathic, there is a

Table 1 Nutritional value of select sweeteners and fruits per 100 g

Item	Kcal	Protein (g)	Dietary Fiber (g)	Calcium (mg)	Iron (mg)	Antioxidants (ORAC)
White sugar	375	0	0	0	0	0
Maple Syrup[a]	333	0	0	102	0	590
Blackstrap molasses[a]	400	6.67	0	667	24	5333
Plums	45	0.65	1.3	6	0.12	6100
Strawberries	36	0.71	2.1	14	0.77	4302
Blueberries	50	0.71	2.9	0	0	4669
Red cherries	300	0	3.3	100	0	3747
Raspberries	50	1.43	6.4	29	0.51	19,220
Gala apples	52	0	2.2	5	0.1	2828

[a]Maple syrup and blackstrap molasses values are for 100 ml rather than 100 g. Data from USDA (2018c) for basic nutrition. Antioxidant data from McBride (2018) and Superfoody (2018), Oxygen Radical Absorbance Capacity (ORAC) units reported as micro mol TE/100 g. TE stands for Trolox equivalent, where Trolox is a water soluble analog of vitamin E

strong dietary component, showing links to dairy consumption in infancy (Åkerblom et al. 1993; Nielsen et al. 2014). A mycobacterium common in cattle and their milk, produces a protein very similar to one in the beta cells. The immune response to the foreign protein also attacks the very similar protein in the beta cells. This kills the beta cells, eliminating the pancreatic production of insulin (Atreya et al. 2014).

Type 2 diabetes begins with a condition called *prediabetes*. Dietary lipids and those leaked from adipose stores in the body penetrate the membranes of muscle cells and interfere with the enzyme cascade that enables the functioning of insulin receptors. The insensitivity to insulin means blood glucose levels remain high and the pancreas produces additional insulin in an effort to reduce the concentration of circulating sugar. That extra insulin stimulates the uptake of glucose by the pancreas and liver. There it is converted to fat, resulting in lipid accumulation in both of those organs.

The condition transitions from prediabetes to type 2 diabetes, as the now fatty liver becomes resistant to insulin. This results in the release of more glucose from glycogen stores, which drives the pancreas to produce evermore insulin. The fatty liver unloads a portion of its lipid store into the blood stream in the form of Very Low Density Lipoprotein (VLDL). The VLDL is a mixture of cholesterol, triglycerides and proteins that accumulate in the pancreas. The VLDL is toxic to the insulin-producing beta cells. At full onset of the disease, a person with type 2 diabetes suffers from both insulin insensitivity and the inability to make sufficient insulin (Kraegen and Cooney 2008; Taylor 2013).

The excess insulin produced by the failing pancreas promotes the growth of cancers (Westley and May 2013). The persistent elevated blood glucose remarkably alters the biochemistry of tissues in various complex ways linked to oxidation (Brownlee 2005). Uncontrolled diabetes leads to an extensive list of complications including increased chances of; dementia, heart disease, stroke, high cholesterol, high blood pressure, chronic kidney disease, nerve damage, loss of vision, loss of hearing, digestive problems, and damage or amputation of toes and feet (CDC 2018c). Diabetes is the fourth leading cause of death in the US for both men and women and is a major contributing factor to the two leading sources of mortality, heart disease and cancer (Heron 2018).

Diabetes Rates Are Higher in the Chesapeake Bay Region and Disproportionately Afflicts People of Color Presently 9.4% of the US population has diabetes, and 26.1% has prediabetes, that if untreated will result in type 2 diabetes within five years (CDC 2018d). The states comprising the Chesapeake Bay region exceed the national average for diabetes; West Virginia (15.2%), Delaware (11.3%), Pennsylvania (10.6%), New York (10.5%), Virginia (10.5%), Maryland (10.4%). The rate of diabetes for the District of Columbia was below the national average at 7.8% (Diabetes US 2018). Statistics show that the rate of diabetes differs by race/ethnicity. The rates by group are; Native Americans (15.1%), non-Hispanic blacks (12.7%), Hispanics (12.1%), Asian Americans (8.0%) and non-Hispanic whites (7.4%) (ADA 2018). These figures are aligned with those for obesity presented in chapter "Introduction: Starting the Journey to a Sustainable Ecosystem and Healthy People".

Preventing and Treating Type 2 Diabetes with Diet Populations that have recently moved from eating mostly plant-based diets, to a version of the Standard America Diet (SAD), have experienced large increases in the incidence of type 2 diabetes. The SAD contains both excessive amounts of added sugar and animal products. Which is more important in promoting diabetes? China conducted the experiment for us. The spectacular prosperity that brought China from a developing nation in the 1970s to the world's second largest economy in the following four decades meant sufficient affluence for the population to increase the amount of animal products and processed foods in their diets. Chinese annual per capita consumption of meat went from 8.0 kg in 1970 to 13 kg in 1980 and hit 52.5 kg in 2018. For

comparison, 2018 annual per capita meat consumption in the US was 98.3 kg (OECD 2018). In the 30 years between 1980 and 2010, the rate of type 2 diabetes in China increased 17-fold, to 11.6% of the population. This follows the four-fold increase in per capita consumption of animal food in the diet, a reduction in dietary fiber, as well as other risk factors, such as increased air pollution (Shen et al. 2016).
During the 1980–2010 interval when Chinese per capita annual meat consumption quadrupled, per capita annual sugar consumption in that country increased only 23% from 4.5 kg to 5.8 kg (HelgiLibrary 2018). *While added sugar is an important part of the obesity and diabetes crisis, the Chinese experience points to meat consumption as the primary driver.*

Cohort and prospective studies that included thousands of participants demonstrated that those eating plant-based diets reduced their chances of developing type 2 diabetes by about 60%, and as much as 75%. In clinical trials, a whole-plant-based diet was tested against the animal food—containing American Diabetes Association (ADA) recommendations for patients with diabetes. The vegan diet group reduced their need for medication by 43%, versus 26% for the ADA diet followers. Moreover, the lower the amount of meat, seafood and dairy consumed, the lower the risk of type 2 diabetes (McMacken and Shah 2017).

To address the crisis of obesity, type 2 diabetes, and related diseases, the World Health Organization calls for a reduction of added sugar to less than 5% of total caloric intake (WHO 2018). The US Center for Disease Control goes only half the way there, considering a healthy diet to include less than 10% of calories from added sugar (CDC 2018d). Neither organization takes a strong stand on the reduction of animal products for reducing obesity and diabetes. However, the United States Department of Agriculture (USDA) weighs in on the issue with recommending a modest reduction in animal protein for middle-aged adults, but this is canceled by its strong advocacy for increased consumption of dairy products for all ages (USDA 2018e, f).

Overweight and obese individuals need to reduce caloric consumption. Yet the US food system is doing the opposite, providing increasing amounts of calories to consumers. Average daily energy consumption remained nearly constant at about 2500 calories between 1909 and about 1975, and thereafter increased until leveling-off at about 3100 calories in the year 2000 (Fig. 8).

A look at how US consumers spend their money on food helps explain the high calorie diet that is making them obese and ill. One comprehensive study found that animal products (30%) and sweetened beverages (7%) combine to consume nearly 40% of the money spent on food (Table 2). The high lipid content of most animal foods renders them energy dense

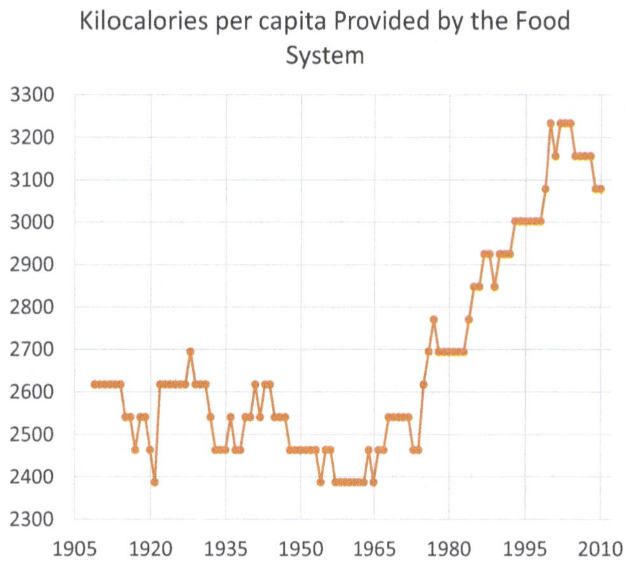

Fig. 8 The annual average daily kilocalories (calories) per capita provided by the US food system since 1909. Note the sharp increase beginning in the late 1970s. This corresponds to the increasing role of added sugars and animal products in the diet. This data is from the USDA and is an estimate of calories delivered to consumers based upon sale by retailers (Gerrior et al. 2004). In 2010, about 21% of calories from food was lost by consumers due to spoilage and non-consumption (Buzby et al. 2014). We applied that estimate for loss to yield the estimate for consumption

in comparison to whole plant- based foods. The combination of energy dense animal food and nutrient deficient sweeteners sets up the high-calorie diet that leads to obesity and type 2 diabetes.

While the total calories delivered by the US diet remained relatively steady until starting an upward trend in the 1970's, consumers began to shift what they eat several decades earlier (Figs. 8 and 9). Beginning in the 1920's, the US diet moved to proportionately less carbohydrates and fiber, replacing those macronutrients with increases in protein and fats. Despite the decline in percentage of diet, the absolute daily per capita consumption of carbohydrate in recent years is nearly identical to the 500 g recorded for 1909. This contrasts with daily per capita fiber consumption, which dropped from 30 g in 1909 to 25 g in 2010. Per capita daily consumption of protein was about 90 grams in the early twentieth century and began to rise in the 1980s to 120 g in 2010. The source of the protein also shifted, going from being evenly split between animal and plant food in 1909, to 63% from animals in 2000. Total daily per capita fat in the US diet went from 120 grams in the early twentieth century to 190 grams by 2010 (Gerrior et al. 2004; USDAe 2018).

The increase in dietary fat certainly contributed to the upward trend in total caloric consumption, as lipids generate over twice the energy of carbohydrates and proteins. The decline in fiber also means an average higher caloric content

Table 2 Expenditures on food by general category for recipients of the Supplemental Nutrition Assistance Program (SNAP) and non-SNAP recipients from 2011 USDA study (USDA 2018d)

Category	SNAP%	Non-Snap%
Meat, Poultry, Eggs, Dairy	31.4	30.5
Sweetened Beverages	9.3	7.1
Vegetables	7.2	9.1
Breads & Crackers	5.4	6.3
Fruits	4.7	7.2
Salty Snacks	3.4	3.1
Prepared Foods	3.1	2.2
Breakfast Cereal	2.8	3
Condiments & Seasonings	2.7	2.8
Fats & Oils	2.4	2.4
Candy	2.1	2.2
Baby Food	1.9	0.6
Pasta, Grains, Flours	1.8	1.6
Juices	1.7	1.9
Coffee & Tea	1.3	1.8
Bottled Water	1.2	1.2
Soups	1	1.3
Sugars	0.9	0.8
Nuts & Seeds	0.8	1.4
Beans	0.6	0.7
Jams & Jellies	0.4	0.4

SNAP (previously the USDA Food Stamp Program) supports lower-income individuals. SNAP and Non-SNAP shoppers spent their money in very similar ways. Animal products and sweetened beverages combined to constitute about 40% of money spent on food for both SNAP and Non-SNAP customers. Non-SNAP shoppers spend more on fruit and vegetables than SNAP purchasers do. SNAP shoppers spend three times more on baby food than non-SNAP recipients do

of the diet, since dietary fiber contributes essentially no calories to human metabolism. The reduced fiber, added protein, fat and sugars means the current US Diet contains about 9% more calories per gram of food then it did in the early twentieth century (Figs. 9 and 10).

Animal Foods Increase the Risk of Type 2 Diabetes As noted earlier, animal foods are rich sources of the type of lipids associated with NAFLD and type 2 diabetes. Prospective studies show significant relationships between meat consumption and type 2 diabetes, particularly for poultry and processed meats. Overall, consumption of meat increases the risk of type 2 diabetes by 32%. In addition to the diabetogenic lipids, animal products contain a host of other compounds that promote diabetes. These include; highly oxidative advanced glycation end products, nitrosamines, excess sodium and heme-iron (Peppa et al. 2002; InterAct Consortium et al. 2013; Feskens et al. 2013). Animal proteins are high in the amino acid leucine, which signals production of mammalian target of rapamycin complex 1 (mTORC1), which induces insulin resistance (Melnik 2012). Animal tissues concentrate persistent organic pollutants that disrupt the endocrine system and increase the prevalence of type 2 diabetes (Magliano et al. 2014).

Glycemic Load Versus Calories From above, it is clear, that diets with excessive calories lead to obesity, NAFL disease and type 2 diabetes. However, it is not just the total calories in the diet that count, it is also how those calories are delivered over time. After a meal, the beta cells in the pancreas releases insulin in proportion to the perceived level of

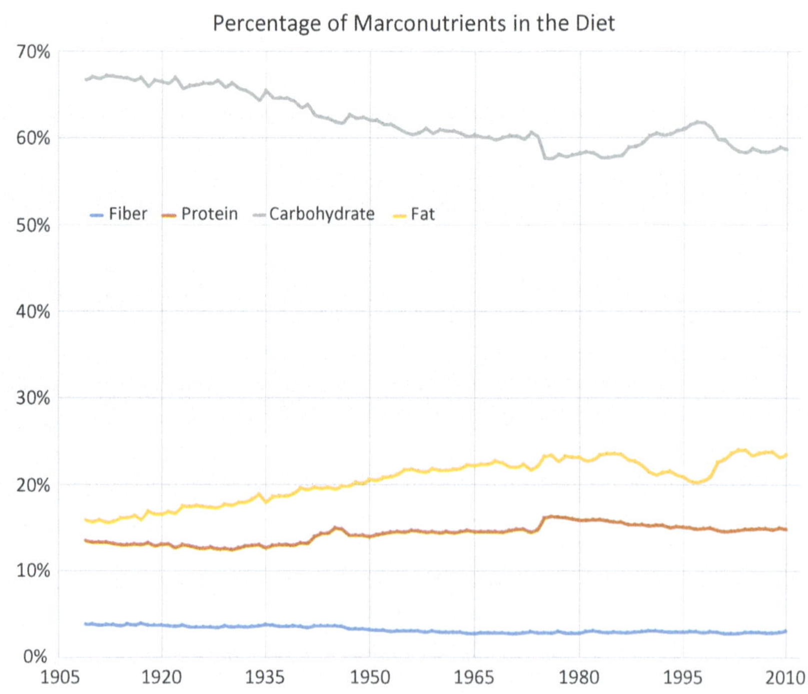

Fig. 9 Distribution of mass of food between the four macronutrient groups of fiber, protein, carbohydrate, and fat calculated from data in Gerrior et al. (2004) and Hiza and Bente (2011). Note the relative decline of fiber and carbohydrates and their replacement by protein and fats

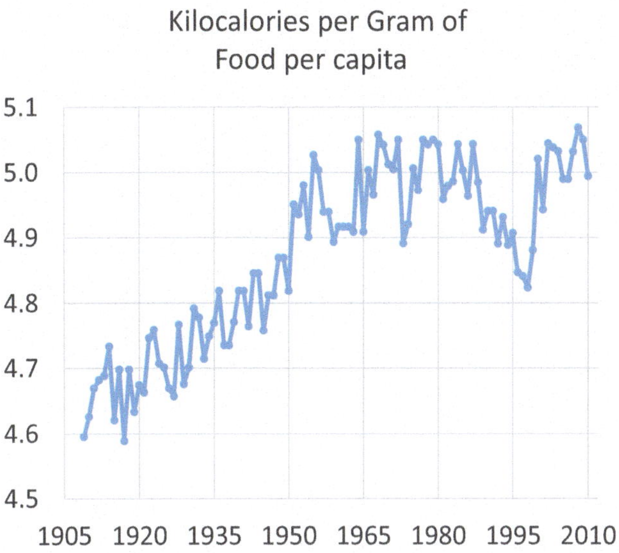

Fig. 10 Increasing caloric content of the food supply over time. The estimate is based upon food sales at the retail level with the assumption that about 89% of the purchased food was consumed. This reflects the larger role of calorie dense fats and added sugars in the diet. Calculation is based upon data from Gerrior et al. (2004) and Hiza and Bente (2011)

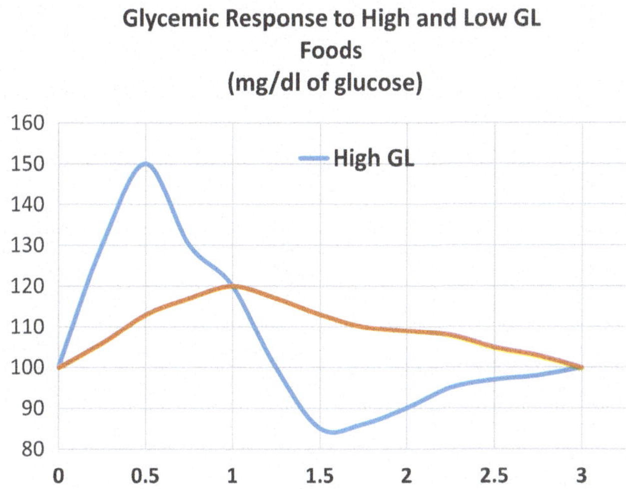

Fig. 11 Typical glycemic response curves to a high glycemic (High GL) and low glycemic (Low GL) load food. The high GL food floods the blood stream with glucose, causing a large release of insulin. The excessive insulin causes a fast drop, into the region of hypoglycemia (below 100 mg/dl). This then triggers glucagon release, to reestablish normal blood sugar concentrations. The subject suffers low blood sugar during the next two hours while glucagon is doing its job. The low GL food causes a lesser post-meal blood sugar increase, and that induces a smaller release of insulin. This returns the subject to pre-meal concentrations of glucose, saving them from a period of hypoglycemia

glucose and other products of digestion absorbed in the blood stream from the intestines. A meal rich in simple carbohydrates leads to rapid digestion and uptake of sugars, triggering a quick and intense release of insulin. The sugar high from circulating excess glucose (hyperglycemia), is followed by a sugar low (hypoglycemia), as the spike of insulin temporarily reduces blood sugar concentrations below optimal levels (Fig. 11). Glucagon release from the alpha cells eventually returns the post-meal blood sugar to normal. To compensate for the temporary hypoglycemia, the liver also dumps fatty acids into the blood stream. The cycle of intense increases and decreases of the blood glucose creates two problems. First, the extra insulin promotes fat storage, a first step to obesity, NAFLD, and type 2 diabetes. Second, the intense cycling damages the beta cells and their ability to produce insulin.

The *glycemic index* and *glycemic load* provide two measures for characterizing the effect of different foods on blood glucose regulation. The glycemic index is a measure of the capacity of the carbohydrate in a food to raise blood sugar in comparison to a dose of glucose, with 100 being the same as glucose (or sometimes white bread), and a score of 0 reflecting no effect on blood sugar. The *glycemic load* takes into account how much carbohydrate is in a serving of that food item, as it is the product of the glycemic index times the mass of the carbohydrate in that food (LPI 2018). For example, raw carrots score 47 on the glycemic index, but only 3 for glycemic load (Table 3). That's because, in addition to the 4.7 g of sugar and 9.6 of starch per 100 g serving, the carrot also offers 2.8 g of dietary fiber, 0.9 g of protein, and 0.2 g of fat (USDA 2018c). Carrots, as a whole food, produce a blood glucose response much lower than predicted by simply looking at its load of carbohydrates.

For glycemic index, scores $\geq$70 are considered high, and $\leq$ 55 low. For glycemic load, scores $\geq$20 are considered high, and $\leq$ 10 considered low (Foster-Powell et al. 2002). Processed carbohydrates, such as white rice and sugary breakfast foods raise blood sugar levels more than whole foods such as fruits, vegetables, legumes, nuts and whole grains (Table 3). The whole plant foods deliver their load of simple carbohydrates in a package of dietary fiber and other phytochemicals that reduce blood sugars (Törrönen et al. 2012). Adding nuts, legumes, whole grains, berries, vegetables and other low glycemic-load foods to dishes prepared with high glycemic load items effectively reduces glycemic spikes (O'Keefe et al. 2008; Törrönen et al. 2013). To reduce the chances of NAFLD, obesity, and type 2 diabetes, consumers should construct diets around whole plant foods with low glycemic load scores. Surprisingly, adding animal protein and fat to the meal may increase the glycemic spike associated with carbohydrate rich foods, such as mashed potatoes (Hätönen et al. 2011).

Sugar Addiction If reducing the consumption of sweeteners will improve human health, why do so many people in the Chesapeake region and the US in general, continue to

Table 3 Glycemic Index and Glycemic Load for some foods commonly consumed in the Chesapeake Bay region

Item	Serving Size (g)	Available Carbohydrate (g)	Glycemic Index	Glycemic Load
Waffles	35	13	76	10
Fruit Froot Loops cereal	30	26	69	18
Pop Tarts, double chocolate	50	36	70	25
Bagel (white flour)	70	35	72	25
Whole wheat bread	30	20	52	10
White rice	150	43	68	36
Brown rice	150	33	66	21
Spaghetti (white flour)	180	48	38	18
Sweet corn on the cob	150	33	60	20
Yogurt, low fat with fruit	200	30	31	9
Apple	120	16	40	12
Banana	120	25	51	13
Cherries	120	12	22	3
Kidney beans	150	25	28	7
Lentils	150	17	30	5
Beetroot	80	7	64	5
Carrots, raw	80	6	47	3
Sweet potato	150	28	61	17
White potato	150	30	85	26
Walnuts	28	4	15	1
Cashews	28	9	25	2
Peanuts, raw	28	6	18	1
Broccoli, raw	80	5	0	0

Glycemic Index is based upon glucose as the standard. Data from Foster-Powell et al. (2002)

consume them? Some people may be unaware of the amount of sugar they eat, some may not care, and others who might like to quit the sugar habit may be physiologically addicted to sweeteners, just as other people develop addictions to opioids, stimulants, and alcohol. Such drug addictions seem to apply only to extracts of plants, or synthetic molecules that mimic them, and not to the whole plants from which they are concentrated and purified. The evolutionary basis for addiction is found in the neural circuits to motivate and reward foraging and food intake. Rodents demonstrate addiction to offerings of sugar in patterns identical to other addictive substances. This includes craving, tolerance, withdrawal and sensitization. Sugar causes the release of endogenous opioids and dopamine. These molecules bind to receptor sites in the brain, providing the sensation of reward (Avena et al. 2008). Alcohol and drug addicted patients, and the children of alcoholic fathers demonstrate heightened sweet preference. This is consistent with a genetic component for potential sugar addiction (Fortuna 2010). Areas of the human brain that react to addictive drugs are also involved in response to food and may be hyper responsive in obese individuals (Iozzo et al. 2012). Mounting evidence suggests that extracted sugar is addictive for a portion of the human population (Lennerz et al. 2013; Wiss et al. 2018).

Taxing Sugary Beverages Added sugar primarily enters the US diet through sweetened beverages (Fig. 7). Public health advocates urge the adoption of special taxes on sugary beverages to discourage their consumption (Roache and Gostin 2017). Wang et al. (2012) calculated that a US wide penny per ounce soda tax would reduce the consumption of soda by adults by 15%, and over a 10 year interval, prevent 2.4 million diabetes person-years, 95,000 coronary heart events, and 26,000 premature deaths. A 2015 initiated penny per ounce tax on sweetened beverages in Berkeley, CA resulted in a 21% reduction in consumption, better than predicted by Wang et al. (Falbe et al. 2016). In 2010, Washington, DC became the first and only jurisdiction in the Chesapeake Bay Watershed to adopt a soda tax. The rate is 5.75% and contrasts to zero tax on other food items. Expanding adoption of the soda tax to more jurisdictions faces stiff opposition from the beverage and sugar industry who fund multimillion dollar campaigns to counter the public health efforts (Roache and Gostin 2017).

Artificial Sweeteners By the late nineteenth century, the public understood the link between the consumption of caloric sweeteners and excessive weight gain. Caloric sweeteners also caused major problems for persons with diabetes in the era before artificial insulin. Constantine Fahlberg, a Chesapeake Bay area scientist and professor at Johns Hopkins University, accidentally discovered Saccharin in 1879. Saccharin is a coal tar derivative that is 300 times sweeter than sucrose. The high intensity of the sweetness meant it essentially provided no calories when added to a food. Saccharin remained a specialty item for persons with diabetes until the advent of World War II, when sugar was in short supply, promoting others to use this artificial sweetner. After the war, dieters embraced Saccharin in their efforts to shed unwanted weight and it received FDA approval for use in food in 1958. The next artificial non-caloric sweetener, cyclamate, came out of the laboratory in 1937, and gained wide use in the US in the 1960s, until it was banned in 1969 as a carcinogen. Other non-caloric sweeteners came to market; Aspartame (1980), Acesulfame K (1988), Sucralose (1998) and neotame (2001). The use of artificial sweeteners went from 3.3% of consumes in 1965 to 15.1% by 2005. Despite the increased use of non-caloric artificial sweeteners, the public continued to gain weight. A host of prospective studies showed the use of artificial sweeteners to ironically be associated with weight gain (Yang 2010).

Non-caloric artificial sweeteners short-circuit the food reward pathways. The initial sweetness isn't followed by the expected post-ingestive component. This only serves to increase appetite as the body searches for the calories from its promised meal. Rather than facilitating the weaning away from sucrose, the sweetness of the artificial molecule reinforces the desire for sugar. Artificial sweeteners also alter the gut microbiome. This induces glucose intolerance and promotes excessive insulin response. Aspartame reaches the large intestine where it is broken down into toxic methanol and formaldehyde (Pretorius 2012; Pepino 2015). The expanded use of artificial sweeteners appears to be driving a 5–12 fold increase in inflammatory bowel disease, including ulcerative colitis and Crohn's disease (Qin 2012). It seems that artificial non-caloric sweeteners do more harm than good and promote the outcomes that their users are seeking to avoid (Swithers 2013).

Retraining the Pallet The ability to perceive sweetness and fat is blunted for obese and diabetic subjects (Wasalathanthri et al. 2014; Stewart et al. 2011). This insensitivity of taste promotes a desire to consume added sugars and fatty foods, which comes from years of eating the sweet-fat Standard American Diet (SAD). Avoiding the health damaging consequences of sweeteners and excess fats means stepping away from the SAD. The key to reducing the ingestion of added sugar is satisfying the natural urge for sweetness by increasing the consumption of fruit, choosing an apple for desert rather than apple pie. Overcoming the expectation of sweetness in other foods and beverages and increasing sensitivity to the taste of fat requires retraining the pallet. This can be achieved over the course of several weeks by switching to a whole-food plant-based diet (HHL 2011). Just as Capt. John Smith and his band of colonists needed to learn from the Algonquin how to eat the food offered by the Chesapeake Bay ecosystem, today's residents need to discover the benefits of rebuilding their diets around nutrient dense plant-based foods.

References

ADA (2018) Statistics about diabetes. In: American Diabetes Association. http://www.diabetes.org/diabetes-basics/statistics/. Accessed 16 Dec 2018

Åkerblom H, Saukkonen T, Vaarala O et al (1993) The case for elimination of cow's milk in early infancy in the prevention of type 1 diabetes: the finnish experience. Diabetes Metab Rev 9:269–278. https://doi.org/10.1002/dmr.5610090407

Ananthakrishnan A, Khalili H, Konijeti G et al (2013) A prospective study of long-term intake of dietary fiber and risk of Crohn's disease and ulcerative colitis. Gastroenterology 145:970–977. https://doi.org/10.1053/j.gastro.2013.07.050

Anon (2018) How do salt and sugar prevent microbial spoilage?. In: Scientific American. https://www.scientificamerican.com/article/how-do-salt-and-sugar-pre/. Accessed 31 Dec 2018

ASA (2018) Where is sugar produced? - American Sugar Alliance. In: American Sugar

Atreya R, Bülte M, Gerlach G et al (2014) Facts, myths and hypotheses on the zoonotic nature of *Mycobacterium avium* subspecies paratuberculosis. Int J Med Microbiol 304:858–867. https://doi.org/10.1016/j.ijmm.2014.07.006

Austin H (1928) History and development of the beet sugar industry. In: Mountainscholar.org. https://mountainscholar.org/bitstream/handle/10217/5540/COAB_13218511.pdf?sequence=1&isAllowed=y. Accessed 2 Dec 2018

Avena N, Rada P, Hoebel B (2008) Evidence for sugar addiction: behavioral and neurochemical effects of intermittent, excessive sugar intake. Neurosci Biobehav Rev 32:20–39. https://doi.org/10.1016/j.neubiorev.2007.04.019

Bowman S (2017) Added sugars: definition and estimation in the USDA Food Patterns Equivalents Databases. J Food Compos Anal 64:64–67. https://doi.org/10.1016/j.jfca.2017.07.013

Bray G, Popkin B (2014) Dietary sugar and body weight: have we reached a crisis in the epidemic of obesity and diabetes? Diabetes Care 37:950–956. https://doi.org/10.2337/dc13-2085

Brownlee M (2005) The pathobiology of diabetic complications: a unifying mechanism. Diabetes 54:1615–1625. https://doi.org/10.2337/diabetes.54.6.1615

Buzby J, Wells H, Hyman J (2014) The estimated amount, value, and calories of postharvest food losses at the retail and consumer levels in the United States. In: Endhunger.org. https://www.endhunger.org/PDFs/2014/USDA-FoodLoss-2014.pdf. Accessed 18 Dec 2018

CDC (2017) In: Diabetes.org. http://www.diabetes.org/assets/pdfs/basics/cdc-statistics-report-2017.pdf. Accessed 31 Dec 2018

CDC (2018a) Consumption of added sugars among U.S. adults, 2005–2010. In: Cdc.gov. https://www.cdc.gov/nchs/products/databriefs/db122.htm. Accessed 17 Dec 2018

CDC (2018b) National diabetes statistics report, 2017. In: Cdc.gov. https://www.cdc.gov/diabetes/pdfs/data/statistics/national-diabetes-statistics-report.pdf. Accessed 16 Dec 2018

CDC (2018c) Prevent complications | living with diabetes | diabetes | CDC. In: Cdc.gov. https://www.cdc.gov/diabetes/managing/problems.html. Accessed 16 Dec 2018

CDC (2018d) New CDC report: more than 100 million Americans have diabetes or prediabetes. In: CDC. https://www.cdc.gov/media/releases/2017/p0718-diabetes-report.html. Accessed 16 Dec 2018

Chiba M, Tsuda H, Abe T et al (2011) Missing environmental factor in inflammatory bowel disease. Inflamm Bowel Dis 17:E82–E83. https://doi.org/10.1002/ibd.21745

Ciesla W (2002) Non-wood forest products from temperate broad-leaved trees. Food and Agriculture Organization of the United Nations, Rome

de Oliveira M, Sichieri R, Venturim Mozzer R (2008) A low-energy-dense diet adding fruit reduces weight and energy intake in women. Appetite 51:291–295. https://doi.org/10.1016/j.appet.2008.03.001

Diabetes in US (2018) Diabetes in the United States. In: The State of Obesity. https://stateofobesity.org/diabetes/. Accessed 16 Dec 2018

Dinsmore C (2017) Baltimore's Domino refinery to receive largest sugar shipment ever. In: Baltimoresun.com. https://www.baltimoresun.com/business/bs-bz-domino-sugar-shipment-20171019-story.html. Accessed 1 Dec 2018

Dunn R (2012) Human ancestors were nearly all vegetarians. In: Scientific American Blog Network. https://blogs.scientificamerican.com/guest-blog/human-ancestors-were-nearly-all-vegetarians/. Accessed 1 Dec 2018

Falbe J, Thompson H, Becker C et al (2016) Impact of the Berkeley excise tax on sugar-sweetened beverage consumption. Am J Public Health 106:1865–1871. https://doi.org/10.2105/ajph.2016.303362

Feskens E, Sluik D, van Woudenbergh G (2013) Meat consumption, diabetes, and its complications. Curr Diab Rep 13:298–306. https://doi.org/10.1007/s11892-013-0365-0

Fortuna J (2010) Sweet preference, sugar addiction and the familial history of alcohol dependence: shared neural pathways and genes. J Psychoactive Drugs 42:147–151. https://doi.org/10.1080/02791072.2010.10400687

Foster-Powell K, Holt S, Brand-Miller J (2002) International table of glycemic index and glycemic load values: 2002. Am J Clin Nutr 76:5–56. https://doi.org/10.1093/ajcn/76.1.5

Gallivan M (2016) The Powhatan landscape: an archaeological history of the algonquin chesapeake. University Press of Florida, Gainesville

Gerrior S, Bente L, Hiza H (2004) nutrient content of the U.S. food supply, 1909–2000. In: Cnpp.usda.gov. https://www.cnpp.usda.gov/sites/default/files/nutrient_content_of_the_us_food_supply/FoodSupply1909-2000.pdf. Accessed 18 Dec 2018

Gupta P, Gupta N, Pawar A et al (2013) Role of sugar and sugar substitutes in dental caries: a review. ISRN Dent 2013:1–5. https://doi.org/10.1155/2013/519421

Gutman D, Golub J, Tulp C, Cioffi C (2012) The last of its kind: domino sugar perseveres in Baltimore's inner harbor. In: CNS Maryland. https://cnsmaryland.org/2012/10/24/the-last-of-its-kind-domino-sugar-perseveres-in-baltimores-inner-harbor/. Accessed 1 Dec 2018

Guyenet S (2012) By 2606, the US diet will be 100 percent sugar. In: Wholehealthsource.blogspot.com. https://wholehealthsource.blogspot.com/2012/02/by-2606-us-diet-will-be-100-percent.html. Accessed 6 Dec 2018

Hätönen K, Virtamo J, Eriksson J et al (2011) Protein and fat modify the glycaemic and insulinaemic responses to a mashed potato-based meal. Br J Nutr 106:248–253. https://doi.org/10.1017/s0007114511000080

HelgiLibrary (2018) HelgiLibrary – sugar consumption per capita in China. In: Helgilibrary.com. https://www.helgilibrary.com/indicators/sugar-consumption-per-capita/china/. Accessed 17 Dec 2018

Heron M (2018) Deaths: leading causes for 2016. In: Cdc.gov. https://www.cdc.gov/nchs/data/nvsr/nvsr67/nvsr67_06.pdf. Accessed 18 Dec 2018

HHL (2011) Controlling what – and how much –we eat. Harvard Health Letter. Oct. 2011. https://www.health.harvard.edu/staying-healthy/controlling-what-and-how-much-we-eat. Accessed 21 Dec 2018

Hiza H, Bente L (2011) Nutrient content of the U.S. food supply: developments between 2000 and 2006. Home Economics Research Report No. 59 US Department of Agriculture. In: Cnpp.usda.gov. https://www.cnpp.usda.gov/sites/default/files/nutrient_content_of_the_us_food_supply/Final_FoodSupplyReport_2006.pdf. Accessed 18 Dec 2018

InterAct Consortium Bendinelli B, Palli D et al (2013) Association between dietary meat consumption and incident type 2 diabetes: the EPIC-InterAct study. Diabetologia 56:47–59. https://doi.org/10.1007/s00125-012-2718-7

Iozzo P, Guiducci L, Guzzardi M et al (2012) Brain PET imaging in obesity and food addiction: current evidence and hypothesis. Obes Facts 5:155–164. https://doi.org/10.1159/000338328

Jensen T, Abdelmalek M, Sullivan S et al (2018) Fructose and sugar: a major mediator of non-alcoholic fatty liver disease. J Hepatol 68:1063–1075. https://doi.org/10.1016/j.jhep.2018.01.019

Kahaly G, Hansen M (2016) Type 1 diabetes associated autoimmunity. Autoimmun Rev 15:644–648. https://doi.org/10.1016/j.autrev.2016.02.017

Kitson M, Kemp W, Roberts S et al (2014) Industrial, not fruit fructose intake is associated with the severity of liver fibrosis in genotype 1 chronic hepatitis C patients. J Hepatol 60:676–677. https://doi.org/10.1016/j.jhep.2013.10.013

Kraegen E, Cooney G (2008) Free fatty acids and skeletal muscle insulin resistance. Curr Opin Lipidol 19:235–241. https://doi.org/10.1097/01.mol.0000319118.44995.9a

Le M, Devaki P, Ha N et al (2017) Prevalence of non-alcoholic fatty liver disease and risk factors for advanced fibrosis and mortality in the United States. PLoS One 12:e0173499. https://doi.org/10.1371/journal.pone.0173499

Lennerz B, Alsop D, Stern E et al (2013) Effects of dietary glycemic index on brain regions related to reward and craving in men. Br Dent J 215:225–225. https://doi.org/10.1038/sj.bdj.2013.856

Lozano R, Naghavi M, Foreman K et al (2012) Global and regional mortality from 235 causes of death for 20 age groups in 1990 and 2010: a systematic analysis for the Global Burden of Disease Study 2010. Lancet 380:2095–2128. https://doi.org/10.1016/s0140-6736(12)61728-0

LPI (2018) Glycemic index and glycemic load. In: Linus Pauling Institute. https://lpi.oregonstate.edu/mic/food-beverages/glycemic-index-glycemic-load. Accessed 20 Dec 2018

Lustig R (2013) Fructose: it's "Alcohol without the Buzz". Adv Nutr 4:226–235. https://doi.org/10.3945/an.112.002998

Magliano D, Loh V, Harding J et al (2014) Persistent organic pollutants and diabetes: a review of the epidemiological evidence. Diabetes Metab 40:1–14. https://doi.org/10.1016/j.diabet.2013.09.006

Maple Festival (2018) Highland maple festival - Highland County Virginia. In: Highland County Virginia. http://www.highlandcounty.org/events/maple-festival/. Accessed 6 Dec 2018

Mayo Clinic (2018) Nonalcoholic fatty liver disease – Symptoms and causes. In: Mayo clinic. https://www.mayoclinic.org/diseases-conditions/nonalcoholic-fatty-liver-disease/symptoms-causes/syc-20354567. Accessed 16 Dec 2018

McBride J (2018) High-ORAC foods may slow aging: USDA ARS. In: Ars.usda.gov. https://www.ars.usda.gov/news-events/news/research-news/1999/high-orac-foods-may-slow-aging/. Accessed 9 Dec 2018

McMacken M, Shah S (2017) A plant-based diet for the prevention and treatment of type 2 diabetes. J Geriatric Cardiol JGC 14(5):342–354

Melnik B (2012) Leucine signaling in the pathogenesis of type 2 diabetes and obesity. World J Diabetes 3:38. https://doi.org/10.4239/wjd.v3.i3.38

Mintz S (1985) Sweetness and power, 1st edn. Penguin Books, New York

Mountford K (2003) Bay journal – article: history behind sugar trade, Chesapeake not always sweet. In: Bayjournal.com. https://www.bayjournal.com/article/history_behind_sugar_trade_chesapeake_not_always_sweet. Accessed 1 Dec 2018

Myfooddiary (2018) BBQ sauce nutrition facts. Calories, fat, carbs & 12 other nutrients. In: Myfooddiary.com. https://www.myfooddiary.com/foods/search?q=BBQ+sauce. Accessed 11 Dec 2018

Nielsen D, Krych Ł, Buschard K et al (2014) Beyond genetics. Influence of dietary factors and gut microbiota on type 1 diabetes. FEBS Lett 588:4234–4243. https://doi.org/10.1016/j.febslet.2014.04.010

Northrup L (2016) In: Consumerist.com. https://consumerist.com/2016/12/09/we-have-sugar-to-thank-for-the-american-tobacco-industry/. Accessed 31 Dec 2018

NPS (2018) African Americans at Jamestown – Historic Jamestowne Part of Colonial National Historical Park (U.S. National Park Service). In: Nps.gov. https://www.nps.gov/jame/learn/historyculture/african-americans-at-jamestown.htm. Accessed 2 Dec 2018

O'Keefe J, Gheewala N, O'Keefe J (2008) Dietary strategies for improving post-prandial glucose, lipids, inflammation, and cardiovascular health. J Am Coll Cardiol 51:249–255. https://doi.org/10.1016/j.jacc.2007.10.016

OECD (2018) Agricultural output – Meat consumption – OECD Data. In: Organisation for Economic Co-operation and Development. https://data.oecd.org/agroutput/meat-consumption.htm. Accessed 17 Dec 2018

Pedder J (1836) Report made to the Beet sugar society of Philadelphia, on the culture, in France, of the beet root. In: Biodiversitylibrary.org. https://www.biodiversitylibrary.org/item/80521#page/7/mode/1up. Accessed 3 Dec 2018

Pepino MY (2015) Metabolic effects of non-nutritive sweeteners. Physiol Behav 152:450–455

Peppa M, Goldberg T, Cai W et al (2002) Glycotoxins: a missing link in the "relationship of dietary fat and meat intake in relation to risk of type 2 diabetes in men". Diabetes Care 25:1898–1899. https://doi.org/10.2337/diacare.25.10.1898

Pretorius E (2012) GUT bacteria and aspartame: why are we surprised? Eur J Clin Nutr 66:972–972. https://doi.org/10.1038/ejcn.2012.47

Qin X (2012) Etiology of inflammatory bowel disease: a unified hypothesis. World J Gastroenterol 18:1708. https://doi.org/10.3748/wjg.v18.i15.1708

Roache S, Gostin L (2017) The untapped power of soda taxes: incentivizing consumers, generating revenue, and altering corporate behavior. Int J Health Policy Manag 6:489–493. https://doi.org/10.15171/ijhpm.2017.69

Samford P (2015) One sweet tale: sugar molds from the Shutt and tool sugar refinery. In: Maryland history by the object. https://jeffersonpatterson.wordpress.com/2015/10/22/one-sweet-tale-sugar-molds-from-the-shutt-and-tool-sugar-refinery/#more-887. Accessed 1 Dec 2018

Shen X, Vaidya A, Wu S et al (2016) The diabetes epidemic in China: an integrated review of national surveys. Endocr Pract 22:1119–1129. https://doi.org/10.4158/ep161199.ra

StarchEU (2013) Factsheet on glucose fructose syrup and isoglucose – starch Europe. In: Starch Europe. https://www.starch.eu/blog/2013/06/10/factsheet-on-glucose-fructose-syrups-and-isoglucose/?redirect=true. Accessed 3 Dec 2018

Statistica (2018) Major sugar consumers worldwide by country, 2017/2018 | Statistic. In: Statista. https://www.statista.com/statistics/496002/sugar-consumption-worldwide/. Accessed 1 Dec 2018

Stewart J, Newman L, Keast R (2011) Oral sensitivity to oleic acid is associated with fat intake and body mass index. Clin Nutr 30:838–844. https://doi.org/10.1016/j.clnu.2011.06.007

Sugar Association (2018) Sources of added sugars. In: The Sugar Association. https://www.sugar.org/diet/sources-of-added-sugars/. Accessed 10 Dec 2018

Superfoody (2018) Antioxidant ORAC value: plums, raw. In: Superfoodly.com. https://www.superfoodly.com/orac-value/plums-raw/. Accessed 9 Dec 2018

Swithers S (2013) Artificial sweeteners produce the counterintuitive effect of inducing metabolic derangements. Trends Endocrinol Metab 24:431–441. https://doi.org/10.1016/j.tem.2013.05.005

Taylor R (2013) Banting memorial lecture 2012 reversing the twin cycles of type 2 diabetes. Diabet Med 30:267–275. https://doi.org/10.1111/dme.12039

Theobald M (2012) Thomas Jefferson and the maple sugar scheme. In: History.org. http://www.history.org/foundation/journal/autumn12/maplesugar.cfm. Accessed 2 Dec 2018

Törrönen R, Kolehmainen M, Sarkkinen E et al (2012) Postprandial glucose, insulin, and free fatty acid responses to sucrose consumed with blackcurrants and lingonberries in healthy women. Am J Clin Nutr 96:527–533

Törrönen R, Kolehmainen M, Sarkkinen E et al (2013) Berries reduce postprandial insulin responses to wheat and rye breads in healthy women. J Nutr 143:430–436. https://doi.org/10.3945/jn.112.169771

USDA (2018a) Food composition databases show foods – pure cane sugar, UPC: 074175820107. In: Ndb.nal.usda.gov. https://ndb.nal.usda.gov/ndb/foods/show/45047292?fgcd=&manu=&format=&count=&max=25&offset

USDA (2018b) Estimated calories needs per day. In: Cnpp.usda.gov. https://www.cnpp.usda.gov/sites/default/files/usda_food_patterns/EstimatedCalorieNeedsPerDay.pdf. Accessed 9 Dec 2018

USDA (2018c) Food composition databases. In: Ndb.nal.usda.gov. https://ndb.nal.usda.gov/ndb/search/list?SYNCHRONIZER_TOKEN=91e139fe-a5b7-4084-9344-f3186004dd82&SYNCHRONIZER_URI=%2Fndb%2Fsearch%2Flist&qt=&qlookup=water+melon&ds=&manu=. Accessed 9 Dec 2018

USDA (2018d) Food and nutrition services. In: Fns-prod.azureedge.net. https://fns-prod.azureedge.net/sites/default/files/ops/SNAPFoodsTypicallyPurchased-Summary.pdf. Accessed 11 Dec 2018

USDA (2018e) Nutrient content of the U.S. food supply, 1909–2010 | Center for Nutrition Policy and Promotion. In: Cnpp.usda.gov. https://www.cnpp.usda.gov/USFoodSupply-1909-2010. Accessed 19 Dec 2018

USDA (2018f) A closer look at current intakes and recommended shifts – 2015–2020 dietary guidelines – health.gov. In: Health.gov. https://health.gov/dietaryguidelines/2015/guidelines/chapter-2/a-closer-look-at-current-intakes-and-recommended-shifts/. Accessed 18 Dec 2018

USDA ERS (2018) USDA ERS – sugar and sweeteners yearbook tables. In: Ers.usda.gov. https://www.ers.usda.gov/data-products/sugar-and-sweeteners-yearbook-tables.aspx. Accessed 3 Dec 2018

Walsh L (2003) The transatlantic slave trade and colonial chesapeake slavery. OAH Mag Hist 17:11–15. https://doi.org/10.1093/maghis/17.3.11

Wang Y, Coxson P, Shen Y et al (2012) A penny-per-ounce tax on sugar-sweetened beverages would cut health and cost burdens of diabetes. Health Aff 31:199–207. https://doi.org/10.1377/hlthaff.2011.0410

Wasalathanthri S, Hettiarachchi P, Prathapan S (2014) Sweet taste sensitivity in pre-diabetics, diabetics and normoglycemic controls: a comparative cross sectional study. BMC Endocr Disord. https://doi.org/10.1186/1472-6823-14-67

Westley R, May F (2013) A twenty-first century cancer epidemic caused by obesity: the involvement of insulin, diabetes, and insulin-like growth factors. Int J Endocrinol 2013:1–37. https://doi.org/10.1155/2013/632461

WHO (2018) Guideline: sugar intake for adults and children. In: Apps. who.int. http://apps.who.int/iris/bitstream/handle/10665/149782/9789241549028_eng.pdf;jsessionid=58D00FDA0FE6F753B61DE0CE5F33C5D2?sequence=1. Accessed 9 Dec 2018

Wiss D, Avena N, Rada P (2018) Sugar addiction: from evolution to revolution. Front Psych. https://doi.org/10.3389/fpsyt.2018.00545

Wood L, Garg M, Smart J et al (2012) Manipulating antioxidant intake in asthma: a randomized controlled trial. Am J Clin Nutr 96:534–543. https://doi.org/10.3945/ajcn.111.032623

Yang Q (2010) Gain weight by "going diet?" Artificial sweeteners and the neurobiology of sugar cravings: Neuroscience 2010. Yale J Biol Med 83(2):101–108

Reversing the Eutrophication of the Chesapeake Bay and Its People

Benjamin E. Cuker

How do we know when more is less?
Too many of the wrong nutrients in the Bay leads to ecosystem distress.
And for people isn't it also true?
For our health, eating too much of the wrong foods is something we shouldn't do.

Abstract

Eutrophication overtook the Chesapeake Bay beginning in the mid twentieth century, and continues to be a major problem. The human food system produces excess inorganic nutrients and oxygen-demanding pollution that are primary drivers in the eutrophication of the Chesapeake Bay. The bulk of that pollution emanates from an agricultural system designed to produce the elements of the Standard American Diet (SAD), including meat and highly refined carbohydrates. Inputs of N and P from modern agriculture's artificial fertilizers and animal manure, overwhelm the natural, high-fiber inputs of detritus that used to predominate. This results in frequent algal blooms and attendant periods of hypoxia in the bottom waters of the Bay during seasonal stratification. Humans respond to the SAD in a fashion similar to how the Bay responds to the agricultural system that produces the foods for that diet. The excessive biomass of algal blooms is equivalent to the unnecessary and dangerous deposits of adipose that create obesity in people. Hypoxia in the bottom waters of the Bay is analogous to that in the brain and adipose tissues of obese and over-weight people (75% of the Bay's human population), causing various health problems. Hypoxic bottom waters promote populations of benthic bacteria that release toxic H_2S. The high animal component of the SAD produces dysbiosis of the human biome, including bacterial production of disease-promoting H_2S in the lower intestines. Both the Bay and humans living there suffer from a lack of fiber-rich organic matter. Reversing eutrophication of the bay and its people requires a shift away from the SAD to a whole food plant based diet.

Keywords

Standard American diet · Hypoxia · Obesity · Diabetes · Coronary heat disease

On April 26, 1607 three ships from London sailed through the Virginia Capes to establish the first permanent English colony in the New World. The *Susan Constant*, *Godspeed*, and *Discovery* easily navigated the clear waters of the Chesapeake Bay. The ship's lookouts could see the extensive beds of underwater grasses and oyster reefs that told of shoals to avoid. Capt. John Smith, leader of the expedition, noted the clarity of the water in the Bay, and its tributaries that he subsequently explored over the next 2 years. He also recorded seeing seemingly endless schools of fish, and a great diversity of birds and other animals making their home in the this pristine estuary (Smith 1612).

It's been many years since the Chesapeake Bay looked much like what greeted Capt. John Smith. Abundant populations of suspended microscopic algae (phytoplankton) now color the water shades of green or brown for most of the year. After storms, suspended sediment washed in from agricultural fields add to the murkiness. The mix of phytoplankton, sediments, and small particles of detritus (dead material) scatter the sunlight penetrating the water column. This turbidity greatly reduces the illumination beyond depths of 2 m

B. E. Cuker (✉)
Department of Marine and Environmental Science, Hampton University, Hampton, Virginia, USA
e-mail: Benjamin.cuker@hamptonu.edu

B. E. Cuker (ed.), *Diet for a Sustainable Ecosystem*, Estuaries of the World, https://doi.org/10.1007/978-3-030-45481-4_15

(6.5 ft) for much of the Bay. That affords little remaining light to support the growth of the once luxuriant beds of underwater grasses. Long gone are the clear waters that once floated Capt. John Smith's small fleet of ships.

The story of the decline of the Chesapeake Bay's ecosystem begins with the decedents of those first English Colonists and the many others who immigrated from various places to the region in subsequent years. On the surface, the decline simply traces to the 99% increase in the number of people living in the Bay's watershed, going from 20,000 in 1607 to 18.2 million in 2017. More people means more pollution and more habitat disruption. Turning back the clock and removing all but 1% of the current residents from the watershed would certainly help to recover the health of the ecosystem, but clearly isn't a sustainable or practical solution. However, changing the way the residents live in the watershed can and must be done to regain the health of the Bay. While various human activities combined to diminish the state of the Bay, one stands out above all others as the main culprit, the *food system*. And that same food system that robbed the Chesapeake Bay of its health, also sickened the people that live there.

When called to action in 2010 to help restore the health of the estuary, the US Environmental Protection Agency prescribed what they called a *pollution diet* for the Bay. EPA spelled out specific reductions in *Total Maximum Daily Load (TMDL)* of nitrogen (25%), phosphorus (24%), and sediment (20%) (Pimental 2010). That the agency chose the term *diet* reflects the understanding that the Bay, like an obese or overweight person, is getting *too many nutrients* and in *an unbalanced* way. As will be seen, the food system generates these excess nutrients from farming practices and how it mishandles the end products of urine and feces. This chapter addresses the ways the food system diminished the water quality of the Chesapeake Bay and draws parallels to how the Standard American Diet compromises the health of the residents of the area.

Eutrophication of the Bay

Excess Nutrients from the Food System and Other Sources Causes Cultural Eutrophication The abundant algae coloring and clouding the water of today's Chesapeake Bay are a product of the process called *cultural eutrophication* (Fig. 1). This occurs when human activity causes excess nutrients to drain from the land and over-stimulate plant growth (Kemp et al. 2005; Smith et al. 2006; Testa et al. 2017). In the Bay phosphorous (P) tends to stimulate algal productivity in the oligohaline (less salty) reaches, while nitrogen (N) does that job in the polyhaline (saltier) portions (Conley et al. 2009a). The food system is responsible for nearly 60% of the excess N entering the Bay; 43% directly from agriculture, and 16.8% from human urine and feces (CBP 2018, see chapter "The Journey from Peruvian Guano to Artificial Fertilizer Ends with Too Much Nitrogen in the Chesapeake Bay" on fertilizers).

The excess nutrients unbalance the ecosystem, causing the phytoplankton to grow so fast and numerous that natural populations of algae-eaters can't keep up with the increase (Malone et al. 1996). In the time of Capt. John Smith, the once extensive beds of oysters, grazing zooplankton, and other suspension-feeding animals consumed much of the crop of algae in the Bay. That's no longer possible. The oysters are mostly gone, and the algal growth is just too quick (Cerco and Noel 2007). Stimulated by the excessive supply of two nutrients, nitrogen and phosphorus, the algae can easily double their population in the course of a day (Brush 2008; Mulholland et al. 2009). The accumulation of these fast growing phytoplankton is called an *algal bloom*.

In the Bay proper, the algal blooms tend to occur in the oligohaline to mesohaline portions of the estuary, corresponding to the upper and middle section of that waterbody. Algal blooms also appear in the major tributaries in zones of similar salinity. This reflects the nutrients flushing off of farm lands, concentrated animal feeding operations (CAFOs), and other sources (Figs. 2 and 3).

Chlorophyll *a* Concentration Is a Marker for Eutrophication Cloudy green, red or brown tinged waters signal eutrophication. Direct quantification of the algal biomass in the water column is best done with laborious microscope work, identifying, counting and measuring the size of the cells present in a set of samples. Much quicker is the indirect estimate of mass by measuring the concentration of chlorophyll *a* in the water. Chlorophyll *a* is the central molecule for the light reaction of photosynthesis. It can be measured after filtering and extracting a sample with a solvent such as acetone. Or approximated by in situ measurement with a fluorometer. Although variable, chlorophyll *a* typically constitutes about 1–3% of phytoplankton total carbon (Cloern et al. 1995).

While phytoplankton is essential to the food chain of the Bay, eutrophication creates too much of it. How much is too much for a healthy Bay? Current maximum chlorophyll *a* criteria for the James River, a major tributary of the Bay, range from 10 to 23 μg/l depending on the season and salinity (USEPA 2017). However, Buchanan (2016) suggests a range of 3.4–9.5 μg/l to achieve desired improvements of water quality. Yet, since the mid twentieth century chlorophyll concentrations typically exceed the 10–20 μg/l range from spring through fall. Figure 2 displays the distribution of chlorophyll *a* on June 21–22 in 2007 for the mainstem of the Bay between its mouth and the Chesapeake Bay Bridge

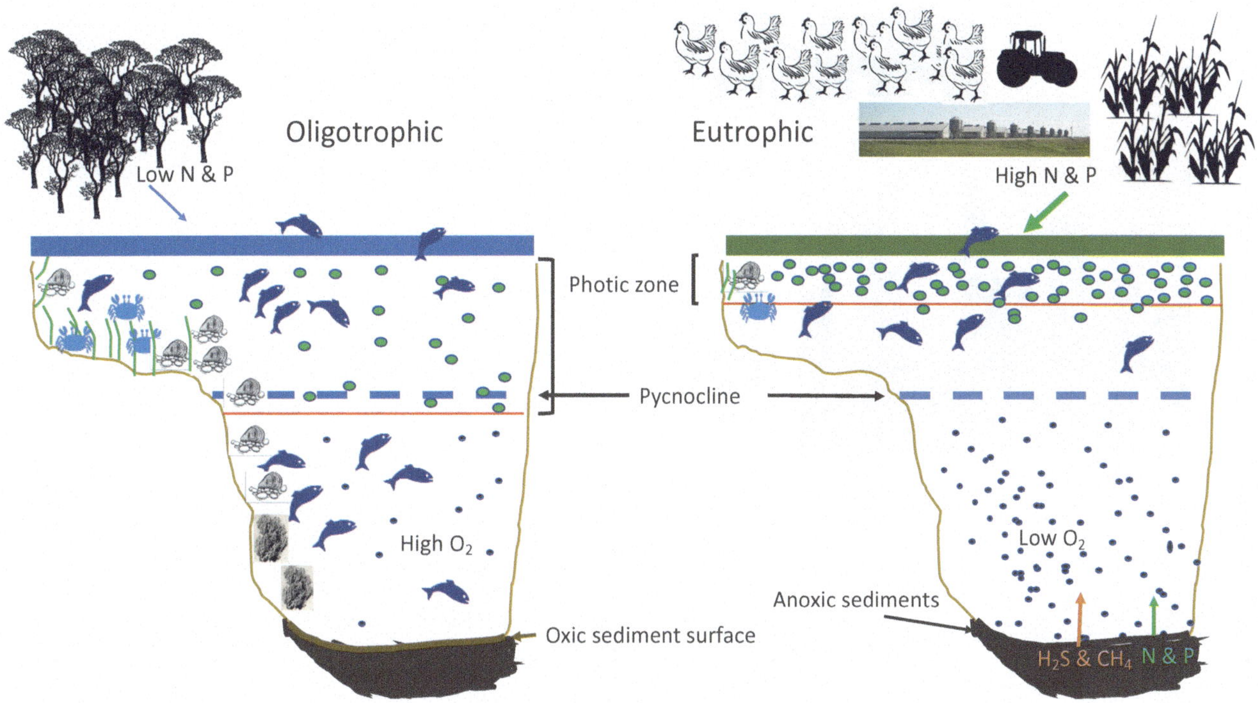

Fig. 1 Excess nutrients, primarily from the food system causes cultural eutrophication of the Chesapeake Bay. The image on the left depicts the conditions of the Bay prior to European colonization. At that time the Bay was oligotrophic, with few nutrients entering from the mostly forested watershed. The deep photic zone facilitated extensive grass beds. Fish, crabs and other metazoans extended to the deep waters all year long. The image on the right depicts the current eutrophic state of the Bay. The high loading of N & P from animal agriculture and the crops that support it causes frequent algal blooms (green dots) that reduce light penetration, making for a shallow photic zone, reducing habitat for submerged aquatic vegetation. Much of the phytoplanktonic algae settles to the bottom waters and sediments, where bacterial decomposition (black dots) reduces or eliminates free oxygen. The anoxic sediments produce toxic H_2S and regenerated N & P to sustain eutrophication

north of Annapolis. In this example concentrations equaled or exceeded 10 μg/l for most of the surface waters, and reached 20–45 μg/l in the upper reaches of the Bay, clearly denoting eutrophic conditions.

Eutrophication Causes Phytoplankton Blooms That Diminish SAVs Biological productivity in the Chesapeake Bay of Capt. John Smith's time was concentrated in the benthos, with extensive beds of underwater grasses harboring a diverse food web. Subsequent eutrophication meant super abundant phytoplankton that shaded much of the grass beds out of existence. Phytoplankton and colored organic material they release reduces the penetration of light into the Bay (Testa et al. 2019). Eutrophication changed the Bay from a primarily benthic to a mostly planktonic system (Brush 2008).

There's three broad groups of primary producers in the Chesapeake Bay system; *emergent* wetland plants, *submerged aquatic vegetation* (SAV), and *planktonic algae* (phytoplankton). The first two groups consist of macroscopic vascular plants rooted in the sediments.

Emergent plants make their home in marshes and along the shores of waterways. They feature strong stems that support the plant as it reaches above the surface of the water. Examples from the freshwater reaches of the Bay include; Borad-leaved cattail (*Typha latifolia*), narrow-leaved cattail (*T. angustifolia*), sweet flag (*Acorus calamus*), soft rush (*Juncus effuses*), great brush (*Scirpus validus*), river bulrush (*S. fluviatilis*), sedge (*Carex lurida*), common reed (*Phragmites australis*), and wild rice (*Zizania aquatica*). Examples from saltier waters are; Saltmarsh cordgrass (*Spartina alterniflora*) saltmeadow cordgrass (*S. patens*), saltgrass (*Distichlis spicata*), and black needlerush (*Juncus roemerianus*). The excess nutrients of eutrophication may stimulate the growth these plants and favor some species over others, but the overall effect is minor. Emergent plants grow in sediments and soils rich in organic matter that detains excess nutrients. These marsh communities filter water draining from the adjacent land on its way to the Bay and its tributaries, removing excess nutrients and organic matter (Mitsch et al. 2000). So, emergent plants aren't much affected by eutrophication, but play a role in mitigating it.

In contrast to emergent plants, SAVs suffer significantly from eutrophication (Orth et al. 2010). SAV's are submerged vascular plants rooted in the sediments. The surrounding water supports their soft stems and leaves that bend with

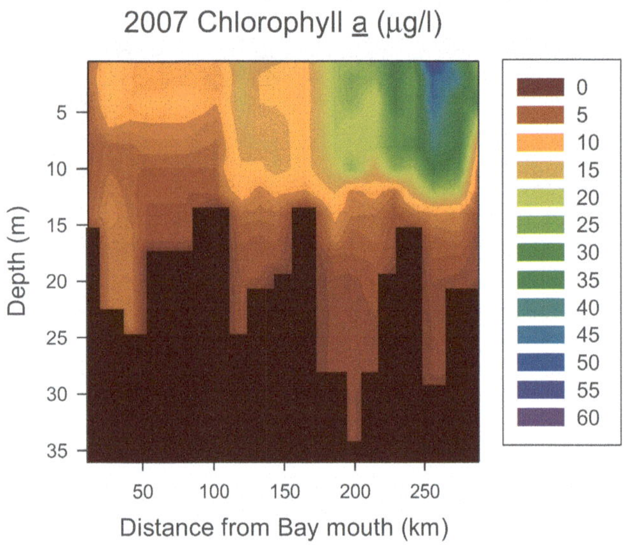

Fig. 2 The distribution of algae in the Chesapeake Bay on June 21–22 of 2007, as indicated by concentration of the pigment chlorophyll *a*. This is a side view of the Bay that exaggerates the depth relative to the length of the estuary. The mouth of the Bay is to the left. The black rectangles along the bottom indicate the depth contour. Note the algal bloom evident in the upper-right of the figure, which corresponds to the freshwater input to the Bay coming from the Susquehanna River. The bloom often extends further toward the mouth over the course of the season. Data is from 20 stations along the main stem of the bay, the last just south of the Chesapeake Bay Bridge. Data collected with a Seabird CTD that used a fluorometer to determine in situ chlorophyll concentrations. This is unpublished data collected by B. Cuker

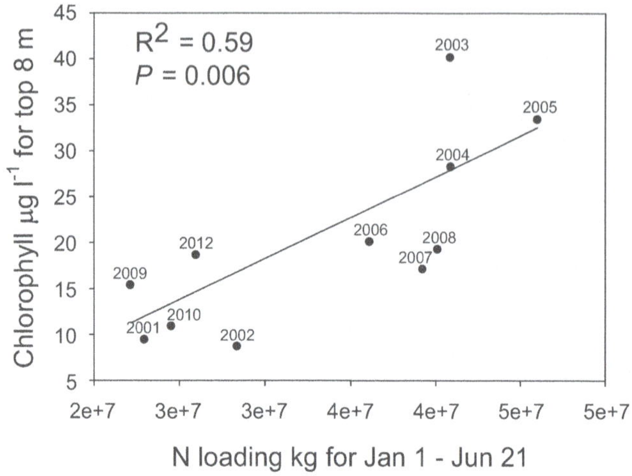

Fig. 3 Loading of nitrogen from the Susquehanna River between January and June predict mean chlorophyll *a* concentration in the upper 8 m of the water column of the mainstem of the Chesapeake Bay. Data are from June 21–22 for the years 2001–2012. Susquehanna River flow accounts for about half of the freshwater entering the Bay. It drains well fertilized farmlands in Pennsylvania and New York. Unpublished data from a study conducted between 2001 and 2012. Chlorophyll estimates come from in situ measurements with a WETstar fluorometer on a Seabird 25 CTD, deployed at 20 stations along the mainstem of the Bay. Nutrient loading calculated from data from the Conowingo Dam supplied by the USGS (2019)

waves and current. Because the plants live entirely submerged they require fairly clear waters to permit sufficient light for photosynthesis. They suffer from eutrophication because the excess nutrients stimulates the growth of two kinds of algae that compete for sunlight. Phytoplanktonic algae clouds the water, intercepting and scattering the light, leaving less for SAVs to do photosynthesis. The other kind of algae stimulated by the excess nutrients are the *epiphytic* species that grow attached to the SAV leaves and stems. These epiphytes form a coating of mucus, algal cells and bits of sediment. This layer reduces the amount of light getting through to the leaf (Stankelis et al. 2003; Kemp et al. 2004). Examples of SAVs from the fresher waters of the bay; wild celery (*Vallisneria americana*), common water weed (*Elodea canadensis*), sago pondweed (*Potamogeton pectinatus*), and redhead pondweed (*Potomogeton perfoliatus*). Saltier waters of the Bay lack the biodiversity of SAVs shown in the fresher reaches. The large grass beds noted by Capt. John Smith in the southern and saltier part of the Bay proper were undoubtedly eel grass (*Zostera mariana*).

SAVs played a central role in the ecology of a once healthy Chesapeake Bay. The beds slow the movement of currents and waves and that promotes the settling of suspended particles, making for clearer waters. Juveniles of blue crabs and many species of fish find refuge from predators amongst the leaves and stems of the SAVs. Various water fowl dine on SAVs. The scientific name of the canvasback duck, the once favorite of Chesapeake Bay hunters, is *Aythya valisineria*. It's species epitaph is but one letter different from the genus of its favorite food, *Vallisneria americana* (wild celery). The epiphytes on the leaves and stems of SAVs provide food for grazing snails and a variety of crustaceans. Small fish in the grass beds prey upon those invertebrates. The SAVs grow from rhizomes and sometimes seeds in the spring. They dieback in winter. But even in death the SAVs feed the Bay. The leaves breakup and decay, providing food in the form of plant *detritus* (dead organic matter) for suspension and deposit feeding animals. The depletion of SAVs and the nursery ground they provide for economically important species in the Chesapeake Bay causes significant loss to the fisheries industry (Kahn and Kemp 1985).

SAVs in the Bay began to decline circa 1900, when most of the watershed was cleared for farming. The decline dramatically increased in the mid twentieth century following the advent of industrial-chemical agriculture in the watershed. By the 1984 only remnants of the original beds remained (Orth et al. 2010). Fortunately, recent improvements in practices in agriculture, land management, sewage treatment and storm water control are driving increases in the extent and location of SAVs beds. A 2018 study found a 23% reduction in nitrogen loading to the Bay since 1984 is associated with the highest coverage of SAVs since the 1970s. Most of the SAV recovery is in the lower salinity areas. The authors of that study point out the role of the food system in suppressing SAVs; "We further find that high nutrient loads result from fertilizer application in all salinity zones, with an additional clear linkage to manure application in the mesohaline zone only (Lefcheck et al. 2018)."

In addition to eutrophication, several other food system practices caused the decline of SAVs in the Bay. Suspended mineral particles washed from cultivated fields adds to the turbidity that shades the SAV beds. Fishers dredging for oysters carved through the SAV beds. Herbicides sprayed on crop lands to kill weeds wash into the water along with those sediments. Although diluted in the water of the Bay, these chemicals still stress the SAVs (Powell et al. 2017).

Besides the food system problems, other factors helped drive the decline in SAVs. Hurricanes in 1933 and 1972 dropped the salinity so much as to kill most of the ell grass beds. Various pathogens weakened the SAVs. Dredging operations to deepen navigation channels dumped sediment on the beds (USFWS 2011). Invasive introduced species also threaten SAVs. The freshwater plant *Hydrilla verticillata* displaces native SAVs in lower salinity waters (Posey et al. 1993). The introduced mute swan (*Cygnus olor*) reduced % cover by SAVs as much as 79% in field experiments (Tatu et al. 2007).

Harmful Algal Blooms (HABs) Eutrophication not only drastically increases the number of phytoplankton in the Bay, but also changes the composition of the algal community. In contrast, unpolluted estuarine waters typically feature diverse phytoplankton communities consisting of a few numerically dominant species, 10 or so common ones, and may rare taxa. Certain species respond to added nutrients better than others (Buchanan et al. 2005; Marshall et al. 2006). During bloom conditions only one or two species comprises almost all of the community biomass. This can affect the food chain, as the algal species dominating under eutrophic conditions may prove less palatable to zooplankton and other suspension feeders. This is especially true when eutrophication brings on *harmful algal blooms* (HABs). HABs are labeled such for the disruption they cause to the ecosystem, human health and esthetics.

Some of the HAB species cause their harm by producing toxins to dissuade consumers. These toxins may accumulate up the food chain, sickening larger animals including humans. The Pocomoke River drains into the Chesapeake Bay from the Delmarva Peninsula. Twice in 1997 it experienced large fish kills (10,000–15,000 fish) of menhaden (*Brevoortia tyrannus*) that traced to the dinoflagellate *Pfiesteria piscicida*. People exposed to the toxin produced by that HAB experienced cognitive problems. The high levels of nitrogen (N) and phosphorus (P) draining from the agricultural lands and chicken CAFOs (Concentrated Animal Feeding Operations) were blamed for the bloom (Gilbert et al. 2001). In the Chesapeake Bay the number of HAB events nearly doubled between the 1990s and the 2000s, going from about 13 to 23 per year. Dinoflagellates (*Prorocentrum minimum* and *Karlodinium veneficum*) typically comprise the HABs in the saltier reaches of the Bay, while cyanobacteria HAB occur in the lowest salinity areas (Li et al. 2015).

Eutrophication Combines with Annual Stratification to Cause a Dead Zone in the Chesapeake Bay The Chesapeake Bay is a semi-mixed, salt wedge type estuary. Gravity drives the freshwater draining from the land to flow toward the Atlantic Ocean. At the same time salty water from the sea enters the mouth of the Bay in the opposite direction. Since the fresher water is less dense than the water from the ocean, it will float on top of the incoming seawater. Some of the momentum from the outgoing freshwater transfers to the deeper layer of incoming saltwater. It seems counterintuitive, but that force applied to the seawater drives it up the Bay, in the opposite direction. This is called *entrainment*, and the whole process is described as *estuarine circulation* (Dyer 1997). A wedge of freshwater draining from the Bay floats atop a wedge of incoming seawater. Hence the Bay is stratified (Figs. 4, 5, and 6). The intensity of this stratification varies over the course of the year, being strongest from early spring to late autumn (Testa and Kemp 2014). During winter the stratification may breakdown completely. The annual cycle in Bay temperatures explains the seasonal difference in stratification. In the warmer months the surface waters achieve temperatures several or more degrees above the incoming seawater. Since warmer waters are less dense than cooler ones, that adds to the intensity of stratification. The transition in density from one layer to the next is called the *pycnocline*. The pycnocline reflects the density differences due to both temperature and salinity.

From spring to autumn the outgoing wedge of warm freshwater is considerably less dense than the incoming wedge of saltier and cooler seawater. The system stays stratified like that till winter, when the temperature of fresh surface waters drops so low that its density isn't much different from the incoming seawater. Wind from a winter storm mixes the two layers, ending stratification till next spring.

Not Enough Light Penetrates to Deeper Waters in the Bay for Photosynthesis So far we focused on stratification by density, but another type of layering is also at

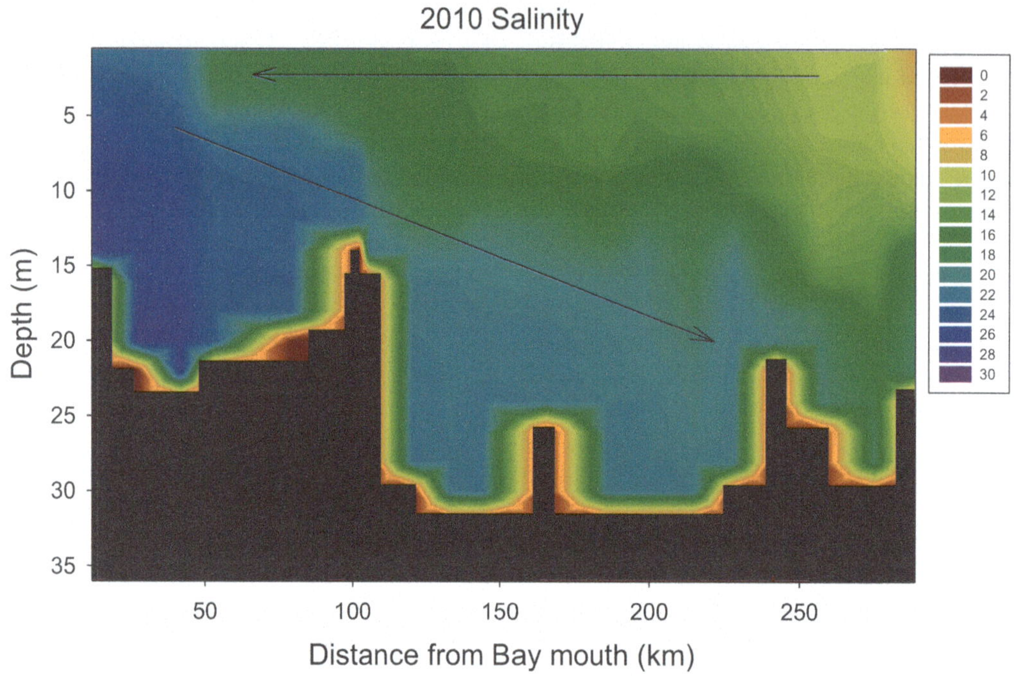

Fig. 4 Distribution of salinity in the Chesapeake Bay during stratification in early summer (June 21–22) of 2010, illustrating a classic salt-wedge. The mouth of the Bay is to the left. This is a side view of the Bay that exaggerates depth. The black rectangles on the bottom of the image indicate the depth contours of the bay. Note the higher salinity (22–24 PSU) of the Atlantic Ocean water intruding at the mouth of the Bay, and the subduction of that water mass with distance up the Bay. Freshwater enters the Bay from river-flow, creating the lower salinity (4–18 PSU) water mass in the upper right portion of the figure. The arrows indicate direction of flow for the two water masses. Unpublished data from a 2-day cruise. Measurements were taken from 20 stations beginning near the mouth of the Bay and with the last one just south of the Chesapeake Bay Bridge, across from Annapolis, MD. A Seabird CTD was used for the measurements

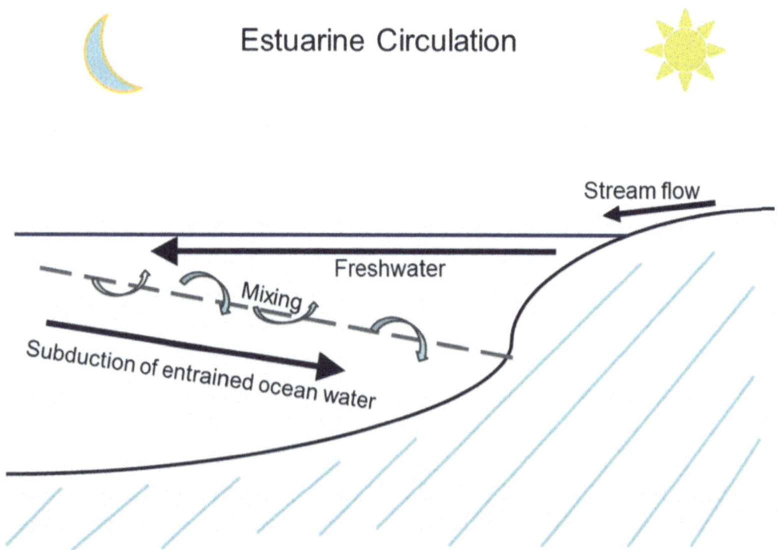

Fig. 5 The momentum of freshwater leaving the Bay transfers to the incoming water from the ocean, with some mixing at the boundary between the two water masses, note the surface waters have sufficient light to support photosynthesis (PS) during the day, but this is not the case for deeper waters

work in the Bay. That's stratification by light. As light enters the Bay it is absorbed or scattered by the water itself and the things dissolved of suspended in it. The more dissolved and suspended mater in the water, the less light can infiltrate. The amount of light penetrating the water decays in a logarithmic fashion with depth, so only 0.5% of the surface light reaches a depth of about 3–4 m during typical summertime conditions (Fig. 6). The high population of phytoplankton provided by eutrophication quickly attenuates the light. Consequently, the Bay is stratified with respect to light, the surface illuminated

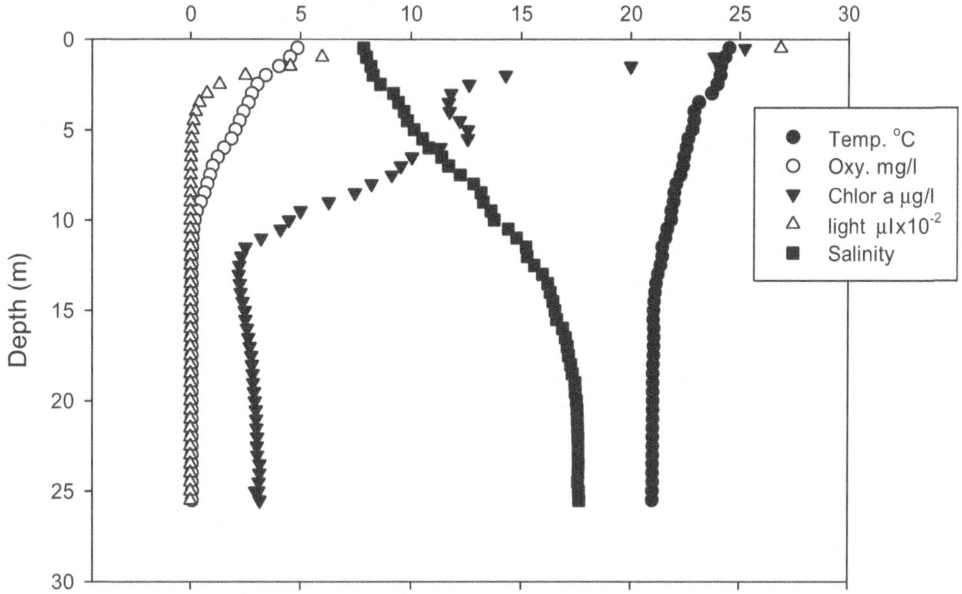

Fig. 6 Vertical distribution of temperature, oxygen concentration, chlorophyll *a*, light and salinity in a stratified section of the Chesapeake Bay at a station south of the Chesapeake Bay bridge and across from the city of Annapolis taken on June 22, 2011. The values for light are µI (microeinstein) divided by 100 to fit the scale of the graph. The Bay is stratified with less dense warmer and less salty water in the upper water column overlying a layer of deeper, saliter and cooler water. The bottom of the pycnocline is about 12 m, as indicated by the inflection in salinity and temperature. The concentration of oxygen drops to near zero at 11 m. Stratification prevents mixing down of oxygen from the surface and the lack of light halts photosynthesis. Note that most light is gone by a depth of 5 m. Chlorophyll *a* concentration is a proxy for phytoplankton and is concentrated in the top 3 m of the water column. Unpublished data collected with a Seabird CTD

layer is called the photic zone. The deeper darker layer is the aphotic zone. Generally the 0.5% of surface light level demarks the bottom of the photic and top of the aphotic zones. This is the *composition point*, or minimum amount of light needed by phytoplankton to balance respiration with photosynthesis. However, SAVs generally require more light than phytoplankton, with a compensation point circa 11% of the surface illumination (Duarte 1991).

Stratification of the Bay Means Restricted Delivery of Free Oxygen to the Bottom Waters During the 8 months or so of stratification, the top and bottom waters don't mix much beyond that which happens along the wedge where the water masses touch each other. This lack of mixing sets up the onset of a dead zone (Figs. 5 and 6). When the water from the Atlantic enters the mouth of the Bay it starts its journey inland near the surface. As it proceeds up the Bay, that dense water mass continues to *subduct* along the way, sinking deeper with each kilometer traveled. The Atlantic Ocean water is saturated with oxygen when it first enters the Bay, about 8 mg/l depending upon the temperature. Since it comes from near the ocean surface, that water benefited from both mixing with the atmosphere and plenty of sunlight for photosynthetic production of oxygen (explained below). Yet that oxygen begins to disappear as the ocean water sinks deeper and deeper on its passage up the Bay. By the time the mass of ocean water gets a quarter of the way up the Bay, it has sunk to below the photic zone. The lack of light in the bottom layer turns off photosynthetic production of oxygen. Additionally, stratification isolates that deeper water layer of Atlantic origin from the atmosphere, the other source of free oxygen (Schindler 2006). However, wind events sometimes disrupts stratification in the Bay, providing temporary reoxygenation of deeper waters (Scully 2016).

Bacteria and zooplankton living in the deeper layer of water, and in the sediments below it, continue to use aerobic respiration (explained below) to fuel their metabolism. This begins to exhaust the restricted supply of oxygen. But where is the food coming from to feed the bacteria and requiring them to use-up all of the oxygen? That food arrives as sinking algae and detritus the upper layer of the water column.

Much of the Phytoplankton, But Not the Oxygen It Produced Sinks to the Bottom Layer Phytoplankton in the upper layer of the Bay get their energy primarily from the light dependent process of photosynthesis. The algae uses sunlight to produce the chemical energy need to fix carbon dioxide into sugars and a host of other compounds. Importantly, photosynthesis requires the splitting of water molecules, separating the two hydrogen atoms from the one oxygen atom. The liberated hydrogen ions, play a role in the production of ATP (Adenosine Tri-Phosphate), the energy currency of the cell. Stripped of their hydrogen partners, the oxygen atoms join in pairs to produce the waste product of free oxygen (O_2). During times of algal blooms the photosynthesis is so intense as to super-saturate the water with the free oxygen. In such conditions oxygen concentrations can reach the level of 200–300% saturation (Cuker and Watson 2002). Because the oxygen concentrations exceed saturation, any amount over 100% will in a few hours combine to make bubbles that float to the surface, burst and diffuse into the atmosphere. Remaining in the surface layer of the water column is only a portion of the oxygen produced in photosynthesis. But also in the water are the energy-rich molecules packaged into the phytoplankton cells.

The phytoplankton cells are denser than the water, thus slowly sink and accumulate in the bottom waters and sediments at the bottom of the Bay. As the senescent (aging) phytoplankton cells sink into the darker waters below, they degrade and are attacked by bacteria. Bacteria use the process of aerobic respiration to release energy stored in the biological molecules of the decaying algae. This requires free oxygen to collect the low-energy electrons produced in the final step of aerobic respiration. Now here's the problem. Long gone is all of that oxygen released by the phytoplankton during the time of intense photosynthesis when it floated near the surface of the water column. So the bacteria quickly deplete the oxygen in the bottom waters, creating first hypoxic (oxygen concentration <2 mg/l) and then anoxic (no oxygen) conditions (Diaz and Rosenberg 2008).

While the saltier ocean water is making its way up the bay, the fresher surface water is traveling in the opposite direction on its way out of the Chesapeake. Loaded with excess nutrients (N & P) from the food system and other sources, the phytoplankton populations quickly grow beyond the capacity of their potential enemies to eat them. The algal cells are denser than the water and sink through the water column. Within a few days they pass through the pycnocline and enter the deeper salty layer. These senescent algal cells feed the aerobic respiration of the bacteria, causing those decomposers to exhaust the supply of oxygen. In addition, detritus from the watershed (*allocanthous* input, from outside the system) and the bay itself (*autochanthous* input, from within the system) also sinks to the bottom waters, providing more food for the community of bacteria and other decomposers. As the waters from the ocean progress up the Bay, they continuously lose oxygen to bacterial respiration (Fig. 7). This results in an extensive zone of hypoxia in the bottom waters of the Bay during the warmer months of the year (Fig. 8).

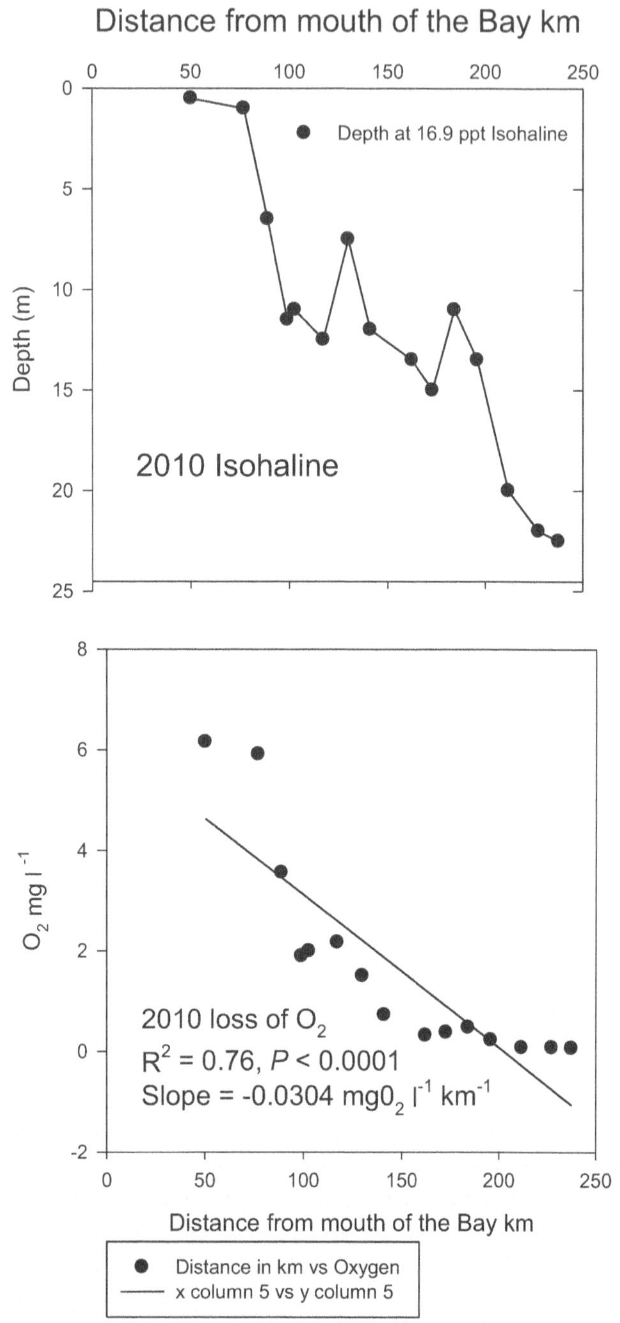

Fig. 7 The loss of oxygen in the waters entering the Chesapeake Bay from the Atlantic Ocean along an isohaline. The top graph traces water from near the mouth of the bay using its salinity of 16.9. Salinity of the ocean water is about 33, so in this example there was some prior mixing with fresher water from the Bay near the mouth of the estuary. As this water mass moves from the mouth (on the left) up estuary, it sinks below the outgoing fresher water mass. This excludes the sources of reoxygenation of the water, photosynthesis (too dark) and mixing with the atmosphere. The bottom graph traces the decline of the oxygen in the lower water mass. From unpublished data collected on June 21–22, 2010 with a Seabird CTD

Estuarine Circulation Causes the Bay to Retain Much of Its Organic Matter in Bottom Waters and the Sediments The nature of estuarine circulation makes the situation even worse in terms of oxygen depletion

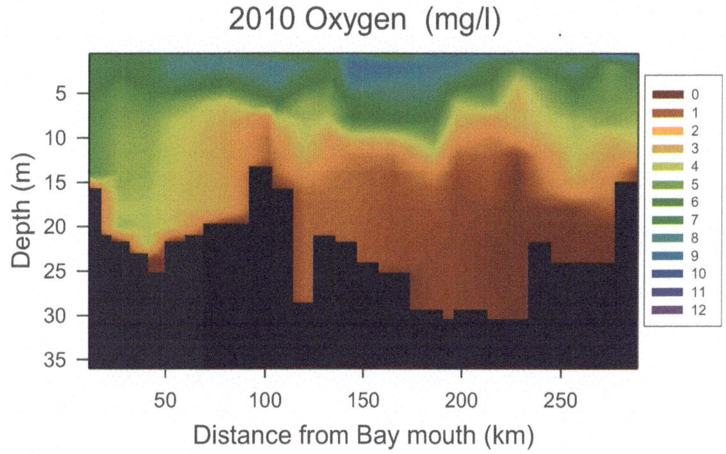

Fig. 8 The distribution of oxygen in the Chesapeake Bay on June 21–22, 2010. The mouth of the Bay is to the left. The black rectangles indicate the bottom contour. The blue area indicates high oxygen concentrations possibly associated with algal blooms in those areas raising concentrations above saturation. Oxygen in the bottom waters is hypoxic beginning about 100 km from the Bay's mouth. Unpublished data from a study sampling 20 stations along the axis of the main channel with a Seabird CTD

of bottom waters. The constant rain of senescent algal cells and detritus arriving in the deeper layer of water doesn't just stay there. Since that deep layer is moving up the estuary, the load of bacteria-feeding algae and detritus is also transported with it (Malone et al. 1988). So algae that sank into the deep layer near the mouth of the Potomac River, may in a few days be transported to the vicinity of Annapolis, about 100 km to the north.

This constant up-estuary transport of detritus in the bottom waters during the months of stratification means that the Bay retains its load of extra organic matter produced by eutrophication for many months. That extends the time for it to feed the bacteria that deplete oxygen. After cooling of surface waters breaks down stratification in late autumn, some of the organic matter deposited on the sediments may be flushed out of the Bay.

The Dead Zone Is Alive with Bacteria It's not just the bacteria in the bottom waters that needs oxygen for aerobic respiration. All of the animals living there require it for that same purpose. The absence of oxygen halts their ability to produce ATP to run the biochemical machinery of their cells. Anoxia in the Chesapeake Bay excludes most animals (metazoans) from those bottom waters. Sponges, worms, oyster, crabs, sea jellies, fish and most other animals either die in place or seek refuge in other areas (Sturdivant et al. 2014).

Although popularly called dead zones, the hypoxic and anoxic waters and underlying sediment produced in eutrophication are still quite alive. Not with the familiar marine animals mentioned above, but with trillions of anaerobic bacteria evolved to live in such anaerobic conditions. During the warm months of the year bacterial concentrations can exceed 20 billion cells/l, and produce 7 billon new cells/l daily (Jonas 1997; Heidelberg et al. 2002). Despite their tiny size, the sheer numbers of bacteria accounts for 60% or more of the plankton biomass. Even more numerous than the bacteria in the Bay are viruses (mostly bacteriophages) that feed upon them (Wommack et al. 1992). In just 1 day the bacterioplankton can remove 9 mg/l of dissolved oxygen from the water. That's 130% of the potential free oxygen typically dissolved in the water during the summer (Jonas 1997).

Since bacteria quickly reduce oxygen concentrations in the bottom waters during stratification, the system switches to one dominated by anaerobic metabolism. Bacteria use a variety of biochemical pathways to produce ATP in the absence of free oxygen. All of these alternative anaerobic processes lack the efficiency of aerobic respiration, releasing much less energy to be used by the bacteria for a given amount of food (decaying algae and detritus). Where is that energy that would have gotten released in aerobic respiration if oxygen was present? Its packaged into the energy-rich waste products of anaerobic respiration. That is, anaerobic respiration results in only a partial release of the energy stored in organic matter. The rest remains in the waste product. That's because free oxygen can accept low-energy electrons better than any other molecule in the natural system. These energy-rich waste products of anaerobic respiration include methane and hydrogen sulfide (Dai et al. 2017).

Bacterial Decomposition of Organic Matter in the Sediments Creates Oxygen Demand and the Release of Reduced Compounds A portion of the senescent algae and detritus settling to the bottom is decomposed along the way. The rest accumulates on the sediment surface. Under aerobic conditions a host of suspension and deposit feeding animals (worms, bivalves, etc.) living in or on the sediments use this as food. The anaerobic conditions associated with eutrophication eliminates these metazoans. The rich supply of decomposing organic matter in the sediments adds to the demand for oxygen in the bottom waters (Moore and Cuker 2018).

The switch from aerobic to anaerobic respiration in the sediments of the Chesapeake Bay involves the following electron acceptors and reduction products; NO_3 to N_2, Mn (IV) to Mn(II), Fe(III) to Fe(II), SO_4 to H_2S, and organic compounds to CH_4. The reduction of nitrate (NO_3) to elemental nitrogen (N_2) is called denitrification because it takes reactive nitrogen out of the system. The N_2 bubbles into the atmosphere. That means less NO_3 to stimulate algal growth.

The other reduced products contain much energy and fuel various bacteria (Middelburg and Levin 2009).

For example, methanogenic bacteria in anaerobic water and sediments produce methane as their waste product when they consume organic matter (Ferry 2011). Methane is the primary component of natural gas, one of the main fossil fuels used to make electricity in power plants. This illustrates how much energy remains in the waste product of that type of anaerobic respiration. And another group of marine microbes in the Chesapeake Bay called methylotrophic archaebacteria takes advantage of the energy in that methane as it bubbles in to the overlying aerobic waters (Sieburth 1993; Hinrichs et al. 1999). But much of the methane produced in the anaerobic sediments of the Chesapeake Bay escapes to the atmosphere. This is a problem since methane is a potent and long-lived greenhouse gas that drives climate change (Gelesh et al. 2016).

The hydrogen sulfide (H_2S) produced by the reduction of sulfate (SO_4) is fuel for sulfide bacteria in the presence of oxygen, but a deadly toxin for many organisms, including humans. It binds to iron (Fe) in the aerobic respiratory pathway, shutting down that process. Its presence in anaerobic sediments helps exclude many animals, reducing survival time in hypoxic environments by 30% (Vaquer-Sunyer and Duarte 2010). The build-up of H_2S in the bottom waters and the sediments of the Chesapeake Bay over the summer teams with low oxygen to eliminate most animals.

Release of N & P from Sediments Helps Sustain Eutrophication Much of the N & P taken up by phytoplankton is transported to the sediments as constituents of the sinking algae. Decomposition of those cells potentially releases these nutrients back to the water column. Regular tidal currents wash along the sediments, creating eddy-diffusion of remineralized nutrients and other metabolites back into the overlying waters. Wind-driven mixing events occasionally disrupts stratification for short periods, allowing the regenerated N & P from the bottom waters and the sediments to stimulate new generations of phytoplankton (Cowan and Boynton 1996). The biogeochemistry of the interaction between the sediments and overlying water column is complex. Generally, under aerobic conditions phosphate (PO_4) will bind with iron in the sediments, restricting its ability to return back to the water column. Anaerobic conditions of eutrophication reverses the flux of P (Moore and Cuker 2018). H_2S and other metabolites released from anaerobic bacteria drops the pH of the sediments, promoting the release of P back into the water. Although eutrophication started with allochanthous inputs of nutrients from the watershed and atmosphere, it may be sustained by autochanthous regeneration P and N from the sediments. Once this internal recycling is set in motion it moves the system to a new stable-state of self-sustaining eutrophication. This creates disappointment for environmental management agencies. Despite significant reductions in allocanthous nutrient loading, the eutrophic conditions often persist in managed estuaries (Conley et al. 2009a,b).

The situation for reactive nitrogen is more complicated. Anaerobic conditions promotes denitrification, where nitrate (NO_3) is used as an electron acceptor by certain bacteria. This converts the NO_3 to N_2 ($2\ NO_3^- + 10\ e^- + 12\ H^+ \rightarrow N_2 + 6\ H_2O$, Seitzinger et al. 2006). Denitrification helps reduce the load of reactive nitrogen streaming back into the water as NH_4 or NO_3.

Industrialized Chemical-Agriculture for the Production of Meat Brings on the Dead Zone In previous centuries localized hypoxia and anoxia in the bottom waters of the Chesapeake Bay occurred intermittently. These trace to extensive clearing and cultivation of the watershed during the late nineteenth century, the application of phosphate fertilizers at that time, and initial use of artificial N fertilizers in the first half of the twentieth century (Zimmerman and Canuel 2000; Testa et al. 2017). However, extensive and persistent periods of hypoxia first became a feature of the Bay in the second half of the twentieth century, concomitant with the widescale adoption of chemical-industrial agriculture for the production of meat.

Chesapeake Bay area farmers used plant and animal manures (including imported guano) to fertilizer their fields since the seventeenth century. Artificially produced N-fertilizers came on the scene after WWI (1918), and became the standard practice after WWII (1945) with transformation of munitions factories to agricultural chemical plants (see chapter "The Journey from Peruvian Guano to Artificial Fertilizer Ends with Too Much Nitrogen in the Chesapeake Bay"). As the Delmarva Peninsula developed a large-scale poultry industry that produced copious quantities of nutrient-rich manure (see chapter "Livestock and Poultry: The Other Colonists Who Changed the Food System of the Chesapeake Bay"). The vast majority of the corn and soy grown in the Bay's watershed is for the production of animal feeds, most of that going to the chicken industry. Layton Land points to animal agriculture as the primary source of nutrient loading in the Bay. "The reason water quality has not improved significantly is simple. The largest source of

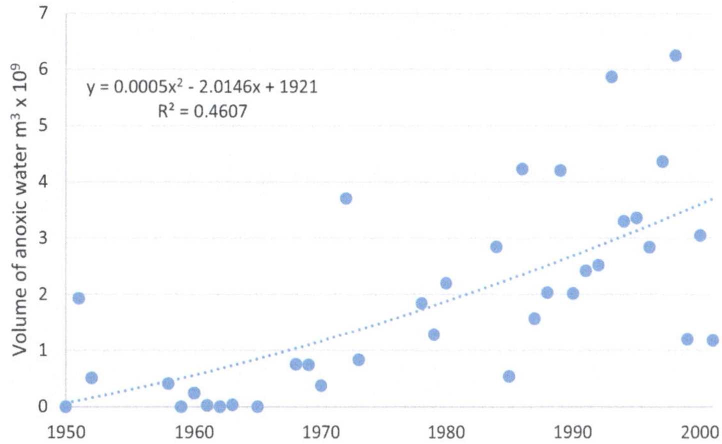

Fig. 9 The increasing volume of anoxic ($[O_2] < 0.2$ mg/l) water during stratification in the Chesapeake Bay between 1950 and 2001. Data from Hagy et al. (2004). A fairly steady increase in anoxic volume took place between 1950 and about 1986. After that, 5 years displayed anoxia well above the trend line (second order polynomial)

pollution, inefficient crop fertilization, has never been meaningfully addressed. That conclusion applies to water bodies worldwide, including the Great Lakes and the Gulf of Mexico. Farmers, supported by the powerful agricultural lobby, seek to maximize their harvest (profit) without paying for the pollution they cause. Society benefits from the cheap food they produce. Most of the grain produced in the Bay watershed feeds animals raised for meat, not humans (Land 2018)."

Farms in Pennsylvania and New York account for much of the agricultural production in the Chesapeake Bay Watershed. These drain to the Susquehanna River, which delivers about half of the load of freshwater to the Bay. Sampling stations in Harrisburg, PA and at the Conowingo Dam in Maryland monitor the pollution flowing down the Susquehanna River and into the Bay. Hagy et al. (2004) studied the relationship between nitrate $\left(NO_3^-\right)$ loading from the Susquehanna River and the appearance of hypoxia and anoxia in the bottom waters of the Chesapeake Bay between 1950 and 2001. This corresponds to the era of transition from essentially organic to chemical-based agriculture for most of the watershed. They also considered differing rates of river-flow in their analysis. The volume of anoxic ($[O_2] < 0.2$ mg/l) water during stratification went from zero to as high as 6.2 $m^3 \times 10^9$ by the end of the twentieth century (Fig. 9). They attributed this primarily to increased loading of total N into the estuary.

The annual average volume of hypoxic water in the Bay peaked between 1986 and 2000, as did N loading. Both features exhibit extensive inter-annual variability mostly driven by differences in precipitation and stream flow. Higher precipitation promotes additional runoff with its load of displaced nutrients. Sporadic wind events also promote aeration of the Bay. Adoption of better farming practices and reductions in air pollution from autos and power plants in the 1980s stemmed the increase of N loading to the Bay. Consequently, hypoxic volumes decreased, but still persist (Fig. 10). Hagy et al. (2004) suggested reducing total N loading by 40%, to 50 Gkg per year to reduce the extent and duration of hypoxia to an acceptable level. They also note that climate change will increase the likelihood of hypoxia. Increased precipitation will flush more nutrients from the land in to the Bay and warmer temperatures will intensify stratification, reducing reoxygenation of bottom waters.

Phosphorus As noted earlier, P also stimulates phytoplankton growth, but mostly in the lower salinity regions of the Bay. P is the most common limiting factor to algal growth in freshwater systems, as N can usually be supplied by nitrogen-fixing cyanobacteria. However salinities of 8–10 and higher inhibit N-fixation in estuaries. While N-fixation does take place in full salinity ocean water, it is diminished in estuarine conditions, possibly due to a combination of factors including sulfate inhibition and grazing by zooplankton (Conley et al. 2009b). Since farmers usually use mixed formulations of N and P as fertilizer, management practices to reduce runoff of N also helps keep P on the land. The story is different for sewage treatment plants. Those emptying into freshwater systems must meet requirements for reducing P in their effluent. However, those installations use biological decomposition to treat the organic matter in the sewage and this releases reactive nitrogen into receiving waters. For the Chesapeake Bay this means the freshwater tributaries may be low in P but will deliver substantial loads of additional N to the system as a consequence wastewater treatment. So the wastewater industry may soon need to add denitrification to their process in order for jurisdictions to meet their TMDLs (see chapter "Finishing the Journey: Wastewater as a Misplaced Resource" on wastewater treatment).

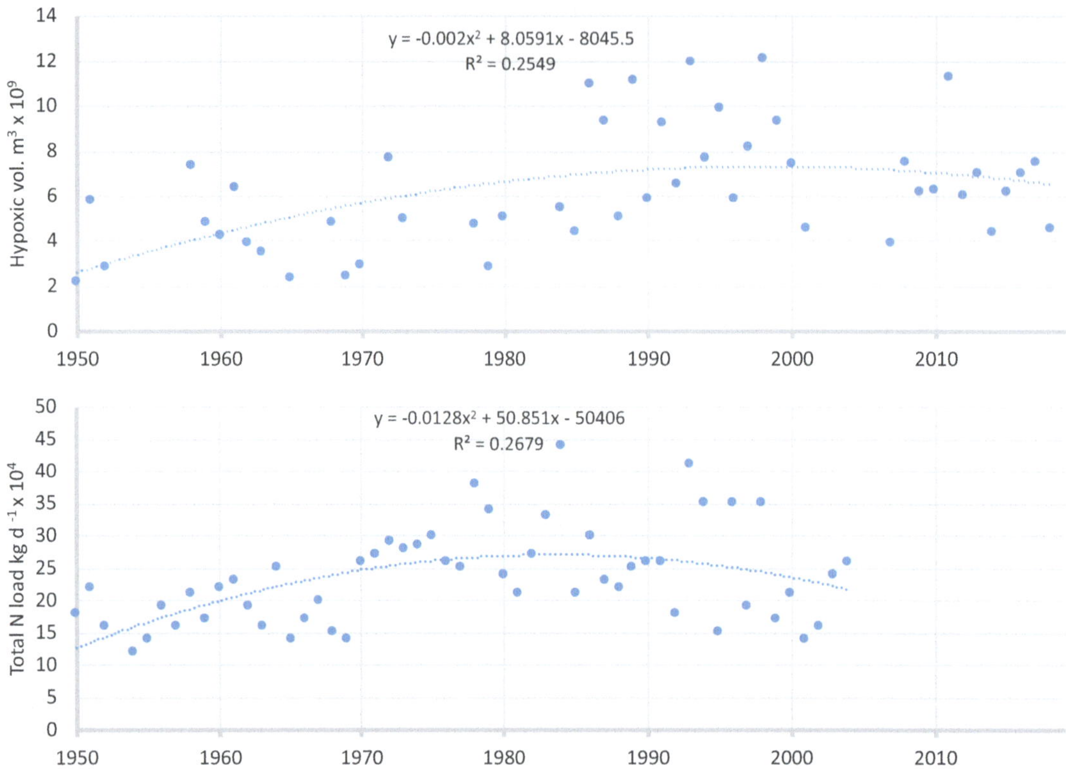

Fig. 10 Top graph, volume of hypoxic ($[O_2] < 2.0$ mg/l) water in the Chesapeake Bay during stratification between 1950 and 2017. Data combined from Hagy et al. (2004) and Scavia et al. (2018) to produce the graph. The second-order polynomial suggest that volume of hypoxic water peaked between 1980 and 2000, with a subsequent decline. Bottom graph, total daily load of total N (Jan.–May) entering the Chesapeake Bay from the Susquehanna River. Total N loads peaked in the 1980s with subsequent decline as suggested by the second-order polynomial line of best fit. Graph recreated from Scavia et al. (2018)

Easily Decomposed Organic Pollution Can Contribute to Eutrophication As detailed above, high loading rates of inorganic nutrients stimulate the excess growth of algae as the first step in the process of eutrophication. Readily decomposable organic matter added to the Bay also causes eutrophication. It does so in two ways. First, the organic matter flowing in from the watershed feeds the bacterioplankton. The well-fed bacteria deplete dissolved oxygen as they respire. Second, the bacteria mineralize the inorganic nutrients from their meal of organic matter. Those newly-regenerated nutrients go on to stimulate excess phytoplankton growth.

Not all organic matter has the same effect. Detritus naturally flowing in from the watershed consists of mostly dead plant material. Leaves and pieces of wood from forested lands and decomposing marsh plants get washed in with receding tides and rainstorms. This natural plant detritus contains mostly cellulose and other digestion-resistant fibers. Digestion of such natural detritus requires microbes that produce cellulase and other enzymes specific to the various fibers (Marinucci et al. 1983). The process is slow and often incomplete, leading to build-up of resistant organic residue in marsh soils and bottom sediments. Being slow and incomplete, decomposition of natural detritus tends not to exhaust the supply of dissolved oxygen. Since the detritus is mostly plant fiber with little N or P, remineralization of eutrophication-causing nutrients isn't a problem. The detritus washed into the Bay in late winter and early spring decays slowly during the subsequent months. Before European Colonization, this provided sufficient nutrients to support a diverse and productive food chain. The high C:N ratio of detritus proved healthy to the Bay.

Organic matter that arrives to the Bay as waste from the food system is generally more quickly decomposed than natural detritus, as it has a lower C:N ratio. The most obvious example is human fecal matter transmitted to the Bay from flush toilets. Prior to the mid-twentieth century human feces and urine flushed down toilets went directly to the Bay and its tributaries (see chapter "Finishing the Journey: Wastewater as a Misplaced Resource"). Today wastewater treatment installations remove most of the organic matter from sewage. Animals manures from chicken, pig and cattle operations are rich in easily decomposed organic matter and can quickly deplete receiving waters of oxygen. Slaughter houses and other animal-based food processing plants produce oxygen-demanding effluents comprised of blood and bits of other animal tissue. Smithfield Foods had to pay the largest fine ever levied at the time for violating the Clean Water Act for polluting the Pagan River (tributary of the James River) with organic waste from its slaughter house (USEPA 1999). Plant-based food processing plants, such as beer breweries must also treat their waste to prevent pollution from organic matter (Westbrook 2018).

Organic matter arrives to the Bay in two basic forms, dissolved and particulate. Dissolved organic matter (DOM) is produced by decomposition of terrestrial and aquatic plants and animals. It fuels the *microbial loop food chain*; organic matter to bacteria to protozoans to zooplankton (Malone et al. 1991; Anderson and Ducklow 2001). In parts of the Bay enriched with DOM, the microbial loop may rival the standard phytoplankton-fed food chain (Jonas 1997). DOM is such an important part of natural waters that in most aquatic systems it exceeds the mass of organic matter residing in plankton, nekton, and benthos. That is, for a given area of the Bay, there is more organic carbon dissolved in the water than there is in all the bacteria, algae, fish, crabs, etc. (Wetzel 2001). For the Bay, DOM accounts for 60–100% of bacterial respiration (Jonas 1997).

Particulate organic matter is anything too large to be dissolved (>0.5 μm). It ranges in size from tiny bits of leaves and algal cells to logs and large fish. But it usually is used to refer to the smaller end of the size spectrum. The particles may float at the surface, remain suspended in the water column, or sink to the bottom. Fiber-rich particles from plants are slow to decompose, as noted above, while those consisting of soft animal tissue quickly reenter the food chain. Detritivores (scavengers such as crabs) eat the larger particles, and microbes feast on the smaller ones.

Modeling The Chesapeake Bay's Dead Zone

Jeremy M. Testa

He is an authority on eutrophication of the Chesapeake Bay with many publications on the topic, including *Long-term nutrient reductions lead to the unprecedented recovery of a temperate coastal region* (Lefcheck et al. 2018), which brought him the Cozzarelli Prize from the National Academy of Sciences.

Modeling Eutrophication in the Chesapeake Bay

Dr. Testa explains the importance of using models to understand the outcomes of the eutrophication process and the effectiveness of potential interventions:

"Scientists have used a variety of observational methods to pinpoint the causes of eutrophication in coastal waters, including controlled 'enrichment' experiments, evaluations of water conditions across different tidal ecosystems relative to local land use, and evaluation of time trends in biological and chemical variables. Modeling has helped inform these data-based investigations by allowing scientists to isolate the different controlling variables in model 'experiments' to see which one is more important in driving those processes we associate with eutrophication. For example, oxygen concentrations are driven by a rich interplay between algal growth and death, water temperature, the physical movement of water, and mixing between the water and air (that is enhanced by wind). At any given time and place, each of these different processes may control oxygen levels—creating a particular challenge in associating low oxygen concentrations with any one variable—especially with the algal growth and death we associate with eutrophication. Thus, we can run the model with and without the elevated nutrient inputs that drive eutrophication-linked oxygen depletion (while holding wind, temperature, etc. constant) and quantify how important the elevated loading is. In some ways, the role of modeling in remediating eutrophication is much more straightforward. There are two primary questions that we are asked to help decide what level of pollution reduction is needed to achieve a specific improvement in oxygen conditions; (1) how much pollution reduction is needed? and (2) where should that pollution reduction occur? In each of these cases, we can't easily peer into the future or use existing data to quantitatively answer those questions, but we need some sort of prediction system that allows for different scenarios. Thus, using models of the watershed, the estuary, and of the coupled watershed-estuary system, we can iteratively reduce the hypothetical nutrients in different places and at different levels and see what combination leads to the oxygen improvement desired by the management agency."

Eutrophication of the Bay's People

From Medicine to Ecology and Back to Medicine: When More Is Less The concept of eutrophication first developed in the early twentieth century to describe the changes in freshwater systems in response to increased levels of inorganic nutrients (Smith et al. 2006). Aquatic scientists embraced a somewhat obscure medical term, *eutrophia*, a Greek word meaning well fed, to describe lakes. Pristine lakes with clear waters and relatively little nutrient input from the watershed were called *oligotrophic*. With a continuum of increasing nutrient inputs the lakes progressed to *mesotrophic*, *eutrophic*, and *hypereutrophic* status.

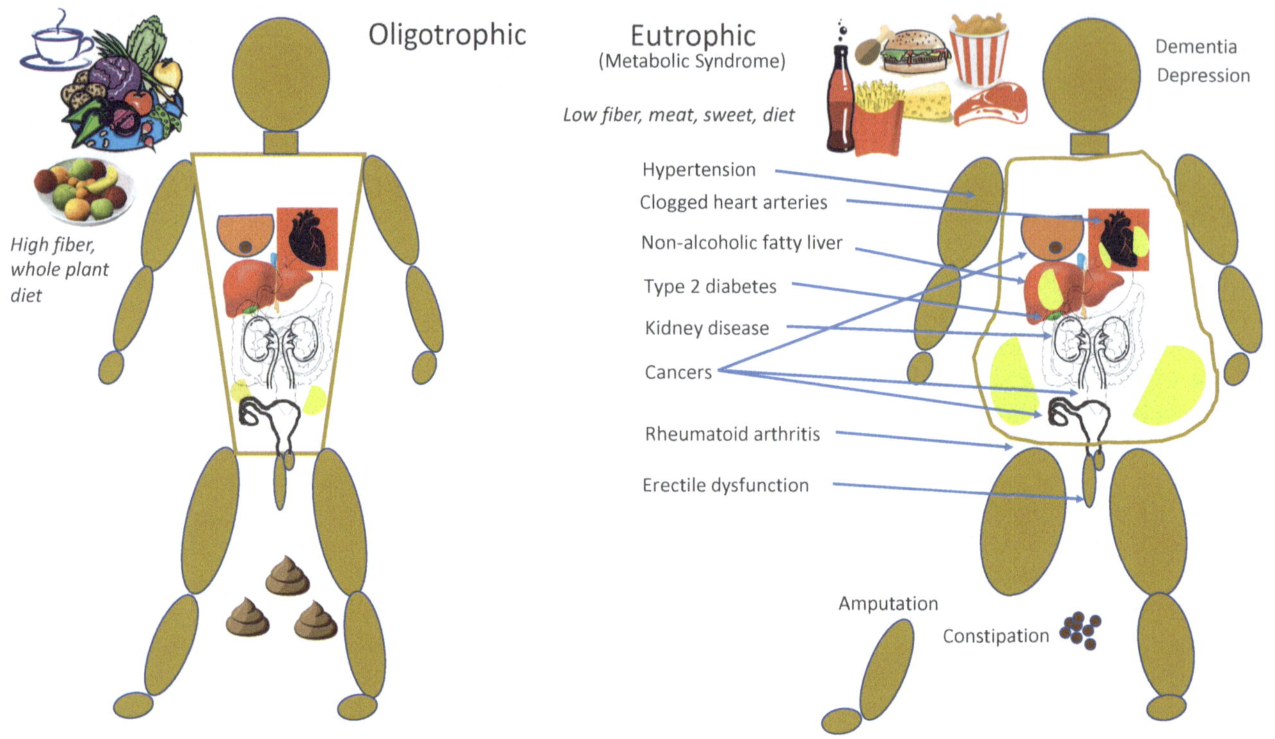

Fig. 11 Eutrophication of humans by the Standard American Diet (SAD). The combination of animal-based and highly processed foods results in numerous maladies. These includes dementia, depression, hypertension, coronary artery disease, non-alcoholic fatty liver, type 2 diabetes, kidney failure, cancers (breast, uterine, colon), rheumatoid arthritis, erectile dysfunction, diabetes-related amputation, and constipation. The SAD is missing sufficient fiber, protective antioxidants and other phytonutrients and is high in caloric density. The healthful oligotrophic diet includes whole plant foods featuring fruits, vegetables, legumes, nuts, and whole grains. These low caloric density foods provide fiber and a variety of phytonutrients

Eventually scientists working in estuarine and marine systems adopted the system, as did some terrestrial ecologists (Smith et al. 1999). Although the term *eutrophia* suggested a state of good health for humans, eutrophication means otherwise for aquatic systems. As detailed above, the eutrophic Chesapeake Bay suffers from excess nutrients that promote algal blooms, inadequate oxygen in bottom waters, and loss of biodiversity.

This is a problem of when more means *less*. The sequence form oligotrophic to hypereutrophic is based upon the positive relationship between nutrients and productivity. More nutrients means more fixation of carbon by the water body, up until the point when light becomes limiting to any further increase of photosynthesis. While a more productive body of water features plenty of algae to feed the planktonic food chain that leads to fish and birds, the outcome isn't usually desirable. The species of algae responding to the nutrients my not be edible, the loss of oxygen in the bottom waters reduces useable habitat, and the fish promoted by those conditions are usually considered of lower value. So from the perspective of the human value-system, the extra productivity of eutrophic systems means more biomass but less desirability.

And more nutrients means *less* for most of the human residents of today's Chesapeake Bay ecosystem. For almost all of human evolution the problem with food was its scarcity. Being well fed, in a state of *eutrophia*, was essential to good health. But the human diaspora out of tropical Africa and to all of the other continents of the world save Antarctica, meant that the gathering-and-hunting people often lacked sufficient nutrition, particularly in seasonal environments. The invention of agriculture (8000 years BCE) eventually meant the ability to produce and store more food than was needed for good health. In the modern era agriculture is so productive that less than 2% of the US population produces food to feed the rest, with a substantial amount left for export (USDA 2019). Exit undernutrition and enter the diseases of affluence; coronary artery disease, obesity, diabetes, and a host of other maladies stemming from eating too much of the wrong kinds of foods. In many ways the human *body* responds poorly to the oversupply of nutrients, similar to what's seen for a *body* of water, like the Chesapeake Bay. Hence, *the contemporary food system causes eutrophication of both the Chesapeake Bay and its people* (Figs. 11 and 12).

The Eutrophic State of the People of the Bay The eutrophication of Chesapeake Bay is representative of how the food system impaired many of the nations' waterways (Michalak et al. 2013). And the eutrophication of its human residents is consistent with the degradation of the health of the US population in general (Figs. 11 and 12). Much of the

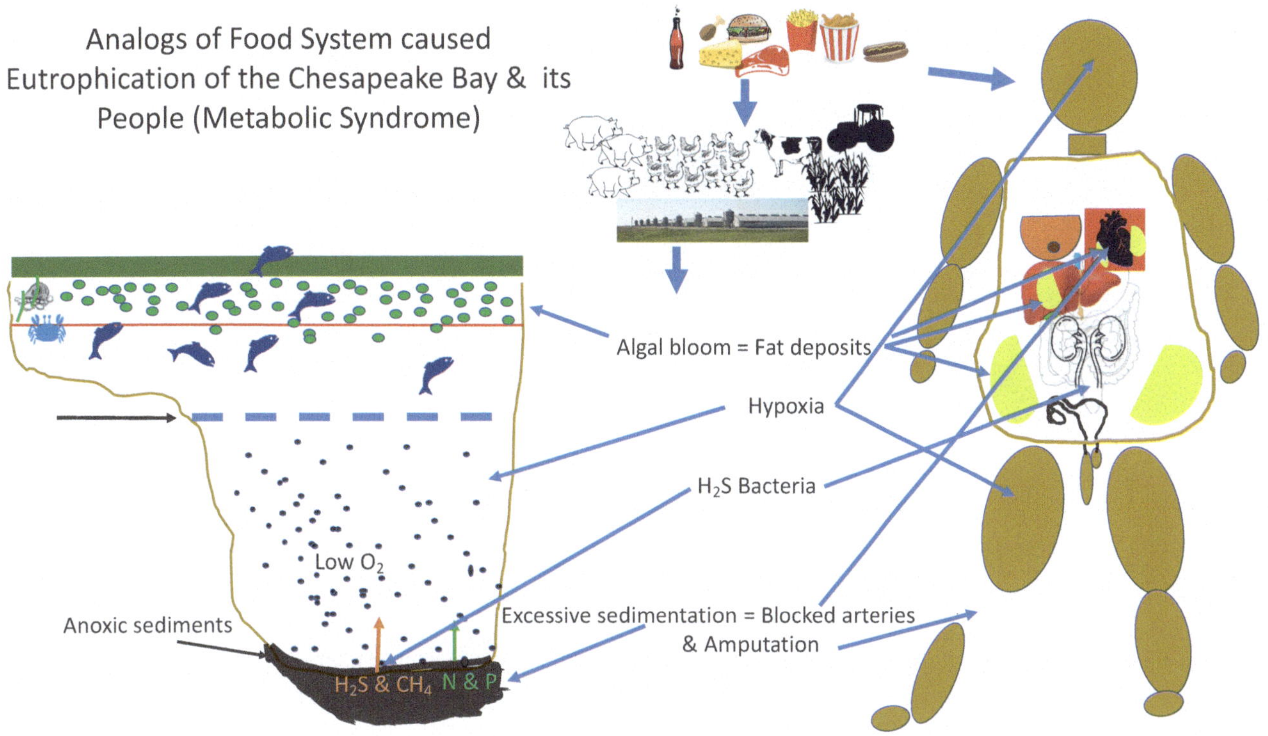

Fig. 12 Using analogy to understand the consequences of the Standard American Diet (SAD) and the chemical-industrial agricultural system that supports it. The SAD sickens people in a way analogous to how chemical agriculture degrades the Chesapeake Bay. N & P from chemical fertilizers and animal manures promote algal blooms. This added biomass to the Bay is analogous to the deposits of fat on humans caused by the SAD. Hypoxia appears in both systems, from decomposition of algae in the Bay and from poor blood circulation in the human. H_2S releasing bacteria form in anoxic sediments and in the lower intestines of meat eaters. The erosion of soil associated with conventional agriculture causes sedimentation that fills channels in the Bay. Deposition of cholesterol from the SAD clogs arteries, and complications of type 2-diabetes may result in amputation of toes or limbs

following discussion is based upon data drawn from the entire nation, and when available, for the Chesapeake Bay Region.

The US Department of Health and Human Services and US Department of Agriculture team to publish recommended dietary guidelines and investigate adherence by the US population. The guidelines for the 2015–2020 opens with this statement, "Today, about half of all American adults—117 million people—have one or more preventable, chronic diseases, many of which are related to poor quality eating patterns and physical inactivity. Rates of these chronic, diet-related diseases continue to rise, and they come not only with increased health risks, but also at high cost. In 2008, the medical costs linked to obesity were estimated to be $147 billion. In 2012, the total estimated cost of diagnosed diabetes was $245 billion, including $176 billion in direct medical costs and $69 billion in decreased productivity (DHHS 2015)."

The Standard American Diet (SAD) consists primarily of animal-based foods, refined carbohydrates, and processed foods with added sugars and fats. "About three-fourths of the population has an eating pattern that is low in vegetables, fruits, dairy, and oils. More than half of the population is meeting or exceeding total grain and total protein foods recommendations, … are not meeting the recommendations for the subgroups within each of these food groups. Most Americans exceed the recommendations for added sugars, saturated fats, and sodium. In addition, the eating patterns of many are too high in calories (DHHS 2015; Fig. 13)."

The USDA provides very general guidance in its dietary recommendations. It sets no prohibitions on any type of food, but does suggest moderation in consumption of some of the most health-damaging ones, such as added sugars, meats, and sodium. It essentially recommends a lite version of the SAD with the addition of more fruits, vegetables, legumes and whole grains. It also suggests increased intake of fats and dairy products (DHHS 2015). This is a controversial stance arising from the agency's duel missions of promoting US agricultural goods and recommending healthy diets. Some agricultural products aren't at all healthy, such as tobacco. Yet the USDA still facilitates tobacco marketing (USDA 2018b). Dairy provides no nutritional benefits over plant-based foods, and its consumption is associated with a variety of diseases. These include gastroesophageal reflux, acne, cancers, and coronary heart disease (Farahmand et al. 2011; Adebamowo et al. 2008; Aune et al. 2014; Campbell and Campbell 2016; Willett 2017). Although dairy is famous for calcium (Ca), it contains less than green leafy vegetables.

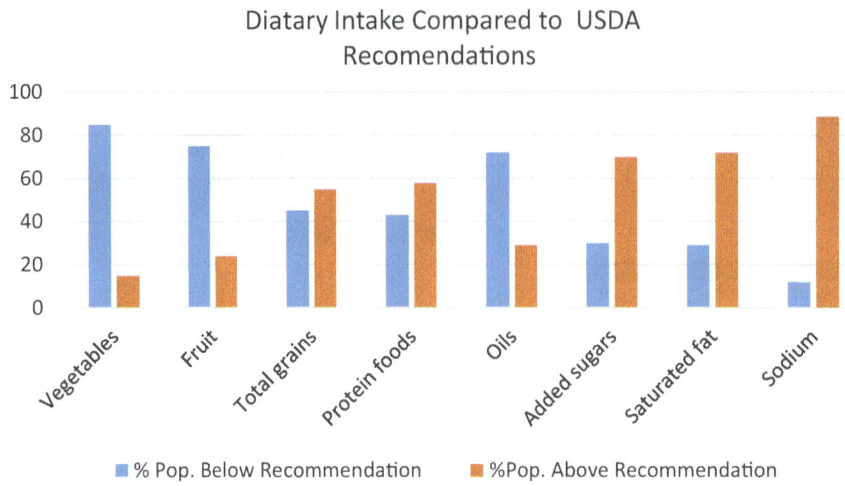

Fig. 13 Comparison of dietary intake and USDA recommendations for individuals age 1 year and older in the US population. According to the USDA, the Standard American Diet is deficient in vegetables, fruits, (whole) grains, and for some, protein. Most people eat too many simple carbohydrates, added sugars, saturated fat and sodium. Graph redrawn from DHHS (2015)

Cows get their calcium from eating those plants, as humans can too. For example, swiss cheese has only 0.27 mg of Ca per g, compared to 0.87 mg Ca per g for spinach. Dairy is also high in saturated fat and cholesterol, both contributing to the leading cause of death in the US, coronary artery disease (USDA 2018a; CDC 2019a).

Algal Blooms and Obesity For the casual observer, the persistent algal blooms that color the water of the Chesapeake Bay present the most obvious indicator of degradation of the ecosystem caused by the food system. Looking at the people that live there, the high incidence of obesity and overweight provide the same sort of visual evidence of ill health. The Obesity Medicine Association defines obesity as, "A chronic, relapsing, multifactorial, neurobehavioral disease, wherein an increase in body fat promotes adipose tissue dysfunction and abnormal fat mass physical forces, resulting in adverse metabolic, biomechanical, and psychosocial health consequences (Welcome 2017)." This definition emphasizes the adverse health consequences of obesity. In addition, obese people suffer discrimination of various forms (O'Brien et al. 2012).

Most epidemiological studies rely on the Body Mass Index (BMI), weight in kilograms divided by the square of height in meters. The metric is imperfect, over-scoring muscular and heavy-boned people, so requires some caution in its application, particularly to individuals. This is the scale; BMI < 18.5 = underweight, BMI of 18.5–24.9 = normal, BMI of 25–29.9 = overweight, BMI > 29.9 = obese (CDC 2019e). Percent of body in fat provides a better, but more difficult to obtain measure of obesity; $\geq$25% for males, and $\geq$32% for females (Welcome 2017).

Consider the analogy of excess algae for the Bay and too much adipose tissue in the human body. Both represent more biomass (organic matter) than is healthy for each system. The ills of algal blooms and eutrophication are detailed above. Obesity is correlated with; "all-causes of death (mortality), high blood pressure (Hypertension), high LDL cholesterol, low HDL cholesterol, or high levels of triglycerides (dyslipidemia), type 2-diabetes, coronary heart disease, stroke, gallbladder disease, osteoarthritis (a breakdown of cartilage and bone within a joint), sleep apnea and breathing problems, some cancers (endometrial, breast, colon, kidney, gallbladder, and liver), low quality of life, mental illness such as clinical depression, anxiety, and other mental disorders, body pain and difficulty with physical functioning (CDC 2019f)."

Add dementia (Alzheimer's disease) to the list of obesity linked diseases. Midlife obesity predicts the onset of dementia (Profenno et al. 2010). Various mechanisms seem at play. Blood volume doesn't increase proportionately to match the added biomass of adipose tissue of an obese person. The greater mass of tissues serviced by the vascular system stress the heart, causing enlargement of the left ventricle (Adams and Murphy 2000). These changes result in deficient delivery of oxygen to the brain and general inflammation promoting formation of tangles of microtubules of tau protein, the morphological marker for Alzheimer's disease (Zhang and Le 2010; Gao et al. 2013). Obesity promotes type-2 diabetes that results in insulin insensitivity in the brain and general inflammation. Dysbiosis (microbial imbalance) in the gut flora associated with obesity drives general inflammation that reaches the brain. Fat deposits harbor organochlorine pesticides from the food system. Release of these toxins also promote dementia (Anjum et al. 2018).

Obesity and Hypoxia Hypoxia in the deep waters of the eutrophic Chesapeake Bay is mirrored in the brain (as noted

Table 1 Incidence of obesity in the US and the Chesapeake Bay region

Jurisdiction	Adult obesity % 1990	Adult obesity % 2017	% Children 10–17 obese	Adult obesity % 2017 Black	Adult obesity % 2017 Latino	Adult obesity % 2017 White
U.S.	11.1	39.6	16.1	46.8	47.0	37.9
Delaware	15.8	35.0	18.5	41.1	35.1	32.7
District of Columbia	15.8	25.3	17.9	39.8	21.7	11.4
Maryland	11.9	34.4	18.6	43.0	30.4	30.9
New York	10.2	28.3	16.3	36.7	31.6	27.2
Pennsylvania	15.1	34.8	15.6	40.5	38.2	33.1
Virginia	12.4	33.1	15.5	45.1	32.9	30.7
West Virginia	15.1	41.9	21.9	48.0	31.9	40.7
Mean Ches. Bay	13.8	33.3	17.8	42.0	31.7	29.5

Rates of adult obesity increased three to fourfold between 1990 and 2017. African American and Latino residents suffer obesity at higher rates than whites. The individual state data is based upon self-reporting of height and weight, while the US data comes from clinical measurement. To make the two data sets compatible, the state values were increased by 10% to correct for documented self-under-reporting (Yun et al. 2005). Data from RWJF (2018)

Table 2 The prevalence of mortality from coronary artery disease and stroke, and morbidity rates of type-2 diabetes and hypertension in the Chesapeake Bay region and the US

Jurisdiction	Heart disease deaths per 100,000	Cancer deaths per 100,000	Stroke deaths per 100,000	Adults % diabetes	% Pop. hypertension
U.S.	165.5	152	36.6	9.4	33.2
Delaware	158.4	160.4	38.8	11.3	34.9
District of Columbia	211.7	189.5	35.5	7.8	26.7
Maryland	164.5	151.25	38.0	10.4	32.4
New York	171.2	141.2	26.1	10.5	29.4
Pennsylvania	176.0	161	36.7	10.6	32.6
Virginia	154.5	152.6	37.0	10.5	32.4
West Virginia	192.0	179.4	45.3	15.2	43.5
Mean Ches. Bay	175.5	162.2	36.8	10.9	33.1

Data from CDC (2019b, c, d)
Residents of the Bay Region suffer higher rates of coronary artery disease, cancer and diabetes compared to the general US population

above) and other tissues in people suffering from obesity (Fig. 13). Restricted angiogenesis (production of capillaries) in adipose tissues creates hypoxia. This leads to a complex set of health-damaging responses. "…many biologically active molecules produced by fat have been identified, evoking a role of this tissue in previously unsuspected processes, including inflammation, insulin, and glucose homeostasis, angiogenesis, and hemostasis. Hormonal factors produced by adipose cells have been collectively called "adipokines", and include about a hundred of known molecules, such as leptin, adiponectin, resistin, apelin, visfatin, VEGF, PAI-1, and many other, not yet identified factors, that have recently emerged by proteomic strategies. It has been postulated that the adipose tissue of the obese, becoming dysfunctional by changes in the secretion pattern of adipokines, may sustain inflammation, insulin resistance, and a pro-thrombotic state with endothelial dysfunction, all conditions underpinning obesity-related comorbidities (Foti and Brunetti 2017)."

Obesity is on the rise. Its incidence in the US adult population quadrupled between 1990 and 2017, going from 11.1% to 39.6%. Adults of the Chesapeake Bay region suffer at the rate of 33.3%, with the 41.9% of West Virginia topping the list for the area and the entire US. The 17.8% rate of obesity for children and teenagers (ages 10–17 years) in the region tops the 16.1% rate for the US. Obesity is highest for Latino and African American residents in the Chesapeake Region and the rest of the US (Table 1). In addition to the nearly 40% of adults in the US being obese, another 32.5% are overweight, bringing the combined portion to 72.5%. The extent of this food system pandemic (similar rates are now seen world-wide) should promote all thoughtful citizens to reconsider the role of diet in their lives.

The growing incidence of obesity means increasing rates of coronary artery disease, stroke, type-2 diabetes and hypertension (Table 2) and other SAD related diseases. Due to the decline in smoking (now 17% of US population), cancer is no longer the number one cause of death. As of 2017, coronary heart disease was first, followed by the other food system related causes: cancer (2), stroke (5), Alzheimer's Disease (6), Diabetes (7), kidney disease (9) (Murphy et al. 2018). Coronary heart disease, the leading cause of death in the US and world-wide is virtually non-existent in populations that eat plant-based (Higginson and Pepler 1954; Thomas et al. 1960).

H_2S Producing Bacteria in the Chesapeake Bay and People's Intestines One consequence of seasonal anoxia in the Chesapeake Bay is the production of H_2S by bacteria. As discussed above, this helps render inhospitable the deeper waters of the Bay. Additionally, H_2S contributes to the acidification of the system, adding to the effect of elevated CO_2 levels associated with fossil fuel consumption (Cai et al. 2017). Bacteria also produce toxic H_2S gas in the anoxic intestines of people eating the SAD. It arises mostly from bacteria digesting the sulfur-containing amino acids, methionine and cysteine. These sulfur-amino acids typically concentrate 4–12 times higher in animal-based than plant-based foods, and production of H_2S gas in the large intestine is associated with consumption of meat (Magee et al. 2000). The H_2S produced by bacteria in the colon promotes ulcerative colitis and the proliferation of cancer cells in the intestinal wall (Pitcher and Cummings 1996; Cai et al. 2010).

In the US, colorectal cancer is the second leading cause of cancer-deaths that afflict both men and women (CDC 2019g). Yet, this disease is rare in populations that eat plant-based diets. The rate of colon cancer for African Americans is 65:100,000, 65 times the 1:100,000 for rural South Africans. African Americans, as others in the US, mostly eat the meat-rich SAD, while rural South Africans eat little meat and mostly plant based foods. The protection afforded by eating plants includes reduced production of H_2S, as well as benefits from increased dietary fiber (O'Keefe et al. 2015). The connection between H_2S in the Chesapeake Bay and the colon's of its residents is clear, as W. Cai served as the lead author on both the ecosystem and human biome studies cited above.

Sedimentation in the Bay, Amputation, and Blocked Blood Vessels In addition to enriching the Bay with nutrients, industrial agriculture for the support of the SAD causes soil erosion that increases rates of sedimentation in the Bay. The sediments settle in channels, causing them to shoal and in some cases completely fill. The agricultural-erosion and sedimentation process certainly pre-dates the advent of industrial agriculture. However, the extensive cultivation needed to produce animal food, and in some places high densities of livestock, promote continued soil erosion. Expensive dredging operations keeps open the most important shipping channels and harbors, but is rarely done for sections of the Bay system with less economic importance. This means reduced circulation of water in some places and in others, the complete loss of portions of the Bay and its tributaries to sedimentation.

The sedimentation of the Conowingo Dam on the Susquehanna River presents a clear and urgent example. The Dam was erected on the river in 1928 to produce hydroelectricity. The structure produced two food system impacts. It blocked the migration of shad and other fish (see chapter "A Fishing Trip: Exploiting and Managing the Commons of the Chesapeake Bay"). It also stilled the waters of the Susquehanna, allowing about half of the suspended sediments eroding from agricultural lands to settle behind the dam. Trapping the sediments mitigated poor agricultural practices. However, that feature of the dam is gone, as nearly a century of sedimentation has filled to capacity the reservoir behind the dam (and two smaller upstream structures). Intense storms produce extreme flows that scour and resuspends the sediments, sending them down to the Bay. Dredging the sediments and using the spoils to replenish depleted food-crop fields is out, since they are contaminated with pesticides and other toxins (PCB's and mercury) that might taint the harvest. A small scale study is underway, relocating spoils to eroding shorelines, landfill covers, and other non-food system land. So for now, the ecosystem services once provided by the stretch of the Susquehanna River behind the dam have been functionally amputated (Langland 2015; Staff 2019). Just as the cultivation of foods for the SAD is clogging the channels of the Chesapeake Bay, the products of that system are clogging the blood vessels of people that consume them. The reduced circulation of blood addressed above and blood vessel dysfunction related to diet induced type 2 diabetes may result in the death of the tissues of extremeties, requiring amputation (Davis et al. 2006).

Lack of Fiber in the Diets of the Bay and Its People Damages the Health of Both When the USEPA put the Chesapeake Bay on a pollution diet, they focused on the inorganic nutrients of N and P, mostly coming from the food system. Under natural circumstances (pre European colonization) the Bay received little inorganic N or P, as those nutrients were mostly locked-up in the organic molecules of life. Those nutrients largely entered the Bay in fibrous plant detritus washed from the watershed. Slowly decomposing detritus would release the nutrients over many months. This contrasts with high flows of N and P from fields amended with artificial fertilizer, leaching from stores of animal manure, and effluent from wastewater treatment plants. Curtailing N and P pollution from those food system sources is essentially returning the Bay to the high-fiber diet that once prevailed.

As with the Bay, humans do better with a high-fiber diet. According to the USDA, 97% of people in the country eat diets deficient in fiber, most consuming half or less than the 25–38 g/day recommendation (King et al. 2012). High fiber diets offer multiple advantages. Foods featuring high fiber content provide low caloric and high nutritional density. Fruits, vegetables, legumes, and whole grains provide, antioxidants, vitamins and other phytonutrients packaged with many fewer calories than animal products. For example, 100 g of skinless chicken breast provides 107 Kcal, compared to 80 Kcal for navy beans, 50 Kcal for apple, or 35 Kcal for kale of the same mass. The chicken provides no fiber, while respectively the plant items provided 5.4, 2.4, and 4 g of fiber per 100 g of mass. The chicken is void of the vitamins and antioxidants provided by the beans, apple and kale (USDA 2018a).

The benefits of fiber in the human diet goes beyond the company it keeps in the foods that provide it (Burkitt 1979). Its more than those important associated phytonutrients. Fiber itself proves health promoting. Fiber extracts cholesterol from bile salts and foods, reducing its concentration in the body and the associated potential for heart disease (Jenkins et al. 2003). A healthy microbiome in the intestines requires dietary fiber to produce short-chained fatty acids that prove anti-inflammatory and protective against various diseases (Mudgil and Barak 2013; Sonnenburg and Sonnenburg 2014). See chapter "Instead of Eating Fish: The Health Consequences of Eating Seafood from the Chesapeake Bay Compared to Other Choices" for a detailed discussion of dietary fiber.

Reversing Eutrophication of the Bay and Its People Converting the food system to one of whole plant based foods could solve the dual problems of ecosystem degradation and human population ridden with diet-caused disease. Eliminating the production of animal foods for the SAD will reduce the amount of land under cultivation needed for the production of animal feeds. The reduced pressure on land for feed stock production will facilitate a shift to organic techniques of somewhat lower intensity agriculture that eliminates artificial fertilizers, and keeps nutrients and soil from running off into the Bay. The plant based diet would virtually eliminate or significantly reduce many of the diseases plaguing the residents of the Bay. The list includes; coronary artery disease, obesity, diabetes, many types of cancer, kidney disease, and dementia.

Chapters "The Journey from Peruvian Guano to Artificial Fertilizer Ends with Too Much Nitrogen in the Chesapeake Bay" and "A New Food System for The Chesapeake Bay Region and a Changing Climate" further explore the environmental and health consequences of moving the Bay's food system to one based upon whole plants.

References

Adams J, Murphy P (2000) Obesity in anaesthesia and intensive care. Br J Anaesth 85:91–108. https://doi.org/10.1093/bja/85.1.91

Adebamowo C, Spiegelman D, Berkey C et al (2008) Milk consumption and acne in teenaged boys. J Am Acad Dermatol 58:787–793. https://doi.org/10.1016/j.jaad.2007.08.049

Anderson T, Ducklow H (2001) Microbial loop carbon cycling in ocean environments studied using a simple steady-state model. Aquat Microb Ecol 26:37–49. https://doi.org/10.3354/ame026037

Anjum I, Fayyaz M, Wajid A et al (2018) Does obesity increase the risk of dementia: a literature review. Cureus. https://doi.org/10.7759/cureus.2660

Aune D, Navarro Rosenblatt D, Chan D et al (2014) Dairy products, calcium, and prostate cancer risk: a systematic review and meta-analysis of cohort studies. Am J Clin Nutr 101:87–117. https://doi.org/10.3945/ajcn.113.067157

Brush G (2008) Historical land use, nitrogen, and coastal eutrophication: a paleoecological perspective. Estuar Coasts 32:18–28

Buchanan C (2016) From programmatic goals to criteria for phytoplankton chlorophyll a. In: Potomacriver.org. https://www.potomacriver.org/wp-content/uploads/2016/02/ICP16-1_Buchanan.pdf. Accessed 10 July 2019

Buchanan C, Lacouture R, Marshall H et al (2005) Phytoplankton reference communities for Chesapeake Bay and its tidal tributaries. Estuaries 28:138–159. https://doi.org/10.1007/bf02732760

Burkitt D (1979) The protective value of plant fibre against many modern western diseases. Qual Plant Plant Foods Hum Nutr 29 (1–2):39–48

Cai W, Wang M, Ju L et al (2010) Hydrogen sulfide induces human colon cancer cell proliferation: role of Akt, ERK and p21. Cell Biol Int 34:565–572. https://doi.org/10.1042/cbi20090368

Cai W, Huang W, Luther G et al (2017) Redox reactions and weak buffering capacity lead to acidification in the Chesapeake Bay. Nat Commun. https://doi.org/10.1038/s41467-017-00417-7

Campbell TC, Campbell TM (2016) The China study. BenBella Books, Dallas

CBP (Chesapeake Bay Program) (2018) ChesapeakeProgress—2017 and 2025 Watershed Implementation Plans (WIPs). In: Chesapeakeprogress.com. https://www.chesapeakeprogress.com/clean-water/watershed-implementation-plans. Accessed 9 Sept 2018

CDC (2019a) Products—data briefs—number 328—November 2018. In: Cdc.gov. https://www.cdc.gov/nchs/products/databriefs/db328.htm. Accessed 18 July 2019

CDC (2019b) Stats of the states—heart disease mortality. In: Cdc.gov. https://www.cdc.gov/nchs/pressroom/sosmap/heart_disease_mortality/heart_disease.htm. Accessed 18 July 2019

CDC (2019c) New CDC report: more than 100 million Americans have diabetes or prediabetes. In: CDC. https://www.cdc.gov/media/releases/2017/p0718-diabetes-report.html. Accessed 18 July 2019

CDC (2019d) Stats of the states—cancer mortality. In: Cdc.gov. https://www.cdc.gov/nchs/pressroom/sosmap/cancer_mortality/cancer.htm. Accessed 18 July 2019

CDC (2019e) Defining adult overweight and obesity. Overweight & Obesity. CDC. In: Cdc.gov. https://www.cdc.gov/obesity/adult/defining.html. Accessed 19 July 2019

CDC (2019f) The health effects of overweight and obesity. Healthy Weight. CDC. In: Cdc.gov. https://www.cdc.gov/healthyweight/effects/index.html. Accessed 19 July 2019

CDC (2019g) Basic information about colorectal cancer. CDC. In: Cdc.gov. https://www.cdc.gov/cancer/colorectal/basic_info/index.htm. Accessed 19 July 2019

Cerco C, Noel M (2007) Can oyster restoration reverse cultural eutrophication in Chesapeake Bay? Estuar Coasts 30:331–343. https://doi.org/10.1007/bf02700175

Cloern J, Grenz C, Vidergar-Lucas L (1995) An empirical model of the phytoplankton chlorophyll: carbon ratio-the conversion factor between productivity and growth rate. Limnol Oceanogr 40:1313–1321. https://doi.org/10.4319/lo.1995.40.7.1313

Conley D, Carstensen J, Vaquer-Sunyer R et al (2009a) Ecosystem thresholds with hypoxia. Hydrobiologia 629:21–29. https://doi.org/10.1007/s10750-009-9764-2

Conley D, Paerl H, Howarth R et al (2009b) ECOLOGY: controlling eutrophication: nitrogen and phosphorus. Science 323:1014–1015. https://doi.org/10.1126/science.1167755

Cowan J, Boynton W (1996) Sediment-water oxygen and nutrient exchanges along the longitudinal axis of Chesapeake Bay: seasonal patterns, controlling factors and ecological significance. Estuaries 19:562. https://doi.org/10.2307/1352518

Cuker B, Watson M (2002) Diel vertical migration of zooplankton in contrasting habitats of the Chesapeake Bay. Estuaries 25:296–307. https://doi.org/10.1007/bf02691317

Dai X, Hu C, Zhang D et al (2017) A new method for the simultaneous enhancement of methane yield and reduction of hydrogen sulfide production in the anaerobic digestion of waste activated sludge. Bioresour Technol 243:914–921. https://doi.org/10.1016/j.biortech.2017.07.036

Davis WA, Norman PE, Bruce DG, Davis TME (2006) Predictors, consequences and costs of diabetes-related lower extremity amputation complicating type 2 diabetes: the fremantle diabetes study. Diabetologia 49(11):2634–2641

DHHS (2015) In: Health.gov. https://health.gov/dietaryguidelines/2015/resources/2015-2020_Dietary_Guidelines.pdf. Accessed 16 July 2019

Diaz R, Rosenberg R (2008) Spreading dead zones and consequences for marine ecosystems. Science 321:926–929. https://doi.org/10.1126/science.1156401

Duarte C (1991) Seagrass depth limits. Aquat Bot 40:363–377. https://doi.org/10.1016/0304-3770(91)90081-f

Dyer K (1997) Estuaries: a physical introduction, 2nd edn. Wiley, New York

Farahmand F, Najafi M, Ataee P et al (2011) Cow's Milk Allergy among Children with Gastroesophageal Reflux Disease. Gut and Liver 5:298–301. https://doi.org/10.5009/gnl.2011.5.3.298

Ferry J (2011) Fundamentals of methanogenic pathways that are key to the biomethanation of complex biomass. Curr Opin Biotechnol 22:351–357. https://doi.org/10.1016/j.copbio.2011.04.011

Foti D, Brunetti A (2017) Editorial: "linking hypoxia to obesity". Front Endocrinol. https://doi.org/10.3389/fendo.2017.00034

Gao L, Tian S, Gao H et al (2013) Hypoxia increases Aβ-induced tau phosphorylation by calpain and promotes behavioral consequences in AD transgenic mice. J Mol Neurosci 51:138–147. https://doi.org/10.1007/s12031-013-9966-y

Gelesh L, Marshall K, Boicourt W et al (2016) Methane concentrations increase in bottom waters during summertime anoxia in the highly eutrophic estuary, Chesapeake Bay, U.S.A. Limnol Oceanogr 61:S253–S266. https://doi.org/10.1002/lno.10272

Gilbert PM, Magnien R, Lomas MW, Alexander J, Fan C, Haramoto E, Trice M, Kana TM, (2001) Harmful algal blooms in the Chesapeake and coastal bays of Maryland, USA: comparison of 1997, 1998, and 1999 events. Estuaries 24(6):875

Hagy J, Boynton W, Keefe C et al (2004) Hypoxia in Chesapeake Bay, 1950–2001: long-term change in relation to nutrient loading and river flow. Estuaries 27:634–658. https://doi.org/10.1007/bf02907650

Heidelberg J, Heidelberg K, Colwell R (2002) Seasonality of Chesapeake Bay bacterioplankton species. Appl Environ Microbiol 68:5488–5497. https://doi.org/10.1128/aem.68.11.5488-5497.2002

Higginson J, Pepler W (1954) Fat intake, serum cholesterol concentration, and atherosclerosis in the South African Bantu. Part II. Atherosclerosis and coronary artery disease. J Clin Invest 33:1366–1371. https://doi.org/10.1172/jci103013

Hinrichs K, Hayes J, Sylva S et al (1999) Methane-consuming archaebacteria in marine sediments. Nature 398:802–805. https://doi.org/10.1038/19751

Jenkins D, Kendall C, Marchie A, Jenkins A, Connelly P, Jones P, Vuksan V (2003) The Garden of Eden—plant based diets, the genetic drive to conserve cholesterol and its implications for heart disease in the 21st century. Comp Biochem Physiol A Mol Integr Physiol 136:141–151. https://doi.org/10.1016/s1095-6433(02)00345-8

Jonas R (1997) Bacteria, dissolved organics and oxygen consumption in salinity stratified Chesapeake Bay, an anoxia paradigm. Am Zool 37:612–620. https://doi.org/10.1093/icb/37.6.612

Kahn J, Kemp W (1985) Economic losses associated with the degradation of an ecosystem: the case of submerged aquatic vegetation in Chesapeake Bay. J Environ Econ Manag 12:246–263. https://doi.org/10.1016/0095-0696(85)90033-6

Kemp W, Batleson R, Bergstrom P et al (2004) Habitat requirements for submerged aquatic vegetation in Chesapeake Bay: water quality, light regime, and physical-chemical factors. Estuaries 27:363–377. https://doi.org/10.1007/bf02803529

Kemp W, Boynton W, Adolf J et al (2005) Eutrophication of Chesapeake Bay: historical trends and ecological interactions. Mar Ecol Prog Ser 303:1–29. https://doi.org/10.3354/meps303001

King D, Mainous A, Lambourne C (2012) Trends in dietary fiber intake in the United States, 1999–2008. J Acad Nutr Diet 112(5):642–648

Land L (2018) Bay journal—article: we can't improve the Bay's water quality without addressing manure. In: Bayjournal.com. https://www.bayjournal.com/article/we_cant_improve_the_bays_water_quality_without_addressing_manure. Accessed 15 July 2019

Langland M (2015) Sediment transport and capacity change in three reservoirs, Lower Susquehanna River Basin, Pennsylvania and Maryland, 1900–2012: U.S. Geological Survey Open-File Report 2014–1235, 18 p. https://doi.org/10.3133/ofr20141235

Lefcheck J, Orth R, Dennison W et al (2018) Long-term nutrient reductions lead to the unprecedented recovery of a temperate coastal region. Proc Natl Acad Sci USA 115:3658–3662. https://doi.org/10.1073/pnas.1715798115

Li J, Glibert P, Gao Y (2015) Temporal and spatial changes in Chesapeake Bay water quality and relationships to Prorocentrum minimum, Karlodinium veneficum, and CyanoHAB events, 1991–2008. Harmful Algae 42:1–14. https://doi.org/10.1016/j.hal.2014.11.003

Magee E, Richardson C, Hughes R et al (2000) Contribution of dietary protein to sulfide production in the large intestine: an in vitro and a controlled feeding study in humans. Am J Clin Nutr 72:1488–1494. https://doi.org/10.1093/ajcn/72.6.1488

Malone T, Crocker L, Pike S, Wendler B (1988) Influences of river flow on the dynamics of phytoplankton production in a partially stratified estuary. Mar Ecol Prog Ser 48:235–249. https://doi.org/10.3354/meps048235

Malone T, Ducklow H, Peele E et al (1991) Picoplankton carbon flux in Chesapeake Bay. Mar Ecol Prog Ser 78:11–22. https://doi.org/10.3354/meps078011

Malone T, Conley D, Fisher T et al (1996) Scales of nutrient-limited phytoplankton productivity in Chesapeake Bay. Estuaries 19:371. https://doi.org/10.2307/1352457

Marinucci A, Hobbie J, Helfrich J (1983) Effect of litter nitrogen on decomposition and microbial biomass in *Spartina alterniflora*. Microb Ecol 9:27–40. https://doi.org/10.1007/bf02011578

Marshall H, Lacouture R, Buchanan C et al (2006) Phytoplankton assemblages associated with water quality and salinity regions in Chesapeake Bay, USA. Estuar Coast Shelf Sci 69:10–18. https://doi.org/10.1016/j.ecss.2006.03.019

Michalak A, Anderson E, Beletsky D et al (2013) Record-setting algal bloom in Lake Erie caused by agricultural and meteorological trends consistent with expected future conditions. Proc Natl Acad Sci USA 110:6448–6452. https://doi.org/10.1073/pnas.1216006110

Middelburg J, Levin L (2009) Coastal hypoxia and sediment biogeochemistry. Biogeosciences 6:1273–1293. https://doi.org/10.5194/bg-6-1273-2009

Mitsch W, Horne A, Nairn R (2000) Nitrogen and phosphorus retention in wetlands—ecological approaches to solving excess nutrient problems. Ecol Eng 14:1–7

Moore T, Cuker B (2018) Sedimentary oxygen demand and orthophosphate release: sustaining eutrophication in a tributary of the Chesapeake Bay. J Water Resour Ocean Sci 7:42. https://doi.org/10.11648/j.wros.20180703.13

Mudgil D, Barak S (2013) Composition, properties and health benefits of indigestible carbohydrate polymers as dietary fiber: a review. Int J Biol Macromol 61:1–6

Mulholland M, Morse R, Boneillo G et al (2009) Understanding causes and impacts of the dinoflagellate, Cochlodinium polykrikoides, blooms in the Chesapeake Bay. Estuar Coasts 32:734–747. https://doi.org/10.1007/s12237-009-9169-5

Murphy S, Xu J, Kochanek K et al. (2018) Mortality in the United States, 2017. In: Cdc.gov. https://www.cdc.gov/nchs/products/databriefs/db328.htm. Accessed 19 July 2019

O'Brien K, Latner J, Ebneter D et al (2012) Obesity discrimination: the role of physical appearance, personal ideology, and anti-fat prejudice. Int J Obes 37:455–460. https://doi.org/10.1038/ijo.2012.52

O'Keefe S, Li J, Lahti L et al (2015) Fat, fiber and cancer risk in African Americans and rural Africans. Nat Commun. https://doi.org/10.1038/ncomms7342

Orth R, Williams M, Marion S et al (2010) Long-term trends in submersed aquatic vegetation (SAV) in Chesapeake Bay, USA, related to water quality. Estuar Coasts 33:1144–1163. https://doi.org/10.1007/s12237-010-9311-4

Pimental A (2010) EPA establishes Chesapeake Bay "pollution diet". Chesapeake Bay Program. In: Chesapeakebay.net. https://www.chesapeakebay.net/news/blog/epa_establishes_chesapeake_bay_pollution_diet. Accessed 10 July 2019

Pitcher M, Cummings J (1996) Hydrogen sulphide: a bacterial toxin in ulcerative colitis? Gut 39:1–4. https://doi.org/10.1136/gut.39.1.1

Posey M, Wigand C, Stevenson J (1993) Effects of an introduced aquatic plant, Hydrilla verticillata, on benthic communities in the Upper Chesapeake Bay. Estuar Coast Shelf Sci 37:539–555. https://doi.org/10.1006/ecss.1993.1072

Powell K, Cope W, LePrevost C et al (2017) A retrospective analysis of agricultural herbicides in surface water reveals risk plausibility for declines in submerged aquatic vegetation. Toxics 5:21. https://doi.org/10.3390/toxics5030021

Profenno L, Porsteinsson A, Faraone S (2010) Meta-analysis of Alzheimer's disease risk with obesity, diabetes, and related disorders. Biol Psychiatry 67:505–512. https://doi.org/10.1016/j.biopsych.2009.02.013

RWJF (2018) The state of obesity 2018. Robert Wood Johnson Foundation. In: Media.stateofobesity.org. https://media.stateofobesity.org/wp-content/uploads/2019/02/19162010/stateofobesity2018.pdf. Accessed 17 July 2019

Scavia D, Bertani I, Long C et al (2018) In: Chesapeake Bay Hypoxic Volume Forecasts. http://scavia.seas.umich.edu/wp-content/uploads/2018/06/2018-Chesapeake-Bay-forecast_Final.pdf. Accessed 12 July 2019

Schindler D (2006) Recent advances in the understanding and management of eutrophication. Limnol Oceanogr 51:356–363. https://doi.org/10.4319/lo.2006.51.1_part_2.0356

Scully M (2016) Mixing of dissolved oxygen in Chesapeake Bay driven by the interaction between wind-driven circulation and estuarine bathymetry. J Geophys Res Oceans 121:5639–5654. https://doi.org/10.1002/2016jc011924

Seitzinger S, Harrison J, Böhlke J et al (2006) Denitrification across landscapes and waterscapes: a synthesis. Ecol Appl 16:2064–2090. https://doi.org/10.1890/1051-0761(2006)016[2064:dalawa]2.0.co;2

Sieburth J (1993) C1 bacteria in the water column of Chesapeake Bay USA. I. Distribution of sub-populations of O2-tolerant, obligately anaerobic, methylotrophic methanogens that occur in microniches reduced by their bacterial consorts. Mar Ecol Prog Ser 95:67–80. https://doi.org/10.3354/meps095067

Smith J (1612). A map of Virginia. Oxford: Joseph Barnes, pp 1–3. In: The Complete Works of Captain John Smith (1580–1631), edited by Philip L. Barbour (1986) 3 vols., 1:119–190. Chapel Hill: University of North Carolina Press. Available in Capt. John Smith's complete works. http://www.virtualjamestown.org/exist/cocoon/jamestown/fha-js/SmiWorks1

Smith V, Tilman G, Nekola J (1999) Eutrophication: impacts of excess nutrient inputs on freshwater, marine, and terrestrial ecosystems. Environ Pollut 100:179–196. https://doi.org/10.1016/s0269-7491(99)00091-3

Smith V, Joye S, Howarth R (2006) Eutrophication of freshwater and marine ecosystems. Limnol Oceanogr 51:351–355. https://doi.org/10.4319/lo.2006.51.1_part_2.0351

Sonnenburg E, Sonnenburg J (2014) Starving our microbial self: the deleterious consequences of a diet deficient in microbiota-accessible carbohydrates. Cell Metab 20(5):779–786

Staff (2019) State says pilot project to collect, test dredge material from Conowingo Dam will be completed this year. In: Baltimoresun.com Feb 25, 2019. https://www.baltimoresun.com/maryland/harford/aegis/bs-md-ha-conowingo-dredge-pilot-20190225-story.html. Accessed 19 July 2019

Stankelis R, Naylor M, Boynton W (2003) Submerged aquatic vegetation in the mesohaline region of the Patuxent estuary: past, present, and future status. Estuaries 26:186–195. https://doi.org/10.1007/bf02695961

Sturdivant S, Díaz R, Llansó R, Dauer D (2014) Relationship between hypoxia and macrobenthic production in Chesapeake Bay. Estuar Coasts 37:1219–1232. https://doi.org/10.1007/s12237-013-9763-4

Tatu K, Anderson J, Hindman L et al (2007) Mute Swans' impact on submerged aquatic vegetation in Chesapeake Bay. J Wildl Manag 71:1431–1439. https://doi.org/10.2193/2006-130

Testa J, Kemp W (2014) Spatial and temporal patterns of winter–spring oxygen depletion in Chesapeake Bay bottom water. Estuar Coasts 37:1432–1448. https://doi.org/10.1007/s12237-014-9775-8

Testa J, Clark J, Dennison W et al (2017) Ecological forecasting and the science of hypoxia in Chesapeake Bay. Bioscience 67:614–626. https://doi.org/10.1093/biosci/bix048

Testa J, Lyubchich V, Zhang Q (2019) Patterns and trends in Secchi disk depth over three decades in the Chesapeake Bay estuarine complex. Estuar Coasts 42:927–943. https://doi.org/10.1007/s12237-019-00547-9

Thomas W, Davies J, O'Neal R (1960) Incidence of myocardial infarction correlated with venous and pulmonary thrombosis and embolism. Am J Cardiol 5:41–47. https://doi.org/10.1016/0002-9149(60)90007-2

USDA (2018a) Food composition databases show foods list. [Online] Ndb.nal.usda.gov. https://ndb.nal.usda.gov/ndb/search/list. Accessed 12 June 2018

USDA (2018b) Cotton & tobacco. Agricultural Marketing Service. In: Ams.usda.gov. https://www.ams.usda.gov/about-ams/programs-offices/cotton-tobacco. Accessed 31 July 2018

USDA (2019) USDA ERS—farm labor. In: Ers.usda.gov. https://www.ers.usda.gov/topics/farm-economy/farm-labor. Accessed 16 Jul 2019

USEPA (1999) 09/15/99: appeals court upholds ruling against Smithfield Foods for polluting Virginia River. In: Archive.epa.gov.

https://archive.epa.gov/epapages/newsroom_archive/newsreleases/c7a68726816ff7b3852567ef0053e790.html. Accessed 11 Feb 2019
USEPA (2017) Ambient water quality criteria for dissolved oxygen, water clarity and chlorophyll a for the Chesapeake Bay and its tidal tributaries: 2017 technical addendum. In: Chesapeakebay.net. https://www.chesapeakebay.net/documents/2017_Nov_ChesBayWQ_Criteria_Addendum_Final.pdf. Accessed 10 July 2019
USFWS (2011) Decline of submerged plants. Chesapeake Bay Field Office. Northeast Region, U.S. Fish and Wildlife Service. In: Fws.gov. https://www.fws.gov/chesapeakebay/savpage.html. Accessed 4 July 2019
USGS (2019) USGS Chesapeake Bay trends summary—monthly loads query. In: Cbrim.er.usgs.gov. https://cbrim.er.usgs.gov/monthly_loads_query.html?sorts%5Bstaid%5D=1&sorts%5Bpcode%5D=1&sorts%5Byear%5D=1&sorts%5Bmonth%5D=1&sorts%5Bq%5D=1&sorts%5Bconc%5D=1&sorts%5Bload%5D=1&sorts%5Bfnconc%5D=1&sorts%5Bfnload%5D=1&sorts%5Bstart_Yr%5D=1&sorts%5Bend_Yr%5D=1. Accessed 11 July 2019
Vaquer-Sunyer R, Duarte C (2010) Sulfide exposure accelerates hypoxia-driven mortality. Limnol Oceanogr 55:1075–1082. https://doi.org/10.4319/lo.2010.55.3.1075
Welcome A (2017) Definition of obesity—obesity medicine association. In: Obesity Medicine Association. https://obesitymedicine.org/definition-of-obesity/. Accessed 19 July 2019
Westbrook G (2018) Case study: wastewater treatment helps craft brewery meet discharge limits. In: Water Technology. https://www.watertechonline.com/case-study-craft-brewery-wastewater-treatment-discharge-limits/. Accessed 16 July 2019
Wetzel R (2001) Limnology. Academic, San Diego
Willett W (2017) Will it be cheese, bologna, or peanut butter? Eur J Epidemiol 32:257–259. https://doi.org/10.1007/s10654-017-0257-8
Wommack K, Hill R, Kessel M, et al. (1992) Distribution of viruses in the Chesapeake Bay. Appl Environ Microbiol 58: 2965–2970
Yun S, Zhu B, Black W et al (2005) A comparison of national estimates of obesity prevalence from the behavioral risk factor surveillance system and the national health and nutrition examination survey. Int J Obes 30:164–170. https://doi.org/10.1038/sj.ijo.0803125
Zhang X, Le W (2010) Pathological role of hypoxia in Alzheimer's disease. Exp Neurol 223:299–303. https://doi.org/10.1016/j.expneurol.2009.07.033
Zimmerman A, Canuel E (2000) A geochemical record of eutrophication and anoxia in Chesapeake Bay sediments: anthropogenic influence on organic matter composition. Mar Chem 69:117–137. https://doi.org/10.1016/s0304-4203(99)00100-0

Finishing the Journey: Wastewater, a Misplaced Resource in the Environment

Carie Curry-Cutter

You are what you eat,
What you eat is what you excrete,
In, out, steady state.

Abstract

Urban metabolism releases water, materials, and nutrients into the environment. When these fluxes become too great and out of balance with the natural ecosystem, an estuary like the Chesapeake Bay can become off balance. Among the inputs of urban metabolism to the Chesapeake Bay is human waste. The goal of effective waste treatment is to produce clean, safe water that is returned to the environment with little impact as possible, recover resources within the waste stream, and responsibly dispose of non-biodegradable materials. Within the Chesapeake Bay watershed two types of sewage treatment systems are used: centralized sewage treatment and septic systems. Approximately three quarters of domestic sewage is treated through centralized sewage treatment, and septic systems treat the remaining quarter. Over a 32-year period (1985–2017), treated wastewater and untreated wastewater (as combined sewage overflows) have decreased nutrient inputs to the Chesapeake Bay by 63% for nitrogen and 77% for phosphorus. Septic systems could potentially be a problem for the Chesapeake Bay if their use is continued as it has increased nitrogen inputs by 125,000 pounds per year over the 32-year period. Nutrient reductions can be made with the implementation of advance treatment technologies to both centralized sewage systems and septic systems.

C. Curry-Cutter (✉)
Department of Water Quality, Central Environmental Laboratory, Hampton Roads Sanitation District, Virginia Beach, VA, USA
e-mail: ccurry@HRSD.com

Keywords

Wastewater · Septic systems · Nitrogen · Phosphorus · Urban metabolism

Introduction

"Tell me what you eat and I will tell you what you are," said French author Anthelme Brillat-Savarin in his 1826 work, *Physiologie du Gout, ou Medetations de Gastronomie Transcendante*. In other words, you are what you eat. Then naturally, "*what you eat is what you excrete*." Indeed, it has been demonstrated that a given diet has a matching waste profile (Rose et al. 2015), showing that diet has an impact on waste. The human body strives to maintain a steady state balance of input and outputs; quantity and quality of food intake is equal the quantity and quality of excretion output minus that what which remains in the body for growth and repair. The food system produces what we eat and has wide ranging implications or the Chesapeake Bay ecosystem, as explored in many of the other chapters in this book. This chapter addresses a topic that most people take for granted when thinking about food systems, the fate of the residue of what we eat after it leaves our bodies. In our modern world we view human waste not as a renewable resource, but as a displaced resource we call pollution.

From the perspective of the environment, the currency of pollution, or the measurable constituents, are primarily nitrogen (N) and phosphorus (P). Excessive levels of these nutrients promote eutrophication as addressed in chapter "A Fishing Trip: Exploiting and Managing the Commons of the Chesapeake Bay". Modern diets can be generally separated

B. E. Cuker (ed.), *Diet for a Sustainable Ecosystem*, Estuaries of the World, https://doi.org/10.1007/978-3-030-45481-4_16

into higher and lower protein categories (Rose et al. 2015). The nitrogen and phosphorus profiles of these two categories are significantly different. Phosphorus in feces is directly proportional to protein intake (Calloway and Margen 1971). Diets higher in protein have increased amounts of ammonia and double the amounts of nitrite in fecal waste (Silvester et al. 1997). Diets lower in protein have increased fiber, which has higher percent moisture in feces as well as increased amounts of sodium and phosphorus (Silvester et al. 1997). Inputs such as heavy metals from contaminated farmlands or lead from industrially polluted air are also important factors of quality food that impacts waste (Rose et al. 2015).

Natural, healthy ecosystems depend on the interaction of primary autotrophic producers and heterotrophic consumers. In a sustainable ecosystem the production of organic material is balanced by its consumption. This basic balance is the ecosystem that sustains the life of the Chesapeake Bay. No doubt, increasing population and industries are making substantial impacts to this environment. *Urban metabolism* has been used to broadly quantify resources needed from the environment for the human biome. Resources such as an adequate water supply, effective sewage disposal, and air pollution control are necessary for a healthy city or region (Wolman 1965). Moreover, urban metabolism has been further refined to include the flow of water, materials, and nutrients as mass fluxes into the environment (Kennedy et al. 2011). The use of these natural resources affects the surrounding natural ecosystems. When the fluxes from urban metabolism become too great and out of balance with the natural ecosystem, nature struggles to survive. How long an estuary can be off balance or how resilient is the ecosystem to recovery are difficult questions. This delicate balance has likely changed over time and could possibly create an intermediate stable state at a diminished quality. This chapter's focus is on the ultimate repository of food resources through wastewater management and resource recovery. Historical to modern practices of waste management will be discussed, along with sanitation for a healthy population and environment, untreatable components within the waste stream, and closing the nutrient resource loop with advanced wastewater treatment technologies.

Historical Perspective on Sewage When Capt. John Smith arrived on the Chesapeake Bay in 1607, the human population was small and dispersed over wide areas of the watershed (see chapter "The Bay and Its Watershed: A Voyage Back in Time"). The ecosystem readily absorbed waste produced by the Algonquin and other indigenous tribes due to their nomadic way of life, and respect for the land and its resources. The remains of Algonquin meals and their waste products were returned to the land, decomposed through natural cycles, and added fertility back to the soil. However, decomposition of waste takes time, and even the small villages received a probably welcome "airing out" every fall and spring as their populations relocated to new hunting and fishing camps (Rountree et al. 2008).

It was not until the introduction of the first permanent, colonial settlements along the Chesapeake Bay that an imbalance of resources began. The Jamestown colonists (1607) suffered from poor waste management. Not only was traveling outside the fort walls hazardous when Native Americans were hostile, but the woods surrounding the fort also became increasingly loaded with infectious microorganisms including dysentery, and typhoid (Sewage Treatment in Virginia 2019). By 1610, the 3-year-old Jamestown settlement had its first set of waste management rules and regulations which "forbade washing soiled clothing and cooking utensils or doing the necessitates of nature within a quarter mile of the new well" (Tarter 2012).

As settlements eventually began to prosper, colonies grew from few hundreds to populations well into the thousands. The wealthy built large houses with indoor chamber pots and then tossed the contents into the environment. Battalions of soldiers marched across the countryside and dug pits and privies at troop encampments. Waste management was still no more than out of sight, out of mind. The mismanagement of human waste was no better than the sanitary dark ages left after the collapse of the Roman Empire (Lofrano and Brown 2010).

When rural communities became towns, concentrations of human waste accumulated in the streets. "Look out below" was quite a relevant expression as residents tossed out the contents of chamber pots out into the streets. Towns began growing into cities, inundated with human and animal waste accumulating in the streets. Waste was primarily conveyed by stormwater into cesspools which leached into soils and groundwater. Eventually rivers diluted waste runoff, and transformed by bacteria, sunlight and oxygen, waters would become rejuvenated by their natural ecosystems further downstream.

The North American Industrial Revolution of the mid-1800s had towns and cities growing and thriving. Industries sprang up along the Chesapeake Bay shores taking advantage of trade routes on and off the water. Shipbuilding, iron smelting, coal mining, textile mills, food processing, and canning, stripped and depleted the Chesapeake Bay watershed of its natural resources. The potent, sometimes toxic, industrial wastes were returned to the Chesapeake Bay's rivers and streams untreated. During this period, indoor plumbing including flushing toilets became more frequent in use (George 2001). Outbreaks of cholera, typhoid and dysentery became more frequent as more wastes seeped into drinking water sources (Grumet 2000).

Cities outside the Chesapeake Bay watershed, such as Boston and Chicago, had established sewer systems and sanitary practices in the 1700s (Lofrano and Brown 2010). The cities and towns surrounding the Chesapeake Bay lagged behind. Baltimore, Maryland, population about 50,000 in 1814, grew to over 300,000 by 1875. Pipes and conveyances were laid to divert waste out of the city into Jones Falls, a tributary of the Patapsco River. By 1890 Baltimore Harbor was seriously polluted, and the malodorous gases were "horrendously ripe" under the hot, humid Baltimore summers. In 1905 the City of Baltimore approved a comprehensive sewage system as part of rebuilding the city after the Great Baltimore Fire (Baltimore History 2019).

The Potomac River is both the water supply and sewerage system for the District of Columbia's (now called Washington, DC) population. In 1859 the Washington Aqueduct was constructed to supply river water to the District. Culverts and conveyances discharged waste into small canals within the city that ultimately drained into the Potomac River below the city. A surge in population during the Civil War quickly increased pollution, which led to epidemics of smallpox, typhoid and malaria (DC Water History 2019). Conditions in the canals of the District remained fouled until 1880 when construction of sewers that diverted both waste and stormwater further away to the marshes of the Potomac and Anacostia Rivers. In 1938 the Blue Plains treatment plant was constructed along with a sewage system that served approximately 650,000 people and operated at the furthest point south along the Potomac River (DC Water History 2019).

About the same time just a little over 200 miles to the south, a thriving seafood industry in Hampton Roads, Virginia (today the cities of Hampton, Newport News and Norfolk) was evolving. By 1881, perhaps one of the largest oyster companies in the world harvested over 200,000 bushels a season from the southern Chesapeake Bay and eastern Virginia shore areas (see chapter "The Chesapeake Bay Oyster: Cobblestone to Keystone"). As the towns and cities of Hampton Roads grew, pipes and conveyances channeled raw sewage and stormwaters directly into the James and Elizabeth Rivers. By 1925, authorities closed approximately 10,000 acres (4047 ha) of Hampton Roads waters due to over harvesting and the remaining oysters contaminated with human pathogens. The area's shellfish industry was devastated along with the livelihood of oystermen, boat builders, ice providers, wholesalers, and retailers. In 1940 the Hampton Roads Sanitation District Commission was created, as a result of passage of public referendum, which joined 13 cities, towns and counties to regionally treat sewage (HRSD History 2019).

As the population of the Chesapeake Bay has increased, so has waste production and its impacts to the waterways. Through grassroots efforts, local laws, and federal regulations, the management of waste has vastly improved over the last 300 years. Rudimentary wastewater practices mitigated some organic and inorganic loading into the receiving waters, but still at a level detrimental to overall Bay health. Technological advances have elevated best practices but require updating wastewater infrastructure. Ultimately, recovering valuable resources from the waste stream and cleaning water using advanced wastewater technologies is the goal. The Chesapeake Bay is a very special estuary where life on land and in the water can do more than just co-exist, but thrive together. Water, no matter how or where it's form, is a precious resource that each and every one of us are responsible and accountable for its use.

Wastewater Methods

Within the Chesapeake Bay watershed two types of sewage treatment systems are largely used, centralized sewage treatment systems and septic systems. There are 472 municipal and industrial sewage treatment systems within the Chesapeake Bay that have been designated as significant sources of nutrient pollution to the Bay (Rath 2019). Publicly owned treatment works (POTW) are centralized sewage treatment systems owned by a government agency or a locality. POTW's treat approximately three quarters of the domestic sewage within the Chesapeake Bay watershed. Septic systems treat the remaining one quarter of domestic sewage (EPA 1997).

Septic Systems Septic systems, or decentralized wastewater systems, are underground waste treatment structures generally used in areas without centralized sewer systems. Septic systems use modern technology in conjunction with natural processes to clarify domestic wastewater.

Typical septic system consists of an underground septic tank and a drainfield or soil absorption field (Fig. 1). Waste is conveyed from the home to the septic tank where organic matter and solids (sludge) are separated from floatable matter like oil and grease (scum; Fig. 2). The liquid discharge, or effluent, from the septic tank flows through a distribution box and into a series of perforated pipes buried in the drainfield. The effluent is slowly released into the soil. Septic system design and size can vary widely, due to factors such as household size, soil type, slope, lot size, proximity to sensitive water bodies, weather conditions, and local regulations (USEPA 2019).

Within the septic tank (Fig. 1), microorganisms break down organic matter in an environment with little to no oxygen. This process is referred to as anoxic digestion. Pathogens are collected within the underground tank, settled

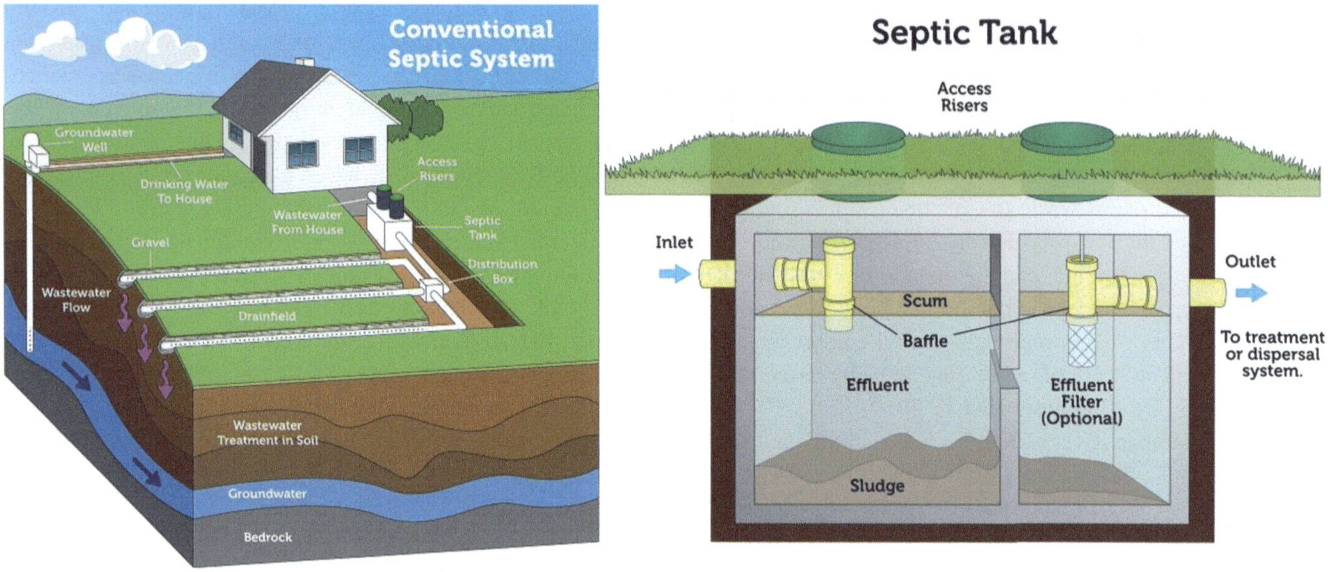

Fig. 1 Septic system—https://www.epa.gov/sites/production/files/styles/large/public/2018-11/conventional_septic_system.jpg (EPA 2019—Conventional septic Systems)

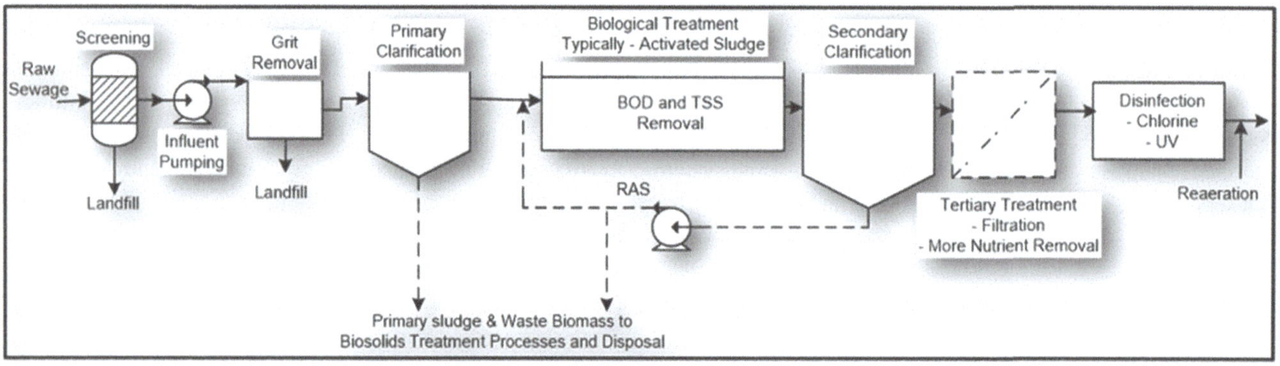

Fig. 2 Image: Waste treatment flow diagram. Where "BOD" is biochemical oxygen demand, "TSS" is total suspended soilds, and "RAS" is return activated sludge. From Charles Bott, Hampton Road Sanitation District (HRSD) with permission

by gravity into the sludge, and decayed or digested (Adegoke and Stenstrom 2019). When the effluent is passed through the distribution system and into the soils below the drainfield, the effluent becomes more oxygenated. As the effluent percolates through the soils, microorganisms continue to breakdown organic matter in an environment with more oxygen. This process is referred to as aerobic digestion. This combination of anoxic and aerobic digestion effectively removes organic matter, reduces nitrogen, phosphorus and further reduces pathogens to safe levels before reaching groundwater or adjacent surface waters.

Periodic servicing is required for septic systems to remove accumulated solids and maintain optimum working conditions. Improperly maintained septic systems can contaminate groundwater, surface waters, spread pollution into the environment, and pose risk to the health of the Chesapeake Bay and its people (EPA 1997). In addition, emergency repairs can be costly to homeowners and may subject to local penalties and fines. Septic systems located in environmentally sensitive areas, proximate to Chesapeake Bay's surface waters, tidally-influenced water, or recreational waters, may be subject to more stringent state and local regulations. For example, the state of Virginia requires septic systems to be pumped out every 5 years (Chesapeake Bay Preservation Act of 2014) and the state of Maryland requires septic systems to be upgraded to best available technology (Bay Restoration Fund Best Available Technology for Removing Nitrogen from Onsite Systems of 2004). The resistant solids removed from septic systems are hauled to POTW's for further treatment.

In most cases phosphorus discharged from septic systems is retained in the soils below drainfields. Phosphorus is quickly scavenged in surrounding sands and soils by precipitation into minerals or absorbed onto particles (Taylor et al. 2013). This natural process prevents phosphorus from being released to ground or surface waters. However, nitrogen is not removed as well as phosphorus within the septic system and drainfields. Excessive amounts of nitrogen can be transported through groundwater to adjacent surface water ponds, streams, or groundwater. Groundwater wells can be negatively impacted by increased levels of nitrogen and pose a health risk (Laws 2000). In anoxic conditions, microbes reduce nitrate $\left(NO_3^-\right)$, a stable form of combined nitrogen in aerobic conditions, into nitrite $\left(NO_2^-\right)$ a relatively unstable and reactive oxidative state of nitrogen. Recharging of groundwater with sewage effluent high in nitrite can contaminate groundwater and potentially create a health problem (Fewtrell 2004). When nitrite-contaminated water is consumed, the ferrous iron in blood hemoglobin is oxidized to a ferric form, resulting in methemoglobin (Laws 2000). While hemoglobin readily transports oxygen throughout the body, methemoglobin cannot bind oxygen. The resulting condition is methemoglobinemia characterized by unusual bluish or brownish skin color, irritability, drowsiness and lethargy. Methemoglobinemia, largely affecting infants less than 3 months old, is also known as 'blue-baby syndrome'.

Septic systems rely on the hydraulic downward movement of water, and microbial and chemical processes. The effectiveness of septic waste treatment is dependent on soil type, water residence time within the soil, and depth of soil between waste infiltration and the water table. Optimal treatment takes place in conditions where waste percolates through soils that are aerobic and have relatively low moisture content, allowing organic matter and nutrients to be removed by microbial degradation. The effectiveness of treatment can be impacted with changes in soil moisture and temperature. Climate change, through the combined effects of increasing temperature and sea-level rise, directly impacts the effectiveness of septic waste treatment within the soils (Cooper et al. 2016). In coastal areas of the Chesapeake Bay, rising water tables due to rising sea levels and saltwater intrusion have diminished the volume of unsaturated soils available for waste treatment via septic systems (Vogelsong 2019). Furthermore, higher groundwater levels result in overall wetter conditions that can enhance the survival and transport of bacteria and viruses, lower phosphorus removal within the soils, and decrease the degradation of organic matter (Cooper et al. 2016). Thus, septic systems may work well when properly maintained, but may not be a reliable method of waste treatment in the Chesapeake Bay in the future.

Centralized Sewage Treatment Though the design flow and complexity of modern sewage treatment plant designs may differ greatly, there are several basic features that are similar. When raw sewage reaches a POTW, it usually passes through a series of screening devices to separate large objects that have entered the sewage collection system which could clog or damage pumps and pipes (Fig. 2). The solid waste material collected during screening, such as rags, paper, plastic, wipes, metal and other large objects, are collected and transferred into landfills. Screened sewage is pumped through grit removal process (Fig. 2), where particles about the size of sand or course grit are removed. Removal of course grit is again necessary to protect pumps and pipes within the POTW. The solid waste materials collected during grit removal are also disposed of in landfills. Screened and grit-separated sewage is then pumped into a primary clarifier (Fig. 2). The primary clarifier is a large quiescent tank where floatable solids are separated from settable solids. Material skimmed from surface and solids pumped and removed from the bottom of the tank are called primary sludge. At this stage, removal of primary sludge is called primary sewage treatment.

Two metrics used to quantify the effectiveness of sewage treatment are the reduction of suspended solids and the reduction of oxygen consuming wastes. Total Suspended Solids (TSS) are measured by filtering a known volume of water through a pre-weighed filter, drying the filter and reweighing it. Oxygen consuming waste can be measured by biochemical oxygen demand (BOD). BOD is defined as the amount of oxygen consumed in a given volume of wastewater incubated in a stoppered bottle in the dark at 20 °C for 5 days. Primary sewage treatment typically removes 40–75% suspended solids and 30–40% BOD (Laws 2000).

Secondary sewage treatment is the ongoing treatment of sewage where biological processes are used to further reduce organic matter, nitrogen, and phosphorus. Effluent from the primary clarifier flows into a second tank, or a series of tanks where the sewage is vigorously aerated and turbulently mixed. The aeration and mixing of sewage facilitate the rapid growth of bacteria and other microorganism. During this activated sludge process, resident organisms are thriving and consuming organic matter, nitrogen, and phosphorus. The effluent from aeration tanks is transferred to a secondary clarifier. The secondary clarifier, another quiescent settling tank, allows biological matter to flocculate and settle. The accumulated biomass in the secondary clarifier is removed, and a small portion is returned to the aeration tank as "Return Activated Sludge" (RAS), to ensure an adequate population of resident organisms is maintained within the aeration tank. The remaining activated sludge is moved to biosolids treatment processes and disposal. As a result of secondary treatment, suspended solids and BOD are reduced by approximately 85% (Laws 2000).

Secondary sewage treatment continues with enhanced Biological Nutrient Removal (eBNR) process. This process takes advantage of microbes that thrive on waste constituents under conditions unique to their metabolism. The microbial consumption of nitrogen, phosphorus and carbonaceous organic matter is facilitated through a series of tanks, each with varying amounts of oxygen. Much like the activated

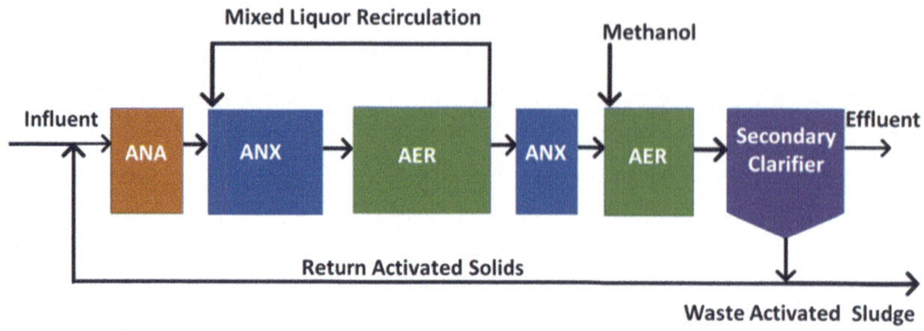

Fig. 3 5-Stage Bardenpho process. The 5-Stage Bardenpho process cycles effluent from secondary wastewater treatment through reaction zones with different levels of free oxygen, facilitating a suite of biological processes to reduce concentrations of nitrogen, phosphorus, and carbon

sludge process where aerobic organisms thrive in oxygenated conditions, anaerobic organisms thrive in anoxic conditions where no oxygen is present. In conditions where oxygen is present but at low levels, facultative anaerobes that can switch from aerobic to anaerobic metabolism to thrive. Nitrification is the biological process of oxidizing ammonia to nitrite, then nitrite to nitrate in aerobic conditions (Eq. 1). Denitrification is the biological process of reducing nitrate to nitrite, to nitric oxide, to nitrous oxide, and finally to gaseous nitrogen in conditions with very little oxygen (Eq. 2).

$$\text{Nitrification}: \ NH_4^+ \rightarrow NO_{2-} \rightarrow NO_3^- \quad (1)$$

$$\text{Denitrification}: NO_{3-} \rightarrow NO_{2-} \rightarrow NO \rightarrow N_2O \rightarrow N_2 \quad (2)$$

The 5-stage Bardenpho process (Fig. 3) involves an anaerobic zone, then an anoxic zone, and an aerobic zone, followed by second, smaller anoxic and aerobic zones. Each zone plays an essential role by providing appropriate conditions for the activity and growth of microbes with different metabolisms. The combination of these five reaction zones provides excellent removal of carbonaceous organic matter, nitrogen and good removal of phosphorus within wastewater (Esfahani et al. 2017). The 5-stage Bardenpho process can remove approximately 99% of suspended solids, BOD, nitrogen and approximately 90% removal of phosphorus (Emara et al. 2014). Due to its importance in modern sewage treatment, the following section expands on the description of the 5-stage Bardenpho process.

1. Anaerobic (ANA)—Waste from the primary clarifier begins the 5-stage Bardenpho process in anaerobic conditions where facultative anaerobes begin the waste digestion process.
   - Dissolved oxygen is low
   - Nitrate and nitrite concentrations are low
   - Ammonia and phosphorus concentrations are high
   - Phosphorus accumulating organisms thrive and bioaccumulate phosphorus
2. Anoxic (ANX)—The waste stream is next moved into anoxic conditions where dentification occurs.
   - Dissolved oxygen is very low
   - Denitrification reduces nitrate and nitrite to gaseous nitrogen
   - Ammonia and phosphorus concentrations are decreasing
   - BOD is substantially reduced
3. Aerobic (AER)—From anoxia, waste is moved to aerobic conditions where nitrification occurs. This is the heart of the of the Bardenpho process. Greater nitrogen removal within the Bardenpho process is achieved through mixed liquor circulation, where a portion of the waste stream is recycled back into the anoxic zone.
   - Dissolved oxygen is higher
   - Nitrification oxidizes ammonia and nitrite to nitrate
   - Nitrate is at its highest concentration
   - Phosphorus and BOD are significantly reduced
4. Anoxic (ANX)—The waste stream is moved into anoxic conditions where a second step of denitrification occurs. Organic carbon can become a limiting resource within the waste stream at this stage of the Bardenpho process. The addition of methanol or other supplemental carbon source can be used for organisms to continue nitrogen and phosphorus removal.
   - Dissolved oxygen is very low
   - Denitrification reduces nitrate to nitrite to gaseous nitrogen
   - BOD is very low
   - Phosphorus is low
5. Aerobic (AER)—The waste stream moves to the final step of the Bardenpho process, aerobic conditions where a second step of nitrification occurs. The reduction of nitrogen, phosphorus, and BOD will have reached their desired levels.
   - DO is higher
   - Nitrification of any remaining nitrite back to nitrate
   - Phosphorus is low

Effluent from the 5-stage Bardenpho process is moved to a secondary clarifier, a large settling tank where biomass flocculates, settles, and is removed. A portion of the waste-activated sludge is recycled back to the primary anaerobic tank for treatment as it is highly concentrated in phosphorus (Esfahani et al. 2017). The remaining activated sludge is moved to biosolids treatment for further processing or disposal.

Disinfection is one of the primary mechanisms for the destruction or deactivation of pathogens within the waste stream. Suitable disinfection depends on many factors including safe handling and storage, effectiveness under normal operating conditions, and absence of toxic by-products from the disinfection process. Chlorine is widely used for disinfection in POTW's. Chlorine can be delivered to treated waste streams in many forms which include chlorine gas, hypochlorite solutions or chlorine salts. Effective disinfection can also be performed by ozonation or exposure to ultraviolet light. After disinfection, residual chlorine can persist and could have detrimental effects on receiving waters and aquatic life. To minimize negative impacts, chlorinated wastewater must often be dechlorinated before discharged into the environment. Dechlorination, or the process of removing excess chlorine, is typically performed by the chemical addition of sodium bisulfite, sodium metabisulfite, or sulfur dioxide. Alternate methods of disinfection, such as ozonation or ultraviolet light, generally do not have residual effects that could impact receiving waters and aquatic life. While disinfection is a standard part of modern wastewater treatment, its efficacy requires monitoring at the plant and ensures the protection of the receiving waters. While POTW's in the Chesapeake Bay watershed rarely have issues with disinfection, inadequate disinfection of wastewater can substantially impact the environment and human health as discussed below.

Wastewater is cleaned and discharged into the environment with as little of impact as possible. The Clean Water Act of 1972 prohibits the discharge of pollution through a point source into the waters of the United States. The National Pollution Discharge Elimination System (NPDES) is a federal program that regulates point sources pollution, providing limits on what type and how much pollution can be discharged. The Environmental Protection Agency (EPA), authorizes state governments for the administration, permitting of pollution discharge, and the enforcement of regulations. In addition to limiting pollution, requirements for monitoring and reporting requirements are mandated to ensure protection of the environment and human health. State authorities adapt NPDES requirements and provisions to the unique characteristic of each point source discharge as well as the receiving bodies of water. NPDES permits are required for other point sources, such as animal feed operations, and aquaculture (NPDES 2019).

Preventing Human Disease

Part of a healthy human diet is growing and maintaining a diverse population of microbes in the gut. More than 1000 species of bacteria play important roles in the human body such as participating in the immune system, the central nervous system or "second brain" (Forsythe et al. 2014), a healthy heart (Aron-Wisnewsky and Clément 2015) and maintaining a healthy weight (Patterson et al. 2016). Just as these microbiomes have evolved in humans, so too has the diversity of life in the oceans, rivers, and estuaries. Chesapeake Bay's ecosystem evolved with a healthily balance of diverse organisms such as algae, bacteria, clams, worms, sea grass and pelagic fish. While this chapter's focus is on the recovery and repository of food resources in the environment, it is difficult to separate these concerns from adequate sanitation. In this context, inadequate sanitation of wastewater could cause harm through direct exposure to contaminated water or indirectly through contaminated food. All vested parties such as consumers, fisheries regulators, and sanitation authorities are aware of the risk to human health. Thus, recreational waters are rigorously monitored by local health departments, impaired bodies of water are restricted from public assess when deemed unsafe, and food safety regulations are enforced. According to the World Health Organization (WHO 2019a):

> Sanitation refers to the provision of facilities and services for the safe management of human excreta from the toilet to containment and storage and treatment onsite or conveyance, treatment and eventual safe end use or disposal. More broadly sanitation also included the safe management of solid waste and animal waste. Inadequate sanitation is a major cause of infectious diseases such as cholera, typhoid and dysentery world-wide. It also contributes to stunting and impaired cognitive function and impacts on well-being through school attendance, anxiety and safety with lifelong consequences, especially for women and girls. Improving sanitation in households, health facilities and schools underpins progress on a wide range of health and economic development issues including universal health coverage and combatting antimicrobial resistance.

The diverse population of microorganisms in the human intestines poses a significant source for infection. Unhealthy individuals excrete feces infected with pathogens, or disease-causing microorganisms that can become mobilized in the environment. Cholera (*Vibrio cholerae*) provides an instructive example throughout history, often referred to as "Ancient Scourge", with six world-wide pandemics between 1817 and 1923. John Snow, a London-based physician, first proposed that cholera was a communicable disease transmitted through infected stool (Snow 1855; Harris et al. 2012). Today, cholera still affects 47 counties across the globe and is a stark marker of the inequalities that exist for clean drinking water sources and basic sanitation facilities (WHO 2019b).

Ideal Indicators Indicator organisms, or a target microorganism, are used as markers of fecal pollution in the environment. Research and analysis of indicator organisms remain at the forefront of water and wastewater microbiology and the foremost technique of detecting and quantifying pollution. Typical bacterial and viral enteric pathogens will be discussed below. Many more types of enteric protozoa,

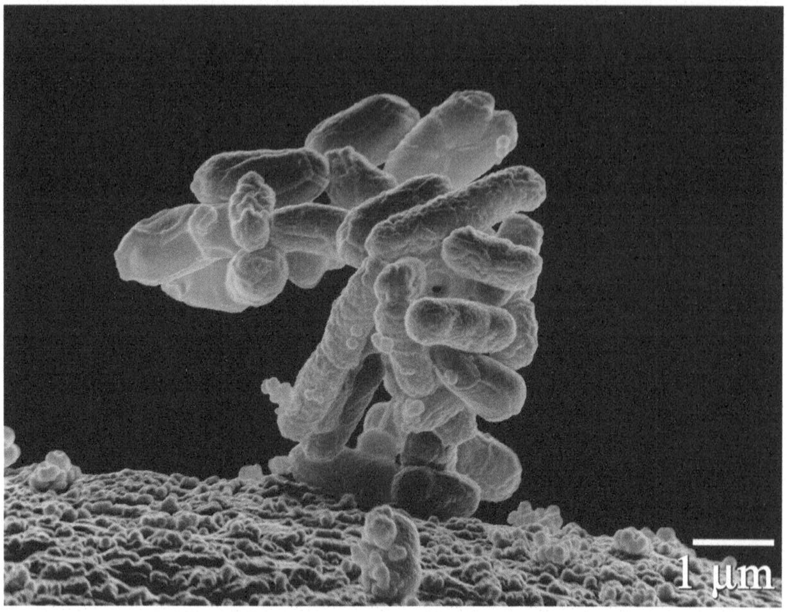

Fig. 4 E. Coli at 10,000 times magnification. Photo by Eric Erbe, digital colorization by Christopher Pooley, both of USDA, ARS, EMU. ARS Image Gallery Image Number K11077-1 (highres), Public Domain, https://commons.wikimedia.org/w/index.php?curid=130129

fungi, parasites and other opportunistic pathogens that are equally important part of wastewater sanitation and protection of human health will not be discussed.

Bacterial characteristics of the colon have been profiled and analytical methods have been developed for the analysis of coliform and enterococcus bacteria (Standard Methods 2018). The coliform group includes a multitude of species, often referred as total coliforms or fecal coliforms. The principal indicator organism of the fecal coliform group is *Escherichia coli* (Fig. 4). Fecal enterococcus species naturally occur in the intestines of humans and many other animals, including cattle, pigs and birds (Wheeler et al. 2002). Recently, risk to public health from large quantities of dog feces on streets and urban areas have been identified as an environmental concern (Cinquepalmi et al. 2012). Pet waste has been proposed as a significant source of excess nutrients and harmful bacterial loading to the Chesapeake Bay (Chesapeake Stormwater Network 2017). In 2012, the state Maryland identified pet waste as contributing to 68% of bacterial loading in the Severn River watershed and 87% in the Forked Creek tributary of the Magothy River (Vaughn 2012). Hence, grass roots efforts such as "Stop the Poo-llution" and "Scoop the Poop" have a meaningful impact on reducing the amount of untreated waste that enters the Chesapeake Bay.

As with bacteria indicator organisms, viral indicator organisms provide evidence of fecal contamination in the environment. The principal indicator organisms of enteric viruses are enteroviruses and rotaviruses (Duncan and Horan 2003). Enteroviruses include the Coxsackieviruses and Polioviruses, and there are nine species of Rotaviruses. Analytical methods have been developed for these pathogenic viruses (Ramírez-Castillo et al. 2015). In contrast to the examination of bacteria, which is easily and safely performed in the laboratory, examination of viruses is performed in specialized laboratories with highly technical instrumentation, increased level of containment and specially trained personnel. Pathogenic human, enteric viruses in marine waters like the Chesapeake Bay are not well quantified (Griffin et al. 2003). Over 100 types of pathogenic viruses from humans and animals end up in municipal wastewater (Qiu et al. 2015). Work has been done to assess the treatability and fate of viruses in wastewater and demonstrated that virus removal is variable and dependent on the treatment process (Worley-Morse et al. 2019). Contamination of marine waters with viruses is and will continue to be an important public health issue.

Indicators at Work in the Chesapeake Bay Virginia Department of Health's (VDH) Beach Monitoring program (VDH 2019) and Maryland's Healthy Beaches Program (MHBP 2019) combined have more than 200 locations throughout the Chesapeake Bay where local health departments monitor and track water quality and safety. The VDH Beach Monitoring Program encompasses 21 locations in Norfolk, Newport News, Hampton, Yorktown, Middle Peninsula, and the Eastern Shore. Maryland's Healthy Beaches Program encompasses 182 locations throughout the northern Chesapeake Bay, Eastern Shore, Potomac River and up to the Susquehanna River. Protected beaches are monitored for enterococcus weekly from May through September (Memorial Day to Labor Day). The monitoring period occurs during the Chesapeake Bay's seasonal recreation timeframe and when the Bay's water temperature is at its maximum. When high levels of the indicator organism enterococcus is detected, advisories are issued and the public is notified. Swimming advisories remain in effect until sampling show enterococcus meet safe levels.

Monitoring the Chesapeake Bay's waters for pathogens is important for recreation and is critical for food safety. To date, the Chesapeake Bay has been fortunate that no outbreaks of food illness from pathogenic human enteric bacteria or viruses have been reported. However, the potential for illness is possible, as *Cryptosporidium* spores and *Giardia* cysts have been observed within oysters of the Chesapeake Bay (Fayer et al. 1998). Food-borne illnesses from Chesapeake Bay's shellfish are more routinely associated with natural types of bacteria and algae. The bacterium *Vibrio parahaemolyticus* that naturally occurs in coastal waters, is regulated in the shellfish industry to reduce consumer risk. Harmful algal blooms (HAB's) or sometime called 'Red Tide', are photosynthetic algae that can produce potent toxins and bioaccumulate in shellfish (Griffith and Gobler 2019). Degraded water quality from excessive nutrient concentrations promotes the development and persistence of HAB's (Heisler et al. 2008, see chapter "Reversing the Eutrophication of the Chesapeake Bay and Its People"). Reducing nutrient loading to the Chesapeake Bay can make a positive impact by reducing the frequency and severity of HAB's.

Stewardship of the Environment

Stewardship is the responsibility of overseeing and protecting of something worth caring for and preserving. In 1880, untreated sewage and stormwater from the DC area was diverted into the Anacostia River by culverts and other conveyances. Basic sewage treatment in the DC area began in 1938 but polluted conditions remained. By 1971 the water quality of the Anacostia River was so bad that local laws were passed making swimming or wading in the river punishable by a \$300 fine or 10 days in jail (Fenston 2019). Still, many people continued to recreate and swim in the river, mainly African Americans, as public swimming pools were segregated and off-limits.

Proper sewage treatment was implemented over 50 years ago and the water quality of the Anacostia River has incrementally improved but is still impaired. The main source of pollution to the river is the antiquated combined sewer system. In a combined sewer system (Fig. 5), stormwater is diverted into a shared system of pipes and treated as wastewater at a publicly owned treatment works (POTW) or sewage treatment plant. Heavy rainfall adds a large pulse of water into the sewer system and can overwhelm the system and spill untreated sewage into the environment.

In order to reduce combined sewer overflows (CSO's) and improve water quality of the Anacostia River, the District of Columbia Water and Sewer Authority (also known as DC Water) has invested over \$2.7 billion dollars in a massive infrastructure program designed to capture and treat wastewater during rainfall events before it reaches the rivers. DC Water's Clean Rivers Project is comprised of a system of deep tunnels, sewers and diversion facilities to capture CSO's and deliver it to DC Water's Blue Plains Advanced Wastewater Treatment Plant. The Anacostia and Potomac River systems include more than 29 km (18 miles) of tunnels (DC Water). Within the first year of operation, the Anacostia River Tunnel has captured and treated 4.5 billion gallons (14.5 billion L) of sewage (Mooring 2019). The Anacostia River water quality has been routinely monitored through a grassroots, citizen-science effort, Anacostia Riverkeeper, since 2015 and results show a reduction in nutrients and a decrease in severity of summer algal blooms (Solomon et al. 2019). Within the near future the Anacostia River should be healthy and safe enough to lift swimming restrictions and open for recreation. Plans for community swimming facilities in the restored Anacostia River by 2025 are envisioned (District of Columbia, Office of Planning 2003).

Stewardship of the environment also applies to the handling of the waste products from the treatment of wastewater. Resources that enter the treatment process through the waste stream leave in one way or another. The goal of effective waste treatment is to produce clean, safe water that can be returned to the environment with as little of impact as possible, recover and reuse resources within the waste stream, and responsible disposal of non-biodegradable materials and refuse.

Untreatable refuse is removed "as is" and permanently disposed of in landfills. Solid waste like rags, wipes and grits removed from the headworks of a centralized treatment plant are non-biodegradable and cannot be reused or further treated. The Resource Conservation and Recovery Act of 1976 (RCRA) governs waste management and materials recovery and reuse, including the disposal of both hazardous and non-hazardous solid waste. Before non-biodegradable waste can be disposed in landfills, it must be evaluated for toxic contamination. In support of RCRA, the EPA has developed solids waste test methods for the analysis of various environmental media, like refuse from the treatment of wastewater (EPA 1993).

Contaminants of Emerging Concern (CEC's) and Pharmaceuticals and Personal Care Products (PPCP's) originate from many different sources and are more frequently being detected at low levels in POTW influents, effluents, biosolids and adjacent surface waters (EPA 2010a). CEC's are a class of chemicals that include but are not limited to flame-retardants, per-fluorinated alkyl substances 1 (PFAS), surfactants, phthalates, bis-phenol A, steroids, and hormones. PPCP's are a class of both prescription and over-the-counter medications as well as non-medicinal chemicals that include

Fig. 5 Combined sewer overflows. By U.S. Environmental Protection Agency (EPA)—U.S. Environmental Protection Agency, Washington, D.C. "Report to Congress: Impacts and Control of CSOs and SSOs." Document No. EPA 833-R-04-001, Public Domain, https://commons.wikimedia.org/w/index.php?curid=3474776

but not limited to hand sanitizers, fragrances in lotions and soaps, and the ultraviolet filters in sunscreens. There are many CECs and PPCPs that also act as so-called endocrine disruptors that alter the normal functions of hormones resulting in a variety of health effects (EPA 2010a).

Many personal care products are washed down the drain from sinks and showers and enter the waste stream. Pharmaceuticals enter the waste stream through direct disposal, though medication should never be discarded in the sink or toilet, and through the ingestion and excretion of medications and their metabolites. Though POTW's are highly effective at removing biodegradable materials and pathogens, they are not designed to remove CEC's or PPCP's and may only incidentally reduce their concentrations. There is concern that these compounds may have a negative impact on aquatic life and it is important to monitor and evaluate their potential impact (EPA 2010a). New developments in technology have led to improvements in detecting and quantifying CEC's and PPCPs in wastewater effluents and biosolids (EPA 2010b). However, despite recent advances in CEC and PPCP research, the full extent and consequences of their presence in aquatic environments is still largely unknown (EPA 2013).

Organic matter, nitrogen and phosphorus within the waste stream are transformed into biological matter or biosolids by POTW's. The processes in which biosolids are handled vary greatly. Biosolids can be dewatered and thickened to reduce the volume of waste. Air drying or thermally drying of biosolids reduces moisture and reduces pathogens in biosolids. After processing, biosolids can be disposed of in several ways. Biosolids can be transported to landfills for disposal. Biosolids are frequently conveyed directly from solids handling to onsite incineration. Ash generated from the incineration of biosolids is placed in landfills. Biosolids and biosolids ash disposed in landfill is also regulated under RCRA. Instead of landfill disposal, biosolids can be transformed from a waste product back into a usable resource of organic matter, nitrogen and phosphorus.

Closing the Resource Loop

Biosolids and Land Application Benefits Biosolids contain organic matter, nitrogen and phosphorus that can be beneficial for soils and help reclaim disturbed areas of land. The addition of organic matter through the application of biosolids helps improve soil structure, makes it more resistant to erosion, and increases the soil's ability to retain water. The application of biosolids provides mainly nitrate as a source of nitrogen to soils. Nitrate is released slowly from the organic matter within the biosolids and therefore moves through soils more slowly than nitrate from chemical fertilizers. Phosphorus in biosolids is generally less soluble than phosphorus contained in chemical fertilizers and thus less likely to cause surface water enrichment. Proper management of biosolids application and soil monitoring is necessary to prevent elevated levels of phosphorus in soils and excessive nitrogen leaching into groundwater and surface water (WRF 2011).

Biosolids as Energy Source Anaerobic digestion of biosolids generates biogases. Digester biogas is typically 60–70% methane, remainder primarily carbon dioxide, and has particularly unpleasant odor. Whereas digester gases are routinely contained and burned to control methane and odor, it can be captured and use to generate heat and power. Combined Heat and Power (CHP) uses methane as a fuel source for heating tanks and digestors within the treatment process, providing heat for climate control of buildings, and can generate electrical power from turbines. CHP systems can significantly reduce power needs, enhances power reliability for the treatment plant, and reduces overall greenhouse gas emissions and other air pollutants by replacing utility power with a renewable resource (EPA 2011). Each million gallons (3.8 million L) per day of wastewater flow can produce enough biogas to produce 26 kilowatts (kW) of electric power. Average residential electrical use, within the U.S., is approximately 29 kW per day (EIA 2019).

Biosolids and CAMBI Process DC Water's Blue Plains Advanced Wastewater Treatment Plant is one of the largest plants of its kind in the world, treating almost 300 million gallons (1135 million L) of wastewater per day and has the capacity to treat over 1 billion gallons (3.8 billion L) per day during peak flows. Blue Plains Advanced Wastewater Treatment Plant uses advanced solids treatment to manage the vast amounts of biosolids generated each day. The CAMBI Thermal Hydrolysis Process (THP) is used to pretreat biosolids as a sterile carbon source for anaerobic digestion. CAMBI-THP acts like a pressure cooker and applies high-heat and pressure to lyse cells within biosolids, providing a readily available food source for microbes at the next step of treatment process. DC Water is the first thermal hydrolysis facility in North America, the largest in the world, processing over 200,000 tons of biosolids into soils each year. In addition, DC Water's Blue Plains Advanced Wastewater Treatment Plant uses CHP technology to power approximately one third of the plant, and heat is converted to steam which is used for the entire THP process with no external energy sources needed.

Phosphorus Recovery Excessive amount of phosphorus is not only harmful to the environment, it is also very difficult to manage within the waste stream. Pipes within the sewage system buildup with a concrete-hard scale over time, like clogged arteries, with a phosphorus rich mineral deposit of struvite (Le Corre et al. 2009). Advanced wastewater treatment technology has been developed to remove this resource out of the waste stream and protect wastewater infrastructure. Ostara's proprietary Pearl® and WASSTRIP® technologies create an environmentally friendly fertilizer, called Crystal Green®. In June 2010, Hampton Roads Sanitation District's (HRSD) Nansemond Treatment Plant, located in Suffolk, Virginia, implemented the first Ostara facility within the Chesapeake Bay and began producing a commercial grade fertilizer from phosphorous that otherwise was a nuisance in pipes. HRSD's Nansemond Treatment Plant is designed to treat up to 30 million gallons (114 million L) of wastewater per day, removing more than 85% of the phosphorus from the waste streams, and has the capacity to produce more than 1 million pounds (0.45 million kg) of fertilizer annually (HRSD 2010).

Sustainable Water Initiative for Tomorrow HRSD has advanced from wastewater treatment operations to innovative water treatment and creating a sustainable water supply. HRSD's Sustainable Water Initiative for Tomorrow (SWIFT) is a long-term project to protect southeast Virginia's environment, sustain the region's groundwater supply, mitigate saltwater intrusion, alleviate environmental stress to the Chesapeake Bay, and potentially slow or reverse land subsidence. The SWIFT process takes highly treated wastewater effluent that would be otherwise discharged into the Elizabeth, James or York Rivers of Virginia, and puts it through an advanced multi-step disinfection process, amends it to match the aquifer geochemistry, and returns the water back to the Potomac Aquifer. In southeast Virginia, groundwater withdraws for domestic and industrial use are faster than the Potomac Aquifer can recharge. The addition of SWIFT Water® will offset groundwater withdraws and protect the aquifer from further overuse damage and saltwater intrusion. In 2019, the SWIFT Research Center is operational and recharging the Potomac Aquifer with up to 1 million gallons (3.8 million L) of drinking water quality SWIFT Water® per day. Within the first year of operation, the SWIFT Research Center has recharged over 100 million gallons (380 million L) of water back into the aquifer. The SWIFT Research Center is providing invaluable information and data that will be critical in designing and implementing five full-scale SWIFT facilities. HRSD plans to begin construction of the first full scale SWIFT facility by 2020 and potentially recharging the Potomac Aquifer with 100 million gallons (380 million L) per day when all five facilities are operational in 2032.

Synopsis

As discussed throughout this and other chapters, the currency of pollution in estuaries is primarily nitrogen and phosphorus. These elements can come from many different sources including, agriculture, forests, urban runoff, wastewater and combined sewer overflows, septic systems, and atmospheric deposition. Though not all these categories directly pertain to the food system or resource management of wastewater, it is valuable to understand the significant impact food system or resource management of wastewater have in relation to these other sources of nitrogen and phosphorus. The following section examines nitrogen and phosphorus sources and changes with time.

Table 1 shows the changing amounts of nitrogen and phosphorus inputs to the Chesapeake Bay. This starts with 1985 baseline data for nutrient inputs to the Chesapeake Bay. Nutrient sources in 2009 are an important benchmark, as the implementation of the Chesapeake Bay's Total Daily

Table 1 Data for nitrogen and phosphorus loadings

Nitrogen by source	1985 (million lb/year)	2009 (million lb/year)	2017 (million lb/year)
Agriculture	142	114	106
Urban runoff	34	40	41
Wastewater + combined sewer overflow	89	52	33
Septic	5	8	9
Forest + non-tidal water atmospheric deposition	47	46	46
Atmospheric deposition to watershed	26	3	1
Atmospheric deposition to tidal water	26	19	16
Total basinwide	370	283	253
Phosphorus by source	1985 (million lb/year)	2009 (million lb/year)	2017 (million lb/year)
Agriculture	11	11	8.4
Urban runoff	2.9	3.0	2.7
Wastewater + combined sewer overflow	10	4.0	2.3
Forest	1.7	1.7	1.7
Total basinwide	26	19	15

Data obtained from the Chesapeake Bay Program's Chesapeake Progress website (https://www.chesapeakeprogress.com/clean-water/watershed-implementation-plans) with downloadable data and interactive charts. Accessed 14 September 2019

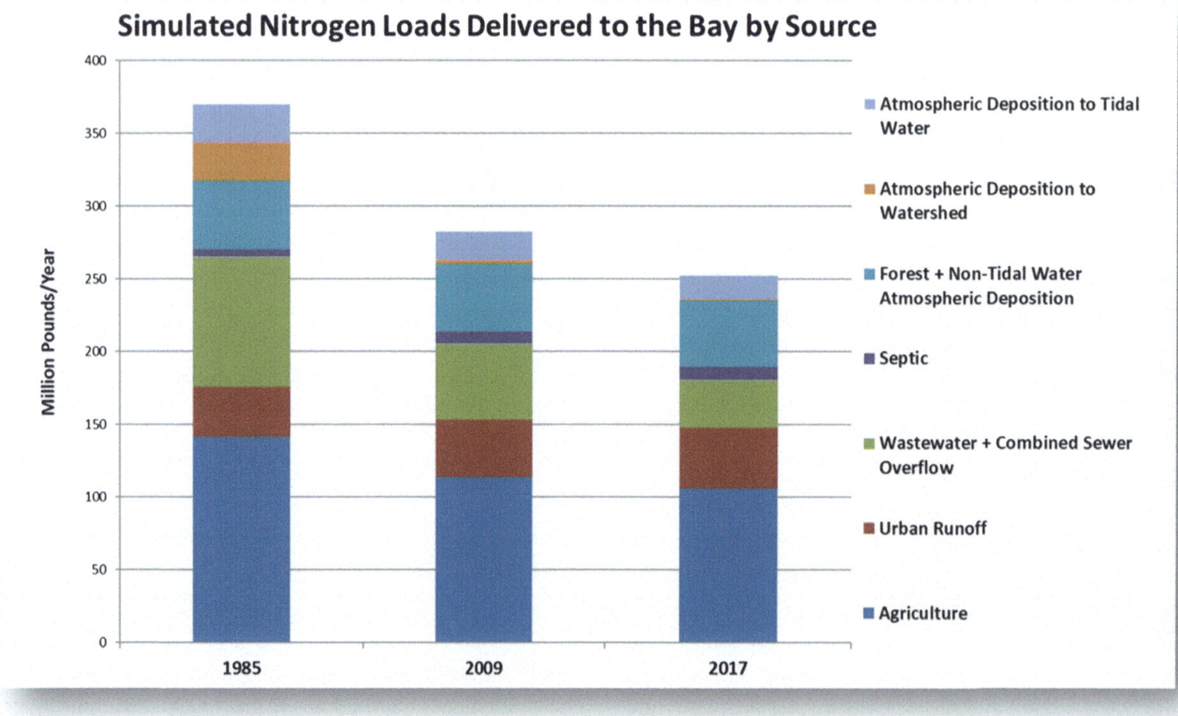

Fig. 6 Nitrogen loading to the Chesapeake Bay by category. Loads simulated using 5.3.2 version of Watershed Model and wastewater discharge data reported by Bay jurisdictions. Data obtained from the Chesapeake Bay Program's Chesapeake Progress website (https://www.chesapeakeprogress.com/clean-water/watershed-implementation-plans) with downloadable data and interactive charts. Accessed 14 September 2019

Maximum Load (TMDL), or a comprehensive pollution diet for the Bay began in 2009. Nutrient sources in 2017 are a critical benchmark as it was the mid-point assessment after the implementation of the Bay's TMDL's. The next benchmark will be in 2025 when all pollution control measures needed to fully restore the Bay and its tidal rivers are in place (Chesapeake TMDL 2019).

Nitrogen Loading Baseline data for the Chesapeake Bay in 1985 estimated 24% of nitrogen loading was from wastewater + CSO's. In 2009, 18% of nitrogen loading was from wastewater + CSO's. During this time, nitrogen loading from overall basin wide loadings decreased 24% and wastewater + CSO's decreased 42%. In 2017, 13% of nitrogen loading to the Chesapeake Bay was from wastewater + CSO's. Since 1985, nitrogen inputs to the Chesapeake Bay from wastewater + CSO's sources have decreased by 63% (Fig. 6). In fact, wastewater + CSO's have already met and exceeded year 2025 target goals for nitrogen reduction (38 million pounds or 17.2 million kg per year; Chesapeake TMDL 2019).

Baseline data for the Chesapeake Bay in 1985 estimates 1% of nitrogen loading to the Bay was from septic sources, approximately 5 million pounds (2.3 million kg) per year. In 2009, 3% of nitrogen loading to the Bay was from septic sources, increasing to 8 million pounds (3.6 million kg) per year. During this time, nitrogen loading from overall basinwide loadings decreased 24% as septic sources increased 60%. In 2017, 9% of nitrogen loading to the Bay was attributed to septic sources, increasing 125,000 (57,000 kg) pounds per year. While most all other nitrogen sources are decreasing and basinwide loadings continue to decrease, septic sources continue to show increases. The target 2025 goals for nitrogen loading from septic sources are 6 million pounds (2.7 million kg) per year (Chesapeake TMDL 2019). Septic systems could potentially be a problem for the Chesapeake Bay if its use is continued for sewage treatment.

Phosphorus Loading Baseline data (Table 1) for the Chesapeake Bay in 1985 estimates 38% of phosphorus loading from wastewater + CSO's. In 2009, 21% of phosphorus loading was from wastewater + CSO's. During this time, phosphorus loading from wastewater + CSO's decreased 60% as well as overall basinwide loadings decreased 27%. In 2017, 15% of phosphorus loading to the Chesapeake Bay was from wastewater + CSO's. Since 1985, phosphorus from wastewater + CSO's has decreased by 77% (Fig. 7). In fact,

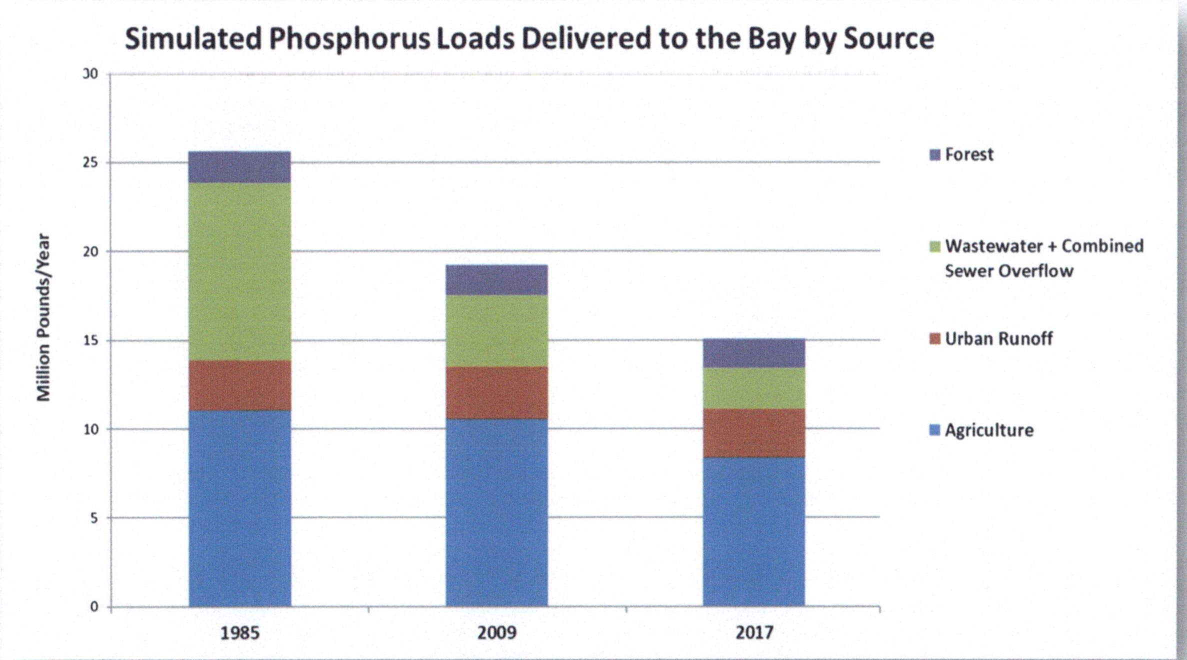

Fig. 7 Phosphorus loading to the Chesapeake Bay by category. Loads simulated using 5.3.2 version of Watershed Model and wastewater discharge data reported by Bay jurisdictions. Data obtained from the Chesapeake Bay Program's Chesapeake Progress website (https://www.chesapeakeprogress.com/clean-water/watershed-implementation-plans) with downloadable data and interactive charts. Accessed 14 September 2019

wastewater + CSO's have already met and exceeded 2025 target goals for phosphorus reduction (2.9 million pounds or 1.3 million kg per year; Chesapeake TMDL 2019).

The human population is the driving force of nutrient pollution in the Chesapeake Bay. Population of the Chesapeake Bay watershed in 1985 was approximately 14 million, increasing about 1% each year to 18.1 million over the 32-year period covered in Table 1. Despite a growing population, the aggregate loads of nitrogen and phosphorus to the Chesapeake Bay have decreased over the 32-year period shown in Table 1 and Figs. 6 and 7. Agriculture inputs have decreased by 7% for nitrogen and 24% for phosphorus. Inputs from urban runoff are essentially the same. Nutrient pollution from agriculture and urban runoff sources are significant and their reductions to date have been less than remarkable. Nutrient pollution from septic sources has not significantly changed over the 32-year time period, nor has septic treatment technology. However, over the 32-year period wastewater + CSO's have decreased their inputs to the Bay by 63% for nitrogen and 77% for phosphorus. As population continues to increase within the Bay watershed, additional nutrient reductions will be needed through the implementation of source controls and best management practices in agriculture, stormwater runoff and through application of advanced wastewater treatment technologies throughout the Chesapeake Bay watershed described above.

Disclosure Statement Views of author only, work not commissioned by HRSD.

References

Adegoke A, Stenstrom T (2019) Septic systems. Global Water Pathogen Project

Aron-Wisnewsky J, Clément K (2015) The gut microbiome, diet, and links to cardiometabolic and chronic disorders. Nat Rev Nephrol 12:169–181. https://doi.org/10.1038/nrneph.2015.191

Baltimore History (2019) History of wastewater removal and treatment. In: Baltimore City Department of Public Works. https://publicworks.baltimorecity.gov/pw-bureaus/water-wastewater/wastewater/history. Accessed 9 July 2019

Calloway D, Margen S (1971) Variation in endogenous nitrogen excretion and dietary nitrogen utilization as determinants of human protein requirement. J Nutr 101:205–216. https://doi.org/10.1093/jn/101.2.205

Chesapeake Stormwater Network (2017) Be a Chesapeake Bay retriever: designing effective outreach programs to reduce pet waste

Chesapeake TMDL (2019) Chesapeake Bay TMDL. US EPA. In: US EPA https://www.epa.gov/chesapeake-bay-tmdl. Accessed 20 Sept 2019

Cinquepalmi V, Monno R, Fumarola L et al (2012) Environmental contamination by dog's faeces: a public health problem? Int J Environ Res Public Health 10:72–84. https://doi.org/10.3390/ijerph10010072

Cooper J, Loomis G, Amador J (2016) Hell and high water: diminished septic system performance in coastal regions due to climate change. PLoS One 11:e0162104. https://doi.org/10.1371/journal.pone.0162104

DC Water History (2019) History—water system. DCWater.com. In: Dcwater.com. https://www.dcwater.com/history-water-system. Accessed 10 July 2019

District of Columbia, Office of Planning (2003) The Anacostia waterfront framework plan. District of Columbia

Duncan M, Horan N (2003) Handbook of water and wastewater microbiology. Academic, London

EIA (2019) How much electricity does an American home use? FAQ—U.S. Energy Information Administration (EIA). In: Eia.gov. https://www.eia.gov/tools/faqs/faq.php?id=97&t=3. Accessed 20 Sept 2019

Emara M, Ahmed F, Adb El-Aziz F, Adb El-Razek A (2014) Biological nutrient removal in Bardenpho process. J Am Sci 10:1–9

Environmental Protection Agency (EPA) (2019) Conventional Septic System

EPA (1993) Test methods for evaluating solid waste, physical/chemical methods SW-846

EPA (1997) The potential for nutrient loadings from septic systems to ground and surface water resources and the Chesapeake Bay 4

EPA (2010a) Stability of pharmaceuticals, personal care products, steroids, and hormones in aqueous samples, POTW effluents, and biosolids. U.S. Environmental Protection Agency Office of Water, Washington, DC

EPA (2010b) Guidelines for deriving numerical National Water Quality Criteria for the protection of aquatic organisms and their uses. Office of Research and Development environmental research laboratories. Duluth, Minnesota

EPA (2011) Opportunities for combine heat and power at wastewater treatment facilities: market analysis and lessons from the field. U.S. EPA, Combined Heat and Power Partnership, Washington, DC

EPA (2013) Contaminants of emerging concern (CECs) in fish: pharmaceuticals and personal care products (PPCPs). EPA Office of Water 4305T, Washington, DC

Esfahani E, Zeidabadi F, Bazargan A, McKay G (2017) The modified Bardenpho process. Handbook of environmental materials management 1-50. https://doi.org/10.1007/978-3-319-58538-3_87-1

Fayer R, Graczyk T, Lewis E et al (1998) Survival of infectious Cryptosporidium parvum oocysts in seawater and eastern oysters (Crassostrea virginica) in the Chesapeake Bay. Appl Environ Microbiol 64:1070–1074

Fenston J (2019) Could we be swimming in D.C.'s rivers soon? DCist. In: DCist. https://dcist.com/story/19/07/23/could-we-be-swimming-in-d-c-s-rivers-soon/. Accessed 14 Aug 2019

Fewtrell L (2004) Drinking-water nitrate, methemoglobinemia, and global burden of disease: a discussion. Environ Health Perspect 112:1371–1374. https://doi.org/10.1289/ehp.7216

Forsythe P, Bienenstock J, Kunze W (2014) Vagal pathways for microbiome-brain-gut axis communication. Adv Exp Med Biol:115–133. https://doi.org/10.1007/978-1-4939-0897-4_5

George R (2001) The history of plumbing. Plumb Eng 3:45–53

Griffin D, Donaldson K, Paul J, Rose J (2003) Pathogenic human viruses in coastal waters. Clin Microbiol Rev 16:129–143. https://doi.org/10.1128/cmr.16.1.129-143.2003

Griffith A, Gobler C (2019) Harmful algal blooms: a climate change co-stressor in marine and freshwater ecosystems. Harmful Algae. https://doi.org/10.1016/j.hal.2019.03.008

Grumet R (2000) Bay, plain, and piedmont a landscape history of the Chesapeake heartland from 1.3 billion years ago to 2000 the Chesapeake Bay heritage context project. U.S. Department of the Interior, National Park Service, Annapolis, MD

Harris J, LaRocque R, Qadri F et al (2012) Cholera. Lancet 379:2466–2476. https://doi.org/10.1016/s0140-6736(12)60436-x

Heisler J, Glibert P, Burkholder J et al (2008) Eutrophication and harmful algal blooms: a scientific consensus. Harmful Algae. https://doi.org/10.1016/j.hal.2008.08.006

HRSD (2010) HRSD, Ostara partner to recover nutrients and make fertilizer—environmental protection. In: Environmental Protection. https://eponline.com/Articles/2010/06/01/HRSD-Ostara-Partner-to-Recover-Nutrients-and-Make-Fertilizer.aspx?Page=1. Accessed 20 Sept 2019

HRSD History (2019) History. HRSD.com. In: Hrsd.com. https://www.hrsd.com/history. Accessed 9 July 2019

Kennedy C, Pincetl S, Bunje P (2011) The study of urban metabolism and its applications to urban planning and design. Environ Pollut 159:1965–1973. https://doi.org/10.1016/j.envpol.2010.10.022

Laws E (2000) Aquatic pollution, 3rd edn. Wiley, New York, NY

Le Corre K, Valsami-Jones E, Hobbs P, Parsons S (2009) Phosphorus recovery from wastewater by struvite crystallization: a review. Crit Rev Environ Sci Technol 39:433–477. https://doi.org/10.1080/10643380701640573

Lofrano G, Brown J (2010) Wastewater management through the ages: a history of mankind. Sci Total Environ 408:5254–5264. https://doi.org/10.1016/j.scitotenv.2010.07.062

MHBP (2019) Maryland healthy beaches program provides information about the condition of public beaches. In: Marylandhealthybeaches.org. http://www.marylandhealthybeaches.org/current-conditions.php. Accessed 25 Sept 2019

Mooring P (2019) DC water's Anacostia River tunnel saves river from 4.5 billion gallons of sewage overflow in 2018. DCWater.com. In: Dcwater.com. https://www.dcwater.com/whats-going-on/news/dc-water%E2%80%99s-anacostia-river-tunnel-saves-river-45-billion-gallons-sewage-overflow. Accessed 18 Aug 2019

NPDES (2019) NPDES permit basics. US EPA. In: US EPA. https://www.epa.gov/npdes/npdes-permit-basics. Accessed 23 Oct 2019

Patterson E, Ryan P, Cryan J et al (2016) Gut microbiota, obesity and diabetes. Postgrad Med J 92:286–300. https://doi.org/10.1136/postgradmedj-2015-133285

Qiu Y, Lee B, Neumann N et al (2015) Assessment of human virus removal during municipal wastewater treatment in Edmonton, Canada. J Appl Microbiol 119:1729–1739. https://doi.org/10.1111/jam.12971

Ramírez-Castillo F, Loera-Muro A, Jacques M et al (2015) Waterborne pathogens: detection methods and challenges. Pathogens 4:307–334. https://doi.org/10.3390/pathogens4020307

Rath M (2019) Wastewater. Chesapeake Bay Program. In: Chesapeakebay.net. https://www.chesapeakebay.net/issues/wastewater. Accessed 20 Sept 2019

Rose C, Parker A, Jefferson B, Cartmell E (2015) The characterization of feces and urine: a review of the literature to inform advanced treatment technology. Crit Rev Environ Sci Technol 45:1827–1879. https://doi.org/10.1080/10643389.2014.1000761

Rountree H, Clark W, Mountford K, Barber M (2008) John Smith's Chesapeake voyages, 1607–1609. University of Virginia Press, Charlottesville

Sewage Treatment in Virginia (2019) Sewage treatment in Virginia. In: Virginiaplaces.org. http://www.virginiaplaces.org/waste/sewer.html. Accessed 9 July 2019

Silvester K, Bingham S, Pollock J et al (1997) Effect of meat and resistant starch on fecal excretion of apparent N-nitroso compounds and ammonia from the human large bowel. Nutr Cancer 29:13–23. https://doi.org/10.1080/01635589709514596

Snow J (1855) On the mode of communication of cholera, 2nd edn. John Churchill, London

Solomon C, Jackson M, Glibert P (2019) Chesapeake Bay's "forgotten" Anacostia River: eutrophication and nutrient reduction measures. Environ Monit Assess. https://doi.org/10.1007/s10661-019-7437-9

Standard Methods (2018) Standard methods for the examination of water and wastewater, 23rd edn. American Public Health Association, Washington, DC

Tarter B (2012) Lawes divine, Morall and Martiall. In: Encyclopediavirginia.org. http://www.EncyclopediaVirginia.org/Lawes_Divine_Morall_and_Martiall. Accessed 9 July 2019

Taylor J, Iwan G, Mains C, Fekete J (2013) Phosphorus and onsite wastewater systems. Pipeline 24:1–9. http://www.nesc.wvu.edu/pdf/WW/publications/pipline/PiL_SU13.pdf

USEPA (2019) How your septic system works. US EPA. In: US EPA. https://www.epa.gov/septic/how-your-septic-system-works. Accessed 23 Oct 2019

Vaughn C (2012) Baltimore sun the scoop on poop: pet waste a major polluter of urban waterways, studies show. In: Capitalgazette.com. https://www.capitalgazette.com/cg2-arc-4c73d98e-32bb-5f73-b727-fa48d2113e31-20121028-story.html. Accessed 7 Aug 2019

VDH (2019) Swimming advisories and monitored beaches map—environmental epidemiology. In: Vdh.virginia.gov. http://www.vdh.virginia.gov/environmental-epidemiology/beach-monitoring/swimming-advisories-and-monitored-beaches-map. Accessed 25 Sept 2019

Vogelsong S (2019) Septic system failures expected to increase in coastal Virginia. Bay Journal

Wheeler A, Hartel P, Godfrey D et al (2002) Potential of as a human fecal Indicator for microbial source tracking. J Environ Qual 31:1286. https://doi.org/10.2134/jeq2002.1286

Wolman A (1965) The metabolism of cities. Sci Am 213:179–190

World Health Organization (2019a) Sanitation. In: World Health Organization. https://www.who.int/topics/sanitation/en/. Accessed 14 Aug 2019

World Health Organization (2019b) Global task force on cholera control. Ending cholera: a global roadmap to 2030. World Health Organization

Worley-Morse T, Mann M, Khunjar W et al (2019) Evaluating the fate of bacterial indicators, viral indicators, and viruses in water resource recovery facilities. Water Environ Res 91:830–842. https://doi.org/10.1002/wer.1096

WRF (2011) National manual of good practice for biosolids. The Water Research Foundation

Plastic Pollution and the Chesapeake Bay: The Food System and Beyond

Robert C. Hale, Meredith Evans Seeley, and Benjamin E. Cuker

Once it was corn in the husk, a walnut in a shell, a glass of water from the tap
Now it's all processed and packaged in plastic, leaving eternal trash all over the map.

Abstract

Plastic pollution is a widespread problem across the world's oceans, bays, estuaries and rivers. Much of this waste is the residue of food system packaging. Excessive production of these durable polymers and inappropriate disposal lead to their accumulation in aquatic environments. Large pieces of plastic debris are a regular feature of shorelines in the Chesapeake Bay, as is true around the world. Over time such plastic debris may be partially degraded, primarily by UV light and abrasion, and fragmented. In fact, a majority of plastic debris falls into the category of microplastics, i.e. debris smaller than 5 mm. The abundance of microplastics in the tributaries of the Chesapeake Bay has been linked to the population density of the surrounding area. Importantly, both large plastic debris items and microplastics threaten marine organisms through ingestion or entanglement. Another threat may accompany the plastic; the surface of the debris may carry sorbed chemical contaminants or pathogens that harm aquatic life. Some plastic polymers also contain chemical additives. These additives may be toxic to aquatic organisms and concentrate at higher trophic levels of the food chain. These phenomena are still being explored by researchers. At present, there is inadequate research detailing the distribution of plastics in the Chesapeake Bay. We explore here the plastic pollution issue, how it may impact the health of the Chesapeake Bay, and possible solutions, with particular attention to food system packaging.

Keywords

Polymers · Food packaging · Water bottles · Solid waste · Microplastic · Plastic

Introduction The modern food system of the Chesapeake Bay owes much of its origin to distant lands on the other side of the continent. If you could travel back 3000 years and about 3000 miles (4828 km) southwest of the Chesapeake Bay, you would meet early Mesoamericans. The corn (*Zea maize*) they cultivated eventually reached the Algonquins of the Bay and transformed their agriculture, and went on to be the dominant food crop in the modern food system. The ancient Mesoamericans also developed another product that has become central to the delivery and storage of food: plastics. Mesoamericans used natural plant-derived polymers to create objects that were integral to their civilization. They used latex from the rubber tree (*Castilla elastica*) and other ingredients to make balls for ceremonial games, footwear and figurines (Hosler et al. 1999). More than entertainment, the ball games had religious and political overtones and were on occasion used to settle disputes. Losers were sometimes beheaded! Ball courts are common features of Mayan and Aztec ruins in Mexico and Central America (Fig. 1). Societies centuries later also combined natural polymers and other ingredients to create new materials, the predecessors of what we now call plastics.

At the start of the twentieth century, no one could imagine the transformation of a Society established on wood, natural

R. C. Hale (✉) · M. E. Seeley
Department of Aquatic Health Sciences, Virginia Institute of Marine Science, Gloucester Point, VA, USA
e-mail: hale@vims.edu; meseeley@vims.edu

B. E. Cuker
Department of Marine and Environmental Science, Hampton University, Hampton, Virginia, USA
e-mail: Benjamin.cuker@hamptonu.edu

B. E. Cuker (ed.), *Diet for a Sustainable Ecosystem*, Estuaries of the World, https://doi.org/10.1007/978-3-030-45481-4_17

Fig. 1 (Left) A Mayan rubber ball (8 cm in diameter) used in Mesoamerican ballgames, from Kaminaljuyu (Guatemala: 300 BC to 250 AD). Also shown is a handstone used by players to strike the ball, from the same area (900 BC to 250 AD). Traces of rubber still adhere to the handstone's surface. Wikimedia Commons CC BY-SA 3.0 (Right) A ball court: Mayan ruins of Coba (Quintana Roo, Mexico). Wikimedia Commons—public domain (image by Ken Thomas)

fibers and animal products (ivory, horn, tortoise shell, bone, tendon) to one dependent on synthetic polymers derived from fossil fuels—that is, plastics. Prior to the commercial production of plastics, circa 1950, people packaged and stored their goods in paper, cardboard or reusable containers made from ceramics, woven fibers, wood, glass or metal. Indeed, archeologists use ceramics and woven baskets to characterize and date ancient societies.

The term "plastic" is derived from the Greek word *plastikos*, meaning a moldable material. Modern plastics came to dominate manufacturing in the twentieth century due to their many useful characteristics. They can be made stiff or flexible, resistant to breakage, light in weight, transparent, translucent, colored or opaque, strengthened by incorporating fibers, formed into any desired shape, non-reactant with many chemicals, and very resistant to environmental break down.

The first synthetic plastics since Mesoamerican times came on the scene in the late nineteenth and early twentieth centuries. In seeking an alternative to ivory for making billiard balls, New Yorker, John Wesley Hyatt in 1870, combined cellulose nitrate, camphor, heat and pressure to make "celluloid" (PHS 2019). The flexible celluloid proved useful in replacing whale bone (baleen) in manufacturing and made possible the roll-film used to make moving pictures. The same ingredients in celluloid were formed as a clear or tinted film called cellophane that became popular as food packaging in the first half of the twentieth century. Moisture and oil resistant cellophane revolutionized the marketing of processed foods. DuPont sold $100 million of cellophane in 1950 (Gordon 2017).

The first fully synthetic plastic, known as "Bakelite" was invented in 1907 in Yonkers, NY by Leo Baekeland, a Belgian immigrant (Kettering 1946; American Chemical Association 1993). He was originally seeking a synthetic alternative to shellac (a scarce commodity made from secretions of a beetle found in Thailand and India), a resin used as a wood finish and as an insulator in some newly emerging electrical applications. Baekeland condensed phenol and formaldehyde in the presence of a catalyst to form a moldable organic polymer (a long chain molecule consisting of a long series of repeating units called monomers). Baekland's polymer has the chemical name "polyoxybenzylmethylenglycolanhydride", so it is obvious why "Bakelite" was a popular alternative. Bakelite entered the food system as handles for pots, pans, and knives; cutting boards, and microwaveable containers.

Bakelite's electrical insulation properties and heat-resistance also made it well suited for radio and telephone casings, inventions that had just taken hold in the U.S. It also was used for jewelry, toys and other products. Bakelite continued in wide use until the 1950s when it was supplanted by other plastics. Articles made of Bakelite are now valuable collectables. Baekeland was already a rich man, due to his previous inventions in the field of photography, by the time he invented Bakelite. However, it added to his wealth and success. He enjoyed sailing his boat immensely and named it the "Ion", combining his love of the water and chemistry. He often sailed from his home in New York past the Chesapeake Bay to his property in Florida. His decedents benefitted from the wealth Leo Baekeland created with his plastic. However, his great grandson, Antony Baekeland lived a troubled life,

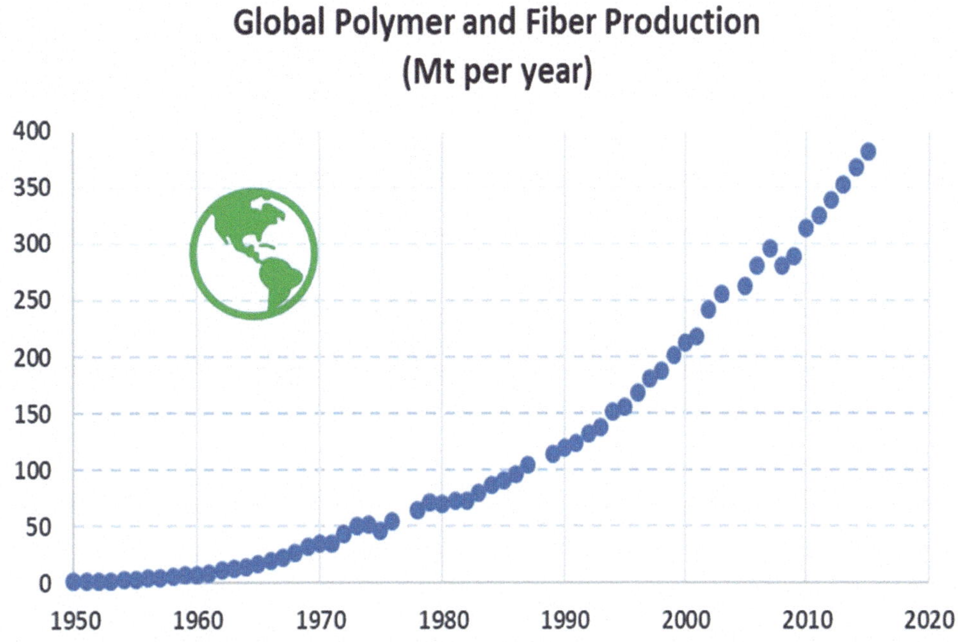

Fig. 2 Global plastics production (*Mt:* millions of metric tons) for various industrial use sectors. Packaging was by far the greatest usage over the interval shown, followed by building and construction materials. Figure prepared from data obtained from Geyer et al. (2017)

being imprisoned for stabbing his mother (to death) and grandmother. While in prison he died by suffocation, ironically from a plastic bag over his head. When Antony's father heard about his son's death, he commented: "It was a beautiful ending—and in plastic too!" (McIver 1990).

Modern, alternative polymers still serve Bakelite's original functions in electronics and electrical devices today. Nonetheless, the manufacture of commercial plastics remained modest until about 1950. Thereafter, their production exploded, spurred by the development of a vast array of new materials, with their attendant low cost, light weight, versatility of form and durability. In the 1967 award winning film "The Graduate" the future of plastics was foretold in an exchange between a recent college graduate (Benjamin: Dustin Hoffman), and a businessman (Mr. McGuire: Walter Brooke):

> Mr. McGuire: Just one word.
> Benjamin: Yes, sir.
> Mr. McGuire: Are you listening?
> Benjamin: Yes, I am.
> Mr. McGuire: Plastics.
> Benjamin: Exactly how do you mean?
> Mr. McGuire: There's a great future in plastics. Think about it. Will you think about it?

Clearly, Mr. McGuire was correct. Plastics now serve a vast range of purposes, including food and beverage packaging, thermal insulation in your home, the chair you sit on, your computer and phone, the carpet beneath your feet and even the cushion beneath it. While traveling, you are surrounded by plastics, including the interior furnishings and controls of your car. On the Chesapeake Bay, the hull of your boat, its sails and rigging, the life preserver you wear and the line you fish with are largely, or entirely, plastic products.

The cumulative total amount of plastic manufactured globally between 1950 and 2015 was about 8.3 billion metric tons (MT), about the weight of 1.7 billion African elephants (National Geographic 2019). A major use (and one that results in massive amounts of discarded plastics entering the environment) is in "single-use" food and beverage containers or packaging, such as bottles, bags and film. This allows food, water, medicine and other supplies to be delivered safely, where and when needed. This advantage has proved critical during disasters such as hurricanes, where supplies must remain stable for extended periods under adverse conditions. These applications save innumerable lives, as well as foster long-distance commerce.

The surge in usage of plastics has resulted in an exponential increase in the mass of waste generated (Fig. 2). As of 2015, the cumulative total of plastic waste generated globally was 5800 metric tons (Geyer et al. 2017), equivalent to 70% of the total plastic manufactured. Plastics are a diverse and complex group of materials, complicating the ease and safety of reuse. Polymers are assigned abbreviations and numbers for convenience and identification: polyethylene terephthalate (PET: #1), high- and low-density polyethylene (HDPE: #2 and LDPE: #4, respectively); polyvinyl chloride (PVC: #3), polypropylene (PP: #5) polystyrene (PS: #6), other plastics (#7). Packaging is the dominant use of plastics (Fig. 3), followed by construction, consumer products and miscellaneous uses. Only a few polymer types (often only the #1 and #2 classifications) are currently widely accepted in the

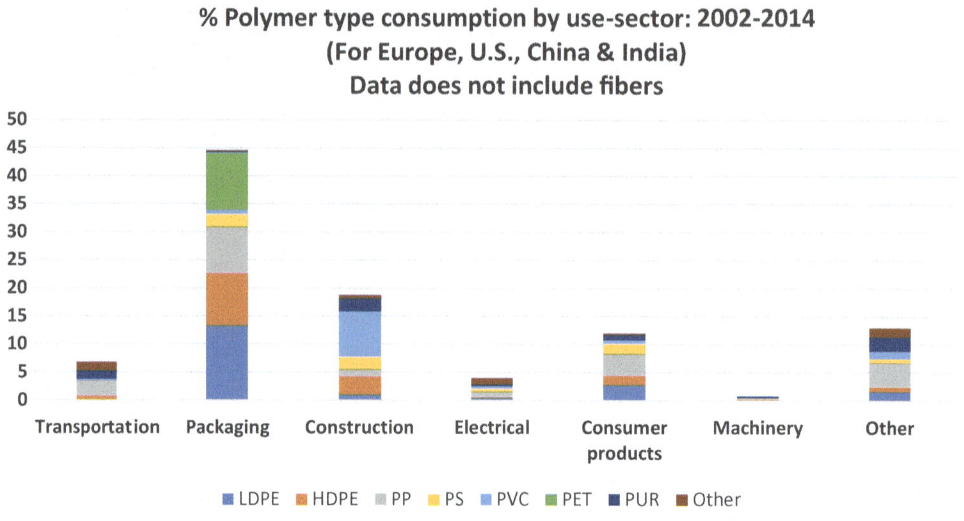

Fig. 3 Global plastic waste generation (in millions of metric tons) and major polymer types consumed for various industrial use sectors for 2002–2014. Figure prepared with data from Geyer et al. (2017)

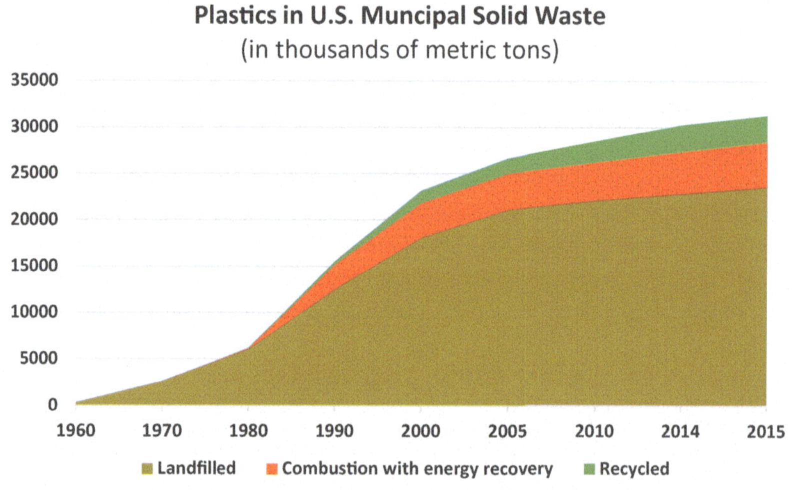

Fig. 4 Illustration of U.S. plastic waste management strategies, from 1960 to 2015. The majority of plastic waste has been disposed of via landfilling. As landfilling costs have increased (and as recognition of negative impacts and energy requirements of plastics were recognized around 1980) recycling and energy recovery have increased modestly over time. Figure prepared from data from the U.S. Environmental Protection Agency (https://www.epa.gov/facts-and-figures-about-materials-waste-and-recycling/plastics-material-specific-data)

U.S. for recycling. At end of their service lives, most plastics are sent to landfills and a small portion is reused beneficially (Fig. 4). A substantial amount is burned for energy recovery. Plastic's low manufacturing cost provides little financial incentive to reuse it. This must change.

The plastics recycling market in the U.S. and elsewhere has declined, since China (the world's largest waste plastic importer) banned imports in 2017 (Brooks et al. 2018). Most of these imports consisted of single-use food and beverage containers. Another smaller stream of waste plastic imported into China has been from the electrical/electronics sector, i.e. "e-waste". These complex products (e.g. cell phones, computers, televisions) have longer service lives than single-use container plastics. However, as technology is advancing rapidly, consumers are discarding electronics at an increasing rate to remain "current". Unfortunately, e-waste plastics contain materials with substantial inherent toxicity. Recycling scavenged valuable metals such as copper from the devices under primitive conditions has been largely done in developing countries where regulations are lax and labor cheap. There, the leftover plastics have typically been haphazardly scattered, shredded or burned. This has released a plethora of toxic chemicals to the local environment, including heavy metals, chlorinated dioxins and flame retardant polymer additives (Zhang et al. 2012; Li et al. 2019).

Globally, less developed countries are increasing their use of plastics on a massive scale. This is enhancing food security by expanding the availability of food and safe drinking water. However, it also has spurred the consumption of processed food, rich in sugars and fat, thus increasing related health consequences, such as diabetes and obesity (Lin et al. 2018). Increased plastic usage also worsens the global debris problem. Poor countries lack advanced waste collection and disposal systems. Most of the plastics discarded near the

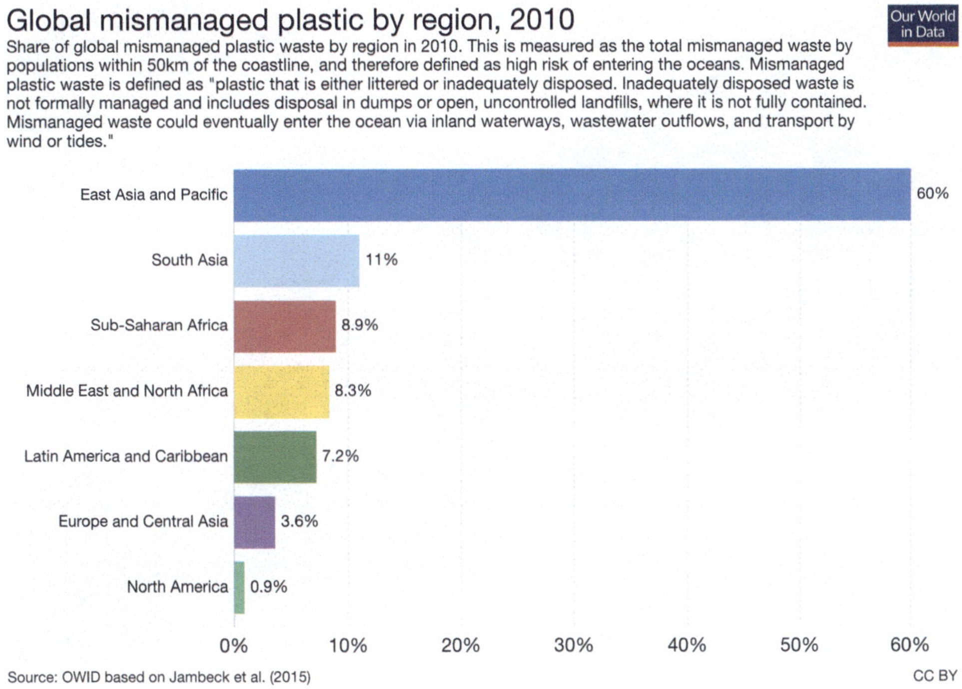

Fig. 5 The distribution of mismanaged plastic waste, by region, for the year 2010. Reprinted with permission from *Our World In Data* (https://ourworldindata.org/grapher/mismanaged-plastic-waste-by-region-2010)

Chesapeake Bay will end up in engineered dumps or shipped as recycling to other countries. Hence, these wastes disappear from the view of most consumers, facilitating our cavalier use of single-use plastics. Landfill space in the U.S. is at a premium. Consequently, fees to dump our wastes have increased. These heightened costs have also incentivized sending of wastes overseas for "recycling," often with undesirable environmental and health consequences. Regions of Southeast Asia, especially China, are the largest release points for plastics to the world's oceans (Fig. 5), dwarfing the input from Europe and North America combined. Ironically, a portion of this waste may be "recycling" from wealthy countries such as the U.S.

Despite a well-developed system of waste management, a significant portion of discarded plastics in the Chesapeake Bay watershed never makes it to landfills or recycling centers. Plastic bottles and food packaging scraps line beaches, deposit in marshes, and float in the open waters of the Bay. Larger pieces colonized by organisms often sink to contaminate the sediments.

One of the benefits of plastic is its greater resistance to breakdown compared to natural products such as plant fibers. Yet such resistance is why plastics are such a problem when they enter the environment. A paper bag or woven basket made from grasses will completely decompose in nature. Not so, for a plastic bag or polystyrene foam food container. However, ultraviolet light and abrasion reduces large pieces of plastic to smaller bits called microplastics.

Microplastics are a growing environmental concern. They are formally defined as fragments less than 5 mm (Fig. 6). They can be primary microplastics (intentionally manufactured smaller than 5 mm) or secondary microplastics (formed from the fragmentation of larger plastic debris). Secondary microplastics are a far more abundant pollutant than primary microplastics (Cózar et al. 2014). In addition, nanoplastics (smaller than 1 nm) can form from microplastics and are also thought to be plentiful in the environment, although they are very difficult to collect and analyze.

Considering the many uses of plastics and routes plastic waste can take, it is easy to see why plastic pollution has become an environmental problem. The media often focuses on a limited number of high-profile topics, such as the Great Pacific Garbage Patch (Parker 2018a). Yet, plastic pollution is a major issue across many scales (large debris to nanoplastics) and locations (terrestrial systems, inland lakes, rivers, bays, estuaries and oceans). Elucidating the source, fate, effects and quantity of plastic in the aquatic environment has become a major research goal. Yet, there are still many lingering questions about plastic pollution, including its biological effects and distribution in waterways. Surprisingly, the distribution of plastic pollution in the Chesapeake Bay is still largely unknown. This chapter will dive into the complexities underlying plastic pollution, specifically in the context of potential effects for the Chesapeake Bay.

Fig. 6 Photograph of microplastics mixed with natural debris, taken at an estuary on Elizabeth Island, Alaska. Plastic pollution from the Pacific Ocean has accumulated and deteriorated in this location over time. Reprinted with permission from Meredith Evans Seeley

What Are Plastics? While plastics share some commonalities, their chemical makeup, properties and environmental behavior often differ in important ways. Plastics are generally distinguished by their polymer composition. Common polymers include: polyethylene (PE), polypropylene (PP), polystyrene (PS), polyvinyl chloride (PVC), polyethylene terephthalate (PET), polycarbonate, acrylonitrile butadiene styrene (ABS), polyamide (nylon), polyester, polyurethane (PU) and polytetrafluoroethylene (trade name: Teflon). Polymers consist of repeating chemical units, or monomers, that themselves are formed from petroleum. In some cases, different polymer types are blended or reacted together.

Natural polymers, synthesized by organisms, are among the most abundant organic materials in the Chesapeake Bay. These include chitin, a component of crustacean exoskeletons (e.g. the blue crab, *Callinectes sapidus*), and cellulose, a major component of many plants (such as eel grass, *Zostera marina*, wood, tree leaves and marsh plants). A new generation of synthetic polymers has begun to be synthesized from renewable starting materials. These harken back to the materials made by our Mesoamerican predecessors. Such "bioplastics" are produced using starch, cellulose, dextrose, hydroxybutyrate or hydroxyalkanoate (bacterial biproducts) precursors. They exhibit greater degrees of degradability in the environment, and some are designed for composting.

Production of plastic products is most commonly achieved by heating polymeric resins or commercially-produced pellets (also known as nurdles), and injecting the melt into a mold or extruding the melt from a die. To achieve specific final product properties (e.g. heat-durability, flexibility, etc.), different polymers may be mixed and additional chemicals mixed in (so-called additives) during this process. Alternatively, the desired shape may be created by cutting the formed plastic after cooling; this technique is frequently applied to PU foams used in furniture. Composite materials are also common (e.g. fiberglass) whereupon a resin is mixed with other chemicals and sets into a solid over time. *Thermoset* plastics are those that cross-link together during curing, hold their shape upon heating and thus cannot be readily remolded. They are chosen for high heat applications such as electronics (e.g. casings and plugs, supplanting Bakelite). In contrast, *thermoplastics* may be reformed by the reapplication of heat and thus are more easily recycled into new products. As a result of the production process applied, plastics may occur in different physical forms (e.g. LD and HDPE, PU foam or coatings, and fibers for synthetic fabrics). The size of the finished product may range from microbeads of less than 100 μm (common in abrasives and personal care products) to articles meters across (such as home siding and insulation, boats and floating docks).

Additional chemicals may be used to catalyze the polymerization of the monomers. These may co-react with the polymers, although some may be left over as residuals within the finished product. Furthermore, products are often imbued during formation with a range of organic or inorganic chemical additives to fine-tune the properties of the material to fit specific applications. Additives include fillers, colorants, anti-oxidants, anti-microbial agents, plasticizers, flame retardants, antistatic and blowing agents, heat and light stabilizers and lubricants. The contribution of these additives to the overall product weight may be exceed 10%. Additives may react and bind with the polymer itself, or be suspended within the polymer matrix (and thus be more likely to migrate out). Additives may have toxic properties and some are known carcinogens or endocrine disruptors. These factors must be considered when evaluating the fate, recycling and toxicological potential of plastics and additives.

The choice of polymer, additive(s) and production technique are dictated by the ultimate purpose that the plastic product will serve. Polyethylene is a good example of the diversity of forms and uses of plastics: available in low, medium, high and ultra-high densities. LDPE is common in flexible applications, such as films and plastic bags. Medium and high densities are often used in more rigid applications such as bottles, plastic lumber and water piping. Ultra-high density PE is a small production volume material, used in critical machine parts subject to wear and in artificial joints. Another example, rigid PVC, is common in a variety of products including credit cards, containers, house siding and water pipes. Flexible PVC is used for tubing, IV (intravascular) fluid bags and electrical wiring. PVC has often contained phthalate additives, some of which have been proven to be toxic endocrine disrupters (Lyche et al. 2009). Polystyrene is commonly used for CD/DVD cases, containers and lids, and disposable cutlery. In foam form, PS is used as insulation, shipping containers, and for flotation in boat hulls, fishing floats and docks. Styrofoam is a trademarked PS foam product. Polyurethane may be used in surface coatings, synthetic fibers and automotive parts. In foam form, PU is used for insulation, flexible furniture cushioning and other applications.

Flame retardant additives are common in PS insulation and PU foam furniture and electrical devices. Polybrominated diphenyl ether (PBDE) flame retardants were detected in high concentrations in filets of Virginia fish (Hale et al. 2001) consumed by local residents. The chemicals were tracked to discharges from a North Carolina wastewater treatment plant which received wastes from a plastics manufacturing operation. The treatment plant released the PBDEs to a tributary of the Dan River where the contaminated fish were collected. This illustrates the persistence, complex transport and fate of plastic additives.

Synthetic rubber, consisting primarily of styrene-butadiene polymer, is widely used in vehicle tires. Tires are complex polymeric products that also contain steel and nylon belts, zinc oxide, sulfur vulcanizing agents, UV blockers (such as carbon black), and other materials. (See Box 1 for further background on tires as an environmental pollutant.)

Box 1 Repurposing Used Tires to Help the Food System?

Since the invention of the automobile, disposal of scrap tires has been a problem. Millions of tires are used and discarded worldwide. The problem is particularly large in the U.S. due to our driving habits. Derelict tires are often stock-piled on land. Beingly largely petroleum-based and combustible, fires occasionally have broken out at tire dump sites, resulting in high volumes of chemical pollution. Abandoned tires also catch water and may serve as mosquito breeding areas. This has become an even greater concern since the arrival of Zika virus-carrying mosquitos in some areas of the U.S. Chipping of tires into "crumb" for use as artificial mulch and on athletic fields or children's playgrounds has also been employed in an attempt to reuse the material. This has been controversial, however, as concerns have arisen regarding possible health impacts on exposed individuals. Weighing down tires with concrete or binding them together with bands for use in constructing artificial reefs has also been attempted (Fig. 7). As an example, approximately 2 million tires were sunk to create a 36-acre (14.6 ha) artificial reef off of Fort Lauderdale, Florida (Florida DEP 2016). Over time, the tires broke free and damaged nearby natural coral reefs. Cost to remove the tires has been estimated in the millions of dollars.

Creation of physical structure in waters such as the Chesapeake Bay has also been applied, in an attempt to increase habitat and populations of commercial fin and shellfish. In the 1990s interest arose regarding the use of tire chips in mesh bags as a substrate for artificial reefs in the Bay. This project was abandoned after it was found that the bags ruptured during storms, releasing their contents and covering local shorelines with debris. At the time, research was also conducted at the Virginia Institute of Marine Science to assess the composition and toxicity of aqueous leachates from the tire chips in the laboratory. No organic contaminants of concern were detected in the leachates. However, analytical methods and the composition of tires have both evolved since then, so further analysis is warranted.

Most waste tires are now reprocessed into fuel, or ground and mixed into roadway asphalt. In fact, in 2017 greater than 80% of scrap tires were repurposed, versus only 11% in 1990—a success story. Nonetheless, tire wear results in the generation of considerable amounts of small plastic fragments. Kole et al. (2017) estimated that each person in the U.S. generated on average 4.7 kg/year of tire wear microplastics, resulting

(continued)

Box 1 (continued)
in the release of 1.8 million metric tons/year to the environment. As such, although repurposing of discarded tires has improved drastically, microplastics from tire wear still pose environmental risks and are a complex ecotoxicology issue.

Durability of Plastics A common misconception is that all plastics last indefinitely in the environment. While durable, plastics show a range of persistence depending on their composition, usage and surrounding environmental conditions. Some polymers (e.g. PU foam) are quite vulnerable to degradation when exposed to UV light and may breakdown into a powder of microplastics in a period of months of exposure. Other plastics, e.g. PE bottles, may last for centuries after being discarded into the environment. Lifetimes are extended in the absence of UV light, frictional forces and microbial (a.k.a. bacterial) action. Interestingly, in beach surveys of litter in the Chesapeake Bay region, cigarette filters were the most abundant debris reported (Table 1). Many assume that cigarette filters are degradable. In reality, they are typically made of plastic (cellulose acetate) fibers, which may persist for years. Plastic food and beverage containers were also common forms of plastic debris removed during coastal cleanups of the Chesapeake Bay.

Fig. 7 Used tires being reutilized as artificial reef. Reprinted with permission from Wikimedia Commons, author *Navy Combat Camera Dive Ex-East* (https://commons.wikimedia.org/wiki/File:Tires540.jpg)

Large plastic debris breaks apart into smaller pieces over time, increasing the number of fragments and obscuring their original use or form. Fragmentation is a continuous process, so for a given piece of plastic debris the number of particles generated will increase over time, while their sizes decrease. Ultraviolet light is one of the more effective degraders of plastics. However, some plastics contain additives to reduce its impact. Plastic submerged below the water's surface or buried in sediments is shielded from UV exposure, extending its lifetime.

Frictional forces also fragment plastic. These include wave action on a beach, tires on roadways or washing of textiles. As they fragment, the actual weight of plastic in a sample may be largely unchanged. Consider a bag of potato chips: if one crushes its contents, the bag contains the same weight of chips, only a far greater number of smaller chips or fragments. Now consider the analytical complexity if each original chip consisted of a different composition. The fragments become increasingly difficult to measure and identify as they get smaller. Nanoplastics may eventually form. Until recently, small plastic fibers were overlooked due to their narrow dimensions and difficulty resolving them from naturally-occurring fibrous materials. As a consequence, these materials often escaped collection. Indeed, a 2014 paper suggested that 99% of plastic that has entered the world's oceans was "missing" (Cózar et al. 2014). Much of this missing plastic was likely present in less obvious compartments, such as sediments or remote areas, was obscured by biofouling, or consumed by organisms. A 2015 study (Enders et al. 2015) observed that smaller microplastics (~10 μm) were actually the most abundant in Atlantic Ocean water column samples. This finding will become even more obvious as plastics continue to fragment and our abilities to detect microplastics improve.

Deficiencies in commonly employed methods for the collection of microplastics have confounded our ability to understand and appreciate their true distribution. For example, researchers have often used relatively coarse (e.g. 300 μm) nets to strain plastics from the water, missing smaller microplastics. In addition, microplastics may coalesce with other materials such as colloidal substances or marine snow, resulting in more complex aggregates (Zhao et al. 2018). This may alter their physical behavior or lead to ingestion by organisms and vulnerability to UV or microbial degradation. The amount of plastic present in a given sample is important, but so are the size range, shape and polymeric composition of the particles. At present, the analytical capabilities we apply are inadequate to determine the abundance and multi-dimensional nature of plastics in the environment. This characterization is an essential first step to fully understand the environmental impact of plastics, and then design solutions.

Finding microplastics in complex environmental samples is an extremely difficult analytical challenge, akin to finding a needle in a haystack. After collection from the field, analysis generally starts with pre-purification of the sample matrix to remove interfering materials. As most polymers are less dense than inorganic particles such as sand, gravity

Table 1 Top 20 items by number reported by volunteers during 20 years of the International Coastal Cleanup Program in Virginia (Register et al. 2016)

Rank	Item
1	Cigarettes/Cigarette Filters
2	Plastic Beverage Bottles
3	Food Wrappers/Containers
4	Plastic Bags
5	Beverage Cans
6	Glass Beverage Bottles
7	Plastic Cups, Plates, Forks, Knives, Spoons
8	Plastic Caps, Lids
9	Plastic Straws, Stirrers
10	Building Materials
11	Tires
12	Balloons
13	Clothing, Shoes
14	Tobacco Packaging/Wrappers
15	Toys
16	Fishing Line
17	Rope
18	Cigar Tips
19	Bait Containers/Packaging
20	Cigarette Lighters

Red indicates smoking-related, blue food and beverage. Reproduced with permission from Kathleen Register

sedimentation of components is often first applied. Since plastics are chemically more stable than many natural biogenic materials, procedures then incorporate some form of oxidation of labile material, using hydrogen peroxide or other caustic chemicals. Many researchers then use visible light microscopy on a subsample of the purified sample and manually count what they believe are microplastic particles. However, this approach is vulnerable to false identifications, as natural polymers (e.g. cellulose and chitin, mentioned previously) and other particulates may still remain and interfere. Common polymers, being largely organic materials, absorb infrared (IR) light in characteristic patterns as a function of their structure, producing signatures that can be matched to those of known plastic standards (Fig. 8; Song et al. 2015). Application of IR micro-spectroscopy, i.e. using a hybrid microscope/IR spectrometer, to identify polymer composition of individual microplastics within a viewing field is increasing. Some systems can automatically scan a viewing field, generate spectra for the particles therein and computer-match those to libraries of polymer standards. However, the reliability of all methods is lower for small or irregular particles and thin fibers. Weathering of the polymer chemically modifies it and will alter the spectrum, complicating identification. Plastics that are composites of more than one polymer type will also complicate identification. Raman scattering micro-spectroscopy has also been applied, especially for microplastics smaller than 20 μm (Schymanski et al. 2018). Unfortunately, IR- and Raman micro-spectroscopy systems capable of automated scanning are complex and expensive, typically exceeding $200,000. This limits their wide availability. Researchers are still refining analytical procedures for determining microplastics in environmental samples.

Behavior and Properties of Plastics Another common misconception is that all plastics float on the water's surface. While the densities of common polymers such as PE and PP (commonly used in plastic films, bags, wrappers and containers) are less than water, many others (e.g. acrylonitrile butadiene styrene (ABS), PVC, nylon, polyester and polycarbonate (PC) are greater and sink. In addition, while the specific gravities of PS and PU are greater than 1.0, foam versions contain air pockets that provide buoyancy. If void spaces become saturated with water, these materials will eventually sink. Sunken or sediment entrained plastics are less apparent to surface-oriented human observers; thus, their contributions are underestimated. Water surface tension (affecting small particles, as it does for insects such as water striders, *Gerridae*) will alter expectations. That is, denser plastic may reside at the surface for a period, but later sink. Indeed, only recently have scientists begun to sample for microplastics in marine and freshwater sediments. The early results are astounding. Macro- and microplastics have been commonly detected, including in the deepest zones of the ocean such as the Mariana Trench (11 km deep). In 2019, Jamieson et al. reported on the discovery of microplastics and fibers in the digestive tracks of 72% of amphipods (small shrimp-like crustaceans) living in all six abyssal Pacific Ocean trenches they sampled. This, in combination with recent studies in the Arctic (Cózar et al. 2017), Antarctic (Waller et al. 2017) and isolated oceanic islands (Barnes et al. 2018), proves that waste plastics have now invaded the most remote areas of the planet.

In shallow, biologically productive areas such as Chesapeake Bay and its tributaries, exposed solid surfaces such as plastic debris, are rapidly covered by an organic film and colonized by fouling organisms. These initially form microbial biofilms, which are then colonized by larger organisms (Fig. 9). The biofilms increase the density of particles, causing them to sink, akin to the plastic/organic matter aggregates described above. The Chesapeake Bay is an estuarine system. Hence, circulation is complex and a function of bathymetry, seasonality, tidal action and freshwater input. In general, there is an upstream movement of saline water at depth and a downstream movement of overlying, fresher surface water (Guo and Valle-Levinson 2007). Surface-associated plastics are likely carried downstream in tributaries and the Bay. However, denser plastics and those extensively biofouled sink, and may be moved upstream

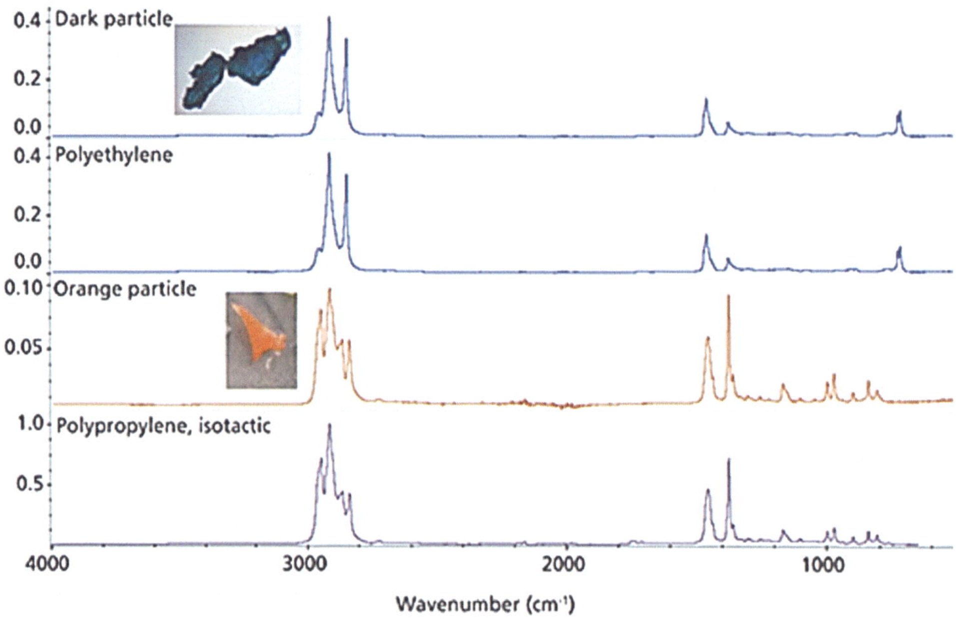

Fig. 8 Infrared absorption spectra of two microplastics matched to spectra of polyethylene and isotactic polypropylene standards. Reprinted with permission from Spectroscopy Online, Bradley et al. (2017)

Fig. 9 Biofilms that formed on three polyethylene plates that were submerged in the York River, VA for 30 days near the sediment, midwater column and in the intertidal zone (from left to right). Photo printed with permission from Corinne Audemond, Virginia Institute of Marine Science

before sedimentation. However, detailed studies on their fate in estuaries such as the Bay have not yet been conducted.

Microbial degradation of plastics varies as a function of composition, but is generally modest. Some researchers have reported pitting of surfaces of plastic debris collected from the mid-Atlantic Ocean gyre (Zettler et al. 2013). This suggests some microbial species can degrade the polymer to a limited extent. Different microbial communities have also been observed colonizing plastic surfaces, compared to those freely living in the water column or on natural substrates such as wood (McCormick et al. 2014; Miao et al. 2019). In a few cases, members of the pathogenic genus *Vibrio* appeared to be over-represented on plastic debris. Some *Vibrio* species/strains may infect open wounds, causing food poisoning or even cholera. Finally, plastic debris/microplastics, along with biological "hitchhikers" may be transported long distances (Keswani et al. 2016). For example, following the catastrophic 2011 Japanese tsunami, considerable foreign debris accumulated on Northwest U.S. beaches. A number of other catastrophic events, including hurricanes, have transported substantial amounts of marine debris to distant shores. These movements may have implications in regard to disease transmission and introduction of non-native species into new habitats—an extremely damaging form of biological pollution with potential profound effects for coastal food systems.

Another concern relates to possible broader ecosystem effects of plastics in the environment. Large amounts of plastics may impact processes such as microbially-mediated nutrient cycling. If such bottom-up effects do indeed occur, ramifications on the health of the Chesapeake Bay and other aquatic systems may be severe.

Occurrence and Pathways of Plastics to the Chesapeake Bay PE, PP, PS, PVC and PET polymers are commonly used for food and beverage containers, as well as films and sheeting. Consequently, they are major components of the solid waste generated by our single-use society. Such debris on the water surface and shorelines of the Chesapeake Bay are obvious, even to casual observers. It can arise directly by losses of fishing gear, dumping or littering. Yet, indirect routes (i.e. initial release to land or via wastewater treatment plants, followed by transfer to surface waters) may be even more substantial.

While beach and shoreline surveys have been conducted, these capture predominantly large, buoyant materials—not microplastics. Beach surveys in the Commonwealth of Virginia have shown that a majority of this debris is derived from food and beverage uses, with smoking-related products also making a sizeable contribution (Table 1). Figure 10 shows plastic debris on a section of beach near Portsmouth, VA. A variety of plastics are visible, including intact bottles, caps, as well and weathered, discolored PS and PU fragments.

In the only currently published scientific study of microplastics to date conducted on Chesapeake Bay tributaries, Yonkos et al. (2014) evaluated the amounts of relatively large microplastics (300–5000 μm) in surface waters of four estuarine tributaries. Water samples were collected between July and December 2011. Sites included those exhibiting a range of human population in the watershed, as well as land-use characteristics (degree of urbanization, agriculture and forests). The scientists observed microplastics in all but one water sample, with levels varying 1000-fold. Concentrations correlated with human population densities and level of urban/suburban development within the watersheds sampled. The authors indicated that levels peaked following rain events and increased surface runoff, supporting the hypothesis that land-based sources were major contributors.

Additionally, storm drains contribute to plastic pollution in the Chesapeake Bay following rainfall events. Storm drains are intended to remove excess water from the streets in extreme rain to prevent flooding. In the Bay region, these drains typically lead straight to the local river or tributary, ultimately reaching the Bay. The rain water will carry with it any plastic pollution that may have been littered on the ground, or that may be caught by the rain or flood waters in spilled garbage bins. As such, storm drains present an

Fig. 10 Plastic debris on the shoreline near the mouth of the Chesapeake Bay (Portsmouth, VA.). Plastic bottles were the most frequently collected item during the associated Clean the Bay Day effort. Fragments of PS and PU are also apparent in the photo. Photo credit: Pamela K. Spaugy (U.S. Army Corp. of Engineers). Creative Commons Attribution 2.0 Generic license

opportunity for large pulses of debris to enter the Bay from many locations around the greater watershed area.

Another source of plastic debris to the Bay and tributary waters may be "plasticulture" (plastic mulching). Here, the soil around the plants is covered with plastic sheeting, commonly LDPE. This serves as a weed block, temperature moderator and reduces water loss. It is used in eggplant, strawberries, green peppers, and cucumber farming. Tomato plasticulture is especially common on the Eastern Shore of Virginia and Maryland. Globally, plasticulture has increased dramatically as it enhances crop growth rate and yield, and lowers labor costs. However, around the Bay concern over increased runoff of applied pesticides into adjacent waterways where plasticulture is used has been previously voiced (Dietrich and Gallagher 2002). In addition, over time the plastic deteriorates, fragments and must be replaced. This leads to an accumulation of plastic debris in these soils (Brodhagen et al. 2017) and likely associated waterways.

Household activities occurring in the Chesapeake Bay watershed also contribute to microplastic pollution, notably from washing of synthetic textiles and clothes. Released fibers (PE, acrylic and PP or other polymers) will first enter wastewater treatment plants and septic systems. The use of laundry detergents during machine-washing has been reported to increase the release of fibers tenfold (De Falco et al. 2018). The average number of fibers generated from a 5 kg wash load of polyester fabrics has been estimated to exceed 6 million. Use of a fabric softener reportedly reduces fiber release by about 35%. Ironically, clothing composed of synthetic fibers are particularly popular with those who frequent the outdoor environment due to their insulative, water-resisting or water wicking characteristics. Yet, many do not make the connection between their clothing and plastic pollution, when at home or on the water. In addition, Schreder and La Guardia (2014) recently observed that laundry wastewater contained high levels of flame retardant polymer additives.

The use of intentionally manufactured microbeads is an additional source of microplastics. Their release pathway is similar to that of textile fibers, i.e. initially entering wastewater treatment plants and domestic septic systems. Microbeads are typically between 10 and 500 μm. These may be round or irregular in shape. For comparison, the diameter of a human hair is about 80 μm. Microbeads have been used in rinse-off personal care products to exfoliate or cleanse the skin, as well as in toothpaste. The U.S. Microbead-Free Waters Act of 2015 ended these uses, but did not affect non-rinse off applications, such as some cosmetics (McDevitt et al. 2017). The Act also did not ban use of microbeads in applications such as plastic-media surface cleaning (akin to sand-blasting), surface treatments such as paints and coatings, or in domestic and industrial cleaners and oil drilling fluids.

Finally, an additional route of plastic pollution is via atmospheric deposition. Plastics are lightweight and therefore, may become airborne and transported into surface waters and deposited in watersheds. Microplastics and microfibers are essentially dust, and are more easily transported than macro-debris. Measurements of associated atmospheric plastic transport and deposition are rare. However, Dris et al. (2016) calculated substantial atmospheric deposition of fibers, amounting to 3–10 metric tons each year on 2500 km^2 (corresponding to the size of the greater Paris metropolitan area). They estimated that 29% of all fibers observed were synthetic in origin. Small fibers were more abundant than larger ones. Deposition was higher during rain events. Most were fibers 7–15 μm in diameter. Deposition was greater at urban than suburban sites. The substantial human population surrounding the Chesapeake and more distant states in the airshed, may thus contribute substantial amounts of microplastics to the Bay via the airborne deposition route.

Wastewater Wastewaters from homes, institutions, industries and stormwater runoff from urban areas are increasingly being routed to advanced treatment facilities. The treated effluents are eventually discharged to nearby surface waters. Raw wastewaters may bypass normal treatment during large rainfall events and enter receiving waters untreated. Large human populations adjoin the Chesapeake Bay and its tributaries. These include the cities of Baltimore, Annapolis, Washington (DC), Richmond, Norfolk and Virginia Beach, as well as many associated suburbs and small communities. Some municipalities have antiquated combined sewage overflows (CSOs). During storms, these divert a portion of untreated wastewater, storm drain water and associated plastic debris directly to local water bodies, in order to avoid flooding the treatment plants.

Wastewater treatment plants were not designed to remove microplastics, and their effectiveness at doing so is only now being investigated. Fate is influenced by the specific gravity of the plastics and any biofilms that form on their surfaces, which may alter their overall buoyancy. Modern treatment involves an initial crude screening to eliminate macro-debris and settling of influent to separate dense material such as sand and grit. These materials are normally disposed of as solid waste and sent to a landfill. The remaining wastewater is passed to *primary treatment*, which consists of additional settling and removal of high organic solids ("sludge"). Surface skimming to remove floating solids and films is typically applied here. The bulk of plastics that escaped removal in the initial step may be captured by this settling and skimming.

Secondary treatment subjects the effluent from the primary step to microbially-mediated, activated sludge treatment designed to oxidize readily degraded organic matter. This has minimal impact on recalcitrant plastics. To improve solids removal, a chemical flocculant is often added, followed by additional surface skimming and sludge removal steps. Ironically, this flocculent is often a water-soluble polymer that may be resistant to subsequent breakdown.

Solids removed from the wastewater during *primary* and *secondary* treatment may be combined, stabilized to reduce odor, volume and pathogens, and then land-applied as a soil conditioner/fertilizer (sludge, or "biosolids"). Stabilization of this sludge to reduce odor and volume may include anaerobic degradation, composting, heat treatment/drying or lime addition. A new technique, thermal hydrolysis, has been instituted by the Blue Plains wastewater facility near Washington, DC. There the sludge, which is mostly water, is heated to 130 °C–180 °C under pressure. The fate of plastics under these conditions has not been studied. Biosolid applications occur most commonly on agricultural fields and represents about 60% of the total wastewater sludge generated in the U.S. Land application may reintroduce a substantial percentage of the plastics removed from the wastewater into the terrestrial environment (Nizzetto et al. 2016). Indeed, Zubris and Richards (2005) suggested that plastic fibers may serve as a tracer of land-applied sludge in soils and that their distribution therein tracked preferential water flow patterns, i.e. voids in the soil created by worm tunneling and other processes. The microplastics and fibers entering soils from land-applied biosolids may eventually reach surface waters.

Tertiary treatment (or advanced treatment) is increasingly applied to the secondarily treated wastewater to remove residual nutrients or for other purposes. Thereafter, the effluent may be chlorinated or UV-treated to kill pathogens, and discharged to surface waters. Removal efficiency of microplastics from the final effluent released to surface waters may exceed 90% of the original influent loads (Raju et al. 2018). Nonetheless, the large volumes of water discharged contribute substantial amounts of microplastics to receiving waters. Some Chesapeake Bay-associated treatment facilities (e.g. Hampton Roads Sanitation District) currently discharging to Chesapeake Bay tributaries have initiated deep aquifer injection of treated wastewaters under a program entitled SWIFT: Sustainable Water Initiative for Tomorrow (HRSD 2019). They are implementing additional steps above and beyond typical treatment protocols to remove contaminants in the effluent injected. The fate of microplastics in aquifers is unknown. However, aquifer flow rates are low, and microplastics therein would be expected to follow available preferential flow paths, as seen in soils. Degradation from UV, abrasion and microbial processes would be expected to be nil.

Leaching of Additives from Plastics Chemical additive concentrations in plastics are often high, in some cases representing several percent of the entire product's weight. Plastic debris is thus a substantial potential source of chemical pollutants. Relocation of the debris will relocate any contained additives. Chemical concentrations toxic to organisms (sometimes in the parts per million or less) are orders of magnitude less than what is contained within some polymers. The extent of bioavailability or release of these additives from the plastic to the environment is a critical variable controlling toxic potential. Different plastic products contain different additive mixtures, depending on their intended use. Highly engineered plastics, such as those used in electronics or home furnishings, may contain substantial loads of potentially toxic additives, such as flame retardants or phthalates (Hermabessiere et al. 2017). Historically, it was believed that plastics remain intact and additives were retained therein indefinitely. However, we now know that additives are released to varying degrees (Hahladakis et al. 2018) to the environment. Once released, polymer additives are considered pollutants, and likely cannot be traced back to the original polymer from which they leached.

Plastics eventually fragment into microplastics, the rate depending on environmental conditions and polymer composition. As particles get smaller, their surface area to volume ratio increases. This enhances their contact and interaction with the surrounding environment. It also reduces the distance within the plastic particle that additives must travel to reach its surface and escape. Some additives, e.g. phosphate flame retardants, exhibit substantial (g/L) water solubilities, and therefore readily leach from the plastic. Additives with lower water solubilities (e.g. brominated flame retardants) exhibit much slower migration from the polymer. However, such hydrophobic (literally water-fearing) chemicals also tend to bioaccumulate in lipid-rich tissues of exposed aquatic organisms and humans that may consume them (Hale et al. 2001). If these chemicals are resistant to degradation their levels may magnify up the food chain. Warmer water temperatures, within the range encountered in the Chesapeake Bay, modestly enhance the release of some hydrophobic additives from plastics. The presence of dissolved organic matter (e.g. humic acids found in Bay waters and surfactants that constitute organismal digestive fluids) enhances hydrophobic additive release from plastics (Fig. 11). Dissolved organic matter concentrations are especially high in productive waters, marshes and the interstitial spaces of sediments, all hallmarks of the Chesapeake Bay. Hence, release of additives from plastics to the surrounding

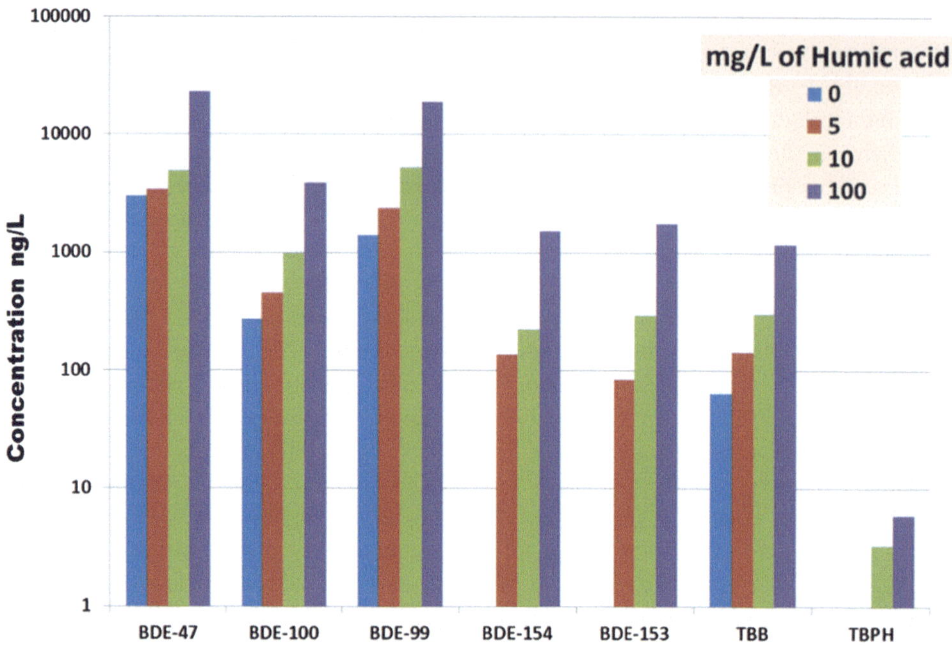

Fig. 11 In a lab experiment at VIMS, increasing humic acid concentrations (from 0 to 100 mg/L) in water greatly enhanced leaching of several brominated flame retardant additives from microplastics produced from commercial polyurethane foam. Note that the *y*-axis is on a log scale. BDE compounds are constituents of the pentaBDE mixture, now banned in production of new products. TBB and TBPH are constituents of a replacement mixture flame retardant, in current use. Figure printed with permission from Robert C. Hale, Virginia Institute of Marine Science

water in such circumstances will be greater. Dilution factors within sediments will also be lower than in open water, resulting in greater chemical exposure of organisms living therein. Thus, recent findings of substantial amounts of plastics in sediments are worrisome (Kazmiruk et al. 2018).

Sorption of Pollutants to Plastics Most plastic surfaces readily sorb hydrophobic contaminants such as chlorinated pesticides, polychlorinated biphenyls (PCBs), brominated flame retardants and polycyclic aromatic hydrocarbons (PAHs) from the water. Indeed, scientists have used polymer strips for decades to monitor concentrations of some water-borne pollutants. Researchers (Browne et al. 2013) have proposed that ingestion by biota of plastics with surface-sorbed pollutants could represent a pathway for increased contaminant exposure. Wardrop et al. (2016) demonstrated this in the laboratory with microbeads, PBDEs and fish. Others have suggested that exposure to non-contaminated plastics could even reduce body burdens by tying up pollutants from the organism onto ingested plastics (Herzke et al. 2016). Isolation of contaminated plastics away from vulnerable organisms, e.g. in sediments, might also reduce pollutant exposure. Biofilm formation on plastics in the water (effectively coating them with more polar material, see Fig. 9) would likely alter the sorptive capacity of the plastic for nonpolar pollutants under real world conditions. Additional investigation to fully understand likely outcomes is needed.

Biological Consequences of Plastics and Microplastics in the Environment We lack adequate understanding of the toxicological consequences of organismal contact with plastics/microplastics. Beyond esthetic concerns associated with careless discard, plastics were long deemed inert and harmless. Ironically, this same flawed logic was applied in the past to some very harmful chemical pollutants such as PCBs and perfluorinated chemicals (e.g. PFOS: perfluorooctane sulfonate used in fabric protectors such as Scotchgard™).

Unfortunately, data on plastics/microplastic ingestion in and consequences to organisms within the Chesapeake Bay and contiguous waters are rare. Therefore, some findings associated with representative organisms similar to those in the Bay, but performed elsewhere, are included in this section.

Entanglement Originally, humans living adjacent to the Chesapeake Bay used natural woven materials to construct fishing nets (Fig. 12). Today, we rely heavily on synthetic polymers for fishing nets, lines and associated gear. These are stronger and more weather-resistant than most natural polymers, which weaken after long-duration exposure to the

Fig. 12 A printed, hand-colored engraving from ca. 1590, titled: *Incolarum Virginiae piscandi ratio (The Method of Fishing of the Inhabitants of Virginia)*. Reprinted with permission from The Mariners' Museum and Park in Newport News, Virginia. Original Author: Theodor de Bry after John White. Created: ca. 1590

elements. In contrast, fishing gear, plastic bottles and bags, diapers, and other products made with synthetic polymers may persist for hundreds of years, depending on the conditions (Fig. 13). This stability is valuable while the material is in use, but is hazardous if improperly discarded into the environment, where it may compromise animals and ecosystems.

Fishing nets may be towed through the water (trawling) or set in fixed locations (e.g. gill nets, pound nets, fike nets). Cast and drift nets are also used. While fishers seek certain species, most gear lacks specificity. Hence, large numbers of unwanted organisms are caught. Even if this by-catch is returned to the water promptly, individuals often do not survive. Especially conspicuous are marine mammals, birds and turtles. These organisms may be drowned or injured if entangled. (An example of bird entanglement from the Chesapeake Bay is provided in Box 2.) So-called *excluder* devices are incorporated in some nets to allow non-target species to escape. Regardless, much fishing gear is lost, but its ability to "ghost fish" continues. Globally over 6 million tons of gear are lost each year (Wilcox et al. 2015a). Not only does this material trap mobile organisms, it also snares on

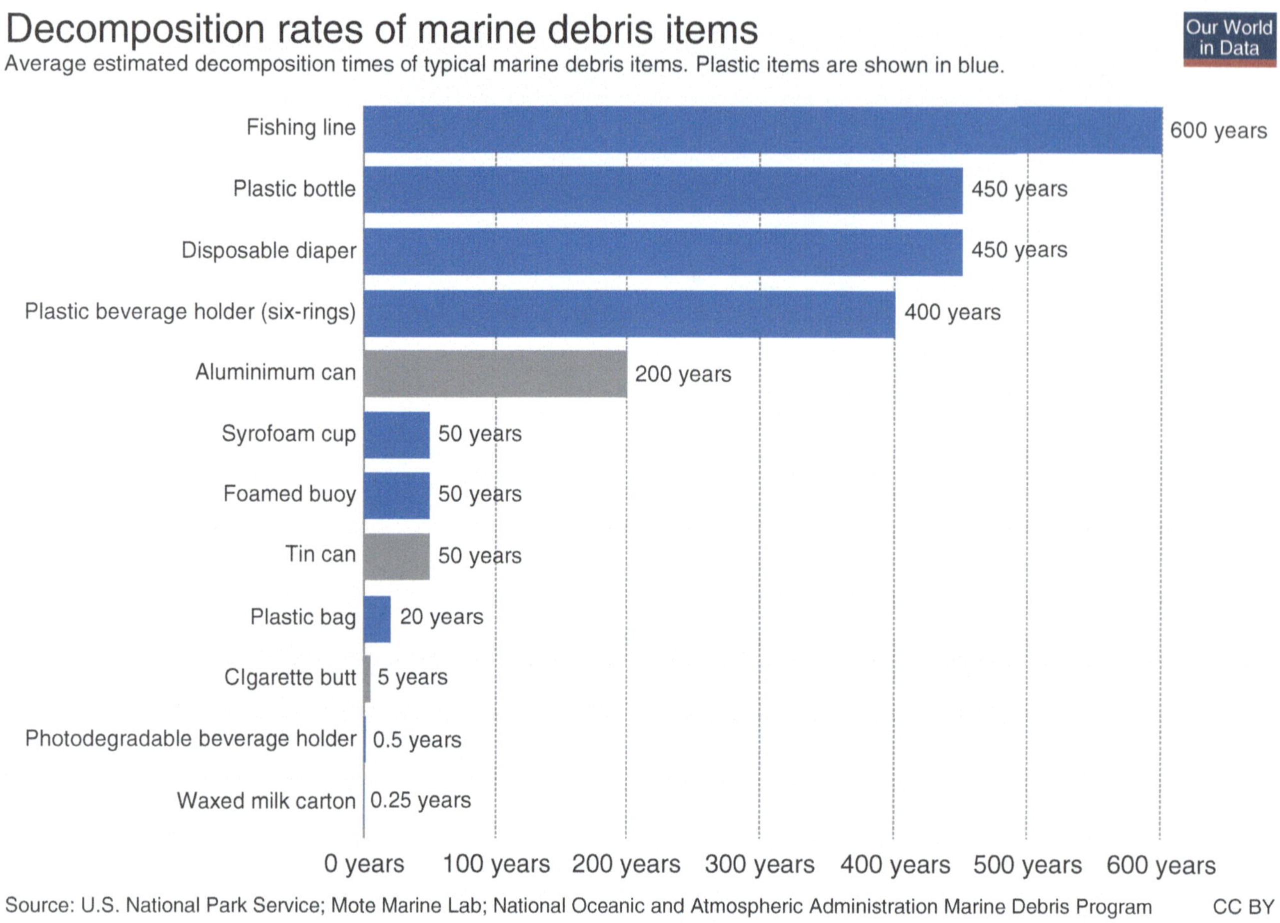

Fig. 13 The lifetimes of different types of plastics. Reprinted with permission from *Our World In Data* (https://ourworldindata.org/grapher/decomposition-rates-marine-debris)

submerged structures such as reefs that serve as oases of aquatic life. Thereon, damage may be physical (as discussed in Box 1) or less obvious. For example, a 2018 study reported disease incidence on Asian-Pacific reefs increased from 4 to 89% for corals in contact with plastic debris (Lamb et al. 2018).

In the Chesapeake Bay, the blue crab is an iconic species and supports a major fishery. Metal wire mesh traps (called crab pots) are common, although plastic and plastic-coated metal traps are also used. Pots are commonly tethered to PS floats, typically with synthetic polymer lines. If a line fails or a float detaches, the pot is generally lost. Between 72,000 and 120,000 pots are lost annually (Bilkovic et al. 2016) in Bay waters. These derelict crab pots kill about 3.3 million crabs per year, or about 20 crabs per lost pot. Diamond back terrapin turtles, a variety of finfish (e.g. toadfish, black sea bass, croaker, eels, white perch and catfish), diving ducks and muskrats also die in lost pots. The carcasses of the trapped animals then act as bait for others. Major efforts to locate and remove derelict pots have been undertaken. This saves the lives of organisms and increases the available harvest—a positive outcome for fishermen and the Bay. In addition, pots are equipped with an operable or removable panel to allow extraction of trapped animals. Scientists at the Virginia Institute of Marine Science developed a biodegradable escape panel made of polyhydroxyalkanoate (PHA). If the pot is lost, this PHA panel will degrade over time, providing an avenue of escape (VIMS 2019). Such degradable panels can be installed in other fishing gear such as lobster traps.

Box 2 A Case Study: Osprey Entanglement

Ospreys (*Pandion haliaetus*) are a symbol and health indicator of Chesapeake Bay wildlife. The Bay's osprey population, similar to those of other raptors such as the bald eagle, has rebounded from historical lows caused by exposure to persistent pollutants, e.g. the pesticide DDT. These fish hawks are often seen soaring overhead, carrying prey or building materials back to their nests. The wing span of adult

(continued)

Box 2 (continued)
birds may reach 2 m. The Chesapeake watershed is home to about 25% of all North American ospreys. Some local landowners build nesting platforms to attract breeding pairs, which typically mate for life. The Virginia Institute of Marine Science maintains a nesting platform in the York River near its Gloucester Point campus. Staff installed an "Osprey-cam" so interested persons can watch a live video stream of the nest. In 2018, a pair successfully nested and began raising two chicks. The attentive parents maintained the nest, bringing food regularly. However, one day a parent returned to the nest with a section of discarded monofilament fishing line. According to Bryan Watts, Director of the William & Mary Center for Wildlife Biology, ospreys often retrieve discarded plastics and return these to their nests due to their physical similarities with natural materials. These have included items as large as flip-flop sandals. A 7–8 week-old fledgling became entangled in the fishing line. Plastic fishing line is designed to be extremely strong and long-lasting, increasing its risk to wildlife. While the line was carefully removed by rescuers, the damage inflicted on the fledgling's wing was too severe (Fig. 14). Regrettably, the young bird had to be euthanized. This illustrates just one avenue for food system plastics to impact wildlife and ecosystems.

Fig. 14 In the photo to the left, Dr. Bryan Watts (Director of the Center for Conservation Biology, right) holds an osprey chick while Jim Goins (Virginia Institute of Marine Science Marine Safety Officer, left) removes the monofilament line that was wrapped around its wing. The young bird is pictured in the photo to the right, after which it regrettably had to be euthanized due to the damage inflicted by carelessly discarded plastic fishing line. Photos reprinted with permission from Dave Malmquist, News & Media Director at the Virginia Institute of Marine Science

Ingestion Plastic macro-debris and microplastics may be ingested during feeding. Particles/fibers may lodge in the digestive track or in respiratory surfaces of organisms of all sizes: from zooplankton and filter feeders (e.g. oysters) to large cetaceans (e.g. whale, dolphins and porpoises). The previously mentioned microplastic aggregate study (Zhao et al. 2018) also observed the presence of plastics in filter-feeding mussel feces and pseudo-feces, evidence of their ingestion. Such uptake of plastic certainly applies to many species that can be found in the Chesapeake Bay.

Sea and shore birds are common inhabitants of the Chesapeake Bay. While no Bay-based studies of plastics in local birds are yet available, seabirds are the most well-studied globally in regard to plastic ingestion. The plight of albatrosses on the remote island of Midway, in the middle of the Pacific Ocean, was recently documented in the film "Albatross" (https://www.albatrossthefilm.com/) by Chris Jordan, which is offered as a free public artwork.

Sea birds exhibit a number of feeding strategies that make them vulnerable to microplastic ingestion. These include foraging of debris on beaches, sediments and the water surface. It has been predicted by 2050 that 99% of all seabird species will regularly ingest plastics (Wilcox et al. 2015b). A recent study found that the biological growth on the surface of plastic may even release dimethyl sulfide (DMS), which is a key "info-chemical" attracting foraging birds to their prey (Savoca et al. 2016). Provencher et al. (2018) found that most of the plastics in the gut of Northern fulmars collected from the Labrador Sea were fragments, while those in fecal precursors (material collected from dissected specimens just before excretion) were fibers. They suggested that the fibers were not retained for long, but the fragments remained and likely fragmented further before being excreted. Birds are also vulnerable to entanglement, as described in Box 2.

Whales Whales are impressive and charismatic marine mammals. While normally inhabiting open water, some species are occasionally seen in the Chesapeake Bay. In 2014, a 15 m-long sei whale (*Balaenoptera borealis*) was observed in the Elizabeth River, a small tributary near the mouth of the Chesapeake Bay (Groc 2015). The species is endangered. Its appearance here was unusual as sei whales frequent deep water, feeding on small organisms such as krill and fish.

The whale was swimming erratically for a week in the area and appeared malnourished. It was later found dead in nearby St. Julien's Creek. An autopsy by the Virginia Aquarium Stranding Response Team revealed the presence of a broken plastic compact disk case in its stomach. The plastic case had lacerated the whale's stomach lining, likely contributing to its death. The carcass also showed evidence of a boat strike.

Ingestion of plastic debris is commonly observed during autopsies of whale strandings. In 2018, a dead, 10-m sperm whale (*Physeter macrocephalus*) in Indonesia was found to contain 6 kg of plastic, including: 115 cups, 15 bags, bottles, flip-flops and a bag with 1000 pieces of string (Parker 2018b). In 2017 a pilot whale (genus *Globicephala*) was found to contain 8 kg of plastic, including 80 bags. Plastic has been documented in 56% of necropsied cetacean species. Occurrences of ingestion were estimated to be as high as 31% in some populations (Baulch and Perry 2014). Impacts can range from satiation due to the presence of the plastic, blockage of the digestive track, or associated damage of the tissue lining.

Sea Turtles Loggerhead turtles (*Caretta caretta*) are commonly observed in the Chesapeake Bay between May and November. This species is designated as *threatened*. They are migratory, some traveling over 3000 miles in a lifetime. The turtles may reach 150 kg and live up to 60 years. The species feeds on crustaceans, sponges, jellyfish and sea grasses. Some marine debris (e.g. plastic bags) resemble these food items and are accidentally ingested. Turtles are also commonly trapped in lost nets and traps.

There are no studies yet regarding the impact of plastic debris on sea turtles in the Chesapeake Bay region. However, research on loggerheads in the southwest Indian Ocean indicated 51% exhibited macroplastics (>0.5 cm) in either their feces or digestive tracks. The frequency is likely much higher, as microplastics were not included. Incidence of plastics was higher in dead than surviving turtles. Examination of feces of organisms represents a nondestructive approach for evaluating plastic ingestion. In one study, a large percentage of turtles ingested plastic bottle caps (Hoarau et al. 2014). A Mediterranean study (Duncan et al. 2019) focusing on microplastics in the digestive tracks of seven different sea turtle species found that all specimens examined contained microplastics. Fibers were the most common form encountered; much of those consisted of elastomers, likely derived from tire wear. As discussed above, this might be a particular problem in the Chesapeake Bay due to the high density of vehicle traffic in surrounding areas.

Another point in the interaction between turtles and plastic debris does not involve ingestion, but reproduction. Females lay eggs on sand beaches and these are zones for some of the highest plastic debris accumulations. Carson et al. (2011) observed plastic contributions to mass of sediment as high as 30.2%, with an average of 3.3% in the upper 5 cm of Kamilo Beach (HI). These are extraordinary levels of plastic. Data on Chesapeake Bay region beaches are currently not available. However, plastic burdens will likely increase over time as production and discard rates increase. Further, Carson et al. (2011) also found that the presence of plastics in sediments (0–29.4% by weight) had an insulative effect, i.e. resulted in the substrate warming and cooling more slowly. Their 1.5% plastic treatment reduced the maximum temperature by 0.75 °C. The authors suggested this might affect the sex ratio of sea turtle hatchlings, citing data that an incubation temperature difference of 4 °C changed the hatching outcome from 100% males to 100% females in loggerhead sea turtles. For hawksbills (*Eretmochelys imbricata*) only a 1.8 °C temperature difference was needed to generate a similar outcome.

Oysters Decades of over-harvesting, habitat destruction, and disease has reduced the population of Chesapeake Bay oyster (*Crassostrea virginica*) to 3% of that believed to be witnessed by the first English colonists. As a result of conservation efforts and other factors, populations are increasing. As these organisms are prodigious filter-feeders, accumulation of particles is to be expected. Indeed, the species has been promoted for reducing contaminants and turbidity in the Chesapeake. No data are currently available regarding impacts of microplastics on Bay oysters; however, existing research on related species elsewhere indicates their uptake. Some researchers have evaluated the toxicological consequences of microplastic exposure to oysters under laboratory conditions. It must be noted that effects may vary depending on environmental factors, particle size, dosage and shape. One recent study (Sussarellu et al. 2016) examined the consequences of exposure to small PS microspheres on the physiology of the related Pacific oyster (*Crassostrea gigas*) in-laboratory. Adult oysters were exposed for 2 months during their reproductive cycle. Oysters preferentially ingested 6-μm over 2-μm diameter particles. Consumption of algae and absorption efficiency were significantly higher in exposed oysters. Exposed oysters showed decreased oocyte numbers, diameters and sperm motility, suggesting physiological effects. The larval development of offspring from microplastic-exposed oysters was lower than unexposed controls. Data suggested elevated maintenance costs in exposed oysters, hypothesized to be due to compromised energy uptake and a shift of energy allocation from reproduction to growth. Therefore, the consequences of microplastic pollution in the Chesapeake Bay to wild oyster populations should be further evaluated.

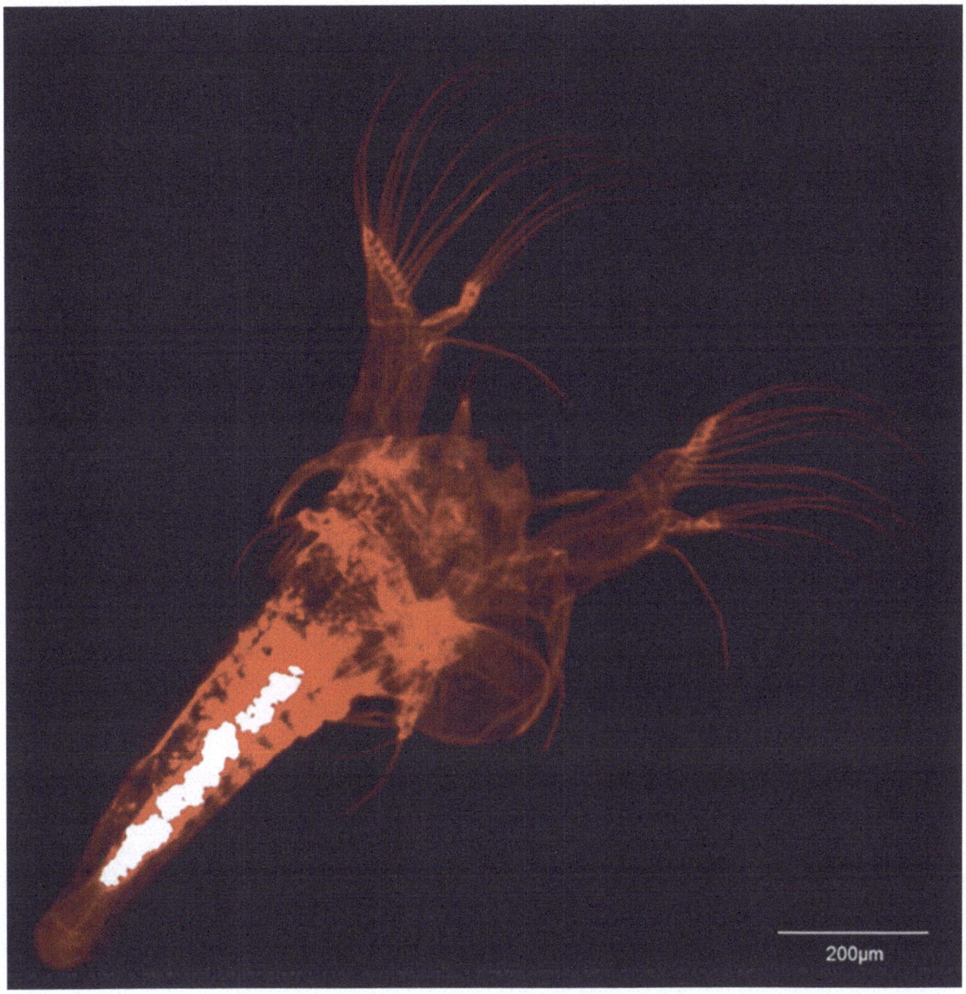

Fig. 15 Image of PU foam microplastics (fluorescing in bright white), ingested by brine shrimp nauplii, in red, in a controlled laboratory experiment. Imaged on an Olympus FluoView FV1200 laser scanning confocal microscope using GFP-uv and Alexa-Fluor 633 filters for imaging the host *Artemia* and microplastic, respectively. Photo credit to Hamish Small, VIMS

Zooplankton Due to the minute dimensions of microplastics, small organisms are potentially vulnerable to impacts associated with their uptake by lodging in their digestive tracks or respiratory surfaces. These organisms (e.g. holoplanktonic zooplankton and meroplanktonic larvae of species such as the blue crab) are abundant in the Bay and serve as food for many larger species. Plastics within the bodies of prey will subsequently be ingested by predators and transferred up the food chain, potentially to humans, with unknown consequences. Figure 15 depicts a brine shrimp nauplius that consumed PU foam microplastics in a laboratory experiment. The microplastics were cleared from the gut within 48 h after cessation of exposure.

While little work has been done to evaluate effects of microplastics in the Chesapeake Bay or on its biota, lab and field studies have been conducted elsewhere and those findings are likely transferable. Laboratory studies of zooplankton show reductions in feeding rate, reproduction, growth, development and lifespan (Botterell et al. 2019). Observations of interactions in the field generally have focused on occurrence of microplastics in small organisms, rather than demonstrated effects. The size, shape, color, density, extent of weathering, polymeric composition and biofouling of the microplastic may affect its likelihood of ingestion. This underlines the need for improved analytical methods.

Humans Considering the ingestion of microplastics across all trophic levels, from zooplankton to shellfish and finfish, plastic may find its way to humans through seafood. We, of course, typically remove the digestive tracts of finfish before consuming them. However, exceptions exist. Shellfish such as oysters, mussels and shrimp may be eaten whole or at least with intact digestive tracts. Studies have detected low levels of microplastics in the digestive tract of bivalves (Van Cauwenberghe and Janssen 2014). These authors reported higher burdens in farmed vs. wild shellfish. Smith et al. (2018) published a review of microplastics in seafood and potential human health consequences.

Exposure to microplastics is not just limited to seafood. They have also been detected in a variety of food items, including honey, sea salt and beer. They have also been observed in drinking water (Koelmans et al. 2019), which is particularly problematic as we consume high volumes to survive. The water source plays a role in determining the extent of contamination. For example, much "pure" drinking water is stored in plastic bottles, so shedding of microplastics from the container and closure/cap may occur. Leaching of additives is also likely, especially as the storage temperature and duration increases. Beverages that contain higher levels of dissolved organic matter are also more likely to leach additives from the container (see previous discussion of leaching). Collected rainfall will scavenge suspended microplastics, particularly fibers, from the atmosphere. If surface waters are used as a source, then water burdens of microplastics will reflect those of the source, e.g. lakes and rivers. Filtration of water destined for human consumption is often done. However, the filters themselves are often composed of plastics, such as woven fibers, which may shed during use. Not surprising, microplastics were recently detected in human feces in a Chinese study (Zhang et al. 2018), as well as those of domestic animals consumed by humans, such as chickens from Mexico (Lwanga et al. 2017). The human health ramifications of ingestion of microplastics are currently unknown.

Strategies for Reducing the Environmental Impacts of Plastics No single option will be adequate to solve the Chesapeake Bay plastics problem. The issue is global in scope. A successful strategy for reducing their effects depends upon an adequate knowledge of the extent of the problem. We are not yet there. However, in the interim we can learn from past experiences with another major pollution problem, oil spills. There, we learned that it is far better to first prevent spills. The next best strategy is to contain spills at their source than to wait and pursue piecemeal cleanups, e.g. on distant beaches. In some cases, spill cleanups may cause more ecological damage than the original release. We also must be cognizant of all the ecosystems at risk. For example, with oil spills use of chemical dispersants at sea might protect beaches, but increase exposure of open ocean organisms to oil, as well as toxic chemical dispersants. We should apply this "do no harm" framework to devising effective strategies to the plastics issue.

First, we should focus attention on preventing release of plastics to the environment. This is particularly true and poignant for single-use plastics, which are designed for convenience and not with the ultimate fate of the waste in mind. Imagine your bathtub is overflowing; your first decision to fix the problem would *not* be to wipe up the mess on the floor, but to first turn off the spout. The reduction of our use of plastic (particularly single-use plastic) is an appropriate first step in reducing marine debris. Likewise, collection approaches of debris and microplastics in the environment are inadequate, at best. Schemes to remove small plastic debris from open water, e.g. the gyres in the middle of the major ocean basins (The Ocean Cleanup 2019) are unlikely to be successful. They may actually harm organisms, such as the neuston that inhabit the surface waters targeted for cleanup (Helm 2019), and the funding for such projects could instead be redirected to stop plastic pollution at the source. While our society is currently dependent on (and addicted to) plastics, reduction in their use is a clear need.

In addition, green design of plastics should be pursued to reduce their negative ecological impact, and improve their recyclability and environmental degradability. The presence of toxic additives in many plastics is incompatible with just enhancing fragmentation and rapid polymer degradation, as these will increase additive release and bioavailability—and the minute dimensions of microplastics will put smaller organisms at greater risk. Also, the more a piece of plastic has fragmented or been dispersed into the environment, the more difficult it will be to physically remove. Investments in improving waste plastic management in developing countries would likely be far more impactful on a global and financial basis, as these nations have become the dominant debris contributors. Improvements in prevention strategies locally in Chesapeake Bay, its tributaries and the surrounding watershed would likewise bear the most fruit. Additional improvements would include reducing indirect transport routes, such as combined sewage overflows and storm drains.

In regards to the Chesapeake, the simplest remedy is for local residents to choose to decrease their use of single-use plastics, thus reducing the problem at its origin. Most of those single-use items emanate from the food system: beverage containers, plates, cups, bowls, utensils, grocery bags, packaging, and drinking straws. The European Union parliament voted overwhelmingly to outlaw many single-use plastics by 2021 (Yeginsu 2018). The jurisdictions in the Bay watershed could move to do the same. Municipalities and entire countries around the world have banned plastic bags shopping bags. These include the Chesapeake Bay's Washington, DC, San Francisco, CA, Seattle, WA, Boston, MA, Karnataka, India, Kenya, Chile, United Kingdom, Australia, and China (Strand and Kerr 2019). In the U.S., however, several states have issued preemptions on such bans, which essentially prohibits municipalities from enforcing a plastic bag ban, or other similar initiatives (Gibbens 2019). Financial incentives (or disincentives), such as fees on disposables, are a market tool and may be more appealing in a capitalist society.

Perhaps the most frequently encountered macroplastic on the Bay and its shores is a discarded single-use water bottle. This product essentially did not exist until the 1980s. In the 1970s people began to reduce consumption of soda to reduce their intake of sugar and artificial sweeteners. The bottlers turned to selling water instead, convincing consumers that it was better than what came out of their tap. However, tap water often is cleaner than bottled, with lower bacterial counts and concentrations of contaminating chemicals (Hu et al. 2011; Felton 2019). Around the world, one million plastic bottles are sold every minute, putting us on track for using a half a trillion bottles per year by 2020. The overall rate of plastic recycling was 9.1% in 2015, although rates for PET and HDPE containers were 29.9 and 30.5%, respectively (US EPA 2019). In 2017, U.S. consumers spent $16 billon for 13.7 billion gallons (51.8 billion/L) of bottled water (Karg 2018). As some of this water is sold in larger containers, we estimate that this resulted in the sale of about 60 billion 500 mL single-use bottles, or about 200 bottles per year per person in the US. That suggests the 18 million residents of the Chesapeake Bay Watershed may have consumed about 3.6 billion single-use plastic water bottles in 2017. If only 0.01% of those ended up in the Bay, it would mean 3.6 million plastic bottles entered the system in that year. It is estimated that 2% of U.S. waste is mismanaged, meaning this value is likely much higher (Jambeck et al. 2015).

It is not just the discarded bottles that taxes the environment. It takes 5.6–10.2 MJ/L of energy to bring a chilled water bottle to a consumer. This includes the energy cost of turning petroleum into PET, treatment of the water and transportation. By comparison, it takes only 0.005 MJ/L for tap water (Gleick and Cooley 2009). Bottled water also costs the consumer, as its price is 560–10,000 times higher than tap water (Bosque 2011). This disproportionality affects minority communities, as African Americans and Hispanics consume more bottled water than whites. African Americans and Latinos spent about $20 per month on bottle water, compared to $12 per month for non-Hispanic whites (Gorelick et al. 2011). In 2017 bottled water became the number one beverage consumed in US. The annual 6.2% increase in consumption is in part due to consumers turning away from sugared beverages (Karg 2018), but also traces to them abandoning tap water due to contaminant concerns. Per capita bottled water consumption nearly tripled between 1999 and 2017, going from 60 to 160 L (Fig. 16).

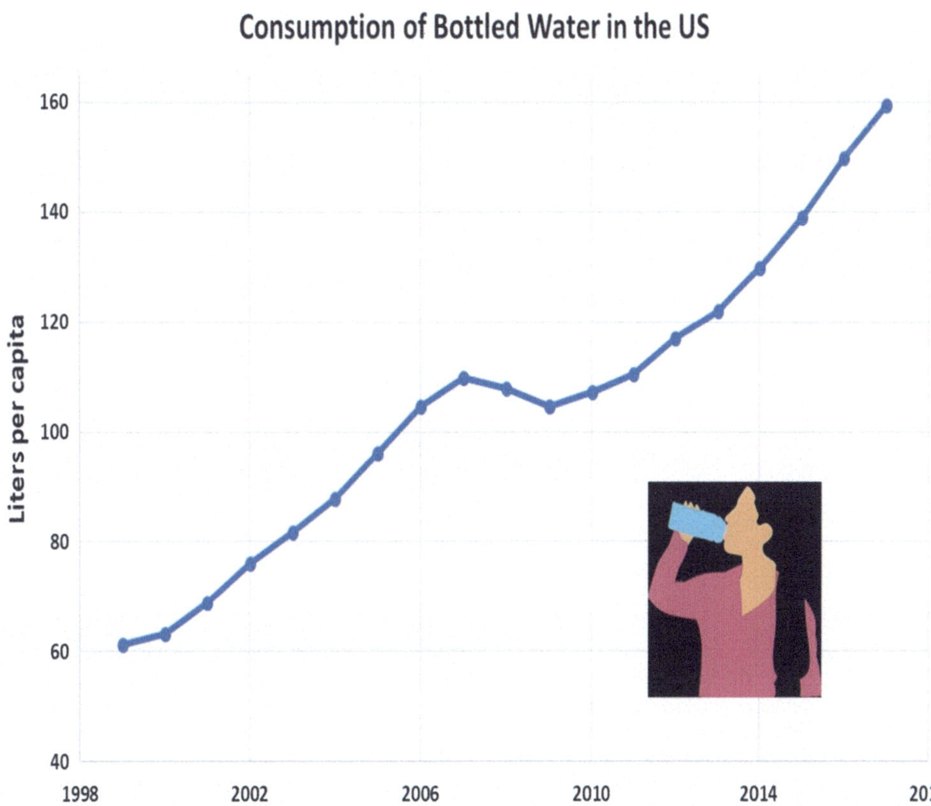

Fig. 16 The per capita consumption of bottled water in the US between 1999 and 2017. Note the slump in consumption during the Great Recession. Data from Statistica (2019)

No Deposit, No Return, Means No Sustainability From about 1885 to the early 1960s bottling companies sold soft drinks and beer in robust glass bottles meant for return and reuse. Consumers returned the bottles to vendors to claim a deposit that was about 40% of the product cost. Even if discarded in litter by a thoughtless consumer, enterprising low-income people and children would scour parks and other public venues to collect the valuable bottles.

That changed in the 1960s when bottlers sold their customers on the convenience of no deposit, no return products as a way to compete with metal cans. By 1970 the market moved to 33% canned, 27% in one-way bottles and 40% in returnable bottles. Returnable glass bottles left the market with the advent of plastic beverage containers in the 1980s. This created a littering problem and 11 states eventually instituted laws requiring deposit beverage containers (Friedel 2014). The states and territories with beverage container laws are: California, Connecticut, Hawaii, Iowa, Maine, Massachusetts, Michigan, New York, Oregon, Vermont, and Guam. Note that New York is the only Chesapeake Bay Watershed state to require deposits on beverage containers (NCSL 2019). The last state to institute a beverage container deposit law did so in 1986, as opposition from the industry is well funded. States with such laws see 35–83% reduction in littering (BottleBill 2019). It is ironic that the beverage industry that once managed a sustainable system of refillable, deposit-bearing bottles, now lobbies against laws requiring deposits.

Two solutions to plastics pollution from beverage containers could help resolve the problem for the Chesapeake Bay and beyond. First, people could simply quit using single-use plastic bottles, as these containers are completely unnecessary and are easily replaced by a permanent refillable bottle. Second, the jurisdictions of the Bay watershed could institute significant deposit requirements on all beverage containers. The easily recyclable PET could be used for more bottles or many other products, saving much wasted energy. Although this is a small portion of the Bay's plastic pollution problem, we can see that feasible solutions are possible. Many of these solutions are in the hand of the consumer.

References

American Chemical Association (1993) "The Bakelizer". Commemorative booklet produced by the National Historic Chemical Landmarks program of the American Chemical Society, 6 p. https://www.acs.org/content/dam/acsorg/education/whatischemistry/landmarks/bakelite/the-bakelizer-commemorative-booklet.pdf. Accessed 12 Mar 2019

Barnes D, Morley S, Bell J et al (2018) Marine plastics threaten giant Atlantic marine protected areas. Curr Biol 28(19):R1137–R1138. https://doi.org/10.1016/j.cub.2018.08.064

Baulch S, Perry C (2014) Evaluating the impacts of marine debris on cetaceans. Mar Pollut Bull 80:210–221

Bilkovic D, Slacum H Jr, Havens K et al (2016) Ecological and economic effects of derelict fishing gear in the Chesapeake Bay 2015/2016 final assessment report, 45 p. https://marinedebris.noaa.gov/sites/default/files/publications-files/DFG_Effects_Chesapeake_Bay_Final_Report_2016.pdf

Bosque T (2011) 15 Outrageous facts about the bottled water industry. In: Business Insider. https://www.businessinsider.com/facts-bottled-water-industry-2011-10. Accessed 8 Apr 2019

Botterell Z, Beaumont N, Dorrington T et al (2019) Bioavailability and effects of microplastics on marine zooplankton: a review. Environ Pollut 245:98–110

BottleBill (2019) BottleBill.org—litter studies in bottle bill states. In: Bottlebill.org. http://www.bottlebill.org/about/benefits/litter/bbstates.htm. Accessed 8 Apr 2019

Bradley M, Sukumaran S, Lowry S et al (2017) Tracking microplastics in the environment via FT-IR microscopy. Spectroscopy 32 (8):17–23

Brodhagen M, Goldberger J, Hayes D et al (2017) Policy considerations for limiting unintended residual plastic in agricultural soils. Environ Sci Pol 69:81–84

Brooks A, Wang S, Jambeck J (2018) The Chinese import ban and its impact on global plastic waste trade. Sci Adv 4(6). https://doi.org/10.1126/sciadv.aat0131

Browne M, Niven S, Galloway T et al (2013) Microplastic moves pollutants and additives to worms, reducing functions linked to health and biodiversity. Curr Biol 23(23):2388–2392

Carson H, Colbert S, Kaylor M et al (2011) Small plastic debris changes water movement and heat transfer through beach sediments. Mar Pollut Bull 62:1708–1713

Cózar A, Echevarría F, González-Gordillo J et al (2014) Plastic debris in the open ocean. Proc Natl Acad Sci USA 111(28):10239–10244

Cózar A, Marti E, Duarte C et al (2017) The Arctic Ocean as a dead end for floating plastics in the North Atlantic branch of the thermohaline circulation. Sci Adv 3(4):e1600582. https://doi.org/10.1126/sciadv.1600582

De Falco F, Gullo M, Gennaro G et al (2018) Evaluation of microplastic release caused by textile washing processes of synthetic fabrics. Environ Pollut 236:916–925

Dietrich A, Gallagher D (2002) Fate and environmental impact of pesticides in plastic mulch production runoff: field and laboratory studies. J Agric Food Chem 50(15):4409–4416

Dris R, Gasperi J, Saad M et al (2016) Synthetic fibers in atmospheric fallout: a source of microplastics in the environment? Mar Pollut Bull 104(1–2):290–293

Duncan E, Broderick A, Fuller W et al (2019) Microplastic ingestion ubiquitous in marine turtles. Glob Chang Biol 25(2):744–752

Enders K, Lenz R, Stedmon C et al (2015) Abundance, size and polymer composition of marine microplastics $\geq$10 μm in the Atlantic Ocean and their modelled vertical distribution. Mar Pollut Bull 100 (1):70–81

Felton R (2019) Arsenic in some bottled water brands at unsafe levels, consumer reports says. In: Consumer Reports. https://www.consumerreports.org/water-quality/arsenic-in-some-bottled-water-brands-at-unsafe-levels/. Accessed 21 Apr 2019

Florida Department of Environmental Protection (FDEP) website (2016) History and overview of the osborne reef waste tire removal project updated August 8, 2016. https://floridadep.gov/waste/permitting-compliance-assistance/content/osborne-reef-waste-tire-removal-project. Accessed 12 Mar 2019

Friedel R (2014) American bottles: the road to no return. Environ Hist 19:505–527. https://doi.org/10.1093/envhis/emu061

Geyer R, Jambeck J, Law K (2017) Production, use, and fate of all plastics ever made. Sci Adv 3(7):e1700782. https://doi.org/10.1126/sciadv.1700782

Gibbens S (2019) See the complicated landscape of plastic bans in the U.S. In: National Geographic. https://www.nationalgeographic.com/environment/2019/08/map-shows-the-complicated-landscape-of-plastic-bans/. Accessed 24 Sept 2019

Gleick P, Cooley H (2009) Energy implications of bottled water. Environ Res Lett 4:014009. https://doi.org/10.1088/1748-9326/4/1/014009

Gordon J (2017) Cellophane: an early advance in chemistry. In: Barrons.com. https://www.barrons.com/articles/cellophane-an-early-advance-in-chemistry-1501907245. Accessed 6 Apr 2019

Gorelick M, Gould L, Nimmer M et al (2011) Perceptions about water and increased use of bottled water in minority children. Arch Pediatr Adolesc Med 165:928–932 https://doi.org/10.1001/archpediatrics.2011.83

Groc I (2015) How a DVD case killed a whale: the number of whales and dolphins harmed by floating marine debris seems to be on the rise. National Geographic. https://news.nationalgeographic.com/news/2015/01/150107-sea-trash-whales-dolphins-marine-mammals/. Accessed 19 Mar 2019

Guo X, Valle-Levinson A (2007) Tidal effects on estuarine circulation and outflow plume in the Chesapeake Bay. Cont Shelf Res 27 (1):20–42

Hahladakis J, Velis C, Weber R et al (2018) An overview of chemical additives present in plastics: migration, release, fate and environmental impact during their use, disposal and recycling. J Hazard Mater 344:179–199. https://doi.org/10.1016/j.jhazmat.2017.10.014

Hale R, La Guardia M, Harvey E et al (2001) Polybrominated diphenyl ether flame retardants in Virginia freshwater fishes (USA). Environ Sci Technol 35:4585–4591

Hampton Roads Sanitation District (HRSD) (2019) SWIFT factsheet. http://swiftva.com/factsheets/. Accessed 15 Mar 2019

Helm R (2019) How plastic cleanup threatens the Ocean's living islands. The Atlantic. https://www.theatlantic.com/science/archive/2019/01/ocean-cleanup-project-could-destroy-neuston/580693/. Accessed 15 Mar 2019

Hermabessiere L, Dehaut A, Paul-Pont I et al (2017) Occurrence and effects of plastic additives on marine environments and organisms: a review. Chemosphere 182:781–793

Herzke D, Anker-Nilssen T, Nøst T et al (2016) Negligible impact of ingested microplastics on tissue concentrations of persistent organic pollutants in Northern Fulmars off coastal Norway. Environ Sci Technol 50(4):1924–1933

Hoarau L, Ainley L, Jean C et al (2014) Ingestion and defecation of marine debris by loggerhead sea turtles. Caretta caretta, from by-catches in the South-West Indian Ocean. Mar Pollut Bull 84:90–96

Hosler D, Burkett S, Tarkanian M (1999) Prehistoric polymers: rubber processing in ancient Mesoamerica. Science 284:1988–1991

Hu Z, Morton L, Mahler R (2011) Bottled water: United States consumers and their perceptions of water quality. Int J Environ Res Public Health 8:565–578. https://doi.org/10.3390/ijerph8020565

Jambeck J, Andrady A, Geyer R et al (2015) Plastic waste inputs from land into the ocean. Science 347:768–771

Jamieson A, Brooks L, Reid W et al (2019) Microplastics and synthetic particles ingested by deep-sea amphipods in six of the deepest marine ecosystems on earth. R Soc Open Sci 6(2):180667. https://doi.org/10.1098/rsos.180667

Karg T (2018) Bevindustry.com. https://www.bevindustry.com/articles/91268-2018-state-of-the-beverage-industry-bottled-water-remains-the-no-1-beverage-in-the-united-states? Accessed 8 Apr 2019

Kazmiruk T, Kazmiruk V, Bendell L (2018) Abundance and distribution of microplastics within surface sediments of a key shellfish growing region of Canada. PLoS One 13:e0196005. https://doi.org/10.1371/journal.pone.0196005

Keswani A, Oliver D, Gutierrez T et al (2016) Microbial hitchhikers on marine plastic debris: human exposure risks at bathing waters and beach environments. Mar Environ Res 118:10–19

Kettering C (1946) Biographical memoir of Leo Hendrik Baekeland 1863–1944. National Academy of Sciences Biographical Memoirs Vol. XXIV-Eighth Memoir. http://www.nasonline.org/publications/biographical-memoirs/memoir-pdfs/baekeland-leo-h.pdf

Koelmans A, Mohamed Nor N, Hermsen E et al (2019) Microplastics in freshwaters and drinking water: critical review and assessment of data quality. Water Res 155:410–422. https://doi.org/10.1016/j.watres.2019.02.054

Kole P, Löhr A, Van Belleghem F et al (2017) Wear and tear of tyres: a stealthy source of microplastics in the environment. Int J Environ Res Public Health 14(10):1–31

Lamb J, Willis B, Fiorenza E et al (2018) Plastic waste associated with disease on coral reefs. Science 359(6374):460–462

Li H, La Guardia M, Liu H et al (2019) Levels and risks associated with brominated and organophosphate flame retardants in sediments upstream and downstream of the Guiyu, China e-waste recycling zone. Sci Total Environ 646:58–67

Lin T, Teymourian Y, Tursini M (2018) The effect of sugar and processed food imports on the prevalence of overweight and obesity in 172 countries. Glob Health 14(35). https://doi.org/10.1186/s12992-018-0344-y

Lwanga E, Vega J, Quej V et al (2017) Field evidence for transfer of plastic debris along a terrestrial food chain. Nat Sci Rep 7:14071

Lyche J, Gutleb A, Bergman A et al (2009) Reproductive and developmental toxicity of phthalates. J Toxicol Environ Health B Crit Rev 12(4):225–249. https://doi.org/10.1080/10937400903094091

McCormick A, Hoellein T, Mason S et al (2014) Microplastic is an abundant and distinct microbial habitat in an urban river. Environ Sci Technol 48(20):11863–11871

McDevitt J, Criddle C, Morse M et al (2017) Addressing the issue of microplastics in the wake of the microbead-free waters act—a new standard can facilitate improved policy. Environ Sci Technol 51 (12):6611–6617

McIver S (1990) Florida's plastic snowbird. South Florida Sun Sentinel. https://www.sun-sentinel.com/news/fl-xpm-1990-04-08-9001010983-story.html. Accessed 17 Mar 2019

Miao L, Wang P, Hou J et al (2019) Distinct community structure and microbial functions of biofilms colonizing microplastics. Sci Total Environ 650(2):2395–2402

National Geographic (2019) African elephant. https://www.nationalgeographic.com/animals/mammals/a/african-elephant/. Accessed 18 Sept 2019

NCSL (2019) State beverage container deposit laws. In: Ncsl.org. http://www.ncsl.org/research/environment-and-natural-resources/state-beverage-container-laws.aspx. Accessed 8 Apr 2019

Nizzetto L, Futter M, Langaas S (2016) Are agricultural soils dumps for microplastics of urban origin? Environ Sci Technol 50:10777–10779

Parker L (2018a) The Great Pacific garbage patch isn't what you think it is. In: News.nationalgeographic.com. https://news.nationalgeographic.com/2018/03/great-pacific-garbage-patch-plastics-environment/. Accessed 6 Apr 2019

Parker L (2018b) Sperm whale found dead with 13 pounds of plastic in its stomach. National Geographic Environment: Planet or Plastic. https://www.nationalgeographic.com/environment/2018/11/dead-sperm-whale-filled-with-plastic-trash-indonesia. Accessed 19 Mar 2019

PHS (2019) Celluloid. In: The Plastics Historical Society. http://plastiquarian.com/?page_id=14221. Accessed 6 Apr 2019

Provencher J, Vermaire J, Avery-Gomm S et al (2018) Garbage in guano? Microplastic debris found in faecal precursors of seabirds known to ingest plastics. Sci Total Environ 644:1477–1484

Raju N, Carbery M, Kuttykattil A et al (2018) Transport and fate of microplastics in wastewater treatment plants: implications to environmental health. Rev Environ Sci Biotechnol. https://doi.org/10.1007/s11157-018-9480-3

Register K, McKay L, Witmer V (2016) Virginia marine debris reduction plan: summary & look ahead. Prepared for the Virginia Coastal Zone Management Program, 20 p

Savoca M, Wohlfeil M, Ebeler S et al (2016) Marine plastic debris emits a keystone infochemical for olfactory foraging seabirds. Sci Adv 2 (11):1–8

Schreder E, La Guardia M (2014) Flame retardant transfers from U.S. households (dust and laundry wastewater) to the aquatic environment. Environ Sci Technol 48(19):11575–11583

Schymanski D, Goldbeck C, Humpf H et al (2018) Analysis of microplastics in water by micro-Raman spectroscopy: release of plastic particles from different packaging into mineral water. Water Res 129:154–162

Smith M, Love D, Rochman C et al (2018) Microplastics in seafood and the implications for human health. Curr Environ Health Rep 5 (3):375–386

Song Y, Hong S, Jang M et al (2015) A comparison of microscopic and spectroscopic identification methods for analysis of microplastics in environmental samples. Mar Pollut Bull 93 (1–2):202–209

Statistica (2019) Per capita consumption of bottled water in the U.S., 2017. In: Statista. https://www.statista.com/statistics/183377/per-capita-consumption-of-bottled-water-in-the-us-since-1999/. Accessed 8 Apr 2019

Strand B, Kerr C (2019) 10 Cities and countries confronting plastic bag pollution head-on. In: Earth Day Network. https://www.earthday.org/2018/04/20/10-cities-and-countries-confronting-plastic-bag-pollution-head-on/. Accessed 8 Apr 2019

Sussarellu R, Suquet M, Thomas Y et al (2016) Oyster reproduction is affected by exposure to polystyrene microplastics. Proc Natl Acad Sci USA 113(9):2430–2435

The Ocean Cleanup (2019). https://www.theoceancleanup.com/

US EPA (2019) Facts and figures about materials, waste and recycling. https://www.epa.gov/facts-and-figures-about-materials-waste-and-recycling/plastics-material-specific-data. Accessed 21 Sept 2019

Van Cauwenberghe L, Janssen C (2014) Microplastics in bivalves cultured for human consumption. Environ Pollut 193:65–70

VIMS (2019) In: Vims.edu. https://www.vims.edu/ccrm/_docs/marine_debris/biodegradablepanel_factsheet.pdf. Accessed 7 Apr 2019

Waller C, Griffith H, Waluda C et al (2017) Review: microplastics in the Antarctic marine system: an emerging area of research. Sci Total Environ 598:220–227

Wardrop P, Shimeta J, Nugegoda D et al (2016) Chemical pollutants sorbed to ingested microbeads from personal care products accumulate in fish. Environ Sci Technol 50:4037–4044. https://doi.org/10.1021/acs.est.5b06280

Wilcox C, Heathcote G, Goldberg J et al (2015a) Understanding the sources and effects of abandoned, lost, and discarded fishing gear on marine turtles in northern Australia. Conserv Biol 29(1):198–206

Wilcox C, Van Sebille E, Hardesty B (2015b) Threat of plastic pollution to seabirds is global, pervasive, and increasing. Proc Natl Acad Sci USA 112:11899–11904. https://doi.org/10.1073/pnas.1502108112

Yeginsu C (2018) European Parliament approves ban on single-use plastics. In: Nytimes.com. https://www.nytimes.com/2018/10/25/world/europe/european-parliament-plastic-ban.html. Accessed 7 Apr 2019

Yonkos L, Friedeldietri E, Perez-Reyes A et al (2014) Microplastics in four estuarine rivers in the Chesapeake Bay, U.S.A. Environ Sci Technol 48(24):14195–14202

Zettler E, Mincer T, Amaral-Zettler L et al (2013) Life in the "plastisphere": microbial communities on plastic marine debris. Environ Sci Technol 47(13):7137–7146

Zhang W, Wu Y, Simonnot M (2012) Soil contamination due to e-waste disposal and recycling activities: a review with special focus on China. Pedosphere 22(4):434–455

Zhang J, Peng Y, Wang L (2018) Occurrence of microplastics in human faeces of children in Tianjin, China. Abstracts of Papers, 256th ACS National Meeting & Exposition, Boston, MA, United States, August 19–23

Zhao S, Ward J, Danley M et al (2018) Field-based evidence for microplastic in marine aggregates and mussels: implications for trophic transfer. Environ Sci Technol 52(19):11038–11048

Zubris K, Richards B (2005) Synthetic fibers as an indicator of land application of sludge. Environ Pollut 138(2):201–211

Part V

Looking to the Future: Ecology, Economics, Ethics, and Policy for Restoring the Health of the Bay and Its People

A New Food System for the Chesapeake Bay Region and a Changing Climate

Benjamin E. Cuker, Kari St. Laurent, Victoria J. Coles, and Dawn Gerbing

Rising seas drown the islands and shores of the Chesapeake Bay,
the warming climate taking a building toll each day.

Burning the fossils of life long past,
is changing the Bay terribly fast.

Add to that the lust for meat,
with its emissions that increase the heat.

For the sake of the future we can and must do better,
seeking a food system that won't make our feet any wetter.

Food systems have the potential to nurture human health and support environmental sustainability; however, they are currently threatening both. . . . Increasing evidence shows that food production is the largest cause of global environmental change, and a transition to sustainable food production is necessary for global sustainable development (Willett et al. 2019).

Abstract

Climate change is reshaping the Chesapeake Bay ecosystem and its linked food system. Sea level rise and increased precipitation drive regular flooding events and the erosion of islands and coastal features. This results in the loss of habitat for various species of birds and fishes. Increased temperatures extend the growing season and reduce the incidence of frosts. Rising water temperatures accommodate the immigration of warm-adapted species and endanger populations of cold ones. Increased temperature interacts with cultural eutrophication to extend the duration and extent of seasonal hypoxia. Higher regional temperatures threaten human health, particularly populations of poorer residents that lack refuge from heatwaves and flooding events. Acidification of the Bay, which is linked to climate change agents, endangers species that rely upon calcium carbonate for shell and skeleton. Emission of greenhouse gases (GHGs, primarily CO_2, NO_2, CH_4) from human economic activity drives climate change. The food system accounts for 25–30% of human GHGs emissions. Half of these comes from livestock, which contribute 14.5% of the global anthropogenic GHG load. Eliminating animal-based foods, particularly ruminants (cattle) would facilitate building a sustainable food system with improved health outcomes for the residents of the Chesapeake Bay Region.

Keywords

Ruminants · Animal-based-foods · Plant-based-foods · Ocean acidification · Sea level rise · Habitat loss

B. E. Cuker (✉)
Department of Marine and Environmental Science, Hampton University, Hampton, Virginia, USA
e-mail: Benjamin.cuker@hamptonu.edu

K. St. Laurent
Department of Natural Resources and Environmental Control, Delaware National Estuarine Research Reserve, Dover, DE, USA
e-mail: Kari.StLaurent@delaware.gov

V. J. Coles
Horn Point Laboratory, Cambridge, MD, USA
e-mail: vcoles@umces.edu

D. Gerbing
International Certified Lactation Consultant, Hampton, VA, USA

The climate of the Chesapeake Bay region is changing, as is true for the entire world. Three centuries of burning fossil fuels, deforestation, population growth, and an expanding

B. E. Cuker (ed.), *Diet for a Sustainable Ecosystem*, Estuaries of the World, https://doi.org/10.1007/978-3-030-45481-4_18

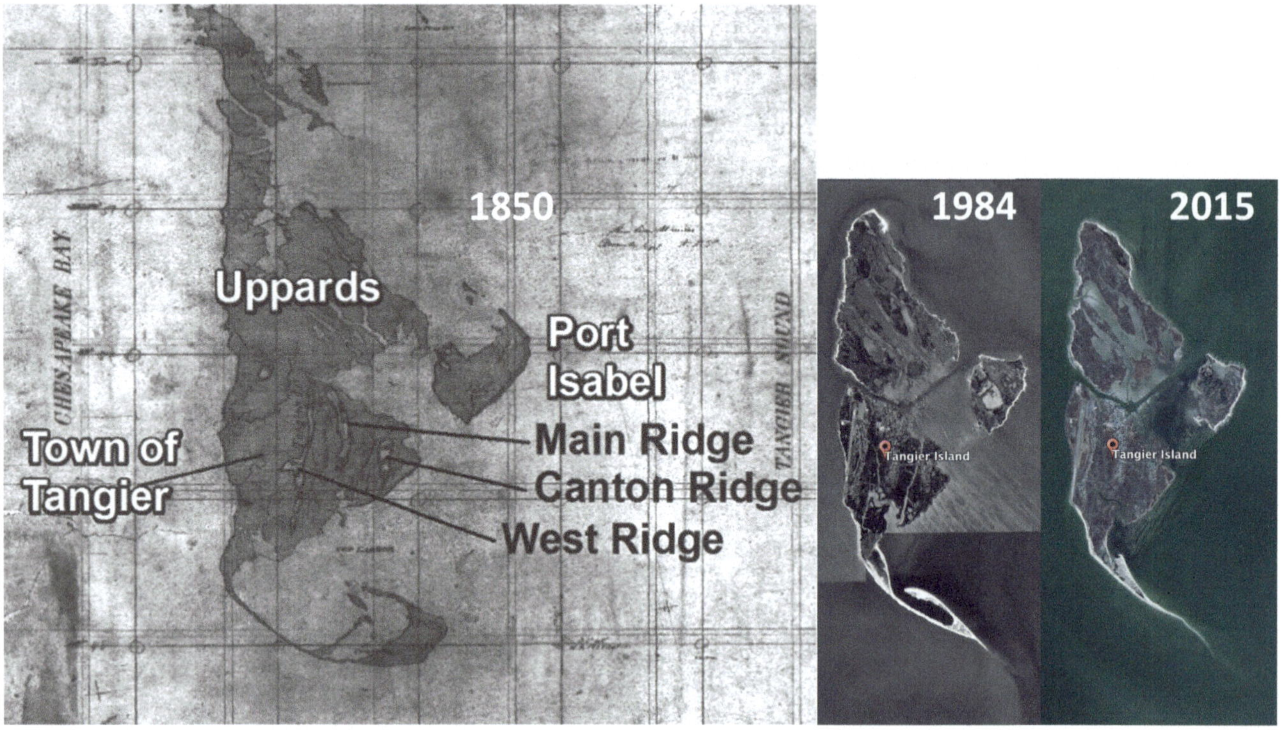

Fig. 1 Tangier and Port Isabel Islands as they appeared in 1850, 1984 and 2015. The extensive loss of land is apparent. The once inhabitable "Uppards" illustrated in the 1850 map has been converted to marsh by the rising sea level. Images from historic map and Google Earth

animal-based food system have combined to increase the concentration of greenhouse gases (CO_2, CH_4, N_2O and others) in the atmosphere. The concentration of CO_2 in the atmosphere went from 310 ppm in 1950 to 415 ppm in June of 2019. Since 1950, earth's global temperature increased 1 °C, and is on track to increase another 0.5° by 2050 (IPCC 2018). The rate of increase in air temperature for the Chesapeake Bay Region is 0.036 °C/year, 30% higher than the global rate of 0.025 °C/year (Irby et al. 2018). About 31% of the CO_2 produced from burning fossil fuels is dissolving in the ocean, having caused a drop in pH from 8.2 to 8.1; a 30% increase in acidity (Gruber et al. 2019; NOAA 2019a). The Chesapeake Bay's ecosystem and food system are responding to climate change in numerous ways. The current food system contributes to the problem by adding to the load of green-house gasses. This chapter explores how climate change has already altered the Chesapeake Bay ecosystem and considers a new food system to meet the challenge of curtailing the upward trajectory of warming and acidification.

The Changing Climate of the Chesapeake Bay: Rising waters Extreme high tides and storms regularly bring the waters of the Chesapeake Bay into the streets and front yards of Tangier Island (Fig. 1). It wasn't always that way. Previously such flooding occurred every few years in association with hurricanes and other severe storms. Now it's expected several times a year (Sweet et al. 2018).

Tangier Island, VA (37° 49′27.48″ N, 75°59′33.64 W′) is about halfway between Cape Henry and Annapolis, MD. It lies just 20 km southwest of Crisfield, MD on the coast of the Delmarva Peninsula. To the northwest and 22 km across the Bay is Smith Point, marking the southern side of the mouth of the Potomac River. Sitting just 1.2 m above mean sea level, Tangier is one of just two remaining inhabited islands in the Chesapeake Bay. The other is Smith Island, which lies 15 km to the north of Tangier. In the sixteenth Century, English cartographers named about 400 islands in the Bay. Most no longer exist, succumbing to rising sea level and sinking land.

The Bay is a dynamic system and islands have always come and gone. But the system is far from equilibrium, no longer do new islands emerge to take the place of those washed away. The process of island loss accelerated starting about 1850. That year marked the end of the little ice age and coincided with intensification of the mostly coal-fired Industrial Revolution. The natural warming cycle, joined by human-produced greenhouse gasses and deforestation, accelerated post-glacial Sea Level Rise (SLR). Melting glaciers and thermal expansion of the sea started to drown the Bay's islands. Add to that the land subsidence from natural settling following the last glacial cycle and withdrawal of water from the aquifer to supply the food system (Kenney and Brainard 2019). The fate facing Tangier is stark evidence of how a changing climate is reshaping the Chesapeake Bay ecosystem (Beringer 2018).

A map from 1850 shows Tangier Island with two contiguous main sections, the more northerly "uppards," and the southerly town of Tangier. The uppards is now all but gone, converted to low marsh and shoals. Presently, about 300 ha (740 acres) comprise the island, but only 34 ha (83 acres) are high enough to support habitation. Tangier Island is a third of the size of what it was in 1850, and projections suggest it will become uninhabitable if not completely washed away by the end of the twenty-first Century (Tangiers Plight 2019).

As their ancestors for generations before them, the less than 800 current residents of Tangier Island make their living primarily from the food system. Most work the surrounding waters to catch crab and fish, raise soft shelled crab, or provide seafood dinners to tourists. Native Americans used Tangier as a fishing camp prior to it being taken by English Colonists in the seventeenth Century (Worrall 2018).

Despite being just an hour's boat ride away from Maryland's Eastern Shore, Tangier Island retains a distinctive culture and unique speech, tracing to Cornwall, England, the ancestral home of most of the current residents. The Island also figured large in the Nation's history. During the War of 1812, it harbored 1000 people who escaped enslavement and sought the protection of the British. It was the primary British base for Royal Navy operations in the Bay, and from its shores, vice-admiral, George Cockburn launched the attack on Baltimore in September of 1814. The successful resistance of Fort McHenry to that invasion is the subject of a poem by Francis Scott Key that became the US National Anthem (NPS 2019). Climate change and the rising waters of the Chesapeake stand poised to wash away what remains of this fishing village and physical evidence of its storied past. And it's not just Tangier and the few other remaining islands of the Bay threatened by rising sea level. Nearby Crisfield, MD is one of a dozen coastal communities facing regular and extensive flooding that will restrict future habitation (Kleinosky et al. 2006).

The long term rate of SLR (Sea Level Rise) for the Chesapeake Bay ranges from 3.17 mm/year in Baltimore over the past century, twice that of the global average of 1.6–2.1 mm/year over the past century, to 4.7 mm/year in Norfolk, VA over the course of a century (VIMS 2019; CSSR 2017; Fig. 2). The region is slowly subsiding from relaxation of the glacial forebulge, tracing back to the Pleistocene (DeJong et al. 2015). But sea level rise has been accelerating. By the year 2100, sea level in the Bay will rise 0.4–1.3 m, depending on the world's ability to reduce the release of greenhouse gases (GHGs) (Boesch et al. 2013). The rising waters won't affect all communities equally. Those in low-lying coastal areas where land forms focus tidal surge will see the greatest damage. Such places are often home to large populations of African Americans. Crisfield is located in Somerset County, which has the lowest income and some of the lowest lying lands in Maryland. Its population is 40% African American, and many residents have suffered from flooding during recent storms (Kobell 2015).

Chesapeake Bay communities face different kinds of flooding, all linked to climate change. Sunny day (nuisance, or high-tide) flooding occurs during high-tide events, often tied to surge from winds and atmospheric pressure differentials. High-tide flooding of the sort that regularly inundates Tangier Island and other coastal communities is on the rise in the Chesapeake, occurring about eight times per

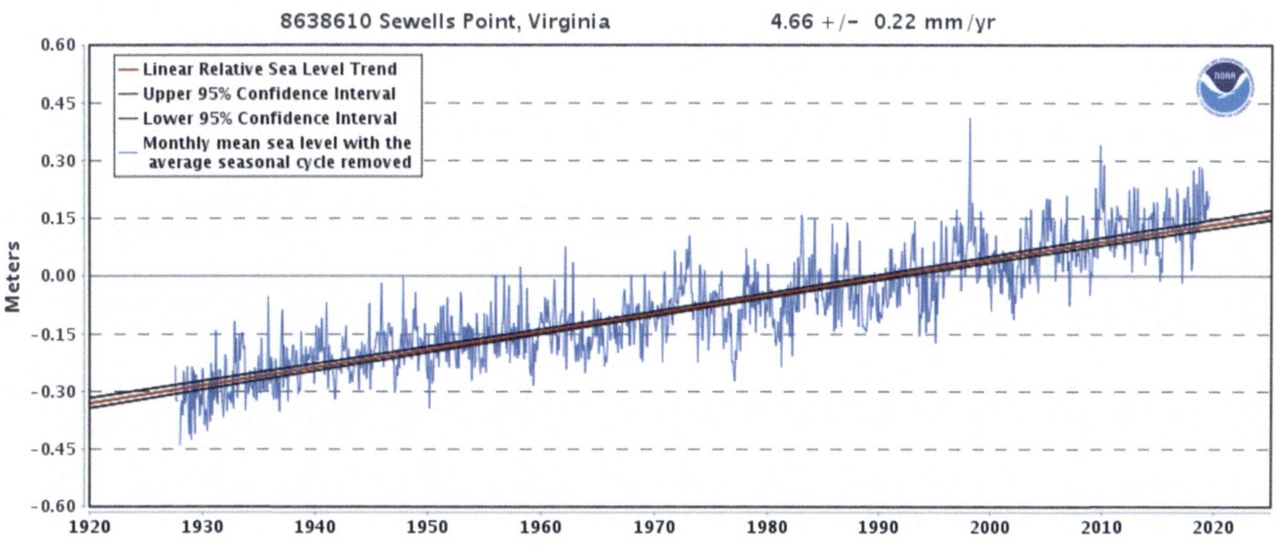

Fig. 2 Rising sea level at Sewells Point, Norfolk, VA as indicted monthly mean sea level data from 1927 to 2018. This is equivalent to a change of 0.47 m in 100 years. This station is located in Hampton Roads, at the southern end of the Chesapeake Bay (NOAA 2019b)

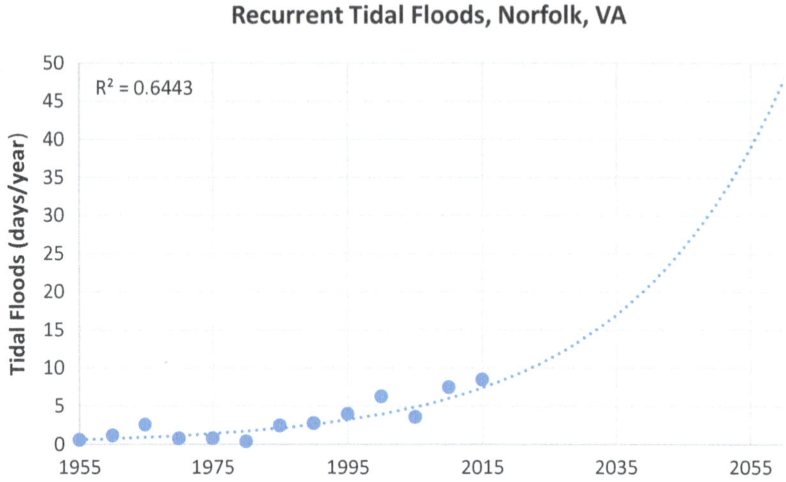

Fig. 3 Five year averages of recurrent tidal flooding in Norfolk, VA between 1955 and 2015. The fitted line is an exponential curve. Figure after Sweet et al. (2017) and modified to show projection to 2060

year in 2015, compared to once per year in 1955. Importantly, the rate of increase is exponential, predicting 40 events per year by 2055 (Fig. 3).

Storm surge associated with tropical storms and nor'easters pushes water from the Atlantic Ocean into the Bay, where it breaches natural shores and seawalls. Intensified rains associated with storms drain from the watershed to cause tributaries to overflow into natural and developed lands along their banks. At times the different drivers of flooding may co-occur, making the flooding more severe.

Flooding is part of natural hydrography and has long informed the food system. For eons before the evolution of humans, periodic storms and seasonal melting caused rivers to escape their banks and deposit rich sediments in floodplains. Indeed, civilization and agriculture evolved on these nutrient-rich soils that provided bountiful harvests (Crawford et al. 1998). Of course floodplains do flood, at times inundating and destroying crops. However, climate change, deforestation and increased urbanization combine to make flooding more common and intense. A warming Earth means faster evaporation, hotter air able to temporarily store more moisture, and more intense storms coming more frequently (Nicholls et al. 2012).

Coastal Flooding Destroys Habitat Flooding of developed lands catches people's attention. Impassable roadways, flooded buildings, and undermined structures reduce human habitability. But excessive flooding also destroys habitability of natural areas for non-human populations in the Chesapeake Region. Rising water levels in the Bay drown the shore-side nests of laughing gulls (*Leucophaeus atricilla*). In 1993 counts found 45,387 nesting pairs in the Bay. That's down to 16,653 in 2019 (Dietrich 2019). To reduce erosion and flooding landowners often harden the shore line with rip-rap (large rocks) or bulkheads. These structures eliminate the natural ecotones (transitions) between aquatic and terrestrial systems. About 63 different species of water birds require those transitions for feeding, mating and nesting along the shores of the Bay's tributaries. As little as 5% shoreline hardening significantly reduces water bird habitat and counts (Prosser et al. 2017). As of 2016, 18% of the shoreline of the Bay and its major tributaries had been hardened, substantially reducing habitat (Mitchell 2016).

Tidal marshes of the Chesapeake Bay come in three basic varieties; *extensive* (large areas adjacent to the shore or independent islands), *embayed* (comprising the sides of creeks), and *fringing* (narrow expanses along the shoreline). All provide habitat for wildlife and important ecosystem services. Marshes slow runoff and filter particulate matter from the water. They also sequester carbon, nitrogen and other nutrients, reducing the effects of water pollution. Marshes dampen wave activity, protecting shores from erosion (Mitchell et al. 2017). SLR is combining with subsidence and land development to cause a net loss in marsh area for the Bay. The York River lost 9% of its marsh area between 1973 and 2009 (Mitchell 2016). Marshes in this area may accrete at a rate of 5 mm/year in the Bay, which roughly matches the combined effects of SLR and subsidence. Consequently, eroded and drowned fringing and extensive marshes aren't being replaced much by shoreward migration. Most of the new marsh comes from flooding of adjacent, low-lying forest in the upper reaches of the tributaries. Stands of dead Loblolly pines (*Pinus taeda*) that succumbed to drowning of their roots, stand as silent witnesses to the encroaching waters of the Bay. The 11% expansion of up-tributary marshes by the process of forest drowning can't make up for the 32% loss of down-tributary fringing and extensive marshes, observed for the interval between 1973 and 2009 (Mitchell et al. 2017).

Rising and More Extreme Temperatures The average temperatures of the air and water are rising. The mean annual air temperature in Baltimore, the largest city on the Bay, went from about 13 °C in the 1890s to 15.5 °C by 2016. The mean

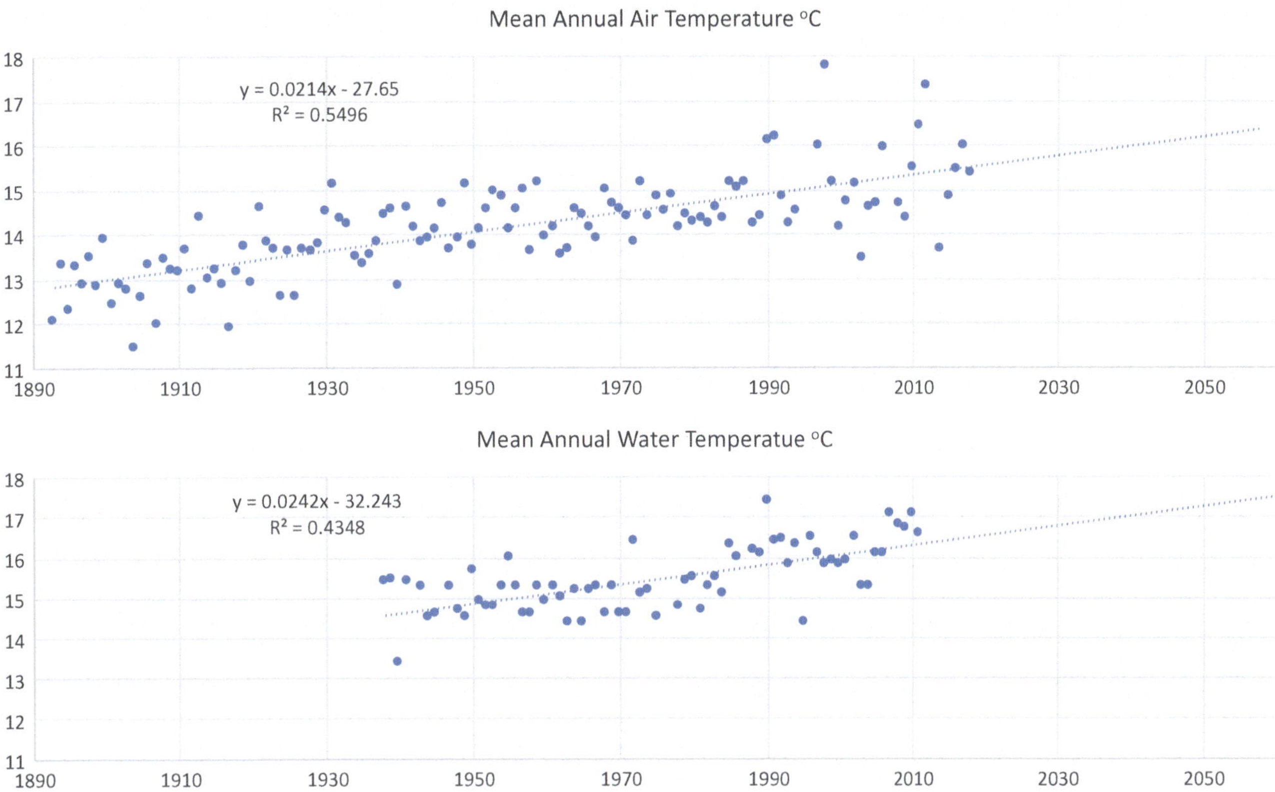

Fig. 4 Mean annual air temperature for Baltimore, MD and mean annual surface water temperature of the Patuxent River. Increasing average air and surface water temperatures indicate a warming Chesapeake Bay climate since the end of the nineteenth century. Mean air temperatures increased from 12.8 °C in the 1890s to 15.5 °C in 2015. If the trend continues, mean annual air temperature will hit 16.4 °C by 2060 and that for water will reach17.5 °C. Data for air temperature from NOAA-NCEI (2019). Data for Patuxent River from Orth et al. (2017)

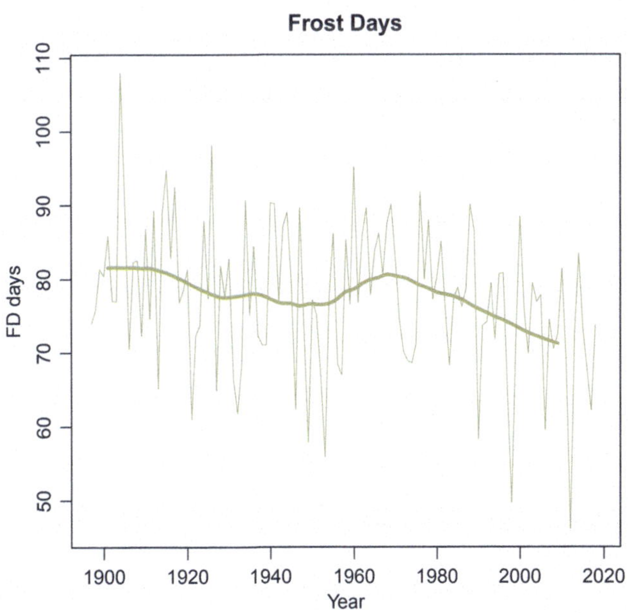

Fig. 5 Number of days each year where daily minimum temperatures dropped below 0 °C (FD = Frost Days). Data is a composite of 146 weather stations located around the Chesapeake Bay. Figure created by V. Coles. Data from the NOAA National Centers for Environmental Information Global Historical Climatology Network Daily database (NCDC 2019)

annual surface water temperature of the Patuxent River rose from 14.5 °C in the 1940s, to 17.0 °C in 2010 (Fig. 4). If these trends continue, we could expect to see mean annual air temperatures of 16.4 °C by 2060 and 17.5 °C for surface waters.

It's not just the average temperature that's rising. The region is experiencing about ten fewer days of frost (temperatures ≤ 0 °C) and ten more tropical (nighttime temperatures ≥ 20 °C) nights per year than it did in at the end of the nineteenth Century (Figs. 5 and 6). This difference is exacerbated in the hottest regions within urban areas which can be as much as 6 °F warmer than the cooler urban regions in Baltimore. These hotspots are correlated with high rates of poverty, and low levels of green space (NPR 2019). Fewer cold days means a growing season length (GSL) about 15 days longer than in 1900 (Fig. 7). GSL is, "…the number of days between the first span of at least 6 days with daily mean temperature warmer than 5 °C and the first span of 6 days after July first with daily mean temperature below 5 °C (Mueller et al. 2015)." Output of crops should be increasing with expansion of the GSL in the Chesapeake Region, as is generally true for mid-latitude farms (Gornall et al. 2010). However, climate change also brings more days of extremely warm temperatures, longer periods of drought and springtime

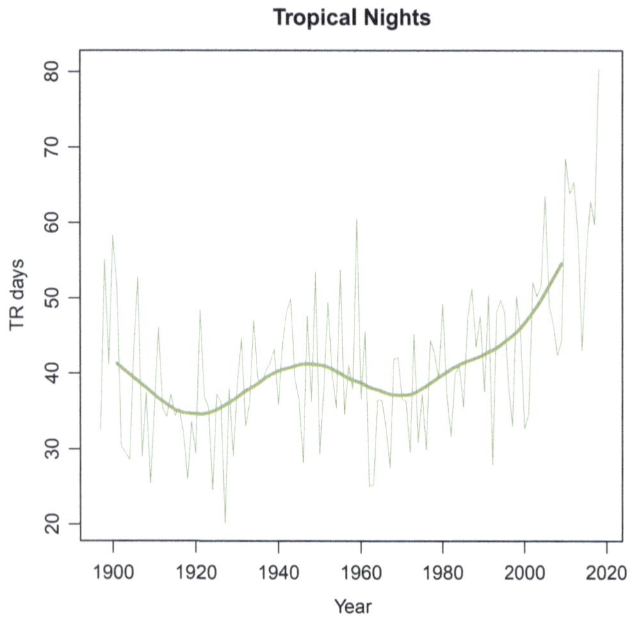

Fig. 6 The number of tropical nights (TR) recorded in the Chesapeake Bay Region increased between the end of the nineteenth Century and the beginning of the twenty-first Century. TR refers to days where minimum temperatures remain above 20 °C. Data integrated for the entire region. Data from the NOAA National Centers for Environmental Information Global Historical Climatology Network Daily database (NCDC 2019)

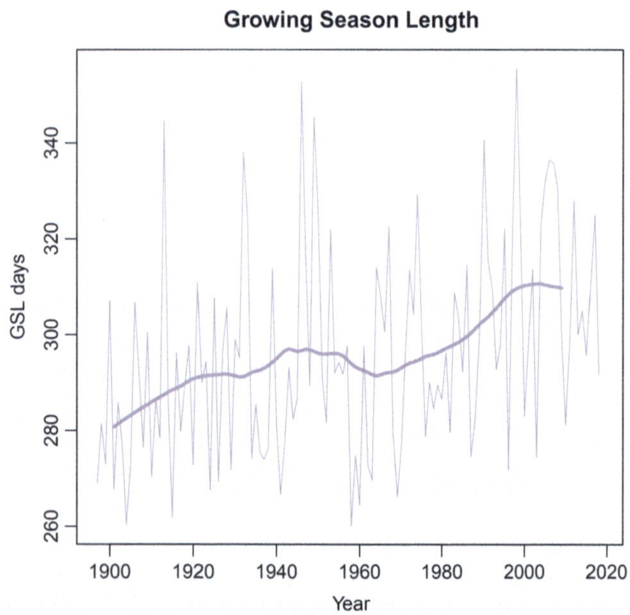

Fig. 7 Increasing growing season length (GSL) in the Chesapeake Bay Region. GSL is defined as the continuous days suitable for growing a crop of wheat (minimum temp ≥5 °C) each year. Data from the NOAA National Centers for Environmental Information Global Historical Climatology Network Daily database (NCDC 2019)

flooding of fields, which may sporadically reduce crop yields. "While change in long-term mean climate will have significance for global food production and may require ongoing adaptation, greater risks to food security may be posed by changes in year-to-year variability and extreme weather events. Historically, many of the largest falls in crop productivity have been attributed to anomalously low precipitation events (Gornall et al. 2010)."

Climate Change Increases the Frequency and Intensity of Rain Events Higher temperatures means more evapotranspiration and the capacity for the atmosphere to hold more moisture. The total amount, frequency, and intensity of precipitation has increased since 1900 in the Chesapeake Bay Region. The rate of increase in annual precipitation is 1 cm per decade. This trend is expected to continue and intensify in coming decades. More extreme rain-events will bring increased flooding. The future appears to hold wetter springs followed by summer droughts (Kunkel et al. 2013).

Increased precipitation means more nutrients (N & P) washed from the watershed into the Bay. Wetter years and higher nutrients yield higher concentrations of phytoplankton. Diatoms, the dominant taxonomic groups of phytoplankton show the biggest response to wet years. Although most species of diatoms serve as excellent food for oysters, menhaden, and zooplankton, their numbers in wet years overwhelm these consumer's ability to eat them (Harding et al. 2015). This results in massive oxygen loss due to bacterial respiration of the decaying algae in the deeper waters of the Bay (see chapter "The Journey from Peruvian Guano to Artificial Fertilizer Ends with too Much Nitrogen in the Chesapeake Bay" on Eutrophication).

SLR, Temperature and Precipitation Interact to Cause a Net Reduction of Oxygen in the Bay Climate change involves a suite of factors that are altering the availability of free oxygen in the Bay. SLR will enhance estuarine circulation, bringing more oxygen rich water from the Atlantic Ocean in to the Bay. However, that effect is being overwhelmed by other factors that lead to reduced oxygen.

As noted above, increased precipitation brings more nutrients to the Bay, promoting excessive algal growth. When those algae die, they sink to the deeper waters where bacteria digest them and in doing so deplete the reserves of free oxygen. While intense storm events produce strong winds that can ventilate the Bay and temporarily increase oxygen concentrations, this is offset by runoff laden with nutrients and oxygen-demanding organic material (Najjar et al. 2010).

Higher temperatures increase the rate of microbial metabolism. This speeds both the rate of algal growth and its subsequent decomposition by oxygen demanding bacteria. In addition, the solubility of oxygen (and all gases) is inversely proportional to temperature, thus reducing the capacity of the warmer waters to supply oxygen to meet the demands of increased bacterial respiration. The warming climate is also extending the duration of the Bay's stratification, moving the onset of hypoxic waters a week earlier in the spring. Of the three factors, modeling suggests that increasing temperature appear to be the most important in determining oxygen availability (Irby et al. 2018).

Efforts to reduce eutrophication in the Bay by curtailing nutrient loading (which comes mostly from the food system) are confounded by climate change. Increased temperature and river flows are projected to reduce the effectiveness of nutrient-reduction by 5–45%, with the most detriment for waters at mid-depths that flank the deep main channel of the Bay. This will reduce habitat for fish, crabs, and other demersal species in mid and lower reaches of the Bay (Irby et al. 2018). Climate change is making the job of restoring the Bay harder.

Projecting Climate Change Impacts on the Bay's Ecosystem Najjar et al. (2010) developed a comprehensive narrative for the changes predicted to be wrought by climate change on the Chesapeake Bay by the end of the twenty-first Century. Let's consider some of these that haven't been addressed above. Increased precipitation in winter and spring will drive more intense blooms of diatoms in the spring followed by aperiodic blooms of dinoflagellates in the summer. The nutrient laden runoff will increase productivity of the phytoplankton over the course of the year.

Increasing concentrations of CO_2 in the water will couple with higher temperatures to alter the composition of the algal community. This will likely result in the increased frequency of harmful algal blooms (HABs). The toxins produced by HABs will discourage zooplankton grazing, resulting in net increases of dissolved organic carbon (DOC), that will help fuel an increasingly heterotrophic food chain.

Increasing temperatures will shift the composition of the fin and shell fish in the Bay, with those adapted for colder waters being replaced by warm-waters species expanding their distributions into the estuary. Brown shrimp (*Farfantepenaeus aztecus*) historically supported a strong commercial fishery off the shores of North Carolina, and significant populations now appear in the southern reaches of the Chesapeake Bay. Warm water species already occurring in the Bay are expected to increase. These include southern flounder (*Paralichthys lethostigma*), cobia (*Rachycentron canadam*), spadefish (*Chaetodipterus faber*), Spanish mackerel (*Scomberomorus maculatus*), mullet (*Mugil curema*), tarpon (*Megalops atlanticus*), pinfish (*Lagodon rhomboids*), black drum (*Pogonias cromis*), red drum (*Sciaenops ocellatus*), weakfish (*Cynoscion regalis*), spotted sea trout (*C. nebulosus*), spot (*Leiostomus xanthurus*), northern kingfish (*Menticirrhus saxatilis*) and southern kingfish (*M. americanus*). Most of these species are part of the Bay's food system and stand to benefit from longer growing seasons and milder winters.

Cold-adapted species in the Chesapeake Bay that occupy the southern portion of their range may fare poorly as temperatures rise and oxygen concentrations drop. This includes; soft clam (*Mya arenaria*), yellow perch (*Perca fulvescens*), white perch (*Morone americana*), striped bass (*Morone saxatilis*), black sea bass (*Centropristis striata*), tautog (*Tautoga onitis*), summer (*Paralichthys dentatus*) and winter flounders (*Pleuronectes americanus*), silver hake (*Merluccius bilinearis*), and scup (*Stenotomus chrysops*). The inclusion of striped bass in this list seems counter-intuitive, since their range extends in to the Gulf of Mexico. Yet, stratification of the Bay often leads to pockets of surface water temperatures >30 °C, sufficient to cause stress (Cook et al. 2006).

Increasing temperatures and changes in salinity may increase the abundance and efficacy of pathogens attacking the Bay's fisheries. Oysters already suffer from *Perkinsus marinus* (Dermo) and *Haplosporidium nelson* (MSK). Effects of these pathogens stand to increase in warmer temperatures (Cook et al. 1998).

As noted above, SLR is reducing marsh habitat in the lower reaches of the Bay's tributaries. This may reduce success of juvenile fish and crabs that depend on those habitats as refugia and feeding grounds. Eel grass (*Zostera marina*) is the most important SAV (Submerged Aquatic Vegetation) in the lower (mesohaline) portion of the Bay, its extensive beds serve as important habitat for fin and shell fish. Eel grass is near the southern extent of its range in the Chesapeake. Periods of elevated temperature have already caused massive diebacks (Moore and Jarvis 2008).

The projections outlined above are those with definitive direction. However, the complexity of the ecosystem makes other predictions about the effects of climate change less certain in both direction and magnitude. For example, comb jellies (*Mnemiopsis leidyi*) are important predators of zooplankton in the Bay, including the larvae of fin and shell fish. A warming Bay may increase their population and effective rate of predation. Yet this ctenophore does poorly at temperatures >30 °C, now a common occurrence in the Bay (Haraldsson et al. 2012). The effect of climate change on the trophic interactions surrounding this key planktonic predator are uncertain, as is true for most of the complex ecology in the Bay ecosystem. So, we can add uncertainty to the list of projected changes (Najjar et al. 2010).

Changing Climate Challenges Human Health The warming climate threatens public health, particularly for vulnerable populations. People of color, the poor, elderly, young children, pregnant women, and those with various illnesses stand to suffer the most from the overheating of the Chesapeake Bay region (USEPA 2017). High temperatures have direct impacts on health. People exposed to high temperatures for extended periods may suffer dehydration and heatstroke, as well as exacerbation of cardiovascular, cerebrovascular, and respiratory disease. This falls hardest on people with outdoor jobs, such as in, agricultural, transportation and construction; and those dwelling in buildings without air-conditioning. People living in urban areas, such as Baltimore, Richmond, and the District of Columbia will suffer from the heat island effect. The abundance of concrete and dearth of trees renders such cities several degrees warmer during heat waves (Imhoff et al. 2010; NPR 2019).

Indirect effects of warming are manifold. Higher temperatures promote the production of ground-level ozone. The secondary air pollutant reduces lung function and exacerbates asthma (Sheffield et al. 2011). In addition to creating GHGs, fossil fuel combustion produces particulate matter, such as black carbon, that stresses pulmonary function. More frequent forest fires associated with climate change will also add such particles to the air (Flannigan et al. 2000).

The extended growing season brought by warmer temperatures isn't just for crops. The spring pollen season is getting longer, posing problems for those with allergies. The higher temperatures extend the infective seasons for many vector-borne disease. A longer season for mosquitoes and ticks means increased chances of infection from West Nile virus and Lyme disease (Khatchikian et al. 2015).

More frequent and more intense storms increases the washing of pathogens from the watershed into the Bay and its tributaries. This exposes swimmers and fishers to a variety of diseases such as *Cryptosporidium* and *Giardia*. Warmer waters also yield higher levels of *in situ* pathogens and toxin producing microbes. These include the bacterium *Vibrio vulnificus* and *Vibrio cholerae*, and harmful algal blooms (Najjar et al. 2010). The Bay staying warmer for a longer period of time each year also increases the chances of encountering these pathogens.

Poultry dominates the animal food industry of the Chesapeake Bay. Most commercial flocks in Concentrated Animal Feeding Operations (CAFOs) and the meat they produce are contaminated with *Salmonella*, and Campylobacter. These cause food poisoning (see chapter "Livestock and Poultry: The Other Colonists Who Changed the Food System of the Chesapeake Bay"), and grow faster at higher temperatures (USEPA 2017).

The mental health of people subjected to the ardors of climate change also suffers. Some psychotherapeutic medications interfere with body temperature regulation, endangering those taking the drugs during heatwaves. People with pre-existing mental illness suffer a tripling of the risk of death during high temperature events (GlobalChange.gov 2016).

Acidification as Part of Climate Change Ocean acidification is a consequence of elevated atmospheric CO_2 concentrations that dissolve into the water. The Chesapeake Bay experiences a second source of acidification, the seasonal production of H_2S in the anaerobic sediments during the warmer months of the year. The combination of CO_2 and H_2S acidification produces pH levels near 7.4 (Cai et al. 2017) in sections of the Bay. This has profound implications for species, such as oysters that use calcium carbonate to build their shells. The eastern oyster (*Crassostrea virginica*) was once the basis for the most economically important fishery in the Chesapeake Bay. It is also a foundational species that provides essential habitat and the ecosystem service of filtering the water (see chapter "The Chesapeake Bay Oyster: Cobblestone to Keystone"). In addition to the devastation wrought by decades of overharvesting and invasive pathogens, oysters now face the new challenge of acidification of the Bay's waters. Levels of pH of 7.5 or less reduce the ability of the oyster to produce its shell and to reproduce (Beniash et al. 2010; Boulais et al. 2017).

Climate Change Challenges the Food System Rising temperatures will influence the food system of the Chesapeake Bay in numerous ways, including some that may not be predicted. The list includes; extended length of growing season, more evapotranspiration increasing sensitivity to drought, expansion of the ranges of plant pests and weeds, stress on pollinators, and changes in migratory patterns of birds and fish.

Higher CO_2 Alters Crop Production and Nutrition Higher concentrations of atmospheric CO_2 will favor some plants over others. About 97% of plant species use C-3 metabolism for collecting CO_2 to be used in photosynthesis. These species suffer in dry conditions when they must keep their stomata closed to reduce loss of water from their leaves. Under such conditions C-3 plants waste much of their energy in the process of photorespiration fueled by high concentrations of oxygen. C-4 plants evolved in dry tropical areas. They use an intermediary molecule with higher affinity for CO_2 combined with special anatomy. This allows them to photosynthesize efficiently with reduced stomatal opening and less loss of water. Despite comprising less than 2% of all plant species, C-4 plants account for 25% of global terrestrial primary production (Ehleringer 2002).

In order, the three dominant crops in the US food system are corn (*Zia maize*), soy (*Glycine max*) and wheat (*Triticum aestivum*). Corn is a C-4 plant, soy and wheat use C-3 metabolism. Both C-3 and C-4 plants increase their yields and growth rates at higher levels of CO_2. But, C-3 plants benefit the most experiencing increases of 20–40%, about two to four times that for C-4 plants. This means that with increasing CO_2 concentrations, corn will suffer more competition from C-3 weeds, such as lambsquarters (*Chenopodium album*). On the other hand, soy will better compete with the C-4 pigweeds (Amaranthus spp.) (Miri et al. 2012).

Increasing CO_2 concentrations changes the macronutrient ratio available to plants. This has two consequences. First, it increases the likelihood of nitrogen (N) limitation for the plant, meaning that more reactive N will be required for the crop to realize its full yield potential. This may encourage farmers to use more N-rich fertilizer on non-leguminous crops, such as wheat. Soy and other legumes use their root nodules that contain mutualistic N-fixing cyanobacteria to fulfil their requirement for that nutrient, so they are expected to increase yields with elevated CO_2. Indeed, a meta-analysis of 111 studies finds a 24% increase in seed yield for soy under elevated CO_2 (Ainsworth et al. 2002).

Second, under elevated CO_2 the relative lack of N will reduce the relative amount of N-rich molecules (protein) in the plant (Ehleringer 2002). One study using CO_2 levels 150 ppm above ambient (expected levels by 2050) found a 10.4% increase in yield of wheat. But, the grain suffered decreased nutritional value; protein dropped 7.4% and iron declined 5.0%. The grain also increased in sugar and lipid content (Högy et al. 2009).

Increasing Temperature Lengthens Growing Season But May Reduce Yields Between 1900 and 2018, Growing Season Length (GSL) in the Chesapeake Bay region increased by about 25 days (Fig. 7). This should be a boon to agriculture. However, the extension of the GSL may be countered by crop damage associated with higher temperatures. The yield for corn increases with rising temperatures up to 29 °C (84.2 °F), and that for soy up to 30 °C (86 °F), but declines with continued warming. Based upon projected temperature increases, yields for these two key crops in the US are expected to decline 30–82% by 2100, with less reduction in yield if the world curtails GHG emissions (Schlenker and Roberts 2009). Temperatures in the region already regularly exceed those damaging to corn and soy. Baltimore recorded an average of 47.5 days with temperatures >32.2 °C between 2010 and 2018 (NOAA-NCEI 2019).

The Animal-Based, Chemical-Based Food System Challenges the Climate, Ecosystem Sustainability and Human Health As addressed above, climate change is altering the food system of the Chesapeake Bay Region. Yet the more important issue is how that food system is responsible for much of the climate change that challenges the sustainability of the local and global ecosystem as well as the degradation of human health (Horrigan et al. 2002; Willett et al. 2019). The current US food system requires 7–10 kcal of energy subsidy from fossil fuel to produce just one kcal of food. That energy subsidy is apportioned as follows; production (14.4%), transportation (3.5%), processing (18.7%), packaging (6.5%), marketing (15.1%), food service (12.9%), household storage and preparation (28.8%). Production of fertilizers and pesticides accounts for 40% of production energy use (Heller and Keoleian 2000). If a 113 g (290 kcal) hamburger seems oily, it should, as it required 3190 kcal of fossil fuel to produce it. That's like drinking 300 ml of diesel fuel!

The food system begins with people's dietary habits. Farmers, fishers and food processors match their products to market demand, as there is no profit in unsold products. Of course, corporate marketing seeks to shape that demand, influencing consumer's decisions. The USDA last published data on food marketing by category in 1997. The food industry spent $7.074 billion in marketing that year. Convenience food, confections, alcoholic beverages, and non-alcoholic beverages accounted for 71.6% of expenditures. Meat of all kinds and dairy combined for 10.1%, while vegetables, fruits, nuts, and beans summed to just 2.2% of the advertising dollar (Gallo 1999). The marketing expenditures seem to work, as

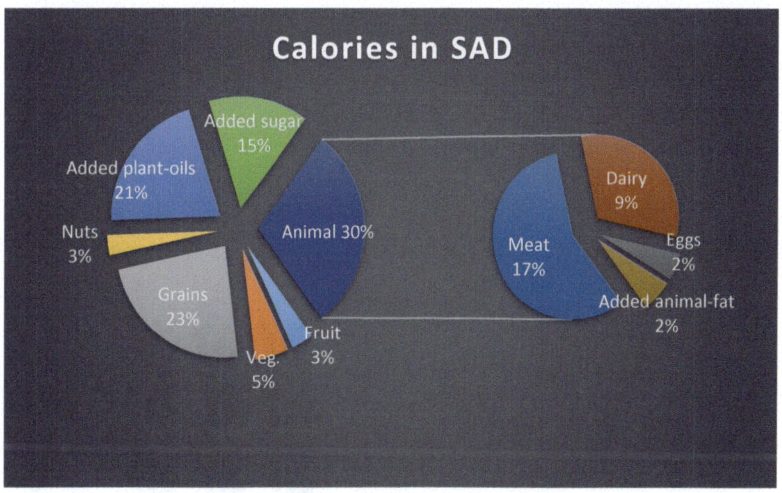

Fig. 8 The percent distribution of calories in the Standard American Diet (SAD) based upon data from the USDA for 2010 (Rehkamp 2016). Meat includes all types of animal flesh. The high proportion of animal products means extensive use of land and water, and the production of greenhouse gases

they roughly reflect how people in the US are getting their calories (Fig. 8). The USDA collects information on the nation's diet, but doesn't parse that out to specific regions or states. So we operate under the assumption that people in the Chesapeake Bay Region eat diets similar to that of most of the nation. What constitutes the Standard American Diet (SAD)?

From the standpoint of mass (weight) of macronutrients, on average people in the US split their diet between carbohydrates (58%), fat (24%), protein (15%) and fiber (3%). This results in a daily average of about 3100 kcal consumed per capita from food that provides an energy density of 5 kcal per gram (Gerrior et al. 2004; Hiza and Bente 2011). The SAD supplies the plurality (30%) of its calories from animal products, with grains, added plant-oils, and added sugar combining for another 59%. That leaves just 8% from vegetables, fruit and nuts (Fig. 8). Note that about 90% of the grains in the diet are highly processed with the elimination of most fiber and the additions of oils, sugars, and sodium (USDA 2019). This diet is unhealthy for both the people eating it and the agroecosystem producing it (see chapter "Eutrophication: Obesity of the Bay and Its People"). It is also unsustainable in terms of GHG production and required subsidies of energy, water, artificial fertilizers and pesticides (see chapters "The Journey from Peruvian Guano to Artificial Fertilizer Ends with too Much Nitrogen in the Chesapeake Bay" and "Pesticides Bring the War on Nature to the Chesapeake Bay").

The food system accounts for 25–30% of all anthropogenic GHG production. Livestock production dominates the food system designed to support the SAD and is responsible for half or more of food system emission. This includes the potent GHGs of methane (CH_4) and nitrous oxide (N_2O) (Barioni et al. 2019, Fig. 9). While CO_2 does promote crop growth, the other carboniferous GHG, methane, causes indirect damage to plants. Photooxidation of methane produces ground level ozone, a potent oxidizing agent that damages crops and that already accounts for about a 5% reduction in agricultural yield (Shindell 2016).

Poore and Nemecek (2018) summarize the environmental problems posed by animal-based diets, "In particular, the impacts of animal products can markedly exceed those of vegetable substitutes . . ., to such a degree that meat, aquaculture, eggs, and dairy use ~83% of the world's farmland and contribute 56–58% of food's different emissions, despite providing only 37% of our protein and 18% of our calories." "We find that the impacts of the lowest-impact animal products exceed average impacts of substitute vegetable proteins across GHG emissions, eutrophication, acidification (excluding nuts), and frequently land use. . . ." "Most strikingly, impacts of the lowest-impact animal products typically exceed those of vegetable substitutes, providing new evidence for the importance of dietary change."

Environmental Footprints of Foods Examining the environmental impacts of specific food-groups reveals the outsized role that animal products play in stressing the environment. Consider the environmental footprint formed by a serving of various foods (Fig. 10). The environmental footprint of a serving of animal product of any kind well exceeds that for almost any plant-based food. The primary reason is a basic ecological principle. Energy is lost with each

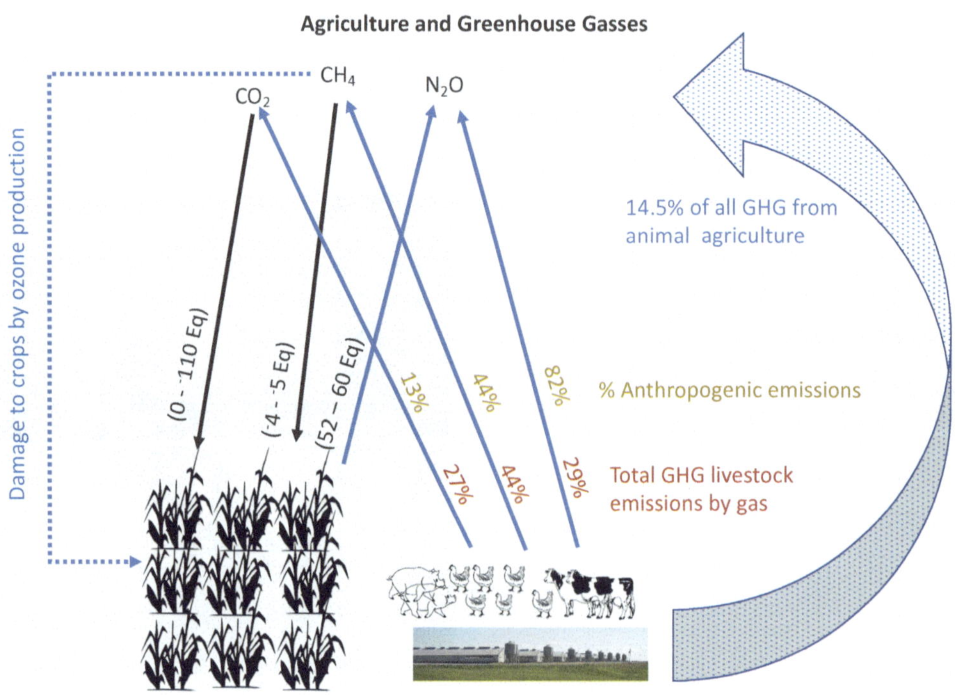

Fig. 9 The relationship between the primary greenhouse gases and agriculture. Crops range from neutral to a net sink for CO_2, are a net sink for CH_4, and a net source for N_2O. Eq = g m^{-2} year^{-1}. The food system accounts for about 25–30% of GHG production, 80–86% of that coming from farming. Farming produces more anthropogenic N_2O than any other activity, and is responsible for 44% of methane release. Livestock stock accounts for 14.5% of all anthropogenic GHG emissions. Although microbes in crop-soils eliminate some CH_4, plants often suffer from methane in the atmosphere that promotes the production of ground-level ozone. Data from Robertson et al. (2000), Vermeulen et al. (2012), Arneth et al. (2019), Barioni et al. (2019) and FAO (2019a)

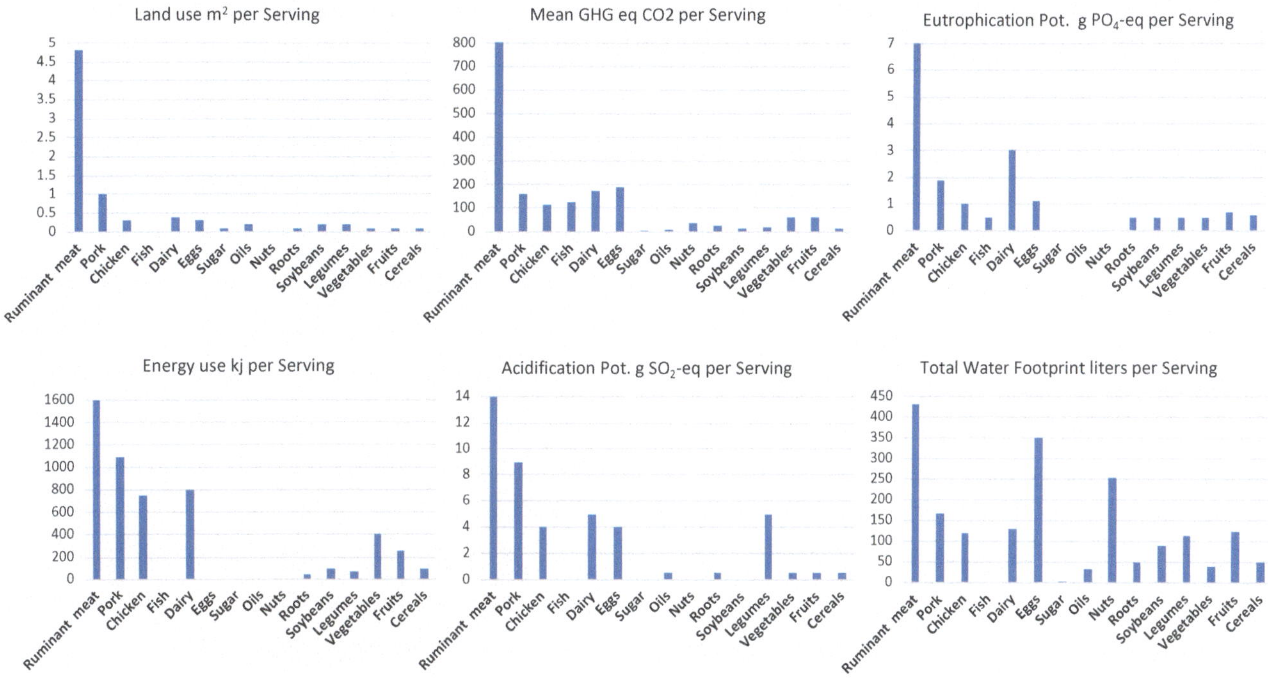

Fig. 10 The environmental footprints based upon portions of common foods consumed in the Chesapeake Bay Region. Portion sizes: For ruminant meat, pork, chicken, fish, nuts, soybeans, legumes, cereals = 28 g; For dairy = 128 g; For eggs = 56 g, For roots, vegetables, fruits = 129 g. Potential is abbreviated as Pot. for eutrophication potential as orthophosphate equivalents and acidification potential as equivalents of sulfate. Data from Willett et al. (2019), Mekonnen and Hoekstr (2010) and Ercin et al. (2011). Note that in all cases ruminant meat (mostly beef) exerts the largest impact on the environment. The absence of bars on the graphs for particular foods indicate a lack of available data. One kj = 0.24 kcal

step up the food chain. In nature, typically only 10% of the energy in primary producers is transferred into the biomass of consumers. That's because the consumers usually can't digest and absorb all of the plant tissue they consume, and what they do absorb is used for various metabolic processes in addition to growth. What consumers burn-up in metabolism won't add to their biomass. Much of the energy absorbed by homeothermic (warm blooded) animals is used to produce heat. All of the major sources of terrestrial animal food comes from homeotherms (cattle, poultry and pigs), meaning much of what they eat is turned into heat rather than flesh. In the US food system, the vast majority of cattle, poultry and pigs are produced in Concentrated Animal Feeding Operations (CAFOs). Rather than grazing from field or forest, these industrially produced animals eat mostly soy and corn. Indeed, 59.5% of US corn goes to animal feed, 30% to ethanol production (engine feed), and 10% for direct human consumption (mostly refined grains and high fructose corn syrup) (NCGA 2018). For soy, 70% goes to animal feed, 15% for human consumption after processing, and 5% for biofuel (USDA 2018; USDAERSA 2019). Feeding the soy and corn to an animal, and then eating that animal means that most of the nutrition from those plants is lost before it gets into one's mouth. Consider pork and soy. It takes 1 m^2 of land area to produce one serving of pork, while it takes 0.02 m^2 of land to produce a serving of soy (Fig. 10).

Traditionally, livestock farmers use a Feed Conversion Ratio (FCR) to determine how much animal product results from a given amount of food. This is typically done as a ratio of weight of the feed to live weight of the animal. FCR is variable, depending on the feed composition and the conditions faced by the livestock. FCR for the most common livestock are; 8–12:1 for beef cows, 5–6.5:1 for pigs, and 2–2.5:1 for chickens (Smil 2000). FCR is deceptive when considering nutrition available to humans, since even the most enthusiastic meat-eaters generally don't consume the entire body of an animal. Processing, butchering, and dining utensils removes uneatable parts from servings of meat. A better way to understand the lack of efficiency associated with livestock is to calculate the transfer of energy from the feed to the edible portions of the animal (Table 1).

For ruminant animals (cows and sheep) only about 1% of the energy in their feed makes it into the edible meat they produce. Protein conservation is only slightly better, in the range of 3–4%. Poultry is the most efficient, transferring 13% of the energy and 20% of the protein from its feed in to meat. Considering that modern intensive animal agriculture is mostly based on very edible (for humans) grain and soy, the practice is clearly wasteful, losing 87–99% of the energy and 75–97% of the protein in the feed.

Globally, livestock produces 14.5% (7.1 gigatons CO_2-eq year^{-1}) of all human induced GHG emissions. To put this in

Table 1 The percent efficiency of edible output of energy and protein from 100 units of feed

Animal	Caloric efficiency %	Protein efficiency %
Sheep	1	3
Beef Cow	1	4
Farmed shrimp	7	15
Dairy Cow	7	16
Pig	10	15
Poultry	11	20
Farmed finfish	12	18
Chicken egg	13	25

The input includes both human-edible (grains, beans, oils) and human-inedible feeds (grasses, crop residues). Note the inefficiency of beef. Data from Searchinger et al. (2019)

perspective consider the magnitude of GHG production by the other main sectors of the global economy; electricity and heat production (35%), industry (21%), transportation (14%), non-animal agriculture (9.5%), buildings (6%) (USEPA 2020). *This means that the portion of the food system devoted to eating animals produces more GHG than comes from autos, trucks, aircraft, trains and ships combined.* All aspects of the food system that produces livestock contribute to GHG production as follow; enteric emissions (39%), feed production (36%), manure (10%), land use change to support animal husbandry (9%), processing and transportation (6%) (Edenhofer et al. 2015). Most of the emissions come from ruminant animals, 41% from beef cattle, and 20% from dairy. Monogastric (simple stomach) animals contribute less; 9% from pigs, 8% from chicken eggs (Gerber et al. 2013).

Meat and dairy from ruminants account for 13 g of protein person^{-1} day^{-1}, about a third of the world's animal protein supply. CAFOs supply 87% of ruminant meat and 6% of dairy production. The balance comes from animals grazed on grasslands. CAFO raised animals receive a grain based diet, generally produced off-site. Some farmers raise animals in a mixed system, combining some grazing with supplemental feed. Proponents of grazing animals on rangeland suggest that it results in increased carbon storage in soil and a smaller GHG hoofprint. Yet, even cattle raised on rangeland are net producers of GHGs, and there is insufficient land available to meet current demand for meat and dairy (Garnett et al. 2017).

Per serving, ruminant meat (mostly beef) exerts an impact on the environment well in excess of any other type of food. Compare a serving of beef to a serving of legumes, another protein rich food. The beef uses 24 times the amount of land, produces 80 times the GHG, yields 14 times the phosphorus water pollution, requires 80 times the amount of energy subsidy, makes 2.8 times the amount of acid-forming pollution, and uses 4 times the amount of water (Fig. 10). The outsized environmental footprints of ruminants traces to the combination of low FCR and the way they digest their food. The low FCR means that ruminants require large swaths of land for the production of feed. This explains much of the environmental footprint of ruminants. The phosphorous pollution comes from runoff of fertilizer used on feed crops. This leads to eutrophication of waterbodies. The energy subsidy is from production of reactive nitrogen fertilizers, transportation, farm equipment, processing, and refrigeration. The acid-forming pollution comes from emissions of CO_2, SO_4, and N_2O. All of these gasses produce acids in the atmosphere, and hence acid precipitation. The water use is from irrigation, CAFO maintenance, and food processing.

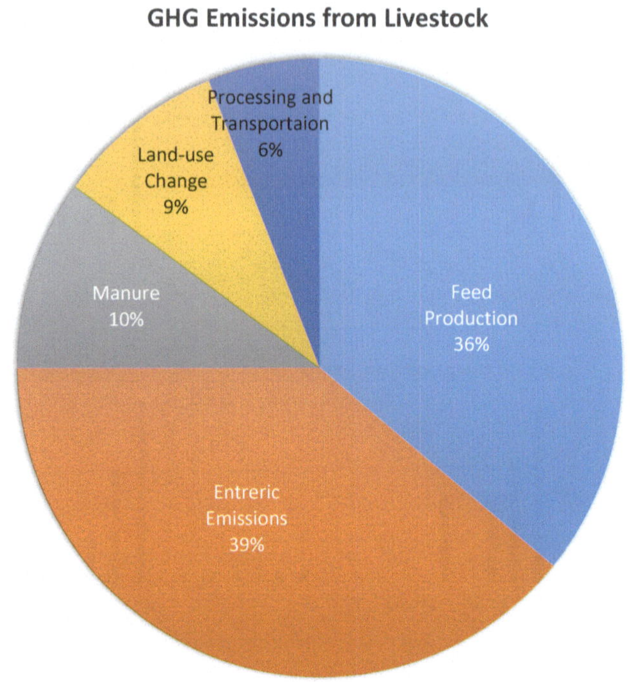

Fig. 11 Greenhouse gas emissions (GHG) from the main sectors associated with the production of livestock. Note that the enteric emissions come from cattle and other ruminants. Land-use change refers to the loss of natural carbon sinks in the natural vegetation and soil of land that is converted to crop production. Data from Garnett et al. (2017)

The low FCR of ruminants also explains a portion of their GHG footprint. The extensive use of fertilizers on feed crops results in NO_2 emissions. The main fraction of GHG comes from the animals themselves in the form of CH_4. Ruminants evolved to eat grasses and other high-fiber plants. Since most of what they ate was cellulose, they evolved a mutualism with microbes that produce the enzyme cellulase, which breaks down the cellulose. Digestion takes place in a specialized four-chamber stomach. One chamber called the rumen is where the mass of digesting fibrous food is fermented, producing copious quantities of methane which is burped out the atmosphere (Garnett et al. 2017). The production of methane by ruminants accounts for 39% of all GHG emissions associated with livestock (Fig. 11).

The extraordinary environmental footprints of beef, dairy and other foods from ruminants surely limits or eliminates their role in a sustainable diet for the Chesapeake Bay Ecosystem. What about non-ruminant animal food? Although carrying half or less the environmental foot print of ruminants, these animals still require more of the environment than almost all plant-based foods (Fig. 10).

Seafood Seafood produced from the Chesapeake Bay spares the land from the production of crops used to feed terrestrial animals. It is instead fed by the Bay's natural food chain, which is highly productive. However, harvesting seafood comes with environmental consequences. The more fish, oysters, crabs, clams and mussels that people catch, the less of these are available for their non-human predators. Some fishing methods destroy important habitats, such as the case with oyster reefs (see chapter "The Chesapeake Bay Oyster: Cobblestone to Keystone"). Destructive overfishing devastated the Bay's oyster population and contributed to the decline of other species. Ecosystem based management of fisheries takes into account the role of these species beyond food for humans, and is discussed in chapter "A Fishing Trip: Exploiting and Managing the Commons of the Chesapeake Bay". In addition to the potential for damaging natural food webs, fishing requires energy subsidies to power; fishing vessels, processing plants, and refrigerated transportation. In general, seafood generates about ten times the GHG of protein rich legumes (Fig. 10). A study of GHG emissions from Chesapeake Bay seafood is needed to quantify the footprint of the industry. It's reasonable to assume that the relatively short distances required by vessels on the Bay to get to fishing grounds means much less fuel consumption compared to coastal and open-sea operations.

Aquaculture is the world's fastest growing food sector. In 1970 it accounted for only 4% of the world's seafood. In 2016 aquaculture produced 80 million metric tons of seafood, nearly as much as the 91 million tons obtained by the world's capture fishery (FAO 2019b). There are essentially three types of aquaculture for human food; (1) farming of seaweeds, (2) un-fed animal systems where nutrition is provided by the environment, and (3) fed animal systems that rely on industrially produced feeds, typically made from wild-caught species of lower economic value fish and sometimes soybeans or other crops. Seaweed farming is fairly benign but not practiced on the Chesapeake Bay. Fed-aquaculture carries a large environmental footprint that includes water pollution, habitat destruction and wasteful harvesting of seafood for the production of feed. The Bay's largest fishery is that for menhaden (*Brevoortia tyrannus*), which is shipped to feed farm raised salmonids and other species in facilities far from the shores of the Chesapeake (see chapter "Menhaden, the Inedible Fish that Most Everyone Eats"). Fortunately, fed-aquaculture farms aren't a significant part of the Bay's current food system. This may change as state authorities sponsor research to promote fed-aquaculture in the region (VATech 2019). Bay area residents intersect with fed-aquaculture when they purchase farm-raised salmon and shrimp. Indeed, a meal of farm-raised seafood may contain mostly protein and oils that originated in the Chesapeake Bay in bodies of menhaden.

Un-fed aquaculture is a growing industry on the Bay. Farmers raise native oysters (*Crassostera virginica*) and clams (hard clam, *Mercenaria mercenaria*). In Virginia these two produced $53.4 million of sales in 2017, 70% going to the more lucrative hard clams (Hudson 2018). Oysters are raised in cages or containers that exclude predators. The cages may be placed on the bottom or suspended in the water column. Hard clams are grown in the sediments and typically protected from predators by a mesh laid down after planting the young bivalves into the bottom. Both species feed on natural populations of plankton, providing the ecosystem service of removing excess algae from the water. Farmers depend on specialized hatcheries to produce the juvenile oysters and clams. Bivalve aquaculture is highly sustainable, providing food and cleaner water with little discernible impact on the environment. However, these suspension feeders may collect pathogens and toxic contaminants, presenting a potential health risk to consumers (see chapter "Instead of Eating Fish: The Health Consequences of Eating Seafood from the Chesapeake Bay Compared to Other Choices"). Certainly, farm-raised oysters and clams present the most sustainable source of animal food for people insistent upon having such in their diet.

The processed food industry offers an alternative for those wanting to abandon consumption of animal products but retain the experience of dining on seafood. Tyson Foods now owns a stake in New Wave, a company that uses seaweed and soy to make a plant-based facsimile of shrimp (Shanker 2019a). The fake shrimp eliminates the environmental damage of fishing for or culturing the real thing. It's also healthier, providing fiber and eliminating the cholesterol found in this crustacean.

Eating Food from Plants and Not Animals Is the Most Sustainable Choice Barioni et al. (2019) summarized numerous studies to determine which diet improvements offer the greatest potential for mitigating GHG production (Fig. 12). Generally, the higher the amount of animal-based food in the diet, the lower the mitigation potential. The vegan diet, which excludes all animal products, proved the best. For example, the "Climate Carnivore" diet that replaces 75% of ruminate meat and dairy with other non-ruminant meat, achieves only 44% of the GHG mitigation created by the vegan diet. Eating only plants also means producing the smallest footprints for land use, energy subsidy, acidification, eutrophication, and water consumption. "Moving from current diets to a diet that excludes animal products has transformative potential, reducing food's land use by 3.1 (2.8–3.3)

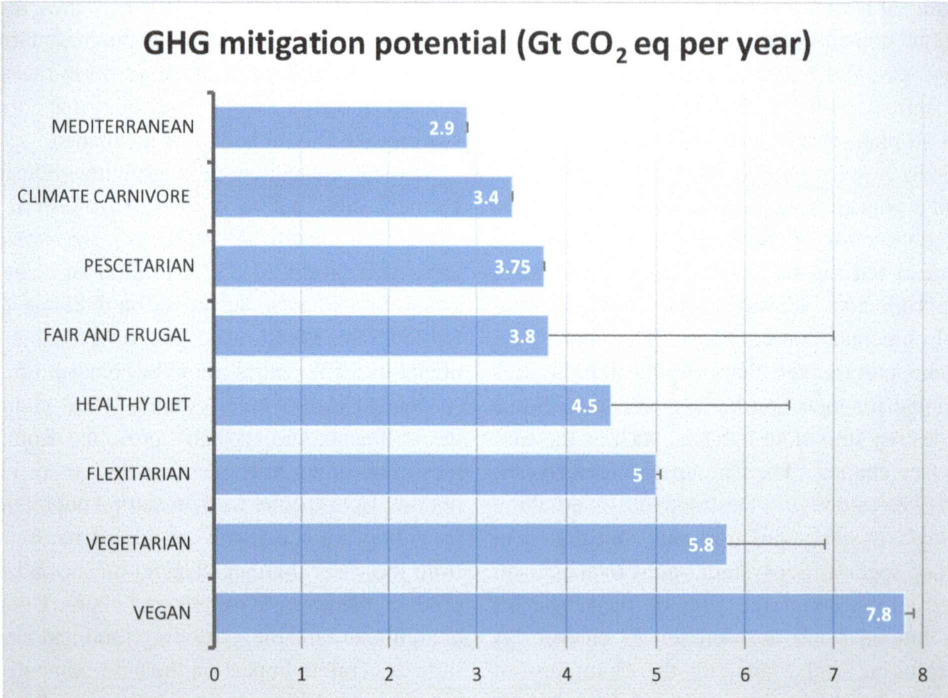

Fig. 12 The mitigation potential for greenhouse gas (GHG) reduction in 2050 for various diets. Estimates include additional effects of carbon sequestration from land-sparing. Data without error bars are from one study only. Vegan = Completely plant-based; Vegetarian = Grains, vegetables, fruits, sugars, oils, eggs and dairy, and generally at most one serving per month of meat or seafood; Flexitarian = 75% of meat and dairy replaced by cereals and pulses; at least 500 g per day fruits and vegetables; at least 100 g per day of plant-based protein sources; modest amounts of animal-based proteins and limited amounts of red meat (one portion per week), refined sugar (less than 5% of total energy), vegetable oils high in saturated fat, and starchy foods with relatively high glycemic index; Healthy diet = Based on global dietary guidelines for consumption of red meat, sugar, fruits and vegetables, and total energy intake; Fair and Frugal = Global daily per-capita calorie intake of 2800 kcal/cap/day (11.7 MJ/cap/day), paired with relatively low level of animal products; Pescetarian = Vegetarian diet that includes seafood; Climate carnivore: 75% of ruminant meat and dairy replaced by other meat; Mediterranean: Vegetables, fruits, grains, sugars, oils, eggs, dairy, seafood, moderate amounts of poultry, pork, lamb and beef. Figure is after Barioni et al. (2019)

billion ha (a 76% reduction), including a 19% reduction in arable land; food's GHG emissions by 6.6 (5.5–7.4) billion metric tons of CO_2eq (a 49% reduction); acidification by 50% (45–54%); eutrophication by 49% (37–56%); and scarcity-weighted freshwater withdrawals by 19% (−5–32%) for a 2010 reference year.... For the United States, where per capita meat consumption is three times the global average, dietary change has the potential for a far greater effect on food's different emissions, reducing them by 61–73% (Poore and Nemecek 2018)."

The USDA doesn't presently include the role of sustainability in figuring their dietary recommendations. Indeed, universal adoption of their ideal diet would result in a 12% increase in GHG emissions in the US. This is due to the high content of meat and dairy (Heller and Keoleian 2014). As discussed in chapter "Ethics and Economics of Building a Food System to Recover the Health of the Chesapeake Bay and Its People", the USDA is conflicted by its mandate to promote agricultural products even when they negatively impact human and environmental health.

Widescale adoption of a plant-based diet means reducing the amount of land under cultivation in the Chesapeake Bay Watershed (CBW), and the attendant environmental foot prints. Pennsylvania, Virginia and Maryland account for most of the crop production in the CBW. For these three states respectively, 75.3%, 56.7% and 68.6% of their agricultural land is devoted to producing animal feed (see Table 4 in chapter "The Journey from Peruvian Guano to Artificial Fertilizer Ends with too Much Nitrogen in the Chesapeake Bay"). Reorienting agriculture in the CBW for a vegan diet will eliminate most, but not all of the cultivation on lands presently devoted to animal-feed. A small portion of these lands (about 10%) will need to stay active to produce more plant-based foods to replace the nutrition from animal products. For the region this means that over half of currently cultivated land will be allowed to revert to forest. The new forests would provide a growing carbon sink and other ecosystem services such as clean water, reduction of air pollution, and wildlife habitat. Agriculture is reponsible for 92% of the global human water foot print, and nearly a third of this traces to the prodution of animal foods (Gerbens-Leens et al. 2013).

Plant-Based Diets Improve Human Health Human health is part of a sustainable ecosystem, as humans are keystone species in this epoch of the Anthropocene. The vegan diet wins the competition for the best human health outcomes (Sabaté and Soret 2014). A 5-year-long study with 96,000 participants found that plant-based diets were associated with; lower body mass index, lower rates of type-2 diabetes, lower prevalence of metabolic syndrome, lower all-cause mortality, and lower cancer rates (Orlich and Fraser 2014). Consumption of animal products is associated with numerous preventable maladies including; coronary artery disease (number one source of mortality in the US and the world), type-2 diabetes, various cancers, obesity, Alzheimer's dementia, hypertension, and ischemic stroke (Godfray et al. 2018; see chapters "Instead of Eating Fish: The Health Consequences of Eating Seafood from the Chesapeake Bay Compared to Other Choices", "Sugar Twice Enslaves: Consequences for the People of the Chesapeake" and "Eutrophication: Obesity of the Bay and Its People"). Consumption of plant protein is associated with increased longevity, while increased animal protein predicts greater mortality (Song et al. 2016). Consequently, vegans live several years longer than omnivores (Fraser and Shavlik 2001). Yet, some authors still allow for animal-based foods in reduced amounts as part of proposed healthy and sustainable diets (Tilman and Clark 2014; Barioni et al. 2019; Willett et al. 2019).

According to Willett et al. (2019); "Healthy diets have an appropriate caloric intake and consist of a diversity of plant-based foods, low amounts of animal source foods, unsaturated rather than saturated fats, and small amounts of refined grains, highly processed foods, and added sugars. Transformation to healthy diets by 2050 will require substantial dietary shifts, including a greater than 50% reduction in global consumption of unhealthy foods, such as red meat and sugar, and a greater than 100% increase in consumption of healthy foods, such as nuts, fruits, vegetables, and legumes. However, the changes needed differ greatly by region. Dietary changes from current diets to healthy diets are likely to substantially benefit human health, averting about 10.8–11.6 million deaths per year, a reduction of 19.0–23.6%." Why not propose complete elimination of animal products from the ideal diet? Most authorities probably view such a change as too radical to be accepted by the population. And in some cases the authors work in the context of funding agencies such as the USDA or FAO that have arms which actively promote the production and consumption of meat and dairy. As this book goes to press in June of 2020 the world is gripped by the Covid-19 pandemic. As with all other human epidemics, the disease jumped from an animal host (Ahmad et al. 2020). Such zoonotic diseases trace to domesticated animals or those gotten as bush meat or in the live animal trade (Greger 2006). The comorbidities associated with the consumption of meat (cardiovasular disease, hypertension, diabetes) increase the risk of Covid-19 (Wang et al. 2020). Avoiding and surviving future pandemics require eliminating or drastically reducing the role of animals in the food system.

Rationalizing the Consumption of Animal Products People of the Chesapeake Bay Region steeped in the SAD may question the efficacy of a plant-based diet in delivering complete nutrition, primarily with concern to protein. Consumers in the US typically eat twice the recommended amount of protein, and this is also true for vegans. Chapter "Instead of Eating Fish: The Health Consequences of Eating Seafood from the Chesapeake Bay Compared to Other Choices" addresses the mythology of protein in detail. In addition, plant-based protein comes at about 2% of the energy costs associated with that from animals (González et al. 2011).

Individuals generally quit eating animal products primarily for one or more of three reasons; damage to the environment, poor health outcomes, and concern for animal welfare (see chapter "Ethics and Economics of Building a Food System to Recover the Health of the Chesapeake Bay and Its People") (Vongkiatkajorn 2019). That consumption of animals remains so widespread in the US and most developed nations begs the question of how omnivores decide to continue this practice in the face of evidence speaking against it? Joy (2010) introduced the three N's to explain why people continue to eat meat. Since most people are raised in families that eat meat, it appears *Normal*. Society teaches that humans are superior animals and it is *Natural* for them to eat lesser ones. The mythology of protein also convinces many consumers that eating meat is *Necessary* for their health, despite abundant evidence otherwise. Omnivores use the three N's to defend consumption of animal products to their own conscious and to their social network. Piazza et al. (2015) added a fourth N for *Niceness* to capture the enjoyment derived from consuming meat, and tested the efficacy of the four N's in a series of studies. They found the four N's encompassed 83–91% or reasons given for eating meat. The four N's differed in frequency of response; *Necessary* 36–42%, *Natural* 17–23%, *Nice* 16–18%, *Normal* 10–12%. Omnivores endorsed the four N's, while they were mostly rejected by vegetarians and vegans. Those embracing the four N's tended to objectify the animals they ate, attributed little mental capacity to cows, were suppurative of social inequality, unmotivated by ethics in making food choices, and frequently ate animal products. "…it would seem that the 4Ns are a powerful, pervasive tool employed by individuals to diffuse the guilt one might otherwise experience when consuming animal products (Piazza et al. 2015)."

Reducing Consumption of Animal Products Attempts to reduce the consumption of animal products in the Chesapeake Bay Region may benefit from adopting a dual-process model of decision making. Food labeling, education,

Table 2 Nutritional value, GHG CO_2-eq, and liters of gasoline equivalents of three non-meat burgers as compares to one typical fast-food meat brand

	Earth Grown™ soy protein burger	Beyond Burger™	Impossible Burger™	Burger King™ Burger
Kcal	129.0	270	240	261.0
Protein g	20.6	20	19	14.1
Total fat	1.3	20	14	10.4
Saturated fat	0	5	8	3.8
Fiber	5.2	3	3	1.0
Kg GHG Eq-CO_2/100 g	0.07	0.07	0.07	2.9
Liter gasoline-Eq	0.02	0.02	0.02	1.2

Values are per 100 g. Data from the nutrition facts labels of the products and Clune et al. (2017) for GHG emissions. The estimates of GHG is based upon values for legumes (assigned to the non-meat burgers) and beef. Gasoline contains 2.35 kg CO_2 per liter (USEPA 2019b). Absent the energy cost of food processing, these values may be conservative. The two imitation meat burgers present nutritional profiles similar to a traditional hamburger, including high caloric and saturated fat content. The soy burger also provides the highest content of health-promoting fiber

and taxes on animal products would appeal to the consumer's reflective decisional system. These would arm citizens with rational reasons to reduce or eliminate consumption of animal products. However, rational thought doesn't motivate everyone to the same degree. Many people might be swayed to reduce meat consumption by appeals to their automatic decisional system which is driven by immediate perceptions and short-term impulses. Examples include; restructuring restaurant menus to include vegan items first, increased marketing of non-meat alternatives, and decreasing the portion size of meat in restaurants (Godfray et al. 2018).

Eliminating or reducing consumption of animal products requires that consumers reshape the way they eat. Advocates of whole food plant based (WFPB) diets suggest eating a variety of; fruits, vegetables, legumes, whole grains, nuts, spices and herbs. This diet eschews highly processed foods and draws upon traditional ethnic recipes (ex. Asian, Mexican and African cuisines) as well as novel efforts to produce appealing dishes (Greger 2017). However, people transitioning from the SAD to a healthier and more sustainable diet may find it difficult to leave behind familiar forms of food. For many this means replacing animal foods with plant-based facsimiles. For example, lightly processed veggie burgers made from beans, ground flaxseed, onions and peppers may not satisfy those used to eating hamburgers made from beef. Food corporations, including some heavily vested in the meat industry, now produce plant-based burgers designed to provide the taste and feel of eating ground beef. These alternative products pose two problems. First, their ingredients may be unhealthy, including levels of saturated fats above that for beef (Table 2). One brand (Impossible Burger™) uses genetically modified yeast to produce heme-iron to simulate the taste of eating muscle. Heme-iron is pro-inflammatory, carcinogenic, linked to coronary artery disease and other maladies (Ward et al. 2012; Fang et al. 2015). That brand uses highly processed GMO soy, coconut oil, and a host of other ingredients to make an earnest imitation of ground beef. This product eliminates most the environmental footprint of beef, but isn't as healthy as a simple traditional veggie burger (Table 2). Second, by so closely imitating meat, this burger is unlikely to help a consumer to transition their palate away from animal-based products. Indeed, most committed vegans, particularly those embracing animal welfare, aren't looking to pretend they are eating meat. Vandana Shiva decries the rise of fake foods as contributing to the industrialization of the world food system, resulting in proliferation of chemical agriculture, loss of biodiversity, the commodification of life and extending corporate control (Shiva 2019).

Despite presenting a nutritional profile not much healthier than traditional meat-based burgers, demand for the Impossible Berger™, overwhelmed the supply during the first half of 2019, shortly after its introduction to the market. Increased manufacturing capacity has placed the burger in over 9000 restaurants across the US, including major fast-food chains (Shanker 2019b). Regardless of the down-sides of fake meat, the quick acceptance of such products indicates that the public is ready to experiment with significant changes to their diet. Will embracing fake meat be a gateway to increasing the level of processed food in the diet? Or will it enable more people to make the transition to a whole food plant based way of eating?

Breastfeeding Improves the Health of People and the Environment

Dawn Gerbing

The story of food, the health of the people of the Chesapeake Bay and the health of the Bay wouldn't be complete without including the smallest members of the community; infants in their first years of life. Breastfeeding was the healthy and only option for all of the native and immigrant peoples of the region until the development of rudimentary cows' milk and soy-based formulas in the early 1900s. There have

(continued)

always been a few infants and mothers needing medical nutrition intervention. However, with the introduction and marketing of commercially produced formulas in the early 1950s, breastfeeding rates plummeted to an all-time low in the early 1970s with fewer than 25% of mothers initiating breastfeeding their baby at all (Gordon 2019).

Through the combined efforts of the National Institute of Health and Human Services, Food and Drug Administration, World Health Organization, Surgeon General's Office and many health professionals' organizations, breastfeeding has made a comeback with 75% of mothers in the United States breastfeeding in the hospital after delivery. However, the USDA estimates that approximately 28 billion ounces (83 million liters) of infant formula are sold in the US each year.

The breastfeeding rates for the states of the Chesapeake Watershed are similar to the national average with the exception of West Virginia (Table 3). However, racial disparities mean that African American women are about 20% less likely to initiate breastfeeding than their white counterparts throughout the watershed (Table 4).

The deleterious health impacts of an infant denied breastmilk are lifelong. The Surgeon General points out that formula fed infants are at much greater risk for developing; ear, upper respiratory and gastrointestinal infections, childhood obesity, asthma, diabetes mellitus, leukemia and sudden infant death syndrome. Also, formula feeding mothers are at greater risk for breast and ovarian cancer (Office of Surgeon General 2019).

The environmental impacts of formula production include not only the production of the cow's milk and transporting that milk to processing plants, but the bottles, packaging, and transport to market. For every one million formula-fed babies, 150 million containers of formula are consumed with most containers ending up in landfills (Torgus and Gotsch 2004). Add to these factors the amount of medical waste for each hospitalized child that may have been protected by breastmilk and the environmental impacts are significant.

When examining how the youngest people of the Chesapeake Bay Watershed are fed, it is important to consider that for a small amount of additional food each day, a mother can provide nourishing, protective milk at the right temperature from the perfect container with no harmful impacts known.

Turning Wasted Food into a Resource North Americans waste 42% of the available food supply. This is about twice the rate for the rest of the world. On a global basis, if wasted food were a nation it would be the third largest emitter of GHG. Its annual 4.4 GT of CO_2-eq is behind the national totals for China's (10.7 GT of CO_2-eq) and the US (5.8 GT of CO_2-eq). In the US, food is lost at all stages of the food system. In terms of calories it's; 17% during production, 6% in storage and handling, 9% in processing, 7% in distribution and marketing, and 61% during consumption. The consumption stage refers to waste once food is bought by consumers, restaurants or caterers. This includes loss due to spoilage, uneaten portions of meals, and discarded products (Searchinger et al. 2019). Perhaps due to their short shelf-live, animal products account for 73% of all the calories of food loss at retail plus consumer levels (Heller and Keoleian 2014). Since most food waste in N. America occurs with the consumer, the responsibility for correcting the problem lies at the intersection of people and the food system.

Just as the residents of the Chesapeake Bay Region will need to alter their diet to protect their health and that of the environment, they will need to address their own wasting of food. This includes; not over-buying groceries, asking for smaller portions, not leaving food on their plate, and pressing restaurants and caterers to reduce food waste. For example, cafeterias that eliminate trays for carrying plates and

Table 3 Breastfeeding rates for the jurisdictions of the Chesapeake Bay Watershed Compared to the National Average (CDC 2019a)

Jurisdictions	Ever breastfed	Breastfeeding at 6 months	Breastfeeding at 12 months	Exclusive breastfeeding at 3 months	Exclusive breastfeeding at 6 months
US National	83.2	57.6	35.9	46.9	24.9
Delaware	77.4	55.6	33.4	47.2	23.6
District of Columbia	83.0	65.5	43.6	52.6	29.1
Maryland	91.0	66.8	41.1	50.1	26.2
Pennsylvania	83.8	59.2	39.0	48.9	25.6
Virginia	81.7	62.5	39.3	45.6	26.6
West Virginia	68.6	40.1	24.3	36.3	20.2

Table 4 Comparison of breastfeeding initiation rates by race for the jurisdictions of the Chesapeake Bay Watershed Compared

Jurisdictions	Breastfeeding initiated White (Non-Hispanic)	Breastfeeding Black (Non-Hispanic)
Delaware	73.7	59.5
District of Columbia	96.3	65.5
Maryland	85.4	75.4
Pennsylvania	77.8	64.4
Virginia	84.2	67.0
West Virginia	61.1	–

West Virginia didn't report breastfeeding rates for African Americans (CDC 2019b)

beverages can achieve a 32% reduction in food waste (Kim and Morawski 2012).

Even the best conservation efforts will result in some food loss, and that material should be thought of as resource rather than a waste product. Uneaten food should be composted for amending soil rather than sent to a landfills or through garbage disposals to a wastewater treatment plant. In the US, only 5.3% of food waste is composted, the rest going to landfill or incineration. Discarded food accounts for 22% of municipal waste (USEPA 2019a). The composting process should be aerobic to avoid production of methane and oxides of sulfur and nitrogen that occur during anaerobic decomposition. However, anaerobic digestion in specialized digesters is useful for producing contained biogas from both food waste and manures. This is a common practice at rural home-settings in Asia and Africa where the fuel is used for cooking (IRENA 2019), but mostly restricted to larger industrial scale operations in the US. Smithfield Foods, on the Pagan River of the lower Chesapeake Bay, is building a large capacity anaerobic digester for biogas production from slaughterhouse waste and manure (Goldstein 2019). Freestate Farms runs a facility in Manassas, VA that produces biogas from food and farm waste (Waste Advantage 2018).

People living in housing that includes some space for gardens can easily compost their kitchen wastes and use the residue to enrich the soil. This isn't an option for apartment dwellers, but some municipalities in the Chesapeake Bay Region offer voluntary curbside pick-up of kitchen wastes. This comes with a cost, about $30 per month in the District of Columbia area. Falls Church, VA subsidizes this service, providing it for $6 per month (FallsChurch 2019). Prohibiting placing kitchen wastes in regular trash receptacles and providing free curbside pickup for compostables would help turn discarded food back into a useful resource.

Although the distribution sector (retail and wholesale) is responsible for only 7% of food waste in the US, much of what they discard can still be welcome nutrition for needy consumers. Various non-profit organizations coordinate the collection and delivery of soon-out-of-date processed food and aged produce to food-insecure clients. One example in the Bay Region is Community Food Rescue (2019) that operates in Montgomery County (between Baltimore and D.C.). Legislation to require food-distributors to participate in such programs could further reduce waste and feed more people.

Features of a Sustainable Food System for the Chesapeake Bay Region Moving to an entirely plant-based food system will provide the best health and environmental outcomes for the Chesapeake Bay region (Fig. 13), and for the planet in general (Alexsandrowicz et al. 2016). However, doing such stands in direct contradiction to the longstanding relationship between residents and animal seafood. Fishing is a dominant form of outdoor recreation in the Middle Atlantic, engaging 11.6% of the region's residents (Lock 2019). The obituary pages of newspapers in the region often feature photos and testimonials documenting the deceased's devotion to fishing. Automobile license plate frames proclaim, "I'd rather be fishing." Sportfishing organizations often form alliances with organizations devoted to restoring the environment. Fishers harvest 227 million kg (500 million pounds) of seafood from the Bay annually. This means $4.4 billion of sales and 34,000 jobs in the local economy (CBF 2019). So, proposing to avid fishers that they forgo their passion for the sport doesn't appear to be a winning strategy for effecting food system change in the region. As it goes, well-regulated local fisheries, along with oyster and clam aquaculture represents a more sustainable source of animal flesh than terrestrial options. However, its difficult to justify the Bay's main fishery, the very large menhaden reduction-effort that produces animal feed for aquaculture production outside of the region (see chapter "Menhaden, the Inedible Fish that Most Everyone Eats").

Eliminating terrestrial animal foods will also encounter resistance, as noted above in the discussion of the four N's. This requires people to learn about the advantages of plant-based foods, and how to prepare them to produce tasty and satisfying dishes. Retraining the pallet also means overcoming addictions to certain foods, such as sugars (see chapter "Sugar Twice Enslaves: Consequences for the People of the Chesapeake") and dairy. Cow's milk contains Beta-casomorphins. These opioid peptides promote bonding between calf and cow. They bind to receptors in the brain, leading to the release of dopamine, that confers the sense of pleasure and reward. So when people say they "love cheese," they mean it (Nguyen et al. 2014). Meat also contains opioid receptor ligands; albumin and hemoglobin, that may form the basis for addiction. Yet, people can also learn to love some plant food in the same way, as spinach and wheat contain similar promoters (Teschemacher 2003).

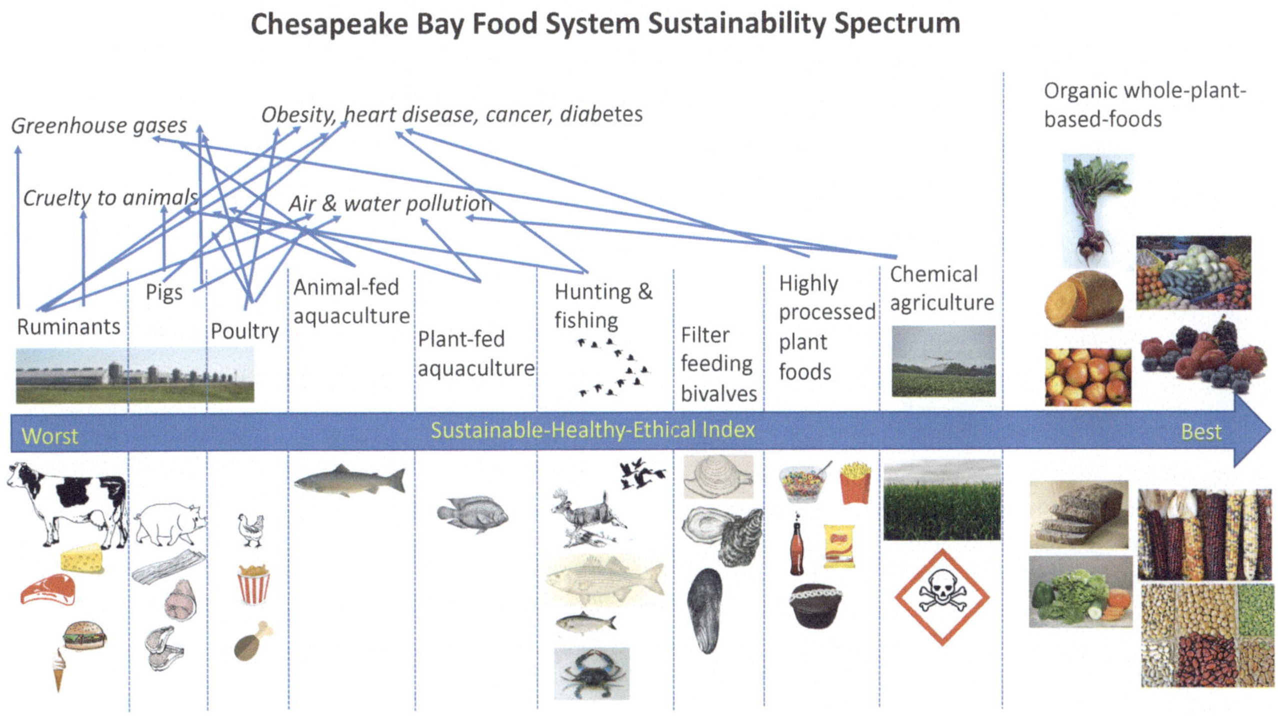

Fig. 13 The sustainability spectrum for the Chesapeake Bay food system. Foods and the way they are produced are represented along a gradient that combines sustainability, human health and animal welfare considerations. Foods from cattle and other ruminant animals are the worst in all regards. Other live stock are less polluting but still compromise health and the environment, as well as unethical treatment of animals. Well-regulated hunting and fishing reduces environmental damage. Wild-caught and cultured filter feeding bivalves may benefit the environment. Highly refined plant foods (mostly grain based) are often raised with chemical agricultural practices. The high caloric and low nutrient density of these junk-foods endangers health. Chemical agriculture damages the environment. The best choice in all regards is a whole-plant-based-diet base with organically grown foods. This produces the best health outcomes, protects the environment, and spares animals from systematic cruelty

The sustainable food system of the future will use organic farming practices, eliminating environmentally damaging pesticides and artificial fertilizers. The drastic reduction of land devoted to producing animal-feed will also mean a deintensification of agriculture, reducing energy and water subsidies. Greater emphasis on producing crops for local consumption will reduce the environmental cost of transporting food over great distances.

To address the changing climate, farmers will need to reintroduce biodiversity of crops, finding those cultivars that do best in their particular microclimates and that can thrive in the regime of higher temperatures and different hydroperiods. This goes against the grain of the current industrial-chemical-agribusiness model that thrives on a few genetically uniform crops sustained by massive inputs of pesticides and fertilizers. "…of more than 14,000 edible plant species, only 150–200 are used by humans with only three (rice, maize, and wheat) contributing 60% of the calories consumed by humans. Many underused plant species have excellent nutritional profiles, as well as traits of interest for adapting food production to climate change (i.e., quinoa, millet, sorghum, or teff for grains, or zapote, chaya, or chenapodes for fruits and legumes). These qualities are especially important considering the increasing risk that climate change will pose on crop yields and the nutritional content of foods. However, food system simplification drives loss of these plant species and varieties, reducing options that support healthy diets from sustainable food systems (Willett et al. 2019)." Agriculture scientists and farmers of the Chesapeake Region will need to meet the challenge of developing a variety of crops that provide consumers with diets rich in whole grains, legumes, vegetables, fruits and nuts.

Even the most sustainable organic farm requires energy subsidies for powering its various operations. The farms can potentially create all the energy they need. Some farms in portions of the watershed (coastal and higher elevation areas) may have sufficient average windspeeds to drive turbines for producing electricity. Photovoltaic solar panels on the roofs of farm structures can provide clean energy. And certain shade-loving crops produce higher yields under arrays of solar panels placed high enough to allow for tractors to pass below (Barron-Gafford et al. 2019). Such *agrivoltaic* installations mean that solar farms need not displace food producing fields.

Although half of the US corn crop goes to producing ethanol, such plant to biofuel operations are barely sustainable, requiring land use change, large inputs of fossil fuels, water, pesticides and fertilizer (Searchinger et al. 2008).

Modern photovoltaic panels can convert 22% of the incident energy from sunlight into electricity (Green et al. 2016). Farm-based biofuel operations can't come close to matching that, as photosynthesis at best converts 6% of incident light energy in to biomass (Zhu et al. 2010) and that would only be during the nine-month-long growing season in Chesapeake Bay area. Perhaps 10% of the biomass of a corn plant attributes to its kernels that are used for ethanol production. And only 49% of the energy in those kernels makes into ethanol (Huang and Zhang 2011). This calculates out to about 0.3% efficiency of converting sunlight into biofuel energy.

Ultimately, consumers will drive the changes to produce a sustainable food system. This includes reducing their role in wasting food, as well as demanding healthier plant-based choices with their dollars. The government can promote this change by subsidizing healthy and sustainable foods and taxing those that aren't (Wirsenius et al. 2010). "Agricultural and nutritional policies that lead to the adoption of plant-based diets at the global level will simultaneously optimize the food supply, health, environmental, and social justice outcomes for the world's population. Implementing such policies is not free of political challenges but is perhaps the most rational, scientific, and moral path for a sustainable future of the human race and other living creatures of the biosphere that we share (Sabaté and Soret 2014)."

References

Ahmad T, Khan M, Haroon, THM, Nasir S, Hui J, Bonilla-Aldana DK, Rodriguez-Morales AJ (2020) COVID-19: zoonotic aspects. Travel Med Infect Dis:101607

Ainsworth E, Davey P, Bernacchi C et al (2002) A meta-analysis of elevated [CO_2] effects on soybean (*Glycine max*) physiology, growth and yield. Glob Chang Biol 8:695–709. https://doi.org/10.1046/j.1365-2486.2002.00498.x

Alexsandrowicz L, Green R, Joy EJ et al (2016) The impacts of dietary change on greenhouse gas emissions, land use, water use, and health: a systematic review. PLoS One 11:e0165797

Arneth A, Barbosa H, Benton T (2019) IPCC special report on climate change, desertification, land degradation, sustainable land management, food security, and greenhouse gas fluxes in terrestrial ecosystems. Ipcc.ch. https://www.ipcc.ch/site/assets/uploads/2019/08/4.-SPM_Approved_Microsite_FINAL.pdf. Accessed 8 Aug 2019

Barioni M, Benton T, Herrero M et al (2019) Food security. In: Climate change and land, an IPCC special report on climate change, desertification, land degradation, sustainable land management, food security, and greenhouse gas fluxes in terrestrial ecosystems. Ipcc.ch. https://www.ipcc.ch/site/assets/uploads/2019/08/2f.-Chapter-5_FINAL.pdf. Accessed 8 Aug 2019

Barron-Gafford G, Pavao-Zuckerman M, Minor R et al (2019) Agrivoltaics provide mutual benefits across the food–energy–water nexus in drylands. Nat Sustain 2:848–855. https://doi.org/10.1038/s41893-019-0364-5

Beniash E, Ivanina A, Lieb N et al (2010) Elevated level of carbon dioxide affects metabolism and shell formation in oysters *Crassostrea virginica* (Gmelin). Mar Ecol Prog Ser 419:95–108. https://doi.org/10.3354/meps08841

Beringer R (2018) Washing away: the disappearing islands of Chesapeake Bay. Earth Island Journal. http://www.earthisland.org/journal/index.php/articles/entry/washing_away/. Accessed 24 Jul 2019

Boesch D, Atkinson L, Boicourt J et al (2013) Updating Maryland's sea-level rise projections. Mdsg.umd.edu. https://www.mdsg.umd.edu/sites/default/files/files/Sea-Level_Rise_Projections_Final.pdf. Accessed 24 Jul 2019

Boulais M, Chenevert K, Demey A et al (2017) Oyster reproduction is compromised by acidification experienced seasonally in coastal regions. Sci Rep 7:13276. https://doi.org/10.1038/s41598-017-13480-3

Cai W, Huang W, Luther G et al (2017) Redox reactions and weak buffering capacity lead to acidification in the Chesapeake Bay. Nat Commun 8:369. https://doi.org/10.1038/s41467-017-00417-7

CBF (2019) The economic importance of the Bay. Cbf.org. https://www.cbf.org/issues/what-we-have-to-lose/economic-importance-of-the-bay/. Accessed 30 Aug 2019

CDC (2019a) 2018 Breastfeeding report card. [Online] Centers for Disease Control and Prevention. https://www.cdc.gov/breastfeeding/data/reportcard.htm. Accessed 6 Sept 2019

CDC (2019b) [Online] Cdc.gov. https://www.cdc.gov/mmwr/volumes/66/wr/pdfs/mm6627a3.pdf. Accessed 6 Sept 2019

Clune S, Crossin E, Verghese K (2017) Systematic review of greenhouse gas emissions for different fresh food categories. J Clean Prod 140:766–783. https://doi.org/10.1016/j.jclepro.2016.04.082

Community Food Rescue (2019) Community food rescue – feed more, waste less. Community Food Rescue. https://www.communityfoodrescue.org. Accessed 23 Aug 2019

Cook T, Folli M, Klinck J et al (1998) The relationship between increasing sea-surface temperature and the northward spread of Perkinsus marinus (Dermo) disease epizootics in oysters. Estuar Coast Shelf Sci 46:587–597. https://doi.org/10.1006/ecss.1997.0283

Cook A, Duston J, Bradford R (2006) Thermal tolerance of a northern population of striped bass *Morone saxatilis*. J Fish Biol 69:1482–1490. https://doi.org/10.1111/j.1095-8649.2006.01211.x

Crawford G, Smith D, Desloges J, Davis A (1998) Floodplains and agricultural origins: a case study in South-Central Ontario, Canada. J Field Archaeol 25:123. https://doi.org/10.2307/530574

CSSR (2017) Climate science special report. In: Science2017. globalchange.gov. https://science2017.globalchange.gov/chapter/12/. Accessed 18 Sep 2019

DeJong B, Bierman P, Newell W et al (2015) Pleistocene relative sea levels in the Chesapeake Bay region and their implications for the next century. GSA Today:4–10. https://doi.org/10.1130/gsatg223a.1

Dietrich T (2019) Gone gull: laughing gulls in Virginia are ravaged by sea level rise. Dailypress.com. https://www.dailypress.com/news/dp-nws-laughing-gulls-gone-20190709-story.html. Accessed 2 Aug 2019

Edenhofer O, Pichs-Madruga R, Sokona Y (2015) Climate change 2014: mitigation of climate change. Contribution of working group III to the fifth assessment report of the intergovernmental panel on climate change. Cambridge University Press, Cambridge

Ehleringer J (2002) Atmospheric CO_2 as a global change driver influencing plant-animal interactions. Integr Comp Biol 42:424–430. https://doi.org/10.1093/icb/42.3.424

Ercin A, Aldaya M, Hoekstra A (2011) The water footprint of soy milk and soy burger and equivalent animal products. Waterfootprint.org. https://waterfootprint.org/media/downloads/Ercin-et-al-2012-WaterFootprintSoy.pdf. Accessed 23 Jul 2019

FallsChurch (2019) Curbside food waste collection. Falls Church, VA – Official Website. Fallschurchva.gov. https://www.fallschurchva.gov/2022/Compost. Accessed 23 Aug 2019

Fang X, An P, Wang H et al (2015) Dietary intake of heme iron and risk of cardiovascular disease: a dose–response meta-analysis of prospective cohort studies. Nutr Metab Cardiovasc Dis 25:24–35. https://doi.org/10.1016/j.numecd.2014.09.002

FAO (2019a) FAO – key facts and findings, by the numbers: GHG emissions by livestock. Fao.org. http://www.fao.org/news/story/en/item/197623/icode/. Accessed 4 Aug 2019

FAO (2019b) SOFIA 2018 – state of fisheries and aquaculture in the world 2018. www.fao.org. http://www.fao.org/state-of-fisheries-aquaculture. Accessed 23 Aug 2019

Flannigan M, Stocks B, Wotton B (2000) Climate change and forest fires. Sci Total Environ 262:221–229. https://doi.org/10.1016/s0048-9697(00)00524-6

Fraser G, Shavlik D (2001) Ten years of life. Arch Intern Med 161:1645. https://doi.org/10.1001/archinte.161.13.1645

Gallo A (1999) Food advertising in the United States. In: America's eating habits: changes and consequences. USDA Economic Research Service, pp 173–180. Ers.usda.gov. https://www.ers.usda.gov/webdocs/publications/42215/5838_aib750i_1_.pdf?v=41055. Accessed 16 Aug 2019

Garnett T, Godde C, Muller A et al (2017) Grazed and confused? Ruminating on cattle, grazing systems, methane, nitrous oxide, the soil carbon sequestration question – and what it all means for greenhouse gas emissions. Oxford University, Oxford

Gerbens-Leenes P, Mekonnen M, Hoekstra A (2013) The water footprint of poultry, pork and beef: a comparative study in different countries and production systems. Water Resour Ind 1–2:25–36. https://doi.org/10.1016/j.wri.2013.03.001

Gerber P, Steinfeld J, Henderson H et al (2013) Tackling climate change through livestock – a global assessment of emissions and mitigation opportunities. Food and Agriculture Organization of the United Nations (FAO), Rome

Greger M (2006) Review of "Bird flu: a virus of our own hatching" by Michael Greger, 1st ed. Lantern Books, New York

Gerrior S, Bente L, Hiza H (2004) Nutrient content of the U.S. Food Supply, 1909–2000. Cnpp.usda.gov. https://www.cnpp.usda.gov/sites/default/files/nutrient_content_of_the_us_food_supply/FoodSupply1909-2000.pdf. Accessed 18 Dec 2018

GlobalChange.gov (2016) The impacts of climate change on human health in the United States: a scientific assessment. Health2016. globalchange.gov. https://health2016.globalchange.gov/. Accessed 12 Aug 2019

Godfray H, Aveyard P, Garnett T et al (2018) Meat consumption, health, and the environment. Science 361:eaam5324. https://doi.org/10.1126/science.aam5324

Goldstein N (2019) Smithfield foods goes big on biogas. Biocycle 60:44

González A, Frostell B, Carlsson-Kanyama A (2011) Protein efficiency per unit energy and per unit greenhouse gas emissions: potential contribution of diet choices to climate change mitigation. Food Policy 36:562–570. https://doi.org/10.1016/j.foodpol.2011.07.003

Gordon S (2019) Infant formula timeline & history. Parent's choice formula. [Online] Parentschoiceformula.com. http://www.parentschoiceformula.com/articles/Infant-Formula-Timeline-What-Should-we-Feed-Baby.aspx. Accessed 6 Sept 2019

Gornall J, Betts R, Burke E et al (2010) Implications of climate change for agricultural productivity in the early twenty-first century. Philos Trans R Soc B: Biol Sci 365:2973–2989. https://doi.org/10.1098/rstb.2010.0158

Green M, Emery K, Hishikawa Y et al (2016) Solar cell efficiency tables (version 48). Prog Photovolt Res Appl 24:905–913. https://doi.org/10.1002/pip.2788

Greger M (2017) The how not to die cookbook, 1st edn. Flatiron Books, New York

Gruber N, Clement D, Carter B et al (2019) The oceanic sink for anthropogenic CO_2 from 1994 to 2007. Science 363:1193–1199. https://doi.org/10.1126/science.aau5153

Haraldsson M, Jaspers C, Tiselius P et al (2012) Environmental constraints of the invasive *Mnemiopsis leidyi* in Scandinavian waters. Limnol Oceanogr 58:37–48. https://doi.org/10.4319/lo.2013.58.1.0037

Harding L, Adolf J, Mallonee M et al (2015) Climate effects on phytoplankton floral composition in Chesapeake Bay. Estuar Coast Shelf Sci 162:53–68. https://doi.org/10.1016/j.ecss.2014.12.030

Heller M, Keoleian G (2000) Life cycle-based sustainability indicators for assessment of the U.S. Food System. The University of Michigan Center for Sustainable Systems, CSS00-04

Heller M, Keoleian G (2014) Greenhouse gas emission estimates of U.S. dietary choices and food loss. J Ind Ecol 19:391–401. https://doi.org/10.1111/jiec.12174

Hiza H, Bente L (2011) Nutrient content of the U.S. food supply: developments between 2000 and 2006. Home economics research report no. 59 US Department of Agriculture. Cnpp.usda.gov. https://www.cnpp.usda.gov/sites/default/files/nutrient_content_of_the_us_food_supply/Final_FoodSupplyReport_2006.pdf. Accessed 18 Dec 2018

Högy P, Wieser H, Köhler P et al (2009) Effects of elevated CO_2 on grain yield and quality of wheat: results from a 3-year free-air CO_2 enrichment experiment. Plant Biol 11:60–69. https://doi.org/10.1111/j.1438-8677.2009.00230.x

Horrigan L, Lawrence R, Walker P (2002) How sustainable agriculture can address the environmental and human health harms of industrial agriculture. Environ Health Perspect 110:445–456. https://doi.org/10.1289/ehp.02110445

Huang W, Zhang Y (2011) Energy efficiency analysis: biomass-to-wheel efficiency related with biofuels production, fuel distribution, and powertrain systems. PLoS One 6:e22113. https://doi.org/10.1371/journal.pone.0022113

Hudson K (2018) Virginia shellfish aquaculture situation and outlook report. Vims.edu. https://www.vims.edu/research/units/centerspartners/map/aquaculture/docs_aqua/vims_mrr_2018-9.pdf. Accessed 23 Aug 2019

Imhoff M, Zhang P, Wolfe R et al (2010) Remote sensing of the urban heat island effect across biomes in the continental USA. Remote Sens Environ 114:504–513. https://doi.org/10.1016/j.rse.2009.10.008

IPCC (2018) Summary for policymakers of IPCC special report on global warming of 1.5°C approved by governments—IPCC. Ipcc.ch. https://www.ipcc.ch/2018/10/08/summary-for-policymakers-of-ipcc-special-report-on-global-warming-of-1-5c-approved-by-governments/. Accessed 7 Aug 2019

Irby I, Friedrichs M, Da F, Hinson K (2018) The competing impacts of climate change and nutrient reductions on dissolved oxygen in Chesapeake Bay. Biogeosci Discuss 1–41. https://doi.org/10.5194/bg-2017-416

IRENA (2019) Biogas for domestic cooking. Irena.org. https://www.irena.org/-/media/Files/IRENA/Agency/Publication/2017/Dec/IRENA_Biogas_for_domestic_cooking_2017.pdf. Accessed 22 Aug 2019

Joy M (2010) Why we love dogs, eat pigs, and wear cows: an introduction to carnism. Red Wheel Weiser, Cork

Kenney A, Brainard J (2019) Chesapeake Quarterly Volume 13 Numbers 2 & 3: Vanished Chesapeake Islands. Chesapeakequarterly.net. https://www.chesapeakequarterly.net/sealevel/main8/. Accessed 24 Jul 2019

Khatchikian C, Prusinski M, Stone M et al (2015) Recent and rapid population growth and range expansion of the Lyme disease tick vector, Ixodes scapularis, in North America. Evolution 69:1678–1689. https://doi.org/10.1111/evo.12690

Kim K, Morawski S (2012) Quantifying the impact of going trayless in a University Dining Hall. J Hunger Environ Nutr 7:482–486. https://doi.org/10.1080/19320248.2012.732918

Kleinosky L, Yarnal B, Fisher A (2006) Vulnerability of Hampton Roads, Virginia to storm-surge flooding and sea-level rise. Nat Hazards 40:43–70. https://doi.org/10.1007/s11069-006-0004-z

Kobell R (2015) Chesapeake's rising seas place extra burden on communities of color. Mdsg.umd.edu, Bay Journal. https://www.mdsg.umd.edu/sites/default/files/files/Come%20High%20Water-Report-2015.pdf. Accessed 24 Jul 2019

Kunkel K, Stevens L, Stevens S (2013) Regional climate trends and scenarios for the U.S. national climate assessment Part 1. Climate of the Northeast U.S. NOAA Technical Report NESDIS 142-1. Scenarios.globalchange.gov. https://scenarios.globalchange.gov/sites/default/files/NOAA_NESDIS_Tech_Report_142-1-Climate_of_the_Northeast_U.S_1.pdf. Accessed 10 Aug 2019

Lock S (2019) Geographical distribution of fishing participants in the US 2018. Statista. Statista. https://www.statista.com/statistics/315867/geographical-distribution-of-fishing-participants-us/. Accessed 29 Aug 2019

Mekonnen M, Hoekstr A (2010) The green, blue and grey water footprint of farm animals and animal products. Ris.utwente.nl. https://ris.utwente.nl/ws/portalfiles/portal/5146067/Report-48-WaterFootprint-AnimalProducts-Vol1.pdf. Accessed 23 Jul 2019

Miri H, Rastegar A, Bagheri A (2012) The impact of elevated CO_2 on growth and competitiveness of C3 and C4 crops and weeds. Eur J Exp Biol 2:1144–1150

Mitchell M (2016) Loss of coastal marshes to sea level rise. Chesapeakebay.net. https://www.chesapeakebay.net/channel_files/23786/cc_workshop_may_2016-_marsh_change_mitchell.pdf. Accessed 2 Aug 2019

Mitchell M, Herman J, Bilkovic D, Hershner C (2017) Marsh persistence under sea-level rise is controlled by multiple, geologically variable stressors. Ecosyst Health Sustain 3:1379888. https://doi.org/10.1080/20964129.2017.1396009

Moore K, Jarvis J (2008) Environmental factors affecting recent summertime eelgrass diebacks in the lower Chesapeake Bay: implications for long-term persistence. J Coast Res 10055:135–147. https://doi.org/10.2112/si55-014

Mueller B, Hauser M, Iles C et al (2015) Lengthening of the growing season in wheat and maize producing regions. Weather Clim Extremes 9:47–56. https://doi.org/10.1016/j.wace.2015.04.001

Najjar R, Pyke C, Adams M et al (2010) Potential climate-change impacts on the Chesapeake Bay. Estuar Coast Shelf Sci 86:1–20. https://doi.org/10.1016/j.ecss.2009.09.026

NCDC (2019) Global historical climatology network (GHCN). National Centers for Environmental Information (NCEI) formerly known as National Climatic Data Center (NCDC). Ncdc.noaa.gov. https://www.ncdc.noaa.gov/data-access/land-based-station-data/land-based-datasets/global-historical-climatology-network-ghcn. Accessed 18 Sept 2019

NCGA (2018) World of corn 2018. Worldofcorn.com. National Corn Growers Association. http://www.worldofcorn.com/#corn-usage-by-segment. Accessed 17 Sept 2018

Nguyen D, Johnson S, Busetti F et al (2014) Formation and degradation of beta-casomorphins in dairy processing. Crit Rev Food Sci Nutr 55:1955–1967. https://doi.org/10.1080/10408398.2012.740102

Nicholls S, Easterling D, Goodess et al (2012) Changes in climate extremes and their impacts on the natural physical environment. In: Field CB, Barros V, Stocker TF et al (eds) Managing the risks of extreme events and disasters to advance climate change adaptation. A special report of working groups I and II of the intergovernmental panel on climate change (IPCC). Cambridge University Press, Cambridge, pp 109–230

NOAA (2019a) Ocean acidification. National Oceanic and Atmospheric Administration. Noaa.gov. https://www.noaa.gov/education/resource-collections/ocean-coasts-education-resources/ocean-acidifi cation. Accessed 7 Aug 2019

NOAA (2019b) Sea level trends – NOAA tides & currents. Tidesandcurrents.noaa.gov. https://tidesandcurrents.noaa.gov/sltrends/sltrends_station.shtml?id=8638610. Accessed 28 Jul 2019

NOAA-NCEI (2019) Find a station. Data tools. Climate Data Online (CDO). National Climatic Data Center (NCDC). NOAA Environmental Centers for Environmental Information Ncdc.noaa.gov. https://www.ncdc.noaa.gov/cdo-web/datatools/findstation. Accessed 26 Jul 2019

NPR (2019) As rising heat bakes U.S. Cities. The poor often feel it most. Npr.org. https://www.npr.org/2019/09/03/754044732/as-rising-heat-bakes-u-s-cities-the-poor-often-feel-it-most. Accessed 18 Sept 2019

NPS (2019) Tangier Island (U.S. National Park Service). Nps.gov. https://www.nps.gov/places/tangier-island.htm. Accessed 24 Jul 2019

Office of Surgeon General (2019) The importance of breastfeeding. [Online] Ncbi.nlm.nih.gov. https://www.ncbi.nlm.nih.gov/books/NBK52687/. Accessed 6 Sept 2019

Orlich M, Fraser G (2014) Vegetarian diets in the Adventist Health Study 2: a review of initial published findings. Am J Clin Nutr 100:353S–358S. https://doi.org/10.3945/ajcn.113.071233

Orth R, Dennison W, Lefcheck J et al (2017) Submersed aquatic vegetation in Chesapeake Bay: sentinel species in a changing world. BioScience 67:698–712. https://doi.org/10.1093/biosci/bix058

Piazza J, Ruby M, Loughnan S et al (2015) Rationalizing meat consumption. The 4Ns. Appetite 91:114–128. https://doi.org/10.1016/j.appet.2015.04.011

Poore J, Nemecek T (2018) Reducing food's environmental impacts through producers and consumers. Science 360:987–992. https://doi.org/10.1126/science.aaq0216

Prosser D, Nagel J, Howlin S et al (2017) Effects of local shoreline and subestuary watershed condition on waterbird community integrity: influences of geospatial scale and season in the Chesapeake Bay. Estuar Coasts 41:207–222. https://doi.org/10.1007/s12237-017-0288-0

Rehkamp S (2016) USDA ERS – a look at calorie sources in the American Diet. Ers.usda.gov. https://www.ers.usda.gov/amber-waves/2016/december/a-look-at-calorie-sources-in-the-american-diet/. Accessed 15 Aug 2019

Robertson G, Paul E, Harwood R (2000) Greenhouse gases in intensive agriculture: contributions of individual gases to the radiative forcing of the atmosphere. Science 289:1922–1925. https://doi.org/10.1126/science.289.5486.1922

Sabaté J, Soret S (2014) Sustainability of plant-based diets: back to the future. Am J Clin Nutr 100:476S–482S. https://doi.org/10.3945/ajcn.113.071522

Schlenker W, Roberts M (2009) Nonlinear temperature effects indicate severe damages to U.S. crop yields under climate change. Proc Natl Acad Sci 106:15594–15598. https://doi.org/10.1073/pnas.0906865106

Searchinger T, Heimlich R, Houghton R et al (2008) Use of U.S. croplands for biofuels increases greenhouse gases through emissions from land-use change. Science 319:1238–1240. https://doi.org/10.1126/science.1151861

Searchinger T, Waite R, Hanson C (2019) Creating a sustainable food future: a menu of solutions to feed nearly 10 billion people by 2050. Wrr-food.wri.org. https://wrr-food.wri.org/sites/default/files/2019-07/WRR_Food_Full_Report_0.pdf. Accessed 19 Aug 2019

Shanker D (2019a) Tyson embraces Faux Fish by backing maker of plant-based shrimp. Bloomberg.com. https://www.bloomberg.com/news/articles/2019-09-05/tyson-embraces-faux-fish-by-backing-maker-of-plant-based-shrimp. Accessed 8 Sept 2019

Shanker D (2019b) The hottest thing in foods is made of peas, soy and mung beans. In: Bloomberg.com. https://www.bloomberg.com/news/features/2019-08-21/fake-meat-is-hot-led-by-impossible-foods-and-beyondmeat. Accessed 14 May 2020

Sheffield P, Knowlton K, Carr J et al (2011) Modeling of regional climate change effects on ground-level ozone and childhood asthma.

Am J Prev Med 41:251–257. https://doi.org/10.1016/j.amepre.2011.04.017
Shindell D (2016) Crop yield changes induced by emissions of individual climate-altering pollutants. Earth's Future 4:373–380. https://doi.org/10.1002/2016ef000377
Shiva V (2019) The industrialisation of fake food. Ecologist. https://theecologist.org/2019/jun/18/industrialisation-fake-food. Accessed 8 Sept 2019
Smil V (2000) Feeding the world: a challenge for the twenty-first century. MIT, Cambridge
Song M, Fung T, Hu F et al (2016) Association of animal and plant protein intake with all-cause and cause-specific mortality. JAMA Intern Med 176:1453. https://doi.org/10.1001/jamainternmed.2016.4182
Sweet W, Kopp R, Weaver C et al (2017) Global and regional sea level rise scenarios for the United States. Tidesandcurrents.noaa.gov. https://tidesandcurrents.noaa.gov/publications/techrpt83_Global_and_Regional_SLR_Scenarios_for_the_US_final.pdf. Accessed 29 Jul 2019
Sweet W, Dusek G, Obeysekera J et al (2018) Patterns and projections of high tide flooding along the U.S. Coastline using a common impact threshold. NOAA technical report NOS CO-OPS 086. https://tidesandcurrents.noaa.gov/publications/techrpt86_PaP_of_HTFlooding.pdf. Accessed 6 Aug 2019
Tangiers Plight (2019) Erosion threatens Tangier Island. Arcgis.com. https://www.arcgis.com/apps/MapJournal/index.html?appid=1582fda56a8a4e0aa3c31e45b794e029. Accessed 24 Jul 2019
Teschemacher H (2003) Opioid receptor ligands derived from food proteins. Curr Pharm Des 9:1331–1344. https://doi.org/10.2174/1381612033454856
Tilman D, Clark M (2014) Global diets link environmental sustainability and human health. Nature 515:518–522
Torgus J, Gotsch G (2004) The womanly art of breastfeeding. Penguin Group, New York
USDA (2018) USDA factsheet soybeans. Usda.gov. USDA Factsheet Soybeans. US Department of Agriculture. https://www.usda.gov/sites/default/files/documents/coexistence-soybeans-factsheet.pdf. Accessed 17 Sept 2018
USDA (2019) A closer look at current intakes and recommended shifts – 2015-2020 dietary guidelines – health.gov. Health.gov. https://health.gov/dietaryguidelines/2015/guidelines/chapter-2/a-closer-look-at-current-intakes-and-recommended-shifts/. Accessed 15 Aug 2019
USDAERSA (2019) Oil crops Outlook: April 2019. Ers.usda.gov. https://www.ers.usda.gov/publications/pub-details/?pubid=92830. Accessed 13 May 2019
USEPA (2017) Climate impacts on human health. Climate Change Impacts. US EPA. 19january2017snapshot.epa.gov. https://19january2017snapshot.epa.gov/climate-impacts/climate-impacts-human-health_.html. Accessed 9 Aug 2019
USEPA (2019a) Sustainable management of food basics. US EPA. US EPA. https://www.epa.gov/sustainable-management-food/sustainable-management-food-basics. Accessed 22 Aug 2019
USEPA (2019b) Greenhouse gas emissions from a typical passenger vehicle. US EPA. US EPA. https://www.epa.gov/greenvehicles/greenhouse-gas-emissions-typical-passenger-vehicle. Accessed 29 Aug 2019
USEPA (2020) Global greenhouse gas emissions data | US EPA. In: US EPA. https://www.epa.gov/ghgemissions/global-greenhouse-gas-emissions-data. Accessed 14 May 2020
VATech (2019) Virginia Seafood Agricultural Research and Extension Center. Arec.vaes.vt.edu. https://www.arec.vaes.vt.edu/arec/virginia-seafood.html. Accessed 25 Aug 2019
Vermeulen S, Campbell B, Ingram J (2012) Climate change and food systems. Annu Rev Environ Resour 37:195–222. https://doi.org/10.1146/annurev-environ-020411-130608
VIMS (2019) Sea-level report cards. Virginia Institute of Marine Science. Vims.edu. https://www.vims.edu/bayinfo/bay_slrc/index.php. Accessed 18 Sept 2019
Vongkiatkajorn K (2019) We asked why you stopped eating meat. Here were your top 5 reasons. Mother Jones. https://www.motherjones.com/food/2019/04/we-asked-you-why-you-stopped-eating-meat-here-were-your-top-5-reasons/. Accessed 25 Aug 2019
Wang B, Li R, Lu Z, Huang Y (2020) Does comorbidity increase the risk of patients with COVID-19: evidence from meta-analysis. Aging
Ward M, Cross A, Abnet C et al (2012) Heme iron from meat and risk of adenocarcinoma of the esophagus and stomach. Eur J Cancer Prev 21:134–138. https://doi.org/10.1097/cej.0b013e32834c9b6c
Waste Advantage (2018) Freestate farms breaks ground at county compost facility in Virginia – Waste Advantage Magazine. Waste Advantage Magazine. https://wasteadvantagemag.com/freestate-farms-breaks-ground-at-county-compost-facility-in-virginia/. Accessed 22 Aug 2019
Willett W, Rockström J, Loken B et al (2019) Food in the Anthropocene: the EAT–Lancet Commission on healthy diets from sustainable food systems. Lancet 393:447–492. https://doi.org/10.1016/s0140-6736(18)31788-4
Wirsenius S, Hedenus F, Mohlin K (2010) Greenhouse gas taxes on animal food products: rationale, tax scheme and climate mitigation effects. Clim Change 108:159–184. https://doi.org/10.1007/s10584-010-9971-x
Worrall S (2018) Tiny U.S. Island is drowning. Residents Deny the Reason. Nationalgeographic.com. https://www.nationalgeographic.com/environment/2018/09/climate-change-rising-seas-tangier-island-chesapeake-book-talk/. Accessed 24 Jul 2019
Zhu X, Long S, Ort D (2010) Improving photosynthetic efficiency for greater yield. Annu Rev Plant Biol 61:235–261. https://doi.org/10.1146/annurev-arplant-042809-112206

An Organic-Based Food System: A Voyage Back and Forward in Time

Martin Kaufman

The Powhattan taught the English how to grow food next to the Chesapeake Bay,
seeds, soil and toil made for a long day.

Shifting fields and menhaden fish,
kept the soil rich.

But then the colonists grew so much tobacco and cotton,
that soon the fertility of the soil was all but forgotten.

In the 20th Century artificial fertilizers and poisons to kill weeds and insects came on the scene,
this rendered the soil and food anything but clean.

Lord Walter Northbourne told farmers to look to the land,
and rediscover what was always at hand.

To sustain our planet and our health,
the farmer of the future must look to nature to earn her wealth.

Abstract

For several decades, conventional agriculture within the Chesapeake Bay Watershed (CBW) has contributed almost half of the nitrogen pollution to the Bay, and significant amounts of its phosphorus and sediment. Since 2005, organic products have represented the fastest growing segment of the U.S. food market. Factors contributing to this rapid growth include the perceived health and environmental benefits, better taste, concern for animal welfare, and concern for farmworkers. This chapter focuses on the environmental aspects of organic agriculture; specifically, its management strategies for soil, water, weeds, and pests. As these practices differ significantly from those of conventional agriculture, it raises the issue of whether the expansion of organic agriculture within the CBW can help mitigate its pollution problems. A review of the existing literature and a case study of the Clagett and Mattawoman Creek organic farms in the CBW provide strong support for expanding organic farming as a means to improve the quality of the air, land, and water within the CBW. An additional benefit would be the transition away from a food system dependent upon multinational agribusiness corporations to one characterized by accessible small farmers who more directly interact with consumers.

Keywords

Organic farming · Chesapeake Bay · Alternative agriculture

Introduction

A quiet revolution in the way the food system works is taking place near the headwaters of the Western Branch of the Patuxent River in Maryland, and along the shores of Mattawoman Creek on the lower Eastern Shore of Virginia, and at scattered sites in between. This is the revolution of replacing conventional methods of farming with *sustainable* organic agriculture in the Chesapeake Bay Watershed (CBW). Conventional agriculture refers to the practice of using various unnatural inputs in the production of crops,

M. Kaufman (✉)
The University of Michigan, Flint, MI, USA
e-mail: martyk@umich.edu

B. E. Cuker (ed.), *Diet for a Sustainable Ecosystem*, Estuaries of the World, https://doi.org/10.1007/978-3-030-45481-4_19

including fossil fuels, artificial fertilizers, chemical pesticides, synthetic herbicides, and Genetically Modified Organisms (GMOs).

In contrast, organic farming relies upon natural plant and animal manures, coupled with crop rotation to enrich the soil, and uses mechanical or natural biological methods to reduce damage from pests and competition from weeds (USDA 2017). Prior to the advent of chemical farming in the post WWII period (circa 1950), most agriculture in the Chesapeake region was what we today call organic. The native people of the Chesapeake and the settlers who displaced them practiced organic farming for 800 years or more (Rountree 1989), but not without causing environmental problems.

While the original "organic" settler farmers worked in the absence of modern conventional inputs, they often still managed to damage their fields. Poor methods of cultivation, leaving barren fields unplanted between seasons, removing natural windbreaks, and the lack of contour plowing promoted soil erosion. Tobacco monoculture in Maryland led to widespread soil exhaustion and soil erosion by the late eighteenth Century (Costa 1975), and in Virginia, concerns over soil problems associated with tobacco prompted the state to enact a law in 1632 to limit the amount of plants each settler could grow (Arts 1981).

Artificial phosphorus fertilizers came on the scene in the second half of the nineteenth Century (Dixon 2018). Since phosphorus was over-applied to the crop, these impacts to soil from tobacco farming persist today in the watershed, as even in sandy soils phosphorus does not leach (migrate to deeper depths) very well. This characteristic, coupled with tobacco not being a very efficient remover of phosphorus, leaves high legacy levels of available phosphorus in the fields once used for tobacco (NC State 2018).

Today, conventional agriculture contributes 45% of the nitrogen reaching Chesapeake Bay along with large quantities of phosphorus and sediment (USEPA 2010, see chapter "The Journey from Peruvian Guano to Artificial Fertilizer Ends with too Much Nitrogen in the Chesapeake Bay"). This fact alone underscores how the health of the Chesapeake Bay ecosystem and the people that live there depend in part on improved agricultural practice. The modern revolution in organic agriculture embraces sustainable techniques to protect the biodiversity, soil, air, ground water and waterways of the region, while producing food free of contaminants and rich in nutrients.

This chapter examines organic farming within the context of the existing U.S. food system. It begins with the current definition of organic farming in the U.S., and then presents a brief history of organic farming, which serves to correlate and distinguish the ancient forms of this practice with and from the modern organic movement. Focus then turns to the application of modern science for growing organic foods, specifically, the benefits that accrue through the management of the soil food web. The Clagett and Mattawoman organic farms within the Chesapeake Bay Watershed (CBW) provide the case studies for this section. Concluding the chapter is a discussion of how organic agriculture can produce enough food sustainably at different geographic scales, while accounting for the profound changes to the existing food system this outcome might provoke. This is cast in the light of developing a sustainable food system for the Chesapeake Bay ecosystem.

What Is Organic Agriculture? The answer depends on your timeframe. Ancient organic agriculture did not have access to synthetic chemicals, GMOs, and irradiation techniques. It was pesticide-free from its inception around 10,000 ybp (years before present) until about 4500 ybp, when the Sumerians began applying sulfur compounds to crops (Kislev et al. 2004; Penn State 2018). When the modern organic movement began in the 1940s, it arose within the context of the post-WWII proliferation of synthetic chemicals. By the 1980s genetic technology came on the scene. The twentieth Century organic movement developed largely as a rejection of the chemical agriculture and genetic engineering of crops. It is, however, possible to see the strong linkages between ancient and current organic agriculture based upon their respective approaches to soil ecosystem management, crop rotation, and nutrient application. These linkages form an evolutionary process, which becomes evident when we examine the current definition provided by the world's largest organic advocacy organization, examples from ancient farming practice, and two case studies of modern organic agriculture.

The International Federation of Organic Agriculture Movements (IFOAM) is the worldwide umbrella organization for the modern organic agriculture movement. It currently represents close to 800 affiliates in 117 countries. IFOAM traces its origins to an international conference held in Versailles, France on November 5, 1972, organized by the French farmer organization Nature et Progréss (IFOAM 2018). Attending were five founding members representing different organizations: Lady Eve Balfour representing the Soil Association of Great Britain, Kjell Arman representing the Swedish Biodynamic Association, Pauline Raphaely representing the Soil Association of South Africa, Jerome Goldstein representing Rodale Press of the United States, and Roland Chevriot representing Nature et Progrès of France. Their definition of organic agriculture cited below reflects the four principles espoused by the organization: health, ecology, fairness, and care (IFOAM 2018): "*Organic Agriculture is a production system that sustains the health of soils, ecosystems, and people. It relies on ecological processes, biodiversity, and cycles adapted to local conditions, rather than the use of inputs with adverse effects.* ***Organic Agriculture combines*** **tradition, innovation,** ***and*** **science** *to benefit the shared environment and promote fair relationships and a good quality of life for all involved*" (emphasis added by the author).

The highlighting of the phrase: "Organic Agriculture combines tradition, innovation, and science" demonstrates a recognition of the linkages between ancient agricultural practices and the modern organic movement beginning in the 1940s. For example, Maya agriculture illustrates early efforts to support the soil ecosystem and practice crop rotation. Maya crop systems used raised fields for the crops; or *chinampas*, since water covered much of the land during the rainy season. Maya farmers dug drainage canals and built stone terraces as far back as 2200 ybp. Behind the terraces, they added marl to raise the plots about 1 m above the water level. Marl is a calcium carbonate or lime-rich mud or mudstone containing variable amounts of clays and silt. The dominant carbonate mineral in most marls is calcite ($CaCO_3$), but other carbonate minerals such as aragonite (also $CaCO_3$), dolomite ($CaMg(CO_3)_2$), and siderite ($FeCO_3$) may be present. Marl raises soil pH, increasing the availability of certain nutrients (Hurt 1987; Turner and Harrison 2000).

Several advantages attributed to raising the beds and applying the nutrients found in marl were: (1) the moist soil of the plots helped seed germination; (2) soil was pulverized and aerated to allow deep root penetration, thus improving tilth; and, (3) soil was given a good shot of organic matter by the addition of muck from surrounding drainage ditches. With patient observation and creativity, these agriculturalists devised ways to support the soil ecosystem through the addition of organic matter. The Maya also rotated beans, corn, and squash; later to become the foundation crops for the "three sisters" of the Indian tribes inhabiting the CBW (Turner 2000).

Another linkage between ancient and modern organic agricultural systems occurs with nutrient application. To avoid the use of synthetic fertilizer, modern organic agriculture uses green manure (e.g., withered plant stems and leaves) to provide nitrogen to crops. Ancient agriculturalists in Mesoamerica also applied green manure. In the thirteenth Century, the Aztecs in the Xochimilco-Chalco Basin of Mexico, like the Maya, practiced *chinampas* agriculture with raised fields. Farmers would form the ridges by carrying sod into the marshes with canoes. The marsh water irrigated the roots even during the dry season, and muck rich in organic nutrients was applied to the plots from surrounding canals. Fertilization consisted of spreading a compost of aquatic weeds (green manure) and possibly human excrement (Armillas 1971).

As these examples make clear, ancient agriculturalists pioneered the basics of modern organic farming, but their socio-technological environment differed markedly from the post-1940 modern organic agriculture period. Two key distinguishing characteristics of today's organic agriculture are: (1) it originated as a response to environmental degradation brought on by synthetic chemicals applied to crops; and, (2) there is more science to wage the battles, on both sides of the issue (e.g., soil organic matter research, GMOs, climate change, etc.).

The Modern Organic Agriculture Movement English-born Sir Albert Howard (1873–1947) pioneered the modern organic farming concept in the period just prior to 1940. After many years of agricultural research in India (1905–1931), he developed a philosophy and concept of organic farming presented in several books, of which two stand out. In his book *The Waste Products of Agriculture* (1931), he focused on soil fertility and the need to effectively recycle waste materials, including sewage sludge, onto farmland. His well-known *An Agricultural Testament* (1943), addressed fertility by building humus with an emphasis on how soil life connected to the health of crops, livestock, and humans.

Much of Howard's thinking was influenced by F. H. King's book, *Farmers of Forty Centuries* (King 1911; Heckman 2006). In turn, Howard's work inspired many farmers and agricultural scientists who furthered the organic movement, including Lady Eve Balfour and Jerome Rodale. In a 1939 field trial called the Haughley Experiment, Lady Eve Balfour compared organic and non-organic farming and helped to popularize organic farming with her publication of *The Living Soil* (1943). Jerome Rodale, a publisher and an early convert to organic farming, was instrumental in the diffusion and popularization of organic concepts in the U.S. He founded the Rodale Institute in Pennsylvania (on the edge of the CBW watershed), which conducts research on organic agriculture (Heckman 2006).

The system of agriculture advocated by Howard and spread by others was coined 'organic' by Lord Walter Northbourne in his book *Look to the Land* (1940). Northbourne was an accomplished individual; being a farmer, writer, Oxford University lecturer in agriculture, and an Olympic silver medalist in rowing. In *Look to the Land*, he warns, "in the long run, the results of attempting to substitute *chemical farming* for *organic farming* are probably far more deleterious than has yet become clear." With that phrase, he laid down the gauntlet for the ensuing clash of agricultures (Paull 2014).

Note that WWII is typically cited as the beginning of the massive conversion to chemical agriculture. While that is true, by the end of the First World War chemical plants were producing artificial fertilizers, including those located in the CBW (see chapters "The Journey from Peruvian Guano to Artificial Fertilizer Ends with too Much Nitrogen in the Chesapeake Bay" and "Pesticides Bring the War on Nature to the Chesapeake Bay"). So Howard, Balfour, Northbourne and Rodale were responding to the first wave of chemical agriculture even before the post-WWII explosion of those new technologies.

Specific Characteristics of Modern Organic Agriculture in the U.S. The Organic Foods Production Act (OFPA; 7 CFR Chapter I, Subchapter M), enacted under Title 21 of the 1990 Farm Bill, served to establish uniform national standards for the production and handling of foods labeled as "organic." The Act authorized a new United States Department of Agriculture (USDA) National Organic Program (NOP) to set national standards for the production, handling, and processing of organically grown agricultural products. In addition, the Program oversees mandatory certification of organic production. The Act also established the National Organic Standards Board (NOSB), which advises the Secretary of Agriculture in setting NOP-based standards. Under the NOP, farmers must be certified organic if they wish to use the word "organic" in reference to their business and products. This rule exempts growers selling $5000 or less a year, who must still comply and submit to a records' audit if requested, but do not have to formally apply. In order to maintain the organic certification, a producer must pay annual certification costs, including costs for items such as application fees, renewal fees, assessment of annual production or sales, and inspection fees (USDA 2018b).

Public participation is accommodated through the 15-member NOSB, who represent sectors of the organic community such as farmers/growers, environmentalists/resource conservationists, and consumer/public interest advocates. The NOSB meets two times a year in a public forum to discuss issues regarding organic standards and materials review. These meetings invite both advance written and in-person oral public comments to gain perspective on issues. Final recommendations are submitted to the NOP (USDA 2018c).

In operation, "organic" represents foods or other agricultural products produced using cultural, biological, and mechanical practices that support the cycling of on-farm resources, promote ecological balance, and conserve biodiversity in accordance with the USDA organic regulations. Farming operations must maintain or enhance soil and water quality, while also conserving wetlands, woodlands, and wildlife. Organic produce does not contain most conventional pesticides, fertilizers made with synthetic ingredients or sewage sludge, bioengineering, or ionizing radiation. In addition, organic produce must be certified to have grown on soil without prohibited substances applied for 3 years prior to harvest. Prohibited substances include most synthetic fertilizers and pesticides. In instances when a grower has to use a synthetic substance to achieve a specific purpose, the substance must be approved according to criteria that examine its effects on human health and the environment. Certified organic farmers must also document the tillage practices and procedures performed as part of their Organic System Plans. Hand weeding and mechanical cultivation are among the allowed weed control measures, and farm records must document the frequency of tillage applications. Land that is certified as organic must have distinct, defined boundaries and buffer zones to prevent the unintended application of a prohibited substance to the crop or contact with a prohibited substance applied to adjoining land that is not under organic management (AMS 2000; USDA 2018c).

Animals raised on an organic operation must meet animal health and welfare standards, not be fed antibiotics or growth hormones, be fed 100% organic feed, and must be provided access to the outdoors (AMS 2018). Additional regulations prohibit organically processed foods from containing artificial preservatives, colors, or flavors and require organic ingredients, with some minor exceptions. For example, processed organic foods may contain some approved non-agricultural ingredients, like enzymes in yogurt, or baking soda in baked goods (USDA 2018c).

Before any product can be labeled "USDA ORGANIC", a Government-approved certifier inspects the farm where the food is grown to make sure the farmer is following all the rules necessary to meet USDA organic standards. Companies that handle or process organic food before it gets to your local supermarket or restaurant must also be certified (USDA 2007).

Rules are important, as are standards. However, without good soil, any sustainable agricultural effort will fail.

Soil and the Soil Food Web Parts of the molten, flowing, and expanding lithosphere uplift the Earth's surface and eject magma through cracks in the crust. The magma cools and hardens to form igneous rocks, and is immediately subjected to physical stress from wind, rain, flowing water, ice, and temperature changes. Other types of rocks (metamorphic, sedimentary) form from crustal pressure, chemical precipitation, or transport and deposition processes, and these rocks receive the same rough treatment as their igneous brethren. Rocks of all types *weather* (fracture and/or chemically decompose), and begin to form the mineral component of soil.

Soil is a mixture of mineral solids, water, air, and organic matter. Mineral solids range in size from sand (0.06–2 mm), silt (0.002–0.06 mm) and clay (<0.002 mm). They mainly consist of silicon, oxygen, aluminum, potassium, calcium, and magnesium. The soil water, or soil solution, contains dissolved nutrients and functions as the main source of water for plants. Essential nutrients are available to plant roots through the soil solution. The air in the soil, through contact with the air above ground, provides roots with oxygen and helps remove excess carbon dioxide from respiring root cells. When mineral and organic particles clump together, aggregates form. They create a soil containing more spaces, or pores, for storing water and allowing gas exchange (Weil and Brady 2016).

Soil is the product of time, physics, biology and chemistry. Soils form when these factors weather the mineral parent material. Parent material reflects the local geology. It ranges in size from huge sheets of igneous or sedimentary bedrock to deposits of fine mineral particles (clays). Freeze-thaw cycles crack rocks, the natural acidity of rain dissolves various minerals, colonizing plants and their decomposition produce organic acids that further erode the parent material, prying plant roots open stress cracks in rocks, and the residue of slowly decaying plants adds organic content to the once solely mineral mix. Dust collects in cracks and small spaces between larger grains of minerals. Microbes (bacteria and fungi) modify the accumulating organic matter. Burrowing invertebrates and eventually small mammals further shape the developing soil (Weil and Brady 2016). In the CBW alluvial sediments (deposited by water) form much of the parent material, from clays and silts washed down from the eroding Appalachian Mountains, to marine deposits of sand left behind during periods of sea-level retreat. Colonizing plants built forests on the better-drained sediments and swamps or marshes on the wetter sites. The detritus from these natural communities further enriched the soils with organic matter, enabling their eventual use in agriculture.

Positioned at the spatial and functional crossroads of the atmosphere, biosphere, hydrosphere, and lithosphere, Earth's great environmental systems, soil is the outer skin of earth's lithosphere. Above the soil lies most of the atmosphere, and below it a good part of the hydrosphere (groundwater) and lithosphere (crust and top part of the mantle); while at the plane of the earth's surface, all four systems co-exist and interact. Due to its good seat in the grand play of Earth's major environmental systems, soil gets to experience and host many of their processes.

Solar energy promotes the evaporation of water from the ocean and terrestrial areas into the atmosphere. As the water vapor cools at higher altitudes it condenses and forms droplets of precipitation. A portion of this water falls on the land. The soil stores a portion of this precipitation, the rest becomes runoff to replenish surface waters, percolates downward to recharge groundwater, evaporates directly back to the atmosphere, or moves through plants in the process of transpiration to evaporate from the leaves. The amount of water stored in soil at any time depends on its texture (sand, silt, clay), the density and type of vegetation, temperature, slope, antecedent moisture, and the amount of organic matter it contains. Gases other than water vapor also exchange between soil and the atmosphere (e.g., oxygen, carbon dioxide, nitrogen, methane).

Natural and agricultural plants rely on the support provided by the soil, the water and gas from the atmosphere and hydrosphere, and the constant input of nutrients from eroded and transported lithosphere, and sunlight. Countless micro- and macro organisms living in the soil enable the cycling of nutrients upon which the plants also depend. Sustaining the productivity of agricultural soils means ensuring the health and biodiversity of the microbial and invertebrate community dwelling there.

Soil organisms form the base of a complex food web, consisting of primary producers and multiple levels of consumers (Fig. 1). This microbial community influences every aspect of decomposition and nutrient availability. Given these critical soil functions underlying food production, how can agricultural practices support them? Which agricultural methods have the best chance to provide healthy food in a sustainable way, while helping the Chesapeake Bay recover? One path to the answers lies in understanding the basic functions performed by the primary in-situ organisms of the soil food web shown in Fig. 1.

Bacteria are tiny, one-celled organisms, approximately 1 μm (4/100,000 of an inch) wide and a bit longer in length. A teaspoon (5 ml) of healthy soil can contain between 100 million and 1 billion of them. Their services to the soil ecosystem include decomposing organic matter, nutrient retention, degrading pollutants, nitrogen fixation, nitrogen cycling, improving water movement and storage, and disease suppression (Ingham 2019).

During their decomposition of soil organic matter, bacteria convert energy into forms useful to the rest of the organisms in the soil food web. These decomposers also *immobilize*, or retain, nutrients in their cells, and prevent the loss of nutrients such as nitrogen from the rooting zone (Vigil and Kissel 1991). Some decomposers can break down pesticides and pollutants in soil (Alexander 1977).

Other bacteria are nitrogen fixers, with the ability to take nitrogen gas from the atmosphere and convert it into a form plants use to make amino acids and proteins. Nitrogen-fixing bacteria often form mutually beneficial associations with plants. One such symbiotic relationship involves the nitrogen-fixing rhizobia group of bacteria living inside nodules formed on the roots of legumes. These bacteria provide nitrogen in a form that leguminous plants can use. To return the favor, the legume provides the bacteria with sugars for energy through root exudation, a process where plant roots secrete a diverse array of compounds (Flores et al. 1999, see chapter "The Journey from Peruvian Guano to Artificial Fertilizer Ends with too Much Nitrogen in the Chesapeake Bay"). Exuded compounds also serve important

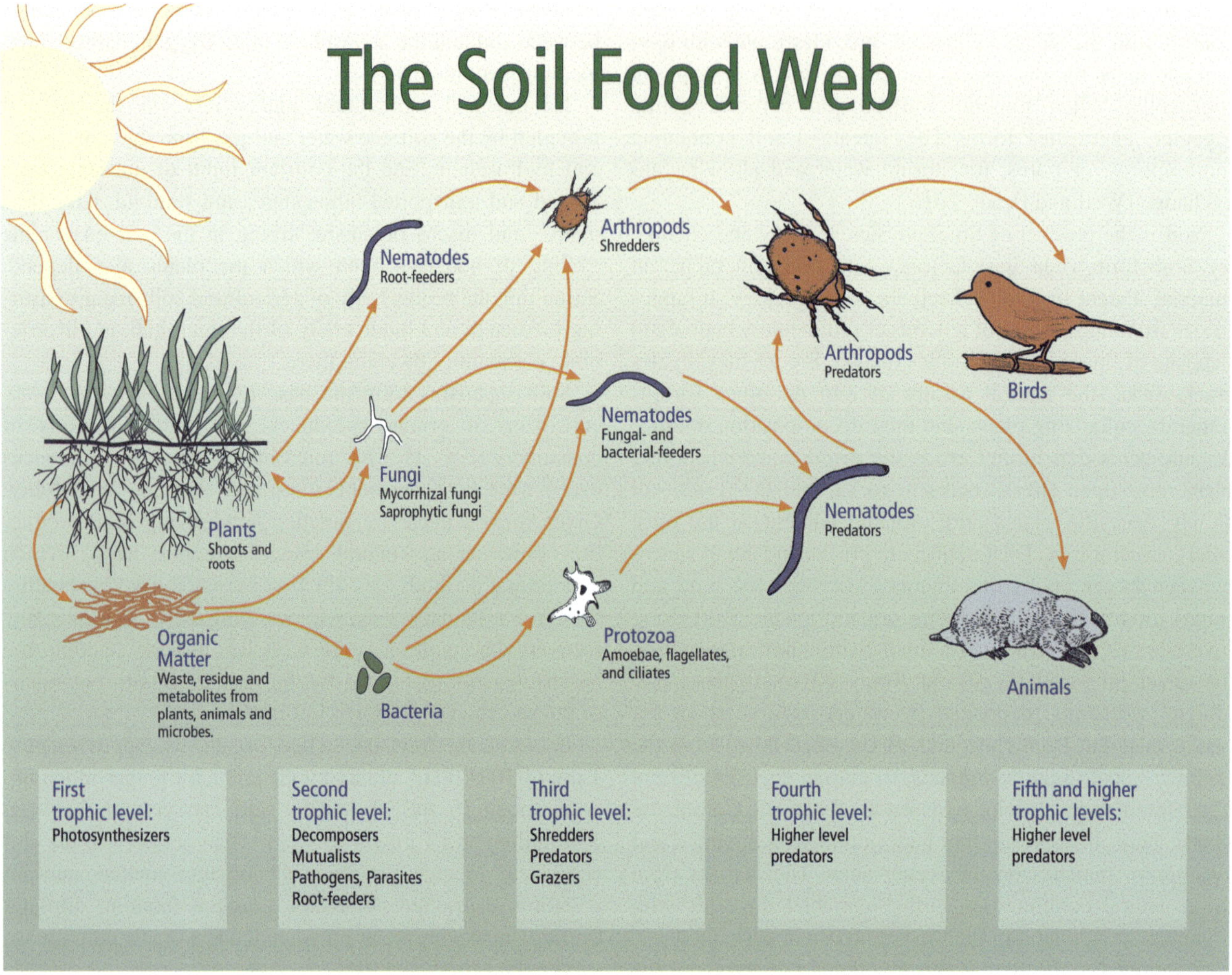

Fig. 1 The soil food web is critical to the cycling of nutrients for agricultural plants. This community decomposes detritus from the plants and manures, releasing nutrient ions in the process. Cyanobacteria and other microbes also fix atmospheric nitrogen (N_2) into ammonia (NH_3^+), a form of this nutrient useable by plants. Source: USDA (2018a)

roles as chemical attractants for beneficial organisms and repellants of detrimental organisms in the rhizosphere, the narrow zone of soil immediately surrounding the root system (Estabrook and Yoder 1998; Bais et al. 2001).

Nitrogen is also made available to plants in a form they can use from nitrifying bacteria. These organisms change the ammonium (NH_4^+) present in fertilizers to nitrite (NO_2^-) then to nitrate (NO_3^-), the preferred form of nitrogen for grasses and most row crops. Denitrifying bacteria convert nitrate to nitrogen (N_2) or nitrous oxide (N_2O) gas. Denitrifiers are anaerobic, meaning they are active where oxygen is absent, such as inside soil aggregates or water-saturated soils.

Soil aggregates are groups of *soil* particles bound to each other more strongly than to adjacent particles. The spaces between the *aggregates* provide pore space for retention and exchange of air and water. Over-application of nitrogen through fertilization can kick-start the denitrifying bacteria, leading to the release of more N_2O, a potent greenhouse gas (Mitsch et al. 2001).

With respect to water movement and storage, certain bacteria affect soil water movement by producing substances that help bind soil particles into small aggregates (those with diameters of 2–200 μm (1/10,000–1/100 of an inch). Stable aggregates improve water infiltration and the soil's water-holding ability (Magdoff and Van Es 2009). Disease suppression benefits from a diverse bacterial community. Then, many beneficial organisms can compete with disease-causing organisms in roots and on aboveground surfaces of plants to limit their potential damage. It seems some soil bacteria have learned their protective role well, as many of the antibiotics used today are manufactured by a group of bacteria called

Streptomyces, belonging to a class of bacteria known as actinomycetes. The actinomycetes are also useful in the soil, helping to degrade resistant compounds, such as chitin and cellulose (Chater 2006).

Fungi are microscopic cells that usually grow as long threads or strands called hyphae, which push their way between soil particles, roots, and rocks. Hyphae are typically only several to a few micrometers in diameter, but they can be long. Single hyphae may span from a few cells to many meters. Mushrooms are the fruiting structures of fungi.

Important soil ecosystem functions of fungi include nutrient cycling, water dynamics, and disease suppression. Along with bacteria, fungi are important as decomposers in the soil food web. Saprophytic fungi feed on dead organic matter and convert it into fungal biomass, carbon dioxide, and organic acids. These organic acids help increase the accumulation of humic acid-rich organic matter which can persist in the soil for hundreds of years (Weil and Brady 2016). Another class of fungi, the mycorrhizal, form close symbiotic relationships with plant roots. Their name derives from the Greek "mukes" meaning fungus, and "rhiza" meaning roots. These fungi colonize the plant roots and draw nutrients from the soil the root system would otherwise not be able to access. In return, the plant provides carbohydrates and other nutrients to the fungi. This is a good trade, since the fungi are able to fuel their growth and to synthesize and excrete glomalin, a glycoprotein that builds better soil structure by binding particles together and adding to organic matter content (Rillig 2004).

Like bacteria, fungi create stable aggregates that help increase soil water infiltration and storage capacity. They also have the capability to convert hard-to-digest organic material such as cellulose and lignin into forms other organisms can use. As for disease suppression, certain fungi parasitize pathogenic nematodes (Alexander 1977). Disturbing the soil through excess tillage can drastically decrease the amount of fungi and prevent them from producing significant benefits on soil structure, plant growth, and plant health (House and Bever 2018).

Protozoa are single-celled animals whose diet consists primarily of bacteria, supplemented by occasional meals of other protozoa, soluble organic matter, and fungi. They are larger than bacteria and range from 5 to 500 μm (1/5000 to 1/50 of an inch) in diameter. After consuming bacteria, protozoa release excess inorganic nitrogen available to plants and other members of the soil food web, a process called mineralization. What causes them to excrete inorganic nitrogen?

The key lies with the ratio of carbon and nitrogen (C:N) in the organism. Protozoa (and nematodes) have a lower concentration of nitrogen in their cells than the bacteria they consume. The ratio of carbon to nitrogen for protozoa is typically 10:1 or higher, but the range for bacteria is only from 3:1 to 10:1. Therefore, the bacteria eaten by protozoa contain too much nitrogen for the amount of carbon protozoa need, and they release the excess nitrogen in the form of ammonium (NH_4^+). This release usually occurs near the root system of a plant. Bacteria and other organisms rapidly take up most of the ammonium, but some is used by the plant (Coleman et al. 2017).

Other important roles of protozoa include the regulation of bacteria populations, serving as a food source for other soil organisms, and the suppression of disease. A single paramecium can consume as many as five million bacteria in 1 day, an example of how protozoa help maintain an ecological balance in the soil. When they graze on bacteria, protozoa stimulate growth of the bacterial population and increase decomposition rates and soil aggregation. Protozoa grazing is equivalent to pruning a tree, where a small amount enhances growth, but too much reduces growth or will modify the mix of species in the bacterial community. Protozoa are also an important food source for other soil organisms and help to suppress disease by competing with or feeding on pathogens (Alexander 1977).

Nematodes are non-segmented worms typically 50 μm (1/500 of an inch) in diameter and 1 mm in length. With their billion-year history on Earth, nematodes are extremely abundant and only insects exceed their diversity (Wang et al. 1999). Many nematode species feed on fungi, bacteria, protozoans, other nematodes, and plants. They can also parasitize insects, humans, and animals. Plant parasitic nematodes feed on plant parts and are widespread in agricultural soils. Two of the most economically damaging nematodes are the root knot nematode (*Meloidogyne* spp.) and soybean cyst nematode (*Heterodera glycines*). These pests enter roots and establish permanent feeding sites where they complete their life cycles without killing the cells around them. As the cyst grows it curtails root function, rendering the plant deficient in nutrients (Collins et al. 2018). Crop rotation with a non-host plant can effectively limit nematode growth. Typically, the cropping system selects plants that nematodes can and cannot grow on. These plants are grown in alternate years and the problematic nematode population decreases dramatically, below damage threshold levels, in the years the non-host is grown (Lambert and Bekal 2002).

The few species of nematodes responsible for plant diseases receive much attention, but the majority of the nematode community play beneficial roles in soil. Many species consume disease causing organisms, such as root feeding nematodes, or prevent their access to roots. Like protozoa, nematodes are important in mineralizing nutrients in plant available forms, especially ammonium (NH_4^+). Other benefits nematodes provide are also important to the soil food web. Nematodes graze to help control the balance between bacteria and fungi. They distribute bacteria and fungi through the soil and along roots by carrying live and dormant microbes on their surfaces and in their digestive systems. In addition, nematodes serve as a food source for higher-level consumers, including predatory nematodes, soil micro-arthropods, and soil insects (Coleman et al. 2017).

Arthropods make their homes in the soil. Their name reflects their jointed (arthros) legs (podos). As invertebrates, they have no backbone, and rely on an external covering called an exoskeleton. Certain row-crop soils contain several dozen species of arthropods in a square kilometer, and although the plant feeders can become pests, most arthropods perform beneficial functions in the soil-plant system. Most soil-dwelling arthropods eat fungi, worms, or other arthropods. Root-feeders and dead plant shredders are less abundant. As they feed, arthropods aerate and mix the soil, regulate the population size of other soil organisms, and shred organic material. Shredding increases the rate of decomposition and helps to return organic material to the soil (Alexander 1977).

Similar to protozoans and nematods, arthropods can stimulate microbial activity by grazing, mineralizing plant nutrients, and enhancing soil aggregation. They also promote the succession of species by consuming the dominant organisms present at a given time, and permitting other species to move in and take their place. This action facilitates the progressive breakdown of soil organic matter (Edwards et al. 2018). Certain arthropods (the generalists), including many species of centipedes, spiders, ground-beetles, rove-beetles, and gamasid mites, feed on a broad range of prey.

Earthworms have segmented bodies, provoking their classification into the phylum *Annelida*, Latin for "little rings". They are the main contributors to enriching and improving soil for plants, animals, and humans. The general understanding of the role earthworms play in improving soils and providing nutrients for plants to flourish traces to the last book Charles Darwin wrote, *The formation of vegetable mould through the action of worms* (Darwin 1892). Earthworms create tunnels in the soil by burrowing, which aerates the soil to allow air, water and nutrients to reach deep within the soil. They also eat the organic matter in the soil, including plant detritus and microbes. After digesting this organic matter, the earthworm releases waste from their bodies called castings, containing many nutrients the plant can use. Some people use earthworm castings as garden fertilizer.

In terms of biomass and overall activity, earthworms dominate the world of soil invertebrates, including arthropods. Earthworms dramatically alter soil structure, water movement, nutrient dynamics, and plant growth. A large amount of soil passes through their guts, and they can turn over the top 15 cm (6 inches) of soil in 10–20 years. This burrowing has several beneficial side effects: it increases soil porosity and water infiltration; the fragmentation of organic matter improves water storage capacity; provides new channels for plant root growth; and gradually buries plant and crop residues deposited on the surface (Edwards 2018).

The desire to grow healthy food for everyone in a sustainable way is common to conventional and organic agriculture, yet the strategies employed for achieving it differ significantly. Conventional agriculture focuses on obtaining higher yields and accomplishes this by using unnatural inputs such as artificial fertilizers, chemical pesticides, synthetic herbicides, and Genetically Modified Organisms (GMOs). As noted in chapters "The Journey from Peruvian Guano to Artificial Fertilizer Ends with too Much Nitrogen in the Chesapeake Bay" and "Pesticides Bring the War on Nature to the Chesapeake Bay", these methods have contributed to the degradation of the land and waters within the Chesapeake Bay Watershed. In contrast, soil health has always been at the core of modern organic agricultural philosophy (Howard 1943).

The Central Role of Organic Matter Organic matter (and secondarily reduced nitrogen) fuels the soil food web. The proper management of soil organic matter (SOM) is the primary means by which sustainable organic agriculture encourages a healthy soil food web (Magdoff and Weil 2004). SOM is the major determinant of soil fertility due to its impact on soil chemical, physical, and biological properties (Romig et al. 1995; Reeves 1997). The benefits of SOM include increased aeration and water holding capacity; the provision of habitat for soil organisms fueling nutrient cycling; and the retention of nutrients critical to productivity (Weil and Brady 2016). Although living organisms are technically not part of SOM, it is now clear their presence is critical to the formation, movement, and breakdown of SOM. Decomposition of organic residues produces nutrients that are available for plant uptake (Fig. 2).

The three main pools of SOM as shown in Fig. 2 are determined by their time for complete decomposition: dissolved organic matter (DOM, 1–2 years), particulate organic matter (POM, 15–100 years) and humus (500–5000 years) (Weil and Brady 2016). Both DOM and POM are biologically active, meaning they are continually being decomposed by microorganisms, and release many previously organically-bound nutrients, such as nitrogen, phosphorus, and other essential ions back to the soil solution. Dissolved organic matter is primarily composed of soluble portions of fresh plant and animal detritus such as simple carbohydrates, amino acids and resistant humic compounds, such as phenols. The remaining residue becomes the POM pool. POM is partially resistant to microbial decomposition and serves as an important long-term supply of nutrients (Wander et al. 1994).

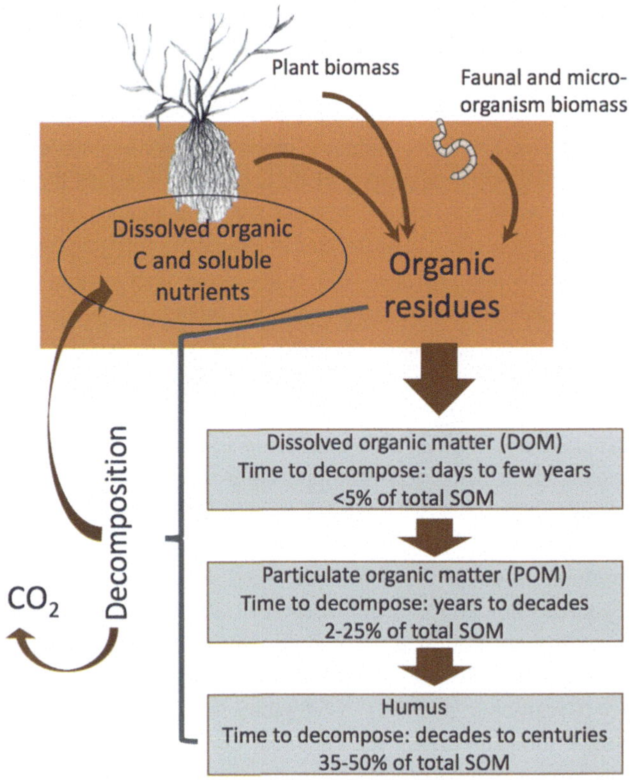

Fig. 2 Organic matter decomposition cycle emphasizing the different resistance to decomposition offered by DOM, POM and humus. Humus is comprised of refractory plant fibers, ligands and a diversity of humic acids. Source: McCauley et al. (2017)

Unlike DOM and POM, humus is not very biologically active and is the pool responsible for many of the soil chemical and physical properties associated with SOM and soil quality. Humus is a dark, complex mixture of organic substances modified from original organic tissue, synthesized by various soil organisms, and resistant to further microbial decomposition. It represents approximately 35–50% of total SOM (Prasad and Power 1997). Because of these qualities, humus breaks down very slowly and may exist in soil for hundreds or even thousands of years. Due to its chemical make-up and reactivity, humus is a large contributor to a soil's ability to retain nutrients on exchange sites. Humus also supplies organic chemicals to the soil solution that can serve as chelators and increase metal availability to plants. Soils high in humus are dark brown or black and help to raise soil temperatures by absorbing more solar radiation.

SOM ranges from about 0.4% to 10% of the dry mass of soils in temperate regions. Although organic matter is a relatively small fraction of the soil, it has large effects on soil structure and soil fertility. In a Maryland experiment, researchers saw an increase of approximately 80 bushels of corn per acre when organic matter increased from 0.8 to 2% (Strickling 1975). Another study demonstrated potential crop yield increases of about 12% for every 1% organic matter increase (Lucas et al. 1977). As soil organic matter decreases, it becomes increasingly difficult to grow plants, because problems with fertility, water availability, compaction, erosion, parasites, diseases, and insects become more common (Magdoff and Van Es 2009).

Strategies Used in Organic Agriculture for Sustainable Crop Production Organic agriculture employs several strategies to effectively manage SOM, produce crops sustainably, and conserve soil. These include minimal tillage, planting cover crops, crop rotation, the use of natural plant and animal manure, and mechanical or natural biological methods to reduce damage from pests and competition from weeds. As demonstrated by research, practical experience, and the case studies presented here, many of these strategies perform multiple functions; for instance, certain crop rotations can reduce weeds and pests.

Tillage is the mechanical modification of soil structure using tools to cut, mill, crush, beat, and rebound soil. The outcomes from tillage depend largely on both the characteristics of the tillage operation (e.g., action, depth and width of disturbance, timing) and the characteristics of the soil tilled (e.g., texture, structure, moisture, friability, plasticity). Tillage suppresses weeds and pests by disrupting their life cycles, manages crop residues, incorporates and distributes fertilizers, manures, and seeds, consolidates rocks, and changes the shape of the soil surface, via leveling, furrowing, or ridging. These uses support the main goal of tillage, which is to provide a proper environment for seed germination and root growth for crop production (UNL 2018).

The three general types of tillage are: (1) *Conventional till* (or intensive tillage), which disturbs all of the soil surface and is performed prior to and/or during planting, and less than 15% of the soil is covered with residue after planting; (2) *Conservation tillage*, is any tillage system that maintains 30% or more of the soil surface covered with crop/cover crop residue, after planting; and, (3) *No till* (or zero tillage, direct drilling), a tillage system maintaining residues (even within crop rows), and plants through these residues using specially designed equipment (CTIC 2018).

Tillage intensity plays a significant role in determining soil organic matter by affecting both soil disturbance and surface residue. Excessive soil aeration oxidizes soil organic matter causing carbon loss as carbon dioxide (Reicosky et al. 1995). Conservation tillage systems reduce the rate of organic matter oxidation compared with the conventional tillage systems, by restricting availability of free oxygen in the soil, thus slowing microbial decomposition of root biomass and below ground organic matter. Residue left on the soil surface slows decay, and helps maximize residue carbon input.

Tilling the soil increases the rate of erosion from the action of water and wind, particularly on sloped fields. Tilling disrupts the aggregates, producing smaller particles easily swept away by either fluid (Mikha et al. 2017).

Increasing awareness of the adverse effects of tillage on soil, combined with the widespread availability of herbicides, has led to the development and adoption of conservation-tillage and no-tillage systems. No-till practices can actually build soil organic matter and soil quality during annual crop production (Sapkota et al. 2012), but the results of no-till in organic agriculture have been mixed, because perennial weeds can become a problem when continuous no-till is attempted without herbicides (Giller et al. 2015). Many organic farmers therefore face the problem of controlling weeds without over-tilling the soil, but recent research performed within the CBW may provide some guidance on this issue.

At the Agricultural Research Service Sustainable Agricultural Systems Laboratory (ARSSASL) in Beltsville, MD, an experiment compared organic farming with no-till. From 1994 to 2002, John Teasdale, the lead scientist for the 9-year study and head of the ARSSASL compared minimal-tillage organic corn, soybean, and wheat with the same crops grown conventionally with no-till. Results indicated organic farming built up soil better than conventional no-till because use of manure and cover crops more than offset the losses from tillage. Teasdale noted, "It takes time for organic matter to build up, so we wouldn't have seen these surprising results had we only looked after a few years" (Schonbeck 2018).

Although soil enrichment improved, there were significant weed problems during the 9 year study that lowered the yields of corn and soybeans (but not wheat) below those of no-till crops. Teasdale persisted, and in another long-term experiment begun in 1996, he learned adding more kinds of crops to the organic rotation helped control weeds. After 10 years, corn yields were higher in diverse organic rotations that included a perennial legume. Other research weighed in on this issue. Liebman and Gallandt (1997) found if tillage and cultivation are utilized in conjunction with other practices (crop rotation, optimizing cash crop growth, cover cropping, intercropping and relay cropping, mulching, mowing, flame weeding, and grazing), the farmer will realize more weed control with less tillage. Reducing weed pressure on crops results from using "many little hammers" employed in a coordinated strategy.

Furthermore, the nonuse of herbicides gives the organic farmers this key advantage: there are no toxic carryovers to constrain crop rotation, cover cropping, or multi-cropping systems. Organic farmers can implement highly diversified rotations of vegetables from many plant families in order to improve weed, pest, and disease management, as well as meet their market objectives. In addition, cover and cash crops can be planted immediately after the previous crop is harvested or terminated without fear of herbicide injury.

Cover crops are plants used primarily to slow erosion, improve soil health, enhance water availability, smother weeds, help control pests and diseases, and increase biodiversity (SARE 2018a). They are typically planted during winter season and fallow periods in annual cropping systems. Experience demonstrates that using more than one offers some benefits. The most common mixtures are a grass and legume, such as winter rye and hairy vetch (a legume). Growing legumes with grasses helps compensate for the decreases in nitrogen availability for the following crop due to uptake by the cover crop grasses. In addition, mixed stands usually do a better job of suppressing weeds than a single species, especially when paired with reduced tillage (Silva and Moore 2017). If planted in rotation with cash crop production, cover crops provide weed suppression through resource competition (Scholberg et al. 2010), niche disruption through modification of light, soil temperature, and/or soil moisture (Abdollahi and Munkholm 2014), and allelopathy (Creamer et al. 1996). Allelopathy is a form of chemical competition, involving the chemical inhibition of one plant (or other organism) by another, due to the release into the environment of substances acting as germination or growth inhibitors.

To keep beneficial insects (pollinators and predators) on the farm it is necessary to create a habitat supporting them all-year-long and at all stages of their life cycle. Most need to feed on something other than insects, so it is necessary to provide alternate food sources as well as shelter and water. Cover crops attracting beneficials include alfalfa, clover, and buckwheat (SARE 2018b). Steps for conserving beneficial insects include recognizing who they are; minimizing insecticide applications; maintaining ground covers and crop residues; and providing pollen and nectar sources or artificial foods (Pearsons and Tooker 2017).

Organic farmers are increasingly using cover crops as a pest management tool. For example, cropping systems providing increased biological diversity over time usually prevent the onset of nematode problems. This outcome may be due to the improvement of the soil ecological balance, and/or a healthier soil structure with a higher amount of organic matter (Litterick et al. 2004). Potato growers reported improved production and less pests after using 2 years of radish in the rotation (MSUE 2018).

It is well established, and known by most farmers, that crop rotation can decrease many disease, insect, nematode, and weed pressures (Magdoff and Van Es 2009). *Crop rotation* is a system where different plants are grown in a recurring, defined sequence. The development and implementation of well-designed crop rotations is central to the success of organic production systems (Lampkin 1990; Stockdale et al. 2001). Crop rotations, including a mixture of leguminous 'fertility building' and cash crops, are the main mechanism for nutrient supply within organic systems. Minimizing the spread of pests, weeds, and diseases is another benefit of planned crop rotations (Altieri 1995).

Soil structure is also modified by crop rotation. The accumulation of organic matter during the fallow period plays a major direct role in soil structure formation (Clement and Williams 1967; Grace et al. 1995). This results from the production of organic binding agents, such as polysaccharides, by microorganisms breaking down organic matter, and the enmeshing effects of the clover and grass roots and fungal hyphae (Breland 1995).

To obtain the benefits of a diverse crop rotation and take advantage of specialty markets, organic farmers usually grow a variety of crops. Thus, organic producers commonly grow 5–10 crop species, and fresh market vegetable growers may grow 30 or more. It is difficult, however, to plan long-range crop rotations because of the large variation in acreage among crops and frequent changes in the crop mix due to weather and shifting market demands. Therefore, many organic farmers don't follow any regular rotation plan, but instead place crops on individual fields (or parts of fields) based on the cropping history of the location and its physical and biological characteristics (e.g., drainage, recent organic matter inputs, weed pressure) (Magdoff and Van Es 2009).

Case Studies of Clagett Farm and Mattawoman Creek Farm: Two Organic Farms in the Chesapeake Bay Watershed

Clagett Farm Clagett Farm is located in Upper Marlboro, Maryland, near the headwaters of the Western Patuxent River on the Western Shore of the Chesapeake Bay. Carrie Vaughn, the farm's vegetable production manager stoops to twine a tomato plant. Her efforts are providing support, which is essential for indeterminate varieties of the plant. Classified by their growth habit, tomatoes fall into the categories of determinate and indeterminate, with growth duration and plant form being the defining traits. Determinate tomato varieties ripen sooner. Indeterminate varieties grow throughout the season, producing taller plants that fruit until the first frost (Ohta and Ikeda 2015).

A good support system will keep the fruit dry and clean, deter pests, provide good air circulation and help prevent disease. Since organic farms use little, if any pesticides (some types are approved for use), these measures, and others such as polyculture (growing multiple crop varieties together) and the use of cover crops, will help promote a healthy soil ecosystem, conserve water and energy, and contribute to the farm's success.

The Chesapeake Bay Foundation (CBF) owns and operates the farm. CBF is a non-governmental organization dedicated to recovering the health of the Bay. They operate the farm to demonstrate agricultural practices that will protect the Bay from pollution. This leading non-profit clearly understands the link between the food system and the health of the Chesapeake Bay ecosystem.

Clagett occupies roughly 120 ha (285 acres) of rolling land, and contains 20 different plots for vegetables, along with pasture areas for sheep and cattle. Potted trees from the farm's native tree nursery are planted throughout the CBW as a buffer against soil erosion. The Maryland Department of Agriculture performs annual inspections and plant tissue tests annually, and they certified the farm's vegetable operation organic in 2011. There are four full-time employees, and three part-timers during the growing season from March to November.

Attention to soil organic matter is evident. When the farm began in 1982, the SOM content was at 1.2%; 36 years later in 2018 it is 3% or more in the vegetable plots, and 4% in the pastures. Organic matter is obtained from summer cover crops (millet, sorghum, cowpeas, and sunn hemp), and the winter cover crop mix which includes rye, hairy vetch, and crimson clover. The legume, sunn hemp (*Crotalaria juncea*) when grown as a summer annual can produce over 2270 kg (5000 pounds) of biomass and over 45 kg (100 pounds) of nitrogen per 0.4 ha (1 acre). Farmers only need 8–12 weeks of frost-free growth conditions to receive these results. Where conditions are favorable, it can provide the benefits of a winter legume prior to a killing frost in the fall, and in the summer after the winter crop harvest. Sunn hemp originated in India at the dawn of agriculture and grows in Brazil and Bangladesh as a soil improving crop (Cook and White 1996).

Within cropped areas at Clagett Farm, animals are not permitted, and animal manure is not used as a source of organic matter and nutrients. These restrictions arise from the excess phosphorus present in the soil from prior tobacco farming and dairying. Instead, blood meal and feather meal provide the phosphorus, and the animal manure is composted and used as potting mix for the Chesapeake Bay Foundation's tree nursery.

The farm uses reduced tillage. Ms. Vaughn indicated no-till is tough to execute with organic practice; it results in more weeds and the soil does not warm up as quickly in the spring, which delays getting produce to the market. To counter the weeds, buckwheat is grown, which also attracts beneficial insects. Other soil management methods include contour plowing, and planting perennials (which have deeper roots than annuals). In areas of higher slope perennials can minimize soil erosion (Kell 2011). Surface runoff is managed

Fig. 3 Tomatoes, straw, and wooded stream buffers at Clagett Farm (photo by the author)

by using grass buffers for each vegetable plot and a forest buffer for the two perennial and two intermittent streams on the property (Fig. 3). The forest buffer at the left of the image serves to protect the surface water nearby by slowing overland flow, increasing infiltration, and removing nitrogen.

Riparian (near-stream) buffer width is positively related to nitrogen removal effectiveness. In data from 45 published studies, wide buffers (>50 m) more consistently removed significant portions of nitrogen entering a riparian zone than narrow buffers (0–25 m). Buffers of various vegetation types were equally effective at removing nitrogen, but buffers composed of herbaceous and forest/herbaceous vegetation were more effective when wider. Subsurface nitrogen is also removed by vegetated buffers (Mayer et al. 2007).

Other efforts to improve the soil on the farm include a current experiment with strip tillage, where a 15–30 cm (6–12 inch) wide strip is planted and the cover crop (clover) left intact. This produces a field with alternating stripes of cover crop and cash crop. There has been some success with strip tillage in Maryland, where it has shown positive economic returns and soil health benefits for organic carbon and nitrogen, and aggregation in organic vegetables (Chen et al. 2015).

Claggett grows a diverse set of vegetables. Straw and mulch surround the plants and help to retain moisture, prevent weeds, and control the Colorado potato beetle (Fig. 3). Mulch is used specifically on squash, cucumbers, and onions. The pediobius wasp (*Pediobius faveolatus*) helps control the Mexican bean beetle (MBB). MBB adults and larvae feed on bean leaves causing a "lacing" effect. Adults chew all the way through the leaves, while larvae scrape the upper surface, leaving a layer of naked leaf tissue where they have fed. The wasp (which does not sting humans) is a larval parasitoid of MBB. Female wasps lay eggs inside of MBB larvae, consuming the host from the inside, and when they hatch the MBB die. Programs controlling MBB on soybeans using releases of *P. foveolatus*, have been successful in the Mid-Atlantic area (MDA 2018). USDA-approved organic pesticides and yellow tanglefoot paint on signs provide additional pest control. Tanglefoot is a sticky substance produced from castor oil and petroleum derivatives that entrap insects that alight on its surface. The pesticides employed are a Bt (*Bacillus thuringiensis*) larval spray, and a bacterium for harlequin bugs and the Colorado potato beetle. On the signs, the yellow color attracts certain bugs, such as the cucumber beetle, and they stick to the paint (Shimoda and Honda 2013).

The attention to soil organic matter, cover cropping and soil nutrients, minimization of erosion, and environmentally sound chemical pest and weed control provide good examples of how organic farming enhances soil and water quality.

After another row of tomatoes was twined, Ms. Vaughn stepped over some drip tape (Fig. 4) used for irrigation to explain the farm's business model. Clagett Farm has a Community Supported Agriculture (CSA) program. In a CSA, which is a marketing tool for farmers, consumers can pre-purchase a "share" of the harvest for the season, creating a mutually supportive relationship between local farmers and community members. Farmers receive financial support from

Fig. 4 Use of drip tape for irrigation (photo by the author)

the community, and the community receives locally grown fresh food at a fair price. For 20 years, the CBF has expanded the traditional concept of the CSA by having Clagett Farm provide free, fresh produce to people living in poverty and near poverty in Washington, D.C. The project, a collaboration with Capital Area Food Bank, blends local, sustainable organic agriculture with social justice. CBF calls the program, "From the Ground Up." Roughly 40–50% of the farm's annual production (about 16,000 kg or 35,000 pounds) is donated to food banks. Over 80% of the farm's operating expenses originate from selling the roughly 200 CSA shares. An extensive group of volunteer weeders and pickers helps to keep the program's operating costs low. These volunteers may have heard about the farm from signs, postcards sent to neighbors, Internet directories; and social media. For low-income households, reduced price shares are available, the new member fee waived, and the same "you-pick" privileges as a full-price membership are included.

Washington D.C. is home to some of the nation's largest food deserts—places where people (mostly poor and African American) don't have ready access to fresh, healthy food. The support lent by this farm to the community, the community's participation in the farm's operations, and locally grown organic food that can reduce environmental impacts comprise the beginnings of an alternative model for agriculture in the CBW (see chapter "Ethics and Economics of Building a Food System to Recover the Health of the Chesapeake Bay and Its People").

Mattawoman Creek Farm Across the Bay, and about 322 km (200 miles) to the southeast of Clagett Farm, you can find Rick Felker near Eastville, Virginia "blind weeding, or "shallow wire cultivating" a plot of land. Rick and his wife Janice own Mattawoman Creek Farms, an 18 ha (44 acres) farm certified organic by Quality Assurance International in 2005. It is situated on the southern Delmarva Peninsula of Chesapeake Bay. Rick and Janice employee three full-time and four part-time workers. Weeds are often the bane of organic farms, so Rick has approached this challenge with shallow (2.5 cm, 1 inch) wire cultivation. He configures his plots with raised beds, and uses a 2-m (6-foot) on center cultivator to allow for a precise fit of the tractor between the rows. Adjusting the wires avoids the food-bearing plants, and if used before weeds emerge, the operation exposes the root hairs of the young weeds, causing them to dry out and die.

A total of 135 permanently raised beds grow the 60 different fruits and vegetables produced by the farm. The beds are formed by disk hillers and bed shapers, steel devices attached to farm tractors for micro excavation. The devices produce a landscape of raised beds situated between troughs the width of a tire tractor. These raised beds are made entirely of earth, different from the lumber sided structures of the same name used in home gardening.

The use of raised beds protects crops from flooding, as the troughs between them expedite drainage. The shape and orientation of the raised beds support crop growth. The local climate experiences a long fall, but also a cold spring due to the adjacent Chesapeake Bay being slow to warm after winter. To optimize solar heat capture, sides of the beds are sloped, and most beds are oriented east to west to maximize exposure to the southern sun. Using raised beds also helps eliminate soil compaction, and no tractor tires have run over the beds in 13 years. This design also enables a year-round growing season and facilitates crop rotation planning. Given this advantage, the farm supports four CSAs, one for each season.

Conservation tillage is practiced through a combination of cover crops and mechanical innovation. The cover crops provide a good share of the nitrogen required for the vegetables, with the balance provided by chicken manure. Winter rye and hairy vetch are planted on beds and traffic alleys at the end of October, the best planting date for weed

Fig. 5 Cover crop of rye and hairy vetch at Mattawoman Creek Farm (photo by the author)

Fig. 6 Soil structure at Mattawoman Creek Farm. Note the dark color typical of high humus content, and the sponginess indicating no compaction, good pore spaces, and high water retention capacity (photo by the author)

suppression, winter coverage, and biomass in the farm's zone 7b/8a climate (Fig. 5). In spring, the cover is flail mowed, followed by a disking to rebuild beds and move residues into the grow zone. Flail mowers are robust devices designed to cut through heavier grasses and scrub. Disk bedders are tools with round disks for slicing through plants or plant residues.

Next, the fields are tilled gently, using a lower gear on a rototiller and driving the tractor at about 4 km/h (2.5 mph). Rick noted that many farmers over till their fields by driving at about 1.6 km/h (1 mph) with the rototiller in high gear. His approach incorporates cover crop residues about 10–15 cm deep (4–6 inches), yet allows visible crumb structure to develop in the farm's sandy soil (Fig. 6). Although it had been 3–4 weeks since cutting the cover, there was very little weed emergence and no soil compaction.

Saving energy, time, and preserving the environment are incorporated into the farm's weed and pest control operations. Rick can perform multiple operations during a single pass with the tractor; e.g., disk bedding and bed shaping; or weeding and planting (Fig. 7). The irrigation method employed also helps with weed control. Drip tape is laid subsurface at a depth of 9 cm (3.5 inches) to allow for shallow

Fig. 7 Mattawoman Creek Farm tractor set up to perform multiple operations during a single pass (photo by the author)

cultivation and promote downward growth of plant roots, which improves root development and drought tolerance.

High density planting is used for most crops to maximize production and harvest efficiency and promote canopy closure and weed suppression. The higher density and depth of crop roots also contributes soil organic matter. Soil pH on the farm is typically between 7.0 and 7.5, essentially neutral, so no lime is required. A buffer of trees and the cover crops provides erosion control, and a better response to rainfall with less runoff is enabled by avoiding the use of plastic soil covers (used extensively in conventional agriculture to suppress weeds).

Several other strategies control weeds and pests. Most crops are transplanted into the field using a water wheel transplanter hooked to a tractor. The wheels poke perfectly sized holes in the soil and fill the holes with water. Several hoop houses (greenhouses formed of plastic sheets) on the farm facilitate germination and initial growth of the young plants. The transplanted crops get a jump on weeds, and do not require as fine a seedbed as direct seeding. Polyculture (growing multiple crops together) fights pests and extends the growing seasons, and timed plantings reduce pest issues. For example, the planting of eggplant is delayed 3 weeks in the spring to minimize flea beetle infestations. Flea beetles emerge in early spring, and attack members of the Solanaceae (nightshade) family of plants (e.g., tomatoes, bell peppers, potatoes, and eggplant) (Brust 2018). The sum of these efforts results in the use of no pesticides and herbicides on the farm.

On the business side, Mattawoman Farms sells some produce wholesale to chain grocers (e.g., Whole Foods), participates in two farmer's markets, and gets a good part of their sales from young military personnel located nearby (Norfolk, Virginia Beach, and Chesapeake) through their CSA. Yet, demand for the farm's products far outstrips the supply. This imbalance at the local scale reflects the severe national shortage of organic farmers and organic farming expertise. Specifically, organic farms comprise only 0.66% of the U.S. land under cultivation, but account for over 5% of the food sales (USDA 2014). In an effort to improve the supply side, the farm actively promotes organic farming methods and works to increase the pool of qualified organic farmers. It offers an apprentice program with the opportunity to learn how to organically grow vegetable crops and pursue a career in organic farming. A summer internship program is also available. To get the word out, the farm appears in food guides, and is on social media. This recruiting effort is sorely needed, and it forms another necessary component for building a new food system within the CBW.

Discussion and Conclusions

The case studies provide some insights into the strategies employed by organic farmers within the CBW to provide fresh, healthy, food while protecting, and improving the environment. Organic farming looks like a promising way to help rehabilitate the CBW. However, some may say that advocating for the expansion of the land area devoted to organic farming may be too adventurous, especially when only two positive samples provide the justification. Why not use Best Management Practices (BMPs) instead? BMPs are essentially methods used to prevent pollution (USEPA 1993). They can reduce erosion and decrease loadings of nitrogen and phosphorus to the Bay. Moreover, BMPs (e.g., contour plowing, stream buffers, no-till, crop rotations, growing cover crops) can be applied to existing conventional agriculture with some success (Montgomery 2007; De Baets et al. 2011; Chase et al. 2016).

True, but with conventional agriculture and the BMPs you will still have GMOs, synthetic chemicals (pesticides, herbicides, and fungicides) and Concentrated Animal Feeding Operations (CAFOs). Chapters "The Journey from Peruvian Guano to Artificial Fertilizer Ends with too Much Nitrogen in the Chesapeake Bay", "Pesticides Bring the War on Nature to The Chesapeake Bay", and "Eutrophication: Obesity of the Bay and Its People" explained the problems with their usage. Upsides to organic agriculture include not only the elimination of GMOs, chemicals and CAFOs, but the growing body of evidence pointing to organic foods' nutritional benefits over non-organic food (e.g., more secondary metabolites, higher amounts of vitamin C, lower pesticide residuals) (Worthington 2001; Baker et al. 2002; Crinnion 2010; Brandt et al. 2011; Barański et al. 2014). In essence, with the USDA's focus on soil health and built-in regulatory requirements for soil and water improvement, organic agriculture *is a BMP*.

Perception of these benefits by U.S. consumers has made organic food the fastest growing segment of the food industry (Lee and Yun 2015; Zhou 2018). In 2017, organic sales in the U.S. totaled a new record of $49.4 billion (5.5% of total food sales), up 6.4% from the previous year (OTA 2018). Further illustration of market penetration is indicated by the 88% of U.S. households who purchased organic products in 2016 (Nielsen Inc. 2018).

These are impressive benefits and economic achievements. However, if organic agriculture is to become a viable strategy for rehabilitation of the CBW, it must also become an integrated component of a sustainable socio-environmental support system within the CBW. Combining watershed and system perspectives is one approach for achieving this goal.

The CBW is extremely diverse, both physically and socially. Its 166,000 km^2 includes parts of New York, Pennsylvania, West Virginia, Delaware, Maryland, Virginia, and the entire District of Columbia (see Fig. 3 of chapter "Introduction: Starting the Journey to a Sustainable Ecosystem and Healthy People" for map). From a physiographic perspective, the CBW spans the Atlantic Coastal Plain, Piedmont, Ridge and Valley, and Appalachian geologic provinces of the Mid-Atlantic region. The Atlantic Coastal Plain is a flat, lowland area with a maximum elevation of about 90 m. The Coastal Plain extends from the edge of the continental shelf, east to a fall line that ranges from 25 to 145 km west of the Chesapeake Bay. "The fall line forms the boundary between the Piedmont Plateau and the Coastal Plain, and is marked by a rapid drop in elevation with waterfalls and rapids, as well as the cities of Baltimore, Washington, D.C., Fredericksburg, and Richmond. Those cities grew by taking advantage of both the potential waterpower generated by the falls and opportunities for tidewater shipping (Shenk and Linker 2013). The entire Eastern Shore is within the Coastal Plain. The Piedmont Plateau sits between the fall line and the Ridge and Valley province to the west. Both the Piedmont and the Coastal Plain are regions of considerable agricultural activity. The Ridge and Valley and the Appalachian provinces cover the western and northern part of the watershed and are typically characterized by forested mountain slopes and agricultural land uses in the valleys (Shenk and Linker 2013).

Presently forests comprise 65% of the CBW. The is rest of the land is agricultural (24%) or developed with buildings, roads, and other features of human occupation (11%) (USEPA 2010). It is home to 17 million people representing a wide variety of ethnicities. The CBW also exhibits a high ratio of land to water (14:1), underscoring an interesting paradox of watershed management: its success hinges on the proper management of different land uses.

Watersheds are physical systems, containing sub-systems of different, but interlinked land uses (e.g., urban, agriculture, forest). As a system, watersheds have a budget consisting of inputs and outputs of matter and energy, and the ability to change themselves through feedback. Feedback occurs in a system when the output from one process influences the rate of overall system operation. When positive feedback occurs, the outcome speeds up or amplifies the system; negative feedback dampens, or retards the system rate. In agriculture, one example of positive feedback may arise when pesticides are applied to crops. The "bad" insects develop immunity, which in turn promotes more pesticide use. Over time, the entire ecosystem suffers; (e.g., beneficial soil organisms die, birds lose food sources), and this damage may spill over to other land uses, such as the nearby forest or urban area. In contrast, negative feedback can remedy this situation by introducing beneficial insects or specific native plants to attract beneficials. Understanding the sequence of events leading to system damage helps to identify where, and when humans can intervene to stabilize them.

Initially, two conditions must be satisfied to create a sustainable food system in the CBW. The first condition is the elimination of the harmful positive feedbacks produced by conventional agriculture. This chapter, and those prior have addressed the efforts needed in this regard (e.g., eliminating synthetic chemicals, reducing fertilizers). The second condition requires the new system (i.e. organic agriculture) to provide beneficial feedbacks (both positive and negative) to not only itself, but to the other land use systems in the watershed. Together, these two outcomes can involve a broader base of people, and provide measurable environmental and economic improvements, which are necessary if organic agriculture in the CBW is to expand.

Combining the principles of systems feedback with the interlinking of land uses in a watershed provides a framework for improving the CBW. Land uses within a watershed, such as urban and agricultural, display a variety of interactions. For example, air and water pollution originating within cities affects exurban areas; or, over-irrigation in agricultural areas may cause water shortages in urban areas. In these cases, an outcome within one land use degrades not only itself, but also others across the watershed. Therefore, to attain sustainable organic agriculture it is imperative to identify the outcomes from activities originating within one or multiple land uses which may be used in a complementary way.

Table 1 presents the outline of a framework providing a simple, and structured method for identifying a broad range of complimentary activities to help achieve sustainable organic farming.

The rows of the table contain the two types of systems feedback. Column headings represent linkages of land uses present within the CBW, and the arrow indicates these may be repeated or new ones specified, such as agriculture/forestry. Each cell represents a step, process, or output from a land use activity with the potential to improve the performance of the food and watershed systems. Deriving the items in each cell is the brainstorming and exploratory part of the process, which can be helped along by referring to a general supply chain of food production (Fig. 8).

Food waste is a component of the "Resource and Waste Recovery" step shown in Fig. 8. If we place "food waste" within the cells (1,1) and (2,1) of Table 1, the next task is to identify complementary activities related to food waste in urban and agricultural land uses. Since 40% all food produced in the United States ends up as waste (Hall et al. 2009), a complimentary use of food waste could occur if urban residents composted their food waste and provided it as a soil amendment to local organic farmers. This exchange may occur at the CSA pickup locations (see chapter "A New Food System for the Chesapeake Bay Region and a Changing Climate"). To generate good positive feedback, the farmer (s) in the CSA may offer a discount on the share price for good compost. This cost incentive could attract more shareholders, who in turn would provide more compost. Negative feedback can be realized environmentally by building soils with more organic content to help minimize erosion.

The second example is "food miles" located in cells 1,2 and 2,2. A food mile is the distance food travels from where it is grown or raised to its point of purchase by the consumer. This concept derives from the "Distribution and Aggregation" circle shown in Fig. 8. In a 1998 study of food miles in Chicago, IL, produce arriving by truck traveled an average distance of 2428 km (1518 miles), a 22% increase over the 1992 km (1245 miles) traveled in 1981. Using similar metrics, the food miles were calculated for three local food projects in Iowa where farmers sold to institutional markets such as hospitals, restaurants, and conference centers. The food traveled an average of 71 km (44.6 miles) to reach its destination, compared with an estimated 2474 km (1546 miles) if these food items had arrived from conventional national sources (Pirog et al. 2001). The constructive positive feedback generated by reducing food miles includes

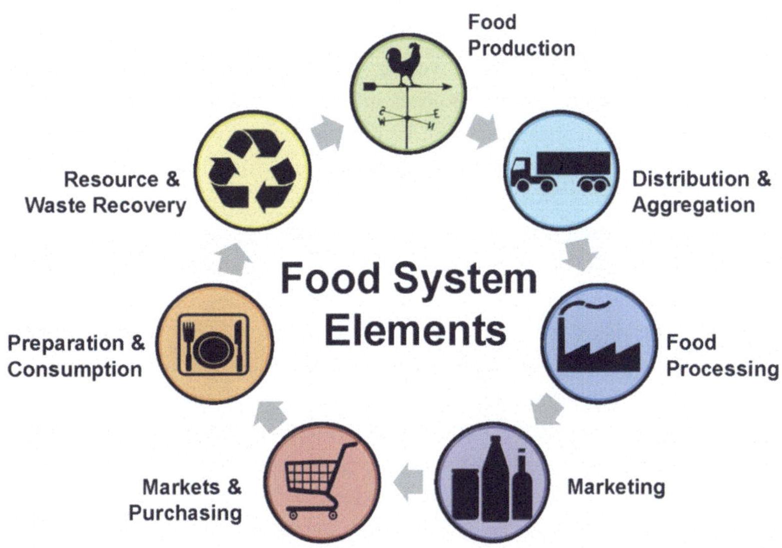

Fig. 8 Food system elements. Source: Eames-Sheavly et al. (2011), and adopted from Shi, C., N.C. State Extension

Table 1 Framework for achieving sustainable organic farming in the CBW (number, number) = row, column position in the table

	Land uses	
Feedback	Agriculture/Urban	Agriculture/Urban (Transportation)
Positive	Food waste (1,1)	Food miles (1,2)
Negative	Food waste (2,1)	Food miles (2,2)

expanding the awareness of locally grown food, which in turn increases farmer's markets and CSA penetration, leading to more citizen awareness. A stabilizing negative environmental feedback occurs with the reductions of fuel use and greenhouse gas emissions. Where could those fuel savings be used?

Sustainable systems sustain themselves by being more efficient, and by creating a general positive feedback loop with those efficiency-related savings. For example, the money saved by reducing transportation costs associated with food distribution could be used to hire more people who teach citizens how to compost.

The conditions are now ripe for a transformation of agriculture across the U.S. and within the CBW. Farm demographics, environmental conditions, and existing expertise support this assertion. Despite the trend to larger farms, small farms in the U.S. are still the dominant farm type, with 90% of all farms classified as "small" (Hoppe and MacDonald 2016). This pattern favors organic agriculture because of the higher labor intensity required than conventional agriculture. In addition, farmers are getting older. Over the last 30 years. Their average age has increased from 50.5 to 58.3 years (USDA 2014). This aging of the labor force will require replacements, especially in the rapidly growing field of organic agriculture.

From an environmental perspective, the climate of the CBW is favorable to agriculture, especially in those areas surrounding the Bay, where a humid subtropical climate exists (Kottek et al. 2006). An inventory of the topography, landforms, and land cover would help identify locations where organic agriculture can significantly reduce environmental impacts. For example, in the Coastal Plain groundwater exists at depths close to the surface. Chemicals used in conventional agriculture threaten this important water resource (USGS 2006). Transitioning to organic agriculture in areas prone to high infiltration can reduce the contamination risk.

At the ecosystem level, maintaining natural areas within and around organic fields and the absence of chemical inputs create suitable habitats for wildlife. The frequent use of under-utilized species (often as rotation crops to build soil fertility) reduces erosion of agro-biodiversity, creating a healthier gene pool—the basis for future adaptation. The provision of structures providing food and shelter, and the lack of pesticide use, attract new or re-colonizing species to the organic area (both permanent and migratory), including wild flora and fauna (e.g. birds) and organisms beneficial to the organic system such as pollinators and pest predators. A study synthesizing the results from 766 scientific papers concluded that organic farming produces more biodiversity than other farming systems (Rahmann 2011).

Although there is a general shortage of organic expertise in the U.S. (Middendorf 2007), the expertise of the current practitioners of organic farming in the CBW is impressive. The staffs at Clagett and Mattawoman Creek Farms are not only accomplished farmers; they have good policy recommendations. Carrie Vaughn at Clagett Farm suggested we do less to encourage commodity crop production and provide more research for sustainable growing practices. Michael Heller, Manager of Clagett Farms noted that much of monoculture land is rented, which may affect the commitment of the owner to long-term sustainability and soil maintenance. He also proposed a way to incentivize young people to become sustainable farmers—forgive their student loans. This recruitment effort would also require incorporating agriculture into the mainstream education system, something from which all of our students could benefit. Rick Felker, owner of Mattawoman Farms suggested we end subsidies for fast food and manage crop risk by starting small.

The case studies also provide a window into how the expansion of organic agriculture can improve democracy, increase environmental stewardship, and reduce a variety of significant health threats to millions, especially from cancer-causing chemicals. Sustainable organic farming will improve democracy because a food production and distribution system consisting of small family farms, co-ops, local markets, and Community Supported Agriculture represents a significant expansion of the people involved with making decisions about our food. This new structure counteracts (and could replace) the few large agribusiness firms who control large market shares in the food industry. More heads in the game will ultimately produce better ideas.

There is a pressing need in the CBW and the U.S. to end the insults to the environment and improve our public health. In the process, we can reconnect towns with country, urbanites with farmers, and people with the environment. Transitioning to organic agriculture can help us realize those goals.

References

Abdollahi L, Munkholm L (2014) Tillage system and cover crop effects on soil quality: chemical, mechanical, and biological properties. Soil Sci Soc Am J 78(1):262–270

Alexander M (1977) Introduction to soil microbiology, 2nd edn. Wiley, New York

Altieri MA (1995) Agroecology: the science of sustainable agriculture, 2nd edn. Westview, Boulder

AMS (2018) National organic program organic livestock and poultry practices. 7 CFR 205. https://www.federalregister.gov/documents/2018/03/13/2018-05029/national-organic-program-nop-organic-livestock-and-poultry-practices

AMS (Agricultural Marketing Service) (2000) The National organic program final rule. 7 CFR 205. https://www.ams.usda.gov/sites/default/files/media/NOP%20Preamble%20Introduction.pdf

Armillas P (1971) Gardens on swamps. Science 174(4010):653–661

Arts J (1981) Private property and soil loss regulations. J Soil Water Conserv 36(6):317–319

Bais H, Loyola-Vargas V, Flores H et al (2001) Root-specific metabolism: the biology and biochemistry of underground organs. Vitro Cell Dev Biol-Plant 37(6):730–741

Baker B, Benbrook C, Groth E III et al (2002) Pesticide residues in conventional, integrated pest management (IPM)-grown and organic foods: insights from three US data sets. Food Addit Contam 19 (5):427–446

Barański M, Średnicka-Tober D, Volakakis N et al (2014) Higher antioxidant and lower cadmium concentrations and lower incidence of pesticide residues in organically grown crops: a systematic literature review and meta-analyses. Br J Nutr 112(5):794–811

Brandt K, Leifert C, Sanderson R et al (2011) Agroecosystem management and nutritional quality of plant foods: the case of organic fruits and vegetables. Crit Rev Plant Sci 30(1–2):177–197

Breland T (1995) Green manuring with clover and ryegrass catch crops undersown in spring wheat: effects on soil structure. Soil Use Manag 11(4):163–167

Brust G (2018) Insect pest management in eggplant. University of Maryland Extension. https://extension.umd.edu/learn/insect-pest-management-eggplant. Accessed 15 Aug 2018

Chase J, Benoy G, Hann S et al (2016) Small differences in riparian vegetation significantly reduce land use impacts on stream flow and water quality in small agricultural watersheds. J Soil Water Conserv 71(3):194–205

Chater K (2006) *Streptomyces* inside-out: a new perspective on the bacteria that provide us with antibiotics. Philos Trans R Soc B: Biol Sci 361(1469):761–768

Chen G, Hooks C, Lekveishvili M (2015) Cover crop and tillage impact on soil quality, greenhouse gas emission, pests, and economics of fields transitioning to organic farming. Final report for project ORG 2011-04944 CRIS Abstracts

Clement C, Williams T (1967) Leys and soil organic matter: II. The accumulation of nitrogen in soils under different leys. J Agric Sci 69 (1):133–138

Coleman D, Callaham M, Crossley D Jr (2017) Fundamentals of soil ecology, 3rd edn. Academic, London

Collins A, Esker P, Murillo-Williams A (2018) Plant parasitic nematodes explained. Pennsylvania State University Extension. https://extension.psu.edu/plant-parasitic-nematodes-explained. Accessed 11 Aug 2018

Cook C, White G (1996) *Crotalaria juncea*: a potential multi-purpose fiber crop. In: Janick J (ed) Progress in new crops. ASHS, Arlington, pp 389–394

Costa J (1975) Effects of agriculture on erosion and sedimentation in the Piedmont Province, Maryland. Geol Soc Am Bull 86(9):1281–1286

Creamer N, Bennett M, Stinner B, Cardina J (1996) A comparison of four processing tomato production systems differing in cover crop and chemical inputs. J Am Soc Hortic Sci 121(3):559–568

Crinnion W (2010) Organic foods contain higher levels of certain nutrients, lower levels of pesticides, and may provide health benefits for the consumer. Altern Med Rev 15(1)

CTIC Conservation Technology Information Center (2018) Tillage type definitions. Purdue University. http://www.ctic.purdue.edu/resourcedisplay/322/. Accessed 20 Aug 2018

Darwin C (1892) The formation of vegetable mould through the action of worms: with observations on their habits, vol 37. John Murray, London

De Baets S, Poesen J, Meersmans J et al (2011) Cover crops and their erosion-reducing effects during concentrated flow erosion. Catena 85(3):237–244

Dixon M (2018) Chemical fertilizer in transformations in world agriculture and the state system, 1870 to interwar period. J Agrar Chang 18 (4):768–786

Eames-Sheavly M, Hadekel C, Hedstrom A et al (2011) Discovering our food system: experiential learning & action for youth and their communities. Cornell University Cooperative Extension, Ithaca, NY

Edwards CA (2018) Earthworms, USDA-NRCS Soils. https://www.nrcs.usda.gov/wps/portal/nrcs/detailfull/soils/health/biology/?cid=nrcs142p2 053863, Accessed 27 August 2018

Estabrook E, Yoder J (1998) Plant-plant communications: rhizosphere signaling between parasitic angiosperms and their hosts. Plant Physiol 116(1):1–7

Flores H, Vivanco J, Loyola-Vargas V (1999) 'Radicle' biochemistry: the biology of root-specific metabolism. Trends Plant Sci 4 (6):220–226

Giller K, Andersson J, Corbeels M et al (2015) Beyond conservation agriculture. Front Plant Sci 6:870

Grace P, Oades J, Keith H et al (1995) Trends in wheat yields and soil organic carbon in the permanent rotation trial at the Waite Agricultural Research Institute, South Australia. Aust J Exp Agric 35 (7):857–864

Hall K, Guo J, Dore M et al (2009) The progressive increase of food waste in America and its environmental impact. PLoS One 4(11): e7940

Heckman J (2006) A history of organic farming: transitions from Sir Albert Howard's war in the soil to USDA National Organic Program. Renewable Agric Food Syst 21(3):143–150

Hoppe R, MacDonald J (2016) America's diverse family farms (no. 262135). United States Department of Agriculture, Economic Research Service, Washington, DC

House G, Bever J (2018) Disturbance reduces the differentiation of mycorrhizal fungal communities in grasslands along a precipitation gradient. Ecol Appl 28(3):736–748

Howard A (1943) An agricultural testament. Oxford University Press, London

Howard A, Wad Y (1931) The waste products of agriculture. Oxford University Press, London

Hurt R (1987) Indian agriculture in America: prehistory to the present. University Press of Kansas, Lawrence

IFOAM (International Federation of Organic Agriculture Movements) (2018). https://www.ifoam.bio/en/about-us/history. Accessed 20 Aug 2018

Ingham E (2019) Soil bacteria. Nrcs.usda.gov. https://www.nrcs.usda.gov/wps/portal/nrcs/detailfull/soils/health/biology/?cid=nrcs142p2_053862. Accessed 4 Sep 2019

Kell D (2011) Breeding crop plants with deep roots: their role in sustainable carbon, nutrient and water sequestration. Ann Bot 108 (3):407–418

King F (1911) Farmers of forty centuries. Mrs. F.H King, Madison, WI

Kislev M, Weiss E, Hartmann A (2004) Impetus for sowing and the beginning of agriculture: ground collecting of wild cereals. Proc Natl Acad Sci 101(9):2692–2694

Kottek M, Grieser J, Beck C et al (2006) World map of the Köppen-Geiger climate classification updated. Meteorol Z 15:259–263

Lambert K, Bekal S (2002) Introduction to plant-parasitic nematodes. Plant Health Instructor 10:1094–1218

Lampkin N (1990) Organic farming. Farming Press Books, Ipswich, MA

Lee H, Yun Z (2015) Consumers' perceptions of organic food attributes and cognitive and affective attitudes as determinants of their purchase intentions toward organic food. Food Qual Prefer 39:259–267

Liebman M, Gallandt E (1997) Many little hammers: ecological approaches for management of crop–weed interactions. In: Jackson LE (ed) Ecology in agriculture. Academic, New York, pp 291–543

Litterick A, Harrier L, Wallace P et al (2004) The role of uncomposted materials, composts, manures, and compost extracts in reducing pest and disease incidence and severity in sustainable temperate agricultural and horticultural crop production—a review. Crit Rev Plant Sci 23(6):453–479

Lucas R, Holtman J, Connor J (1977) Soil carbon dynamics and cropping practices. In: Lockeretz W (ed) Agriculture and energy. Academic, New York, pp 333–451

Magdoff F, Van Es H (2009) Building soils for better crops, 3rd edn. Sustainable Agriculture, Waldorf, MD

Magdoff F, Weil RR (eds) (2004) Soil organic matter in sustainable agriculture. CRC, Boca Raton

Mayer P, Reynolds S, McCutchen M et al (2007) Meta-analysis of nitrogen removal in riparian buffers. J Environ Qual 36 (4):1172–1180

McCauley A, Jones C, Olson-Rutz K (2017) Soil pH and organic matter. Nutrient management module 8. Montana State University Extension. http://landresources.montana.edu/nm/documents/NM8.pdf. Accessed 20 Aug 2018

MDA, Maryland Department of Agriculture (2018) Using the parasitic wasp pediobius foveolatus to suppress Mexican bean beetles in home gardens. https://mda.maryland.gov/plants-pests/Documents/mbbbroch.pdf. Accessed 15 Aug 2018

Middendorf G (2007) Challenges and information needs of organic growers and retailers. J Ext 45(4)

Mikha M, Benjamin J, Vigil M et al (2017) Manure and tillage use in remediation of eroded land and impacts on soil chemical properties. PLoS One 12:e0175533

Mitsch W, Day J, Gilliam J et al (2001) Reducing nitrogen loading to the Gulf of Mexico from the Mississippi River basin: strategies to counter a persistent ecological problem: ecotechnology—the use of natural ecosystems to solve environmental problems—should be a part of efforts to shrink the zone of hypoxia in the Gulf of Mexico. AIBS Bull 51(5):373–388

Montgomery D (2007) Soil erosion and agricultural sustainability. Proc Natl Acad Sci 104(33):13268–13272

MSUE (Michigan State University Extension) (2018) Cover crops program. East Lansing Michigan. wwwcovercropsmsuedu. Accessed 20 Aug 2018

NC State (2018) Tobacco grower's information. North Carolina State Cooperative University Extension. https://tobacco.ces.ncsu.edu/tobacco-fertility-nutrients/. Accessed 8 Aug 2018

Nielsen Inc. (2018) Organics products are showing up in more places-and for less money. http://wwwnielsencom/us/en/insights/news/2017/organic-products-are-showing-up-in-more-places-and-for-less-moneyhtml. Accessed 19 Aug 2018

Northbourne C (1940) Look to the land. J.M Dent, London

Ohta K, Ikeda D (2015) Differences in branch formation in indeterminate and determinate tomato types. Environ Control Biol 53 (4):189–198

OTA (Organic Trade Association) (2018) US organic industry survey 2018. https://ota.com/resources/organic-industry-survey. Accessed 19 Aug 2018

Paull J (2014) Lord Northbourne, the man who invented organic farming, a biography. J Org Syst 9(1):31–53

Pearsons K, Tooker J (2017) In-field habitat management to optimize pest control of novel soil communities in agroecosystems. Insects 8 (3):82

Penn State (2018) A short history of pest management. Pennsylvania State University Agricultural Extension. https://extension.psu.edu/a-short-history-of-pest-management. Accessed 8 Aug 2018

Pirog R, van Pelt T, Enshayan K et al (2001) Food, fuel, and freeways: an Iowa perspective on how far food travels, fuel usage, and greenhouse gas emissions, Leopold Center for sustainable agriculture. Iowa State University, Ames

Prasad R, Power J (1997) Soil fertility management for sustainable agriculture. CRC, Boca Raton

Rahmann G (2011) Biodiversity and organic farming: what do we know? vTI Agric Forest Res 3:189–208

Reeves D (1997) The role of soil organic matter in maintaining soil quality in continuous cropping systems. Soil Tillage Res 43 (1–2):131–167

Reicosky D, Kemper W, Langdale G et al (1995) Soil organic matter changes resulting from tillage and biomass production. J Soil Water Conserv 50(3):253–261

Rillig M (2004) Arbuscular mycorrhizae, glomalin, and soil aggregation. Can J Soil Sci 84(4):355–363

Romig D, Garlynd M, Harris R et al (1995) How farmers assess soil health and quality. J Soil Water Conserv 50(3):229–236

Rountree H (1989) The Powhatan Indians of Virginia. University of Oklahoma Press, Norman

Sapkota T, Mazzoncini M, Bàrberi P et al (2012) Fifteen years of no till increase soil organic matter, microbial biomass and arthropod diversity in cover crop-based arable cropping systems. Agron Sustain Dev 32(4):853–863

SARE (Sustainable Agriculture and Research) (2018a) Cover crops. https://wwwsareorg/Learning-Center/Topic-Rooms/Cover-Crops. Accessed 20 Aug 2018

SARE (Sustainable Agriculture and Research) (2018b) Cover cropping for pollinators and beneficial insects. https://www.sare.org/cover-cropping-for-pollinators. Accessed 20 Aug 2018

Scholberg J, Dogliotti S, Leoni C et al (2010) Cover crops for sustainable agrosystems in the Americas. In: Lichtfouse E (ed) Genetic engineering, biofertilisation, soil quality and organic farming. Springer, Dordrecht, pp 23–58

Schonbeck M (2018) The organic farmer's dilemma: how to manage weeds effectively without compromising soil quality. Extension.org. http://articles.extension.org/pages/18537/the-organic-growers-dilemma:-how-to-manage-weeds-effectively-without-compromising-soil-quality. Accessed 20 Aug 2018

Shenk G, Linker L (2013) Development and application of the 2010 Chesapeake Bay watershed total maximum daily load model. J Am Water Resour Assoc 49(5):1042–1056

Shimoda M, Honda K (2013) Insect reactions to light and its applications to pest management. Appl Entomol Zool 48 (4):413–421

Silva EM, Moore VM (2017) Cover crops as an agroecological practice on organic vegetable farms in Wisconsin, USA. Sustainability 9 (1):55

Stockdale E, Lampkin N, Hovi M et al (2001) Agronomic and environmental implications of organic farming systems. Adv Agron 70:262–329

Strickling E (1975) Crop sequences and tillage in efficient crop production. Abstracts of the 1975 Northeast Branch American Society Agronomy Meetings, pp 20–29

Turner BL (2000) Ancient agricultural land use. In: Turner, BL, Harrison PD (eds) Pulltrouser Swamp: ancient Maya habitat, agriculture, and settlement in northern Belize. University of Utah Press, Salt Lake City, pp 169–171

Turner BL, Harrison PD (eds) (2000) Pulltrouser Swamp: ancient Maya habitat, agriculture, and settlement in northern Belize. University of Utah Press, Salt Lake City, p 247

UNL (University of Nebraska-Lincoln) (2018) Tillage and no-till systems. Cropwatch-Institute of Agriculture and Natural Resources. https://cropwatch.unl.edu/tillage. Accessed 20 Aug 2018

USDA (2007) Consumer brochure. https://www.nal.usda.gov/afsic/organic-productionorganic-food-information-access-tools#define. Accessed 20 Aug 2018

USDA (2014) 2012 Census of agriculture. National Agricultural Statistics Service, Washington, DC

USDA (2017) Certified organic survey 2016 summary. United States Department of Agriculture National Agricultural Statistics Service, Washington, DC

USDA (United States Department of Agriculture) (2018a) Soil biology primer photo gallery. Https://www.nrcs.usda.gov/wps/portal/nrcs/photogallery/soils/health/biology/gallery/?cid=1788&position=Promo. Accessed 20 Aug 2018

USDA (United States Department of Agriculture) (2018b) Becoming a certified operation. USDA Agricultural Marketing Service. https://www.ams.usda.gov/services/organic-certification/becoming-certified. Accessed 19 Aug 2018

USDA (United States Department of Agriculture) (2018c) Organic 101: What the USDA organic label means. https://www.usda.gov/media/blog/2012/03/22/organic-101-what-usda-organic-label-means. Accessed 20 Aug 2018

USEPA (2010) Chesapeake Bay TMDL. https://www.epa.gov/chesapeake-bay-tmdl/chesapeake-bay-tmdl-document. Accessed 19 Aug 2018

USEPA, United States Environmental Protection Agency (1993) Guidelines manual for developing best management practices (BMPs). EPA 833-B-93-004, USEPA Office of Water, Washington, DC

USGS (United States Geologic Survey) (2006) Sustainability of the ground-water resources in the Atlantic Coastal Plain of Maryland, USGS Fact Sheet FS 2006-3009. USGS, Baltimore

Vigil M, Kissel D (1991) Equations for estimating the amount of nitrogen mineralized from crop residues. Soil Sci Soc Am J 55:757–761

Wander M, Traina S, Stinner B et al (1994) Organic and conventional management effects on biologically active soil organic matter pools. Soil Sci Soc Am J 58:1130–1139

Wang D, Kumar S, Hedges B (1999) Divergence time estimates for the early history of animal phyla and the origin of plants, animals and fungi. Proc R Soc Lond B 266:163–171

Weil R, Brady C (2016) The nature and properties of soils, 15th edn. Pearson, Boston

Worthington V (2001) Nutritional quality of organic versus conventional fruits, vegetables, and grains. J Altern Complement Med 7:161–173

Zhou J (2018) Consumers' perception of organic foods. Bus Manage Rev 9(3):43

What Nature, Politics and Policy Demand of the Chesapeake Bay and Its Food System

Bill Matuszeski

Once was the Bay so fair.
That one could swim and fish without a care.
Then too many people moved there.
They farmed with chemicals & put animals in factories to improve their profit share.
The quality of the water declined everywhere.
So folks got together to dare.
That government make better the water and the air.
New laws, new agencies and now officials made aware.
That gave citizens hope to end their despair.
The Bay is getting better, but it will be a long-time affair.

Abstract

By the middle of the twentieth Century the Chesapeake Bay showed clear signs of declining water quality and attendant depletion of seafood harvests. The loss of water quality traced directly to excessive nutrients released into the Bay by industrial agriculture, and poorly functioning wastewater treatment plants. Overuse of artificial fertilizers and drainage from unregulated concentrated animal feeding operations polluted the Bay, resulting in cultural eutrophication that displaced oxygen-loving species from their habitats. Citizen organizations pressed the state and Federal governments to tackle the problem. The creation of the US Environmental Protection Agency in 1970, and passage of the Clean Water Act of 1972, as well as other legislation led to a coordinated Federal and state response that gave rise to the Chesapeake Bay Program. A series of Chesapeake Bay Agreements, beginning in 1983, developed the infrastructure for investigating problems in the Bay, articulating solutions to those problems, and setting in motion actions to restore the health of the ecosystem. The Chesapeake Bay Program and a host of citizen groups have successfully worked to improve the health of the Bay, with much reaming to be done. Reforms in the food system, including reductions in chemical agriculture, adopting soil-conserving organic techniques, and turning away from animal-based foods show promise in helping to recover the Bay. Recovery efforts for the Chesapeake serve as a model for other estuaries facing similar problems.

Keywords

Estuary · Restoration · USEPA · Clean Water Act · Nutrient Pollution

An Ecosystem Both Unique and Representative The Chesapeake Bay is known by the general US public as a nice place along the East Coast that grows oysters, clams, and fish. That is an accurate view of one element of the Bay; in a good year it produces over 500 million pounds (227 million kg) of seafood of all types (CBP 2019a).

However, the Bay is known by those who live near as much more—as a place for recreation and beauty around the Bay and up and down its many rivers, which drain a rich and lovely land of plenty.

But what is not well known by the public is that the Bay and its watershed are in many respects unique on this Earth in terms of size, character, vulnerability and potential.

B. Matuszeski (✉)
Chesapeake Bay Program, Washington, DC, USA
e-mail: bmat@olg.com

B. E. Cuker (ed.), *Diet for a Sustainable Ecosystem*, Estuaries of the World, https://doi.org/10.1007/978-3-030-45481-4_20

Table 1 The distribution of land by area and percent of watershed for the jurisdictions draining into the Chesapeake Bay

Jurisdiction	Area km^2 (miles^2) within the watershed	% of Total watershed area	% of Total jurisdiction in Watershed
Washington, DC	161 (00,062)	0.1	100.0
Delaware	1826 (00,705)	1.0	36.0
Maryland	23,906 (09,230)	14.0	95.0
New York	16,237 (06,269)	10.0	13.0
Pennsylvania	58,560 (2,2610)	35.0	50.5
Virginia	56,363 (21,762)	34.0	55.0
West Virginia	9267 (03,578)	6.0	15.0

Note that New York, Pennsylvania and West Virginia have no land directly contacting the Bay proper, but constitute a major portion of the watershed (Rand McNally 2000; CBP 2019b)

What Makes the Bay Special The massive 64,000 mile^2 (166 km^2) watershed includes all of the nation's capital, Washington DC, 95% of Maryland, 55% of Virginia, 51% Pennsylvania, 33% of Delaware, 20% of West Virginia and 17% of New York State (CBP 2019b, Table 1). Its 11,684 mile (18,803 km) shoreline around the Bay and up the tidal portions of its tributaries is greater than that of the entire US Pacific coast.

Many elements of its ecology are unique. The Chesapeake is the largest coastal estuary in the United States and the third largest in the world. It is 524 miles (843 km) from Cooperstown, New York on Lake Otsego to Norfolk, VA, and 200 of those miles are the Bay itself, which varies from 4 to 30 miles wide (6.4–48.2 km). Even more impressive, the Bay watershed's land to water ratio, the area of land draining into the surface of water, is the highest in the world, five-times the next closest body of water (the Gulf of Finland). This means that there is an enormous amount of runoff from the land into a relatively very small amount of water.

What Made the Bay Sick From the time of the arrival of the first humans to the watershed some 9000 years ago, their activities altered the ecology of the Bay. But it was small at first. The human impact grew as colonizing Europeans left their mark by overharvesting seafood and building of dams that blocked natural migration patterns of fish. They also used poor farming practices which led to sediment pollution in the Bay. Yet the Bay proved resilient to those insults. That changed in the mid-twentieth Century. At the end of World War II, the chemical plants that made explosives were altered to manufacture artificial nitrogen fertilizer and pesticides (see chapters "The Journey from Peruvian Guano to Artificial Fertilizer Ends with too Much Nitrogen in the Chesapeake Bay" and "Pesticides Bring the War on Nature to the Chesapeake Bay"). With that came a new war, one on nature and traditional agricultural practices. This included industrializing the production of chickens in concentrated animal feeding operations (CAFOs) that yield thousands of tons of nutrient rich manure each year (see chapter "Livestock and Poultry: The Other Colonists Who Changed the Food System of the Chesapeake Bay"). Both artificial fertilizers and the CAFOs unleashed a flood of excess nutrients into the Chesapeake Bay and its tributaries. Growing human populations in the watershed made things worse with their largely untreated sewage adding to the nutrient soup. The Bay became eutrophic, sporting extensive blooms of algae that colored the water green and blocked penetration of light to the beds of submerged aquatic vegetation. In summers the algae sinks to the bottom of the Bay. Bacteria exhaust the supply of oxygen in the deeper waters, as they decompose the sunken algae. This excludes fish, crabs, and many other forms of life from large sections of the Bay (see chapter "Reversing the Eutrophication of the Chesapeake Bay and its People").

The challenge we face in the Chesapeake is how to protect and enhance this great resource while recognizing that the lands of the watershed are themselves remarkably valuable sources of agricultural products, *products whose cultivation creates the very conditions that threaten the water quality essential to the Bay's productivity.*

But we are trying to accomplish this clean-up in a place with some major challenges from activities on all this land draining into the Bay. The watershed is home to 18 million people and 84,000 farms and ranches (CBP 2019a). They produce a lot of waste that is carried by streams, rivers and sewer systems into a mostly shallow Bay. And it is the shallowness and the mixing of fresh and salt water that gives the Bay its amazing productivity. The average depth is only 21 feet (6.4 m), and a six-footer could walk through half the acreage of the Bay and not get his or her hat wet. There are 104,843 acres (42,428 ha) of underwater grasses and 284,000 acres (115,000 ha) of tidal wetlands, all essential habitat for the seafood (CBP 2019a).

As we like to say, if we can clean up the Bay, and we are likely to be there within ten years, no one else in the world will have one tenth of an excuse!

Just like the Bay, the agricultural lands of the watershed are in some ways unique. Farmlands in the upper watershed are the most productive non-irrigated acres in the United States. The vegetable and animal food products from this area is within overnight transport of 40% of the US population (CBP 2019a). Most of these farms make extensive use of pesticides and fertilizers that pollute the Bay. Add to that large amounts of manure generated from concentrated animal feeding operations (CAFOs). Due to the high land values for raising fruit, vegetables and meats for the huge nearby markets, the food for the livestock (mostly chickens) is brought in from the Midwest. The high price of shipping means that the animal manure never makes it back to the Midwest to fertilize the fields growing animal feed. Instead, the otherwise resource becomes a pollutant that drains excessive nutrients into the Bay.

The Susquehanna River watershed in which these lands lie provides 50% of the fresh water to the Bay, more than any the other tributaries. Nearly all this area is in Pennsylvania, although small percentages are in Maryland and New York. So while Pennsylvania is responsible for much if not most of the nutrient pollution in the Bay, the shores of the Chesapeake all lie outside its boundaries.

The result is there is no natural lobby within Pennsylvania for protecting the Bay. Fifty percent of the area and well over half the population of the state are not even in the watershed. And even among those who are, Pennsylvania lacks the kind of stakeholders that gain direct income from the Bay, no fishers, no marina operators, no waterfront restaurants. This makes difficult the politics of standing-up to agricultural interests. Since there is no direct Pennsylvania waterfront property on the Bay, obtaining support for measures to clean the Chesapeake requires the focus to shift to show the benefits of improving stream habitat and water quality for the state's waterways. Cleaning the streams running into the Susquehanna River means restoring that river, and ultimately the Bay into which it flows. Pennsylvania residents using these streams for recreation will reap the rewards of improved water quality that is also helping to recover the Bay.

Why National Environmental Laws Are Not Enough for the Chesapeake Bay The vast size and complexity of the ecosystem means that the Bay needs more than the usual environmental legislation at work in the US. While those laws include some good provisions, they alone aren't sufficient to restore the ecosystem. The major nation-wide legislation that applies to the Bay was passed by Congress in the 1970s and includes the following.

The National Environmental Policy Act (NEPA) of 1970 This law requires Federal agencies to evaluate the environmental impact of all their actions. When those actions are significant, the agencies must produce Environmental Impact Statements for public review, seeking improvement to their proposals. Many states have similar laws for their actions which impact the environment. But the bureaucracies of most agencies are reluctant to reveal the known negative effects on the environment of what they are proposing. So the public must be constantly on guard to assure the law is met in fact and in spirit.

The adoption of NEPA also set in motion President Richard Nixon's Reorganization Plan No. 3, which he presented to the US Congress in 1970. This plan consolidated the functions of protecting the environment, previously dispersed in a number of different departments, in a new entity, the US Environmental Protection Agency (USEPA). The USEPA became the responsible body for enforcing NEPA, and for conducting studies related to improving the environment (NEPA 2019). Plan No. 3 also created the National Oceanic and Atmospheric Administration (NOAA), which also plays a role in protecting and recovering the Chesapeake Bay.

The Congress passed the Clean Air Act of 1970 about the same time it adopted NEPA. This act followed a series of previous measures passed in 1955, 1963, and 1967 aimed at reducing air pollution. The Clean Air Act of 1970 put in place major regulatory programs that share responsibilities with states. The USEPA is responsible for enforcing this act (USEPA 2019a).

The Clean Water Act (CWA) of 1972 In 1972 Congress applied the principles of the Clean Air Act to address the problem of widespread water pollution. The Clean Water Act of 1972 is the foundation for water pollution reduction and control nationwide and includes a wide range of provisions to regulate waters, wetlands and similar areas under Federally-funded programs operated for the most part by the states, with oversight by the USEPA. Every few years the CWA is reviewed and revised by Congress. Such updates provide an opportunity to learn from experience how to change programs and provisions for the better. Areas of particular concern for the Chesapeake are provisions to control municipal sewage, protect wetlands and regulate agricultural activities. Section 117 of the CWA established the Chesapeake Bay Program described below (USEPA 2019b). A major failing of the CWA is its lack of authority to regulate non-point discharges, which means essentially no regulation of pollution from agricultural fields. However, concentrated animal feeding operations (CAFOs) and manure applied to fields can be regulated by the USEPA (Table 2).

The Clean Air Act, the Toxic Substances Control Act the Fish and Wildlife Coordination Act and Other laws There is a full set of national regulatory laws focused on environmental issues apart from water, but the effects of the pollutants managed by these laws can impact water bodies like the Chesapeake from such sources as air deposition or leaching from land sites into groundwater, so they need to be managed as part of the solution for clean waters to recover and preserve the Chesapeake. Each of these laws has its own regulations and some of those try to deal with issues such as air deposition of toxics and nutrients on land and water. But they are not well managed

Table 2 Key provisions of the Clean Water Act (CWA) that affect the Chesapeake Bay and why they fall short

Bay needs	CWA limited controls	CWA lack of control
Reduced nutrients	Permits for sewage treatment and urban stormwater	Agriculture
Reduced sediments	Urban stormwater permits	Agriculture
Reduced toxics	Urban and suburban point source permits	Agricultural pesticides
Reduced peak loads	Urban stormwater combined sewer discharges	Agricultural runoff

The CWA fails to protect the Bay from agricultural activities due to the legislation lacking any authority over non-point sources of water pollution

with respect to overlap with other laws or other areas of the environment. The term used in the bureaucracy is that they are managed with a "stovepipe" mentality that ignores impacts on other media.

In the case of the Bay, it became clear by the mid-1970s that the new national laws wouldn't be enough to clean up the varied and inter-related sources of pollution. Even today, the laws do little to protect the Bay from the pollution produced by agriculture. It all came to a head in 1973. Senator "Mac" Mathias of Maryland took a 450-mile (724 km) boat ride on the Bay with scientists. He saw the declining fisheries and water quality and talked to 150 local residents about the condition of the system. Senator Mathias took that information back to Congress, which enacted a $27 million 5-year study of the Bay (Blankenship 2010). Upon its completion in the early 1980s, the study made clear that significant reductions in loads of nutrients, sediments and toxins were required to recover the Bay.

The First Chesapeake Bay Agreement, 1983 This Agreement was a one-page pledge signed by the Governors of Maryland, Pennsylvania and Virginia, the Mayor of DC, the Administrator of EPA, and the Chair of the Chesapeake Bay Commission, a body comprised of legislators from the three states. The group also established itself as the Chesapeake Executive Council and committed to state/Federal cooperation. In recent years, Delaware, New York and West Virginia also joined the pact. Many other Federal agencies joined EPA in working with the states (Table 3). The Executive Council meets annually and adopts new agreements and directives as appropriate. The basic Agreement committed to use Clean Water Act standards to achieve reductions, and established the Bay Program Office under EPA in Annapolis, MD (CBP 2019c).

The Second Chesapeake Bay Agreement, 1987 This second agreement marked a substantial expansion with detailed actions and specific numeric goals. Most significant was adoption of the 40% nutrient reduction goal to be achieved by the year 2000. The Goal applied to both nitrogen and phosphorous, in part because it had been learned that while nitrogen was the controlling pollutant in salt waters, phosphorous was controlling in fresh water systems.

To have a group of states and the Federal Government adopt such a specific and time-limited goal was unprecedented at the time, although it led many other places to do the same in following years. It was also made clear that *the goals were to be applied to agricultural sources as well as point sources*. The modus operandi was for the Federal laws to be followed and Federal funds drawn upon to reach the 40% reduction goals, with maximum flexibility given to the States to develop and publish their strategies for how it would be accomplished.

The 1992 Amendments The Executive Council at its 1992 meeting adopted provisions to improve controls of upstream sources of pollutants and to include a toxics reduction strategy in the Agreement. These were proposed to deal with the emergence of new issues from the data being collected. It had become evident that clean-up technologies for urban waste streams were increasingly well-developed and receiving a good deal of attention from scientists, in part because funds to carry out the measures were available and the benefits from installation were immediate. In contrast, funds to individual farmers and ranchers were more difficult to obtain and the technologies to handle agricultural sources were not developing as quickly. The purpose of the amendments was to have states develop the means to reach out to agricultural landowners with programs of assistance and technology advances.

The Third Chesapeake Bay Agreement, "Chesapeake 2000" While there was mixed progress on a number of goals, it became clear by the late 1990s that the 40% nutrient reduction goal was not going to be met by 2000. Much of the problem was with agricultural sources, and Pennsylvania, in particular, was way behind in reaching the goal. Other states could benefit from upgrades to sewage treatment plants and urban stormwater controls in their cities, but the Pennsylvania watershed was overwhelmingly agricultural.

The intent of "Chesapeake 2000" was to provide the vision and strategy to reach the 40% reduction and other goals by 2010. In all, 102 specific goals were adopted, nearly all numeric. They related to pollution discharges, natural habitat recovery, living resource improvements, sound land use practices, and public engagement, among other goals.

Table 3 Federal partners participating in the Chesapeake Bay Agreement as of 2019 (CBP 2019d)

US Environmental Protection Agency	National Oceanic and Atmospheric Administration
US Army Corps of Engineers	US Fish and Wildlife Service
National Parks Service	US Geological Service
Natural Resources Conservation Service (US Department of Agriculture)	Forest Service (US Department of Agriculture)
Department of Defense	Department of Home Land Security
Department of Transportation	

Each would be monitored and reported on, in the most important cases monthly.

In addition, at this time both Delaware and New York signed on to the water quality goals; they were joined by West Virginia in 2002, completing participation by governments covering the entire watershed for the first time.

The Fourth Chesapeake Bay Agreement This effort began in 2008, when it became clear that despite progress, the 2010 goals were not going to be met. The Chesapeake Bay Program's Principal's Staff Committee (PSC) and Executive Committee requested that USEPA develop a Bay-wide Total Maximum Daily Load (TMDL) allotment for nutrients and sediments. President Obama's Executive Order #13508 issued in 2009 helped set the stage for incorporating the TMDL in a 2014 update of the agreement. For the first time there would be a single Bay-wide Total Maximum Daily Load (TMDL) allotment for nutrients and sediments. The TMDL tool was in the Clean Water Act, but heretofore had been used for single pollutants and in small areas of watersheds. Now it would be combined with individual Watershed Implementation Plans (WIP's) for each state. These WIP's are to include specific steps that needed to be taken to meet the goals for each major pollutant state-wide by 2025. The plan to do all this was upheld in a court challenge.

These documents were developed over a long period with public meetings and involvement, and were signed in 2014 by all six states and DC. This agreement contains 5 themes consisting of 10 broad goals and 31 specific outcomes, which include Bay tributaries and lands (Table 4). Each jurisdiction has specific steps to take to meet the reduction goals by 2025, with workplans and milestones to carry out the Implementation Plans.

The ten goals are interrelated: improvements in water quality can mean healthier fish and shellfish; the conservation of land can mean more habitat for wildlife; and a boost in environmental literacy can mean a rise in stewards of the Bay's resources. Our environment is a system, and these goals will support the health of the public and of the watershed as a whole.

Each goal is linked to a set of outcomes, or time-bound and measurable targets that will directly contribute to its achievement. Each outcome has a management strategy and work plan which offer insight on how the partnership will achieve each outcome, as well as how we will monitor, assess and report progress toward abundant life, clean waters, engaged communities, conserved lands and climate change resilience. Learn more on the goal pages listed below.

Abundant Life Poor water quality and harvest pressure challenge the health of species across the region, while our increasing need for land and resources has fragmented and degraded the habitats they depend on. Supporting sustainable fish and shellfish populations and restoring habitat for native and migratory species will support a strong economy and a balanced ecosystem. This them encompasses two goals; Sustainable *Fisheries* and *Vital Habitats*.

Clean Water Excess nutrients, sediment and toxic contaminants degrade our waterways, harm fish and wildlife and pose risks to human health. Reducing these pollutants is critical to creating safe, healthy waters for animals and people alike. Three goals serve this theme: *Water Quality*, *Toxic Contaminants*, *Healthy Watersheds*.

Conserved Lands Changes in land use and development can impair water quality, degrade habitats and alter culturally significant landscapes. Conserving lands with ecological, historical and community value is integral to maintaining a healthy ecosystem and vibrant culture. This is served by the *Land Conservation* goal.

Engaged Communities The long-term success of the Chesapeake Bay restoration effort depends on the work of individuals and communities living throughout the watershed. Connecting with current environmental stewards and encouraging future local leaders helps build the network that will keep our work moving forward. Three goals address this issue: *Stewardship*, *Public Access*, *Environmental Literacy*.

Climate Change Storms, floods and sea level rise will have big impacts across the watershed. Monitoring, assessing and adapting to these changing environmental conditions will help our living resources, habitats, public infrastructure and communities withstand the adverse effects of climate change. This is addressed by the *Climate Resiliency* goal.

Table 4 The goals set out by the Chesapeake Bay Agreement of 2014 (CBP 2019d)

Goal	Description
Sustainable fisheries	Protect, restore and enhance finfish, shellfish and other living resources, their habitats and ecological relationships to sustain all fisheries and provide for a balanced ecosystem in the watershed and Bay
Vital habitats	Restore, enhance and protect a network of land and water habitats to support fish and wildlife, and to afford other public benefits, including water quality, recreational uses and scenic value across the watershed
Water quality	Reduce pollutants to achieve the water quality necessary to support the aquatic living resources of the Bay and its tributaries and protect human health
Toxic contaminants	Ensure that the Bay and its rivers are free of effects of toxic contaminants on living resources and human health
Healthy watersheds	Sustain state-identified healthy waters and watersheds recognized for their high quality and/or high ecological value
Stewardship	Increase the number and the diversity of local citizen stewards and local governments that actively support and carry out the conservation and restoration activities that achieve healthy local streams, rivers and a vibrant Chesapeake Bay
Land conservation	Conserve landscapes treasured by citizens in order to maintain water quality and habitat; sustain working forests, farms and maritime communities; and conserve lands of cultural, indigenous and community value
Public access	Expand public access to the Bay and its tributaries through existing and new local, state and federal parks, refuges, reserves, trails and partner sites
Environmental literacy	Enable every student in the region to graduate with the knowledge and skills to act responsibly to protect and restore their local watershed
Climate resiliency	Increase the resiliency of the Chesapeake Bay watershed, including its living resources, habitats, public infrastructure and communities, to withstand adverse impacts from changing environmental and climate conditions

The 2018 Review In 2018 a progress review was held of all TMDL's and their WIP's. The conclusion was that there is sufficient progress in meeting pollutant reduction goals to find the effort on schedule or nearly so, with the exception of nitrogen, where the major problem remains in Pennsylvania. The recent change (2018) in political leadership of that state, with the environmental agency appointments of a new Governor in Pennsylvania, as well as the appointment of a high-level Commonwealth official from the state Department of Environmental Protection to be Director of the Bay Program for USEPA in Annapolis, may signal increased commitment to the issue. However, Pennsylvania continues to lag behind the other states in its efforts (CBF 2019a).

Other Important Institutions in the Bay Program The structure of the Bay Program is complicated, with a large number of committees of federal, state and local officials in areas from modeling to fisheries. In addition to those committees and work groups of primarily governmental and academic experts, there are a number of special groups that need to be understood in terms of roles and responsibilities.

The Executive Council, as mentioned above is comprised of the Governors, the Mayor of DC, the Chair of the Chesapeake Bay Commission of state legislators, and the Administrator of EPA. The Principals' Staff Committee is made up of the heads of the environmental agencies of each jurisdiction and the Regional Administrator of EPA. The Management Board is comprised of high-level officials of state and federal agencies. The Goal Implementation Teams have the lead on the ten goals (Fig. 1).

The Federal Leadership Committee is comprised of representatives of 12 Federal agencies with important roles to play in achieving Program goals. Some are key landowners, such as the Department of Defense; some are important regulators such as the US Army Corps of Engineers; Others play multiple roles, such as the National Park Service or the US Forest Service. This Committee attempts to carry out the very complex efforts to coordinate the work of their specific agency in the watershed with that of other Federal agencies, state and local counterparts and other groups in the general public.

There are three standing Advisory Committees—Scientific, Local Government and Citizens. Members are appointed by participating states. Their responsibilities are somewhat self-evident. The Scientific and Technical Advisory Committee (STAC) has been a critical source of expertise in such areas as watershed modeling, fisheries management and emerging technologies. The Local Government Advisory Committee (LGAC) is the single most important way for local governments to participate in the Bay Program and have their views taken into account. The Citizens Advisory Committee has the widest scope and mission, to make the Program aware how it is being perceived by the public and to propose ways to assure continued political support.

In addition to (CAC) providing reports and reviewing Program actions, the three formal Advisory Committees provide their advice directly to the Executive Council (the Governors, EPA Administrator, etc.). Each year they are invited to the private meeting of the Council with no program personnel present, an important element in the Bay Program

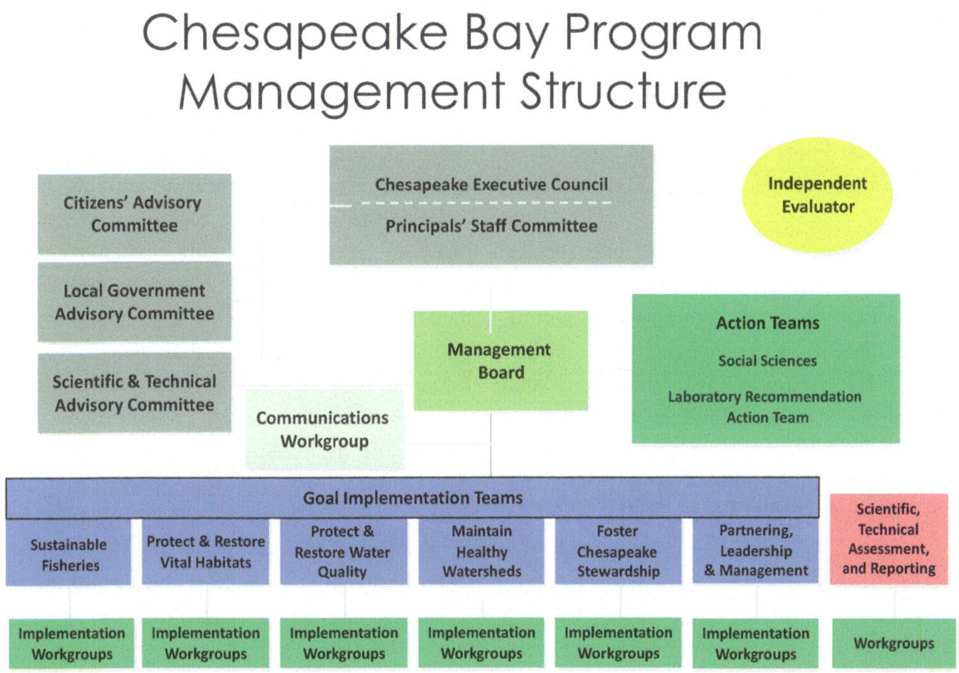

Fig. 1 Organizational chart for the Chesapeake Bay Agreement of 2014 (CBP 2019d)

structure to assure freedom of expression direct from experts to the highest political levels.

Citizen Groups There are literally hundreds of citizen-based groups at work to help restore the lands and waters of the Chesapeake. Many are focused on particular rivers or their tributaries. For example, there are over 40 such groups at work in the Anacostia River in the DC area alone, and it is but one small tributary of the Potomac. In general, all these groups combine citizen engagement in projects with pressuring government agencies to improve conditions in their watersheds.

Taking the Anacostia as an example, the citizen groups there are supported by larger non-profits like the Anacostia Watershed Society (AWS 2019) and the Anacostia Riverkeeper (AR 2019). The latter help organize larger-scale clean-up and restoration events and provide education and technical help to the smaller citizen-based groups. The DC Department of Environment and Energy and the Prince Georges County Department of the Environment also provide technical and financial support to projects undertaken by these groups.

There are also a number of citizen-based organizations that operate Bay-wide in a variety of capacities. Five that bear special mention:

The Alliance for the Chesapeake Bay This organization was founded soon after the Bay Program was launched. The focus of this group is on support through training and expertise of local citizen-based groups. The Alliance also works closely with the Bay Program, even performing under contract such functions as outreach and education (ACB 2019).

The Chesapeake Bay Foundation This group advocates on behalf of the entire Bay and its watershed. It also got an early start in the late 1960s, as interest on the Chesapeake clean-up began. While it operates extensive citizen clean-up and education programs, it's unique role is as a group that lobbies all levels of government, and actively initiates and supports legislative and executive actions. It is also willing to step up and criticize the Bay Program for not acting quickly or smartly enough (CBF 2019b).

The Chesapeake Conservancy This group partners with the National Park Service and related state agencies to support conservation of key public and private land and water resources in the Bay watershed. One important focus is on the Chesapeake Bay Water Trail and its on-shore support facilities. The Water Trail includes signage with historical interpretation to inform boaters and shore visitors of significant sites and events (CC 2019). The Conservancy has created a Conservation Innovation Center that uses high resolution land cover data and other land use data base information to identify the most effective locations to implement pollution control measures to achieve the greatest reductions.

The Chesapeake Legal Alliance By drawing upon the volunteer services of lawyers throughout the watershed, this group assists both regional and local groups who believe they need legal services to achieve their goals through negotiation, legal services of all kinds and proposed administrative and court actions where necessary (CLA 2019).

The Bay Journal An outgrowth of the Alliance for the Chesapeake Bay and with the financial support of the Bay Program, the Bay Journal provides accurate and helpful information both in a monthly news magazine and on-line about all aspects of the Bay clean-up. Over time it has been able to use its growing reputation as a news source to finance itself more and more with both donations and advertising. This has assured its independence and its long-term financial security as a trusted source of information on Bay issues (BJ 2019).

Progress to Date Adding it all up, as noted above the 2018 program evaluation by the Bay Program indicates that progress is on track or nearly so for nearly all the major pollutants requiring reductions, with the exception of nitrogen. Pennsylvania is the only jurisdiction lagging because so much of its nitrogen loadings are coming from its extensive areas of intensive agriculture. And the effects are broadly felt in the Bay because (1) the vast majority of Pennsylvania loadings enter via the Susquehanna River, which enters the very top of the Bay, and (2) nitrogen is the controlling nutrient in salt and mixed salt and fresh waters.

Facing the Challenge of Agriculture While progress in the clean-up of the Bay has been slow, nobody ever thought it would be easy. In particular, the important role of agriculture in the economy of the watershed and its substantial impacts on water quality were always going to be difficult to deal with.

Cities are large sources, but there is technology to handle much of their loads and a broad base of rate-payers to share the cost. Cities produce mostly point source pollution out of pipes. That makes it feasible to intercept, retain, and treat the flow before discharge. The technologies to handle toxic pollutants, many of which originate in urban areas, are also reasonably available and cost-effective.

The situation is more difficult with respect to agriculture. Pollution drains from non-point sources of fields, CAFOs, and contaminated ground water. Clean-up costs for nutrients, sediments and other agricultural pollutants are heavily born by the landowner and can be costly. While states and the US Department of Agriculture provide funding and other assistance, changes in agricultural crops and practices can be difficult for farmers to undertake and accept. The Delmarva Peninsula constitutes the entire Eastern Shore of the Bay, and its home to several thousand CAFOs for the broiler chicken industry (see chapter "Livestock and Poultry: The Other Colonists Who Changed the Food System of the Chesapeake Bay"). These factory farms add pollution to ground water, tributaries draining to the Bay, and the air.

The special case of Amish farmers of Pennsylvania provides some additional challenges and opportunities. They are committed long-term to their lands and most of their farming activity is organic and chemical-free, but their animals produce wastes that are recycled onto the fields. Because they do not accept government funds for most activities they must decide on their own to undertake changes that result in loadings reductions. But this also provides opportunities. For example, while other farmers see forest buffers along streams as a loss of valuable agricultural land, for the Amish the planting of fruit and nut trees is a welcome opportunity to engage their children in the harvest.

And while manure and fertilizer loadings can be controlled by forest buffers and other means, reducing the use of intensive animal agriculture such as feedlots, and eliminating or reducing certain crops can also go a long way to achieving goals. The key is to combine flexibility with clear knowledge of the economic and water quality effects of alternative agricultural practices. The bottom line will probably be some mix of changes in crops, changes in the mix of land use between animals and crops, expanded forest buffer programs, and reductions in fertilizer, pesticide and other chemical use, either through practices that use less or through the development of less polluting alternative products. Developing and convincing farmers to employ this new mix will need to be a major part of Pennsylvania's efforts to meet its goals for the Chesapeake in the next decade.

Part of the solution to the estuaries' problems lies just outside of the Chesapeake Bay Watershed, near Kutztown, PA. J.I. Rodale founded the Rodale Institute experimental farm in 1947 to develop and share organic farming methods free of artificial fertilizers and pesticides (Rodale 2019). Rodale was also one of the first to recognize the link between chemical agriculture and declining public health, and promoted his ideas in a book, *Pay Dirt: Farming and Gardening with Composts* (Rodale 1948). Convincing more farmers to leave the chemical behind and join the organic movement will go a long way in helping to recover the Bay (see chapter "An Organic-Based Food System: A Voyage Back and Forward in Time").

Assuring the Will to Get the Job Done Folks in the Chesapeake watershed have confidence that ultimately the Bay will be restored and the fisheries, grasses and other resources are returned to healthy levels. Much has already been achieved. The submerged aquatic vegetation, for example, is as extensive as they have been since first measured nearly a century ago (VIMS 2018).

The key will be the willingness of agricultural landowners to be flexible as we learn more about what is needed. This will be done by applying the *modus operandi* that has developed over the recent decades of efforts by the citizens and their governments. It requires political will, engagement of all the players, drawing on and pushing the technologies that will get us the needed results, learning and using what we know about the role nature can play in the restoration, and being open to new products and processes. For example, amending soils with biochar (charcoal produced by pyrolysis of organic matter) increases nutrient retention and soil fertility (Lehmann and Joseph 1970). It's also essential to keep everyone at the table. This means working to achieve and show everyone the benefits of what we are all doing, including improvements in local streams and in agricultural profits.

A great deal has been achieved in the Chesapeake watershed already, and people around the world are taking notice of such innovations as the wide use of forest buffers, the daylighting of streams in cities to absorb and cleanse the discharge from upstream storm sewers, and massive reductions of nutrients from point sources. And we are achieving this in a watershed with at least five times the land to water ratio of every other watershed on earth, with amazingly productive agriculture land up the rivers and an equally amazing harvest of fish and shellfish downriver in the shallow Bay. Just remember, if we can do it here, nobody else in the world has one-fifth of an excuse!

References

ACB (2019) Home. Alliance for the Chesapeake Bay. https://www.allianceforthebay.org/. Accessed 18 Apr 2019

AR (2019) Home – Anacostia Riverkeeper. Anacostia Riverkeeper. https://www.anacostiariverkeeper.org/. Accessed 18 Apr 2019

AWS (2019) Anacostia Watershed Society. Anacostia Watershed Society. https://www.anacostiaws.org/. Accessed 18 Apr 2019

BJ (2019) Bay Journal – Home. Bayjournal.com. https://www.bayjournal.com/. Accessed 18 Apr 2019

Blankenship K (2010) Bay Journal – Article: Charles 'Mac' Mathias, founder of Bay cleanup effort, dies. Bayjournal.com. https://www.bayjournal.com/article/charles_mac_mathias_founder_of_bay_cleanup_effort_dies. Accessed 12 Apr 2019

CBF (2019a) 2019 State of the Blueprint. Cbf.org. https://www.cbf.org/how-we-save-the-bay/chesapeake-clean-water-blueprint/2019-state-of-the-blueprint.html. Accessed 6 Sep 2019

CBF (2019b) Chesapeake Bay Foundation. Cbf.org. https://www.cbf.org/. Accessed 18 Apr 2019

CBP (2019a) Watershed. Chesapeake Bay Program. Chesapeakebay.net. https://www.chesapeakebay.net/discover/watershed. Accessed 19 Apr 2019

CBP (2019b) Facts & Figures. Chesapeake Bay Program. Chesapeakebay.net. https://www.chesapeakebay.net/discover/facts. Accessed 10 Apr 2019

CBP (2019c) Bay Program History. Chesapeake Bay Program. Chesapeakebay.net. https://www.chesapeakebay.net/who/bay_program_history. Accessed 12 Apr 2019

CBP (2019d) Chesapeake Watershed Agreement 2014. Chesapeakebay.net. https://www.chesapeakebay.net/documents/FINAL_Ches_Bay_Watershed_Agreement.withsignatures-HIres.pdf. Accessed 16 Apr 2019

CC (2019) Homepage – Chesapeake Conservancy. Chesapeake Conservancy. https://chesapeakeconservancy.org/. Accessed 18 Apr 2019

CLA (2019) Homepage – Chesapeake Legal Alliance. Chesapeake Legal Alliance. https://www.chesapeakelegal.org/. Accessed 18 Apr 2019

Lehmann J, Joseph S (1970) Biochar for environmental management. Routledge, London

NEPA (2019) NEPA. National Environmental Policy Act. Ceq.doe.gov. https://ceq.doe.gov/. Accessed 11 Apr 2019

Rand McNally (2000) Rand McNally premier world atlas. Rand McNally, Chicago

Rodale J (1948) Pay dirt: farming and gardening with composts. Kessinger, Whitefish, MT

Rodale (2019) Rodale Institute – pioneers of organic agriculture research. Rodale Institute. https://rodaleinstitute.org/. Accessed 19 Apr 2019

USEPA (2019a) Evolution of the Clean Air Act. US EPA. US EPA. https://www.epa.gov/clean-air-act-overview/evolution-clean-air-act. Accessed 12 Apr 2019

USEPA (2019b) History of the Clean Water Act. US EPA. US EPA. https://www.epa.gov/laws-regulations/history-clean-water-act. Accessed 12 Apr 2019

VIMS (2018) Underwater grasses in Chesapeake Bay continue record growth. Virginia Institute of Marine Science. Vims.edu. https://www.vims.edu/newsandevents/topstories/2018/sav_2017_report.php. Accessed 19 Apr 2019

Ethics and Economics of Building a Food System to Recover the Health of the Chesapeake Bay and Its People

Benjamin E. Cuker and Karen Davis

The Beatles, 1963: Money
The best things in life are free
But you can keep them for the birds and bees
Now give me money
That's what I want
That's what I want, yeah
That's what I want
...

Songwriters: Janie Bradford/Berry Gordy Jr.

The Beatles, 1964: Can't Buy Me Love
Can't buy me love, love
Can't buy me love
I'll buy you a diamond ring my friend if it makes you feel alright
I'll get you anything my friend if it makes you feel alright
Cos I don't care too much for money, and money can't buy me love.
...

Song writers: John Lennon/Paul McCartney

Abstract

Reimagining a new food system for recovering the health of the Chesapeake Bay and its people requires addressing the ethics of the current way we produce and consume food. The present food system is built on making profit with the aid of government subsidies focused on producton of health-damaging animal agriculture. As is the case for most of the US, the Chesapeake Bay region produces more food than it uses, exporting surpluses elsewhere. Consistent surplus production depresses prices. Wealthier farmers take advantage of a system of price supports to sustain their profits. Despite the excess production of food, most of the workers in the food system earn such low wages that they require government subsidized nutrition, organized through the US Department of Agriculture. Many of the lowest paid food system workers are people of color and migrant laborers from Mexico and Central America. Only 7% of the food dollar goes to pay producers, with most of it going to the food service and food processing industries. About 10% of the residents of the Chesapeake Bay region face regular food insecurity and 14% receive subsidized nutrition benefits. The animal product-centered food system pollutes the Chesapeake Bay and sickens its people, while causing the suffering and premature deaths of industrially farmed animals. Extensive use of pesticides and artificial fertilizers on crops grown for animal feed (corn and soy) further degrade the ecosystem. Elements of a new ethical

B. E. Cuker (✉)
Department of Marine and Environmental Science, Hampton University, Hampton, Virginia, USA
e-mail: Benjamin.cuker@hamptonu.edu

K. Davis
United Poultry Concerns, Machipongo, VA, USA
e-mail: karen@upc-online.org

Two opposing notions of the role of money in structuring one's life as expressed by the 1960s musical group, the Beatles in two of their popular songs. Perhaps they changed their minds between 1963 and 1964. Should we also rethink how we place value on the various aspects of the food system as we reimagine it for the Chesapeake Bay?

B. E. Cuker (ed.), *Diet for a Sustainable Ecosystem*, Estuaries of the World, https://doi.org/10.1007/978-3-030-45481-4_21

food system include organic production for whole plant based diets, imporved compensation for workers, eliminating the exploitation of animals, developing food literacy, increasing the amount of food prepared at home, replacing corporate profit with social and environmental need as the driving force of agriculture, subsidizing healthy foods, and encouraging local farming with direct distribution to consumers.

Keywords

Equity · Animal rights · Income distribution · Racial and gender disparities · Immigration · USDA · Poverty · Food insecurity

A New Ethical Food System A new ethical food system for the Chesapeake Bay region would better compensate workers and shift farm production to support a whole food plant-based diet using sustainable organic techniques. This could include expanding the system of community supported agriculture and farmer's markets. The new system would shift subsidies away from commodity crops to instead support farming fruits and vegetables. The change would also require educating consumers on how to choose and cook sustainable and healthy plant-based foods.

In May of 1607, the English colonists arrived on the shores of the Chesapeake Bay with enough food for only a few months. The Virginia Company expected Capt. John Smith and his party to obtain food from the Algonquin people of the Chesapeake Bay, as well as to learn how to gather, hunt, and grow their own victuals. The colonists proved unable to feed themselves and two thirds of their group perished during the "starving time" of the winter of 1609–1610. That as many survived as did traces in large part to the food they got from the native people of the Bay. They obtained this life-saving food (mostly corn) from their neighbors one of three ways: as a gift, in trade for goods, or by theft. These three ways to obtain victuals from others went on to inform the subsequent development of the food system of the Chesapeake Bay and how it continues to operate today.

Imagining what will restore the health of the Chesapeake Bay and that of the people that live there means contemplating radical changes to the food system. Such changes will require rethinking every aspect of the production, marketing, distribution, and consumption of food. To do this we need to understand the underlying ethical questions of the endeavor. The present food system of industrial agriculture producing for the Standard American Diet (SAD) developed under British colonialism, the institution of slavery, and the US capitalist system. While the American Revolution killed colonialism, and the Civil War did the same for chattel slavery, their ghosts still haunt the Chesapeake Bay, lurking behind the contemporary system of large corporate capitalism.

Government subsidies prop up sectors of the food system that would otherwise fail in their present forms. Government policies also serve to sustain the profitability of the food system, and to constrain the system from doing more harm than it does. Federal food policies in the shape of SNAP (Supplemental Nutrition Assistance Program, formerly called Food Stamps) and WIC (The Special Supplemental Nutrition Program for Women, Infants and Children) assure markets for food producers and basic nutrition for the least advantaged of our citizens.

More than anything, the drive for making money shapes the relationships between land, food, human health, and consequences for the environment. This is the nature of the capitalist system which in full development monetizes all aspects of the way humans relate to each other. Profit seeking is the organizing principle. As such, the current food system primarily operates to maximize profits for those in control of it.

Money and Food Early English colonists bought food from the Algonquin-speaking tribes with barter of manufactured goods. English money didn't translate into a redeemable currency for the Native Americans. They might have valued the coins for their metal and exotic artistry, but certainly not for their symbolic expression of a universal way to accomplish trade. It's not that the concept of money was new to the Powhatan and other indigenous people, as they used seashells woven into belts (wampum) to facilitate a vigorous system of trade between the network of diverse tribes that stretched across the continent (Jacobs 1949). Indians might have also exchanged furs or other goods that they didn't want or need, kept only for their value in future trading. As wampum, British Pounds and local currencies gave way to US Dollars (and for a brief period, Confederate Dollars) as the population of the Chesapeake Bay region moved to a common currency for trade. This meant that a dollar represented some portion of essentially every good, real property, or service exchanged between people. It also meant isolating people from the basis by which the many goods came into being, including food. A person with sufficient money didn't need to farm, hunt, or forage. Importantly, money also limited notions of ethics and morality in commerce.

A dollar acquired in business by a slave trader, pimp, or thief was still a dollar of the same worth as that earned by a farmer, baker, teacher, doctor, mechanic, cleric, or scientist. Immorality doesn't change the color of money. It's still a dollar no matter how much misery or good went into obtaining it. Indeed, for those so liberated from notions of ethics, the only problem with what some might consider tainted money is it t'aint enough. For the food system, this meant that farmers no longer grew their crops just to feed themselves and their neighbors, but to make money. The market and profits came to dictate what they grew as much as did soil conditions, weather, and season. The same dollar used to purchase organically grown crops works just as well in buying meat from concentrated animal feeding operations (CAFOs) or a can of nutrition vacant cola. The care and sustainability of producing organic oats isn't reflected in the money it makes. Nor is the cruelty to chickens, water pollution, or artery clogging cholesterol painted on the dollar traded for a piece of meat. It's all just money, void of any ethical responsibility.

Beyond being void of morality, money influences the food system in another way. Money is managed in the US by the Federal Reserve to promote a robust and stable economy. Functionally that means that the value of money needs to decrease a percentage point or so every year. Such managed inflation means a slight but steady increase in the cost of goods and services. This is ironic in that technological advances coupled with energy subsidies continues to decrease the cost of labor needed to provide goods and services. Indeed, some goods such as electronics do get cheaper, or provide much more functionality for the same price. Industrialization of the broiler industry also made for cheap chicken. But taken on the whole, incremental inflation means an ever-rising cost of living. Why not manage the economy for zero inflation or even deflation to reflect the decreasing cost to produce goods and services? Doing so would mean money invested in savings accounts, stocks, bonds, and real estate would tend to lose value over time. It takes land to produce food. If the value of that land declines over time through deflation, the farmer won't have sufficient collateral to continue to secure the mortgage she used to buy it. The falling price she gets on the food she produces won't pay the loans used for buying equipment, seeds, or soil amendments. The farmer will default on her mortgage and loans. The bank will repossess and auction the land, as was the fate of tens of thousands of farmers during the Great Depression when deflation reached 30% (Alston 1983).

Land for Farming What about that land used for farming in the Chesapeake Bay region, and sustained by incremental inflation? How did the farmer come by it in the first place? The farmer either bought or inherited it as property originally pilfered from its initial owners, the Powhatan and other tribes of the area. Native Americans understood land ownership quite differently from the English. The Indian perspective viewed land as a common, available to all in the tribe. The English replaced that system with set boundaries and a title of individual ownership. To obtain the lands, the colonists violently expatriated the tribes in a series of wars. The English Royal Family then awarded the spoils of colonial expansion to military veterans, members of the Virginia Company Colony, and other favored persons, thereby establishing the lineage of ownership tracing to today's land deeds. King James dismissed the idea of the land belonging to its original inhabitants, but did defer to that claimed by Spain, granting lands "...not now actually possessed by any Christian Prince or People (Virginiaplaces 2019)." Just as today's industrial farmer externalizes the cost of pollution she might produce, the English colonists never paid for the land they acquired from the tribes. The questionable ethics of land acquisition in the Colonies set the tone for how the food system would treat its workers.

The original residents of the Chesapeake Bay Region spent most of their time obtaining food; gathering, hunting, fishing, and cultivating. The English colonists had to do the same, despite the distraction of pursuing money via tobacco agriculture. Beginning in the nineteenth Century, advances in farming technology, industrialization of agriculture, and subsidies of energy, artificial fertilizer, and pesticides freed much of the population from its bond to procuring food. The industrialization of farming also meant a shift from many small family farms to fewer and larger corporate operations. In 1935, the U.S. featured 6.8 million farms, with an average size of 61 ha (150 acres). By 2010 that shifted to two million much larger farms, averaging 178 ha (440 acres) in area. Industrialization reduced the number of workers needed to produce food. Modern farms require only 25% of the labor used in 1940. As of the 2010 Census, less than 2% of the US population now works in farming (USDA 2014). The intensive agricultural model adopted after WWII tripled national yields, and did so with 25% less land under cultivation (Lusk 2016).

Is there an alternative to individual or corporate ownership of farms? Community gardens green the urban landscapes of many of the Chesapeake Bay's towns and cities. They occupy vacant lots owned by local governments or sympathetic landlords. In the US, this movement began in the 1890s in response to urban food shortages. The need for more vegetable production during both World War I and II spurred federal authorities to promote victory gardens in backyards on private land and in community commons. Victory gardens worked, producing 44% of the fruit and vegetable crop in 1942 (Birky 2009). The American Community Gardening Association was founded in 1982 to revive common gardening, and they cite 18,000 efforts across the US and Canada (ACGA 2019). One newspaper article listed 27 community gardens in the cities of South Hampton Roads; the Virginia cities of Norfolk, Virginia Beach, Suffolk, and Chesapeake (Hollingsworth 2016). Additionally, Baltimore offers 22, and the District of Columbia lists 16 community gardens.

From the outset, the issue of owning and controlling land for food production loomed large for African American farmers in the Chesapeake Bay Region. Those enslaved on plantations often sustained themselves with "slave gardens" in which they cultivated numerous crops of African origin (Hatch 2001). The enslaved gardeners sold or bartered their excess produce to obtain tools and other necessities denied them by their forced and unpaid labor. After emancipation, and absent the promised "40 acres and a mule," some freedmen found refuge from sharecropping and tenant farming with a system of collective agriculture (White 2019).

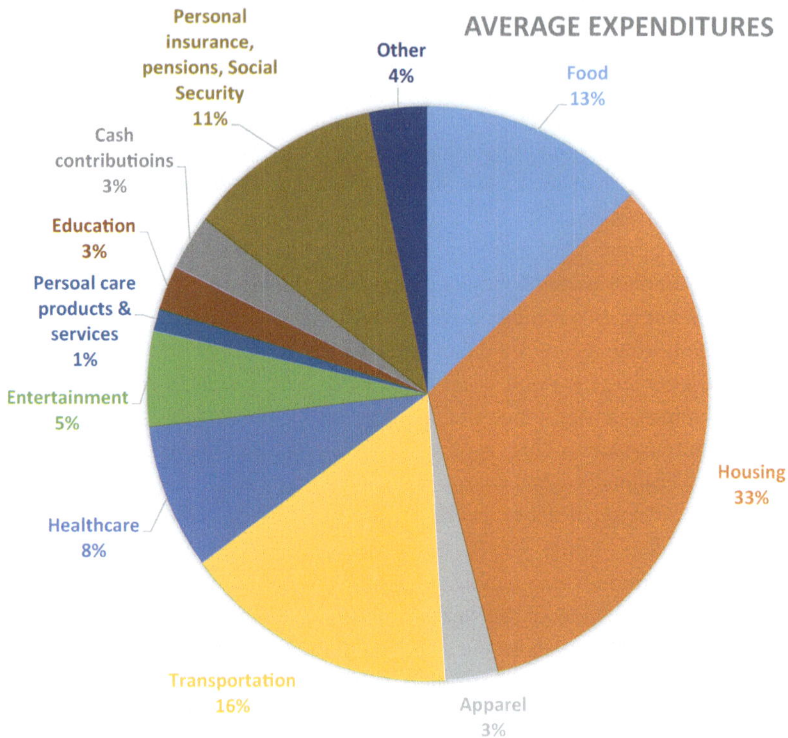

Fig. 1 Average expenditures per consumer unit in the US for 2017. A consumer unit includes families, single persons living alone, or sharing households. This is based upon an average before taxes income of $73,573 and average expenditures of $60,060. The 13% spent on food equals $7729 per year (BLS 2019a)

Many also joined the Colored Farmers' Alliance (CFA), which included chapters in Virginia. Like the Grange, the Agricultural Wheel and National Farmers' Alliance, this late nineteenth Century organization brought farmers and farm workers together to exert political clout. The chapter in Norfolk, VA also served as a cooperative, facilitating the purchase and distribution of farm supplies and crops. (Spriggs 1979; White 2019). In 1891, the CFA of Virginia listed 20,000 members, and about 25% were women. At the time there were somewhat less than 20,000 black-owned farms in the state, so farm laborers must also have been included in the membership count. The CFA eventually collapsed, much because rising racism at the end of the nineteenth Century impeded making common cause with white farmers (Spriggs 1979).

Several civil rights and black power focused organizations of the 1960s and 1970s featured farming and food cooperatives as a means to obtain food security, social justice, and autonomy. This tradition is carrying on as African Americans in urban centers around the US are rediscovering their cooperative farming roots (White 2019).

Money to Buy Food How much money people have determines their relationship to the food system. The United States Department of Agriculture (USDA) publishes estimates of how much money it takes to buy food. For 2019, the USDA projected that costs to feed an adult (age 12 and up) a nutritious diet ranges between $2028 and $4452 per year, depending on age, gender, and thriftiness (USDA 2019a). Yet people are spending much more than that for food, an average of $7729 per year (Fig. 1, BLS 2019a, b). Why are we spending several thousands of dollars more than we need to for food? People use 44% of their food budget on eating out of the home, which is more expensive than preparing food in their own kitchens (BLS 2019a, b). Also, The Standard American Diet (SAD) is built on expensive animal-based foods, lots of refined carbohydrates, and few fiber-rich fruits and vegetables. This diet provides low nutritional value and high caloric intake for the money. The absence of dietary fiber reduces satiation, and prevalence of prediabetes and diabetes means wide swings in blood sugar concentrations. This all leads to overeating, meaning buying more food than is needed.

The 13% of average expenditures devoted to food seems reasonable, if not inexpensive considering it's what keeps us alive. However, for the 29.5% of households in the US with total annual incomes of less than $35,000, the expenditure for food is significant and often prohibitive for maintaining good health. For the poorest 10% of the population, food could easily account for half of what they have to spend (Table 1). The burden of food prices for low income people falls most heavily on people of color. Children in the US and Chesapeake Bay jurisdictions are disproportionally poor, with 21% and 19% respectively living below the poverty level. SNAP and WIC ameliorates food insecurity for 13% of the US population and 14.4% of Chesapeake Bay area

Table 1 The income distribution in the US for households based upon 2017 US Census data (Census 2019)

Race	Under $15,000	$15,000 to $24,999	$25,000 to $34,999	$35,000 to $49,999	$50,000 to $74,999	$75,000 to $99,999	$100,000 to $149,999	$150,000 to $199,999	$200,000 and over	Median $	Mean $
All Races	10.7	9.6	9.2	12.3	16.5	12.5	14.5	7.0	7.7	61,372	86,220
White	9.1	9.2	9.0	12.1	16.7	12.9	15.4	7.5	8.2	65,273	89,632
Black	19.6	12.6	11.6	14.1	15.8	10.1	9.6	3.5	3.1	40,594	58,985
Hispanic	12.0	11.1	11.3	15.0	18.1	12.6	11.5	4.5	3.8	50,486	68,319
Asian	8.9	6.5	6.1	9.7	14.6	12.7	16.4	10.5	14.6	80,961	113,720

About 29.5% of all US households live on less than $35,000 per year, making them vulnerable to food insecurity

residents. This poverty results in 11.5% of Bay residents experiencing food insecurity. Low wages mean that poverty and food insecurity afflicts many working families. On average 44% of SNAP recipients are from working families, and 39.1% for the Bay area (Nchako and Cai 2019, Table 2).

Most Food System Workers Earn Low Wages and Tend to be People of Color Many of those people living in poverty and suffering food insecurity work in the food system. This traces to colonial times. The early English colonists of Virginia introduced a model of underpaid and unpaid labor that still informs conditions for food system workers 400 years later. The colonial planters needed labor to work their fields to produce both food for local consumption and cash crops for export. Impoverished English men and women hungry for a new start in the New World traded the cost of passage on a ship for the promise of 4–7 years of labor owed to a Virginia planter. This system of indentured servitude was well developed in England and easily transported to the Virginia Colony (Ron and Norwood 2018; Wolfe and McCartney 2019). It meant that despite the great shortage of labor in Virginia, wages were kept low.

After serving their required time of labor, these bonded workers were free to build their own lives in the Colony. As such, planters had to look for more workers. They found them in the system of chattel slavery developed by the Caribbean sugar industry, decades earlier. Beginning in 1619, planters bought slaves from importers. The planters owned the labor of their chattel slaves for as long as the slaves lived, *and also owned the labor of the descendants of the slaves*. This was a very different proposition from the temporary system of indentured workers. Virginia planters became so successful at growing their own slaves that the numbers produced exceeded local labor needs. The sale of excess slaves to planters in other states became Virginia's most lucrative "crop" by the 1850s, earning more cash than tobacco, cotton, or other commodities (Freudenberger and Pritchett 1991).

The system of indentured labor was based on class and permitted former white servants to eventually rise in social and economic status. Chattel slavery was based on race. While some slaves managed to buy their freedom, or were freed by sympathetic owners, that change in status didn't alter their racial designation. Such freedmen retained a legal second-class citizenship during antebellum times and extending through the Jim Crow Era (circa 1877–1967) (James 1988). The ideology of racism used as the moral basis for justifying race-based slavery still echoes through society and the food system. This means that into the modern era people of color continue to perform the most dangerous and demanding jobs in the food system while earning the lowest pay (McGwin 2000).

The Civil War ended chattel slavery. Yet many former slaves stayed in place, working for their former owners as sharecroppers. In return for being allowed to retain their meager lodgings, they worked a section of land with a large portion of the crop going to the landlord. However, the sharecroppers lacked funds to capitalize the enterprise and quickly fell into debt to the landlord for seeds and farm implements (Douglas-Bowers 2013). Many poor whites in the Chesapeake Region joined their black counterparts in this newly expanded form of farmworker exploitation.

World War II created a shortage of farm labor. Many young men and women left farms to join the armed forces or work in factories to produce the goods of war. More food was also needed to supply the troops. To meet the demand, the US partnered with Mexico in 1942 to create the Bracero Program of migrant guest workers to tend the crops. However, these guest workers often faced abuse and unpaid wages. The program ended in 1964. One of its legacies was the subsequent waves of documented and undocumented workers crossing the southern border to work in the US food system (Mandeel 2014).

Beginning in the 1940s with the Bracero Program, migrant Mexicans and Central Americans began to replace many African American food system workers in the Chesapeake Bay region (Sills et al. 1994). Although different in hue, skin color still figured into the equation of wages and conditions faced by these new workers of color. Spanish language radio stations now fill the air of seafood processing plants, replacing the work songs of African American women that

Table 2 The percentage of people in the Chesapeake Bay watershed jurisdictions and the US that receive SNAP (Supplemental Nutrition Assistance Program, formerly called Food Stamps), the number of WIC (The Special Supplemental Nutrition Program for Women, Infants and Children) recipients, the percentage of people and children living below the poverty level, and the percentage of people experiencing food insecurity

Jurisdiction	% on SNAP	% SNAP Recipients working	Number on WIC	% Food insecure	% People below poverty line	% Children below poverty line
Delaware	15.0	49.0	16,244	11.1	13.6	18.5
District of Columbia	18.0	21.0	11,154	11.2	16.6	25.6
Maryland	11.0	39.0	123,892	10.4	9.3	12.0
New York	15.0	43.0	372,373	10.9	14.1	19.7
Pennsylvania	14.0	44.0	202,960	12.1	12.5	17.0
Virginia	9.0	42.0	107,736	10.1	10.6	14.0
West Virginia	19.0	36.0	32,167	14.9	19.1	25.9
Average for Bay Jurisdictions	14.4	39.1	–	11.5	13.7	19.0
US overall	13.0	44.0	6,378,885	11.8	12.3	21.0

Data from Nchako and Cai (2019) and Census (2019)

Table 3 Demography of hired US farm labor force for 2016

Demographic	Farm laborers	Farm managers
Percent Female	23.7	16.6
Average Age (years)	38.3	43.7
Percent White, not Hispanic	31.4	64.8
Percent Black, not Hispanic	2.5	1.9
Percent Hispanic	57.8	27.4
Percent US Citizens	49.2	81.3

Note that most hired farm laborers are older, male, and people of color, with half lacking US citizenship. However, hired farm managers are older still, even more male, and mostly white (USDA 2019b)

used to echo from the same walls. Despite the change in language and skin tone, it's still people of color tediously picking crabmeat and shucking oysters.

For the US, Hispanics comprise about 58% of the hired farm labor force, and only 27% of hired farm mangers. Most hired farm workers are middle aged and male, with only half having US Citizenship (Table 3). "Farmworkers are drawn from shifting groups of people whose vulnerability falls along lines of race, ethnicity, and citizenship status (Rodman et al. 2016)."

In 1932, Franklin D. Roosevelt won the Presidency with a promise of a New Deal that would protect working men and women and provide them a decent living. Following much pressure from Roosevelt, Congress passed the Fair Labor Standards Act (FLSA) of 1938, affording new protections to industrial workers. The law covered about 25% of laborers at that time (Grossman 2019). The FLSA provisions set standards for child labor, instituted a federal minimum wage and regulated required overtime work. Members of Congress from southern states made sure the FLSA excluded agricultural and domestic workers. This left farmworker protection to individual states. The Chesapeake Bay states of Virginia and Maryland, and the state of West Virginia still exempt farmworkers from minimum wage standards (Rodman et al. 2016). Note that in the south this meant denying legal protections to mostly African American and Latinx workers.

To address the plight of migrant farm workers, Congress eventually passed the Migrant and Seasonal Agricultural Worker Protection Act (MSAWPA) of 1986 (amended 1995). This act protected agricultural workers from underpayment of wages, requirements of buying food or other items from the employer, and abusive housing practices (USDL 2019a). Agricultural workers generally aren't provided time and a half wage for overtime work (Rodman et al. 2016). And they are often exempt from minimum wage requirements if they are paid on a piece work basis (i.e. paid by number or weight of food items harvested). While violators of the FLSA and MSAWPA potentially face fines and imprisonment, the asymmetrical power status of workers often makes it difficult for them to use these laws to protect their rights. Many of the migrant workers in the Chesapeake Bay Region come from Mexico and Central America, and some lack documentation of legal status. Pressing an issue with an employer might mean not only a loss of a job, but also deportation.

These days documented foreign food system workers arrive under the H-2A Temporary Agricultural Workers program. Agents of industrial farms recruit these workers mostly from Mexico and Central America. They may stay up to 3 years, after which they are required to leave the US for 3 months before applying for another round of work. Employers must pay for transportation between the home country and the US, as well as provide both housing and meals. H-2A workers may bring spouses and non-adult children with them. These workers must continue to work for the employer that recruited them or face charges of abscondment, and possible deportation (USCIS 2019). In that last regard, the system harkens back to the days of indentured servitude. However, unlike the indentured workers of colonial times, these guest workers must return home after their labor is complete and aren't welcomed into society as citizens.

Table 4 Calculated annual wages in dollars for different jobs in the food system

Jurisdiction	Food perp. and serving	Food processer	Farm manager	Meat, poultry, fish trimmer	Farming, fishing, forestry	Wait staff	Farm worker	Food cashier
Delaware	25,230	28,960	28,390	26,430	35,910	23,690	28,390	21,250
District of Columbia	37,480	N/A	N/A	N/A	N/A	N/A	N/A	26,960
Maryland	26,090	25,190	89,660	30,780	35,980	24,160	31,900	22,650
New York	31,000	30,590	N/A	29,090	37,540	34,460	32,720	23,850
Pennsylvania	24,400	27,490	90,650	29,520	33,200	24,850	29,270	20,310
Virginia	24,280	27,560	84,630	27,310	36,230	24,070	N/A	20,690
West Virginia	22,840	N/A	25,430	N/A	32,110	23,140	25,430	20,470
Ches. Bay Average	27,331	27,958	88,313	29,542	28,626	29,590	35,162	22,311
US Average	25,580	27,590	79,940	27,790	30,140	23,540	26,450	22,130

Note that the US Bureau of Labor Statistics determined annual salary by multiplying hourly wage times a full work year of 2080 hours. So these values may substantially overestimate annual income for seasonal laborers, such as farm workers and fishers (USBLS 2019). N/A = no data reported

There is no cap on the number of H-2A workers allowed in the US each year. Since they are paid anywhere between 2% and 33% (depending on the type of position) less than the average for agricultural workers, their presence helps to depress wages for other food system workers (Huennekens 2018). About 24,400 H-2A workers found employment in the jurisdictions of the Chesapeake Bay Watershed in 2016. That's about 10% of the total documented guest workers in the US (USDL 2019b). However, the count of documented H-2A workers reflects only about half of the real number of migrant guest workers in the food system. The other half lack documentation (USDA 2019b).

The mean annual wages compiled by the Bureau of Labor Statistics for food system jobs is based upon the assumption of full-time employment (Table 4). However, many food system jobs are either seasonal or part-time. Farmworkers and fishers often migrate to follow seasonal patterns of planting, harvesting, and catching. Some stay in place during the off-season and switch to different forms of local employment. Food service workers in tourist venues also face seasonal changes in employment. Many workers in the food retail industry can't find full time employment. Large grocery stores purposefully hire part-time workers to avoid paying the additional benefits due to fulltime workers. Thirty hours per week is the minimum for designation as fulltime. Walmart is the largest grocery store in the US (Dudlicek 2018). About 50% of its workers in 2018 were part-time, up from only 20% in 2005 (Bose 2018). Other large groceries in the Chesapeake region use the same model of reduced hours for workers to avoid paying benefits. In practice, many part-time food workers have to work two or three separate parttime jobs to make ends meet. Despite some working 50 h or more combined, these low-wage employees aren't eligible for basic benefits. About half of the 4.3 million US laborers working multiple jobs held two part-time positions in 2018 (BLS 2019b).

The low wages earned by most food system workers makes many of them eligible for assistance from the federal government. The threshold annual income for a family of four to qualify for SNAP (food stamps) is $31,980 (USDA-FNS 2019a). Most food system jobs in the Chesapeake Bay region pay less than that (Table 4). For example, consider a grocery cashier that supports a family of four with an income of $24,000 a year, and a $1000 per month rent. She will receive $380 per month in SNAP benefits (SNAP 2019). This means that food system industries that pay such low wages receive a government subsidy to support their business model. For our example family, that would be $4560 of subsidy per year.

Wastewater treatment plant workers deal with the residue of digested and non-absorbed food. Although at the end of the food production and consumption line, these employees earn some of the highest wages and benefits in the food system. For example, The Hampton Roads Sanitation District (HRSD) that serves the lower Chesapeake Bay starts its workers at $35,000 per year. After 9 years of service and the completion of classes, this grows to $59,000 per year. These wastewater workers also earn excellent benefits, including health insurance and pension (personal communication with HRSD human resources office). What accounts for the discrepancy in pay between wastewater workers and those in the rest of the food system? Local governments operate wastewater treatment plants to protect human health and the environment, not to turn a profit for shareholders. They also operate as a monopoly, so they don't face pressure from competitors to clean sewage at a cheaper price. That protects workers from low wages in the name of corporate survival.

Piecework Any hungry diner who ever sat down to a platter of whole steamed crabs knows how hard it is, and how long it takes to extract the meat from this crustacean. The traditional beer and corn on the cob eases the slow reward of crabmeat that comes from cracking the shell and picking out the muscle tissue. For professional crab pickers its literally no picnic. They spend long tedious days cracking and picking a away to produce expensive crab meat.

A recruitment page posted by one seafood processor in Maryland conveys the sense of what it means to labor in the crab processing industry (Lindy 2019): "*THIS JOB ORDER IS IN CONNECTION WITH A CONCURRENT H-2B LABOR CERTIFICATION APPLICATION.*

The job opportunity is a bona fide, full time temporary position for 105 seafood processors/crab meat pickers from 4/1/2019 to 12/31/19 with Lindy's Seafood, 1548 Taylor's Island Road, Woolford, MD (Dorchester County) c/o Aubrey Vincent; (P) 410-228-5032.

Duties of seafood processor/crab picker: steam, clean and pick whole crabs; weigh, sort, ice, debone, pack, cook, seafood preparation for wholesale/retail sale. General plant labor such as clean and prepare work areas, load/unload, dump, clean, move, dock work, grade, freeze, prepare, seal, lift/carry up to 50 lbs; and any other activities related to seafood processing/crab picking. Must have good work ethic, follow directions and perform manual labor. Workers will assist with unloading raw product (crabs) from boats or trucks and place in processing area or onto tables to be processed. Workers will also assist with loading finished product refrigerators or freezers for storage. Workers will assist with loading finished product onto trucks for shipment.

35hpw; Mon–Fri; 2 shifts 2 am–12 pm and 10 am–8 pm; work extended daily hours and weekends pending product availability.

No experience/no education. Work is supervised. Employer provides 10 days on the job training. No infectious diseases; seafood allergies may be fatal.

$9.60/h; overtime is not guaranteed but if worked rate is paid at time and a half per hour ($14.40) above 40 h per week. Discretionary raise/bonus.

Piece rate is paid at $4.50 per pound shell free meat or guaranteed prevailing wage whichever is higher. Employees must pick 3 pounds picked crab meat per hour after a 10 day training period.

A single workweek will be used in computing the wage due. Frequency of pay is weekly on Friday.

Shared housing available to only seasonal full-time employees. Housing is not offered to non-employees. Employees are not required to live in employer arranged housing, however, if they opt to do so they will be charged $45/week to include utilities. Employees pay $5/month for internet services if requested. Employees may make their own arrangements at their own expense.

The employer will make the following deductions from the worker's wages: all deductions required by law, rent (where applicable), cash advances and repayment of loans, repayment of overpayment of wages to the worker, payment for articles which the worker has voluntarily purchased from the employer, long distance telephone charges, recovery of any loss to the employer due to the worker's damage (beyond normal wear and tear) or loss of equipment or housing items where it is shown that the worker is responsible, and any other reasonable deductions expressly authorized by the worker in writing. No deduction not required by law will be made that brings the workers hourly earnings below the FLSA Federal statutory minimum wage."

From the notice above, this company pays its employees $9.60 per hour, but with the stipulation of producing a minimum of 3 pounds (1.4 kg) of crab meat per hour. Since blue crabs weigh about 10 oz (0.3 kg), and are about 14% meat by mass, that means having to process 21.4 pounds (9.7 kg) of crabs per hour. That works out to processing 32.3 crabs per hour. To earn above the set wage, this means working faster (Seafood 2019).

Harvesting fruit is also paid on the piecework basis. In 2016 The Chesterfield Berry Farm, located near the James River in Virginia, payed strawberry pickers $10.72 hourly or $2.75 per eight quarts (7.6 l) (BerryPicking 2016).

The Food Dollar The USDA (2019c) calculates how the dollar spent by US consumers is apportioned to the various food system industries (Table 5). The food service industry earns 36.7%, food processors 15.0%, retail trade 12.6%, wholesale trade 9.1%, farmers and fishers 7.8%. The remaining 18.8% is divided between agribusiness, packaging, transportation, energy, finance and insurance, advertising, and legal and accounting. Just over half of all costs go to pay wages and benefits. That the food service industry garners the single largest share of the food dollar is consistent with consumers spending 44% of their money for meals by eating out of their homes (BLS 2019a). Considering that Americans obtain most of their calories from highly processed (61%) and moderately processed (16%) food (Poti et al. 2015), it's not surprising that that sector is second in its share of the food dollar. Perhaps what's most striking about the dispensation of the food dollar is that less than 10% of the money finds its way back to the people that grow or capture the food.

Food Security, Farmer's Markets and Community Supported Agriculture The USDA defines food security with four different categories. "High food security: no reported indications of food-access problems or limitations. Marginal food security: one or two reported indications—typically of anxiety over food sufficiency or shortage of food in the house. Little or no indication of changes in diets or food intake. Low food security: reports of reduced quality,

Table 5 How the dollar spent by the US consumer on food is apportioned to the food system components for 2017 (USDA 2019c)

Industry group	Total	Imports	Output taxes	Property income	Salary and benefits
All industries	100.0¢	5.0¢	9.2¢	35.3¢	50.5¢
Agribusiness	2.1¢	0.5¢	0.1¢	1.0¢	0.6¢
Farm production	7.8¢	0.8¢	0.2¢	5.3¢	1.4¢
Food processing	15.0¢	0.8¢	0.8¢	5.8¢	7.7¢
Packaging	2.3¢	0.6¢	0.1¢	0.6¢	1.0¢
Transportation	3.5¢	0.2¢	0.1¢	1.2¢	1.9¢
Wholesale trade	9.1¢	0.0¢	1.6¢	3.1¢	4.5¢
Retail trade	12.6¢	0.3¢	2.2¢	3.5¢	6.6¢
Foodservice	36.7¢	0.6¢	3.5¢	10.2¢	22.4¢
Energy	3.8¢	1.0¢	0.4¢	1.6¢	0.8¢
Finance and Insurance	3.2¢	0.1¢	0.1¢	1.3¢	1.7¢
Advertising	2.6¢	0.1¢	0.1¢	1.2¢	1.1¢
Legal and Accounting	1.3¢	0.0¢	0.1¢	0.5¢	0.7¢

Agribusiness: all establishments producing farm inputs such as seed, fertilizers, farm machinery, and farm services, and all subcontracting establishments. Farm Production: all establishments classified within the agriculture, forestry, fishing, and hunting industry. Food Processing: all establishments classified within the food and beverage manufacturing industries, and all subcontracting establishments. Packaging: all establishments classified within the packaging, container, and print manufacturing industries, and all subcontracting establishments. Transportation: all establishments classified within the freight services industries, and all subcontracting establishments. Wholesale Trade: all non-retail establishments that resell products to other establishments for the purpose of contributing to the U.S. food supply, and all subcontracting establishments. Retail Trade: all food retailing and related establishments, and all subcontracting establishments. Foodservice: all eating, drinking, and related establishments, and all subcontracting establishments. Energy: oil and coal mining, gas and electric utilities, refineries, and related establishments, and all subcontracting establishments. Finance and Insurance: all financial services and insurance carrier establishments, and all subcontracting establishments. Advertising: all advertising services and related establishments, and all subcontracting establishments. Legal and Accounting: establishments providing legal, accounting, and bookkeeping services, and all subcontracting establishments

variety, or desirability of diet. Little or no indication of reduced food intake. Very low food security: reports of multiple indications of disrupted eating patterns and reduced food intake (USDA-ERS 2019a)."

Residents with low incomes often cluster in areas of low availability of nutritious food, called food deserts. Despite the highly productive food system of Chesapeake Bay, many residents live distant from stores with nutritious offerings. This includes low income areas of cities such as Richmond, Newport News, and Baltimore, as well as large swaths of the rural landscape (Fig. 2).

Officially, 11.5% of households in the Chesapeake Bay Region and a similar number for the US in general, experience food insecurity (Table 2). However, the number of families that struggle to put healthy food on the table is certainly larger than that, since nearly 30% earn less than $35,000 per year (Table 1). These low earning families devote 25%–75% of their income to food. The SNAP and WIC programs that help ameliorate food insecurity are one answer to the problem. Another lies in making food cheaper to buy. This means reducing the over 90% of the food dollar that goes to the non-producing portion of the food system. That's best achieved by consumers purchasing food directly from farmers and fishers. Such direct transactions provide the added benefit of eliminating processed food from the equation. Highly processed food is directly associated with poor health outcomes. This is a hidden cost to its consumption (Poti et al. 2015).

From the earliest days of agriculture, farmers used excess production (more than needed by the family or tribe) in trade. The trappings of modernity separated consumers from producers. Food processing, shipping to distant markets, wholesalers, and retailers made for wider distribution of food, but added costs not accrued by the farmers.

The farmer's market movement seeks to reconnect consumers with producers. According to the USDA, the first formal farmer's market in Colonial America opened in the Chesapeake Bay Watershed in the town of Lancaster, PA, in 1730 (Neal 2013). The Lancaster Central Market continues to thrive today, being housed in an 1889, purpose-built structure. The USDA lists 8771 farmer's markets across the US (USDA 2019d). For Bay region jurisdictions, the numbers are; Delaware 35, District of Columbia 56, Maryland 163, New York 675, Pennsylvania 307, Virginia 226, West Virginia 93. These counts don't reflect the numerous informal seasonal roadside fruit and vegetable stands erected along country roads.

Community Supported Agriculture (CSA) refers to a system where clients subscribed to weekly deliveries of produce from a local farm or consortium of growers. The modern version was imported from Europe by Robyn Van En in 1985 (DeMuth 1993). But it also has US roots that trace to American pastoralism (Press and Arnould 2011). Some CSAs deliver a mix of whatever is in season, while others allow their customers to place orders for certain harvested items. The USDA maintains an extensive data base that

Fig. 2 USDA map showing food deserts in green (low income and low availability of nutritious food) in the Chesapeake Bay Region as of 2017. From USDA-ERS (2019d). Limited availability of nutritious food is linked to urban and rural areas of impoverishment

details what listed CSAs offer and many other aspects of their operations (USDA 2019e). As of 2019 USDA listed 845 CSAs in the US. That number may drastically underestimate the actual count. LocalHarvest, an organization that promotes CSAs, shows over 4000 operations across the US in their data base (LocalHarvest 2019). According to the very limited USDA list, the Bay state's offerings of CSAs are; Delaware 3, District of Columbia 2, Maryland 33, New York 67, Pennsylvania 44, Virginia 31, West Virginia 8. Participation in a CSA typically costs a family $300–$500 per season. Most CSAs promote their produce based on its healthfulness for consumers and the environment. Most advertise the absence of genetically modified plants, and organic certification.

While most consumers prefer locally produced foods, obtaining them in the Chesapeake Bay Watershed is often challenging. Despite USDA support for CSAs and local farmer's markets, the general US food policy favors large corporate farms and processors with points of production often remote from the Chesapeake (Tassone et al. 2017).

Health Costs of Food What we pay for in the price of food doesn't include the negative or positive consequences for our health involved in making dietary choices. We eat food to abate hunger and for the enjoyment of the meal, usually with only a remote notion of its nutritional value or potential to cause chronic disease. This is a natural consequence of millions of years of evolution. Humans, as the other great apes, spent most of their evolutionary history foraging for fresh whole-plant foods of low caloric value compared to today's Standard American Diet. The sweeter the fruit, the starchier the root, or the oilier the seed, all the better as that meant more calories to carry on life's functions. To get through lean times of food insufficiency, humans evolved the ability to store fat during times of plenty. It was very much about the calories.

That humans possessed the potential for a highly omnivorous diet enabled their diaspora out of tropical and subtropical Africa and into a wide variety of climates on all of the continents save for Antarctica. This included the extremes of high latitudes that offered little in the way of plant food, but a wealth of energy dense animal flesh. The ability to digest animal flesh also enabled life in seasonal environments. Animal food supplemented scarce supplies of edible plants during winter. The Powhatan neighbors of the English colonists in Virginia made good use of such omnivory, building diets on gathered plant foods, cultivating crops, fishing, and hunting. Despite the diverse set of food resources, the Powhatan still suffered periods of food insufficiency during periods of drought, and years with very cold winters. Survival was very much about the calories. Fish, fowl, oysters, crabs, and deer all provided energy dense food needed to survive at times of deprivation.

While the fat and protein in animal foods provides a dense source of calories, that energy comes with potential health costs that come due later in life. Life expectancy at the time of the founding of Jamestown was about 30 years (Roser 2019). Many people in 1600 did live longer than that, but most succumbed to infectious disease and other sources of mortality earlier in life. Improvements in sanitation, healthcare and other factors reduced rates of mortality. In 2015 life expectancy for men and women in the US was 76.3 and 81.1 years respectively (Arias and Xu 2015). The three or four extra decades granted by modern life expectancy is when the health consequences of the Standard American Diet (SAD aka Western Diet) are most severe. As people age, they get sicker from the diseases of affluence; obesity, coronary heart disease, non-alcoholic fatty liver disease, hypertension, kidney disease, cancer, and others. The leading cause of death in the US and around the world is coronary heart disease (Murphy et al. 2018). This malady is unknown in populations that eat whole plant-based diets (Thomas et al. 1960). The nine next most common causes of morbidity are: cancer, unintentional injuries, chronic lower respiratory diseases, stroke, Alzheimer disease, diabetes, influenza and pneumonia, kidney disease, and suicide. Cancer, stroke, Alzheimer disease, diabetes and kidney disease are also tightly linked to eating animal-based diets (Greger 2015). The global export of the Standard American Diet now means that what we eat and don't eat has replaced smoking as the leading cause of death around the world (Afshin et al. 2019). The SAD includes unhealthy foods such as animal products, hydrogenated plant oils, refined carbohydrates, and excess sodium. Those unhealthy foods displace important healthy choices: dietary fiber and anti-oxidant rich vegetables, fruits, whole grains, legumes, and nuts.

The burden of disease and increased mortality associated with the SAD raises many ethical questions. Should those that profit from producing and selling SAD foods pay a fee to help cover the damage to the public health? Should such unhealthy SAD foods be taxed to raise their price and reduce consumer demand? Should animal and processed foods that contribute to disease be required to display a warning label like that mandated on tobacco products? Should society eliminate subsidies for disease causing foods and shift those funds to reduce the cost of health promoting foods?

The USDA and Farm Subsidies President Lincoln signed legislation to create the US Department of Agriculture in 1862. Isaac Newton, the first leader of this new department of government outlined its mission: "(1) Collecting, arranging, and publishing statistical and other useful agricultural information; (2) Introducing valuable plants and animals; (3) Answering inquiries of farmers regarding agriculture; (4) Testing agricultural implements; (5) Conducting chemical analyses of soils, grains, fruits, plants, vegetables, and manures; (6) Establishing a professorship of botany and entomology; and (7) Establishing an agricultural library and museum (Rasmussen 2019)." Lincoln wanted the USDA to modernize and transform agriculture and it did. During its first four decades the USDA focused on enhancing agricultural productivity by introducing new seeds and promulgating the best farming techniques. The current mission statement reads, "USDA provides leadership on food, agriculture, natural resources, rural development, nutrition, and related issues based on sound public policy, scientific evidence, and efficient management (USDA 2019f)." This captures the original 1862 goals and incorporates the added issues of rural development, food safety and nutrition.

The USDA teamed with farmers to make US agriculture spectacularly successful in producing food. No one has gone hungry in the US over the last 150 years for the lack of food production. Those that do lack food do so for missing the funds to obtain it or living in a food desert. *In fact, since 1900, the most pressing problem for US farmers has been the over production of food and the associated suppression of prices.* Consider corn, the food crop upon which the Nation was founded, and soybean, a protein rich food. *Corn and soy top US crop production, yet little of either is used to directly feed people.* Most is used to feed animals, produce biofuels, or exported.

Food Safety Upton Sinclair's *The Jungle*, published in 1906, revealed the appalling conditions in slaughterhouses that resulted in animal products which sickened consumers. In response, Congress passed two bills to regulate that industry: The Food and Drugs Act and the Meat Inspection Act. Many other bills followed to protect consumers from pesticides, chemical additives, mislabeling, and other issues. It falls to the USDA and the Food and Drug Administration (FDA) to enforce the food safety rules (Lusk 2016; NDSU 2019).

Price Supports The USDA took on another responsibility at the beginning of the Great Depression when President Herbert Hoover established the Food Board. Its purpose was to stabilize the dropping prices of wheat and cotton by having the government buy excess production and store it for later resale. This first attempt to subsidize wheat went awry. "Farmers who grew other crops were now incentivized to grow wheat and cotton because of the price guarantees. The resulting overproduction lowered prices below the floor and resulted in the government's buying the excess crops and quickly exhausting the program's $500 million budget (Lusk 2016)."

When Franklin Delano Roosevelt stepped into the White House, he faced the crisis of falling farm prices left unresolved by Hoover. Roosevelt took a different tack with the Agricultural Adjustment Acts of 1933 and 1936. These acts paid farmers to either not plant crops that were already too abundant to fetch a fair price, or to cultivate alternative crops such as those used as cover for soil restoration. The Agricultural Adjustment Acts established the precedent for the quadrennial Farm Bill that to this day sets food system policy for the US.

Falling crop prices in the early 1950s motivated the creation of the USDA Soil Bank Program that pays farmers not to cultivate portions of their acreage (Helms 1985). It has two components, the Acreage Reserve Program (ARP) and the Conservation Reserve Program (CRP). Both programs intended to increase crop prices by removing land from production, thus removing surplus from the market. The ARP meant to conserve useful land for future production, while the CRP functioned to permanently remove land from future cultivation. The CRP lands are managed to reduce soil erosion, improve water quality, and establish habitat for wildlife and pollinators (USDA 2019g).

Tariffs Support the Sugar Industry In 1789, the US government placed a tariff on imported sugar to provide revenue. The tariffs grew over time and by the passage of the 1934 Sugar Act became a means to subsidize the US sugar industry to protect it from cheaper imports. Along with tariffs, domestic industry protections include import quotas, price support loans, and processor marketing allotments (Wiltgen 2007). Since added sugars damage public health, one could question the ethics of supporting the domestic industry. Yet the effect of the program is to raise sugar prices above the world market level, which in theory should dampen domestic demand for the product.

Crop Insurance Helps Mostly the Wealthiest Farmers In 1938, the federal government began experimenting with selling crop insurance to farmers. Participants would receive payments if their crops failed or if market prices dropped. However, participation remained low, farmers instead counted on federal payments from ad hoc disaster relief packages from Congress after droughts, floods, or other disasters. Beginning in 1980, a series of new acts restructured the crop insurance program. Private companies now provide the insurance, but with the reinsurance backing of the federal government. The government also subsidizes 38–67% of the cost of premiums. By 2008, 89% of planted land that included 128 different crops were insured for $124 billion (ProAg 2019).

Despite covering most of the planted land in the US, only 16% of farmers participate in the program. The discrepancy comes from most of the land being concentrated in the hands of a small portion of farmers. The program is dominated by farmers with gross annual receipts ranging from $150,000 to >$1000,000. Due to the $7 billion of subsidy for premiums, participants garner $2 in indemnities for every $1 they spend on premiums (Faber 2018).

Total Subsidies In 2016, subsidies accounted for $13.9 billion, or 25% of total US farm income (UCS 2016). From 1995 to 2017 the USDA paid out about $370 billion to farmers in various forms of subsidy. The six jurisdictions in the Chesapeake Bay Watershed accounted for only 2.4% of those subsidies (Table 6). This is consistent with the bulk of the most subsidized crops (corn, wheat and soy) being grown in the Midwest (Table 7). Although most subsidies go out of the Bay's watershed, they support crops that are an important part of the food shed. Corn and soy flow into the region as feed to support the massive poultry industry on the Delmarva Peninsula. The subsidies also shape the general US food system, helping to determine what Bay residents eat.

The bulk of the harvest of the three most subsidized and produced crops in the US doesn't go to feed the domestic human population. The most produced US crop is corn. Only 10% of it goes to directly feed humans, with 46% taking an indirect route as animal feed. The rest goes to making biofuel (30%) and exports (13.5%) (NCGA 2018). For soy, the second most produced crop, 46% is exported. Only 15% goes to directly feed humans and most of that in the form of processed food and oils. About 70% of soy becomes animal feed (USDA 2018; USDA-ERS 2019b). Wheat is the third

Table 6 Subsidies payed to farmers 1995–2017 in millions of dollars (EWG 2019)

Jurisdiction	Commodity programs	Crop insurance	Disaster programs	Conservation programs	Total
Delaware	210	98	25	54	386
Maryland	664	299	70	291	1321
New York	1490	367	211	206	2270
Pennsylvania	1270	553	153	491	2470
Virginia	1290	551	265	240	2350
West Virginia	82	26	247	79	211
Ches. Bay States Total	5006	1894	971	1361	9008
US Total	205,400	84,500	31,700	48,100	369,700

Table 7 The number of farmers and total subsidies for the top 13 food crops in the US 1995–2017 (EWG 2019)

Crop	No. recipients	Subsidy in millions USD
Corn	1,886,944	111,174
Wheat	1,574,743	47,696
Soybeans	1,212,252	37,446
Rice	84,098	16,156
Livestock	934,707	10,781
Sorghum	704,262	8371
Dairy	163,032	5582
Peanut	103,781	5436
Barley	388,968	3081
Sunflower	76,790	1270
Oat	715,518	367
Apple	8556	262

most harvested crop. About 93% goes to direct human consumption with 7% for animal feed. About 45% of US wheat is exported (USDA 2019h). This raises the ethical question of having US taxpayers subsidize feed for the domestic and international meat industries and feedstock for the biofuel business.

The Ethics of Reshaping Subsidies to Promote Public Health and Lessen Environmental Impacts At least part of the obesity and type-2 diabetes pandemics sweeping the US traces to subsidies for corn, soy, wheat and sugar (Fields 2004; Jackson et al. 2009). Cheap corn and soy means cheap meat and dairy; cheap wheat and sugar means cheap chips, pastries, and sweetened beverages. Such supports make it easier and cheaper for the population to exist on high caloric and low nutrition density foods that dominate the SAD. It also makes those health-robbing foods appear inexpensive in comparison to the low caloric and high nutrition density of fresh fruits and vegetables. Simply removing subsidies from corn, soy, wheat and sugar may not be enough to turn around dietary preferences shaped by decades of overproduction and consumption of those foods. Promoting a new diet based on whole plants will likely require subsidizing fresh fruits and vegetables as well as whole grains and legumes. As it stands now, only 4% of planted lands in the US is devoted to fruits and vegetables and that is a direct consequence of subsidies for the commodity crops (mostly for animal feed and biofuel) grown on the rest of the land (Jackson et al. 2009).

Jackson et al. (2009) put it well when they wrote of the Farm Bill that funds the USDA, "...it has evolved into a system that creates substantial health impacts, both directly and indirectly. By generating more profit for food producers and less for family farmers; by effectively subsidizing the production of lower cost fats, sugars, and oils that intensify the health destroying obesity epidemic; by amplifying environmentally destructive agricultural practices that impact air, water, and other resources, the Farm Bill influences the health of Americans more than is immediately apparent." They present various recommendations to reshape the food system. These include:

- Develop a life cycle approach with "all cost accounting" to agricultural decisions. Add environmental and health perspectives into all major reports and programming.
- Model "food literate" behavior by purchasing sustainably grown foods.
- Communicate the expense of our current food system in terms of taxes, ill health, and environmental costs.
- Support efforts to maintain strong organic certification qualifications.
- Promote the consumption of locally grown, humane, and sustainably produced, healthy foods at schools and health care institutions.
- Make explicit and public all input in agricultural decisions, including participation from lobbyists.
- Revise the current Farm Bill subsidy structure to equally support naturally grown specialty crops and cap farm payments to large operations to support family farms.
- Increase federal research on structural interventions that reduce the health impact of current food production systems and increase access to healthy calories. Good models are "edible schoolyards," community-supported agriculture, and farmers' markets.
- Support research into the health effects of agricultural production, including antibiotic resistance, pesticides,

Fig. 3 The USDA budget expenditures for 2020. Note that nutrition assistance accounts for the majority of spending. The All Other category includes Rural Development, Research, Food Safety, Marketing and Regulatory, and Departmental Management (USDA 2019f)

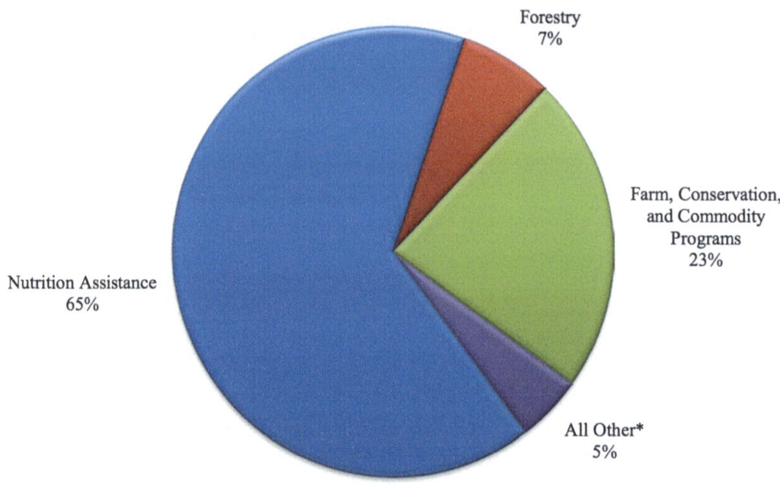

toxins, hormones, corn sweeteners, grain-fed livestock, and the true costs of "traveled" foods.
- Evaluate the costs and benefits of changes to the food system on health care costs related to chronic disease.

Despite the enormous number of dollars going to farm subsidies, those USDA programs account for only 23% of the organization's budget. Most of the money USDA spends goes for providing nutrition assistance to the poorest portion of the population. The 2020 USDA annual budget is about $150 billion, 65% of which goes to food assistance programs such as SNAP and WIC (Fig. 3). SNAP accounts for 10% of the money spent on food in the US, consistent with that portion of the population that it serves (Wolkomir 2018).

The Food and Nutrition Service (FNS) of the USDA administers various nutrition assistance programs (USDA-FNS 2019b, Table 8). This includes SNAP and WIC, addressed above, as well as others that target specific populations such as low-income school children and low-income seniors. FNS also sets dietary guidelines to promote public health. This is a controversial endeavor. While there is near uniform agreement on promoting consumption of fruits, vegetables, legumes, and whole grains, USDA guidelines go beyond that to include recommendations for consumption of dairy and other animal products. Since consumption of animal products are unnecessary and often damage human health, their inclusion is questioned by such groups as the Physicians Committee for Responsible Medicine (Keevican 2017). This points up the ethical dilemma built into the mission of the USDA. The Department is supposed to promote both human health and farm commerce. Yet some farm products it promotes prove detrimental to human health. This includes tobacco, animal products and sugar.

The USDA promotes agricultural products through the Agricultural Marketing Service (USDA-AMS 2019). This includes a variety of activities, such as the Organic Certification and Accreditation Program and setting standards for agricultural commerce. The AMS also facilitates industrial trade groups in promoting the domestic and export sales of their products. This is done through a checkoff system of mandatory assessments paid by the farmers. Examples with annual budgets include the American Egg Board, $20 million; The Cattlemen's Beef Board, $72 million; National Dairy Promotion and Research Program, $112.7 million; The Fluid Milk Processor Promotion Board $92 million; National Pork Board, $76 million; Hass Avocado Board, $56.3 million; Highbush Blueberry Council, $10 million; Mushroom Council, $5.1 million; National Potato Promotion Board, $20.7 million; National Peanut Board, $10.4 million; Popcorn Board, $0.7 million; National Watermelon Promotion Board, $3.2 million; and United Soybean Board, $101.6 million. This list indicates more producer funds for animal products than for plants. Soybean promotion seems to be the exception, but since most of this plant food is destined for animal feed, it is in the end another way of promoting consumption of animals.

Corn and soybeans are the two dominate crops grown in the US. Both overwhelmingly enter the food system as animal feeds. Over 90% of each crop is genetically modified for use with the glyphosate herbicide. This means that the animal product rich SAD is supported by the industrial application of a toxin over much of the cultivated land. This general herbicide is helping to destroy biodiversity while proving less useful each year as the target weeds develop resistance (Schütte et al. 2017).

The system of intensive agriculture separates the raising of livestock from the land, with pigs, cattle, and chickens housed in Concentrated Animal Feeding Operations (CAFOs) often remote from where their feed crops are grown. Its cheaper to use artificial fertilizer to sustain crop growth, rather than returning the animal's manure and load of

Table 8 Programs administered by the USDA Food and Nutrition Services branch

Program	Stated purpose
Supplemental Nutrition Assistance Program (SNAP)	SNAP provides nutrition benefits to supplement the food budget of needy families so they can purchase healthy food and move towards self-sufficiency
Special Supplemental Nutrition Program for Women, Infants, and Children (WIC) (WIC coves 50% of all children born each year)	Provides federal grants to states for supplemental foods, health care referrals, and nutrition education for low-income pregnant, breastfeeding, and non-breastfeeding postpartum women, and to infants and children up to age five who are found to be at nutritional risk
Farmers Market Nutrition Program	Eligible WIC participants are issued FMNP coupons in addition to their regular WIC benefits. These coupons can be used to buy eligible foods from farmers, farmers' markets or roadside stands that have been approved by the state agency to accept FMNP coupons
Senior Farmers' Market Nutrition Program (Over 800,000 seniors receive SFMNP benefits each year)	Provide low-income seniors with access to locally grown fruits, vegetables, honey and herbs
National School Lunch Program (In FY 2018, schools served over 4.8 billion)	Program operating in public and nonprofit private schools and residential child care institutions. It provides nutritionally balanced, low-cost or free lunches to children each school day
Summer Food Service Program (In 2018, the SFSP provided more than 1.5 million nutritious meals and snacks to children during the summer when school was not in session)	SFSP reimburses program operators who serve free healthy meals and snacks to children and teens in low-income areas
School Breakfast Program (In FY 2018, schools served over 2.4 billion breakfasts)	Provides reimbursement to states to operate nonprofit breakfast programs in schools and residential childcare institutions
Special Milk Program (In FY 2017, the Special Milk Program supplied over 41 million half pints of milk to children)	Provides milk to children in schools and childcare institutions who do not participate in other federal meal service programs
Food Distribution Program on Indian Reservations	The Food Distribution Program on Indian Reservations (FDPIR) provides USDA Foods to income-eligible households living on Indian reservations and to Native American households residing in designated areas near reservations or in Oklahoma
Commodity Supplemental Food Program	Works to improve the health of low-income persons at least 60 years of age by supplementing their diets with nutritious USDA Foods
The Emergency Food Assistance Program (TEFAP distributes about half a billion dollars in USDA Foods each year through food banks across the country)	Helps supplement the diets of low-income Americans by providing them with emergency food assistance at no cost
USDA Foods in Schools (USDA Foods comprise about 15–20% of the foods on a lunch tray)	Supports domestic nutrition programs and American agricultural producers through purchases of 100% American-grown and -produced foods for use by schools and institutions participating in the National School Lunch Program (NSLP), the Child and Adult Care Food Program (CACFP), and the Summer Food Service Program (SFSP)
Food Safety	Protects people served by FNS programs from foodborne illness by developing food safety education, training and technical assistance resources to support FNS program operators
Center for Nutrition Policy and Promotion (CNPP)	Works to improve the health and well-being of Americans by developing and promoting dietary guidance that links scientific research to the nutrition needs of consumers
Choose MyPlate	Offers ideas and tips to help create a healthier eating style that meets individual needs and improves health
Dietary Guidelines for Americans	The U.S. Departments of Agriculture (USDA) and Health and Human Services (HHS) work together to update and release the Dietary Guidelines for Americans (Dietary Guidelines) every 5 years. Each edition of the Dietary Guidelines reflects the current body of nutrition science and provides advice on what to eat and drink to promote health and reduce risk of chronic disease

This list leaves out several smaller administrative programs, but it features the most important ones in shaping the US diet (USDA-FNS 2019b)

nutrients to the land. This breaks the natural cycling of nutrients, replacing it with energy-intensive synthetic chemicals. Also, much of the nitrogen-rich artificial fertilizer finds its way into waterways, such as the Chesapeake Bay, causing algal blooms and subsequent seasonal low oxygen in the bottom waters. The unwanted animal manures stored on land, leach additional excess nitrogen and phosphorus into waterways (see chapters "The Journey from Peruvian Guano

Table 9 Some of the top ten food exports from states in the Chesapeake Bay Watershed for 2017, with values in millions of dollars

State	Broiler Meat	Corn	Soy	Other plant products	Fruits	Veg.	Tree Nuts	Dairy	Beef
Dela.	106	20	36	31		6			
Maryland	104	44	112	175				24	
New York	1				5482	3095	7712	931	283
Penn.	59	92	130	75				287	
Virginia	91		116	179				48	50
West Virginia	19		7	32	9				21

Products not in the top ten food exports for a state are not listed. From USDA-ERS (2019c)

to Artificial Fertilizer Ends with too Much Nitrogen in the Chesapeake Bay" and "Livestock and Poultry: The Other Colonists Who Changed the Food System of the Chesapeake Bay" for detailed discussion).

The current food system requires a massive subsidy of fossil fuels, about 17% of the total US consumption. It takes 25 kcal of fossil fuel to produce 1 kcal of animal protein. For plants it is 11 times more efficient, requiring only 2.2 kcal of fossil fuel per 1 kcal of protein (Pimentel and Pimentel 2003). The burning of fossil fuels and greenhouse gas emissions methane and nitrous oxides from animal-based agriculture are also speeding climate change (see chapters "The Journey from Peruvian Guano to Artificial Fertilizer Ends with too Much Nitrogen in the Chesapeake Bay" and "A New Food System for the Chesapeake Bay Region and a Changing Climate" for detailed discussion).

Shifting away from the animal-based SAD and industrial-chemical agriculture would alleviate much of the damage done to the environment by the food system. Including the cost of damage to the environment and human health into the price of animal products would raise their cost significantly and reduced consumer's willingness to purchase them. The price of industrially produced meat, dairy, and corn doesn't include the social cost of contaminated soil, water, air, and food, or the diseases of diet such as obesity and type 2 diabetes (UCS 2019). Redirecting the current subsidies away from animal and industrial agriculture to plant-based organic farming would help farmers grow crops that reduce environmental impact and promote human health (Mader 2002; Pimentel et al. 2005; Rembiałkowska 2007, see chapter "An Organic-Based Food System: A Voyage Back and Forward in Time" for discussion of organic agriculture in the Chesapeake Bay Watershed). For the Chesapeake Bay this means replacing the poultry CAFOs that dominate the food production system of the Delmarva Peninsula with diversified croplands that produce organic fruits, vegetables, grains and legumes (see chapter "Livestock and Poultry: The Other Colonists Who Changed the Food System of the Chesapeake Bay").

Extending Ethical Consideration to Animal Agriculture in a Time of Super Abundance The profit driven food system provides no direct accounting for cruelty experienced by non-human animals. Most consumers are so divorced from the way their food is produced that they have little notion nor concern about the lives of the animals that they eat. Eating animals was essential to the survival of fifteenth Century Native Americans and English Colonists of the Chesapeake Bay region. Hunting deer or raising pigs and chickens to eat made perfect sense at a time when people faced periods of limited food availability that would lead to malnutrition and starvation. Additionally, the hunted or husbanded animals lived essentially natural if not shortened lives. They could mate, nurture their offspring, and engage in long-evolved behaviors. The industrial food system ended that, reducing farmed animals to their parts for consumption by humans.

Since the nineteenth Century, food production in the Chesapeake Bay and most of the rest of the US far exceeded the needs of the population. And so much so that food exportation is an important part of the economies of the states in the Chesapeake Bay Watershed (Table 9). This raises the ethical question of the need to slaughter animals for food. Unlike their ancestors of four centuries earlier, residents of the Chesapeake Bay and the rest of the US certainly don't *need* to eat animal products to fulfil their nutritional requirements. Plant-based diets could easily feed everyone, and with much better outcomes for public health and the environment. So why do animal products still form the center of the SAD?

Society teaches that each meal is built around an animal product and relegates fruits, vegetables, grains and legumes to the status of side dishes. Marketing by the animal food industry reinforces this notion of what constitutes a meal. Dietary tradition of most families is animal-product centered. As such, children develop a palate that favors the meat-sweet-fat SAD. Even when such diets send people to their physicians for treatment of hypertension, coronary artery disease, non-alcoholic fatty liver disease, and type-2 diabetes, their doctors typically prescribe medications rather than changes in what to eat (Bruckert et al. 2012). Eleven of the top 15 most prescribed drugs in the US in 2017 targeted diseases caused by the SAD; hypertension, elevated cholesterol and lipids, type-2 diabetes, and acid reflux (Table 10, Fuentes et al. 2018).

Table 10 The top 15 most prescribed medications in the US in 2017

Drug	Disorder	Disease from diet
Lisinopril	Hypertension	Yes
Levothyroxine	Hypothyroidism	No
Atorvastatin	Hyperlipidemic	Yes
Metformin	Diabetes	Yes
Simvastatin	Hyperlipidemic	Yes
Omeprazole	Acid Reflux	Yes
Amlodipine	Hypertension	Yes
Metoprolol	Hypertension	Yes
Hydrocodone	Pain	No
Hydrochlorothiazide	Hypertension	Yes
Losartan	Hypertension	Yes
Gabapentin	Convulsions	No
Sertaline	Depression	No
Furosemide	Hypertension	Yes
Atenolol	Hypertension	Yes

Eleven of these drugs treat diseases caused by the Standard American Diet (Fuentes et al. 2018)

Physicians know little about diet for disease prevention, as 71% of US medical schools don't meet the recommended requirement of 25 h of nutrition education (Adams et al. 2015). Most physicians continue to eat animal products themselves, so they aren't interested in, nor have the moral authority to recommend their patients do otherwise. Also, the medical industry makes no money from patients who resolve their illness by lifestyle changes. *Is it ethical for people to eat animal products when it means cruelty to the creatures they eat, endangering public health and damage to the environment?*

What is life like for industrially farmed animals? The SAD centers around eating chickens, cows and pigs. These animals are raised with an eye to maximizing profits, and little or no concern for their welfare. Since poultry now dominates the animal product production of the Chesapeake Bay region, and the diet of its residents, let's consider the ethics of the industry that produces it. Dr. Karen Davis, PhD, is President of United Poultry Concerns (UPC), an organization that addresses the plight of these animals in the food system. Davis founded UPC on the Eastern Shore of Virginia, in the midst of the Delmarva broiler industry. She provides the reader with an ethical perspective on the consumption of animal food (see Box).

Ethical Veganism

The plea for ethical veganism, which rejects the treatment of birds and other animals as a food source or other commodity, is sometimes mistaken as a plea for dietary purity, as if formalistic food exercises and barren piety were the point of the desire to get the slaughterhouse out of one's kitchen and out of one's system. Abstractions like "vegetarian" and "vegan" mask the experiential and ethical roots of an animal-free diet. They make the realities of food animal production and consumption seem almost trivial, mere matters of ideological preference and consequence, or of individual taste, like selecting a shirt, or hair color.

The decision that has led millions of people to stop eating animals is not rooted in arid adherence to diet or dogma, but in the desire to eliminate the kinds of experiences that using animals for food confers upon beings with feelings. The human commitment to harmony, justice, peace, and love is meaningless as long as we continue to support the suffering and shame of the slaughterhouse and its attendant operations.

Like most people, I grew up eating animals, and like most people, unless they live on a farm, I did not think of it that way. I ate "meat" like everyone else. My unconsciousness was disrupted in the 1970s when I came across an essay by the great Russian writer of War and Peace, Leo Tolstoy, called "The First Step" (Tolstoy 1883). Tolstoy argued that the first step toward a spiritually compassionate, nonviolent life was to wash the blood of animals out of ourselves, and to stop killing them. He described his visit to a Moscow slaughterhouse. Sitting in my studio apartment in Baltimore, Maryland, I felt the animal terror and pain he described and the effect of slaughtering animals on the workers. That essay woke me up to the meaning of meat, and I became one of those people who, in the words of former Tyson chicken slaughterhouse worker, Virgil Butler, "just couldn't look at a piece of meat anymore without seeing the sad, tortured face that was attached to it sometime in the past" (Butler 2004).

Not only did I stop eating animals, but a decade later a chicken named Viva changed the course of my life and career. When I met her, I was an English teacher at the University of Maryland in College Park completing my doctoral dissertation. I planned to teach English for the rest of my life. But in the 1980s I was increasingly drawn to the plight of animals in human society, particularly farmed animals. The number of animals was staggering. At the bottom of the pile were billions of birds. Farmed animals were generally dismissed as beyond the pale of equal, or even any, moral concern because, it was said, they had been bred to a substandard state of intelligence and biological fitness, and because they were "just food" that was "going to be killed anyway."

My experience with Viva, a crippled and abandoned "broiler" hen, defied these claims. Viva was

(continued)

expressive, responsive, communicative, affectionate, and alert. Though cursed with a disabled body, there was nothing inferior about her personally. She already had a voice, but her soft trills and purrs and peeps needed to be amplified within the oppressive system in which she and her kind were trapped. There were billions of Vivas out there, just as special.

Viva's death was painful to me but knowing her clarified my future. It was not only that Viva had suffered, but that she was a valuable being, somebody worth fighting for. She was not "just a chicken." She was a chicken, a member of Earth's community with a claim equal to anyone else's to justice, compassion, and a life worth living.

Chickens are creatures of the Earth who no longer live on the land. Billions of chickens are confined in huge metal sheds the equivalent of cesspools. If there is such a thing as "earthrights," the right of a creature to experience directly the Earth from which it derived and on which its happiness chiefly depends, then chickens have been stripped of theirs.

Thus far, our responsibility for how we treat chickens and allow them to be treated has been dismissed with rhetoric designed to silence objection: "How the hell can you compare the feelings of a chicken with those of a human being?" Meanwhile, study after study reveals the intelligence and emotional sensitivities of chickens, and farmed animal sanctuaries rout the false rhetoric with observable facts.

Still, people are intimidated. We are told people are capable of knowing just about anything we want to know—except, conveniently, what it feels like to be one of our victims. We are told we are being "emotional" or "anthropomorphic" if we care about a chicken or grieve over a chicken's plight (O'Rourke 2004). In reality, it is not "emotion" as such that is under attack, but the vicarious emotions of pity, sympathy, compassion, sorrow, and indignity on behalf of a fellow creature—emotions that undermine business as usual. By contrast, such "manly" emotions as pride, conquest, subjugation, control, and mastery are lauded and encouraged.

A ploy to silence objection to the mistreatment of chickens is to say that they are "productive"—only "happy" chickens lay lots of eggs and pile on breast muscle tissue. In fact, chickens do not gain weight and lay eggs in inimical surroundings because they are comfortable, content, or well-cared for, but because they are manipulated to do these things through genetics and management techniques that have nothing to do with happiness, except to destroy it. In addition, chickens, like all farmed animals, are slaughtered at extremely young ages, before diseases and death have decimated the flocks as they would otherwise do, even with all the drugs.

Even so, millions of chickens die each year before going to slaughter, and on the way to slaughter, but because the volume of birds is so big—in the billions—the losses are economically negligible. The term "productivity" is an economic measure referring to averages, not the wellbeing of individuals. Michael W. Fox, a veterinary specialist in farmed-animal welfare, states that while chickens and other farmed animals may sometimes appear to be adapted to the adverse conditions in which they are forced to live, "on the basis of their functional and structural 'breakdown,' which is expressed in the form of various production diseases, they are clearly not adapted" (Fox 1983–1984).

The effects of the "human controlled evolution" of chickens bred for the meat industry are described in an article in International Hatchery Practice. Andrew A. Olkowski, DVM and his colleagues state in "Trends in developmental anomalies in contemporary broiler chickens" that chickens with extra legs and wings, missing eyes and beak deformities "can be found in practically every broiler flock," where "a variety of health problems involving muscular, digestive, cardiovascular, integumentary, skeletal, and immune systems" form a complex of debilitating diseases. Poultry personnel, they say, provide "solid evidence that anatomical anomalies have become deep-rooted in the phenotype of contemporary broiler chickens" (Olkowski et al. 2013).

One method we use to appease our guilt about our mistreatment of chickens is to plead that, having no basis for comparison, they cannot know that their lives are desolate. We might as well use this plea to absolve ourselves of responsibility toward anyone in the world who, we decide, because they have never known anything but misery, "cannot know that their lives are desolate." Are we prepared to say that babies who are born addicts as a result of their mother's drug habit, having no basis for comparison, cannot know that their lives are desolate? Is this an acceptable argument for children born into slavery or surrounded by domestic violence? Even if this were so, where does such thinking constructively lead?

As a matter of fact, animals do "know," because knowing is an organic process far deeper than words

(continued)

and concepts can express. Every bodily cell is a repository of experiences including memory and expectation as elements of a cell. The look in a creature's eyes tells us a whole lot about what he or she "knows." Freedom and well-being, as Dr. Michael Fox observes, "are more than intellectual concepts. They are a subjective aspect of being, not exclusive to humanity, but inclusive of all life. This is not an anthropomorphic claim. It is logically probable and empirically verifiable" (Fox 1983–1984).

While every effort must be made to extend legal protection to farmed animals, the question remains how far the law can protect a creature whom it has defined in advance as a piece of property, a thing without rights. It is not that they are going to die anyway that seems to justify our abuse of them, because death is the fate we all share and we do not generalize the argument, but that we are deliberately going to kill them. Henry Spira, who exposed the moral atrocities of Perdue Farms in the 1980s, asked, "What gives us the right to violate the bodies and minds of other feeling beings?"

We have a suspicion that we have no right. People with empathy for animals are reduced to deception and denial. Tolstoy describes this experience in "The First Step," where he writes: "I had wished to visit a slaughterhouse, in order to see with my own eyes the reality of the question raised when vegetarianism is discussed. But at first I felt ashamed to do so, as one is always ashamed of going to look at suffering which one knows is about to take place, but which one cannot avert; and so I kept putting off my visit.

But a little while ago I met on the road a butcher returning to Toúla after a visit to his home. He is not yet an experienced butcher, and his duty is to stab with a knife. I asked him whether he did not feel sorry for the animals that he killed. He gave me the usual answer: "Why should I feel sorry? It is necessary." But when I told him that eating flesh is not necessary, but is only a luxury, he agreed; and then he admitted that he was sorry for the animals.

"But what can I do? I must earn my bread," he said. "At first I was afraid to kill. My father, he never even killed a chicken in all his life." The majority of Russians cannot kill; they feel pity and express the feeling by the word "fear." This man had also been "afraid," but he was so no longer.

Not long ago I also had a talk with a retired soldier, a butcher, and he, too, was surprised at my assertion that it was a pity to kill, and said the usual things about its being ordained; but afterwards he agreed with me: "Especially when they are quiet, tame cattle. They come, poor things! trusting you. It is very pitiful" (Tolstoy 1883).

If we cannot imagine what it is like for an animal to be shoved in a cattle car packed with other terrified animals, if we cannot imagine how chickens feel being grabbed by their legs in the middle of the night by men who are cursing and yelling at them while stuffing them into the crates in which they will travel to the next wave of terror attacks at the slaughterhouse, then let us try to imagine ourselves placed helplessly in the hands of an overpowering extraterrestrial species to whom our pleas for mercy are mere "noise" to the master race in whose minds we are "only animals."

As I write these words in the early morning, I hear our sanctuary hens and roosters outside in the yard beginning their day. I contrast their expressions of exuberant life with my visits to chicken houses on the Eastern Shore where thousands of chickens just a few weeks old sit soundless and motionless in the fetid dark. A "sea of stationary grey objects" is how a journalist described the selfsame scene in a British chicken shed (Purvis 2006).

Chickens represented by the poultry industry as unfeeling beings have rested in my lap with their eyes closed peacefully, and they quickly learn their names. A little white hen from the egg industry named Karla became so friendly, all I had to do was call out "Karla!" and she would break through the other hens and head straight toward me, knowing she'd be scooped off the ground and kissed on her sweet face and over her closed eyes. And I can still see Vicky, our large white hen from a "broiler breeder" operation, whose right eye had been knocked out, peeking around the corner of her house each time I shouted, "Vicky, what are you doing in there?" And there was Henry, who came to our sanctuary dirty and angry after falling from a truck on the way to a slaughter plant. Lavished with attention, Henry, who at first couldn't bear to be touched, became as pliant and lovable as a big shaggy dog. I couldn't resist wrestling him to the ground with bearish hugs, and his joy at being placed in a garden where he could eat all the tomatoes he wanted was expressed in groans of ecstasy. He was like, "Are all these riches of food and affection really for me?"

One of my most poignant memories is of a large black hen named Mavis who'd been dropped off at a shelter. She must have spent her life immobilized in a cage with a keeper who treated her like an object. At first, Mavis could not even stand up without crumbling

(continued)

to the ground, and she was deeply shy and inexpressive. In the chicken yard she sat alone by the fence and poked around a little by herself without showing or attracting interest. I saw no sign that she was ever going to recover from the deprivation of her previous life. During this time, we had three adult broiler hens—Bella Mae, Alice, and Florence. They were the opposite of Mavis. All I had to do was crouch down in the yard, and here comes one of my Three Graces, as I called them, Bella Mae for example, bumping up against me with her ample breast for an embrace. Immediately, Alice and Florence would plod over on their heavy feet to participate in the embracement ceremony. Assertively but without aggression, they would vie with one another, bumping against each other's chests to maneuver the closest possible contact with me, and I would encircle all three of them with my arms. One day as we were doing this, I looked up and saw Mavis just a few feet away, staring at us. The next time, the same thing happened. There was Mavis with her melancholy eyes watching me hugging the three white hens. And then it struck me—Mavis wants to be hugged. I withdrew from the hens, walked over and knelt beside Mavis and pulled her gently toward me. It didn't take much. She rested against me in a completeness of comfort that seemed to include her gratitude that her shy desire had been understood.

Experiences like these have made me an advocate for chickens for more than thirty years. I want to get chickens off of our plates and into our hearts and back on the Earth where they belong. Every day I wake up to work on this project, and the chickens outside in the yard cheer me on.

Note that much of the above discussion was developed in Davis (2009).

KAREN DAVIS, Ph.D. is the President and Founder of United Poultry Concerns, a nonprofit organization that promotes the compassionate and respectful treatment of domestic fowl including a sanctuary for chickens in Virginia. Inducted into the National Animal Rights Hall of Fame for Outstanding Contributions to Animal Liberation, she is the author of *Prisoned Chickens, Poisoned Eggs: An Inside Look at the Modern Poultry Industry*; *More Than a Meal: The Turkey in History, Myth, Ritual, and Reality*; *The Holocaust and the Henmaid's Tale: A Case for Comparing Atrocities*; *For the Birds: From Exploitation to Liberation*, and other works including her children's book; *A Home for Henny* and *Instead of Chicken, Instead of Turkey: A Poultryless "Poultry" Potpourri*, a vegan cookbook.

Hunting and Fishing Is there an ethical argument for consuming animals? Hunting and fishing were important aspects of the Chesapeake Bay food system long before the arrival of Capt. John Smith. The ability of farming to produce abundant plant-based foods eliminates the need to get nutrition from hunting and fishing, yet an ethical argument remains for continuing the practices on a limited scale to regulate wildlife populations. White-tailed deer (*Odocoileus virginianus*) and blue catfish (*Ictalurus furcatus*) provide two prime examples of populations requiring regulation in the Chesapeake Bay region.

White-tailed deer greeted the English Colonists upon their arrival in Jamestown. The Powhatan and other Chesapeake Bay tribes competed with wolves (*Canis lupis*), mountain lions (*Puma concolor*), and black bears (*Ursus americanus*) in hunting the deer. The English colonists essentially eliminated the other predators, leaving only humans, feral dogs and recent coyote-wolf hybrids to regulate deer population size. Most wildlife ecologists argue that without hunting, the deer population would quickly grow, creating conflict with farmers and suburban dwellers, as well as expanding the epidemic of tick-borne Lyme disease caused by *Borrelia burgdorferi* (Telford 2017). Reestablishing sufficient natural predator populations to control the deer is unrealistic given habitat alteration and potential conflict with the expanded human presence. So, hunting by humans seems important for maintaining a population of healthy deer that don't overtax the environment. Yet, deer populations are often managed to increase their numbers by planting "food plots" and restricting the taking of does. Such managing for the benefit of hunters is contributing to the perceived problem of overpopulation (Edwards et al. 2004).

Stocking programs introduced blue catfish to the James River in the 1970s for sport fishers. However, the catfish population size and range expanded rapidly and endangers native species. With few natural predators, fishing pressure is important to regulating this introduced species. Yet, this apex predator accumulates toxins that threaten the health of consumers, challenging the ethics of promoting fishing this species for human consumption (Schloesser et al. 2011).

People engage in fishing often more for the sport than the potential dinner. Many fishers practice catch-and-release, with no intention of keeping their momentary catch. This gets the fisher on to the water to reconnect with nature while sparing the prey from certain death. However, a large portion of the released fish do suffer damage and death from the experience, so this food system related sport still offers ethical challenges (Thompson et al. 2018). Also, the underlying assumption is that fin fish are lesser beings than say marine mammals. Catch and release of Chesapeake Bay dolphins would violate laws protecting marine mammals from harassment. Fish feel pain in their own way (Chandroo et al. 2004). Before being released the fish are essentially suffocating as their gills collapse in the air. Some would argue the ethics of catching fish for sport considering the damage it inflicts.

Should Ethics Get in the Way? Most conservationists perceive sport fishers and hunters as natural allies in reversing the damage wrought by the food industry. After all, it was the duck hunters that pushed through the first US laws for the protection of wildlife (see chapter "Passenger Pigeon and Waterfowl: Flights to Extinction and Not" for discussion of the Migratory Bird Act of 1918). Hunting and fishing licenses help pay the salaries of state wildlife officers and scientists. So, one shouldn't expect organizations like the Chesapeake Bay Foundation or the Virginia Department of Fish and Game to say much about the ethics of killing and eating animals. They instead focus on the most sustainable way to do so. It will take a massive cultural shift for most people involved in the food system to take seriously ethical questions around animal rights and eating healthier diets. Yet societies do change. Several centuries of slavery in the US ended in 1865. Women's suffrage came in 1920. Same-sex marriage became the law of the land in 2015, something unthinkable two decades before. Tobacco smoking rates are 69% lower than in 1965. All of these changes required society to make new decisions informed by evolving ethical notions of what is right and wrong. Raising ethical questions may be uncomfortable but necessary to drive society to a sustainable food system that provides for human and environmental health.

Cooking Is the First Step to Ethical Eating The dietary choices we make influence our health, the environment, and the welfare of animals in the food system (Singer and Mason 2006; Tilman and Clark 2014; Sabaté and Soret 2014). Consumers can reshape the food system by creating demand for healthy, environmentally sustainable, and cruelty-free items, while eschewing less ethical choices. To effectively take control of what one eats requires rediscovering the art of cooking. Preparing one's own meal provides the consumer with more control of their diet than comes from dining out. As noted above, the largest fraction of the food-dollar goes to paying for food prepared outside of one's own kitchen. In 2010 consumers in the US spent half of their dietary budget on Food Away From Home (FAFH). This includes food bought and consumed away from home, and prepared food brought into the home. The percent of dietary calories consumed from FAFH doubled between 1978 and 2012, going from 17 to 34%. FAFH contains more calories, saturated fats, and sodium; with less fiber, calcium and iron than food prepared at home. This is true of school lunches as well as restaurants. FAFH is dominated by animal products and includes few whole grains, vegetable and fruits. Most Americans purchase FAFH meals five times per week (Saksena et al. 2019).

With a shift to FAFH Americans are losing their desire to prepare their own meals. One study revealed that only 10% of respondents "love to cook" and 45% "hate it," with the balance indicating ambivalence (Yoon 2017). In one study 28% of Americans reported not knowing how to cook (Sterling 2011). However, the percentage of people who can cook has increased since 1965, owing to the greater number of men acquiring culinary skills. In 1965, 29% of men cooked. This rose to 46% in 2017. About 70% of American women reported cooking in 2017, up three points from 2016 (Taillie 2018).

In the mid-twentieth Century US public schools taught about 50% of female students to cook in high school home economics (now Family and Consumer Sciences) courses. Presently similar numbers of students take those classes, but with greater gender equity (Werhan 2013). This means that half of all young Americans lack any formal training in cooking, and many don't see their parents cook, owing to high consumption of FAFH. Even when receiving formal cooking classes in school, those students are being trained in preparation of dishes for the SAD, with meals centered around animal products. Adopting an ethical diet means not only learning the basics of cooking, but also how to prepare nutritionally complete plant-based dishes. Implementing a new food system that will help recover the health of the Chesapeake Bay and its people requires educating those people on the basics of whole-food plant-based nutrition and cooking. This means requiring state supported colleges and universities in the Chesapeake Bay region to offer courses in plant-based nutrition, so as to produce the professionals and educators that can inform the general public. It also means promoting informal education to provide science-based advice on ethical eating and food preparation.

Ethics of Economic Disruption, the SAD as the New Tobacco Creating a new food system will displace workers from some jobs and open up other employment opportunities. Considering that the food system employs less than 2% of the workforce, the disruption will impact a relatively small number of people. Jobs will shift away from animal agriculture and commodity crops for animal feed. The production of organically produced grains, legumes, nuts, fruits, and vegetables will create new employment opportunities in the Chesapeake Bay food system. Organic farming produces more jobs and more stable employment than conventional agriculture (Finley et al. 2017). The elimination of pesticides will also create safer working environments.

The disruption of agriculture due to adopting a new ethical food system won't be the first such event experienced in the Chesapeake Bay Region. The area witnessed a similar disruption when sales of tobacco, its founding cash crop, slumped after the public understood the health risks of smoking. From 1965 to 2018 the fraction of adults that smoke in the US went from 45 to 14% (CDC 2019). Consequently, annual production of tobacco in Virginia during that same interval went from 61.4 million kg to 24.0 kg in (USDA/NASS 2019). This forced many tobacco farmers and workers to find other employment. While people consume less tobacco today, on average they eat more meat and sugar than in past decades (see chapters "Livestock and Poultry: The Other Colonists who Changed the Food System of the Chesapeake Bay" and "Sugar Twice Enslaves: Consequences for the People of the Chesapeake"). Until about 2015, smoking was the leading cause of death from a non-communicable disease (NCD) in the US and the rest of the world. However, by 2017 the SAD in the US, and variations of that salty-meat-sweet-fat rich diet elsewhere, replaced smoking as the number one cause of NCD mortality around the world. As of 2017, poor diet kills 11 million people per year globally, compared to 8 million deaths from smoking (Afshin et al. 2019; WHO 2019). Just as lung cancer was an essentially unknown disease before the advent of smoking, coronary artery disease was absent in populations that ate plant-based diets. Due to the SAD, heart disease is now the leading cause of death in the US and around the world.

References

ACGA (2019) American Community Gardening Association – Home Page. American Community Garden Association. https://communitygarden.org/. Accessed 6 May 2019

Adams K, Butsch W, Kohlmeier M (2015) The State of Nutrition Education at US Medical Schools. J Biomed Educ 2015:1–7. https://doi.org/10.1155/2015/357627

Afshin A, Sur P, Fay K et al (2019) Health effects of dietary risks in 195 countries, 1990–2017: a systematic analysis for the Global Burden of Disease Study 2017. Lancet 393:1958–1972. https://doi.org/10.1016/s0140-6736(19)30041-8

Alston L (1983) Farm foreclosures in the United States during the interwar period. J Econ Hist 43:885–903. https://doi.org/10.1017/s0022050700030801

Arias E, Xu J (2015) United States Life Tables, 2015. Cdc.gov. https://www.cdc.gov/nchs/data/nvsr/nvsr67/nvsr67_07-508.pdf. Accessed 9 May 2019

BerryPicking (2016) Chesterfield Berry Farm. Virginia Employment Commission. http://www.vec.virginia.gov/sites/default/files/documents/756771%20Chesterfield%20Berry%20Farm.pdf. Accessed 7 May 2019

Birky J (2009) The modern community garden movement in the United States: its roots, its current condition and its prospects for the future. Scholarcommons.usf.edu. https://scholarcommons.usf.edu/cgi/viewcontent.cgi?referer=https://www.google.com/&httpsredir=1&article=2859&context=etd. Accessed 6 May 2019

BLS (2019a) Bureau of labor statistics consumer expenditures—2017. Bls.gov. https://www.bls.gov/news.release/cesan.nr0.htm. Accessed 21 Apr 2019

BLS (2019b) Multiple jobholders by selected characteristics. Bls.gov. https://www.bls.gov/cps/cpsaat36.htm. Accessed 3 May 2019

Bose N (2018) Half of Walmart's workforce are part-time workers: labor group. U.S. https://www.reuters.com/article/us-walmart-workers/half-of-walmarts-workforce-are-part-time-workers-labor-group-idUSKCN1IQ295. Accessed 3 May 2019

Bruckert E, Pouchain D, Auboiron S, Mulet C (2012) Cross-analysis of dietary prescriptions and adherence in 356 hypercholesterolaemic patients. Arch Cardiovasc Dis 105:557–565. https://doi.org/10.1016/j.acvd.2012.06.005

Butler V (2004) Slaughterhouse worker turned activist. Poultry Press 14.3. http://www.upc-online.org/fall04/virgil.htm. Accessed 21 April 2019

CDC (2019) Current cigarette smoking among adults in the United States. Centers for Disease Control and Prevention. https://www.cdc.gov/tobacco/data_statistics/fact_sheets/adult_data/cig_smoking/index.htm. Accessed 31 May 2019

Census (2019) Income and poverty in the United States: 2017. Census.gov. https://www.census.gov/library/publications/2018/demo/p60-263.html. Accessed 21 Apr 2019

Chandroo K, Duncan I, Moccia R (2004) Can fish suffer?: perspectives on sentience, pain, fear and stress. Appl Anim Behav Sci 86:225–250. https://doi.org/10.1016/j.applanim.2004.02.004

DeMuth S (1993) National agricultural library (U.S.) community supported agriculture (CSA): an annotated bibliography and resource guide. National Agricultural Library, Beltsville, Md

Davis K (2009) Prisoned chickens, poisoned eggs, 2nd ed. Book Publishing: An Inside Look at the Modern Poultry Industry (revised edn), Summertown, TN

Douglas-Bowers D (2013) Debt slavery: the forgotten history of sharecropping. I The Hampton Institute. Hamptoninstitution.org. http://www.hamptoninstitution.org/sharecropping.html#.XM9hfNNKjUI. Accessed 5 May 2019

Dudlicek J (2018) Top 50 Grocers: Amazon in 8th place while rest of industry Restrategizes, Reshuffles. Progressive Grocer. https://progressivegrocer.com/top-50-grocers-amazon-8th-place-while-rest-industry-restrategizes-reshuffles. Accessed 3 May 2019

Edwards S, Demarais S, Watkins B, Strickland B (2004) White-tailed deer forage production in managed and unmanaged pine stands and summer food plots in Mississippi. Wildl Soc Bull 32:739–745. https://doi.org/10.2193/0091-7648(2004)032[0739:wdfpim]2.0.co;2

EWG (2019) EWG's farm subsidy database. Farm.ewg.org. https://farm.ewg.org/region.php?fips=00000&statename=UnitedStates. Accessed 11 May 2019

Faber S (2018) Top 5 reasons to reform crop insurance. EWG. https://www.ewg.org/agmag/2018/06/top-5-reasons-reform-crop-insurance. Accessed 11 May 2019

Fields S (2004) The fat of the land: do agricultural subsidies foster poor health? Environ Health Perspect 112:A820–A823. https://doi.org/10.1289/ehp.112-a820

Finley L, Chappell M, Thiers P et al (2017) Does organic farming present greater opportunities for employment and community development than conventional farming? A survey-based investigation in California and Washington. Agroecol Sustain Food Syst 42:552–572. https://doi.org/10.1080/21683565.2017.1394416

Fox M (1983–1984) Animal freedom and well-being: want or need? Appl Anim Ethol 11:205–209

Freudenberger H, Pritchett J (1991) The domestic United States slave trade: new evidence. J Interdiscip Hist 21:447. https://doi.org/10.2307/204955

Fuentes A, Pineda M, Venkata K (2018) Comprehension of top 200 prescribed drugs in the US as a resource for pharmacy teaching, training and practice. Pharmacy 6:43. https://doi.org/10.3390/pharmacy6020043

Greger M (2015) How not to die. Flatiron, New York

Grossman J (2019) Fair labor Standards Act of 1938: maximum struggle for a minimum wage. U.S. Department of Labor. Dol.gov. https://www.dol.gov/general/aboutdol/history/flsa1938. Accessed 3 May 2019

Hatch P (2001) African-American Gardens at Monticello. https://wwwmonticelloorg/house-gardens/center-for-historic-plants/twinleaf-journal-online/african-american-gardens-at-monticello/. Accessed 28 Aug. 2019

Helms J (1985) Brief history of the USDA Soil Bank Program. Nrcs.usda.gov. https://www.nrcs.usda.gov/Internet/FSE_DOCUMENTS/stelprdb1045666.pdf. Accessed 10 May 2019

Hollingsworth D (2016) Community gardens spreading across South Hampton Roads. The Virginian-Pilot. https://pilotonline.com/life/home/lawn-garden/article_d1625ab5-6158-59e1-8931-30fe30f57080.html. Accessed 6 May 2019

Huennekens P (2018) Unlimited cheap farm labor: evaluating H-2A disclosure data. CIS.org. https://cis.org/Report/Unlimited-Cheap-Farm-Labor-Evaluating-H2A-Disclosure-Data. Accessed 4 May 2019

Jackson R, Minjares R, Naumoff K et al (2009) Agriculture policy is health policy. J Hung Environ Nutr 4:393–408

Jacobs W (1949) Wampum: the protocol of Indian Diplomacy. William Mary Q 6:596. https://doi.org/10.2307/1916752

James D (1988) The transformation of the Southern Racial State: class and race determinants of local-state structures. Am Sociol Rev 53:191. https://doi.org/10.2307/2095687

Keevican M (2017) USDA's myplate is making Americans sick, says doctors group. Physicians Committee for Responsible Medicine. https://www.pcrm.org/news/news-releases/usdas-myplate-making-americans-sick-says-doctors-group. Accessed 14 May 2019

Lindy (2019) Job opportunities. Lindy's Seafood, Inc. – Wholesale Maryland Crabs, Crabmeat & Oysters. lindysseafood. https://www.lindysseafood.com/job-opportunities. Accessed 7 May 2019

LocalHarvest (2019) Community supported agriculture – LocalHarvest. Localharvest.org. https://www.localharvest.org/csa/. Accessed 9 May 2019

Lusk J (2016) The evolving role of the USDA in the Food and Agricultural Economy. Mercatus.org. https://www.mercatus.org/system/files/Lusk-USDA-v1.pdf. Accessed 6 May 2019

Mader P (2002) Soil fertility and biodiversity in organic farming. Science 296:1694–1697. https://doi.org/10.1126/science.1071148

Mandeel E (2014) The Bracero program 1942-1964. Am Int J Contemp Res 4:171–184

McGwin G (2000) Increased risk of agricultural injury among African-American farm workers from Alabama and Mississippi. Am J Epidemiol 152:640–650. https://doi.org/10.1093/aje/152.7.640

Murphy S, Xu J, Kochanek K et al (2018) Mortality in the United States, 2017. NCHS Data Brief, no 328. National Center for Health Statistics, Hyattsville, MD

NCGA (2018) World of corn 2018. Worldofcorn.com. National Corn Growers Association http://www.worldofcorn.com/#corn-usage-by-segment. Accessed 17 Sep 2018

Nchako C, Cai L (2019) A closer look at who benefits from SNAP: state-by-state fact sheets. Center on Budget and Policy Priorities. https://www.cbpp.org/research/food-assistance/a-closer-look-at-who-benefits-from-snap-state-by-state-fact-sheets#Virginia. Accessed 21 Apr 2019

NDSU (2019) Milestones in U.S. Food Law—Food Law. [online] Ag.ndsu.edu. https://www.ag.ndsu.edu/foodlaw/overview/history/milestones. Accessed 15 May 2019

Neal A (2013) "Meet me at the market" – the evolution of a farmers market. Usda.gov. https://www.usda.gov/media/blog/2013/08/07/meet-me-market-evolution-farmers-market. Accessed 8 May 2019

O'Rourke K (2004) The animal rights struggle. Journal of the American Veterinary Medical Association, 15 May, 1567–1568

Olkowski A et al (2013) Trends in developmental anomalies in contemporary broiler chickens. Int Hatchery Pract 28(2). http://www.positiveaction.info/pdfs/articles/hp28.1p7.pdf. Accessed 21 April 2019

Pimentel D, Pimentel M (2003) Sustainability of meat-based and plant-based diets and the environment. Am J Clin Nutr 78:660S–663S. https://doi.org/10.1093/ajcn/78.3.660s

Pimentel D, Hepperly P, Hanson J et al (2005) Environmental, energetic, and economic comparisons of organic and conventional farming systems. Bioscience 55:573. https://doi.org/10.1641/0006-3568(2005)055[0573:eeaeco]2.0.co;2

Poti J, Mendez M, Ng S, Popkin B (2015) Is the degree of food processing and convenience linked with the nutritional quality of foods purchased by US households? Am J Clin Nutr 101:1251–1262. https://doi.org/10.3945/ajcn.114.100925

Press M, Arnould E (2011) Legitimating community supported agriculture through American pastoralist ideology. J Consum Cult 11:168–194. https://doi.org/10.1177/1469540511402450

ProAg (2019) The basics of crop insurance – ProAg. ProAg. https://www.proag.com/basics-of-crop-insurance/. Accessed 11 May 2019

Purvis A (2006) Pecking order. Guardian (UK), 23 September. https://www.theguardian.com/lifeandstyle/2006/sep/24/foodanddrink.features8. Accessed 21 Apr 2019

Rasmussen W (2019) Lincoln's agricultural legacy. National Agricultural Library. Nal.usda.gov. https://www.nal.usda.gov/lincolns-agricultural-legacy. Accessed 10 May 2019

Rembiałkowska E (2007) Quality of plant products from organic agriculture. J Sci Food Agric 87:2757–2762. https://doi.org/10.1002/jsfa.3000

Rodman S, Barry C, Clayton M et al (2016) Agricultural exceptionalism at the state level: characterization of Wage and Hour Laws for U.S. farmworkers. J Agric Food Syst and Commun Dev 1–22. https://doi.org/10.5304/jafscd.2016.062.013

Ron A, Norwood D (2018) America cannot bear to bring back indentured Servitude. Atlantic. https://www.theatlantic.com/politics/archive/2018/03/american-immigration-service-slavery/555824/. Accessed 3 May 2019

Roser M (2019) Life expectancy. Our world in data. https://ourworldindata.org/life-expectancy. Accessed 9 May 2019

Sabaté J, Soret S (2014) Sustainability of plant-based diets: back to the future. Am J Clin Nutr 100:476S–482S. https://doi.org/10.3945/ajcn.113.071522

Saksena M, Okrent A, Anekwe T (2019) America's eating habits: food away from home. Ers.usda.gov. https://www.ers.usda.gov/webdocs/publications/90228/eib-196.pdf. Accessed 29 May 2019

Schloesser R, Fabrizio M, Latour R et al (2011) Ecological role of blue catfish in Chesapeake Bay communities and implications for management. Am Fish Soc Symp 77:369–382

Schütte G, Eckerstorfer M, Rastelli V, Reichenbecher W, Restrepo-Vassalli S, Ruohonen-Lehto M, Saucy AGW, Mertens M (2017) Herbicide resistance and biodiversity: agronomic and environmental aspects of genetically modified herbicide-resistant plants. Environ Sci Eur 29(1)

Seafood Sizes (2019) Seafood sizes, weights, measures – fresh seafood. Fresh-seafood.net. http://www.fresh-seafood.net/fish-shellfish/seafood-sizes-weights-measures/. Accessed 7 May 2019

Sills E, Alwang J, Driscoll P (1994) Migrant farm workers on Virginia's Eastern shore: an analysis of economic impacts. J Agric Appl Econ 26:209–223. https://doi.org/10.1017/s1074070800019313

Singer P, Mason J (2006) The way we eat: why our food choices matter. Rodale, Emmaus

SNAP (2019) SNAP benefit estimator. Department of Health Services. https://aix-xweb1p.state.or.us/caf_xweb/SNAP_Estimate/frmEstimate.cfm. Accessed 7 May 2019

Spriggs W (1979) The Virginia colored farmers' alliance: a case study of race and class identity. J Negro Hist 64:191–204. https://doi.org/10.2307/2717032

Sterling J (2011) Survey finds 28% of Americans can't cook. Delish. https://www.delish.com/food/news/a38951/one-third-of-americans-do-not-know-how-to-cook/. Accessed 29 May 2019

Taillie L (2018) Who's cooking? Trends in US home food preparation by gender, education, and race/ethnicity from 2003 to 2016. Nutr J 17:41. https://doi.org/10.1186/s12937-018-0347-9

Tassone J, Bowen G, Bowen N (2017) Ag policies – production and productivity. AGNR research. Health, safety, and welfare? A report on the factors that favor or hinder the flow of local food in the Chesapeake Bay Region. https://agresearch.umd.edu/agroecol/research/ag-policies-production-and-productivity. Accessed 24 Jun 2019

Telford S (2017) Deer reduction is a cornerstone of integrated deer tick management. J Integr Pest Manage 8(1). https://doi.org/10.1093/jipm/pmx024

Thomas W, Davies J, O'Neal R et al (1960) Incidence of myocardial infarction correlated with venous and pulmonary thrombosis and embolism*. Am J Cardiol 5:41–47. https://doi.org/10.1016/0002-9149(60)90007-2

Thompson M, Van Wassenbergh S, Rogers S et al (2018) Angling-induced injuries have a negative impact on suction feeding performance and hydrodynamics in marine shiner perch, *Cymatogaster aggregata*. J Exp Biol 221:jeb180935. https://doi.org/10.1242/jeb.180935

Tilman D, Clark M (2014) Global diets link environmental sustainability and human health. Nature 515(7528):518–522

Tolstoy L (1883) The first step. Preface to Howard Williams, the ethics of diet. Essays and Letters (trans: Aylmer Maude). H. Frowde, New York (1909) http://www.animal-rights-library.com/texts-c/tolstoy01.htm. Accessed 21 April 2019

UCS (2019) Hidden costs of industrial agriculture. Union of Concerned Scientists. https://www.ucsusa.org/food_and_agriculture/our-failing-food-system/industrial-agriculture/hidden-costs-of-industrial.html. Accessed 18 May 2019

USBLS (2019) Food preparation and serving related occupations. Bls.gov. https://www.bls.gov/oes/current/oes350000.htm. Accessed 6 May 2019

USCIS (2019) H-2A temporary agricultural workers. US Citizenship and Immigration Services (USCIS). https://www.uscis.gov/working-united-states/temporary-workers/h-2a-temporary-agricultural-workers. Accessed 4 May 2019

USDA (2014) 2012 Census of Agriculture – Farm Demographics Highlights. USDA – National Agricultural Statistics Service. Nass.usda.gov. https://www.nass.usda.gov/Publications/Highlights/2014/Farm_Demographics/index.php#how_many. Accessed 6 May 2019

USDA (2018) USDA factsheet soybeans. Usda.gov. USDA Factsheet Soybeans. US Department of Agriculture. https://www.usda.gov/sites/default/files/documents/coexistence-soybeans-factsheet.pdf. Accessed 17 Sep 2018

USDA (2019a) Official USDA Food Plans: Cost of Food at Home at Four Levels, U.S. Average, March 2019. In: Cnpp.usda.gov. https://www.cnpp.usda.gov/sites/default/files/CostofFoodMar2019.pdf. Accessed 21 Apr 2019

USDA (2019b) USDA ERS – farm labor. Ers.usda.gov. https://www.ers.usda.gov/topics/farm-economy/farm-labor/#size. Accessed 7 May 2019

USDA (2019c) USDA ERS – food dollar series documentation. In: Ers.usda.gov. https://www.ers.usda.gov/data-products/food-dollar-series/documentation/. Accessed 7 May 2019

USDA (2019d) Local food directories: national farmers market directory. Agricultural Marketing Service. Ams.usda.gov. https://www.ams.usda.gov/local-food-directories/farmersmarkets. Accessed 8 May 2019

USDA (2019e) Local food directories: community supported agriculture (CSA) Directory. Agricultural Marketing Service. Ams.usda.gov. https://www.ams.usda.gov/local-food-directories/csas. Accessed 9 May 2019

USDA (2019f) USDA 2020 budget summary. Obpa.usda.gov. https://www.obpa.usda.gov/budsum/fy2020budsum.pdf. Accessed 10 May 2019

USDA (2019g) Conservation reserve program. Fsa.usda.gov. https://www.fsa.usda.gov/programs-and-services/conservation-programs/conservation-reserve-program/. Accessed 10 May 2019

USDA (2019h) Wheat Outlook: April 2019. Ers.usda.gov. https://www.ers.usda.gov/publications/pub-details/?pubid=92847. Accessed 13 May 2019

USDA/NASS (2019) USDA/NASS quickstats ad-hoc query tool. Quickstats.nass.usda.gov. https://quickstats.nass.usda.gov/#092A1893-A8CC-3506-9818-613C6937895D. Accessed 1 Jun 2019

USDA-AMS (2019) Agricultural marketing service. Ams.usda.gov. https://www.ams.usda.gov/. Accessed 14 May 2019

USDA-ERS (2019a) USDA ERS – Definitions of Food Security. [Online] Ers.usda.gov. https://www.ers.usda.gov/topics/food-nutrition-assistance/food-security-in-the-us/definitions-of-food-security.aspx. Accessed 15 May 2019

USDA-ERS (2019b) Oil crops outlook: April 2019. Ers.usda.gov. https://www.ers.usda.gov/publications/pub-details/?pubid=92830. Accessed 13 May 2019

USDA-ERS (2019c) USDA ERS – annual state agricultural exports. Ers.usda.gov. https://www.ers.usda.gov/data-products/state-export-data/annual-state-agricultural-exports/. Accessed 19 May 2019

USDA-ERS (2019d) USDA ERS – go to the Atlas. Ers.usda.gov. https://www.ers.usda.gov/data-products/food-access-research-atlas/go-to-the-atlas.aspx. Accessed 30 May 2019

USDA-FNS (2019a) Income eligibility standards. USDA-FNS. Fns.usda.gov. https://www.fns.usda.gov/snap/income-eligibility-standards. Accessed 6 May 2019

USDA-FNS (2019b) FNS nutrition programs. USDA-FNS. Fns.usda.gov. https://www.fns.usda.gov/programs. Accessed 13 May 2019

USDL (2019a) U.S. Department of Labor Wage and Hour Division (WHD) The Migrant & Seasonal Agricultural Worker Protection Act. Dol.gov. https://www.dol.gov/whd/regs/statutes/0001.mspa.htm. Accessed 3 May 2019

USDL (2019b) OFLC performance data. Foreignlaborcert.doleta.gov. https://www.foreignlaborcert.doleta.gov/performancedata.cfm. Accessed 4 May 2019

VirginiaPlaces (2019) How colonists acquired title to land in Virginia. Virginiaplaces.org. http://www.virginiaplaces.org/settleland/headright.html. Accessed 21 Apr 2019

Werhan C (2013) Family and consumer sciences secondary school programs: national survey shows continued demand for FCS teachers. J Fam Consum Sci 105:41–45. https://doi.org/10.14307/jfcs105.4.10

White M (2019) Freedom farmers. University of North Carolina Press, Chapel Hill

WHO (2019) Tobacco. Who.int. https://www.who.int/news-room/fact-sheets/detail/tobacco. Accessed 1 Jun 2019

Wiltgen T (2007) An economic history of the United States sugar program. Scholarworks.montana.edu. https://scholarworks.montana.edu/xmlui/bitstream/handle/1/2554/WiltgenT1207.pdf. Accessed 11 May 2019

Wolfe B, McCartney M (2019) Indentured servants in colonial Virginia. Encyclopediavirginia.org. https://www.encyclopediavirginia.org/indentured_servants_in_colonial_virginia. Accessed 3 May 2019

Wolkomir E (2018) SNAP boosts retailers and local economies. Center on Budget and Policy Priorities. https://www.cbpp.org/research/food-assistance/snap-boosts-retailers-and-local-economies. Accessed 10 May 2019

Yoon E (2017) The grocery industry confronts a new problem: only 10% of Americans love cooking. Harvard Business Review. https://hbr.org/2017/09/the-grocery-industry-confronts-a-new-problem-only-10-of-americans-love-cooking. Accessed 29 May 2017

The manufacturer's authorised representative in the EU is Springer
Nature Customer Service Centre GmbH, Europaplatz 3, 69115 Heidelberg,
Germany. If you have any concerns regarding our products, please
contact ProductSafety@springernature.com

Printed and bound by CPI Group (UK) Ltd, Croydon, CR0 4YY
28/09/2026
02224097-0003